HOUSE CONSTRUCTION DETAILS

by NELSON L. BURBANK

Author of *House Construction Details; Practical Job Pointers; Shopcrafter's Manual*

Co-Author of *Handbook of Building Terms and Definitions*

and ARNOLD B. ROMNEY

as revised by
CHARLES PHELPS

Seventh Edition

McGraw-Hill Book Company

New York St. Louis San Francisco Auckland Bogotá
Hamburg Johannesburg London Madrid Mexico Montreal
New Delhi Panama Paris São Paulo Singapore
Sydney Tokyo Toronto

Library of Congress Cataloging-in Publication Data

Burbank, Nelson Lincoln, date
House construction details.

Includes index.
1. Building — Details. 2. House construction.
I. Romney, Arnold B. II. Title.
TH2025.B85 1986 690'.837 86-2923
ISBN 0-07-008929-9 (pbk.)

1234567890 HALHAL 8932109876

ISBN 0-07-008929-9

Printed and bound by Arcata Graphics/Halliday.

This book was originally published in hardcover by Simmons-Boardman, Inc.

TABLE OF CONTENTS

	Preface	v
	Acknowledgments	vi
1.	Trends in House Design	2
2.	Excavations/Foundation Forms/ Foundations	17
3.	Sills/Girders/Joists/Subflooring	36
4.	Outside Wall Construction	46
5.	Inside Walls/Ceiling Joists	61
6.	Roof Construction/Bay Construction/Roofing	73
7.	Cornices/Porches/Patios/Atriums	101
8.	Windows	111
9.	Doors	128
10.	Exterior Wall Covering	148
11.	Interior Wall Covering/ Interior Trim	169
12.	Insulation	187
13.	Stair Construction	199
14.	Hardware	214
15.	Closets/Shelves/Built-In Equipment	232
16.	Finish Flooring	252
17.	Chimneys and Fireplaces	268
18.	Scaffolds and Hoists	284
19.	Garages and Carports	297
20.	Comfort Conditioning	309
21.	Painting and Finishing Today's House	333
22.	Modern Building Materials	351
23.	Prefabrication	369
24.	Time and Money Saving Details	389
	Glossary	449
	Index	455

PREFACE

This completely revised seventh edition of House Construction Details, in keeping with the aims of the six editions preceding it since 1939, has compiled the latest available information on the design, building and equipping of single family dwellings. It recognizes the trends which have developed since the sixth edition was published – trends toward more efficient design, simpler details to reduce building costs, increased use of pre-cut and factory-made components, and the development of a wide variety of new, durable building materials requiring little, if any, maintenance by the home owner. Today's house is built to live in, not to work on.

While including most of the craftsmanlike hammer-and-saw details that made the earlier editions so useful to both the small contractor and the home builder, this edition also meets the needs of readers who wish to keep abreast of new trends in design and materials, and to anticipate the future direction of the building industry.

Many of the photographs used to illustrate the text were taken specifically for this book. We are also indebted to the many manufacturers and associations serving the building industry for their cooperation in supplying the latest information and illustrations of building products and techniques. A list of these friends is on the following page.

N.L.B.
A.B.R.

ACKNOWLEDGMENTS

Our sincere thanks to the following organizations who contributed so generously toward the preparation of this book:

Armstrong Cork Co.
Amana Refrigeration, Inc.
American Builder Magazine
American Plywood Association
American Standard Co.
Andersen Corp.
Arvin Industries, Inc.
J.H. Baxter & Co.
Better Heating-Cooling Council
Bird & Son
Burnham Corp.
Capital Industries
Clopay Corp.
Condon-King Co.
Dacor Mfg. Co.
Electromode Corp.
Frigidaire Corp.
GAF, Inc.
General Electric Co.
Georgia-Pacific Corp.
Goldblatt Tool Co.
Grumman Energy Systems, Inc.
Heatform Co.
Ida Products Co.
Johns Manville Corp.
Kinkead Industries
Leigh Products, Inc.
Lennox Corp.
The Majestic Co.
Mobay Chemical Co.
Mohawk Door Co.
National Association of Home Builders
National Forest Products Association
National Gypsum Co.
National Homes
National Paint, Varnish & Lacquer Assn.
National Woodwork Mfrs. Assn.
Nutone Corp.
Pease Woodwork Co.
Portland Cement Association
Potlatch Forests, Inc.
RCA Whirlpool Corp.
R.O.W. Window Sales Co.
Reynolds Metals Co.
Rolscreen Co.
Sanford Truss, Inc.
Schlegel Mfg. Co.
Sinclair-Koppers Co.
The Singer Co.
Small Home Council, U. of Ill.
Stanley Works, Inc.
Stanley Vemco Corp.
Tile Council of America
Timber Engineering Co.
U.S. Gypsum Co.
U.S. Plywood Co.
U.S. Steel Co.
Wausau Homes
Westinghouse Electric Corp.

Chapter 1

TRENDS IN HOUSE DESIGN

From the demands of buyers of one-family dwellings, the designs of architects, the technological advances of building materials manufacturers, and from conferences of builders, come the shaping of today's and tomorrow's house for complete and all season living and comfort.

Many advances in the manufacture of materials and components used in present day dwellings have radically changed the shape and appearance of the exterior and interior of our homes.

Basic House Types

The one-story house. Building statistics reveal that the one-story house—with or without a basement—is built in larger numbers each year than any other basic house type. One-story houses have more variation in size, shape and design than any other type. Simplicity of construction provides a base for studies of construction methods, which often result in important cost savings. Most of the building innovations developed during the past ten years have been applied to the one story house most successfully.

The pitched roof and flat ceiling one-level house, **Figure 1,** offers the greatest convenience and livability for all members of a household, regardless of age.

The one-story house designed with a sloping ceiling following the pitch of the roof, **Figure 2,** is relatively simple to construct since ceiling joists are not required. Insulation and ceiling material are applied directly to the rafters.

The one-and-one-half-story house. The Cape Cod house, **Figure 3,** having two living levels and a varying second floor area is the most familiar 1-1/2-story type. This traditional house of New England has been popular for more than two centuries. Its steep sloping roof is also the basis for many outstanding contemporary designs. When dormers are added at the front or rear, more natural light is admitted, providing greater flexibility in planning the second floor living area.

The two-story house. Box-like, but spacious, the two-story house, shown in outline elevation in **Figure 4,** provides the maximum living area for the large family at the least cost per square foot. A wide range of roof types can be used to vary the design characteristics. The upper-level walls may overhang the lower level—garrison style—to gain more floor area and break the high wall appearance.

The addition of a patio, porch, or an attached garage to the two-story aspect will enhance the overall design composition. This house type, as rich in traditional heritage as the 1-1/2-story house, must be designed with respect for proportion and detail, regardless of the particular architectural styling.

The split-level house. In an effort to combine the advantages of the one, one-and-one-half, and two-story dwellings of tradition, structures having three or more living levels have been developed in recent years. **Figure 5** illustrates an outline of a split-level house having three separate living levels, with varying roof and ceiling types. Instead of placing the living levels one above the other, they are connected by segments of stairs, which combine in pairs to total one full stair flight. The multi-levels provide distinct separation of different living functions,

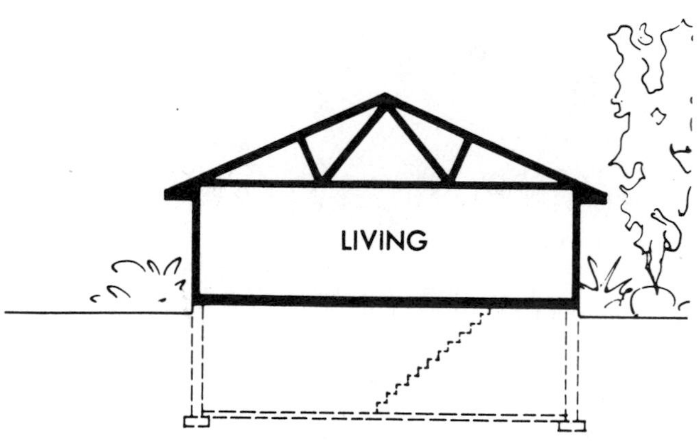

Fig. 1. The pitched roof, flat ceiling, one-story house.

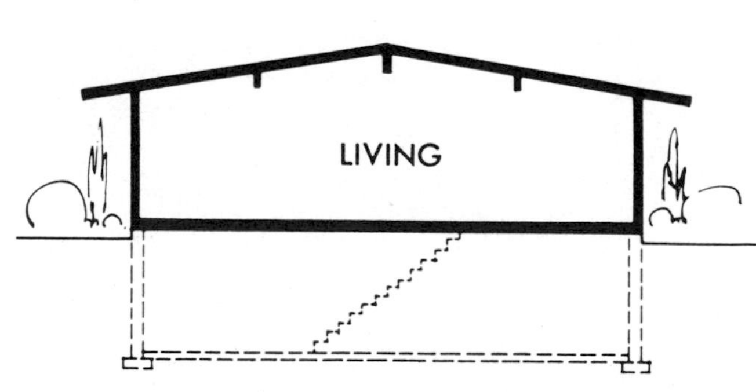

Fig. 2. The pitched roof, sloping ceiling, one-story house.

which can be planned with almost unlimited variety. A split-level house looks best and is most functional when placed on a building site having a distinct slope. This gives the lower-level living areas the advantage of full-story-height exposure instead of being below grade.

A sloping roof construction, which is continuous over two or more floor levels, is outlined in **Figure 6.** A functional design of pleasing proportions results from this construction. Roof slopes of this type split-level are generally 1/6, 1/4, or 1/3 pitch. The popular type split-level with a lower living area exposed to the outdoors, often has a basement under the mid-level portion of the house. It is readily accessible from the lowest living level, and encloses the utility room and storage area.

The bi-level house. Built with all living levels completely or partially above grade, the bi-level house, **Figure 7,** may be identified as a two-story house without a basement, or a raised one-story house with a finished

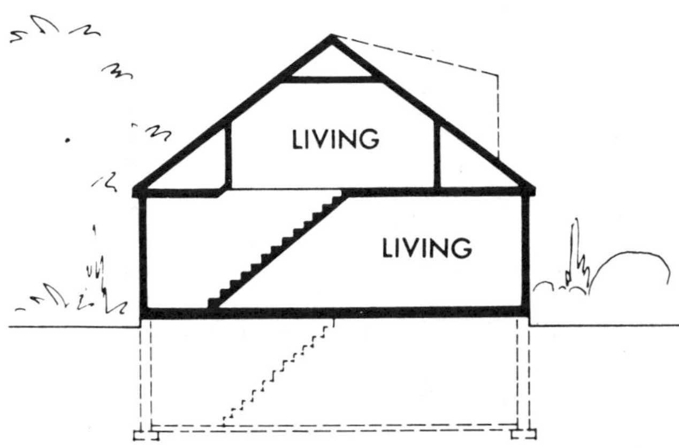

Fig. 3. The Cape Cod, or one-and-one-half-story house.

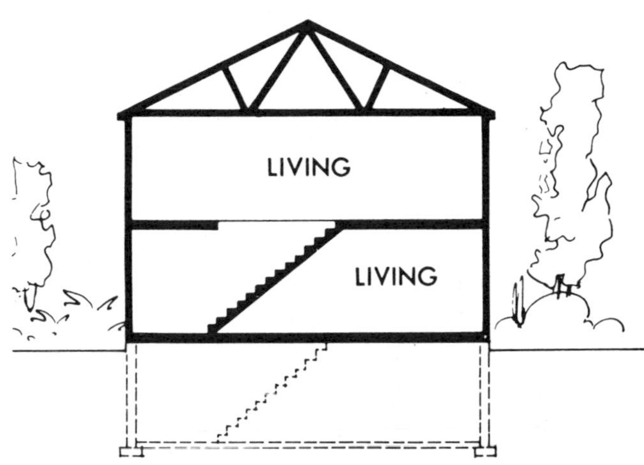

Fig. 4. The two-story house is rich in tradition.

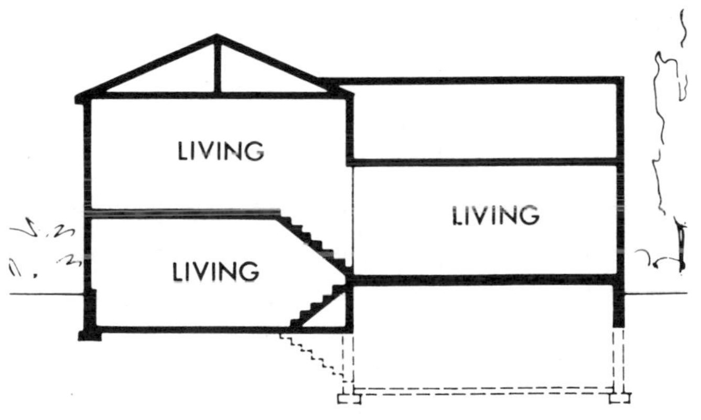

Fig. 5. The split-level house has three living levels.

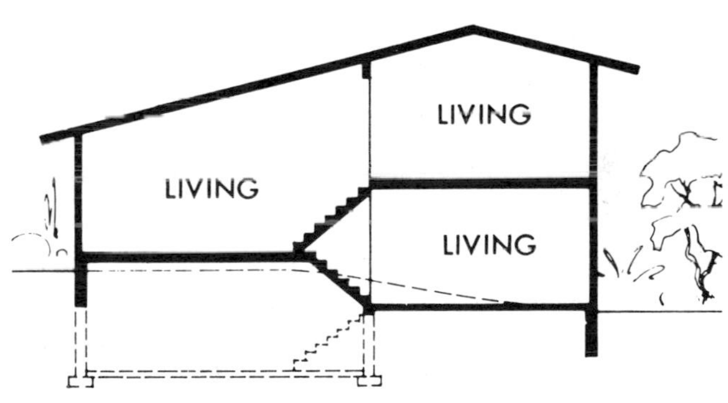

Fig. 6. A sloping roof built over two floor levels.

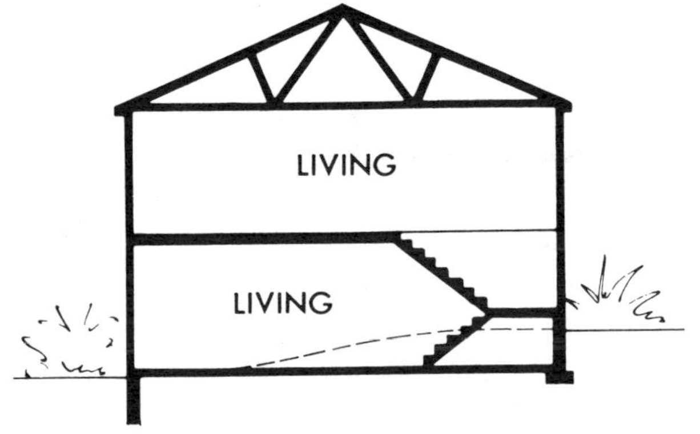

Fig. 7. A bi-level house without a basement.

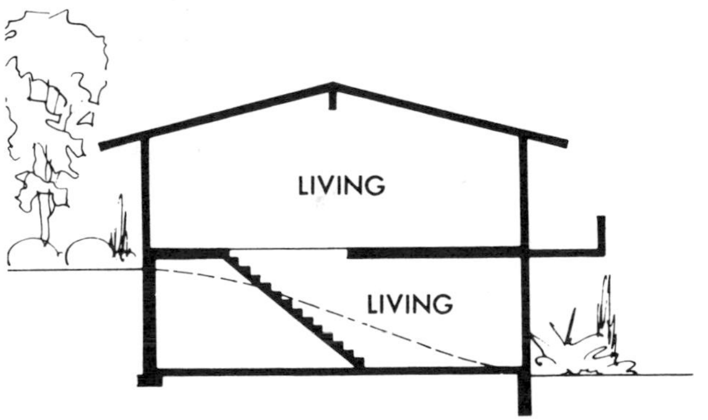

Fig. 8. A bi-level house on a sloping building site.

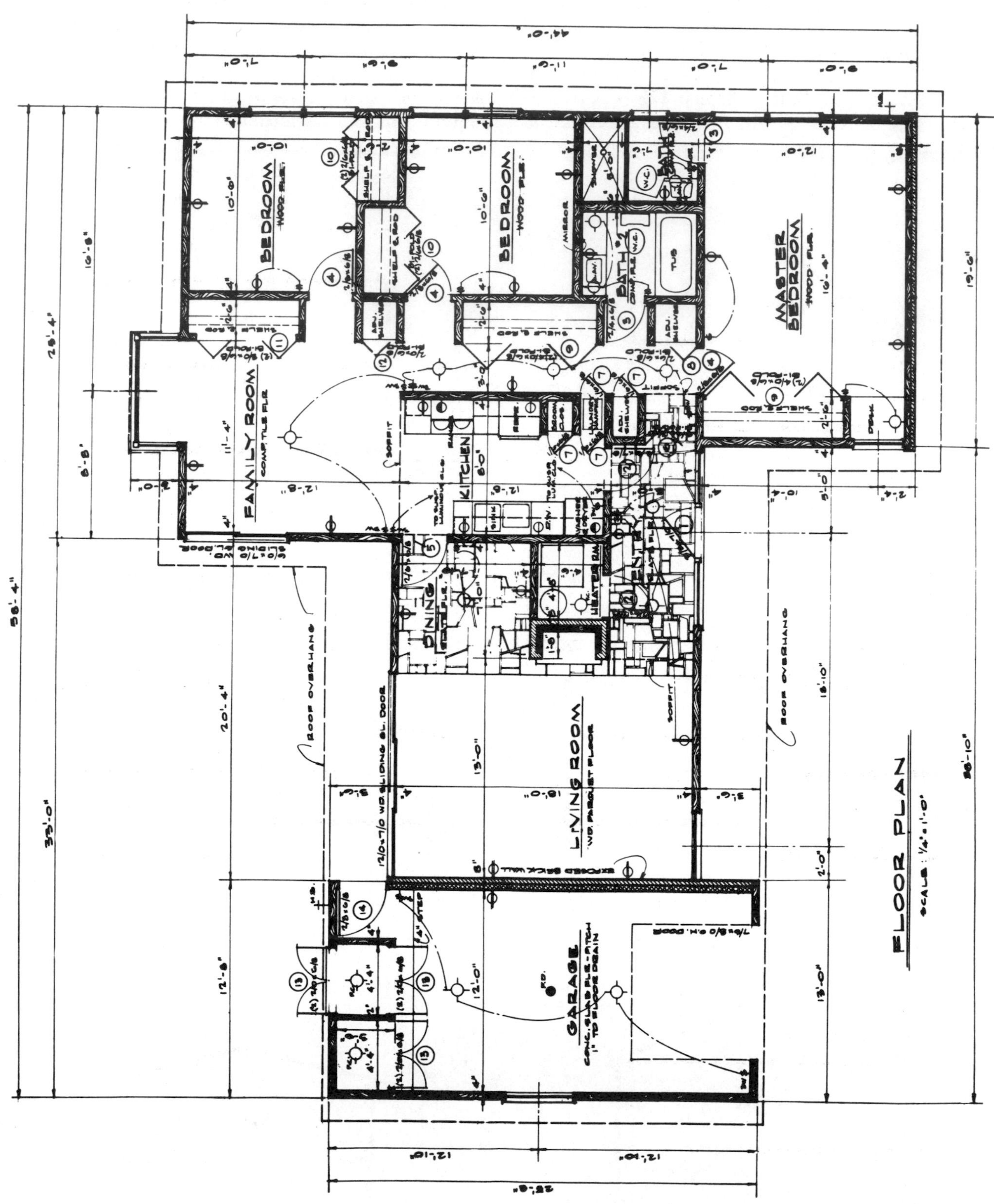

Fig. 9. Floor plan of a one-story house shows good room arrangement, efficient distribution of traffic.

lower level. The split-level entry foyer between the two living levels is a predominating characteristic of this design. The functional arrangement of living areas varies widely in bi-levels. Some have complete living facilities on the upper level, with flexible use of space on the lower level; others combine living and sleeping areas on both levels. The garage may also be a part of the lower level. Roof framing is similar to that used in the two-story house shown in **Figure 4.**

The design of dwellings for hillside plots must be extremely flexible; the house must be properly oriented to the sloping terrain. Note the full-story stairs between living levels in **Figure 8.** The main entrance may be on either the upper or lower grade as determined by the slope of the building site relative to the street, and the identical house may be placed on an uphill or downhill slope with no changes other than the placement of the main entrance. It is, in effect, a reversible house design. The external wall exposure usually determines the area of living space available in the lower level. The remaining space may be used for utilities, storage, hobbies, or other family needs.

House Plans

A one-story floor plan. The arrangement of rooms in any basic house type is a challenge to the owner, the architect and the builder. Individual family requirements and tastes vary widely, and areas of activity must be separated from sleeping areas.

There are more size, shape and design variations in one-story houses than in any other type. A study of the floor plan, **Figure 9,** will reveal some important features of good room layout. Access to the house, either from the main entrance in front, or at the rear, is direct to all parts of the house. An entry foyer distributes traffic, eliminates stepping directly into the living room, and does away with the annoyance of cross traffic and congestion.

The living room is not used as a corridor at the expense of carpeting and furnishings, but provides the privacy for which it was intended. The always busy work center of the house—the kitchen, laundry and family room—is accessible from the patio and main entrance without walking through other rooms. Noisy areas are

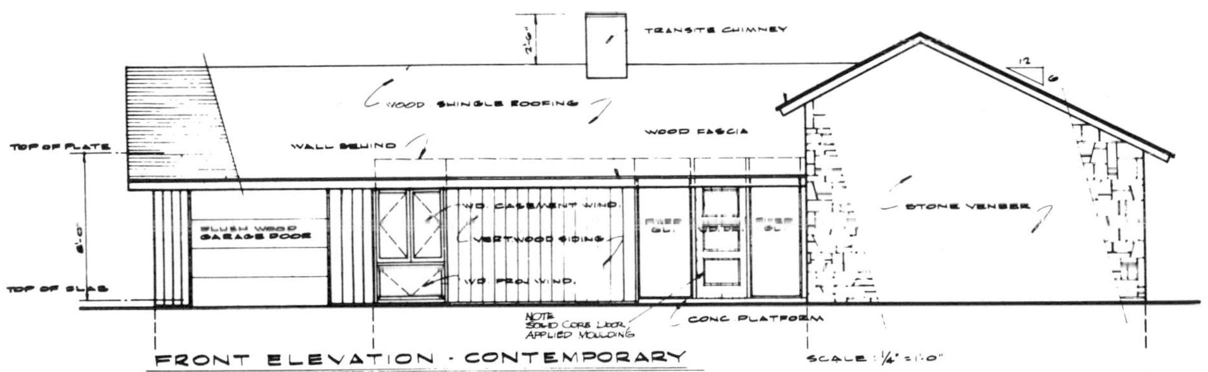

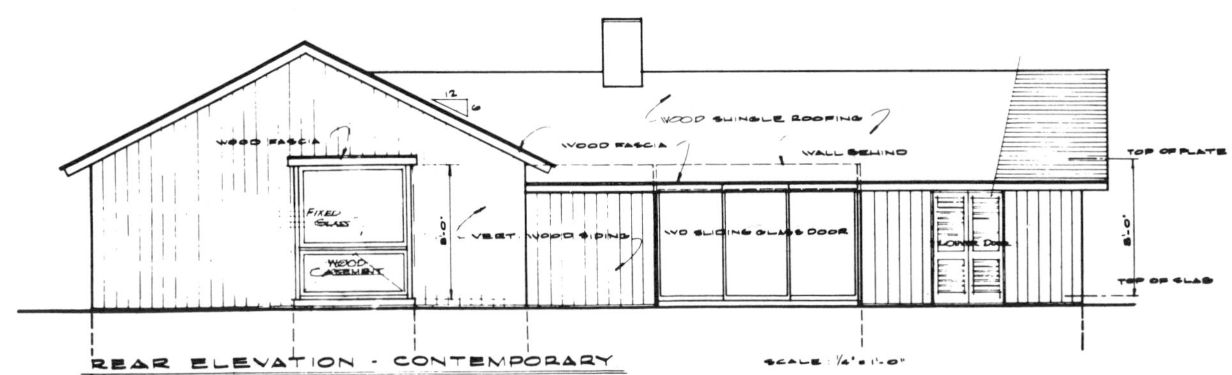

Fig. 10. Attached garage adds length to front and rear elevations of one-story house.

isolated from quiet areas, and children's activities are separated from parent's activities. The master bedroom suite is assured privacy by its remoteness from the family room-kitchen area and the outdoor living space.

The outdoor living area is actually an extension of the family room and living room, providing ample space for leisure activities and entertaining. There is adequate storage and closet space. The rooms are quite generous in size and include a separate dining room, two bathrooms, an attractive entrance hall, and a fireplace featuring crafted woodwork in the living room. The arrangement of the rooms is not complicated, is easy for the builder to frame, and economical to build. The overall area is 1,460 square feet.

The features of this house project an image of modern residential design, while maintaining a link with the past. There is a tasteful combination of building materials, well-placed large glass areas, an interesting intersection of roofs, and careful consideration of mass and proportion. In placing the garage slightly forward of the house, the box-like appearance of a "development house" has been avoided in favor of an interesting array of planes, highlighted by the expanse of stone on the exterior wall of the master bedroom. The selection of building materials avoids stereotype architecture, and provides texture, durability and ease of maintenance.

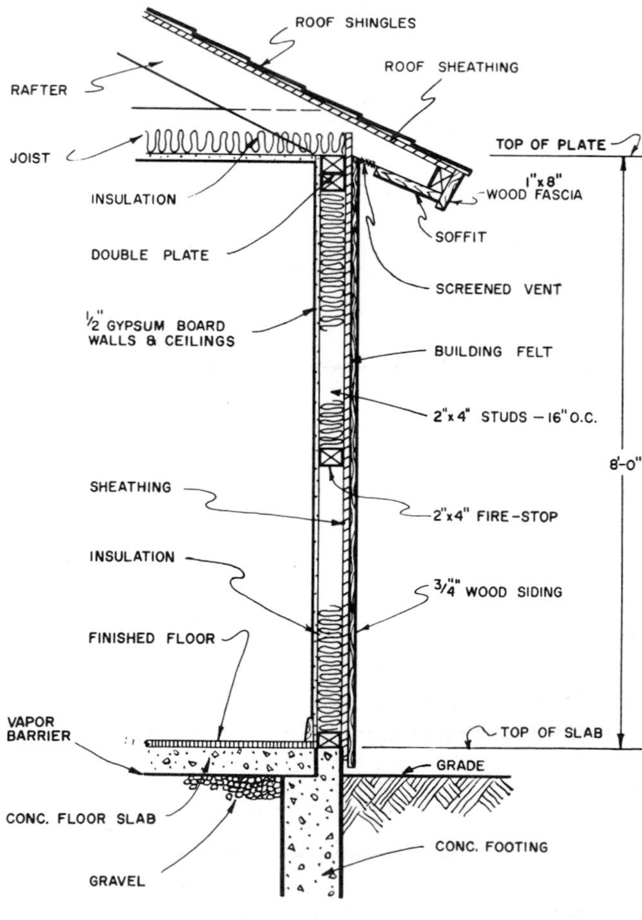

Fig. 11. Sectional view of typical exterior wall construction for one-story house erected on concrete slab-on-grade.

Front and rear elevations for this floor plan are illustrated in **Figure 10.** Walls are surfaced with vertical wood siding, with the exception of the stone veneer laid up on the bedroom wall. The garage is built as an integral part of the dwelling, extending out about a quarter of its length beyond the front entrance. The garage door is made up of four hinged panels, which open overhead.

Exterior wall construction is detailed in the sectional view, **Figure 11.** The entire house is built on a concrete slab. Mineral wool or fiberglass insulation is used in the ceiling as well as in the exterior walls. The exterior wall covering specified is vertical wood siding. Other wall coverings which might be used to finish the exterior include brick or stone veneer, horizontal or vertical aluminum siding in various patterns, horizontal beveled wood siding, steel siding, vinyl lap siding and other materials described in the chapter on exterior wall covering.

Interior wall details are shown in **Figures 12 through 18.** The entire wall of the living room opposite the fireplace wall is faced with brick veneer to add beauty and comfort while separating the living room from the garage as shown in **Figure 9.**

The work center or kitchen walls are adjacent to the fireplace and furnace room, and include the laundry equipment, so that one common chimney can channel out all vapors and fumes of combustion. The sink and dishwasher are on the same wall and spacious storage cabinets are hung over these units. On the opposite wall of the kitchen are placed a laundry hamper, a broom closet, refrigerator and range flanked by additional cabinets at floor level and overhead.

Both baths have ceramic tile surrounding the tub and shower. There are large mirrors above wash basins, and ample closets for linens. The master bath features a large mirrored door.

Walls can be lathed and plastered or surfaced with gypsum panels and given a final finish of paint in pleasing color combinations. Various widths of wood paneling or large panels of veneer plywood can be used to add warmth to walls and ceilings. Structural beams may be stained and left exposed, or dummy beams built up of 1″ lumber can be installed on ceilings to add character to rooms.

Placing the House On the Site

Year 'round comfort requires that a house be properly oriented on the building site to reduce air conditioning and heating loads. Since ideal orientation can seldom be attained in subdivision planning, it is necessary to minimize heat gain by shading the house from the sun. The arrows in **Figure 19** point to the recommended wall treatment and sunshading in the quadrant faced by each side of the house. These suggestions will add to summer comfort even if the house is not equipped with cooling.

In the quadrant 'N' or North, the sun exposure is very brief, if any. Since little sun shading is needed, roof overhang can be kept to a minimum.

In the quadrant 'E' or East, the sun exposure is at a low level during morning hours and is intense, making it necessary to shade the house with tall and medium height trees.

In the quadrant 'S' or South, exposure is overhead and intensified by the day's hottest air. Protection from this exposure should be maximum as indicated, including the construction of an attached carport or garage on the South side of the house.

In the quadrant 'W' or West, the late afternoon exposure to the sun is short, but intensified by its low angle and heated air. Tall trees or a porch are listed among suggestions for shading this side of the house.

Differences in latitude affect the width of the roof overhang needed to minimize heat gain. The nearer to the equator, the greater the overhang required, especially on walls facing South. A house having a roof overhang designed for one arca of the United States will not be effective in another. The southern tip of Florida and Texas are but 25 degrees latitude above the equator.

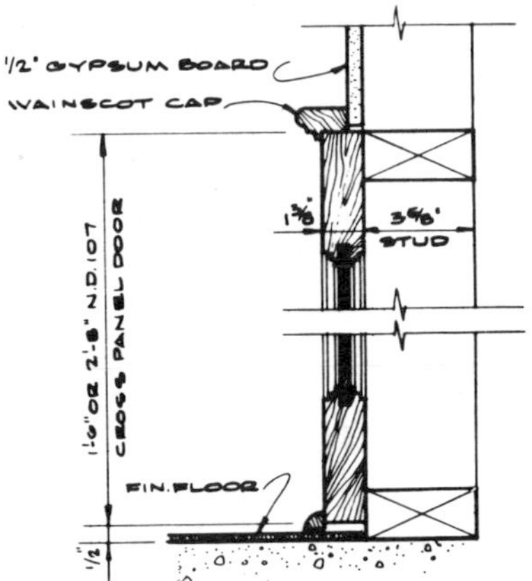

Fig. 12. Sectional view of wainscot paneling.

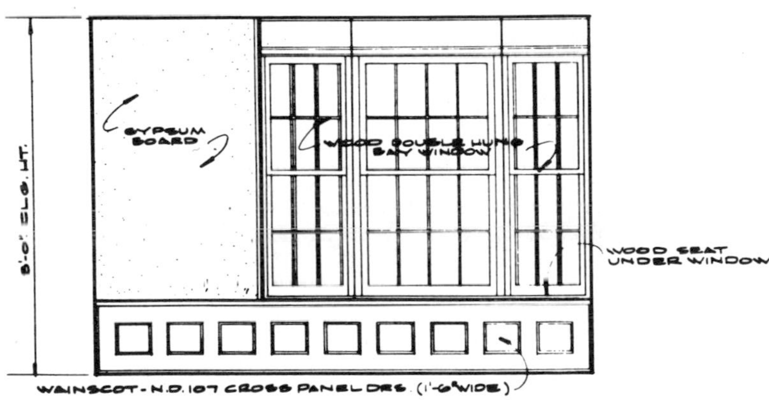

Fig. 13. Wall treatment around bay window in family room.

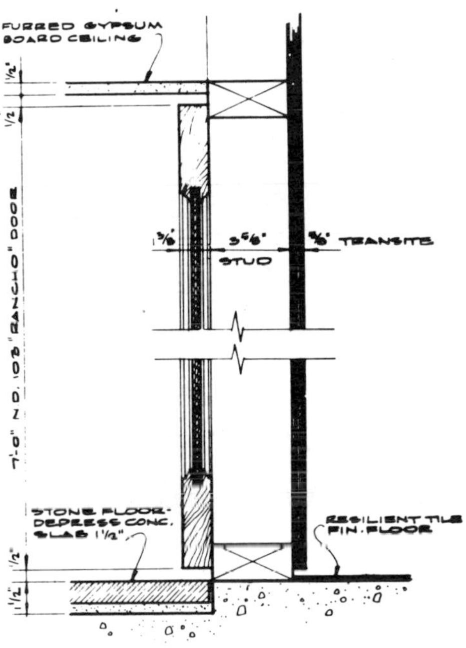

Fig. 14. Sectional view of the entry paneling on the rear wall of the living room shown in the floor plan.

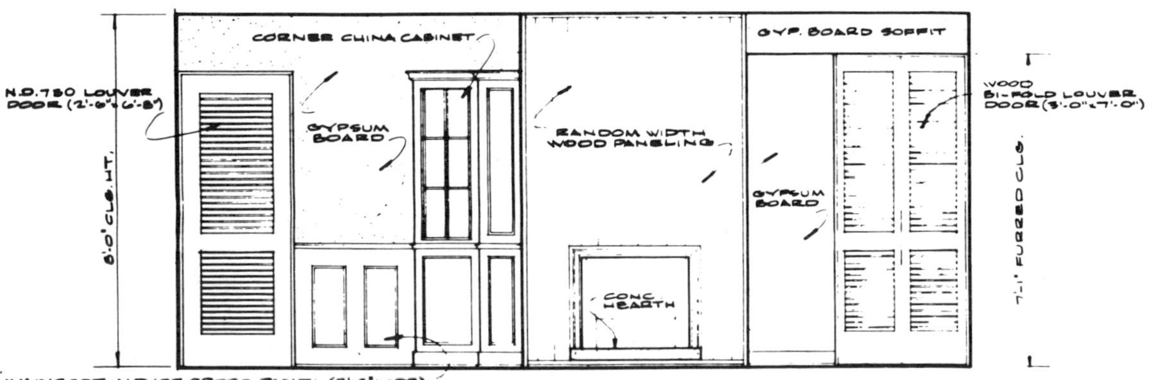

Fig. 15. Louver doors, wainscoted wall, china cabinet and fireplace add interest to living room interior wall.

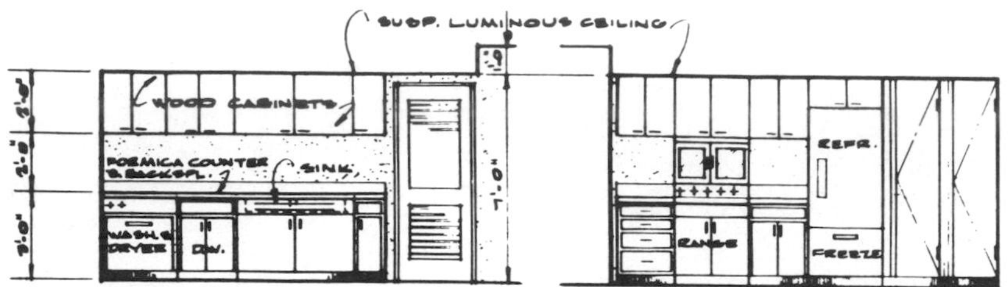

Fig. 16. Kitchen in the one-story house has washing machine, dishwasher and sink grouped on one wall, and range, oven and refrigerator on opposite wall. Overhead storage cabinets are hung on both walls.

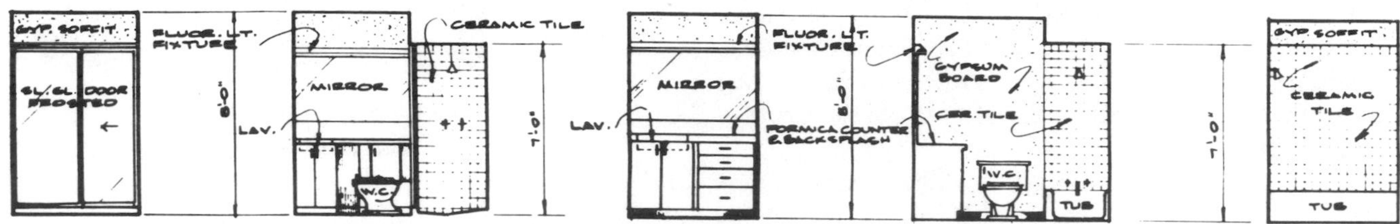

Fig. 17. Master bathroom (left) has stall shower, second bath (right) includes tub and shower combination.

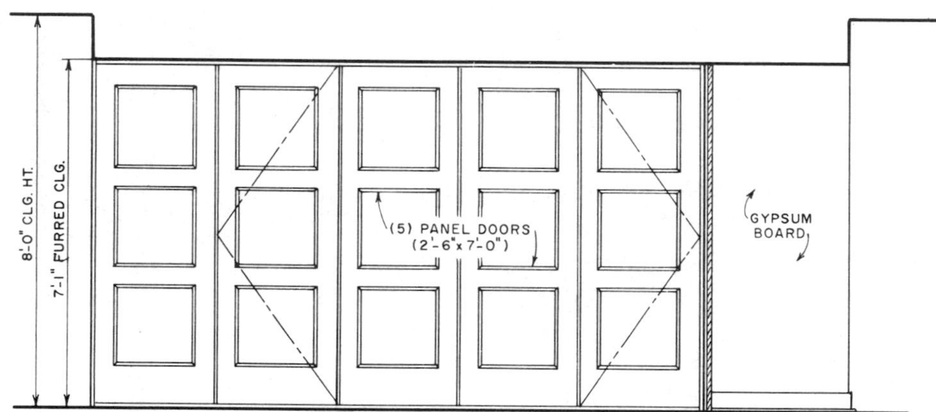

Fig. 18. Suggested use of five panel doors in a wide entry instead of sliding glass doors.

Further North at 30 degrees latitude is New Orleans, La. Charlotte, N.C., and Santa Maria, Calif., are at 35 degrees. Philadelphia, Pa., and Denver, Colo., lie at 40 degrees, and Bangor, Me., and Portland, Ore., are 45 degrees from the equator.

Less protection from the heat of the sun is required for houses built in northern parts of the country, but such houses must be better insulated due to low temperatures during the winter months.

Popular One-Family Dwellings

Three styles account for most single-family construction in the United States: the ranch or rambler, the story-and-a-half house, and the split-level. Each type has several varieties, and regional differences appear within each style. Variations are expressed through the shape and material of the roof, the placement of windows, and features such as carports, garages, porches, verandas, and the covering used on exterior walls.

The ranch style may reproduce the sprawling, informal, high ceiling structures of the open West, or it may be a single level house that once was known as a bungalow. This one-family dwelling may be built over a full or partial basement, a crawl space or a concrete slab foundation. The slab is easily adapted to large scale building, and now most mass developments of small single-level houses are set on slabs, **Figure 20.** This photo shows the front lawn, landscaping and facade of a six-room ranch having exterior walls built with concrete block. The compact design is well balanced by the bedroom ell at the left and the extended garage at the right, with a covered porch entrance between the two.

The floor plan of this house, **Figure 21,** is most economical in its use of floor space and room arrangement. The kitchen is accessible to all rooms. The passage to the bedrooms and bath is used as an access to the utility room. Note the placement of the washer and dryer in the garage at the kitchen wall, convenient to sink plumbing.

The slab foundation plan, **Figure 22,** shows the loca-

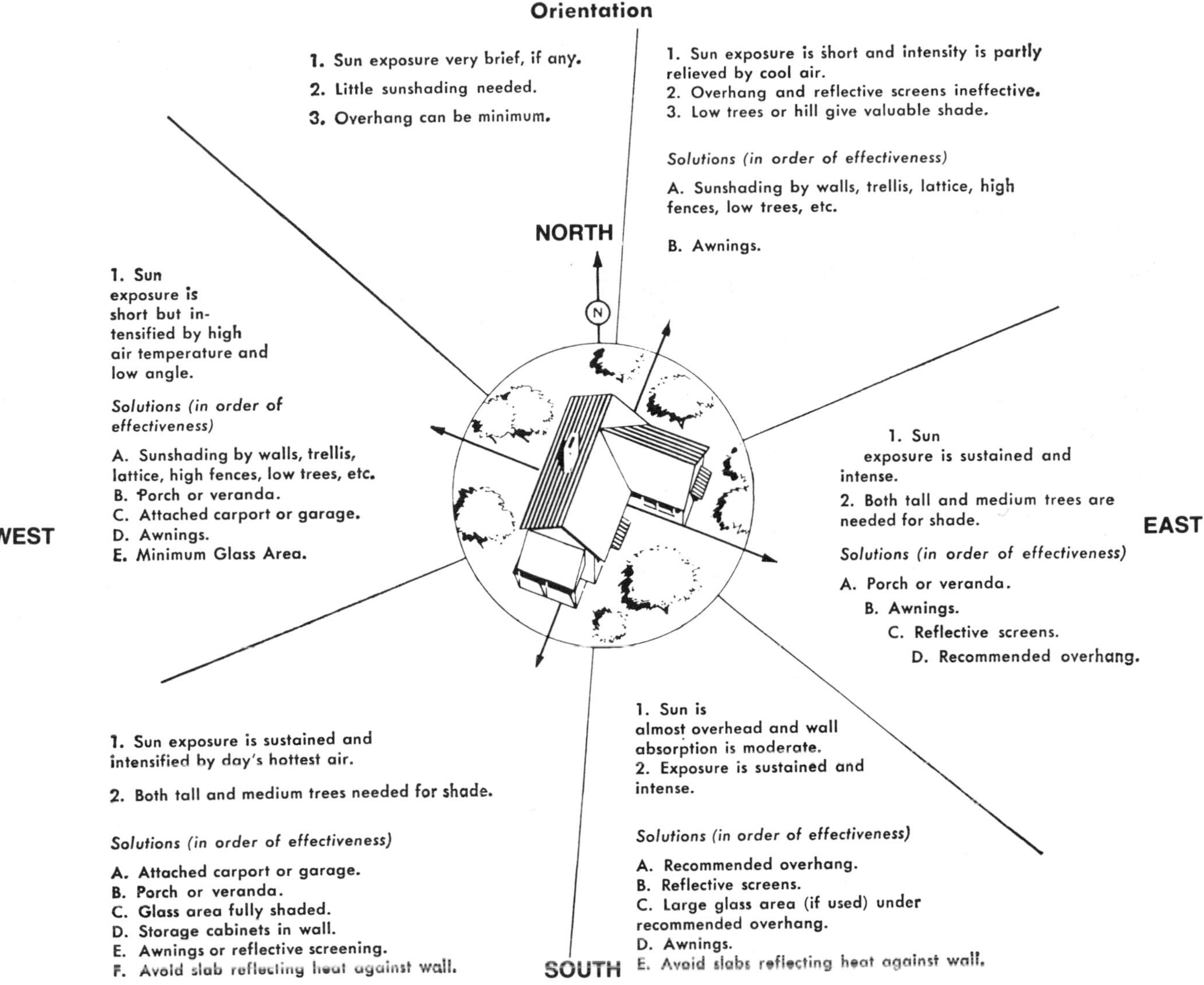

Fig. 19. Orientation of dwelling on a building site, showing solutions to sun exposure on different quadrants.

tion of the heating ducts running from the heating plant in the utility room.

In wood construction, the simple framing of the one-story rambler is handled easily in prefabrication factories, where panelization of parts has been developed to a high degree of craftsmanship and accuracy. When designed with trussed rafters, this style of house can benefit from the popular open planning, in which living, dining, kitchen, and family alcove areas are treated as part of one connected space. Since structural partitions are not needed except for privacy, area functions can be arranged by furniture placement, room dividers, a projecting fireplace, or by folding partitions. Finish flooring can be put down over the entire floor area before framing partitions, reducing labor and minimizing wasted materials.

The front, rear, right and left elevations of the rambler are illustrated in **Figure 23A**. Eight additional construction details are given in **Figure 25**.

Alternate designs of the ranch permit the open or cathedral ceiling with exposed rafters and ridge, eliminating the attic by following the contour of the roof. These designs are marked by long lines, large expanses of glass in the rear wall, **Figure 24,** and a low pitched roof with a wide overhang at the eaves. Note the air conditioning unit mounted on the roof, and the inviting patio accessible through sliding glass doors of the living room and the rear door of the garage.

The split-level design owes much of its popularity to the increasing shortage of level building sites in some sections of the country. The split-level is particularly adapted to sloping building sites, affords a great deal of living space for the plot size, and provides a satisfactory separation of functions at the various floor levels.

One of the first arrangements of rooms in the split-level house was the placement of the living room, dining room and kitchen at the ground level in one section. Adjoining this at a lower level and partially below grade,

Fig. 20. Compact single-level design is made interesting by bedroom ell (left) and garage, with covered porch between.

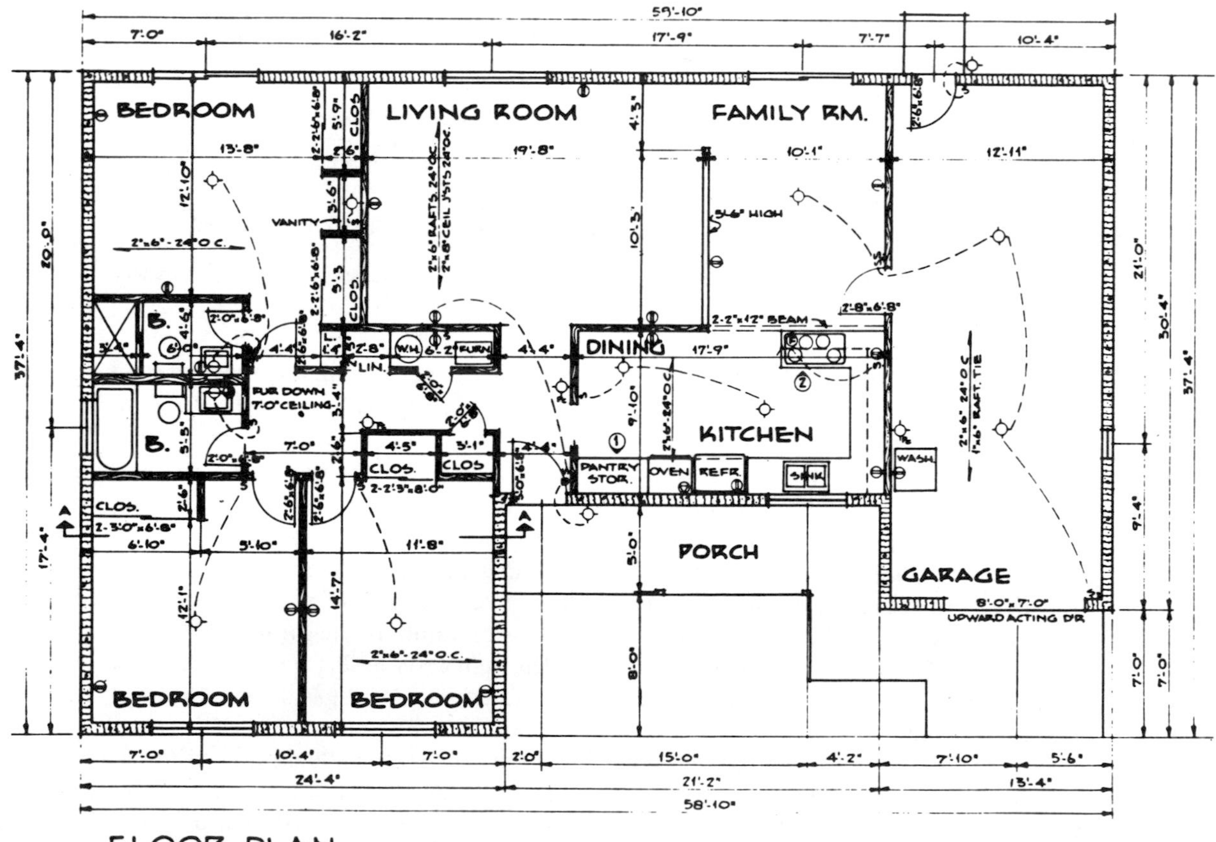

Fig. 21. Floor plan demonstrates economical use of space. From the kitchen one can see through the family room sliding glass doors to the rear patio, as well as to the front of the house. From the family room the eye can travel through the kitchen pass-through and kitchen window, as well as into the living room. Short hall leads to bedroom wing. Oversize garage provides for laundry at side and workshop at rear.

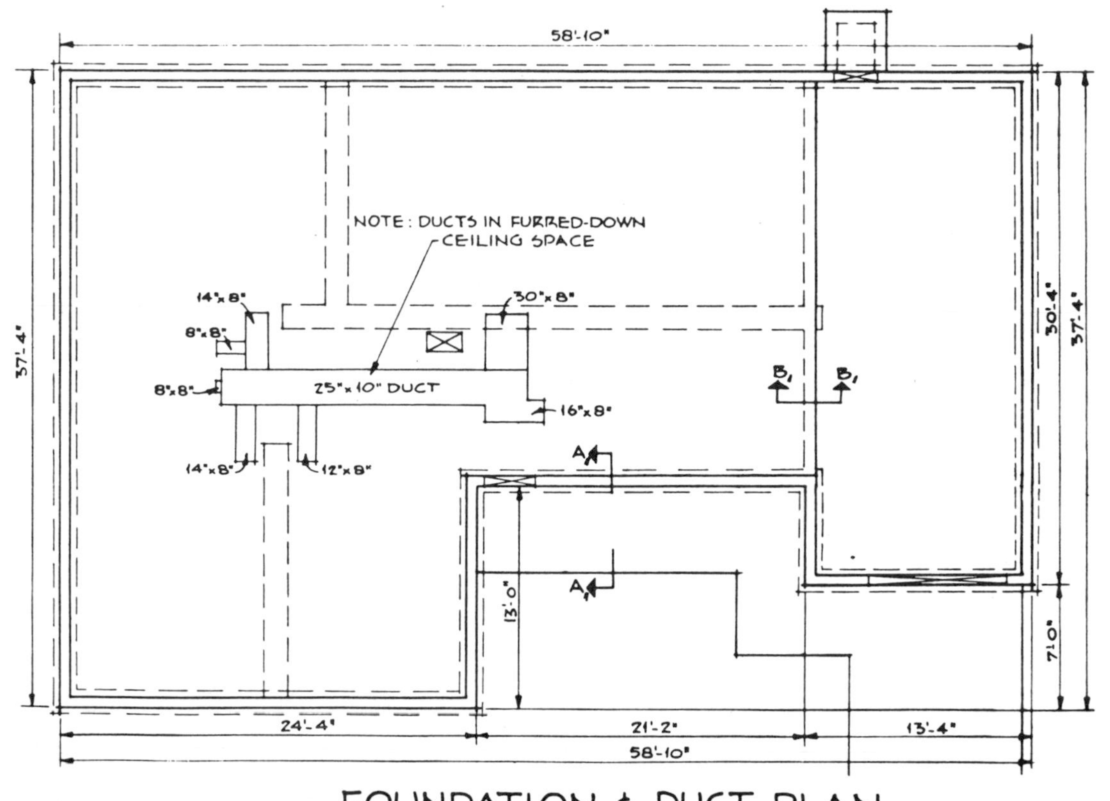

Fig. 22. Slab foundation plan shows location of heating ducts for compact single-level house.

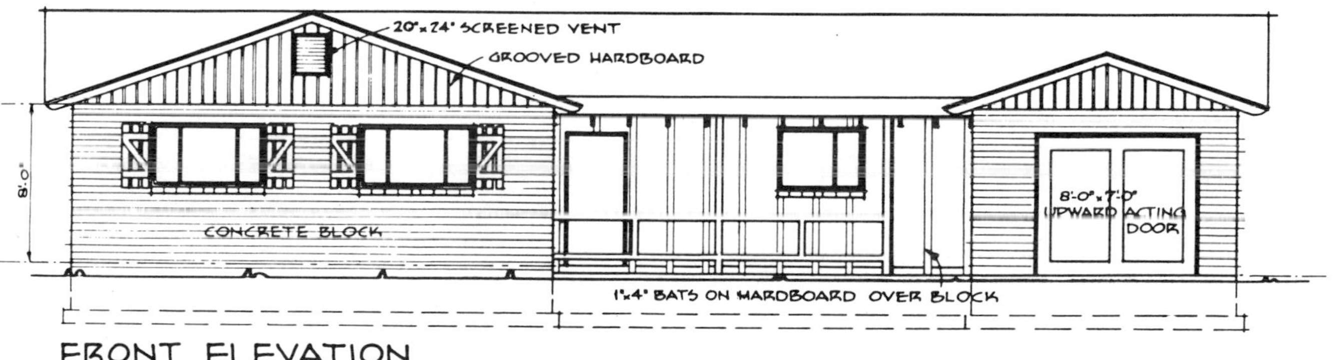

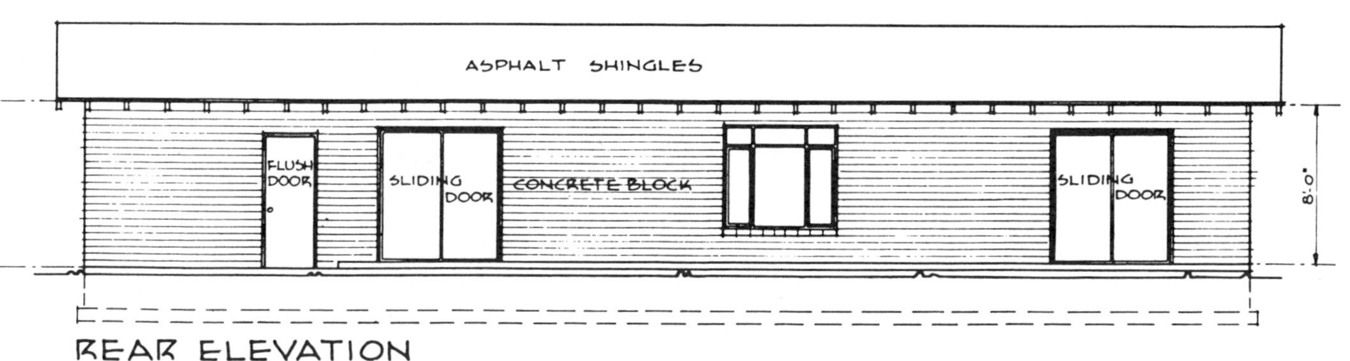

Fig. 23. Front and rear elevations illustrate simplicity of design and use of exterior materials for economy.

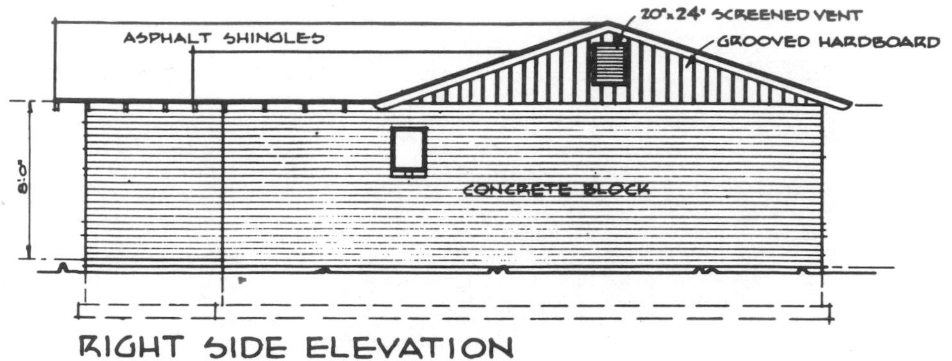

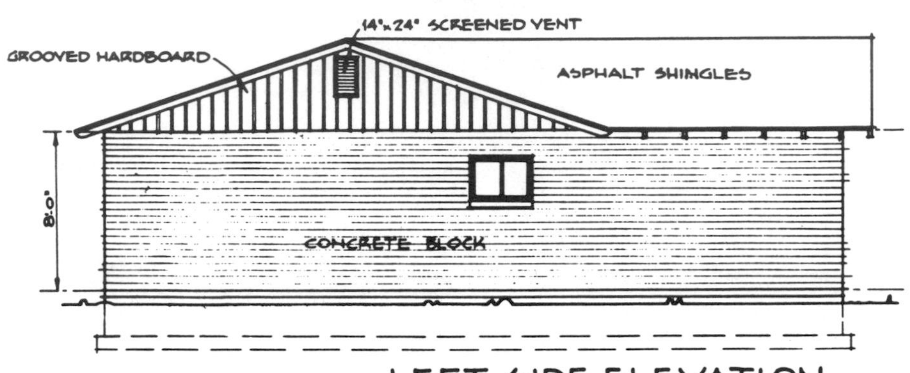

Fig. 23A. Right and left elevations of compact single-level house show interesting roof line from either end.

Fig. 24. Rear view shows garage entry (left), and sliding glass doors to family room and master bedroom. Roof-mounted air-conditioning compressor distributes cool air through ducts running across attic floor.

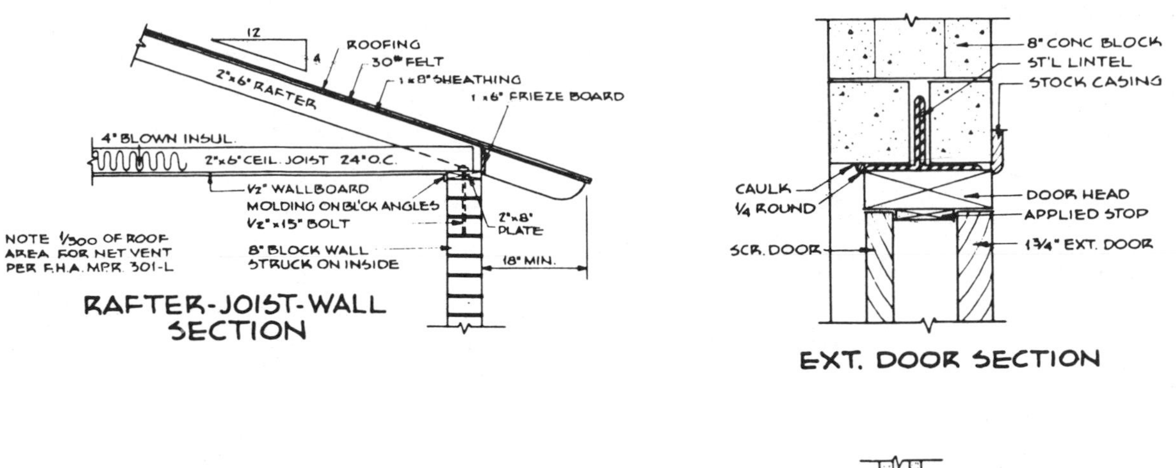

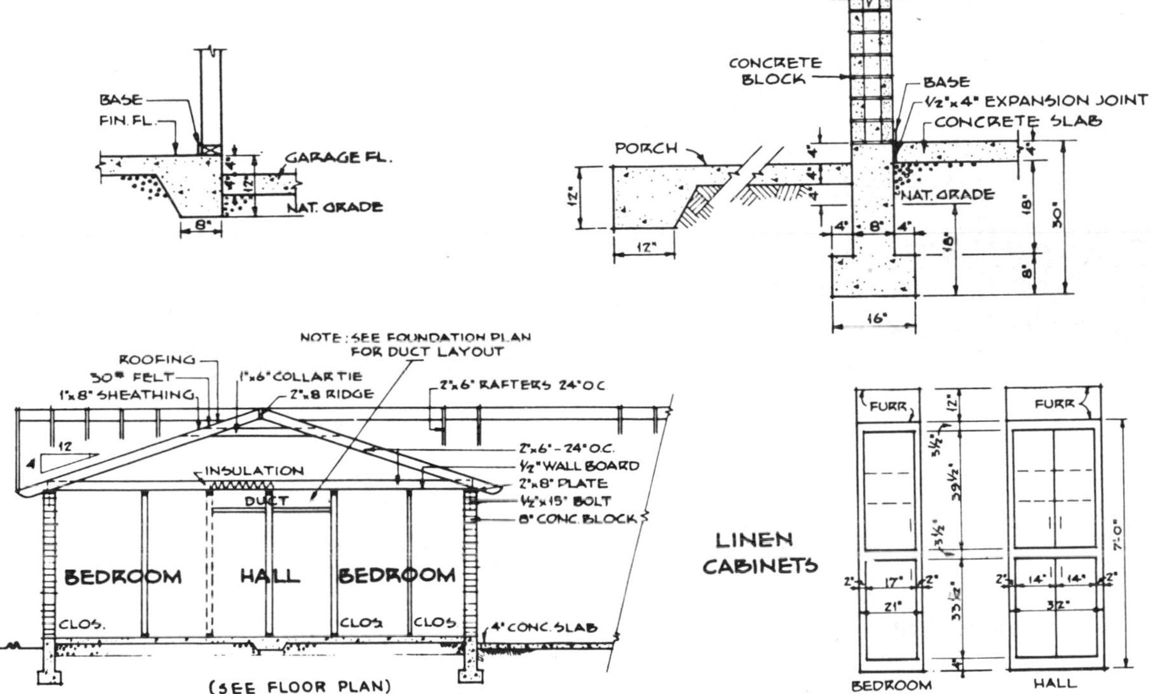

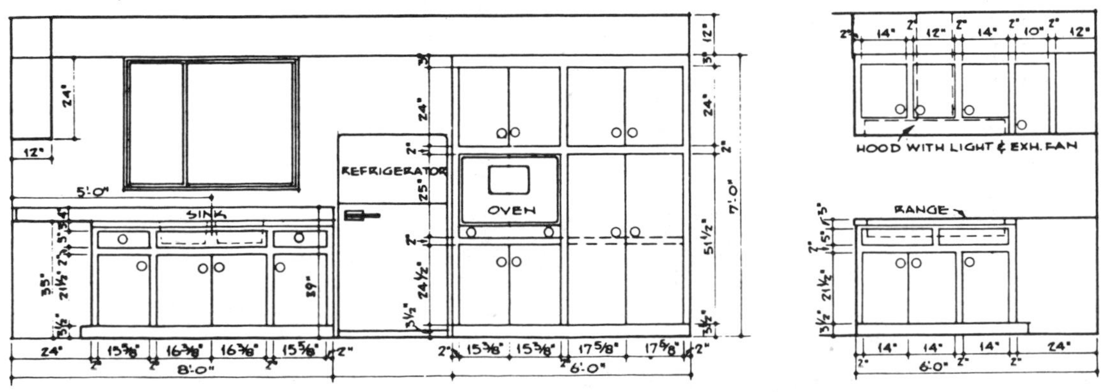

Fig. 25. Miscellaneous construction details used in the compact single-level house on preceding pages.

Fig. 26. A typical side-to-side split-level set properly on slightly sloping plot.

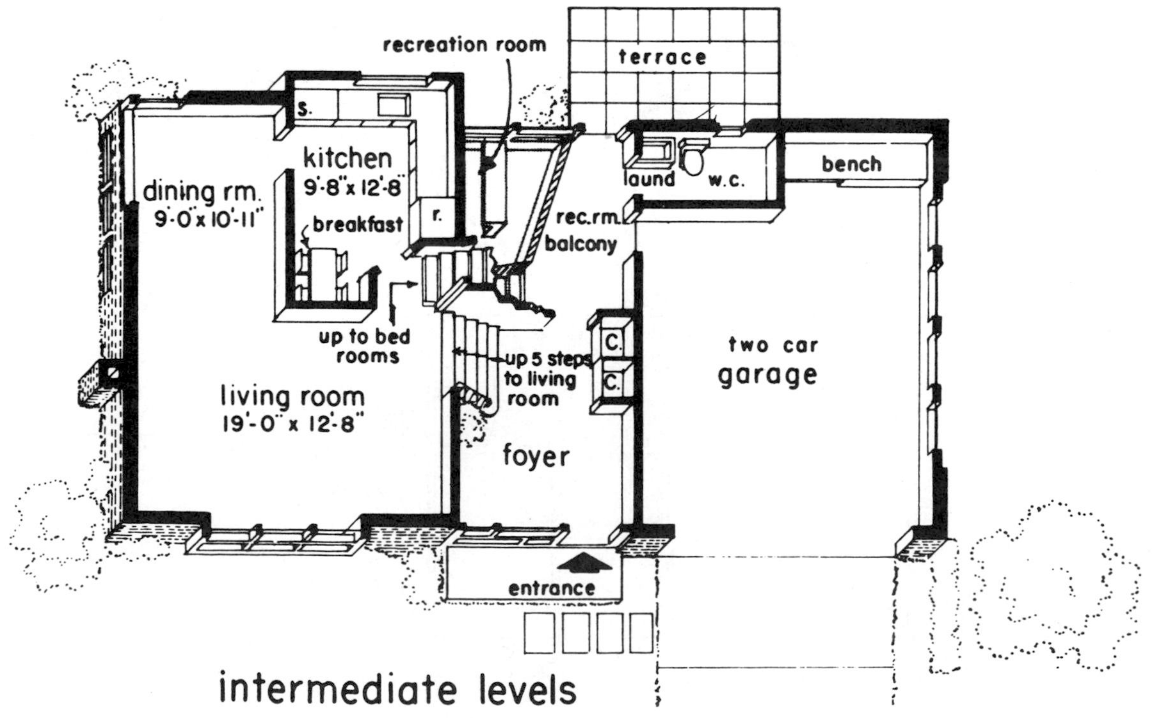

Fig. 27. Intermediate level shows half-flight of steps leading to upper and lower levels.

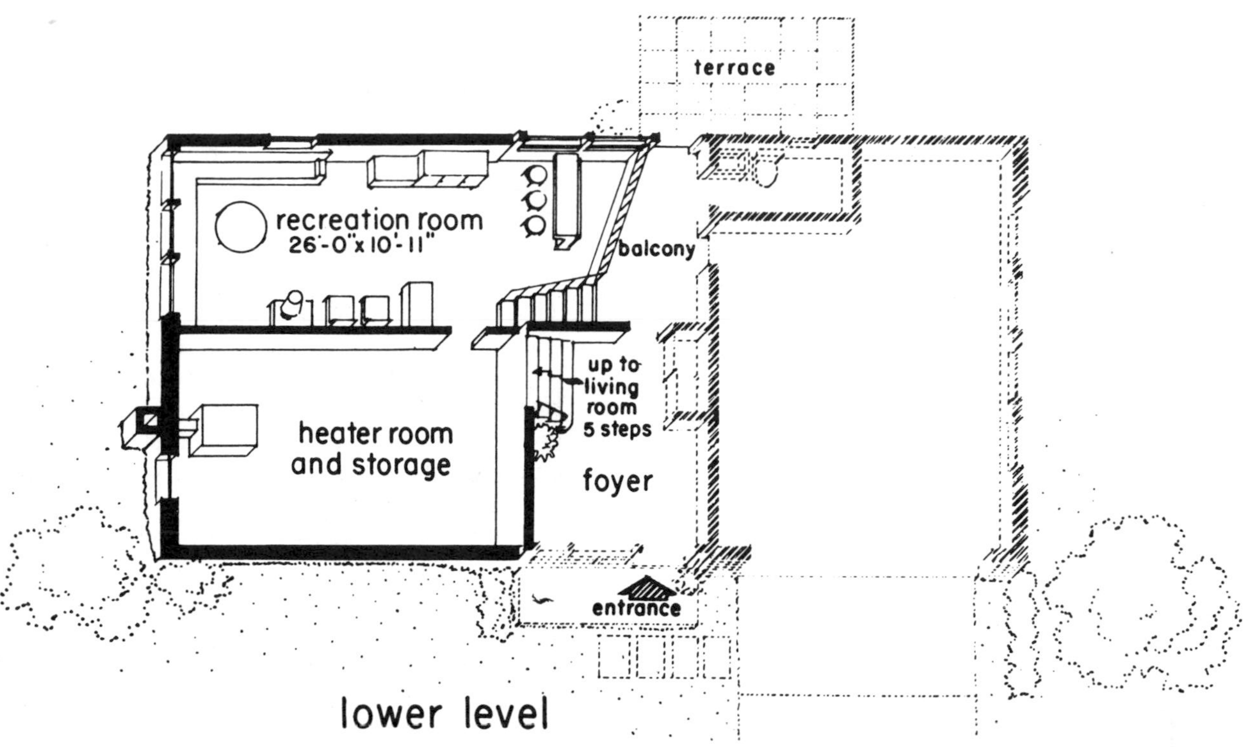

Fig. 28. Lower level beneath living room includes recreation room and storage-utility area.

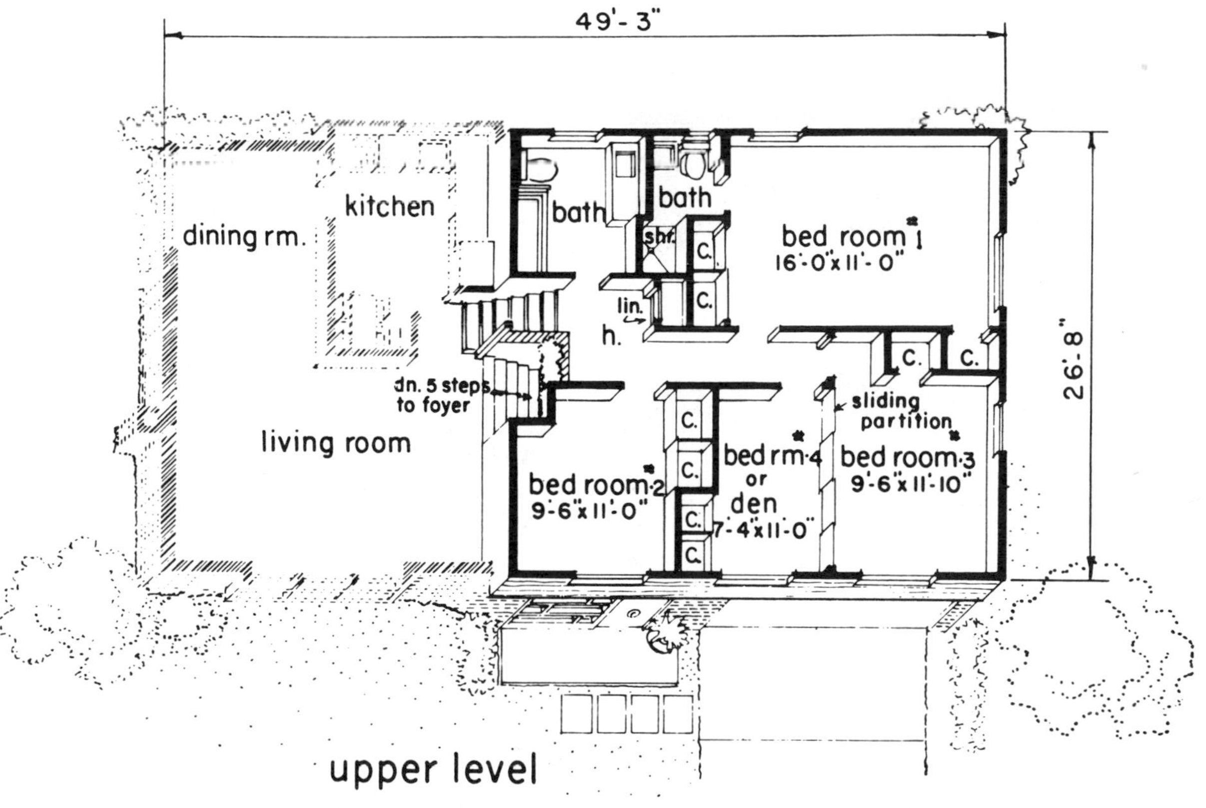

Fig. 29. Bedrooms and baths are on upper level over garage. Shaded areas indicate adjacent levels.

Fig. 30. A side-to-side split-level with a sloping roof over both upper levels to simplify framing.

Fig. 31. A simple Cape Cod style house with full basement and false gable above the entrance.

Fig. 32. Colonial style house for a large family has four bedrooms and two baths on the second floor.

Fig. 33. Contemporary split-level has a double shed roof draining toward leaders at the center of the house.

was the garage, utility room and a main foyer. Three or four bedrooms were placed on an upper level. This arrangement of rooms is commonly known as the side-to-side split, **Figures 26, 27, 28, 29.** It is an ideal dwelling for a large family.

A more attractive arrangement is to place the garage on the ground level next to the living room-kitchen section, putting a large activity or family room in the lower section. Sometimes a four-level, side-to-side split is achieved by placing the family room in an unexcavated section under the living-room-kitchen area. This adapts well on a sharply sloped lot, so that the family room faces out on a patio at the rear.

Usually the two sections of the side-to-side split-level house are framed under distinct roof lines, with the pitch of the higher section running from front to rear, and the pitch of the lower section from side to side. This often creates a problem in design and requires skill in integrating the two sections in an artistic manner to avoid the appearance of two separate houses joined together to make one dwelling.

A contemporary approach to the split-level design is the use of a single roof ridge, running from front to rear, with one long roof section sweeping over the intermediate area, **Figure 30.** This single roof simplifies framing and roof construction.

The story-and-a-half design makes use of the conventional rafter and ridge construction for the upper floor level. The Cape Cod house is the traditional form of this design, having the entrance centered in the front of the house and a central stairway leading to the second floor. Sometimes dormers are constructed in the front or rear of the roof to provide more windows for the bedrooms and gain additional floor space on the upper level. The simple Cape Cod design is often modified by an attached garage with an extra room built above it, or by a false gable over the entrance, **Figure 31.**

Dwelling for Large Families

The two-story house is coming back into favor for large families. The traditional appearance of this design is the brick-faced or wood shingled Colonial, **Figure 32,** having a central entrance and stairway, and a full basement. The classical, two-story Colonial has four bedrooms upstairs. Small panes of glass in windows, false shutters, and paneled wood entry door are typical details.

An interesting contemporary house for the large family is shown in **Figure 33.** This design has the butterfly or double-shed roof leading off from a long clerestory window, which usually runs the length of the living room.

In an effort to escape from the uniformity of mass design that at times has caused monotony in building developments, builders have learned to vary elevations, even though there may be but slight change in the basic floor plans. Often plans are simply reversed and window details altered. Variations can also be made in roof shapes, exterior wall covering and colors, and the addition of fences, patios and planters can create visual differences from house to house.

Chapter 2

EXCAVATIONS/FOUNDATION FORMS/FOUNDATIONS

An excavation is dug in the building site by earth moving equipment for the construction of a foundation to properly support the superstructure of a dwelling. The type and depth of the excavation is determined by the design of the house foundation and the nature of the soil.

If a full basement is specified for the house, the excavation must be of proper width, length and depth for the foundation wall footing, much of the foundation wall, column and chimney stack footings, the basement floor and an adequate drainage system.

An analysis of the soils making up the building site is of great help in determining the placing of the dwelling. Soils vary in color, from red in Hawaii, for example, to black in North Dakota. All soils have some things in common, such as mineral and organic matter, water and air. Various soils have width, length and depth. A specific area of soil has a succession of layers, known as profile, when viewed vertically from the grade downward into the loose weathered rock formation. A profile consists of at least two or more layers of earth lying one below the other and roughly parallel to the land surface, **Figure 1.** The layers of soil of a given area, the profile, or sectional view, are known as horizons. Horizons of soils differ widely in color, texture, structure, porosity, firmness and reaction.

The uppermost or surface layer of soil, 'A', is most abundant in plant roots, bacteria, fungi, and small animal life. Water, in the form of rain or melting snow, reaches this horizon first. The layer 'B', is the horizon known as the sub-soil which is harder when dry, yet stickier when wet, than other layers. The third layer, 'C', is the horizon of parent materials of soil and is the deepest of the three major horizons.

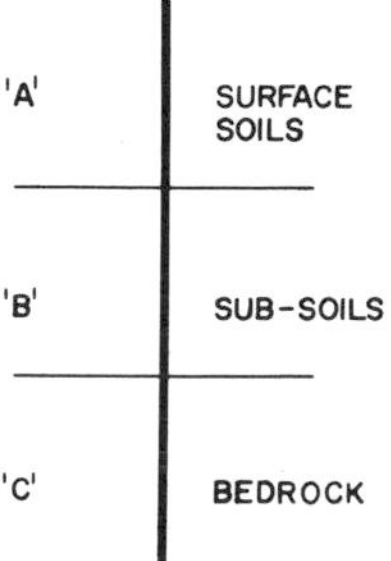

Fig. 1. Diagram showing successive layers of soil or profiles, one below the other and parallel to the surface.

An Examination of the Soil. If a building site or lot can be inspected by the prospective owner, builder or architect before any plans are drawn, or contracts to build are let, much can be learned. If a building site is relatively level an inspection of the soil can be made by boring holes with an earth auger. Sometimes an horizon of solid stone lies near the surface or an horizon of extremely porous or wet soil is found at low depths.

If the building site has a distinct slope it will often determine the style of dwelling to build. If the properties adjacent to the site have structures already built, much can be learned if questions can be asked about any unusual problems of building. An inspection of the surface drainage of water should be made on the site and on land above it if the terrain slopes.

If the soil of the site is to be used for sanitary sewage drainage, contact should be made, if possible, with owners or builders of dwellings in the immediate neighborhood. Often utilities such as electricity, gas and water are available but sanitary sewage is lacking. The absence of sanitary sewers can often alter the placing of the dwelling on the building site to accommodate a septic sewage disposal system.

Types of Foundations

The foundation is that section of a building below the level of beams nearest the grade. It must be strong enough to support the sills and members of the structure resting on it. It usually is made of poured concrete or concrete blocks, but fieldstone or bricks are sometimes used. The foundation must be built on soil firm enough to support the weight of the building and its equipment load.

If the foundation shifts or settles due to unstable soil or heaving caused by frost, the building will be forced out of alignment, and cracks often will appear in plastered walls and masonry.

The full basement foundation, Figure 2, is popular because it provides extra area for living or recreation at minimum cost. It is mostly below grade in one and two-story houses, and includes footings and masonry walls approximately 7′ high.

The crawl space foundation, Figure 3, is similar in construction to the basement foundation, but offers none of the advantages. It requires only a shallow excavation and its 18″ to 24″ high walls provide limited ac-

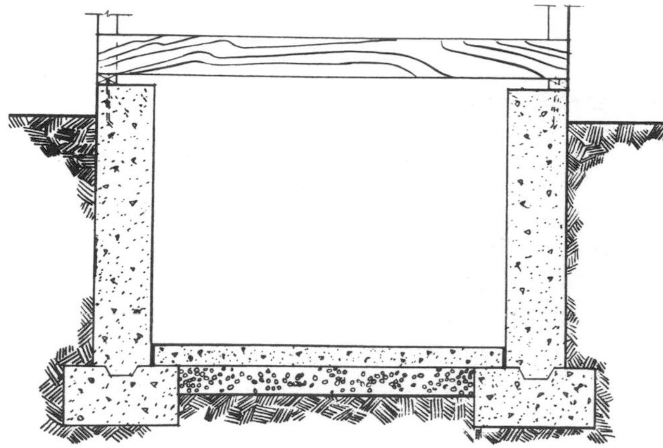

Fig. 2. Cross section of a full basement foundation.

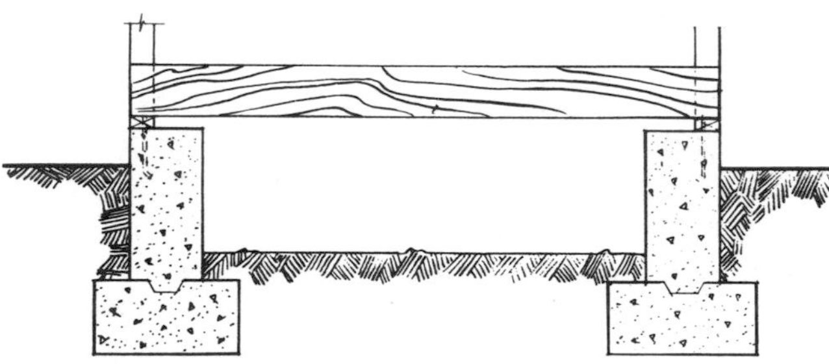

Fig. 3. A crawl space foundation.

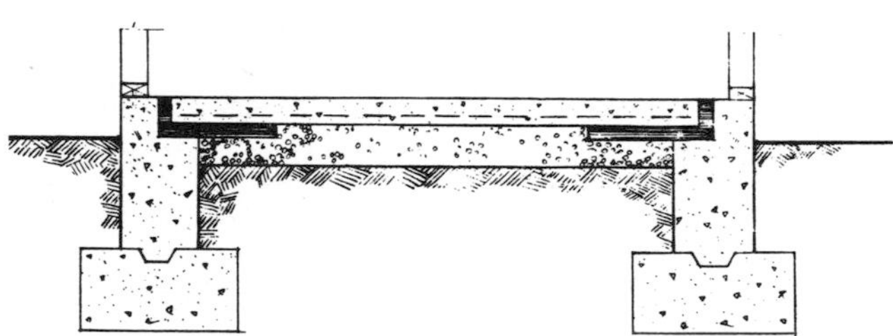

Fig. 4. Slab-on-grade type of house foundation.

cess. This space may, if properly sealed against moisture and insulated, be used as a distribution plenum for warm air heat, eliminating duct work and reducing the cost of heating equipment.

The slab-on-grade foundation, Figure 4, is most economical, since it eliminates framing for the first floor. It is especially suitable for low-cost mass-produced housing and factory-built dwellings.

A combination of these three types of foundations is shown in the sectional view of a concrete block foundation, **Figure 5.** This foundation has a partial basement (foreground), a crawl space foundation between the partial basement and the patio at right, and a slab foundation for the garage floor at rear.

Sixteen important points of foundation construction are numbered as follows: (**1**) The poured concrete slab for the garage floor. (**2**) The water drain in the garage floor connected either to the street sewage drain or a dry well on the building site, but never to the septic sanitary sewage drain. (**3**) The footing for the garage floor slab and the back fill of earth. (**4**) A basement window set below the final grade to be protected by a curved metal guard forming a window well. (**5**)A brick or stucco veneer exterior wall finish placed on a footing in the foundation wall below the finished grade. (**6**) Backfill of earth from foundation excavation returned against the foundation wall by earth-moving machines to establish the finished grade of the yard. It is in these places, or voids, that careless workers drop short pieces of scrap, especially of wood, which must be protected against pest control, particularly termites. An experienced operator can treat these places with an impregnable poison chemical barrier without disrupting the construction schedule. (**7**) The entrance of the main line to the sanitary sewage system in the street is placed here. It is to this line that all drainage from bathroom, lavatories and kitchen fixtures are run. If a septic sewage system is used the drainage from roof gutters and garage floor are run to a dry well. Under no circumstances is rainfall drainage to be piped to a septic tank installation. (**8**) The concrete footing for a hollow steel column, usually called a lally column, which supports a steel I-beam or wood girder carrying the floor joists. (**9**) The basement floor, of poured concrete, placed after the walls are built. Drains are set in place and, in some instances of high ground water levels, a protective waterproofing is used over the sub floor before the finish floor is troweled on. (**10**) The footings for columns supporting the roof over a patio, (**11**) A concrete bed for the patio flagstones. (**12**) The outside crawl space foundation wall supporting the large family room adjacent to the patio. (**13**) The excavated trench and the footing for the crawl space wall. The footings must be adequate and on a firm soil bed, free of attack by termites. This trench-like excavation of soil is done by the use of a backhoe shown in **Figure 6.** This same illustration shows a bulldozer-scarifier used in the removal of earth for a full basement. This is an earth moving machine having a soil loosener or ripper at the rear end and a soil pushing device at the front. (**14**) The single pier of concrete blocks, supporting a central girder or I-beam, which in turn supports the floor joists. (**15**) The inside and outside perimeters of the crawl space should be protected against a termite attack, since the foundation depth is less than half that of the foundation wall of a full basement and does not have the advantage of a basement floor of poured concrete. (**16**) The rear outside entrance to the basement is made up of concrete walls to support an access door and to enclose the steps from the basement to the yard.

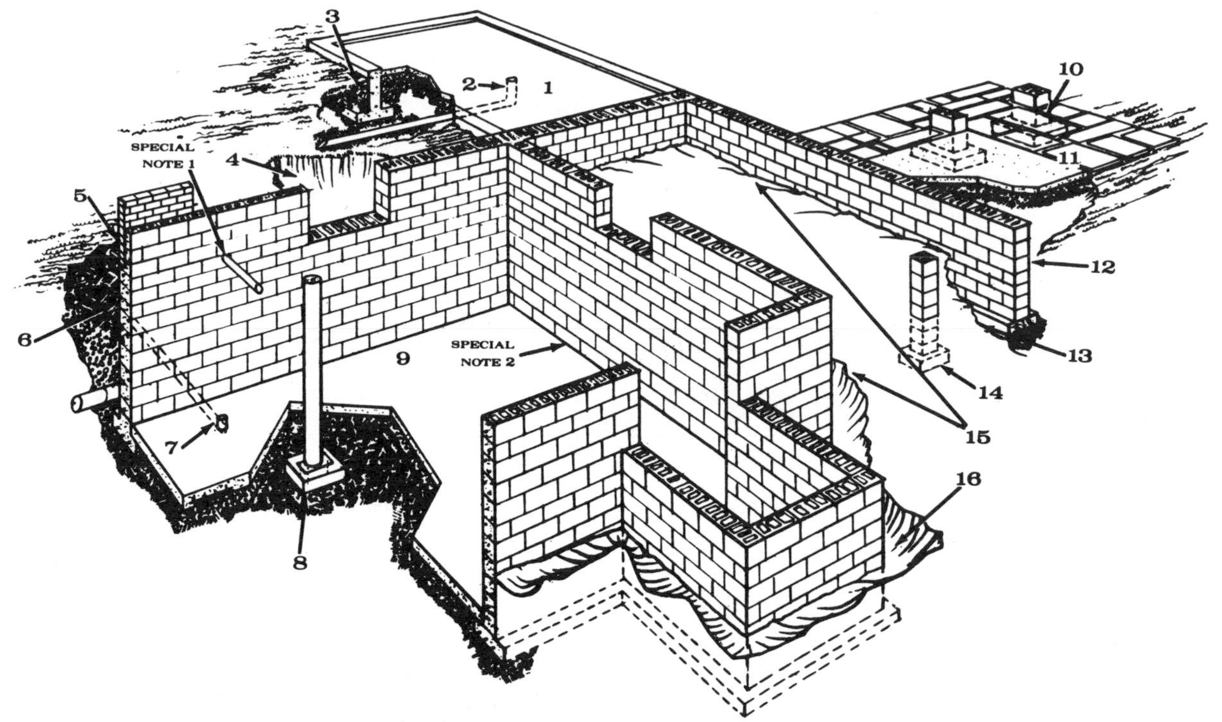

Fig. 5. Sectional view of a combination of three types of excavations and foundations for one house.

In the area, at the Special Note 1, at the front wall, care must be taken if holes are made in foundation walls for utility pipes to have area around the pipes caulked with oakum and finished with cement grouting and the surrounding soils protected against pests.

In the area, at the Special Note 2, where the basement wall joins the slab floor, a bituminous sealer must be installed to prevent seepage of water in the soils outside the foundation.

Laying Out An Excavation. After the boundaries of the building site have been checked and marked off by surveyors, the lot inspected and samples of soils obtained, a decision must be made to locate the dwelling most advantageously on the site. Local building regulations may permit the house to be placed with regard to suggestions of orientation listed in Chapter 1.

The builder lays out the area for the excavation and marks off the exact outside lines of the foundation by erecting a series of four batter boards, **Figure 7,** and cords placed to guide the operator of the earth moving equipment, and the mason building the footings and foundations.

If the house is properly laid out, the diagonals of a square or oblong structure will be equal. The use of the "square of the hypotenuse" principle provides a simple test of whether the angles are an exact 90 degrees. Lay off 6′ along the cord from one side of the stake, and 8′ along the other cord, as shown in the upper corner

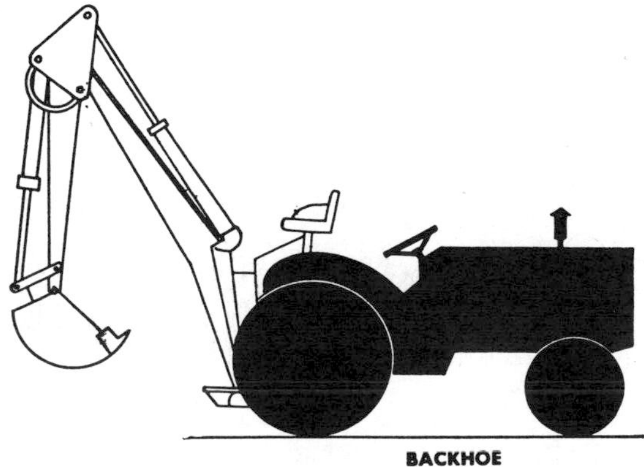

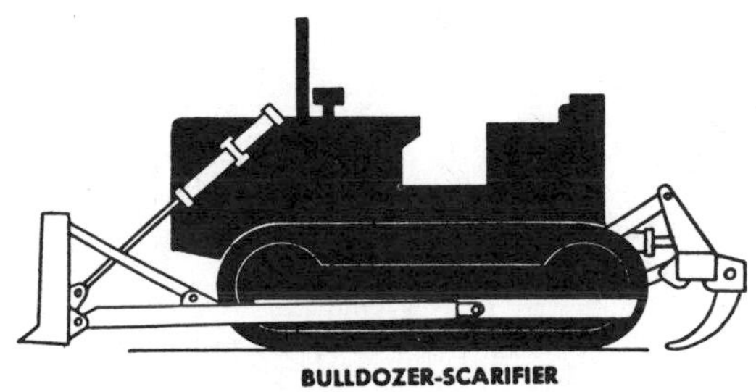

Fig. 6. Typical builder's earth-moving machines, a backhoe (left) for deep digging operations, and a bulldozer-scarifier (right) for pushing earth out of excavations and for grading and leveling operations.

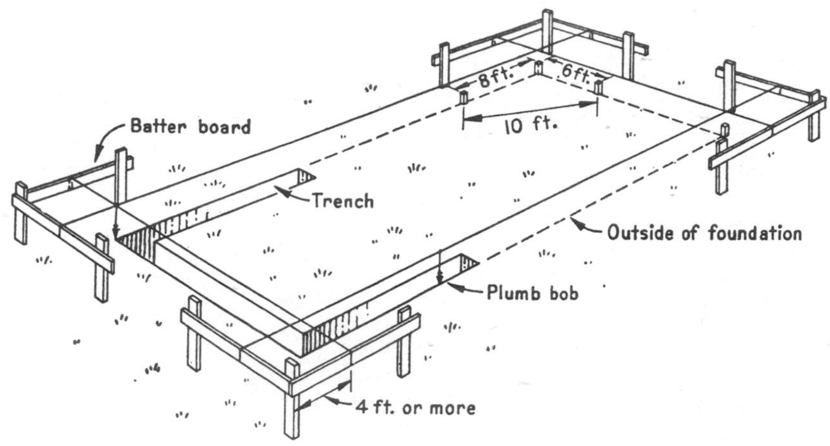

Fig. 7. Laying out an excavation, using batter boards with lines attached to mark the outside of the foundation wall.

of drawing, and mark the two points. The line connecting the two markers is the hypotenuse of the right angle triangle, and should be 10′ in length.

Directly under the cords will be the outside edges of the excavation and the foundation when it is poured or laid up by use of cement blocks. Excavating accurately to within 18″ of these cords will keep costs down by reducing the quantity of soil which must be backfilled.

Earth Moving Equipment In Action. A tractor or bulldozer is used to excavate the soil for a full basement. Often the soil is kept on the building lot to be used as backfill or to modify the grade or terrain. Topsoil is pushed to one side, to be spread over the plot after construction is completed. Sometimes soil is carried away to another building site by truck, **Figure 8,** to fill in low areas. Note the combination bulldozer and backhoe in **Figure 9.** The operator is using a lower seat while operating the scoop lift. He switches to the higher seat when using the backhoe for trenching or excavating for a septic tank.

Fig. 9. Versatile tractor has scoop lift on one end, backhoe on other. Note separate seats for each implement.

International Harvestor

Fig. 8. Scoop lift loads truck to remove earth dug from excavation for a house foundation.

Ford Motor Co.

Fig. 10. Rubber-tired tractor sets up on hydraulically operated legs while digging trenches for footings.

Fig. 11. Fiber form tubes are set in place for piers.

Fig. 13. Completed dwelling supported by concrete piers.

Fig. 12. House framing is erected on concrete piers.

Fig. 14. Portable hole digger digs 12"-diameter holes.

If it is only necessary to dig a footing trench for a slab-on-grade or a crawl space foundation excavation, a back digger is used, **Figure 10,** to lift the soil and deposit it a convenient distance away from the excavation. The combination back digger and scoop shown in this illustration is equipped with rubber tires for good traction while maneuvering rapidly around the site. This device is used in relatively dry soils and for somewhat shallow digging. The earth moving machines shown in **Figures 8 & 9** have two tracks on which the four wheels operate and are used when soils are wet and sticky, or when rock formations are encountered. Another attachment can be mounted on a tractor for service pipe ditches. It is known as a continuous chain trencher.

Concrete Columns and Piers. Some types of dwelling construction permit the use of concrete columns for foundations. Cylindrical fiber forms are used for the foundation columns. Concrete columns are beautiful as well as functional. Steel reinforcing rods set in the forms before pouring concrete, **Figure 11,** add strength to the columns and tie them to the house framing, **Figure 12.**

A completed dwelling supported by round columns

Fig. 15. Pouring concrete into fiber form tubes.

of concrete poured into fiber tube forms is shown in **Figure 13.** This type of support is used where the terrain is very irregular, or slopes sharply downhill.

In some instances where a slab on grade is used or a shallow excavation is made for a basement wall to create a crawl space, the footings may be made of concrete posts placed 8′ or 10′ apart. An earth auger is used to bore holes in the soil for pouring the posts, **Figures 14 & 15.** Pre-cast concrete beams, called grade beams are put in place directly over these concrete posts to provide a continuous bearing for exterior walls of wood or masonry.

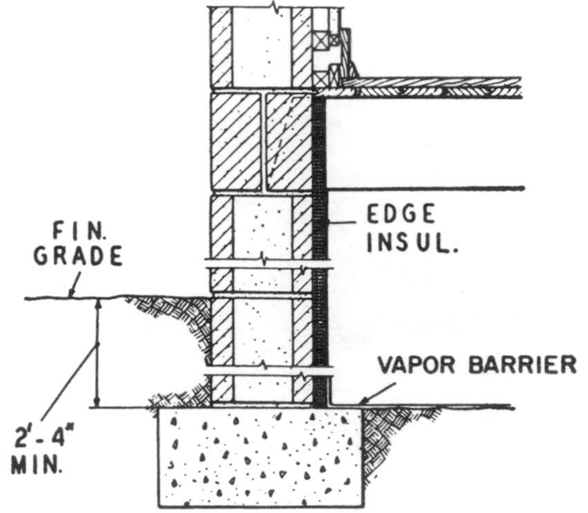

Fig. 16. Detail of foundation wall for crawl space.

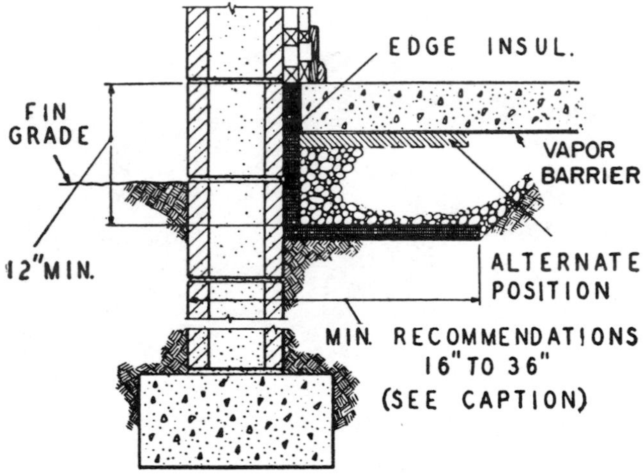

Fig. 17. Slab-on-grade foundation for masonry walls.

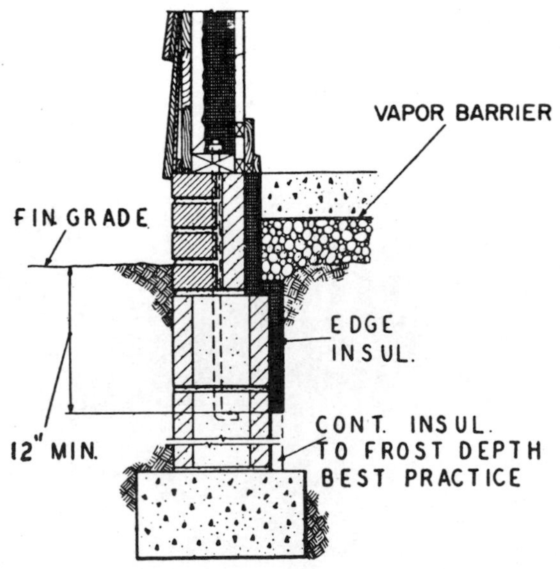

Fig. 18. Slab foundation for wood-framed walls.

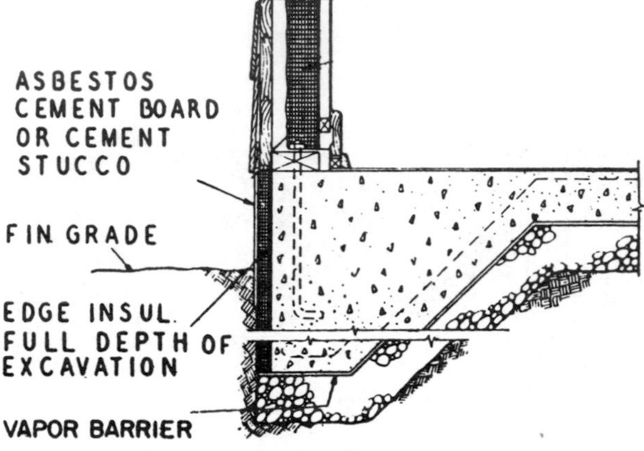

Fig. 19. Floating slab for use on unstable soil.

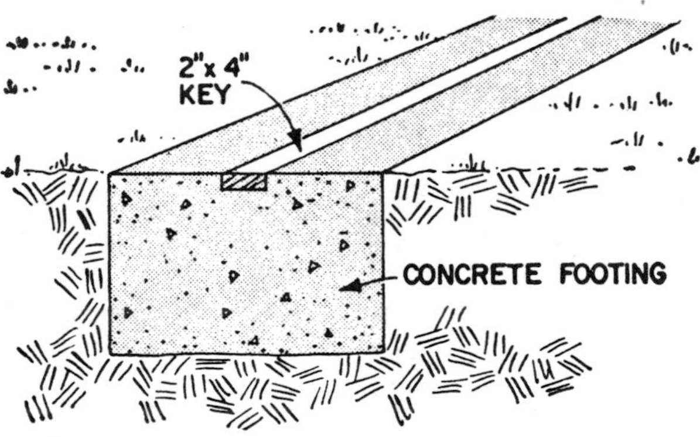

Fig. 20. Trench in firm earth can serve as form for concrete.

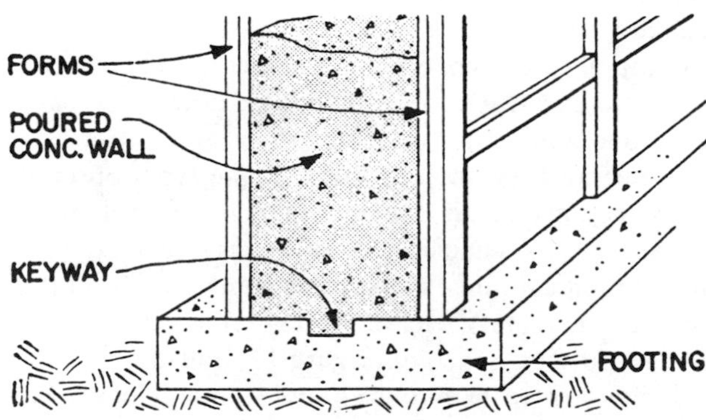

Fig. 21. Keyway bonds foundation wall to footing.

Cross Sections of Sturdy Footings and Foundations. Present day practices require sturdy footing and foundation wall construction for various exterior wall and floor finishes. A shallow excavation for a crawl space is shown in **Figure 16.** The application of a thick edge insulation on the interior of the block foundation reduces heat loss from the living area. A vapor barrier of polyethylene film is placed over the soil, which has been excavated a minimum of 2′-4″. The joists of wood rest on the second course of the concrete block wall. The rough and finished floor is shown in sectional view as it rests on top of the floor joists.

A more heavily insulated concrete slab floor placed over a gravel bed and resting directly on the ground is illustrated in **Figure 17.** The exterior wall is of concrete blocks with the interior furred for gypsum board finish.

A heavily insulated concrete floor slab having a sill of wood and bricks anchored by bolts to the shoe strip is shown in **Figure 18.** The exterior wall of the house is covered with wood sheathing and finished with shingles.

The Floating Slab in **Figure 19** is unique in that the finished concrete floor slab, foundation wall and footing are reinforced with metal rods and poured as one integral part over a gravel bed. An exterior of heavy insulation is surfaced with stucco or asbestos cement board. The shoe plate of the wood-framed exterior wall is anchored at the edge of the foundation and the exterior wall covering is of beveled siding nailed over tongue and grooved sheathing boards. The Floating Slab is used on fill or instable soil, where conventional foundations might settle unevenly and crack.

Foundation Forms. If the soil of the excavation is firm or if the trench will hold its shape without crumbling, no temporary form for holding poured concrete will be needed, **Figure 20.** Note the 2 × 4 placed in the center of the concrete, before it sets, and later removed to form a key as the concrete of the foundation wall is poured.

The foundation wall is placed over the footing, **Figure 21,** in forms erected on top of them. Most builders prefer to have complete control over the footings and foundations and any forms needed. Footing or foundation forms are made of planks, plywood, steel or pressed board braced in various ways. Most forms can be used again and again if handled with reasonable care.

Foundation walls require more bracing than footings. The more concrete poured, the stronger the forms must be. The higher the walls, the more bracing is required. If the forms, which act as a mold until the concrete hardens, do not hold firmly the wall may be irregular in shape. Sometimes the forms may bow or separate and actually fail under the pressure of heavy concrete poured into them.

Forms can be built of tongue and groove stock, using 2 × 4 studding, tied in pairs and 2 × 4 braces and stakes, **Figure 22.** The footing has been poured in a trench without the use of a form in this detail.

A hole or flaw in a form will allow the more fluid portions of the wet concrete, the water and cement, to run out, leaving sand and gravel without cohesive power. All forms must be water tight and all joints tightly fitted together and rigidly secured. When forms are reused for other concrete walls they are scraped smooth after each use, and a special oil applied to the contacting surfaces. This care leaves the set concrete surfaces smooth and prevents bits of the new wall from adhering to the surface of the forms.

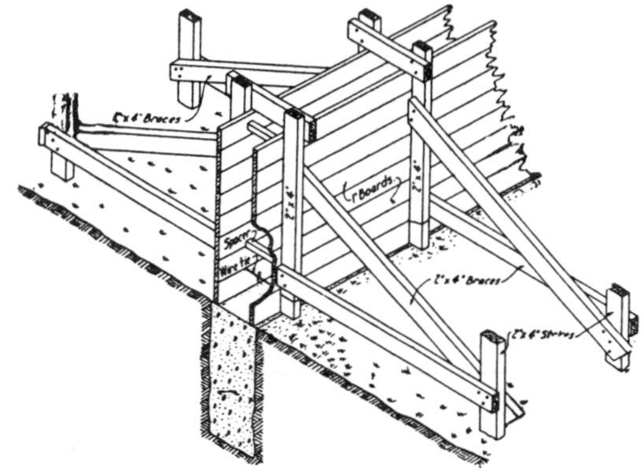

Fig. 22. Wall forms erected on footing poured in trench.

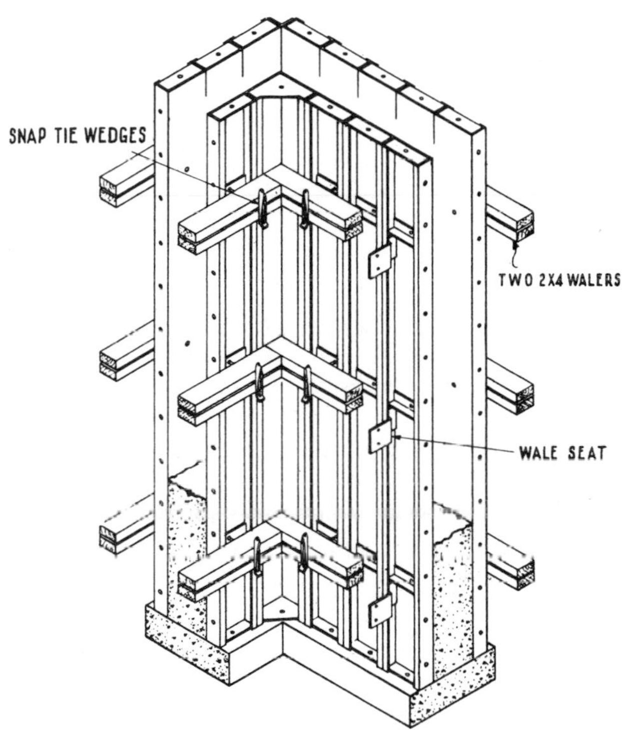

Fig. 23. Full length steel forms for heavier construction.

Quality Concrete Construction

Concrete is a mixture of portland cement, water, and inert materials called aggregates, both fine and coarse. Fine aggregate is sand; coarse aggregate is usually crushed stone, gravel, or some of the lightweight materials such as expanded slag, clay or shale.

When concrete is first mixed, the cement and water form a paste that coats every piece of aggregate. Within a period of 2 to 3 hours a chemical reaction starts to take place between the cement and water. As this chemical reaction progresses, the cement-water paste hardens and binds the aggregates together to form a hard, durable mass.

The quality of the concrete depends upon the binding qualities of the cement-water paste. Concretes of various strengths can be produced by changing the proportion of water to cement. As less water is used, the concrete becomes stronger and more durable and watertight, but more difficult to place. The key to quality concrete is to use the proper amount of cement and water.

Entrained air is an important ingredient in concrete. It is incorporated into concrete by the addition of a chemical called an air-entraining agent. This agent causes the formation of billions of microscopic air bubbles throughout the concrete. Concrete containing these minute air bubbles is called air-entrained concrete. Entraining is a chemical process as in precipitation or distillation. This concrete has superior resistance to destructive scaling that results from freezing and thawing and salt action. Air entrainment also increases workability of freshly mixed concrete, reduces segregation and bleeding in the fresh concrete, and increases sulfate resistance of hardened concrete. Since the freshly mixed concrete is more workable, it requires slightly less mixing water, an added benefit. Many cement manufacturers market portland cement containing an air-entraining agent. Such cements are identified on the bag. Air-entraining agents can also be purchased separately and the proper amount added to the mixer. For general work using 1-1/2″ aggregate, it is desirable to get 5 to 7% air incorporated into the concrete.

Air-entrained concrete is recommended for all concrete exposed to freezing and thawing and to salt action, such as caused by the use of de-icing salts. Sidewalks, patios, and driveways in northern climates should be made of air-entrained concrete. When ordering ready-mixed concrete, specify to the producer that you want air-entrained concrete.

Using Ready-Mixed Concrete. Ready-mixed concrete is sold by the cubic yard (27 cu. ft.). Ready-mix producers will usually deliver any quantity greater than 1 cu. yd. The mix to order depends upon its use. For footings, foundation walls and retaining walls, specify a mix containing at least 5 sacks of portland cement per cubic yard and a maximum of 7 gallons of water per sack of cement. For basement floors, driveways, garage floors, patio slabs, slabs on ground and stairs, order a mix containing at least 6 sacks of portland cement per cubic yard and a maximum of 6 gallons of water per sack of cement. For chimney caps, lintels, septic tanks, the mix should contain at least 6-1/2 sacks of portland cement per cubic yard and a maximum of 6 gallons of water per sack of cement, using a maximum of 3/4″ aggregate.

It is usually wise to order 5% to 10% more concrete than calculated to cover normal wastage and unforeseen factors such as irregular forming or subgrade preparation, or minor miscalculations. Any leftover concrete can be dumped into small forms kept at hand to make patio stepping stones or splash troughs for use beneath downspouts, saving the cost of buying these items when completing the house.

Good site preparation will do much to speed the job. Put forms in place and check for alignment and adequate bracing before the concrete is delivered. Clear access lanes so the truck can get as close as possible to the job site. Have sufficient help available to place and finish the concrete quickly. Assemble the proper tools prior to the delivery of the concrete.

Mixing Concrete On The Job. If it is necessary to mix concrete on the job the mixing site should be as close as possible to where the concrete is to be placed. The mix varies with the job involved. Various suppliers can furnish suggested mix tables. The first step is to make a trial mix using the proportions suggested by the manufacturer of cement. If it is necessary to adjust the mix, do so by changing the aggregate. Do not change the amounts of water and cement. If the trial mix is too wet, add more aggregate. Adjust a mix that is too stony by reducing the amount of coarse aggregate, increasing the sand, or by a combination of these.

Placing Concrete On The Job. Concrete should not be placed on mud or frozen soils. In dry weather the subgrade must be dampened to lessen the absorption of the mixing water by the supporting soils. All forms must be carefully set, plumbed, and leveled. They must be braced adequately and oiled to prevent the bonding of the concrete to them. To keep segregation of the coarse and fine materials to a minimum, the freshly mixed concrete should not be moved farther than necessary. The use of chutes is recommended when the fresh concrete must be dropped more than 3 or 4 feet.

Finishing Concrete. After the concrete has been placed, it is struck off with a straightedge. In flat work it is difficult to strike off sections wider than 10 to 12 feet. The straightedge is usually a 2 × 4 and should be 1 to 2 feet longer than the width of the section being finished. A wood or light metal float is then used to float the concrete. In flat work a bullfloat is very handy; it has a long handle and is easy to use on the wider slabs. Many experienced concrete finishers use a bullfloat first and finish with a hand float. Floating gives a gritty finish that wears well, is attractive, and provides good footing.

If a smooth finish is desired, it will be necessary to steel-trowel the concrete. Steel-troweling must be delayed until the concrete can be finished without bringing an excessive amount of water, cement, and fine sand to the surface. An excessive amount of these materials produces a surface that does not wear well. Power trowels are being widely used to finish concrete. The concrete must be hard enough to hold the power trowel. Here again premature troweling must be avoided. If the operator brings a soupy slurry to the surface, the concrete is too plastic to be finished and troweling should be delayed until it is firm.

A textured finish can be obtained by dragging a broom across the surface. The stiffness of the broom bristles, the amount of pressure applied to the broom, and the plasticity of the concrete will determine how deep the broom scores the surface.

Curing Concrete. Concrete should be cured for at least 5 days to prevent evaporation of the mixing water and to develop all the properties of the concrete. If too much water evaporates, there will not be enough water to react with the cement and the concrete will not gain sufficient strength or durability. The exposed portion of concrete is naturally the first to lose the mixing water. Since the surface will be exposed to the weather, abrasion, and other factors, it must necessarily be durable. Curing is an essential part of every concrete job.

Concrete can be cured by one of several methods. It can be covered with sand, straw, burlap, or other similar material that is dampened and kept damp for the duration of the curing period. Another method is to cover the concrete with a vapor-sealing material such as water-resistant kraft paper or polyethylene film. Commercial curing compounds can also be sprayed on the fresh concrete to form a thin membrane which seals in the mixing water. In vertically formed concrete, cure the concrete by leaving the forms intact for the 5-day period.

Cold-Weather Concreting. Temperature has a considerable effect on the rate of hardening of concrete. The optimum temperature for placing concrete is 70 degrees F. As the temperature drops, the rate of hardening slows. All new concrete must be protected from freezing. In buildings, heat is often supplied with an oil-fired stove called a salamander. The hot air from the salamander should not be allowed to come into direct contact with the concrete as it will dry it out.

When concrete is mixed in cold weather, the water and aggregates are often heated. The water should not be heated to more than 180 degrees F.; overheating is likely to cause a flash set in the concrete.

High-early-strength portland cement is frequently used in winter concrete work because it sets more rapidly than normal portland cement. When normal portland cement is used, calcium chloride can be added to the mix in cold weather. It is not an antifreeze material, but an accelerator that speeds the chemical reaction between the cement and water. The quantity of calcium chloride should not exceed 2%, or two pounds per sack of cement. It should be dissolved in the mixing water, not added in powder form to the mix. The use of calcium chloride is not a substitute for normal cold-weather precautions.

Hot-Weather Concreting. As the temperature rises above 70 degree F., the initial rate of hardening of concrete increases. Evaporation of water from the concrete is also more rapid in hot weather. A combination of wind, high temperature, and low humidity dries concrete rapidly.

In extremely hot weather it may be necessary to reduce the temperature of the freshly mixed concrete. Aggregates should be stockpiled in the shade, if possible, and cool water used to mix the concrete. It is often advisable to delay placing concrete until late in the afternoon to take advantage of lower air temperatures. Curing procedures must be started promptly to prevent the evaporation of water.

Estimating Concrete Needed. The unit of measure of concrete is the cubic yard, which contains 27 cubic feet. To determine the amount of concrete needed, find the volume in cubic feet of the area to be concreted and divide this figure by 27. The following formula can be used to determine the amount of concrete needed for any square or rectangular area:

$$\text{Cubic yards of concrete equals } \frac{\text{Width in feet} \times \text{length in feet} \times \text{thickness in feet}}{27}$$

For example, a 4″ thick floor for a 30′ × 60′ house would require: $\frac{30 \times 60 \times 1/3}{27}$ equals 22.22 cubic yards of concrete. The amount of concrete determined by the above formula does not allow for waste or slight variations in concrete thickness. An additional 5% to 10% will be needed to cover waste and other unforeseen factors.

Mixing Color In Concrete. Colored concrete can be obtained by adding mineral pigments to the concrete mix. The pigments are thus dispersed throughout the concrete. Mineral pigments are inert and do not react with the cement water paste. Mineral pigments are sold by ready-mixed concrete and block producers, lumber yards, and building materials dealers in red, green, yellow, brown, gray and black colors. The common method of packaging is in 5, 9 or 50 pound bags. The maximum amount of pigment to use is 9 pounds per sack of portland cement. The amount to use varies with the intensity of color desired and the color of the sand. Less color pigment is required when the sand is light colored or white. In most cases 5 to 7-1/2 pounds of mineral pigment are added to the mix for each sack of cement used.

The materials must be carefully measured for every batch. Add the cement, sand, coarse aggregate, and pigment to the mixer and mix thoroughly for four minutes. Then add the water and mix for four additional minutes. Under normal conditions this will be sufficient to blend the materials thoroughly and assure uniform coloring. If the concrete has streaked or uneven coloring, it will be necessary to mix it longer.

Ground Water Problems

The time to make a basement watertight is when it is being constructed. A bed of gravel must be laid under and around the planned footing and foundation wall, and a 4″ drain tile pipe laid next to the base to carry away excess water to the sewer line or a sump pit, **Figure 24.** Some tiles are perforated to permit the entrance of water. Some tiles are placed 1/4″ apart and the top half of each joint covered with a strip of building paper to keep silt out of the line, **Figure 25.** A layer of coarse granular material should be placed to a depth of 6 to 8 inches over the drain line to aid water movement into the line.

If water penetrates a block wall, it usually does so through inadequately filled joints. The joints on both the inside and outside of the wall should be tooled with a V-shaped or concave tooling device. The outside of block basement and foundation walls should be covered with a 3/8″ coat of portland cement plaster called parging. The mortar used in laying up the block is often used to plaster the walls. The plastering should be done in one operation and applied in a uniform thickness over the wall. The finished surface should be smooth and even. The area at the bottom should be filled with plaster and rounded to carry the water away from the critical area. The plaster should extend 6″ above the finished grade. The plaster coat should be kept damp for at least 48 hours after application to permit the plaster to cure.

The plaster is then painted with two coats of portland cement base paint. Each coat of paint should be moist-cured for at least 24 hours. An alternate waterproofing with two continuous coatings of hot bituminous material applied at right angles to each other over a suitable priming coat may be used.

Water should not be allowed to accumulate in the excavation behind a free standing wall since it might cause it to collapse. Floor joists give the walls stability and provide a certain amount of bracing, so no backfilling should be done until the first floor framing of the house has been installed. All refuse should be removed from the excavation before the backfilling. One or two feet of earth should be carefully shoveled over the granular material to protect the tile lines from movement. Backfill should be compacted to minimize settling.

Rain gutters should have downspouts connected to the tile lines to carry the water away from the foundation. If the downspouts are not tiled away from the house, much of this roof water will accumulate around

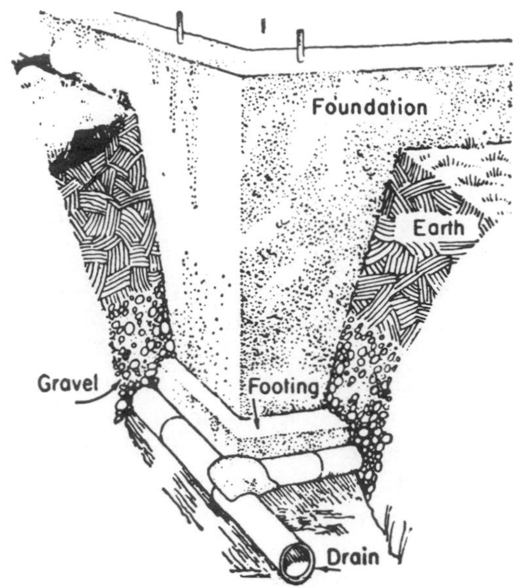

Fig. 24. Cutaway sketch of a footing drain tile laid with tight joints in a bed of gravel. The tile is perforated to permit water to enter and run to an outlet lower than floor.

Fig. 25. Drain tile laid ¼″ apart beside footing. The top half of each joint is covered with a strip of asphalt paper to keep out soil as water seeps in. A layer of coarse gravel is placed over the drain tile.

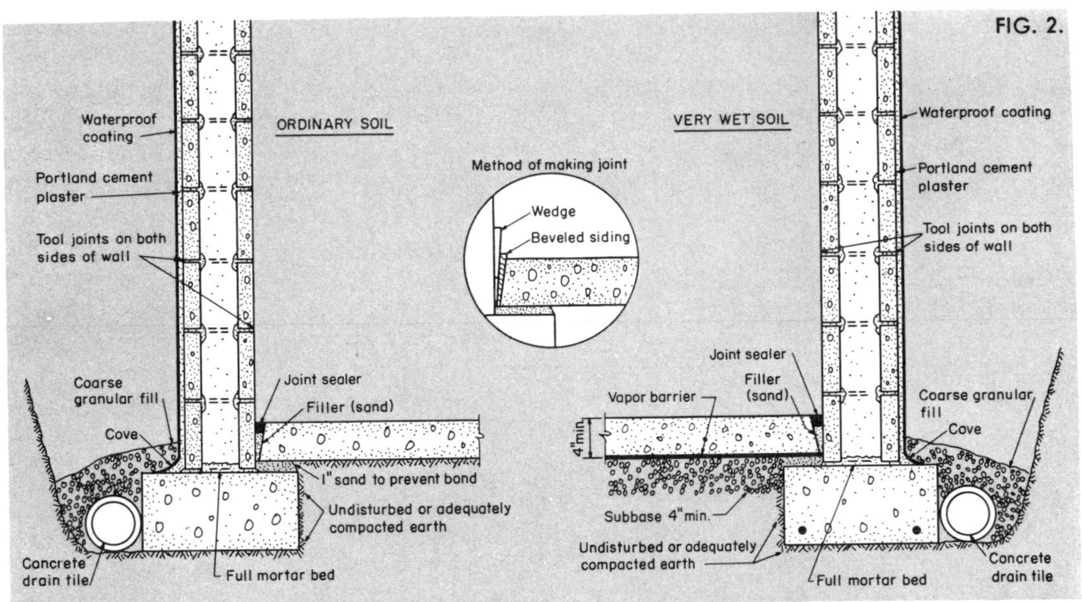

Fig. 26. Foundation at left is constructed for normal soil conditions, one at right for very wet soil.

the foundation. The finish grade should slope away from the house to carry surface water away. **Figure 26** shows sectional views of two outside foundation walls constructed of concrete blocks. On the left is a wall erected in ordinary soils and on the right is a wall erected in very wet soils. Added protection is taken to prevent the seepage of water from the very wet soils.

Good Construction Methods

In the following illustrations are shown good standard construction details and photographs of foundations and foundation forms.

A well planned, insulated slab with foundation wall, moisture barrier, metal reinforcement in the floor and a below grade drain tile placement is shown in **Figure 27.**

Two methods of slab construction for radiant heating are illustrated. The first sectional view, **Figure 28,** is a simple slab construction with radiant heating tubing placed over the sand and gravel fill for passage of hot water heating house. The second sectional view, **Figure 29,** is a more elaborate slab construction containing an 8″ heating duct for passage of forced warm air at the perimeter of the foundation. **Figure 30** shows a slab, reinforced with wire mesh, a shallow integrated footing with an offset to support the exterior brick veneered wall backed by a well insulated stud wall. **Figure 31** shows a slab foundation poured over a tamped

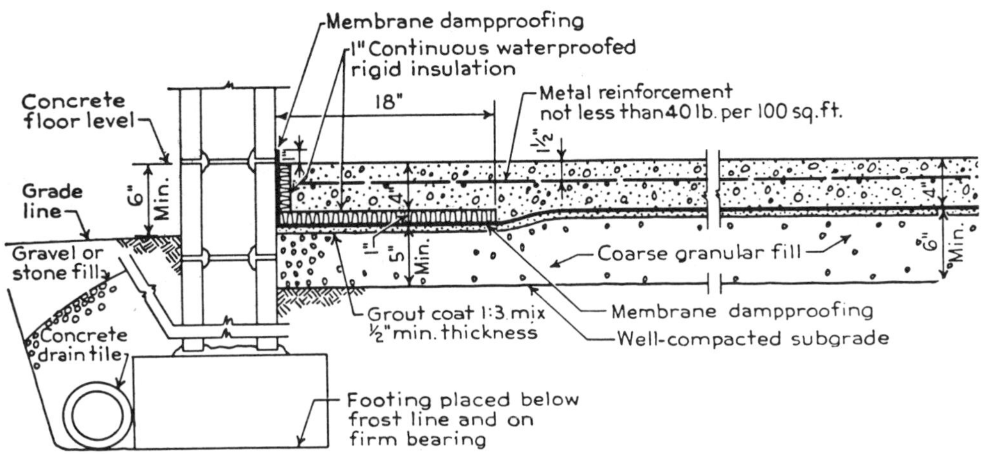

Fig. 27. A well planned, insulated slab, constructed to withstand pressure of heavy ground water.

EXCAVATIONS/FOUNDATION FORMS/FOUNDATIONS

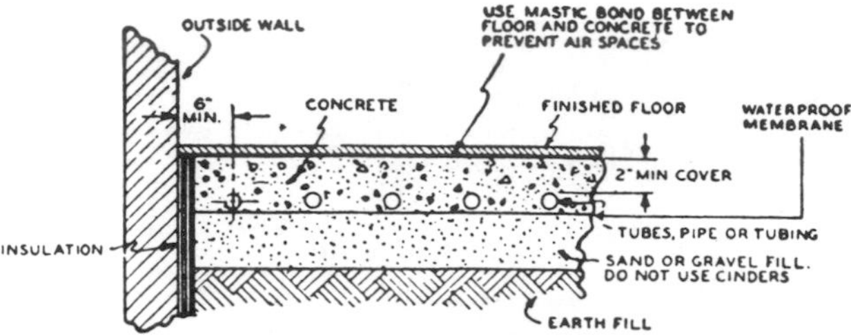

Fig. 28. Slab with radiant heating tubing embedded.

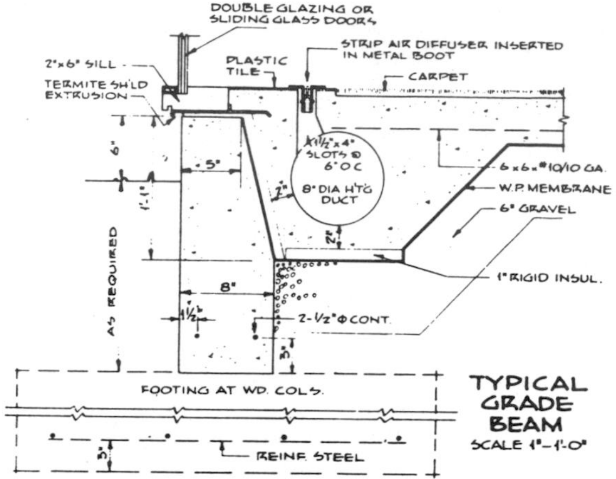

Fig. 29. Slab with perimeter warm air duct embedded is insulated to reduce heat loss at edges.

gravel base and resting on a masonry block foundation wall which rests on a footing poured in firm earth requiring no formwork. The large anchor bolt serves to tie the block and sill of the framed exterior wall together.

Various Types of Forms For Concrete Walls. Forms directly affect the finished appearance of the wall. They must be tight, smooth, defect-free, properly aligned, and well braced to resist lateral pressures created by the plastic concrete. The pressure created by fresh concrete is related to the amount of concrete placed per hour, the outside temperature, and the amount of mechanical vibration. Mechanical vibrators increase the pressure considerably. Walls can be cast using forms built on the job or with commercial prefabricated forms. Forms for walls up to 8′ in height can be made of 1″ matched lumber or 3/4″ form-type plywood. This is nailed to 2 × 4 studs placed 16″ on center. The studs are in turn held by horizontal members called wales, consisting of two 2 × 4s. Wood spacers, cut to a length equal to the width of the wall, should be inserted (but not nailed) in the form to maintain the proper form spread. Wire ties should then be placed near the spacers and drawn tight. As the concrete is placed, the spacers are removed. The wire remains in the concrete and the exposed ends are cut off when the forms are removed.

Commercial forms are sold by a number of companies. In many localities it is possible to rent commercial forms. Builders who erect a number of houses often find it profitable to buy a set of forms. The standard form panel is 2′ wide and is available in heights up to 8′. Filler panels of various sizes are also available to accommodate portions of the wall that cannot be formed with 2′ wide panels. Commercial form ties are used with commercial panels. Form ties are either round or rectangular and act as both a spacer to maintain the proper wall thickness and as a tie to keep the forms from spreading. Form ties are designed to snap off easily far enough into the concrete to permit patching of the tie holes. On basement walls, ties with a rating of 3,000 pounds should be spaced 2′ apart.

It is desirable to place the entire wall in one operation. The concrete should be of medium consistency and placed in 6- to 12-inch layers around the building. When the temperature is 70 degrees F. or above, place concrete at the rate of 6 feet or less per hour. If the temperature is 50 to 70 degrees F. no more than 4 feet of concrete should be placed per hour. If mechanical vibrators are used, the rates should be reduced to 4 feet and 2 feet per hour respectively.

The concrete must be vibrated or tamped to prevent honeycombing and to obtain a smooth dense surface. The concrete should be spaded along both faces of the wall with a flat tool. Mechanical vibrators do an excellent job but should not be used in one spot for more than 30 seconds. In inaccessible areas such as under windows, the forms should be tapped lightly with a heavy hammer to consolidate the concrete.

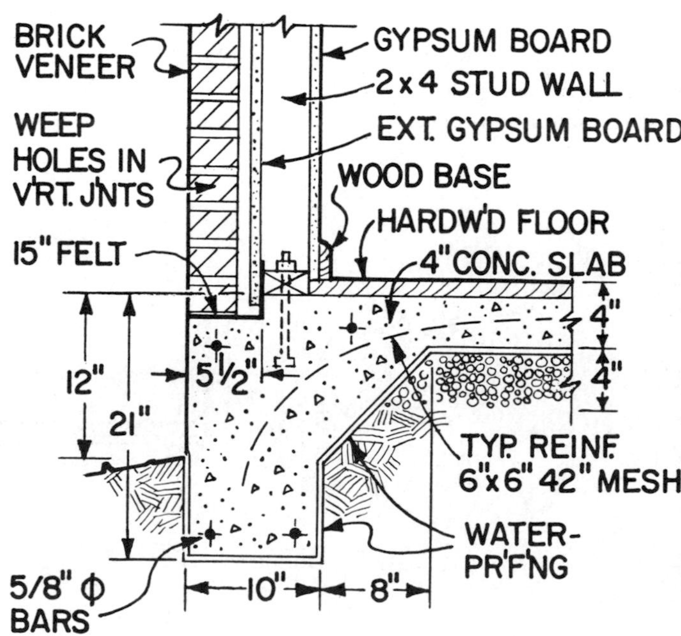

Fig. 30. Diagram shows placement of reinforcing rods and wire mesh in monolithic floating slab.

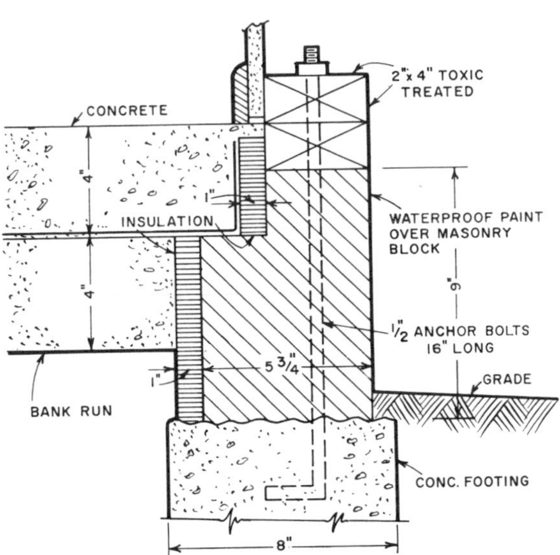

Fig. 31. Anchor bolt set in footing runs through voids in block to tie wood sill to foundation.

Fig. 32. Panel form system with rod walers. Note ties projecting from outside form panels and baffles to divert concrete around window openings.

Figure 32 shows the outside wall of a foundation form of plywood panels tied together with rod wales, sometimes called walers. These forms are placed on foundation footings, already poured and set. In the far corner workmen are setting the first of the inside portion of the form. Forms for basement windows are already set in place. Reinforced baffles at window forms divert the poured concrete around the window forms.

A crane can be used to lower a large section of the inside form in place, **Figure 33.** Note that all outside sections are set in place and properly tied together. The excavation is relatively deep and large enough to provide adequate working space to erect the outside wall of the foundation form.

Fig. 33. Rows of identical foundations can be formed with large prefabbed sections hoisted from one site to another by a crane. Forms are joined at corners.

Small steel plates can be tied together in modules for varied heights of concrete foundation walls, **Figure 34.** Note the half stripped foundation wall on the far side and the protruding tie wires which are eventually cut off at the surface of the wall.

Metal walers brace forms, maintain alignment and tie the corner of an outside foundation form together. A metal wedge is used to tighten the horizontal waler in place, **Figure 35.**

Wooden brackets can straddle the top of the forms to serve as ties for a shallow crawl space wall as shown in **Figure 36.**

Slab On Ground Construction. Many manufactured houses are now assembled in a factory, being produced panel by panel or component by component; shipped by truck to the building site; then placed on a pre-cast slab of concrete on the earth.

The earth under the slab floor is called the subgrade. If a layer of granular materials is placed over the earth,

Fig. 34. Commercial forms consisting of small steel plates can be locked together in modules for varying length and height. These forms can be reused many times.

Fig. 35. Metal walers are locked in place and tightened by triangular wedges driven into slotted brackets.

Fig. 36. Wooden brackets can straddle crawl space walls to brace them and eliminate ties.

it is referred to as the subbase.The water tight membrane between the subbase and the slab is the vapor barrier. Perimeter insulation is placed between the foundation wall or footing and the edge of the slab to minimize heat loss.

The construction area must be cleared of all sod,roots, large stones, and debris. The subgrade should be firm throughout to provide adequate support under the slab. The undisturbed earth should be used when possible. In some instances, it is necessary to fill at least part of the area. The fill should consist of gravel, crushed stone, crushed blast furnace slag, or other suitable material. All holes, irregularities, and trenches for pipes and conduits should be filled first and the material compacted. The balance of fill should be deposited in 6″ layers and adequately compacted by rolling, vibrating, tamping or a combination of these.

The top of the floor slab should be at least 8″ above the finished grade line of the earth at the outside of the foundation wall. The soil around the house must be graded to carry away surface water. The grade should be at least 1′ in 25′ in all directions.

Subbase Preparation. Water moves upward by capillary action through the soil. This action varies with the type of soil and ranges from a low capillary movement in coarse sand soil to severe action in clay soil. A granular subbase drastically reduces capillary action and assures a drier floor slab. A granular subbase should be used under slabs in all living areas and slabs laid in areas where water is a problem, **Figure 37.** It's considered good practice to use a gasoline powered vibratory compactor to consolidate the material of the subbase as shown in **Figure 38.**

Use at least a 4″ thick bed of gravel, crushed stone, or crushed blast-furnace slag under the slab. Other materials may be used if test results prove them to be satisfactory. All fill material should pass through a 2″ screen and be retained on a 1/4″ screen. The subbase material should be tamped or rolled to settle and consolidate it.

Vapor Barrier. A vapor barrier is placed over the subbase to stop the movement of both liquid water and water vapor into the slab. The vapor barrier should have a permeance rating of 0.2 perms or less. Perm is a contraction of the word permeability and is defined as a rating for the amount of water vapor transmitted through a material per square foot, per hour, per inch of mercury vapor difference. Among the materials used as vapor barriers are 55 pound roll roofing, 4 mil polyethylene, and, asphaltic-impregnated kraft papers. A mil is a unit of measure equal to 0.001 part of an inch. Strips should be lapped 6″ to form a complete seal. A vapor barrier is essential under every section of the slab that is in a habitable area. It should extend up along the edges of the slab at the juncture of the slab and footing wall. Care must be taken to prevent puncture of the vapor barrier while placing concrete over it.

Perimeter Construction and Insulation. The floor slab should be constructed so that it does not bond to the foundation wall or to any column footings. In dry areas, bond can be prevented by placing a layer of 55 pound smooth roll roofing between the edge of the slab and the wall. Premolded asphaltic-impregnated strip material is also widely used. The premolded strip material can be bought from most ready-mixed concrete and products producers and building material dealers. This general type of joint does not give positive water control around the slab and is used in areas where ground water levels are not critical.

In wet areas, a watertight joint can be made by placing two pieces of oiled, tapered wood around the perimeter of the slab. When the slab hardens, the wood is removed and the lower 2 to 3 inches of the groove is filled with sand. A joint sealer is then placed over the sand to fill the void and form a tough, watertight seal.

Perimeter insulation is necessary for slabs on gound in living areas, and is used to reduce heat losses from the floor slab to the outside. In hot climates, rigid foamed plastic insulation is used to reduce heat movement into the slab. The insulation material should be non-capillary, stable in the presence of water and wet concrete, and resistant to termites and fungi. The material should also be strong enough to resist crushing. The thickness of the perimeter insulation is usually 1, 1-1/2 or 2 inches, depending upon the outside temperatures and the type of heating used.

When the ground water level is 4′ or more below the outside grade, the insulation is usually placed vertically or between the footing and the slab edge. If the ground water level is less than 2′ below the outside grade the insulation must be placed flat or horizontally around the perimeter of the wall.

Pouring the Slab. All standing water must be removed before pouring the floor slab. Concrete must not be poured on frozen earth. The usual thickness of a slab on ground for a house is 4 to 6 inches, although the slab may be thicker in certain instances. All pipes and reinforcements should have a concrete cover of at least 1″. Any duct that is not crush-resistant, nonabsorbent, or noncorrosive should be encased completely in 2″ of concrete protection.

Reinforcing the Slab With Steel Rods. Reinforcement is often used in floor slabs for houses under the following conditions: 1. In slabs supporting load-bearing partitions if the partitions are more than 4′ from the center axis of the slab. 2. If a slab is placed on fill more than 2′ deep, or more than 10% of the area within the foundation wall has been excavated and backfilled. 3. In slabs in which heat ducts or pipes are embedded. 4. In unheated slabs longer than 30′ in their longest dimension.

The use of reinforcing steel will not assure the prevention of cracks. The steel may, however, reduce the size of the opening if a crack does occur. The steel should be placed 1″ from the top of the slab and it should be held in this position. The practice of placing the steel on the subbase and pulling it up with a rake or hook should not be done; it is virtually impossible to control the location of rods or wire fabric by this method.

If the floor does not contain steel reinforcement but does have warm air ducts, the area over the ducts should be reinforced. Use 6 × 6 × 10 gauge mesh and extend it 18″ past the point where the slab thickness reduces to its normal thickness.

A typical slab floor, **Figure 39,** may include asbestos-cement pipes laid over the vapor barrier. These pipes conduct heated air to the various rooms from a central heating unit. Note the sanitary soil stack in place to be supported by the slab. Note the workmen screeding a poured portion of the slab over the vapor barrier, heating pipes, and the steel mesh.

Figure 40 shows a finished slab on ground for a new house. Note the soil pipes in place, the anchor bolts in place, the footing wall and the completely screeded

Fig. 37. Workman places granular subbase for a slab foundation to reduce capillary action of ground water.

Fig. 38. Gasoline powered vibratory compactor (right) consolidates subbase before placing concrete.

Fig. 39. Section of slab is screeded after concrete is placed over vapor barrier and wire mesh. Note heat duct and soilpipe in place before pouring concrete.

slab in place.

Support of Partition Walls. In many instances the floor slab may have to be thickened to provide adequate interior support for partition walls. A 4″ slab can be used without thickening for loads up to 500 pounds for each linear foot. Thicken the slab 2″ for wall loads up to 1,000 pounds for each linear foot and 4″ for loads up to 1,500 pounds for each linear foot. When

Fig. 40. Troweling a section of a slab with a gasoline powered rotary trowel. Rotating steel blades finish large areas of concrete faster and more evenly than troweling by hand. Note the copper tubing for the radiant heating system and wire mesh ready for the next pour.

Fig. 41. On small jobs a rotary trowel attachment can be driven by an electric drill to speed finishing.

the loads exceed 1,500 pounds per linear foot, a separate footing should be built in to support the wall. In addition, the thickened portion of the slab should be reinforced with two steel bars placed under the partition wall and parallel to it.

Control Joints. To control the possible cracking of concrete slabs, control joints are cut into the slab to create a series of weakened planes. If the concrete cracks, it usually does so at these joints. Control joints are cut to a depth of one-fifth the thickness of the slab when the concrete is being finished. The joints are usually formed with a groover or flat blade. When steel reinforcement is not provided, place control joints approximately 15 to 25 feet apart. Slabs that are not rectangular or square should have a control joint across the slab at the point of juncture of the offset. Reinforcement is not carried across joints; it is brought within 2″ of the joint.

Troweling a Floor Slab. After the pouring of concrete to form a slab it is screeded as shown in **Figure 39.** Further finishing is given the slab by troweling, either by hand, or more rapidly by the use of power driven trowels. Various kinds of rotary trowels are available. **Figure 40** shows a large, self propelled trowel being used even while rain is falling. A second troweling was done when the rain ceased. Note the copper tubing of the radiant heating system and the wire reinforcement ready for the next section of the slab to be poured. A small, power driven trowel may be operated as an attachment to an electric drill, **Figure 41.** This device has a small demountable handle.

Troweling a poured and screeded slab gives it a very smooth finish on which can be placed a finished flooring of the builder's and owner's choosing.

All-Weather Wood Foundation System

General Description. This foundation system is fabricated of pressure treated lumber and plywood. All parts of the supporting element for the house structure are included in the foundation system. Foundation sections of nominal 2″ lumber framing and plywood sheathing may be fabricated in a factory or constructed on the job site. **Figures 42 and 43** show pressure treated framed plywood walls resting on 3″ gravel bed prior to pouring the 4″ slab.

Footing plates for the foundation are nominal 2″ pressure treated wood planks resting on 4″ or thicker leveled bed of gravel or crushed stone. The system can be used for both basement and crawl space construction. The exterior of basement foundation walls is covered with a polyethylene film which is bonded to plywood. Joints are lapped and sealed with a suitable construction adhesive. Concrete slab basement floors are poured over the gravel base, **Figure 44.**

National Forest Products Assn.

Fig. 42. Prefabricated framed plywood walls of pressured treated wood being placed in shallow excavation on a 4" bed of gravel.

National Forest Products Assn.

Fig. 43. Prefabricated plywood wall section being bolted together on gravel bed. A 2" or 3" slab of concrete will be poured on gravel to form a floor.

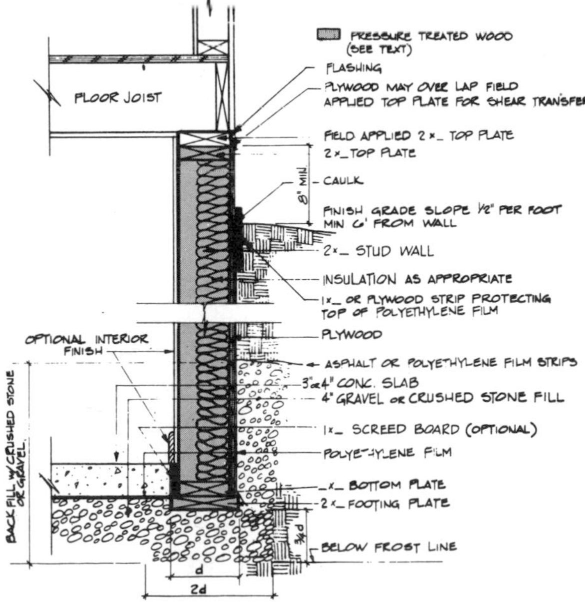

National Forest Products Assn.

Fig. 44. Diagram shows a typical basement wall of pressure treated wood with footing resting on gravel and with a 3" or 4" concrete slab inside the walls.

Site Preparation. For basement construction, the house site is excavated and leveled to the required elevation. Plumbing lines located below the basement floor are then installed and provision made for foundation drainage in accordance with code requirements for the area.

Where there is habitable space below grade, a sump provides assurance of adequate drainage. The sump should be 24" in diameter or 20" square and should extend at least 30" below the bottom of the basement slab. Positive drainage should be provided to channel water away from the sump. This may consist of a drain from the sump to a storm sewer, or where conditions require, a sump pump to keep water accumulation in the sump at a low level should be installed.

After grading and installation of plumbing lines and drainage system, the basement site is covered with a 4" thick layer of porous gravel. A thicker bed of gravel may be required under the foundation walls when 2x8 or wider footing plates are used.

For crawl space construction, trenches are dug to the required depth and gravel placed as a base for the footing plates.

Footings. The width of footing plates is determined by vertical loads on the foundation walls and the bearing capacities of the gravel base and soil. For most house designs and live load conditions, nominal 2" x 6" x 8" footings will be adequate when the bearing capacity of the gravel base and supporting soil are 3,000 pounds per square foot and 2,000 pounds per square foot, respectively. These bearing values are conservative compared to typical allowable values recognized for stable soils by model codes.

Brick Veneer Exterior Construction. When brick veneer construction is used, 2" x 10" or 2" x 12" footing plates are required to provide added width for supporting the veneer. Height of brick should not exceed 16' unless the knee wall footing plate and gravel base are designed to support greater height. To reduce the amount of brick required, the veneer may be supported below grade on a knee wall which rests on and is nailed to the treated wood footing.

The depth of the gravel bed under the footing plate for a continuous wall is ¾ of the required plate width. The bed is extended out from both edges of the plate a distance of ¼ the plate width. For columns or posts, the depth and width of gravel bed should be greater to accommodate the concentrated load. The bearing load is assumed to be distributed outward at a 30 degree angle from the edge of the footing plate to the soil under the gravel bed.

Gravel. Gravel may be crushed stone or bank or river gravel ranging in size from ⅜" to ¾".

Load Bearing Partitions. A load bearing partition or wall may be used as a center support under the first floor.

Foundation Walls

Design method. Lumber framing members in foundation walls enclosing a basement are designed to resist the lateral pressure of fill as well as vertical forces resulting from live and dead loads on the structure. The plywood sheathing is designed to resist maximum inward soil pressures occurring at the bottom of the wall. Soil pressures on the retaining walls are based upon the "fluid pressure" of the retaining earth. For most soils, the assumption of 25 to 30 pounds per cubic foot of equivalent fluid pressure is generally satisfactory. With soils high in clay or fine silt and low permeability, or with poorly drained soils, a higher equivalent fluid pressure should be used.

Crawl Space Construction. Foundation walls for crawl space construction are designed to resist only vertical loads when the difference between the outside grade and level in the crawl space is 12″ or less. Framing is covered with ½″ plywood shield and resistance to wind forces.

Plywood skins are designed as a 12″ wide beam under the uniform earth pressure load acting 6″ from the top of the bottom plate. Plywood structural properties are based on effective section properties and design stresses recommended by the American Plywood Association. Movements and shear due to lateral earth pressure are determined in accordance with methods given in "Concrete Masonry Foundation Walls" published by the National Concrete Masonry Association. Allowable shear capacities of wood foundation and wall construction are based on "Plywood Construction Systems" published by the American Plywood Association.

Construction of Foundation Walls. Treated foundation wall studs are end nailed to the top and bottom plates using two 16 penny nails. Treated plywood panels, ½-inch or thicker are attached to the framing using 8 penny nails spaced 6″ or less on centers along panel edges and 12″ or less on centers at intermediate supports. Six penny nails may be used when height of backfill against the foundation is approximately even on all sides of the building. Panels may be applied with face grain parallel or perpendicular to studs. Vertical joints between panels within wall sections should be located over studs.

Foundation wall sections are attached to wood footing and to adjacent sections using 10 penny nails spaced 12″ on centers and face nailed through the bottom plate and through studs.

Wall panel joints should not occur over joints in the wood footings. A field-installed untreated top plate is face nailed to the treated top plate of the foundation wall with 16 penny nails spaced 8″ on centers. An alternate method of construction is to fabricate the foundation wall sections to the edge of the plywood wall panels extending ¾″ above the top of the treated top plate. The field-applied top plate is then face nailed to the top plate of the wall section using 10 penny nails spaced 16″ on centers and is tied to the overlapping edge of the plywood using 8 penny nails spaced 6″ on centers.

Brick Veneer Walls. Knee walls used to support brick veneer generally are 2x4 studs spaced 16″ on centers, bearing on the footing plate and attached to the foundation wall studs using 16 penny nails spaced 16″ apart.

All nails used in the treated wood foundation must be aluminum, hot-dipped galvanized steel or stainless steel.

Joints in the field-applied top plate are staggered with those in the treated top plate of the wall panels. At corners and wall intersections, the field applied upper plate of one wall is extended across the treated plate of the intersecting wall to tie the building together.

The first floor is then installed following standard platform framing practices.

Height of Backfill. When the height of backfill on a wall parallel to the joists exceeds 4′, solid bridging at intervals of approximately 4′ should be installed between the first four rows of joists nearest the wall. This serves to transfer the lateral load from the top of the foundation studs into the floor system. Any other method which transfers this load may be used.

Plywood panel joints in the foundation wall should provide a space of ⅛″ for caulking with an appropriate sealant. In basement construction, the exterior sides of foundation walls are covered below grade with a 6-mil polyethylene vapor barrier. The film is bonded to the plywood, and joints sealed and lapped 6″ with a suitable construction adhesive.

The top edge of the vapor barrier is completely sealed to the plywood wall with adhesives. Film areas near grade level are protected from mechanical damage and exposure using treated plywood, brick, stucco or other covering appropriate to the architectural design. Where plywood is used for this purpose, the atop edge of the 12″ wide panel of treated material is attached to the wall several inches above the finished grade level. The top edge of the panel is caulked full length before the panel is fastened to the wall.

For habitable basement space, insulation may be installed between studs and an interior vapor barrier and interior finish are applied to wall framing.

Basement Floors

For basement construction a polyethylene film 6 mils thick is applied over the gravel bed, and a concrete slab at least 3″ thick is poured over the film. The slab should be high enough to provide at least 2 square inches of bearing against the bottom of each stud to resist lateral thrust at the bottom of the wall.

A nominal 1″ continuous strip of treated wood nailed to the studs at proper height serves as a guide to leveling concrete. The pouring of the slab can be delayed until after the house is under roof. Backfill should not be placed against the foundation walls until after the concrete slab is in place and set and the top of the wall is adequately braced. Where height of backfill exceeds 4′, gravel should be used for the lower portion.

Center Support

Framing members in the load bearing walls acting as center support for the first floor joist system are designed and constructed in the same manner as exterior foundation walls except that vertical loads only are considered.

In some basement construction, the loadbearing interior wall is made with pressure treated bottom plates and with untreated studs and top plates. A double bottom plate is used to provide clearance between the top of the concrete slab floor and the untreated vertical framing in the wall.

In areas of high termite hazard, pressure treated studs should be used. The interior wall framing is finished both sides as desired. Interior load bearing walls in crawl space construction are made of treated framing and should be braced diagonally.

Interior bearing wall sections are attached to each other by foundation walls using 10 penny nails spaced 12″ apart by face nailing the end studs of the interior wall to the edges of studs in the outside wall, or to blocks nailed between exterior wall studs. The upper top plate of the interior wall should be extended across the treated top plate of the exterior wall.

Interior bearing wall sections are attached to each other by face nailing studs using 10 penny nails spaced 12″ on centers.

Where a post and beam center support system is used, posts should be supported on blocks of sufficient strength that the vertical concentrated loads do not exceed the bearing capacity of the gravel base and soil. Foundation walls supporting the center girder should be designed to carry and distribute the concentrated load from the girder to the wood footing and gravel base, such that allowable gravel and soil bearing capacities are not exceeded.

The Plen-Wood System

An interesting outgrowth of the pressure treated wood foundation system is an underfloor plenum heating/cooling method developed by the wood industry. The system was developed by the American Plywood Assn; American Wood Preservers Institute; Southern Forest Products Assn; and Western Wood Products Assn.

Basically the system makes use of a crawl space under a house which is tightly sealed, insulated and used as a plenum for a conventional down flow heating and air conditioning unit. Registers in the wooden floor permit the flow of warm or cool air. The walls of the plenum may be pressure treated wood or standard concrete blocks or poured concrete. Clearance under joists is 18″ to 24″.

Rough plumbing is brought into the crawl space before the foundation walls go up. Fuel lines and plumbing waste cleanouts are not within the plenum. A hatch in the floor provides for access to the plenum for inspection. An open system is used for return air inside the house. With conventional rectangular-shaped house, uniform air distribution is achieved without stub ducts.

Cost estimates comparing the Plen-Wood System with concrete slab construction by the NAHB Research Foundation show that as much as $216 to $334 can be saved by using the Plen-Wood system, **Figure 45.**

Western Wood Products Assn.

Fig. 45. The Plen-Wood heating/cooling system uses an all-wood or concrete perimeter foundation and a wood floor to create an enclosed air distribution chamber, as a cost-saving alternate to slab construction.

Chapter 3

SILLS/GIRDERS/JOISTS/SUBFLOORING

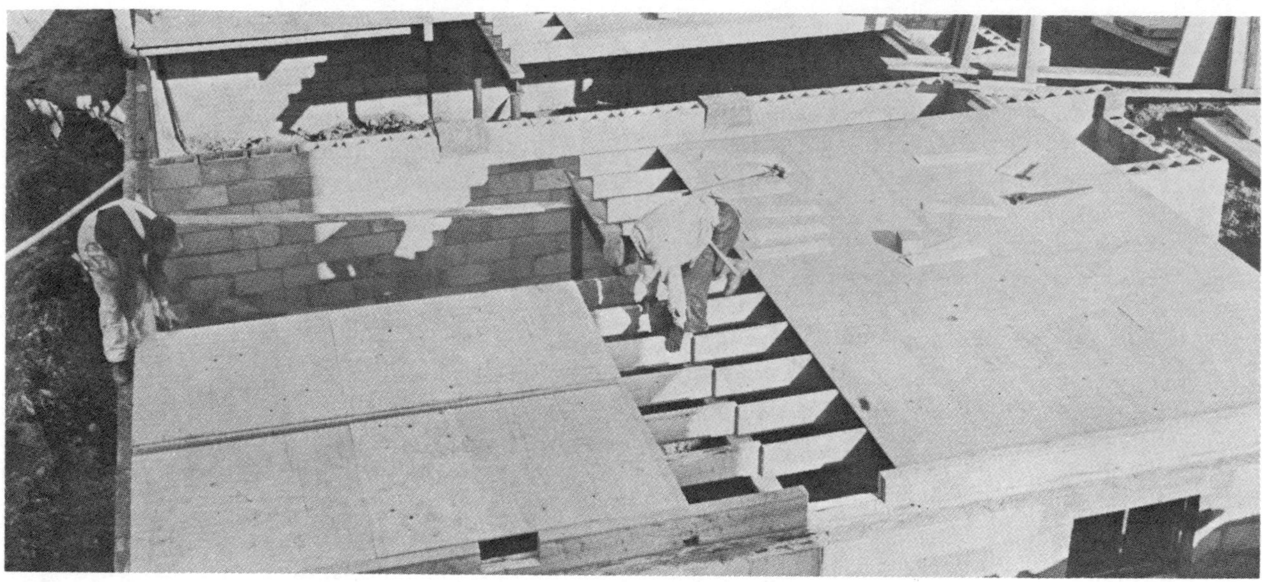

Fig. 1. Joists bear on sills atop foundation walls, and on central girder supported by steel lally columns.

The sills, girders, floor joists and subflooring are the first members of the wood-framed superstructure placed on the foundation walls after the concrete of the foundation walls has set and the forms are removed. When completed this superstructure becomes a platform on which walls may be framed, **Figure 1.**

Figure 2 illustrates a corner of a balloon framed house. This is an assembly of the several members as they are built up. Note the lower sill member embedded in mortar grouting to provide a tight seal and true levelling. The other members shown are the sill header, the floor joists, the subflooring, and the outside wall studding and corner post.

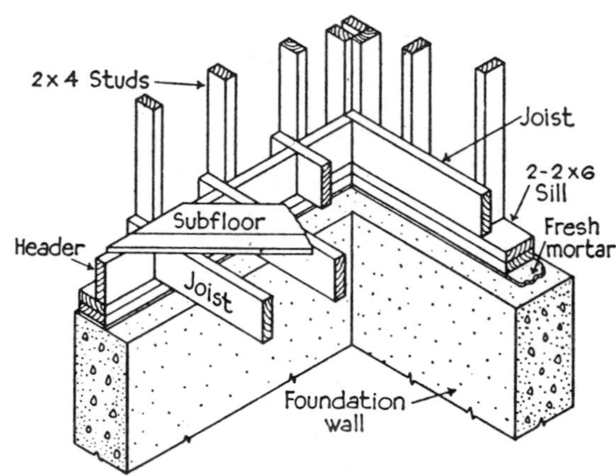

Fig. 2. Sill is embedded in fresh mortar to level it for the joists and framing erected on it.

Sills rest on the foundation walls, carrying all of the dead and live loads of the frame structure except the loads carried by interior girder posts, or by inside bearing walls. Sills are usually made of wood and are held to the foundation wall by anchor bolts. Holes are bored into the sill to fit over the anchor bolts and the sill is secured firmly by washers and nuts.

The drawing at left in **Figure 3** shows the corner construction of a plain sill and the use of corner bolts or anchors embedded in the foundation. At the right is shown the lapping of a solid sill and the anchoring and lapping of a built up sill.

Sills are not needed when the exterior walls are of brick or concrete. Brick veneered walls require a sill and wooden framing.

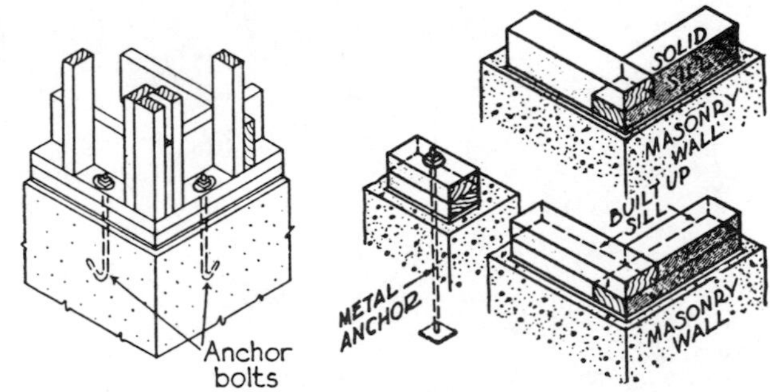

Fig. 3. Anchor bolts secure sill to foundation. Note overlapping corners of solid sill and built-up sill.

Fig. 4. Sill plate bolted to edge of slab foundation.

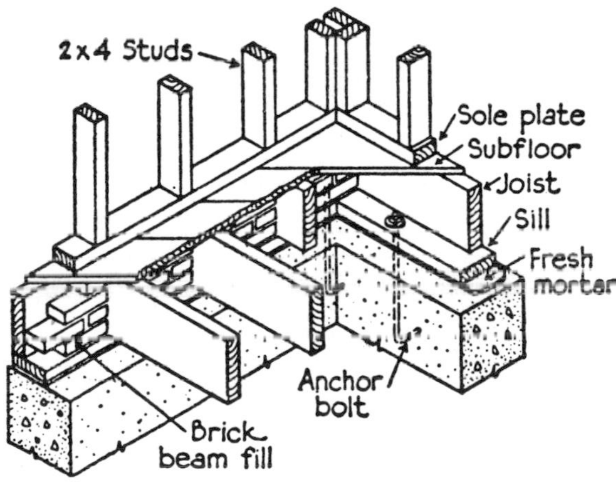

Fig. 5. Box sill assembly with brick fill between joists.

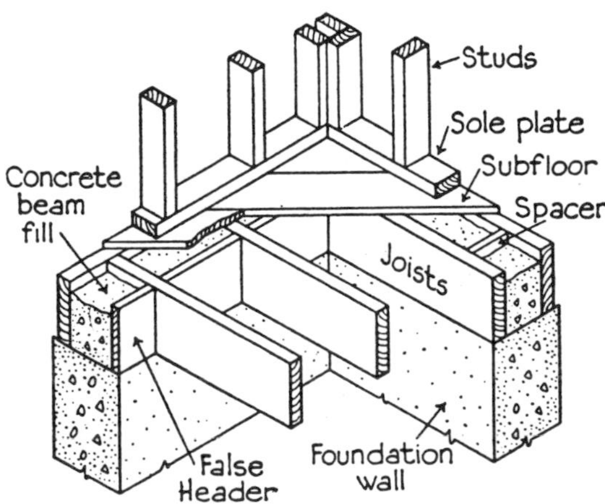

Fig. 6. Box sill assembly without sill plate, concrete fill.

A Sill for a Slab Foundation. Figure 4 shows a wood sill plate bolted to the edge of a slab foundation which is now ready for the placing of factory made panels or components. Note the shoe sills nailed in place for the interior wall partitions. When a house is assembled in this manner, the exterior and interior walls, and possibly bath and utility components are soon under roof.

Types of Sills. In the construction of a box sill a continuous header is placed even with the outside edge of the sill plate which is bolted to the foundation wall. Joists at right angles to the header butt against it and bear on the sill.

Another type of box sill is constructed with a sill plate and sill header at the outside edge of the plate as shown in **Figure 5.** A brick fill between the floor joists is a fire and draft retardant. A box sill assembly of a sill plate is shown in **Figure 6.** Concrete fill is enclosed between the outside header placed directly on the foundation wall, at the outside edge, and short, false headers are nailed between the joists. The inside header running parallel to the joists becomes a joists and an inside member of the sill assembly. The concrete fill acts as a draft stop, fire retardant and termite shield. The box sill type of construction is generally used in the platform frame construction of a house. Often the same sill assembly is repeated at the second floor level.

In the balloon style of framing, **Figure 7,** the exterior studding extends from the sill plate at the foundation to the rafter plate which supports the roof rafters.

The T-sill type of construction, **Figure 8,** has the sole plate resting on the sill rather than on the sub-flooring.

Girder Construction

Girders are load bearing members giving intermediate support to the joists supporting the ground floor. Usually a girder runs midway between opposite parallel foundation walls. In this way joists need not be too long and will support the floor load without deflection. The ends of the girder usually rest on the foundation walls and one or two supporting posts of wood or steel. The steel post is called a lally column and may be used with a girder of wood or a steel I-beam.

If the girder ends rest on the foundation wall a pocket or housing space is provided in the wall, **Figure 9.** This method of framing is designated as platform frame construction. This permits the subsequent joist placement to be at the same height as the sill plate. The girder pocket does not extend through the entire width of the foundation wall. Sometimes a masonry pilaster is built as an integral part of the foundation wall to support the ends of the girder at the same level as the sill plate.

Figure 10 illustrates a steel I-beam girder supported by a steel lally column. The ends of the I-beam may be housed in a recess in the foundation wall or on a pilaster.

Along the length of the girder are one or two columns, depending on the length of the span, which rest on footings in the basement floor. A wood sill or sleeper laid on the I-beam supports the joists, provides for nailing.

Steel Girders and Steel Joists. Figure 11 shows prefabricated plywood floor panels being placed on joists of steel, which are supported, in the center, by a steel girder and columns. The ends of the joists, at the walls, rest on the sill plate and are secured by nailing flanges.

A steel I-beam girder, 8″ deep and weighing 18.4 pounds per linear foot, will carry 19,000 pounds in a 9′ span. The same method of determining the proper size of the beam is used for both steel and wood members. The lightest weight for any given depth is always the strongest for the amount of material used. Steel beams vary in weight, depth, and thickness of web flanges. If the total load on a steel girder is 13,500 pounds of uniformly distributed weight, and the span or distance between basement piers is 9′, then a 7″ by 15.3 pounds per linear foot I-beam is the proper size to use, since it will carry 13, 800 pounds.

A 6″ girder weighing 12.5 pounds per foot will carry 21,800 pounds safely over a span of 4′. A 6″ deep girder weighing 17.25 pounds per foot will carry 10,400 pounds safely over a span of 10′. A 10″ deep girder weighing 25.4 pounds per foot will carry 14,700 pounds safely over a 20′ span.

Girders Made of Wood. A wooden girder, built-up or solid, may vary in cross section size from 2″ × 6″

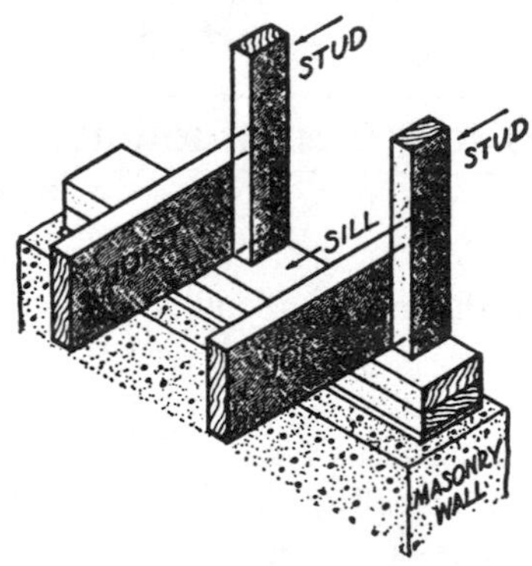

Fig. 7. Sill and joist construction for balloon framing.

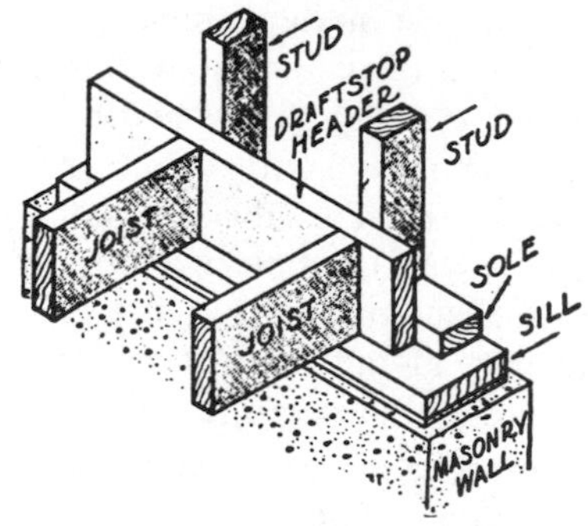

Fig. 8. T-sill construction with sole resting on wide sill.

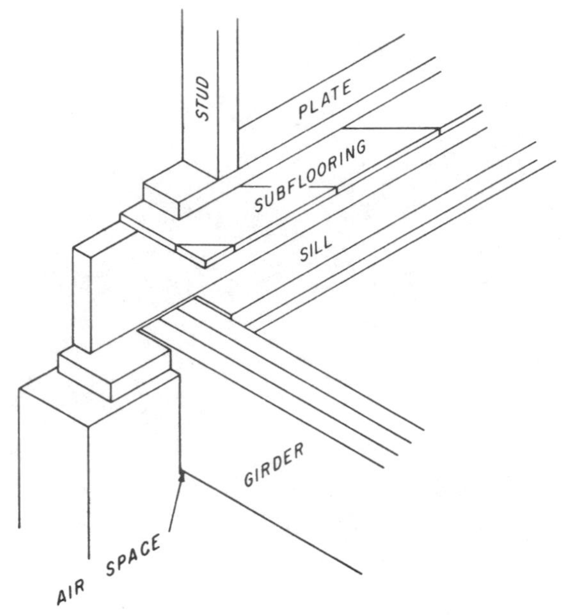

Fig. 9. A built-up wood girder housed in a recess or pocket in the foundation wall. This method is used in platform frame construction. Air space protects girder from moisture.

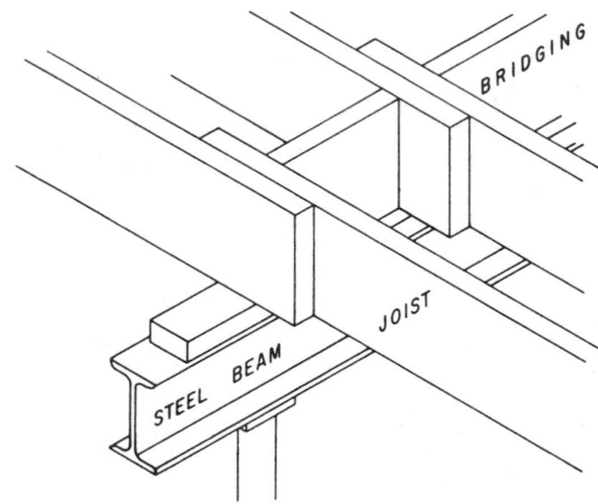

Fig. 10. Steel I-beam girder supported along its length by lally columns. Sometimes a wood sill is placed on top of I-beam to support the overlapping joists and bridging.

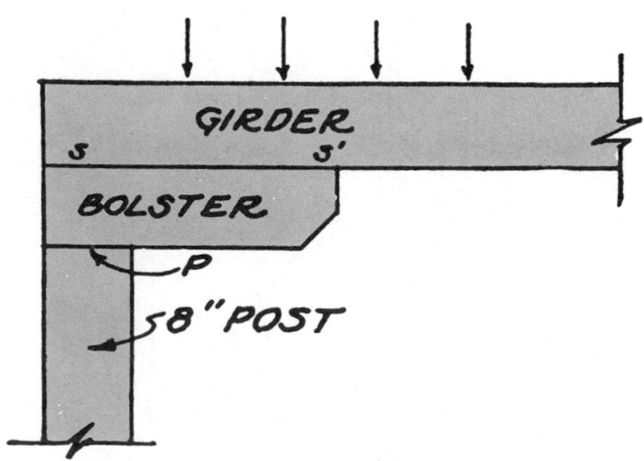

Fig. 11. Bolster plate on post provides broader bearing for girder.

to 14″. The actual size of a 2″ × 6″ piece of framing lumber is 1-5/8″ × 5-5/8″. The actual size of an 8″ × 10″ piece of framing lumber is 7-1/2″ × 9-1/2″. The size of a girder needed will vary with the species and grade used. If a built-up girder is used, special percentage figures applying to built-up members must be used, to allow for the lesser thickness of dressed material. Four 2″ × 10″ members are not as thick as one 8″ × 10″ piece. A built-up girder is as satisfactory as a solid one, and may be better, since it affords an opportunity to select and arrange the material for the best possible results.

A solid beam of dressed material 12″ × 14″ over a span of 8′-5″ will carry a load of 20,140 pounds. A girder 10″ × 10″ over a span of 10′ will carry a load of 8,523 pounds. A #1 common grade of Douglas fir can withstand a load of 1,000 pounds per square inch, with the grain, but only 325 pounds per square inch across the grain.

A wood girder supported at each end and carrying a uniform load of 22,000 pounds, has a force of one half of the weight, or 11,000 pounds, pushing up at each support. This 11,000 force, or reaction, is distributed over only that small girder that rests on the support. If the wood has a safe crushing strength across the grain of 325 pounds per square inch, then enough square inches of girder area must be directly supported to add up to 11,000 pounds or about 34 square inches.

If the girder were 8″ wide (actually 7-1/2″) then it would need a 5″ long bearing on the support to make up the 34 square inches required. The upright wooden post would be pushed along its grain and would require a bearing area of 11,000 divided by 1,000 or only 11 square inches. If a larger bearing is needed for the girder than is provided by the head of the post, a hardwood bolster or metal plate is used, **Figure 11.** The post is strong enough to resist crushing at point 'P' as the bolster distributes the girder bearing area along points (S - S').

Framing Joists and Girders of Wood

To equalize shrinkage of interior supports with that occurring in sills around the building perimeter, joists should be set on ledgers. Construction to equalize shrinkage in balloon framing is shown in **Figure 12.** To equalize shrinkage in braced and western type of framing, use the method illustrated in **Figure 13.** Joist hangers of various shapes and sizes can also be used to frame joists level with the top of girders, **Figure 14.**

A girder can be set into a foundation wall so that the bottom edges of the joists rest level with the top of the sill, **Figure 15.** When a built-up girder bears on a sill, set it on a pillow block to spread the weight over a larger area, **Figure 16.**

Joists are load bearing members of wood, steel, or concrete, the ends of which rest on the sills, if the span is narrow, or on sills and interior girders, if an intermediate support is needed. Joists are placed on edge, and must be of sufficient strength and number to support the floor loads with so little deflection that there need be no fear of plaster cracking. Joists are usually placed 16″ on center.

Joists are most often positioned on top of the sill and the girder. Joists can be notched to a lower position in the framing if supported by a ledger strip, or they can be positioned level with the top of the sill and girder if suspended by metal stirrups. Wooden joists are framed into an I-beam either by fitting the joist flush, or by resting it on a wooden ledger strip bolted into the I-beam or with the lower joist ends notched for a closer fit.

Another way of framing the tops of the joists to the level of the sill is to notch the inside edge of the foundation to the approximate height of the joists. This allows the subflooring to be framed out over the sills and starting the sole plate and studs immediately above the subflooring. In this way the house line is lowered thus saving on exterior wall material and labor.

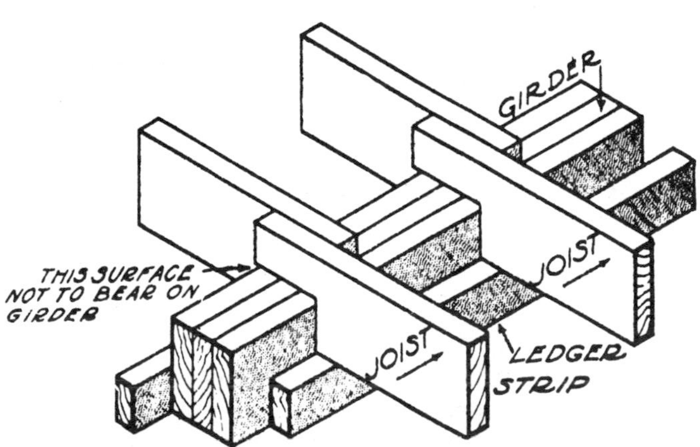

Fig. 12. Girder construction to equalize shrinkage across widths of joists. Used in balloon framing.

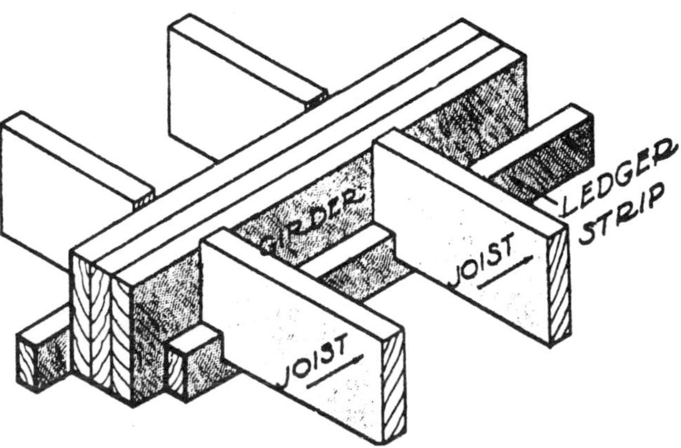

Fig. 13. Ledgers equalize shrinkage, bring tops of joists flush with top of girder. Used in western type framing.

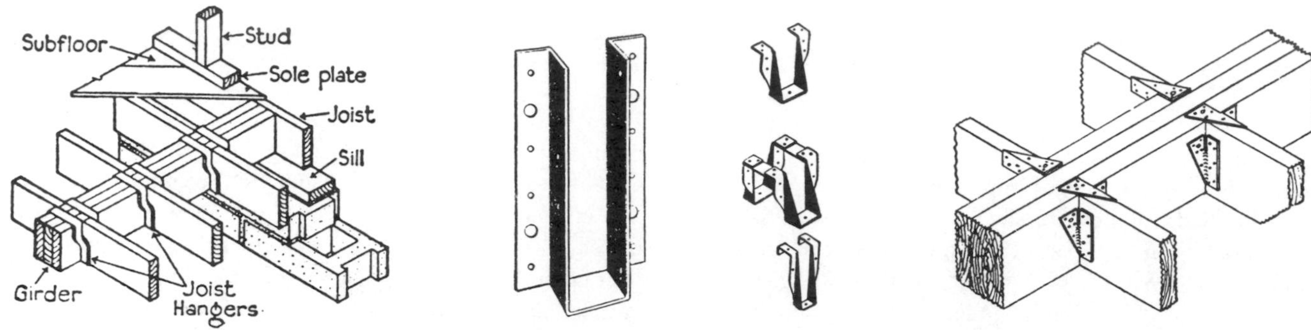

Fig. 14. Metal joist hangers of various shapes and sizes frame joists level with girder, eliminate toenailing.

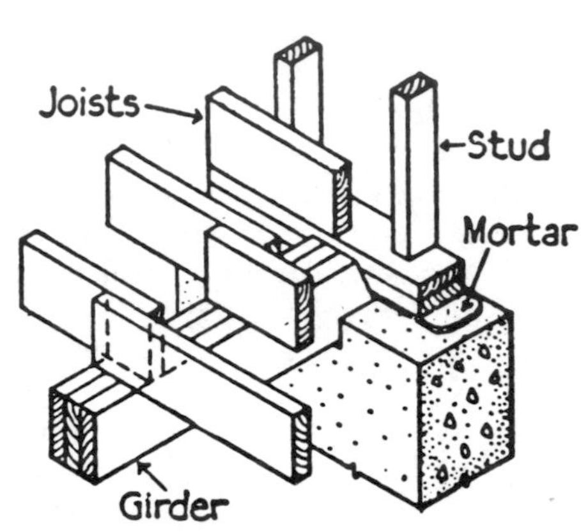

Fig. 15. Girder is set into foundation so its top is flush with top of sill.

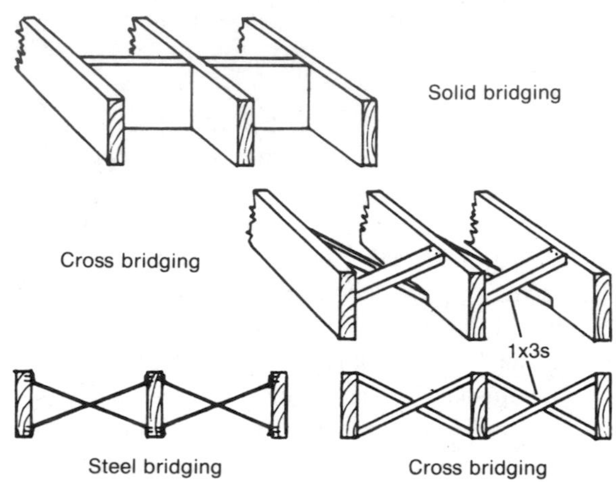

Fig. 17. Types of bridging used to stiffen joists. Space about 7′ apart.

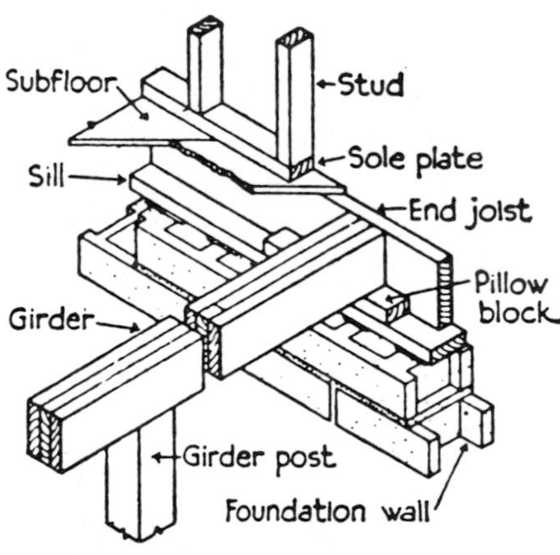

Fig. 16. Built-up girder is set on pillow block to spread weight on sill.

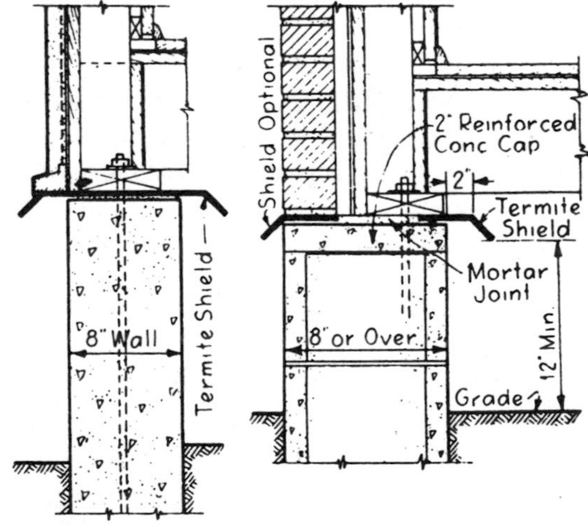

Fig. 18. Continuous metal termite shields have edges turned down 45 degrees.

Preparing Wood Joists for Subflooring

In conventional house construction floor joists of wood will deflect from side to side to some extent even though being firmly secured at places of bearing on the sills and girder. To overcome this tendency to deflect, pieces of wood, usually 1″ × 3″ in size, are nailed crosswise between the top and bottom of adjacent joists at about the center of the span, **Figure 17.**

There are numerous types of metal bridging available, some requiring no nailing. Joists having a span of 12′ need one line of bridging; two lines of bridging are advised for joists having a span of nearly 18′.

Termite Shields

Metal shields are necessary in most sections of the country to protect wood framing against destructive attacks by wood-eating termites. Since termites come from the soil around a house, the common practice is to place the shield, a continuous plate of metal, with edges protruding and turned down about 45 degrees just above the foundation wall, **Figure 18.** The shields are imbedded in cement mortar, just below the mortar bed of the sill. Toxic treatment of the framing lumber is recommended as an added protection against termites. Poisoning the soil around foundations is also recommended.

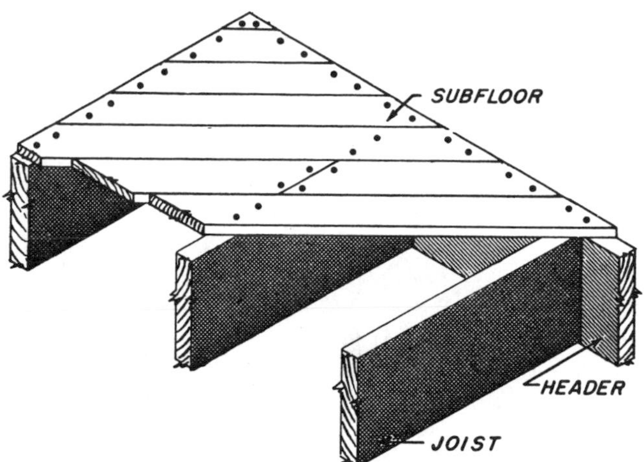

Fig. 19. Diagonal subflooring laid on joists with two 8d nails at each bearing place. End joists occur on joists, are staggered for maximum strength and smooth surface.

Laying Subflooring

Subflooring, when placed over the sills and joists, forms the platform of the first floor on which all subsequent framing rests. The flooring material is made of dressed boards butted edge to edge, usually laid at an angle of 45 degrees to the joists and the walls. Subflooring is placed diagonally to form a firm, rigid platform for a finished flooring of wood which is generally laid parallel with the longest wall in each room. Subflooring boards are either 6″ or 8″ to 12″ wide, oft-time butted edge to edge, or of tongue and groove or shiplapped at the edges. The boards are nailed firmly to resist twisting and warping. Subflooring can be nailed across the joists and at right angles to them, but finished flooring can then be laid in one direction, regardless of room dimensions.

Figure 19 shows the placing of butted diagonal subflooring on joists at the corner of a building with two 8 d. nails at each bearing place. Note the end butting of stock over a joist. Plywood subflooring should be placed with the grain of the surface plies at right angles to the joist and nailed with 8 d. nails spaced 5″ along all edges and 10″ along intermediate members, **Figure 20.** Sole plates for wall framing are nailed directly over the subfloor, **Figure 21.**

Sometimes a layer of pressed board is laid over the rough subflooring to serve as underlayment for mastic and resilient tiles; or heavy plywood serves as both subflooring and base for tiles.

Tongue and Groove Panels Used As Flooring. Developments in the factory manufacture of large panels make possible large single sheets of plywood and of wood fiber panels. Plywood is available in seven ply sheets, 1-1/8″ thick, 4′ wide and 8′ long with precisely engineered tongue and grooved joints to provide solid connections, **Figure 22.** These plywood sheets are stronger, tighter and more economical to use than narrow boards.

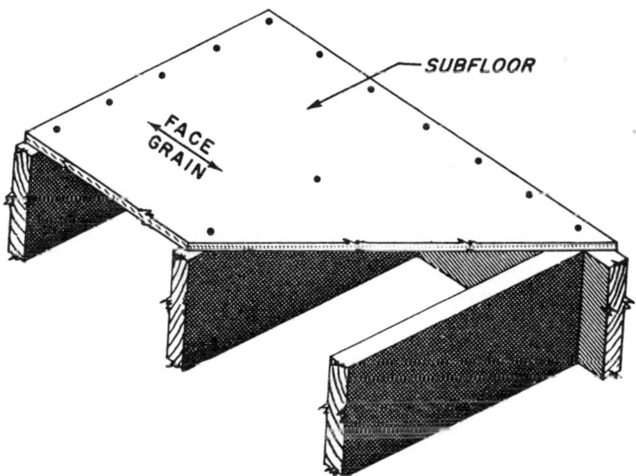

Fig. 20. Plywood subflooring placed with face grain at right angle to joists and fastened with 8d nails spaced 5″ along edges and 10″ along intermediate members.

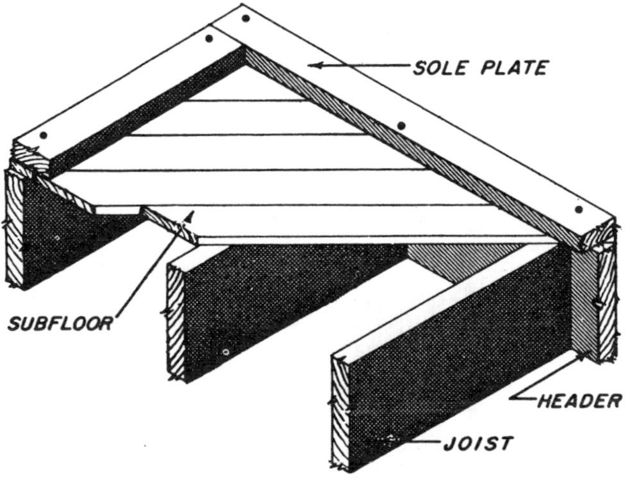

Fig. 21 Sole plates for wall framing are placed over subflooring and nailed, except where door openings occur. Studs are toenailed to sole plates.

Fig. 22. Tongue-and-groove plywood sheets, 1-1/8" thick, can be used as subfloor and underlayment in one layer.

Fig. 23. T&G panels can be laid on beams spaced 4′ o.c. End joints must butt on beam, but T&G edge joints require no blocking or nailing.

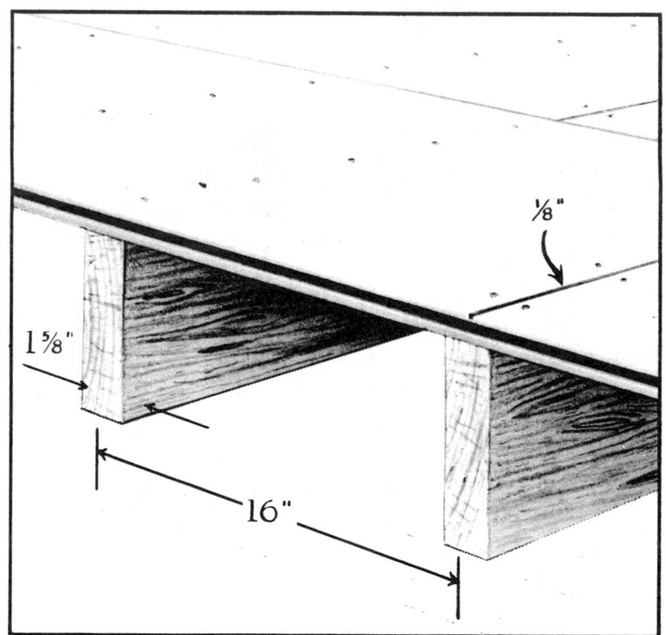

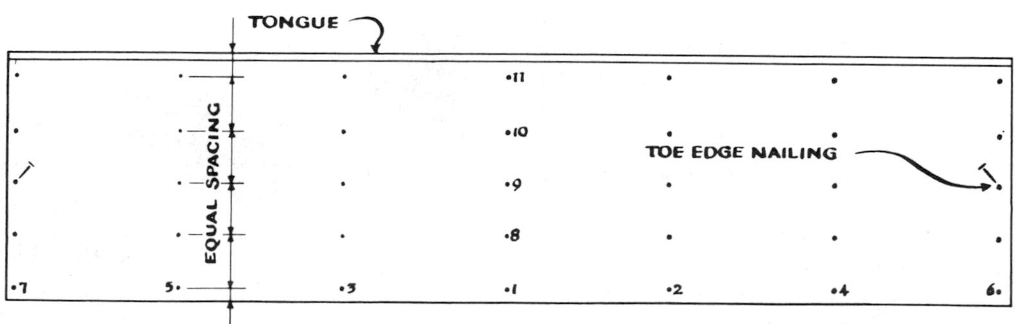

Fig. 24. Tongue-and-groove wood fiber panels, 1-7/8" thick, can also be laid over joists as combination subfloor and underlayment as shown in upper illustration. A diagram for nailing in numerical order is shown above.

Any of the several support systems may be used with the plywood panels. Some suggested methods are box beams, 4'' or two 2'' joists spiked together. Beams can be spaced up to 48" on center, **Figure 23.** When panels are nailed in place the face grain of the panel should run across the main beams, and wherever possible, cover two spans. If the panels are square edged, 2'' × 4'' blocking is required under the edges between the beams. If both sides and ends of the panels are tongue and grooved drive the side joints tight first. Then drive the end joints tight.

When the floor covering is of the thin resilient type, fill any cracks 1/6'' or wider and sand the joints lightly if they are not absolutely flush.

To achieve the greatest resistance to nail popping, and the maximum in withdrawal strength, use 8d. common ring shanked or helically threaded nails spaced 6'' o. c. at all bearing points. Ten penny common smooth shanked nails may be substituted if desired. If the floor covering is of resilient tile, and the beams are not fully seasoned, set all nails 1/8'' in the panels but do not fill. Set the nails just prior to laying the resilient flooring to take advantage of the beam seasoning that has taken place.

Wood fiber panels, 1-7/8'' thick by 2' × 8', are installed as floor decking over conventionally framed 2'' × 10'' or 2'' × 12'' joists of wood spaced 16" on center, **Figure 24.** The diagram shows the order of placing the nails in each panel. Fit the tongue and grooved side edges together snugly and proceed to nail the tongue and groove together, starting from the center of the panel toward each end. Apply the first eleven nails in numercial order shown, using a minimum of five nails per joist, spaced 6" o.c., and countersink the nails approximately 1/6", but do not fill.

The squared end of the panel should be toe nailed into the supporting member, allowing 1/8" separation between squared ends of the panels.

Module Component Construction of Floor Panels. In the factory built floor component, large sections spanning up to 16' are built up as shown in the details at left in **Figure 25.** The panels are set in place on the foundation walls and supporting beams and form the first units of a componentized house. Single panels of plywood with 2 × 4 stringers nailed to them prior to installation can be dropped in place quickly as shown at right in **Figure 25.** The stringers bear on ledgers nailed to beams, and panel edges butt on beam centers.

Glued Construction of Subflooring and Joists. A method of applying plywood subfloors to joists pioneered by the American Plywood Association using both glue and nails eliminates squeaks, nail popping, and increases stiffness of both plywood and beams from 10% to 90%. The bond of the mastic type glue is so strong that joists and plywood behave like integral T-beam units. The strength of the glue bond is greatly increased if the adhesive is also applied to the tongue-and-groove joints of the 4' x 8' panels. The system greatly reduces costs since it makes possible single-layer floors of ½", ⅝", and ¾" plywood over joists ranging from 2x6's to 2x12's.

Application Method. A chalkline is snapped across the joists 4 feet from wall. This serves as a guide for panel align-

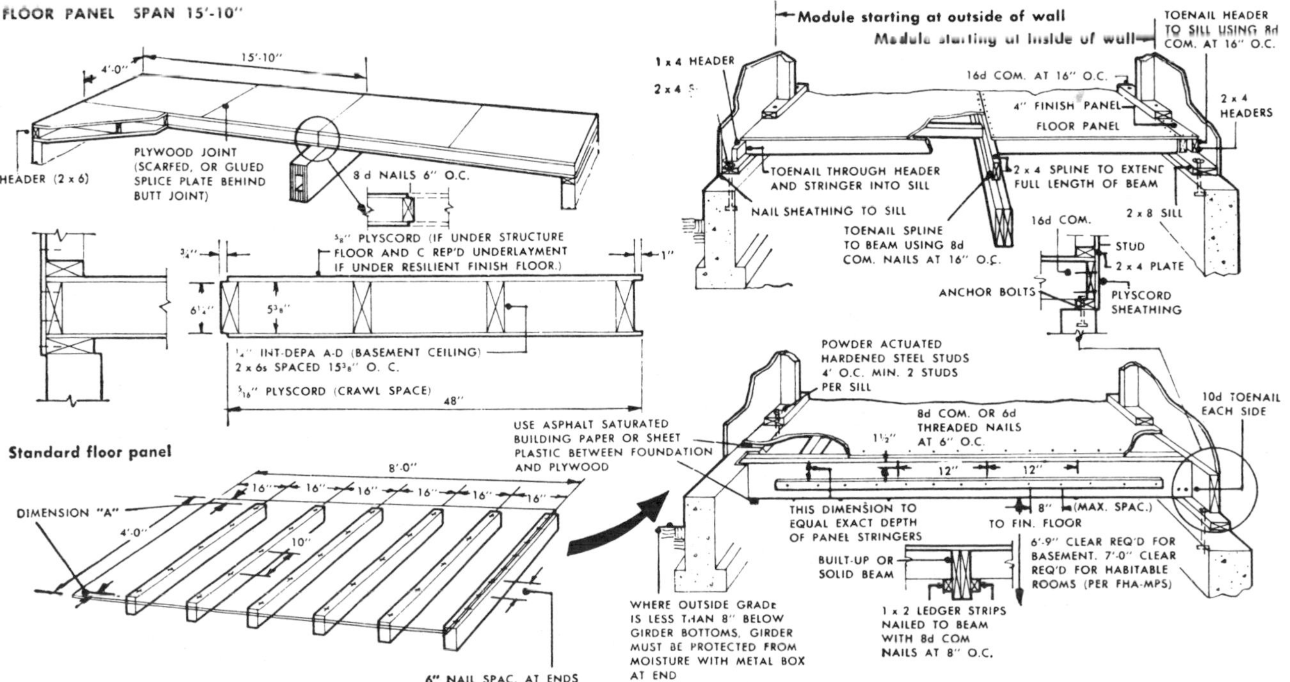

Fig. 25. Construction details of modular component floor panels used in conventional and prefabricated houses.

Fig. 26. Snapping a chalkine across joists for correct alignment of plywood panels.

Fig. 27. Spreading glue on joists using a carbon dioxide gun.

Fig. 28. First panel is laid on glued joists with tongue next to wall.

Fig. 29. Double row of glue is applied where panel ends meet.

Fig. 30. Plywood panel is dropped down on glued joists. T and G edges are also glued for stronger bond.

American Plywood Assn.

Fig. 31. Panels are tapped into position with block and hammer so that tongue and groove fit tightly.

ment and boundary for spreading glue. Glue is spread either with a hand caulking gun from cartridges or with a pneumatic (CO_2) gun. Only enough glue is spread for one or two panels at a time.

First panel is laid with tongue side against wall. This protects tongue of next panel from damage while being tapped into place. Where two panels meet there are double rows of adhesive. Panel joints are offset and nailed before the glue sets. A 1/16″ space is allowed between all end and edge joints. **See Figures 26, 27, 28, 29, 30, 31, 32.**

The single-layer flooring is ideal for resilient flooring since the plywood used is of the underlayment type and is considered excellent for carpets. The precise clear span for joists and the various kinds and thicknesses of plywood used with them can be determined from tables supplied by the American Plywood Association, 1119 A St., Tacoma, Wash. 98401.

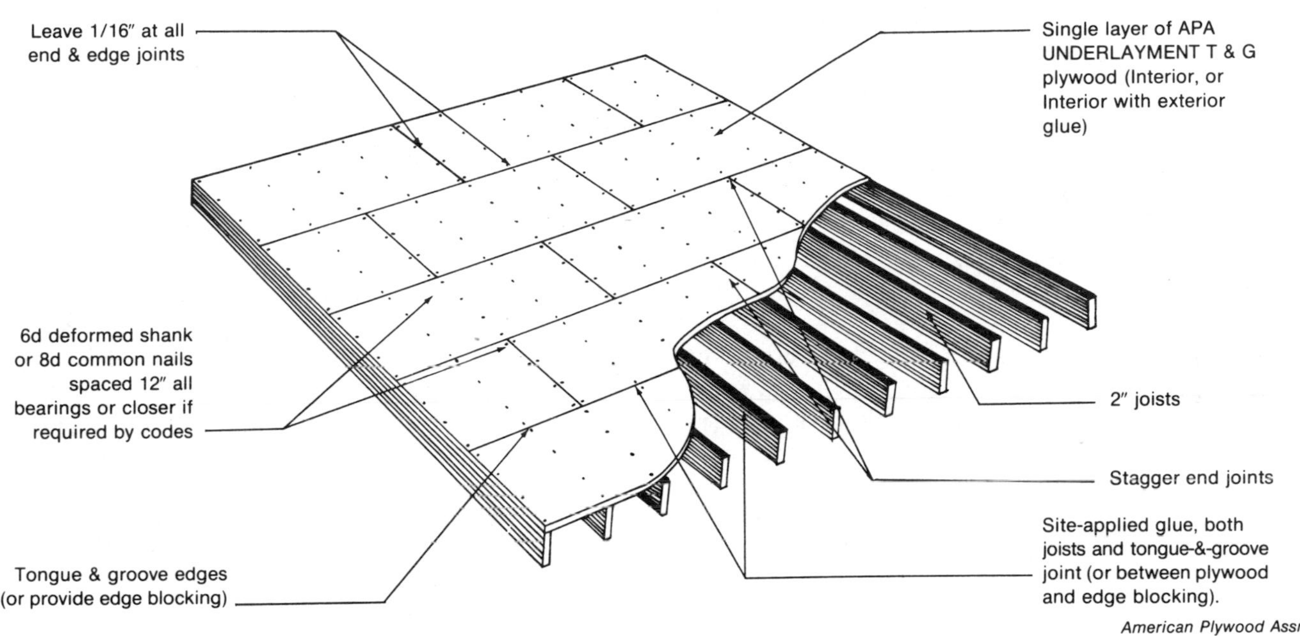

Fig. 32. Nailing system for glued plywood subflooring.

SILLS/GIRDERS/JOISTS/SUBFLOORING

Chapter 4

OUTSIDE WALL CONSTRUCTION

The exterior or outside walls of a house rest directly on the sills or sills and subflooring attached to the foundation walls. These walls are bearing units for the upper floors if any, and the roof of the structure.

Platform Frame Construction. In platform construction, the subflooring extends to the outside edges of the building and provides a platform upon which exterior walls and interior partitions can be erected. Platform construction is the type of framing most generally used for one-story houses **Figure 1.** It is also used alone or in combination with balloon construction for two-story structures. Building techniques in some parts of the country have been developed almost entirely around the platform system.

Platform construction is easier to erect because it provides a flat surface, at each floor level, on which to work. It is also easily adapted to various methods of prefabrication. With a platform framing system, it is common practice to assemble the wall framing on the floor and then tilt the entire unit into place. Exterior walls must resist the lateral loads resulting from winds, and in some areas, from earthquakes. Top plates should be doubled and overlapped at wall and bearing partition intersections. This ties the building together into a strong unit.

Studs in exterior walls are placed with the wide faces perpendicular to the direction of the wall. Studs should be at least nominal 2 × 4 inches for a one or two story building. In three story buildings, studs in the bottom story should be at least a nominal 2″ × 6″ in size. In one story buildings, studs may be spaced 24″ on center, unless otherwise limited by the wall covering. In multi-story buildings, spacing should not exceed 16″ on center. An arrangement of multiple studs is used at the corners to provide for ready attachment of exterior and interior surface materials.

Balloon Frame Construction. In balloon frame construction both studs and first floor joists rest on the anchored sill, **Figure 2.** The second floor joists bear on a 1″ × 4″ ribbon strip which has been let into the inside edges of the studs running continuously from sill to roof. Balloon framing is a preferred type of construction for two story buildings where the exterior covering is of brick or stone veneer or stucco, as there is less likelihood of movement between the wood framing and the masonry veneer.

Where exterior walls are of solid masonry, it is also desirable to use balloon framing for interior bearing partitions. It eliminates variations in settlement which may occur between exterior walls and interior supports. There is less vertical shrinkage in the two-story studs than occurs across the grain of headers and sills.

Openings In Exterior Walls. Where doors or windows occur, a header of adequate size is needed to carry the vertical load across the opening, **Figure 3.** Ends of the header may be supported on studs or framing anchors when the span does not exceed 3′. Where the opening exceeds 6′ in width it is good practice to use triple studs with each end of the header resting on two studs.

A Continuous Header. A continuous header consisting of two 2″ members set on edge, may be used instead of a double top plate, **Figure 4.** The depth of the members will be the same as that required to span the largest opening. Joints in individual members should be staggered at least three stud spaces and should not occur over openings. Members are toe-nailed to studs and corners. Intersections, with bearing partitions, should be lapped or tied with metal straps.

Outside Corners and Studs at Partitions. Two methods of assembly of outside wall studs at corners are illustrated in **Figure 5.** These stud arrangements provide a more rigid corner post and better nailing surface for exterior and interior wall coverings. **Figure 6** shows two methods of framing the outside wall studs to provide adequate nailing surfaces for the interior partition framing. In both instances the lower member or plate is often referred to as the sill or sole plate, and is nailed to the subflooring, to distinguish from the top plate of the outside framing on which the rafters are supported.

Stud Size and Spacing. Studs in exterior walls are placed with the wide faces perpendicular to the direction of the wall. Studs should be at least the nominal 2″ × 4″ in size for the one and for the two story structures. In three story buildings, studs in the bottom story should be at least the nominal 2″ × 6″ in size. In one story buildings, studs may be spaced 24″, on center, unless otherwise limited by the wall covering. In multi-story buildings, spacing should not exceed 16″.

Fastenings. Nails are used most generally for fastening 1″ and 2″ framing lumber. A few recommendations of the quantity and size of common nails to be used in fastening certain members in house framing are shown in the chart on page 49.

Other types of nails, including those with annular or spiral grooves, have demonstrated higher load-carrying capacities, but their use is often limited to special purposes. They have particular value where high withdrawal resistance is required.

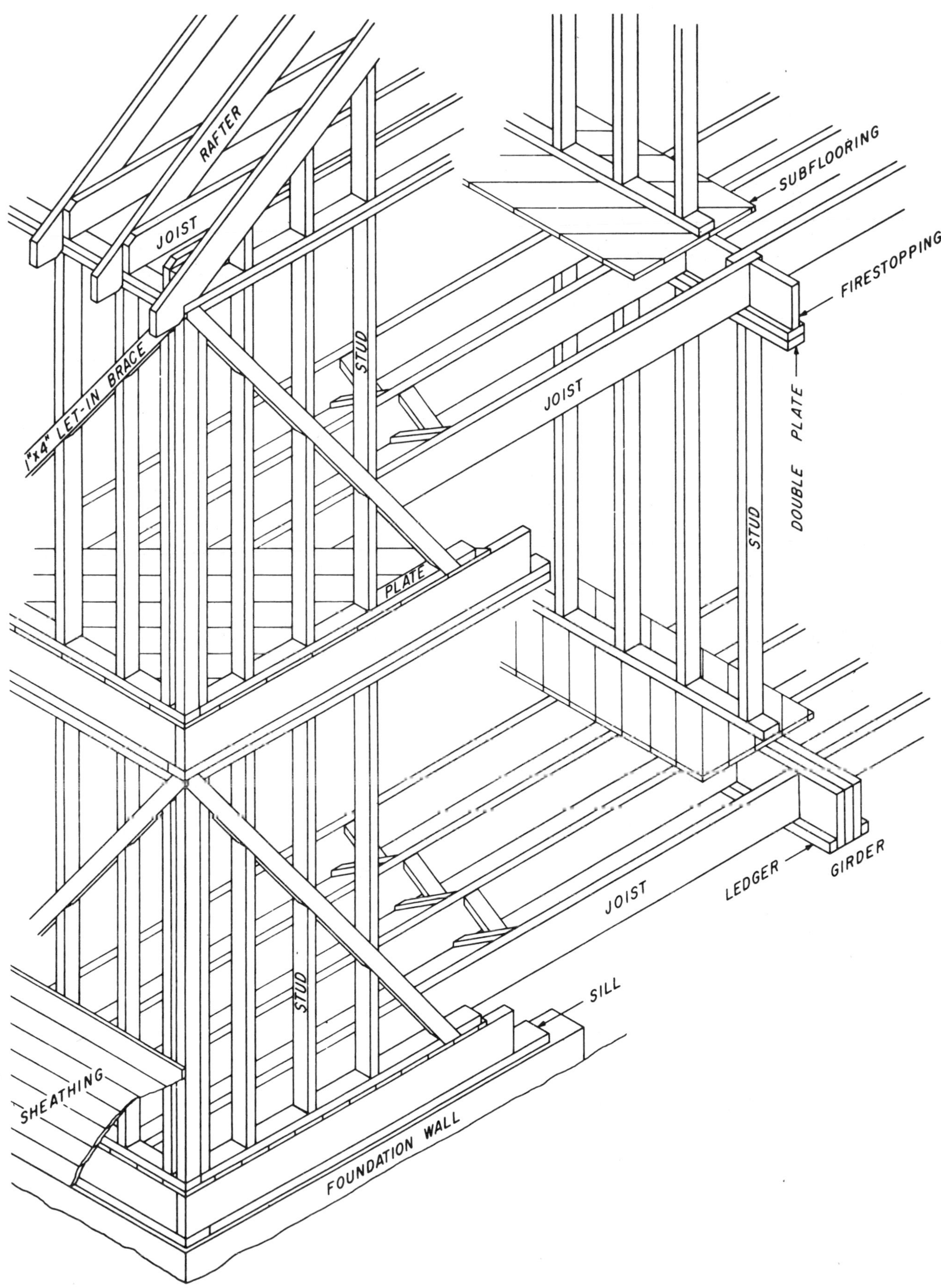

Fig. 1. Structural details of typical platform frame construction.

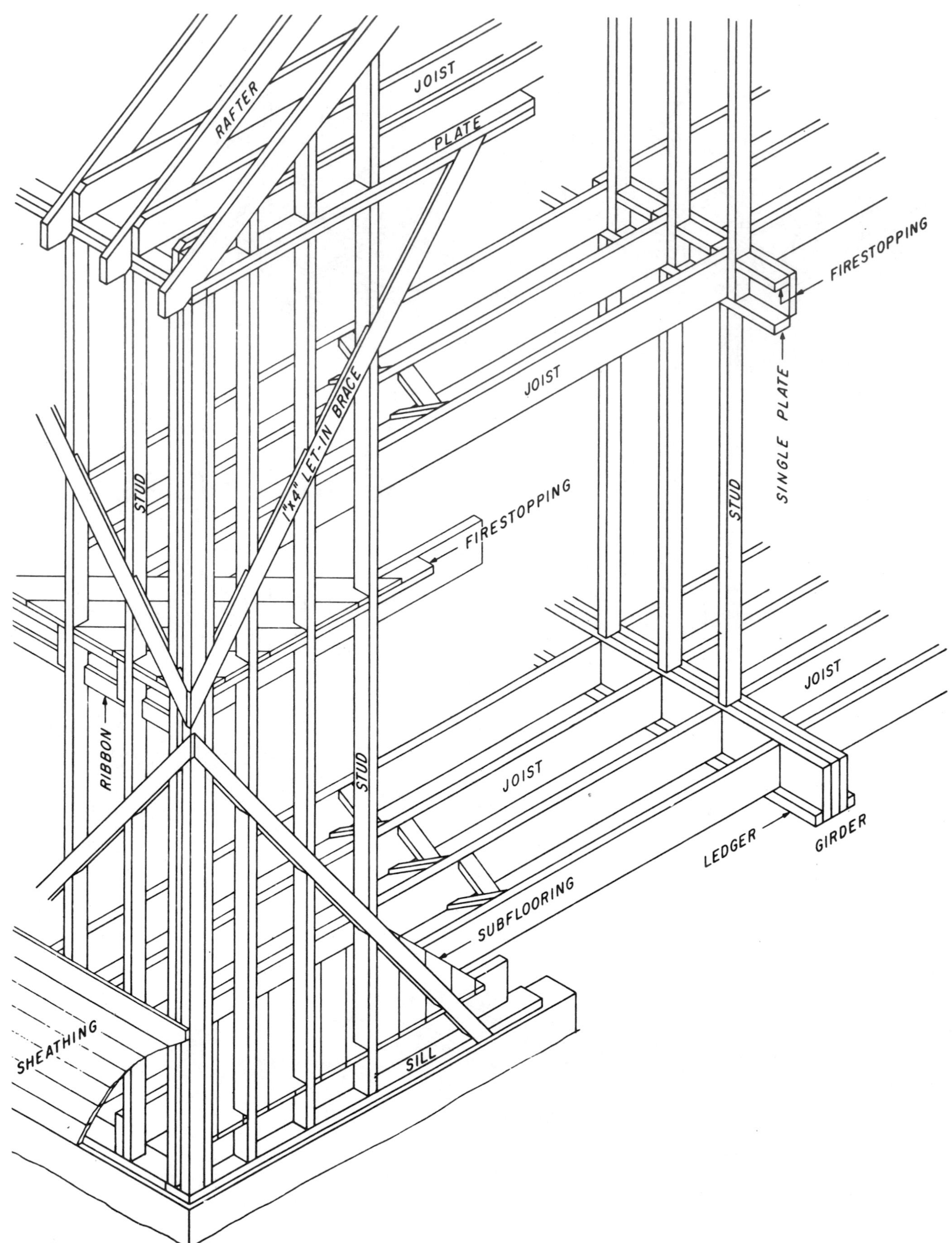

Fig. 2 Structural details of typical balloon frame construction.

Nailed joints are strongest when the load is acting at right angles to the nails. Joints, where the load is applied parallel to the nail in such a way as to cause withdrawal, should be avoided since the nails are weakest when loaded in this manner.

Sometimes, the most practical way to fasten wood members is by toe-nailing, in which nails are driven at an approximate thirty degree angle with the grain. It is a preferred method for nailing studs to sills where wall sheating does not serve to tie these members together. **Figure 7** shows three methods of loading or driving nails into wood.

Outside Wall Studs at the Gable. In the conventional framing of the outside wall the gable studs may be placed above each exterior wall stud, **Figure 8.** Note the cuts required to fit the gable stud to the rafter. Sometimes the gable stud is framed under the rafter with a single bevel cut along its width.

Exterior Wall Framing in Split Level Construction. The wide acceptance of the split level house has introduced complexities in framing. The split level house takes many varied shapes and designs. The cracking of interior finish and the buckling of drywalls may result from uneven expansion of the members of framing meeting at different levels. Many split-level houses have the second floor rooms cantilevered over the first floor rooms and garage, **Figure 9.**

Three simple rules make split level framing more efficient: (1) Have a minimum of bearing partitions; avoid L-shaped plans or irregular perimeters. (2) Keep the "split line" straight; avoid jogs in the line where change of level occurs. (3) Frame the building so that

Common Nail Usage Chart

Members	Number and Size of Nails
Joist to sill or girder, toe nailing	3 - 8d
Bridging to joist, toe nail each end	2 - 8d
Ledger strip (at each joist)	3 - 16d
1″ × 6″ subfloor or less to each joist, face nail	2 - 8d
Over 1″ × 6″ subfloor to each joist, face nail	2 - 16d
Sole plate to joist or blocking, face nail 16″oc	1 - 16d
Top plate to stud, end nail	2 - 16d
Stud to sole plate, toe nail	4 - 8d
Double studs, face nail 24″oc	1 - 16d
Double top plates, face nail 16″oc	1 - 16d
Top plates, laps and intersections, face nail	2 - 16d
Continuous header, two pieces 16″oc along each edge	1 - 16d
Ceiling joists to plate, toe nail	3 - 8d
Continuous header to stud, toe nail	4 - 8d
Ceiling joists, laps over partitions, face nail	3 - 16d
Ceiling joists to parallel rafters, face nail	3 - 16d
Rafter to plate, toe nail	3 - 8d
1″ brace to each stud and plate, face nail	2 - 8d
1″ × 8″ sheathing to each bearing, face nail	2 - 8d
Over 1″ × 8″ sheathing to each bearing, face nail	3 - 8d
Built-up corner studs, nail 24″oc	1 - 16d
Built-up girders and beams, nail 32″oc along each edge	1 - 20d

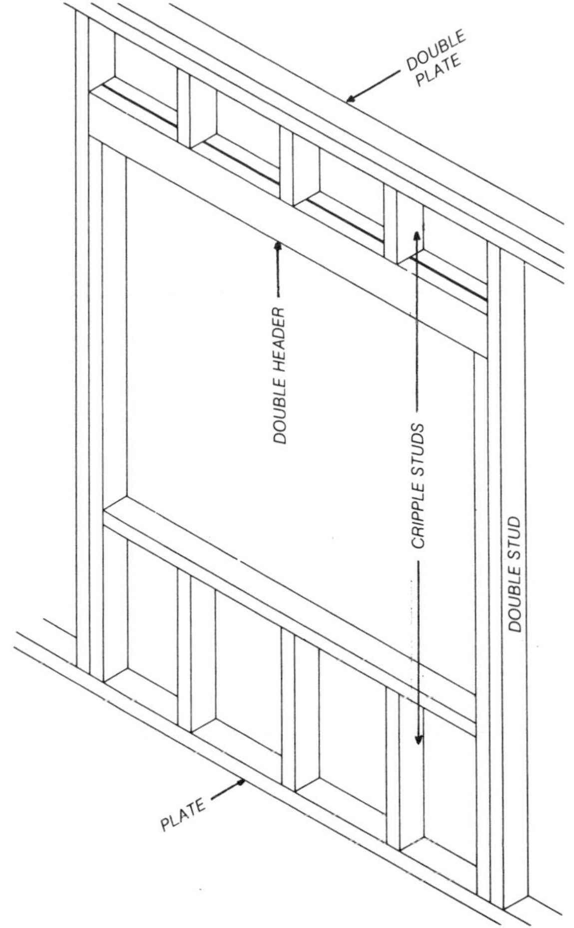

Fig. 3 Framing around exterior wall opening.

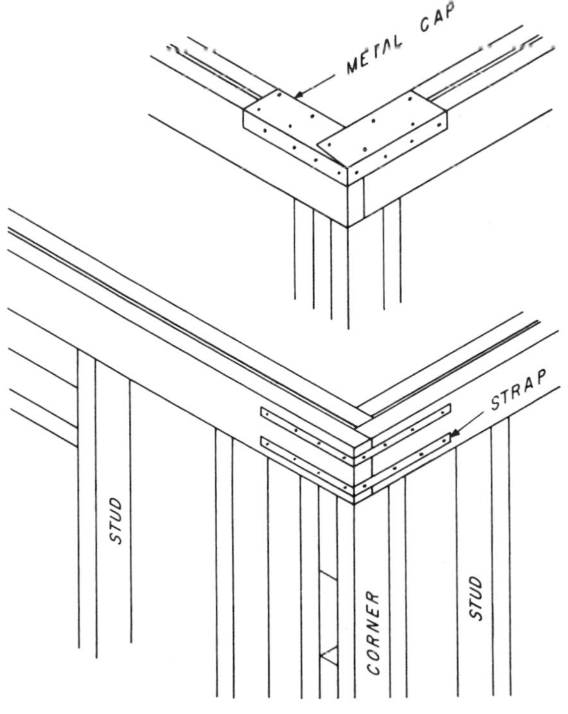

Fig. 4 Methods of joining continuous headers at corners.

all structural members run in the same direction and maintain a regular beam or joist pattern. Framing members at different levels are supported by a center bearing partition, **Figure 10.** This partition runs the length of the house and is the only bearing partition used. **Figure 11** shows a method of framing a front wall with the cantilevered second floor joists supporting 18″ lookout rafters to form a front roof overhang. Note the use of half length (4′) and full length studs (8′) used in the rear wall framing, **Figure 12.** There are two methods of framing an overhanging second floor exterior wall. Lookout joists frame the overhang of the wall when joists run parallel with it **Figure 13.** Joists are extended or cantilevered out beyond the lower wall when they run at right angles to the facade, **Figure 14.**

Firestopping. All concealed spaces in wood framing are firestopped with wood blocking, accurately fitted to fill the opening and arranged to prevent drafts from one space to another. Exterior stud wall spaces should be fire stopped, at each floor level and at the top story ceiling level, with nominal 2″ blocking. In many

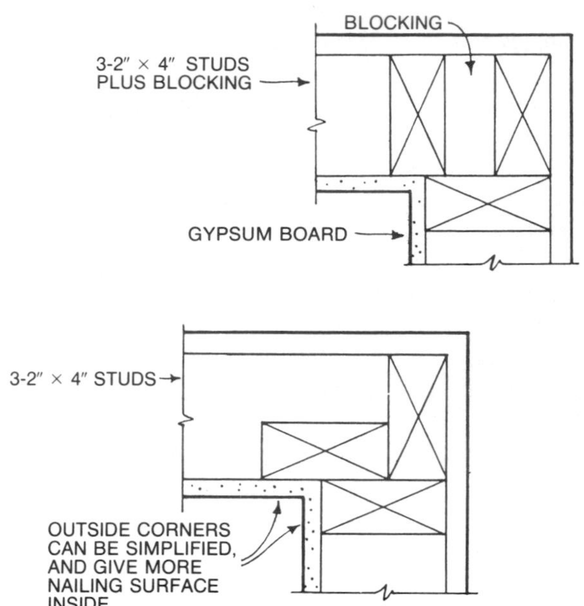

Fig. 5. Arrangement of studs at corners to provide nailing surfaces for wall materials.

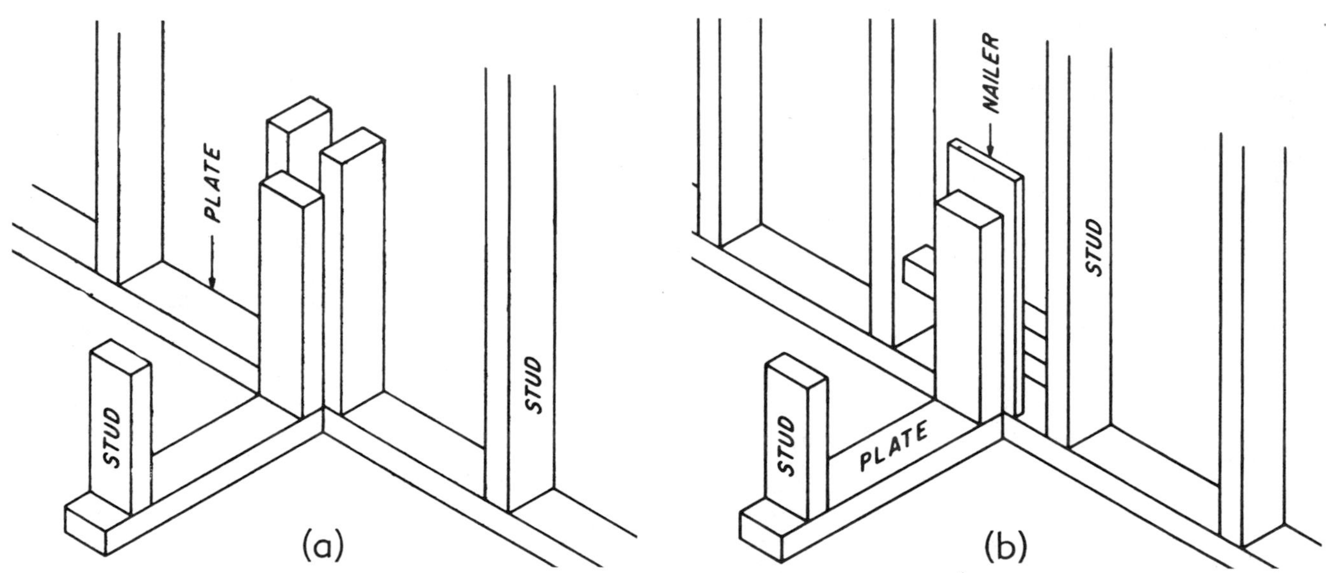

Fig. 6. Two methods of joining interior partition to exterior wall framing, (a) to studs, (b) to blocking.

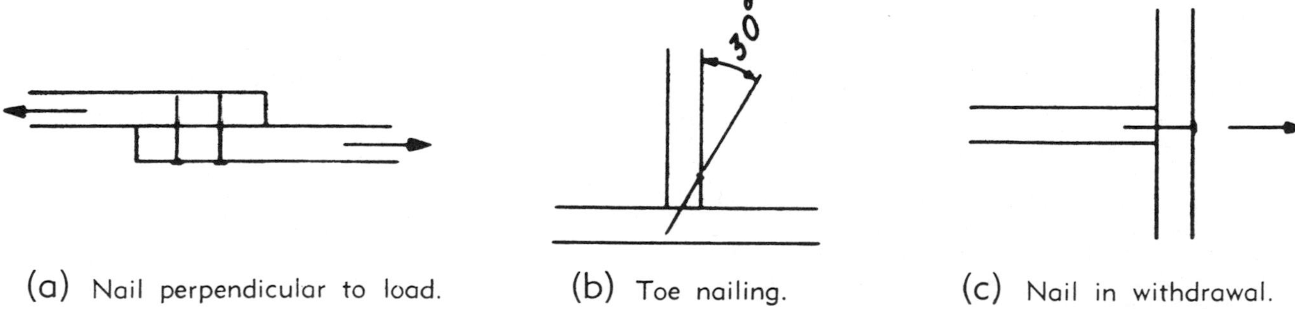

Fig. 7. Three methods of driving nails into wood, as related to assembly of house framing.

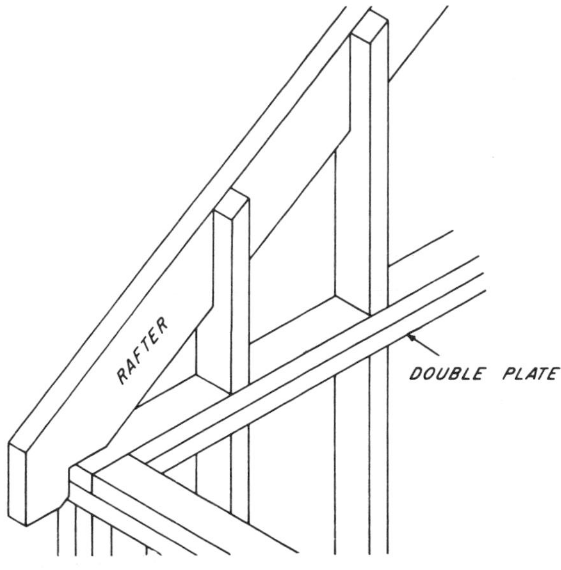

Fig. 8 Cable studs notched to fit end rafter.

Fig. 9 Split level house with second floor rooms cantilevered over first floor rooms and garage.

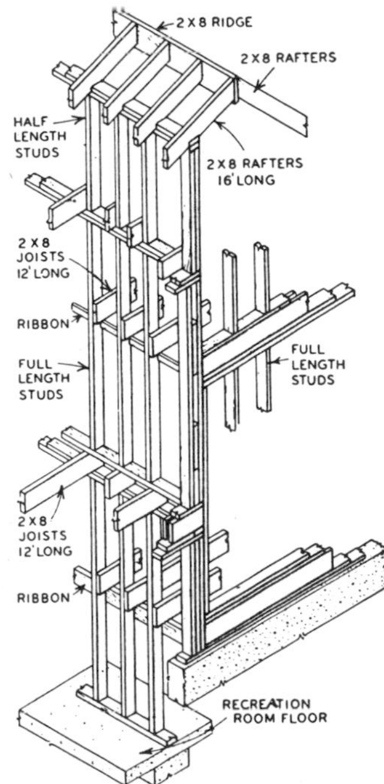

Fig. 10 Center bearing partition in split level runs full length of house.

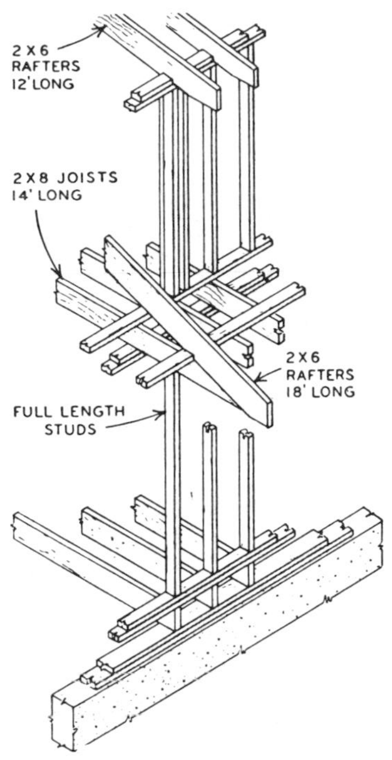

Fig. 11. Front wall framing detail showing joists supporting lookout rafters over porch.

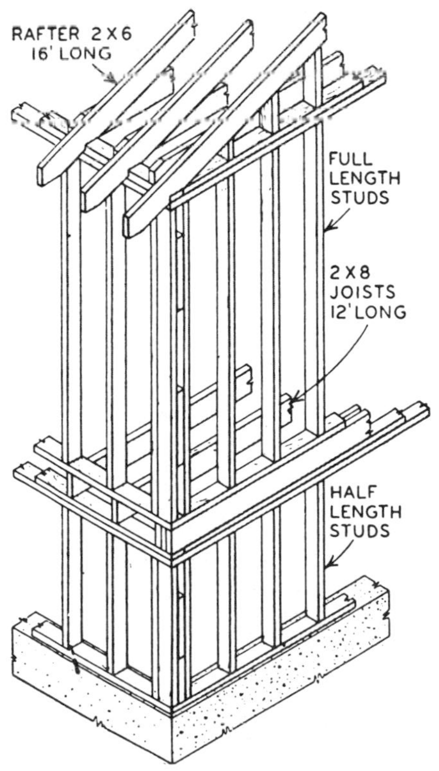

Fig. 12. Rear wall framing. Half length studs are used on half-height foundation.

OUTSIDE WALL CONSTRUCTION

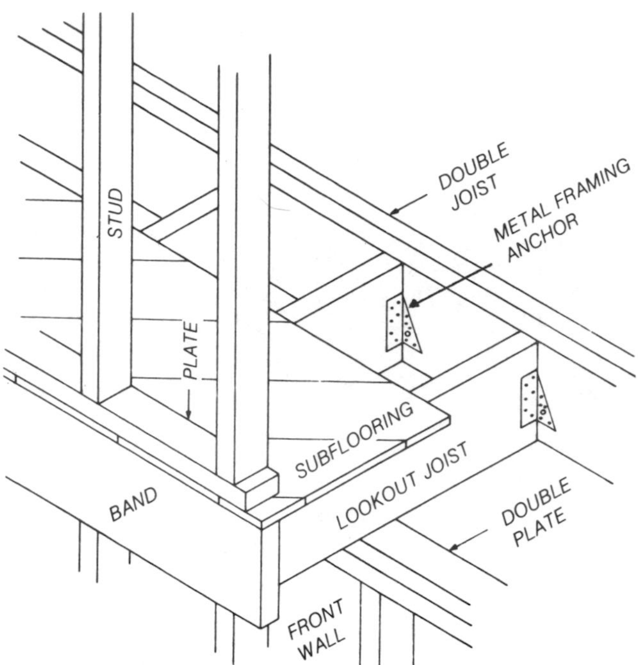

Fig. 13. Lookout joists frame second floor overhang when floor joists run parallel to exterior wall.

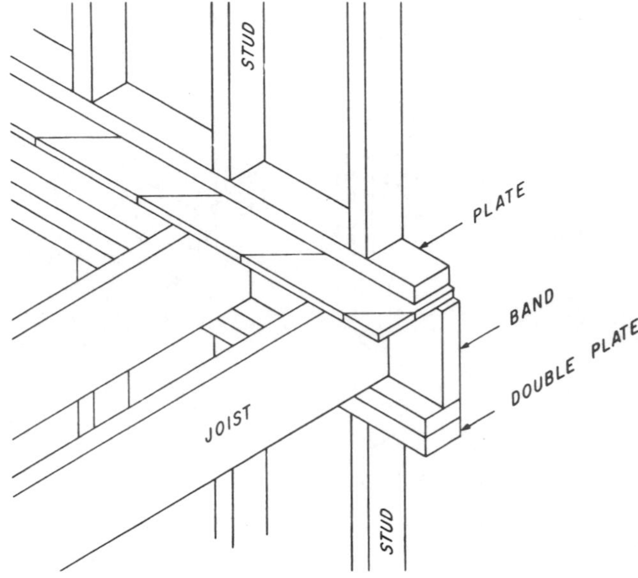

Fig. 15. Second floor framing band and plates serve as firestopping in platform construction.

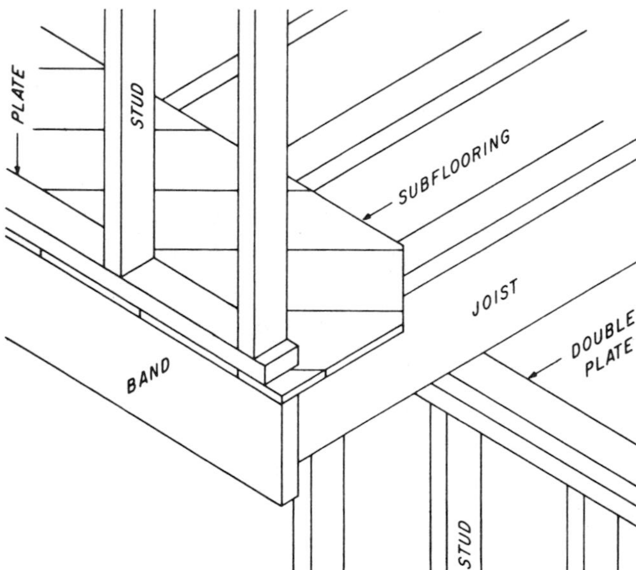

Fig. 14. Floor joists are cantilevered to frame overhang when they run at right angle to exterior wall.

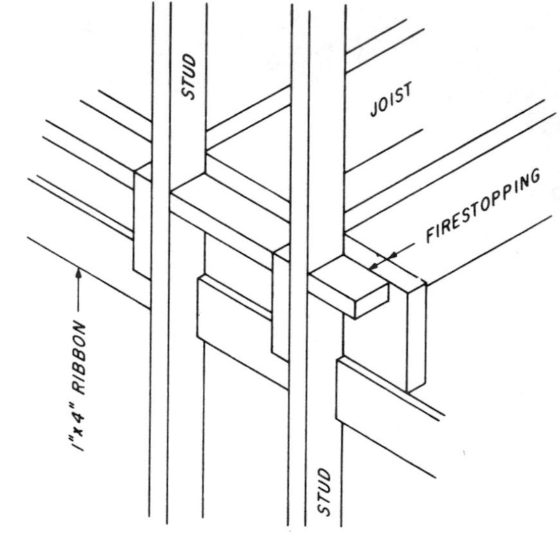

Fig. 16. Blocking nailed between joists and studs serves as firestopping in balloon frame construction.

instances, sills and plates will serve this purpose, **Figure 15**, but where they are not present, additional blocking is necessary. Blocking should be nailed between joists to provide firestopping in balloon frame construction, **Figure 16.**

Simplified Exterior Wall Framing. Of the traditional types of board sheathing, the diagonal method is more effective than the horizontal. Sheathing with 4′ × 8′ panels of various materials has been found to meet racking tests quite satisfactorily.

The development of prefabrication and pre-cutting methods has, however, simplified wall sections in many ways. Complete or partial wall sections are prepared on jigs, usually on a modular system, so that the openings are incorporated in the section with a minimum of time and effort. Windows and doors are generally placed to coincide with the normal placement of the studs.

Another system is to prepare two types of panels: one with simple wall sections, the other with window or door sections; the different types of sections are put up

Fig. 17. Preassembled wall sections are errected side by side in a modular system to make up the final wall size.

Fig. 18. Wall section, complete with exterior siding and its own sole plate, is set on plywood subflooring. Wall section with window openings is shown in background.

Fig. 19. Walls can be assembled on deck with windows installed, then tilted into position as a unit.

side by side according to the design, **Figure 17.** Great flexibility is permitted in the placement of elements and the size of the wall. Extra wall or window sections can be added at will on a modular basis. A complete section of wall may be constructed in one piece, with windows and doors included.

Figure 18 shows workmen setting a complete outside wall in place. The framework, sheathing and outside finish come in single panels. Each panel has its own sole plate, resting on the plywood subflooring. An entire wall section can be tilted up into position at one time **Figure 19.** This section is shipped as a unit from the factory, or it can be assembled flat on the subflooring.

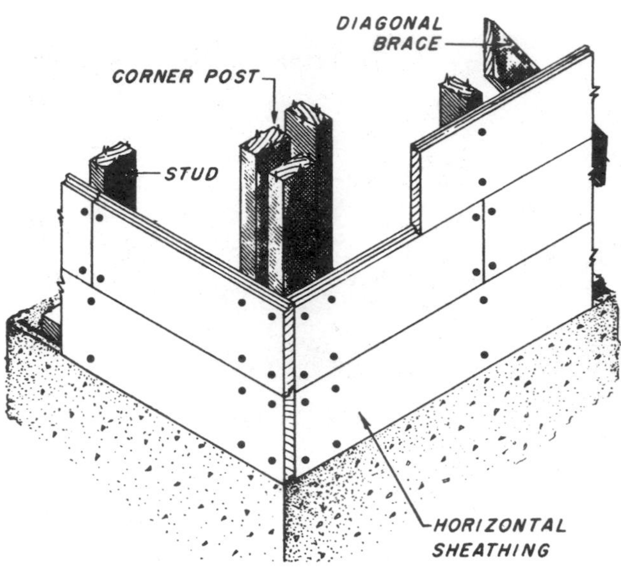

Fig. 20. Horizontal placement of sheathing is easiest.

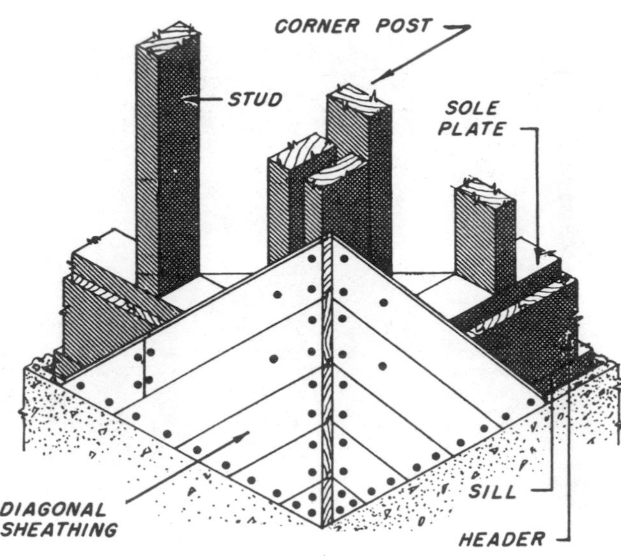

Fig. 21. Diagonal sheathing affords greater rigidity.

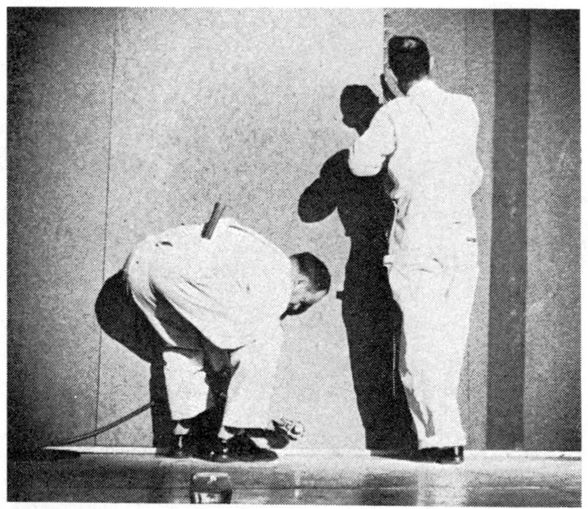

Fig. 22. Workman using power stapling machine fastens panel three times as fast as man with hammer and nails.

Fig. 23. Jig for large wall section is made up of steel angles. Wood members are dropped in place and nailed.

Wall Sheathing. Traditional sheathing boards, shiplapped to aid in watertightness and then covered with building paper, may be used, but mainly in single house construction. The horizontal placement of boards is easiest, **Figure 20.** Diagonal sheathing affords greater rigidity, but requires greater care in cutting and matching and meeting studs, **Figure 21.**

Today, however, outside walls are sheathed with 2′ × 8′ or 4′ × 8′ panels of plywood, gypsum or insulation board. The 2′ × 8′ board is applied horizontally while the 4′ × 8′ board is applied vertically. Large panels can be applied faster than sheathing boards. In pre-cut and prefabricated construction, the sheathing panels are essential to provide rigidity to the wall sections while being transported to the job site.

Another advantage of gypsum or insulation panel sheathing is that it permits the use of a power stapling machine which is about three times faster than hand nailing, as illustrated by workmen in **Figure 22.** The use of a jig for a large wall panel, **Figure 23,** allows fast assembly with power nailers. Sheathing can be stapled on a wall panel much faster as it is assembled on the floor, **Figure 24.**

Sheathing combinations are very commonly used. A house closed in with 2′ × 8′ insulating on the sides and horizontal wood sheathing on the gable end, is shown in **Figure 25.** A prefabricated panel with 4′ × 8′ insulating sheathing attached at the factory to stud frame work with the standard 16″ spacing is shown in **Figure 26.** Plywood is often used on corners to stiffen

Fig. 24. Sheathing is applied to preassembled wall section before tilting it into its final position.

framing covered with gypsum or insulating sheathing.

Masonry Wall Construction. Concrete block walls are the simplest to erect, when compared with brick veneer walls, since they are usually a continuation of the block foundation. The thickness of block walls supporting roofs or floors is usually 8″. The maximum height, including the foundation should not exceed 35′. Faced masonry walls consist of a masonry facing and backing which are of different materials. A typical faced wall would consist of a concrete split unit brick or stone facing with a concrete masonry backing. The two sections of masonry are bonded together to act as one. A cavity wall made of two 4″ wide units is bonded with wire ties to hold sections together, **Figure 27.** When a block partition wall is brought up to an exterior wall, strap metal ties are used to tie walls together. **Figure 28.** Most building codes require that exterior walls of concrete blocks be reinforced with two 1/4″ metal bars or strips of wire mesh in alternate courses, **Figure 29.** Cavity walls usually consist of two 4″ block walls with a 2″ air space, giving a 10″ wall. The two 4″ sections are tied together with 3/16″ wire ties, 36″ on center horizontally, and 16″ on center vertically. The cavity between the walls is filled with a poured insulation of Vermiculite. These are heat expanded minerals with millions of tiny closed air cells which makes them an excellent form of insulation.

A height limitation of 25′ should be placed on 10″ cavity walls. Plaster can be applied directly to the inside face of a cavity wall. Parapet walls should be at least 8″ thick. Unless reinforced, the height of a parapet wall should not exceed three times its thickness. Parapet walls should be carefully laid with fully packed mortar joints. They should be capped with solid units and thoroughly sealed.

There is no substitute for careful workmanship. It affects the performance and appearance of the wall as much as do the materials with which the wall is built. Good workmanship involves proper bedding of the units with adequate mortar, careful placement of the units, and proper tooling of the joints. **Figure 30** shows a stacked bond concrete block exterior wall and a screen wall. Note the clean, sharp joints which result as the workman uses a tooling device to compress the mortar in each joint, **Figure 31.**

Tooling Masonry Joints. Masonry joints should be tooled to compress the mortar between the masonry units and to improve the appearance of the wall. Tooling also makes the wall more watertight and helps locate unfilled or partially filled joints. Any loosely filled areas are repacked at the time of tooling. The tooling instrument should be wider than the joint. If it is not, the mortar may be pushed in too far, disturbing

Fig. 25. House is closed in with insulating sheathing panels on front and rear, and wood sheathing on gable ends.

the bond between the units and creating ledges where water may collect. The tooling device should be 24″ to 36″ long. A long tooling instrument makes a straighter, more uniform joint.

Four types of mortar joints are generally used by masons on brick or concrete block walls, **Figure 32.** Concave and V-shaped joints are the most widely used. They have good weather resistance and can be used both on exterior and interior walls. The flush joint is cut even with the masonry unit. It is then rubbed with a wood float or piece of burlap or carpet to give a texture similar to that of the block. The flush joint's overall weather performance is fair. It should be used only on vertical exterior joints or on interior joints.

Raked joints are made by cutting all mortar out to a depth of approximately 1/4″, giving distinct shadows. The raked joint leaves a ledge that can collect water. Extruded or "ooze" joints are used to give the wall a rustic effect. The mortar is not cut off when the units are laid. Raked and extruded joints should not be used on exteriors except in dry areas.

Anchoring a Framed Roof to a Concrete Block Wall. All roofs must be securely anchored by bolts or nails to the supporting exterior walls of frame or masonry construction. Anchoring, in a concrete block wall, can be done by inserting 18″ bolts, 1/2″ in diameter, into the core spaces at 4′ centers **Figure 33.** A piece of expanded metal or similar material is placed under each core holding a roof anchor bolt. A large washer is

Fig. 27. Masonry cavity wall made up of two 4″ wide units is bonded with wire ties.

Fig. 28. Strap metal tie bonds intersecting partition to masonry exterior wall.

Fig. 26. Factory made wall panels arrive with insulating sheathing already fastened to stud framing laid out on standard 16″ on-center spacing.

Fig. 29. Wire reinforcement is laid on alternate courses of masonry wall.

Fig. 30. Stacked bond concrete block exterior wall and screen wall has joints aligned vertically as well as horizontally.

Fig. 31. Joints are tooled to compress mortar and keep out weather.

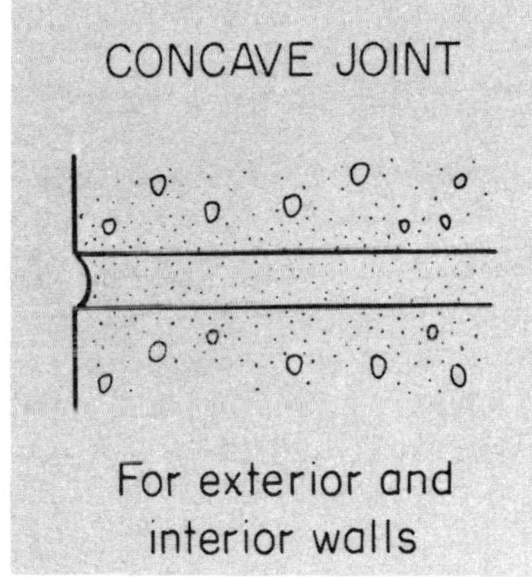

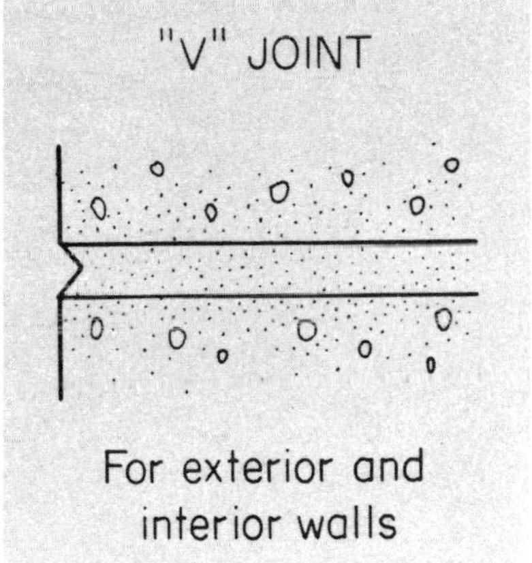

Fig. 32. Four types of mortar joints used by masons for interior and exterior masonry walls.

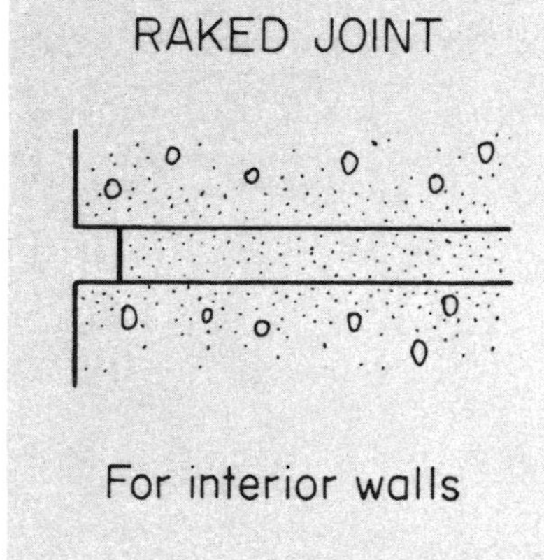

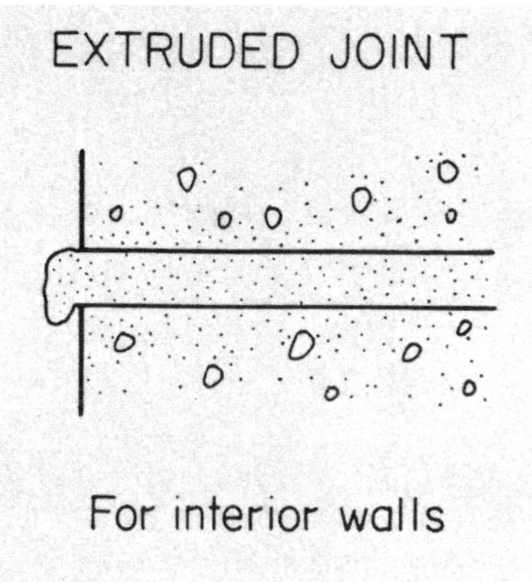

placed on the head of the bolt and the core is then filled with portland cement grouting. When the grout has hardened, the roof plate, generally of two, 2″ × 4″ framing stock, is drilled to accomodate the bolts and secured in place, to hold the rafters. A large washer is used on top of the plate. In high wind areas, metal ties are used to secure the rafters to the top plate.

Plank and Beam Framing of Exterior Walls. The plank and beam or post and beam framing of exterior walls is essentially a skeleton framework. Floor planks are designed on the basis of a moderate, uniformly distributed load that is carried to the beams which, in turn, transmit their loads to columns directly supported by foundation walls or basement column footings, **Figures 34, 35, 36, and 37.** Where heavy, concentrated loads occur in places other than main beams or columns, supplementary beams are needed to carry such loads. In plank and beam framing, exterior walls may be laterally braced by installing solid panels between columns at appropriate intervals. The wall panels and columns are tied together by diagonal bracing or suitable sheathing.

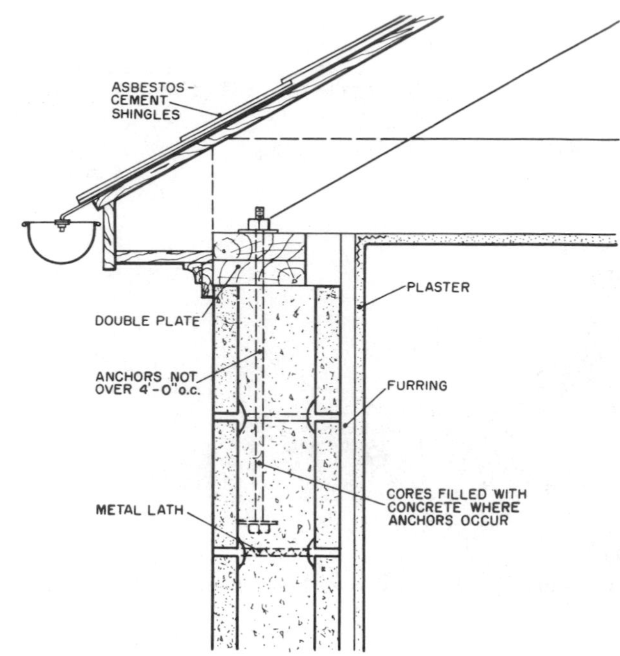

Fig. 33. Sectional view of a sturdy method of anchoring a wood-framed roof to a concrete block wall.

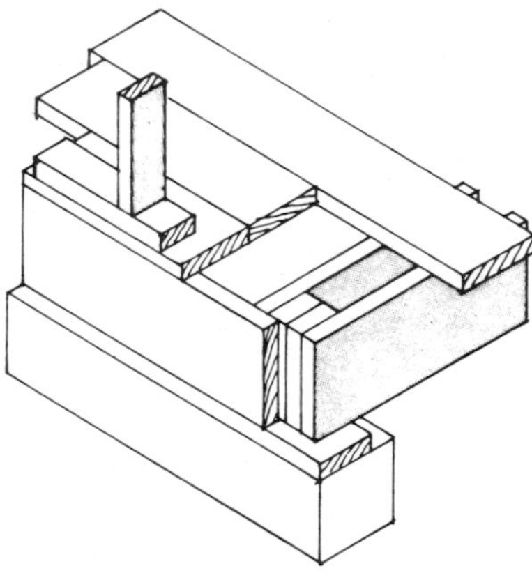

Fig. 34. First floor framing of exterior wall with spaced beam bearing on wood sill plate.

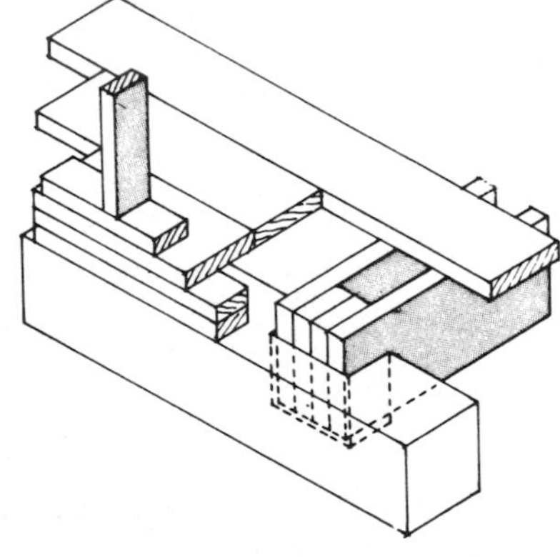

Fig. 35. First floor framing of exterior wall with spaced beam set in masonry foundation wall.

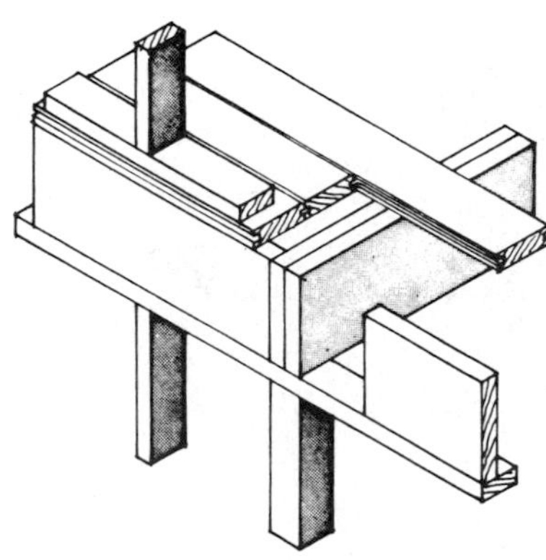

Fig. 36. Laminated beam bearing on second floor exterior wall. Note column under beam.

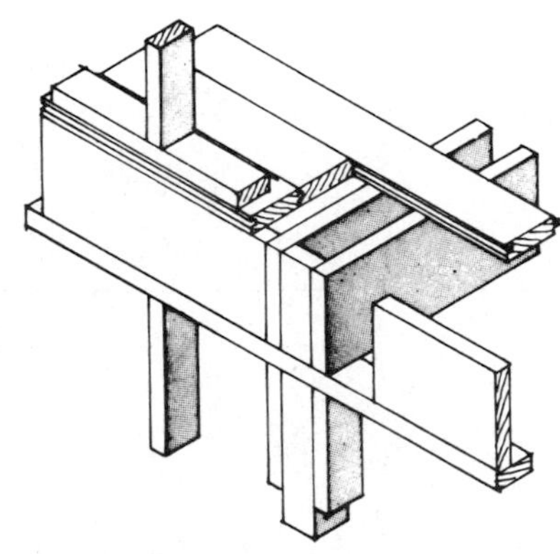

Fig. 37. Spaced beam bearing on second floor exterior wall. Note blocking on column sides.

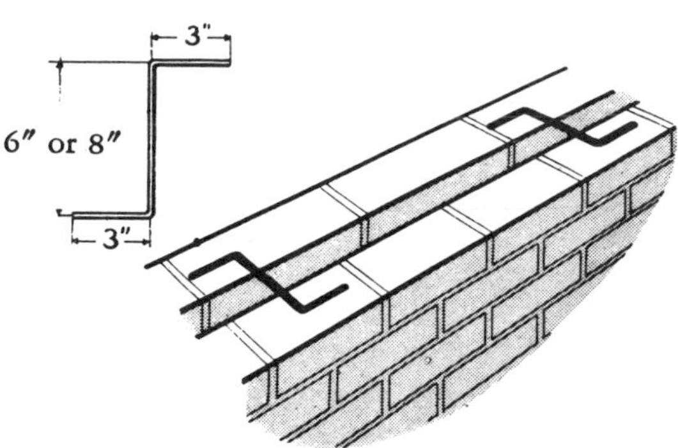

Fig. 38. Metal ties of various shapes are used to bond the wythes of a masonry cavity wall together.

Fig. 40. Interior wall material can be nailed to furring strips fastened to masonry. Insulation is applied between furring.

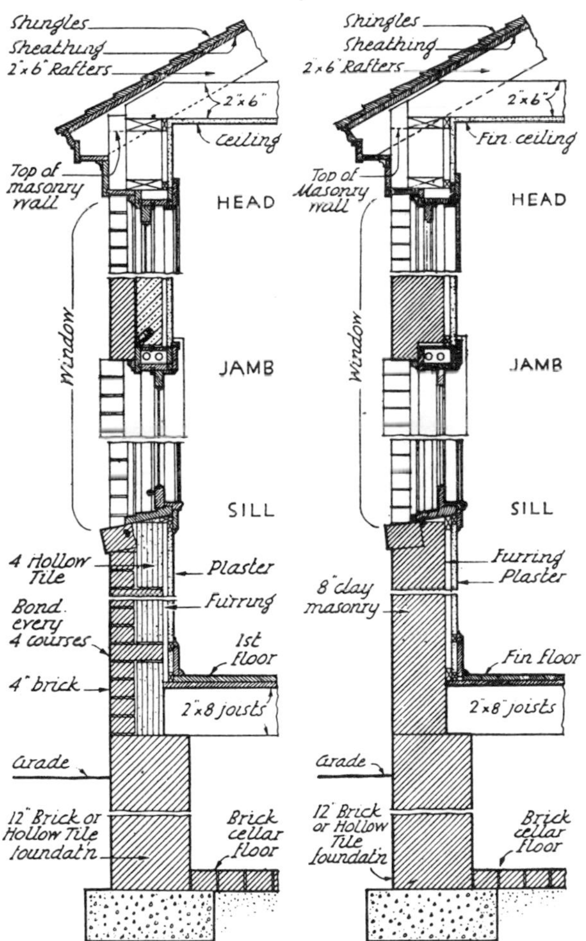

Fig. 39. Solid wall of brick and tile must have bonding courses of brick to tie the two materials together (left). Solid brick wall (right) has less insulating value.

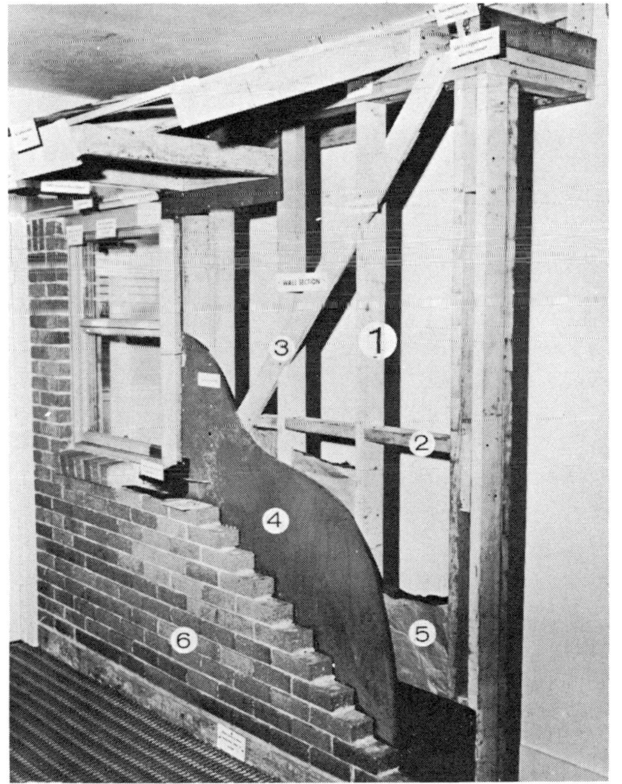

Fig. 41. Cutaway of brick veneer wall: (1) stud (2) blocking (3) bracing (4) sheathing (5) insulation between studs (6) brick veneer held by metal ties 1" away from sheathing.

The most efficient use of a 2″ floor plank occurs when it is continuous over more than one span. Using the same span and uniform load in each case, a plank which is continuous over two spans is nearly two and one half times as stiff as one which extends over a single span. Where standard lengths of plank are used, such as 12, 14 or 16 feet, beam spacings of 6, 7 and 8 feet are used. Beam spacings may affect the house dimensions.

This method of framing exterior walls, the use of plank flooring and box beams is adapted from heavy timber construction. It is also used for floors and roofs together with conventional stud framing, masonry walls, curtain walls, or panels of fixed glass.

Outside Walls of Brick. Solid brick walls of broad brick are used in one or two of the exterior walls of a split level house or in one or two story houses. Cavity walls, whether of brick and brick, brick and tile, or brick and block, require more complicated bonding to the foundation, and the two vertical elements, or wythes, must be tied to each other **Figure 38.** Solid brick walls are backed by tile or concrete and must have bonding courses of brick to integrate the two materials **Figure 39.**

Clips for furring strips can be inserted at proper intervals when a solid brick wall is erected, or the strips can be fastened with cut nails. Insulation can be applied between the furring strips, **Figure 40,** and interior wall material nailed to the furring.

Brick veneer laid over wood-framed walls, **Figure 41,** is most commonly used in residential construction today. Metal ties are nailed to the sheathing to secure the brick veneer. A 1″ air space between sheathing and brick is typical. The foundation must be wide enough to support the wood sill for the framed wall and the brick veneer.

Chapter 5

INSIDE WALLS/CEILING JOISTS

Inside walls form the partitions for the rooms inside the dwelling. One type of inside wall or interior partition is a load bearing wall which supports framing above it. Another type is the non-bearing partition which carries only the weight of its own materials.

Bearing Partitions. In conventional wood framing the studs should be at least 2″ × 4″ in size, set with the wide dimension perpendicular to the partitions and capped with two pieces of nominal 2″ lumber, or by continuous headers which are lapped or tied into exterior walls at places of intersections. Studs supporting floors should be spaced 16″ on centers and those supporting ceilings and roofs may be spaced 24″ on centers. Where openings occur, loads should be carried across the openings by headers similar to those recommended for exterior walls.

Non-Bearing Partitions. In conventional wood framing the studs should be nominal 2″ × 3″ or 2″ × 4″ in size, set with wide faces perpendicular or parallel to the partition. Spacings of studs may be 16″ to 24″, except as limited by the type of wall covering. The interior wall is lighter in weight than the exterior wall, to which one or both of its ends are attached or framed. An interior partition may carry a portion of the weight of the ceiling joists and of the floors above in a multi-story house, but it is not a bearing unit for the roof, the rafters of which rest on the outside wall structure. Interior walls are conventionally of frame construction, but may be of steel, stone, brick, concrete, hollow tile or factory fabricated.

Framing members at the second floor level are built on the subflooring directly over first floor bearing partitions in platform frame construction, **Figure 1.** Framing members at the second floor level bear on the top plate of first floor bearing partitions in balloon frame construction, **Figure 2.**

Non-bearing partitions are framed over the flooring, doubled joists and girder making up the main floor system, **Figure 3,** and are attached to blocking nailed between ceiling joists, **Figure 4.** Blocking should be nailed to studs to support a bathtub against a non-bearing wall partition, **Figure 5.**

Ceiling joists rest on a double plate directly over studs in a bearing partition, **Figure 6.** Note the sole plate of the partition resting on the subflooring; the plumbing lines supported by the framing; the double studding (right) at the bathroom door opening and window (left); and the jack studs over the window header. The roof rafters are supported by the exterior wall framing.

Plumbing presents special problems in framing. The 4″ soil pipe widely used necessitates a 2″ × 6″ studding or at least a wider than usual sole plate. Horizontal plumbing lines are commonly cut through the framing. Sometimes extra studding is needed for

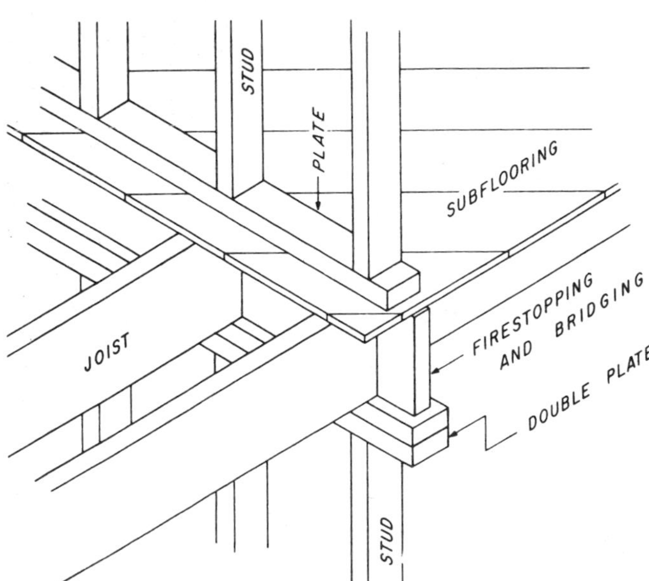

Fig. 1. Loadbearing partition supporting second floor framing is platform frame construction.

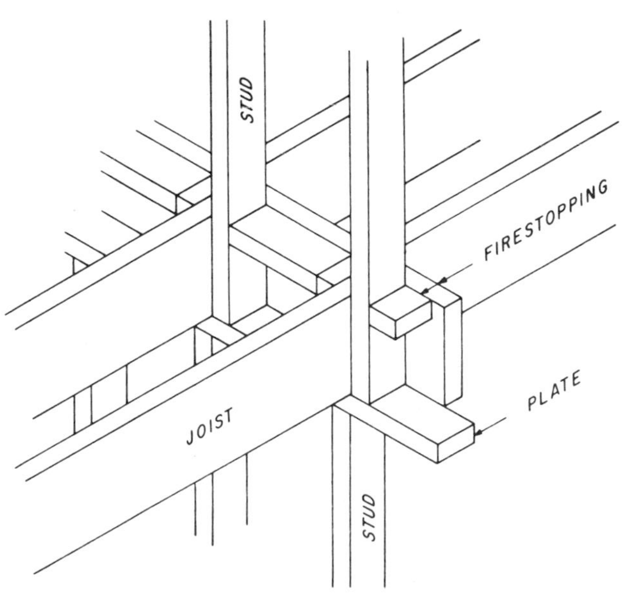

Fig. 2. Loadbearing partition supporting second floor framing in balloon frame construction.

proper support, **Figure 7.** Note the bath tub in place at the right and the base for a shower beyond the double partition.

One solution to placing the plumbing service pipes in a partition is shown in **Figure 8.** In this photograph, the plumbing installations have been simplified by placing rough plumbing in the interior wall between the garage and adjacent kitchen and bath.

Pre-Built Units Installed in Inside Walls. To simplify the construction and to speed up the building of a house, even in conventional construction, many of the techniques of prefabrication are employed. The principle of "parts, no pieces" is applied when an interior partition is set up as a unit, rather than studs nailed to the plate and sole. Some factory built units are available as panels and fitted into the inside wall framing. The prefabricated plumbing panel, **Figure 9,** is built into 2″ × 6″ framing stock and framed into the proper opening. **Figure 10** shows a full length partition wall with the plumbing installed. The soil stack is set in place before the flooring is laid.

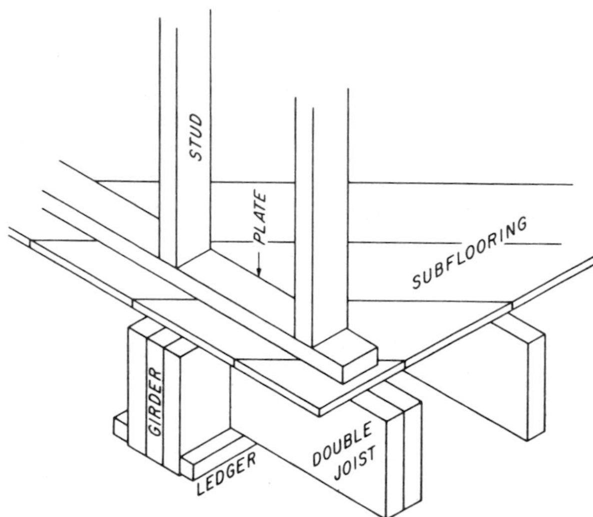

Fig. 3. Doubled joists support non-loadbearing partition.

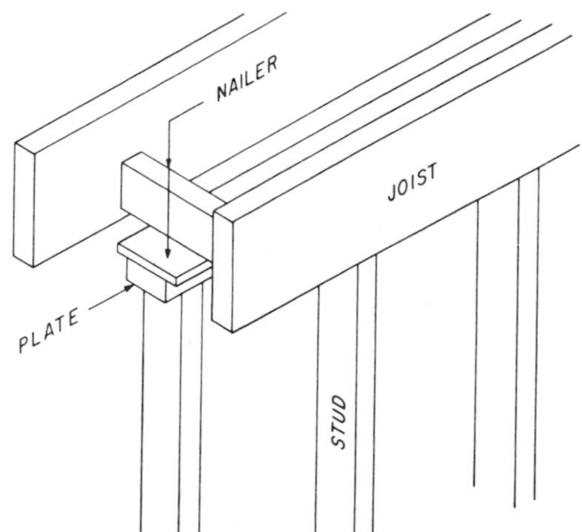

Fig. 4. Blocking between joists secures top of partition.

Mechanical Unit Components. The modular component constructed house, lends itself handily to the installation of the mechanical core, or three dimensional unit, or units, encompassing the most expensive part or parts of the house. Making up any mechanical core involves

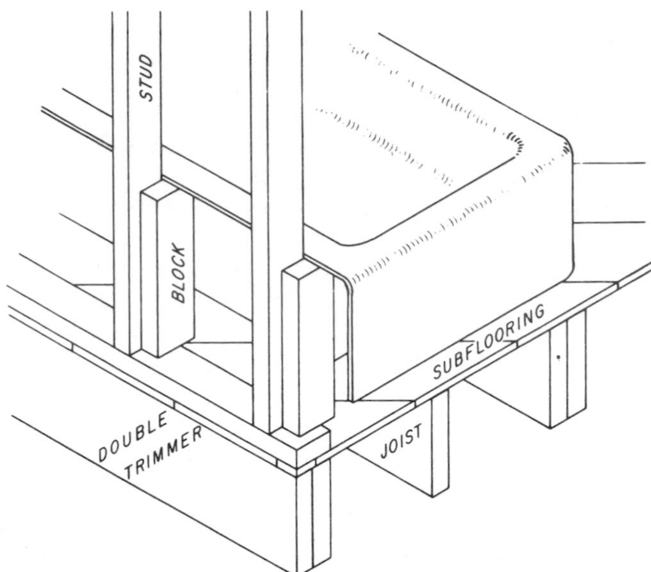

Fig. 5. Blocking nailed to studs supports edge of bathtub.

Fig. 6. Celing joists rest on a double plate directly over studs in a bearing partition. Studs also support plumbing.

the work of many skilled tradesmen. Skilled wood workers and carpenters are involved, in the factory in building the various components, and later erecting these components at the building site.

Skilled electricians, plumbers, tile setters and others are involved in the factory in the making of a mechanical core, which in one unit may include the furnace, the air conditioning system, the water heater and electrical panels. Another core may include the kitchen cabinets, sink and other accessories. Still another core may include bathroom equipment, floor and wall tile, cabinets, doors and decorated walls.

The mechanical core is a pre-planned, finished section or room of a dwelling that is factory made and transported to the house and put in place as a unit. The development of the manufacture of the structural component has now reached a satisfactory and efficient stage in dwelling construction. It is now comparable to the satisfactory and efficient manufacture of the prefabricated house. Mechanical unit component manufacture is now being extended to the making of electrical heating panels, factory wired furnace systems, plumbing wall components, etc. In fact the mechanical parts of a house offer some of the greatest opportunities in component methods of construction.

The core unit shown in **Figure 11** is being set in place so that its floor joists, framed in at the factory, run in the same direction as the floor joists already in place. The ends of the joists rest on the sill plate on the foundation and on the plate attached to the steel I-beam girder in the center of the rough floor platform. The sides of the unit are already framed as any conventional inside partition is framed. The door openings have double studs at the sides and jack studs over the door opening header. The ceiling is completed in this unit. If a second story is to be built to this structure the ceiling joists would rest on top of the pre-framed partition plates.

Storage Unit Components. Inside walls, when serving as non-bearing partitions, can be factory built and moved into place, **Figure 12.** This unit serves as a divider of floor space and provides adequate storage and cabinet space in a bedroom.

Plank and Beam Framing of Inside Walls. One great modification of interior construction in recent years has resulted from the widespread use of trussed rafters. Partitions can be made lighter, since they no longer need support the ceiling joists at an intermediate point between the load bearing outside walls. Partitions may go only part way up to the ceiling or may be omitted completely in open planning. Room separating functions may be filled by non-bearing storage walls, plastic or glass, clear or opaque. Often storage walls can be moved at will, and interior space divisions can be changed as the family's requirements change. A room

Fig. 7. Cast iron soilpipe requires 6" plumbing wall.

Fig. 8. Plumbing wall between garage and kitchen and bath simplifies installation of water lines and drains.

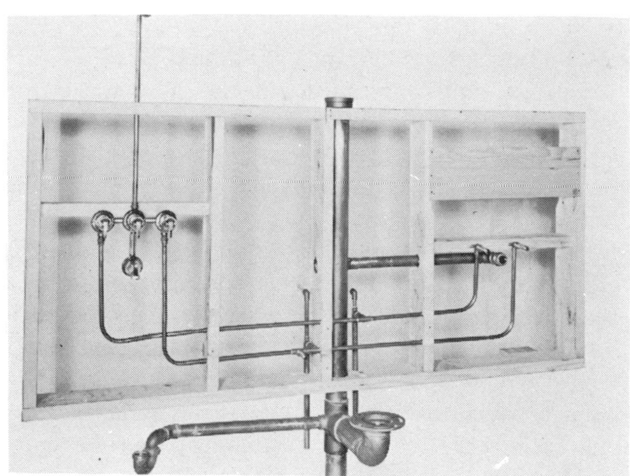

Fig. 9. Prefabricated plumbing panel is built into wall.

Fig. 10. Partition for prefabricated house has plumbing installed, ready to connect to soil pipe and water supply lines.

divider can be composed of various sized drawers, shelves and cabinets, **Figure 13.**

A dwelling interior of post and beam or plank and beam construction permits great freedom in interior partitioning. In this portion of the house the children's play and bed rooms can be closed off from each other and from the other areas of the house by folding doors

Fig. 11. Prefabricated bathroom core with fixtures and heating unit installed is lowered into position.

suspended from tracks attached to the beams, **Figure 14.**

Any wall partition, whether interior or exterior, are mere curtains between the posts, and may be of glass, light framing, or even folding materials, such as plastic or wood slats. Sliding doors, glass window walls, space dividers, and storage walls are common in plank and beam framing.

Figure 15 shows four methods used in framing plank and beam inside walls. The upper left detail shows a solid beam bearing at the second floor over an interior column. The upper right detail shows a spaced beam bearing at the second floor level over an interior column. The lower left detail shows a non-bearing partition parallel to the floor plank, supported by a supplementary beam under the plank floor. The lower right detail shows a non-bearing partition parallel to the floor plank supported by a supplementary beam on the plank floor.

Post and beam or plank and beam designs often require long ceiling framing members not generally covered at the bottom by ceiling material, but closed in on top by a combined ceiling-roof deck, **Figure 16.** Beam type rafters run from outside wall plate to ridge beam. These rafters are covered with tongue and groove plywood roof decking which forms a cathedral style ceiling.

Long beams can also run parallel to the front of the house, **Figure 17.** Note the metal base for a raised hearth in the foreground and the framing for a room divider between the posts. At the far left may be seen the hanging cabinets between posts to form the dining-living room section of the house.

Full-length beams and ridge often are left exposed as a feature of a pre-fabricated house design, **Figure 18.**

Fig. 12. Storage wall, with space for clothes, shoes and linens, serves as non-loadbearing partition between rooms.

Fig. 13. Partial room divider separates activity areas.

Fig. 14. Post and beam construction provides great flexibility of room layout. Note folding partition hung from beam.

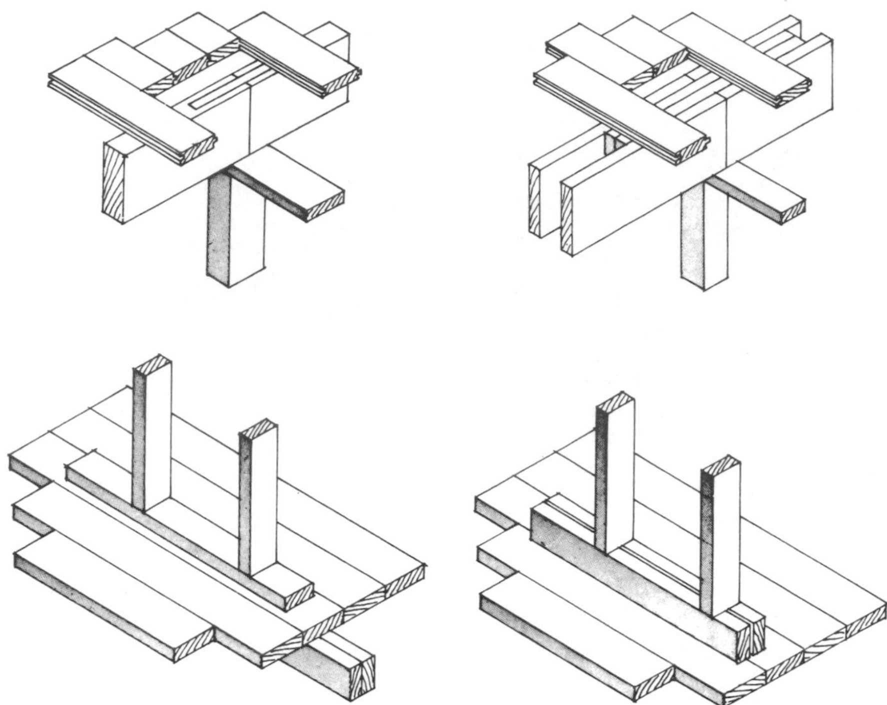

Fig. 15. Framing details for plank and beam interior walls.

Fig. 16. Combination ceiling-roof deck laid over exposed rafters.

Fig. 17. T&G decking spans beams running longitudinally.

Note how the notched, pre-formed interior panels receive the beams and the pre-finished roof panels are set in place. Ceiling joists can be fabricated with a steel plate, called a flitch plate bolted between the joists for added strength and stiffness, **Figure 19.**

An attractive appearance is given the interior of the house in the use of double collar beams on alternate rafters. In **Figure 20,** the vertical planks in the gable end and heavy plank covered rafters present a semi-cathedral style ceiling. The horizontal beams supporting the roof are of heavy laminated stock resting on heavy posts. The vertical wall at the fireplace is built of rough sawn material. The deck is of rough sawn stock. The posts and beams of the exterior wall allow complete fenestration at the gable end of the house.

Inside Walls and Ceiling Joists. Conventional interior framing consists of a combination of sole plate, studding, and top plate (often doubled) to receive the weight of the ceiling joists. The sole plate generally runs across the floor joists or along and over one of the joists depending upon the direction of the partition. The ceiling joists are generally positioned directly over the supporting studs, facing in the same direction.

Where an interior door is to be placed, there must be double studding around the opening to take the place of the missing support, and a header or lintel to support the plate in the space immediately over the opening. Short Jack studs or cripples carry the weight from the plate to the header, **Figure 21.** When the opening is 3′ or less in bearing wall and partition or inside wall, two

Fig. 18. Prefabricated roof panels span beams set in end walls.

Fig. 19. Steel plate is bolted between wood members for strength.

pieces of 2″ × 4″ framing stock are used as indicated, but when the openings are wider and the weight above is heavier than normal, the headers are trussed to carry the main weight to the double studding at the jamb, **Figure 22.** The ceiling joists are well supported in this method of trussing. The use of steel angle in lintels can add support while keeping framing members light, **Figure 23.**

Fig. 20. Laminated beams bear on widely spaced posts, permitting maximum use of glass in end walls.

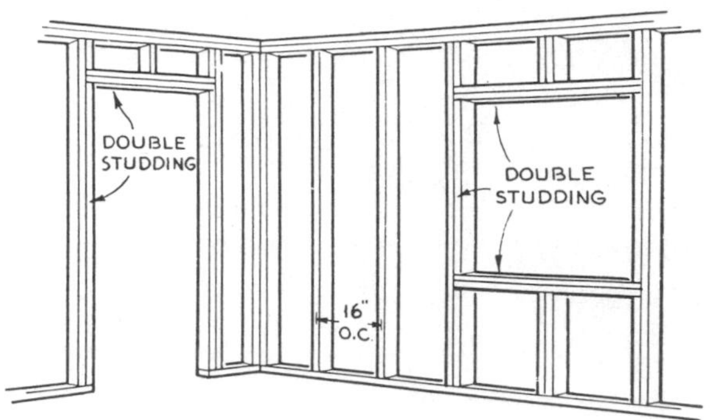

Fig. 21. Jack studs or cripples above openings carry load from top plate to headers bearing on double studding.

Fig. 22. Openings wider than 3′ may have trussed headers to carry floor or roof loads to double studs in jamb.

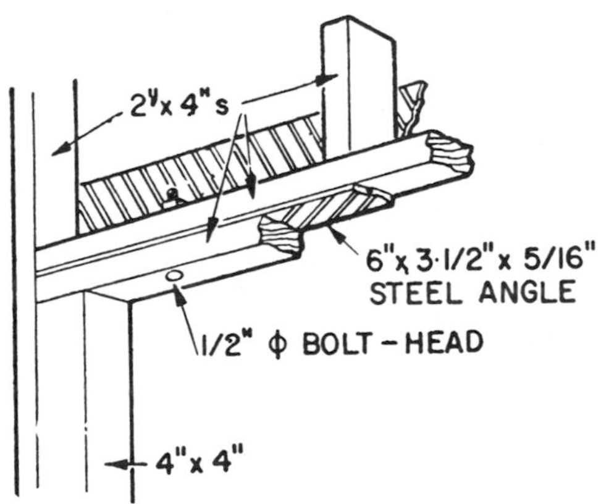

Fig. 23. Lintels of steel angle can be bolted into header framing to add support while keeping framing members light.

In a conventionally framed two-story house, the first floor partition should be placed directly over the girder; the second floor partition whenever possible should be over the first floor partition to carry the weight directly down to the girder, but of course the upstairs is usually different from the down stairs plan, so the second floor partition may rest anywhere on the second floor joists.

When the partitions are non-loadbearing, the framing may be of lighter stock than the traditional 2″ × 4″ material, and may be faced flatwise along the line of the partition in order to take up less floor space, **Figure 24.** In this type of framing the main ceiling weight is carried by the heavy upper beam supported by posts, and the room partition does not run to the ceiling joists.

Fig. 24. Roof load is carried by the post and beam framing, permitting the use of a partial height partition framed with 2″ x 3″ studs set edgewise on 2″ x 2″ top and bottom plates.

Prefabricated construction or factory built methods generally rely on roof trusses to permit the use of easily transported non-loadbearing partitions. Some prefabricated systems use house-length beams to carry the weight of the roof which permits open ceilings, **Figure 25.** Note the panelized roof section in the foreground.

The use of split-level framing provides a variation of interior wall framing, particularly the use of the cathedral ceiling, as well as problems in interior weight distribution and in contraction of plaster and drywall.

In the building of houses many parts, sections or panels are factory built or prefabricated then transported by truck to the building site to be assembled, **Figure 26.** Two of the outside wall panels of the lower level of this split-level dwelling are in place on the slab foundation, and three of the wall panels of the higher level are erected. Note the floor and wall framing of the second level in the background, also the stairs which leads from the lower to the upper level.

Workmen are unloading a non-loadbearing partition from the truck to be set into a notch in the outside wall plate. A portion of the exterior wall of the lower level, right, is in place ready to receive the partition and the remaining section of the exterior wall. An entire section of an inside wall partition is on the floor ready to be raised in place to tie the longer partition with the wall framing of the upper level.

One decisive factor in assembling prefabricated parts is the exact fitting of units as they are set in place. Generally protected from the weather, prefabricated partitions can be butted together at close tolerances without too much complication from unevenly expanded parts. Partitions the entire width of the house can be brought in as a unit and fitted into place in four stages of erection, **Figures 27, 28, 29 and 30.** First, the partition soles are laid out on the sub-floor. Then partitions are raised on the sole plate and attached to the ceiling, which is attached to the rafters of the several trusses of the roof. The framed partition is set tight against the ceiling by shims driven between the partition sole plate on the sub-floor and the shoe plate of this non-loadbearing partition. The shims are later trimmed at the edges of the shoe and sole plates. The partition complete with a door opening, is framed with the width of the studs nailed to the edges of the top plate and shoe plate to save space.

Fig. 25. House length roof beams are supported by posts in end wall and spanned by prefabricated panelized roof sections stacked in foreground. Note screened eave vents in panel.

The roof trusses, built with the ceiling joists as an integral part of the roof rafters, are set in place and covered with roof decking before the non-load bearing walls are moved in.

Split level framing of interior partitions occasionally offers problems of unusual stresses. To accommodate these stresses, a 2″ × 12″ header can be reinforced by a 3/8″ × 11-1/2″ steel plate flitch plate resting on a 4″ × 4″ post and supported by a 2″ × 4″ frame, **Figure 31**. Such a header can be seen in the outside framing with cantilever bracing of ceiling joists, **Figure 32**. Most difficulties come at the line of the meeting of levels, and at the rafters of the cathedral living rooms, **Figure 33.** Ceiling joists may be

Fig. 26. Non-loadbearing partition in prefabricated split-level house is unloaded from truck and set into notch in previously erected exterior wall section. End wall of upper level has siding precut to fit trusses to be placed on lower level.

Fig. 27. With house closed in and walls and ceilings finished, bottom sole plates of partitions are laid out for interior plan.

Fig. 28. Preassembled partition with 1″ shoe on bottom is set in position on the appropriate sole plate.

Fig. 29. Shims are driven between 1″ shoe and sole plate to force partition tightly against ceiling. Shims are trimmed.

Fig. 30. Installation of prehung door (at right above) saves time-consuming fitting of jamb to door at site.

suspended on hangers attached to framed rafters to reduce the angle of the sloping ceiling as it meets the middle partition, **Figure 34.**

In balloon framing the use of the long stud from the girder or foundation to the topmost ceiling joists in exterior walls minimizes shrinkage and drywall buckling, **Figure 35.** Floor joists at the various levels may be supported by let-in ribbon boards or by separate uprights nailed to the framing. Buckling will occur if a ribbon board is not let in to support the floor joist and the ceiling joist. In split level framing, cripple studs or nailers often are used to support joists butting against the studs, **Figure 36.**

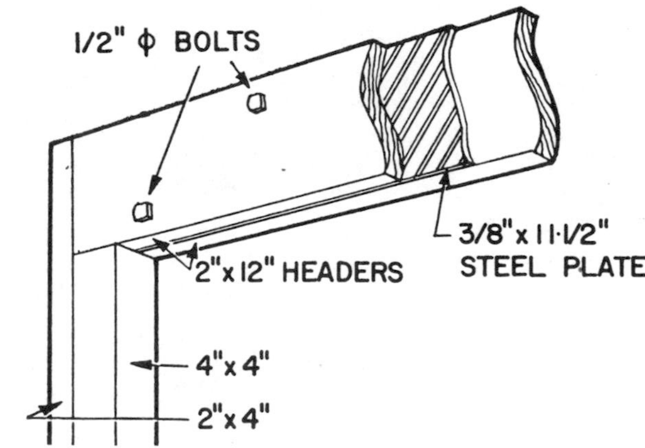

Fig. 31. Steel Flitch plate more than doubles strength of header.

Fig. 32. Flitch plate header in place over wide opening.

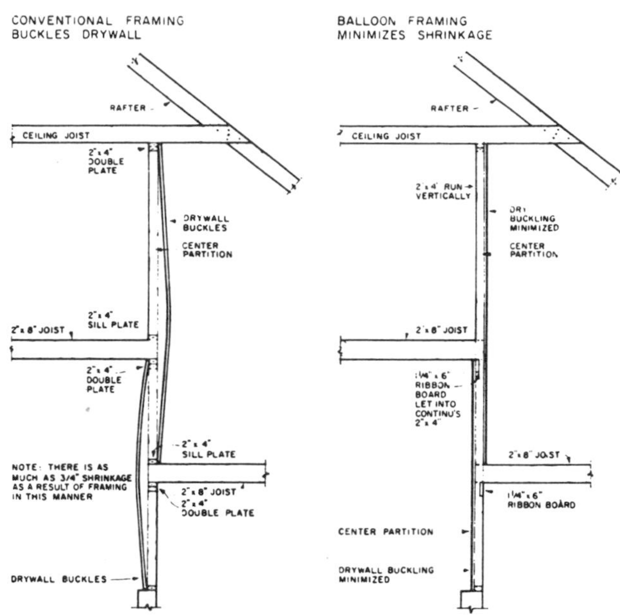

Fig. 35 Long studs in balloon framing minimize shrinkage.

Fig. 33. Rafters in this cathedral ceiling are covered.

Fig. 34 False rafters hung from main rafters reduce slope.

Fig. 36 Cripple studs nailed to framing support floor joists.

Inside Walls of Concrete Masonry

Concrete masonry is used as an inside wall material in rectangular units and in various shapes, sizes and textures. Pleasing patterns, can be achieved in using the conventional block, with molded designs protruding 3/8″ to 1/2″, for inside walls. The minimum thickness of a partition wall, carrying no loads other than its own weight, should be 4″. Partition walls supporting floors or roofs should be 8″ thick.

Partition walls should not be tied to exterior walls with a masonry bond but abut the exterior wall and tied with steel ties. Screen block, usually 4″ thick, are used in dwelling construction to form dividers and low walls.

The molded face and regular rectangular block, **Figure 37,** create an abstract pattern which changes as the natural and artificial light changes, and shadows add accent.

Figure 38 shows a dining room and living room area having a patio to the rear. A food serving counter and china cabinets are on the wall to the right and a molded face block masonry partition wall between the living room and garage. The use of four half-width face blocks alternating with two full-width blocks to form a symmetrical row of clefts near the top of the masonry wall gives a pleasing design to the entire area.

Figure 39 shows a living room having an inside wall of masonry blocks surrounding a fireplace. Portions of the blocks are built with a stretcher bond and a Flemish bond. An extra wide hearth is built out into the room.

Inside Walls of Metal Studs and Gypsum. Because of the high costs of lumber, a limited number of residential builders are making use of steel studs and gypsum wallboard for non-load bearing partition walls. The studs are made of 25 gauge steel and are fastened to square U tracks nailed to the floor and ceiling, **Figure 40.** The ½″ or ⅝″ gypsum panels are nailed into grooves at the edges of the studs. The panels have a special paper which accepts a dense, hard veneer plaster about 1/16″ to 3/32″ thick.

Fig. 37. Inside wall of concrete masonry combines plain block with sculptured block for design effect and economy.

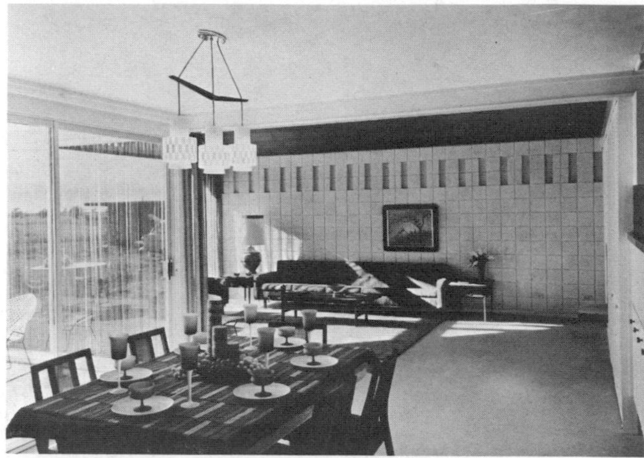

Fig. 38. Non-loadbearing partition of small concrete face blocks is laid up to form symmetrical recesses along top.

Fig. 39. Masonry block wall encloses fireplace and chimney, supports wide cantilevered hearth.

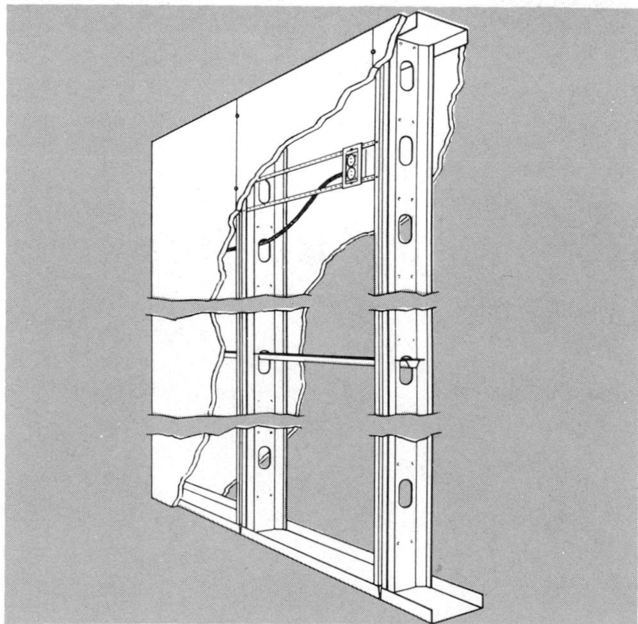

Fig. 40. Nailable steel stud system includes track and bridging. Panels are nailed into grooves formed in stud edges.

Chapter 6

ROOF CONSTRUCTION/BAY CONSTRUCTION/ROOFING

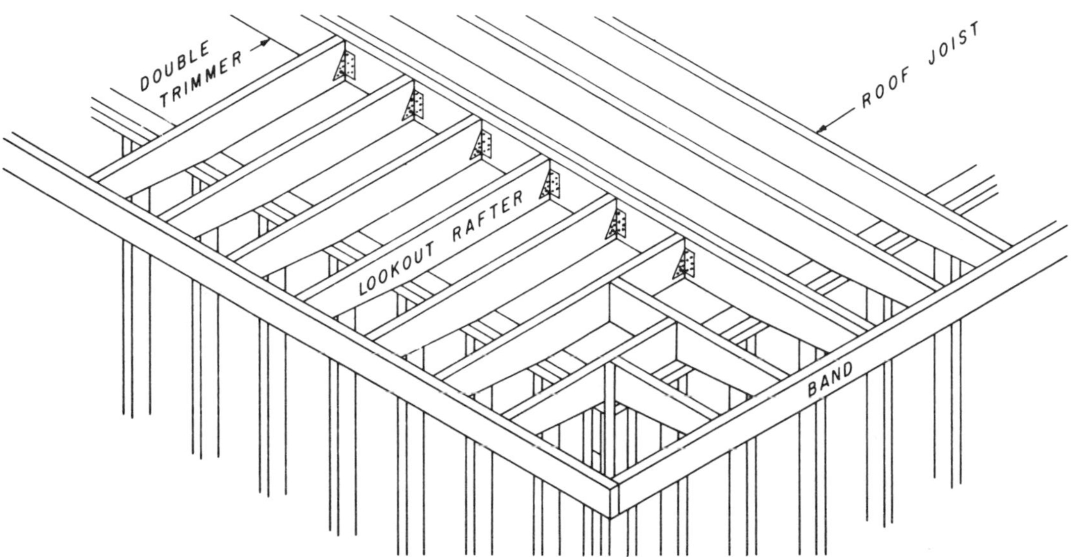

Fig. 1. Structural members used in framing a flat roof. Note lookouts cantilevered over wall running parallel with main joists.

The roof of a house is the uppermost structural assembly supported by the exterior walls. The roof provides protection to the dwelling and its occupants. The shape of the roof helps portray the architectural style of the entire building. The roof is also an integral part of any cornice construction. The roof must be structurally sound and solid and covered with appropriate roofing material.

Roofing is the final covering attached to the structural assembly and exposed to the weather.

Bay construction projects from the exterior wall of a house providing added floor area and windows. Bow and oriel windows also project from an exterior wall. An eyebrow window is built into the slope of a roof.

The Single Surface Flat Roof. Shapes of roof surfaces vary from the single surface flat roof, with a slight pitch to shed water, to a combination of flat, pitched and curved surfaces requiring the utmost skill in framing. In conventional construction, the flat roof is the easiest to frame, **Figure 1.** The horizontal framing is made up of roof joists or rafters overhanging the top plate of the outside wall framing. At right angles to the roof joists are lookout rafters attached by metal fasteners to a double trimmer, or header. The ceiling finish is attached to the roof joists and lookout rafters. A band framing member secures the ends of the joists and rafters. In

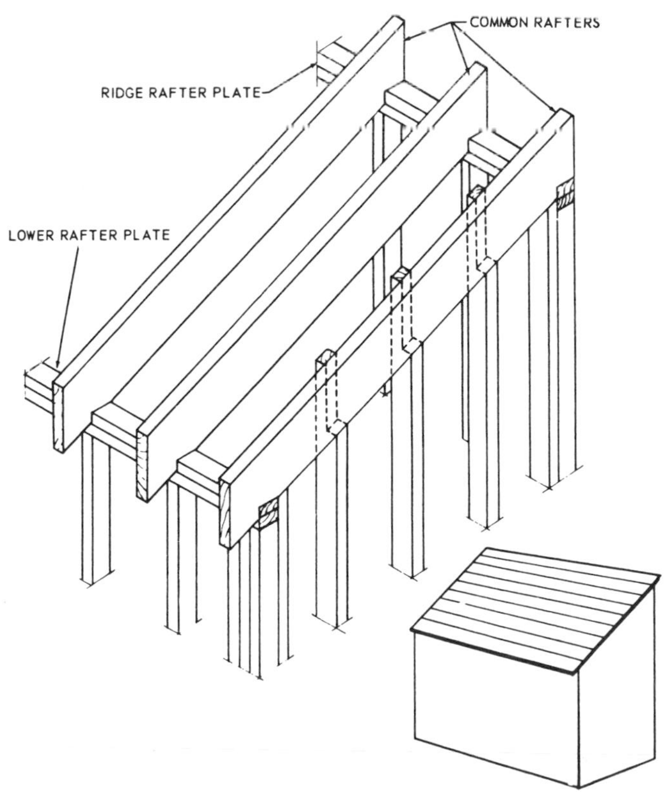

Fig. 2 Shed roof (right) is framed with sloping rafters. Endwall studs are cut and beveled to fit under end rafter.

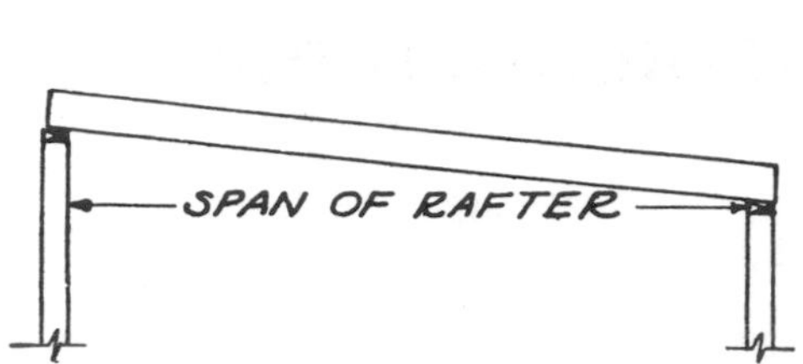

Fig. 3. Rafter span of shed roof measured from wall to wall.

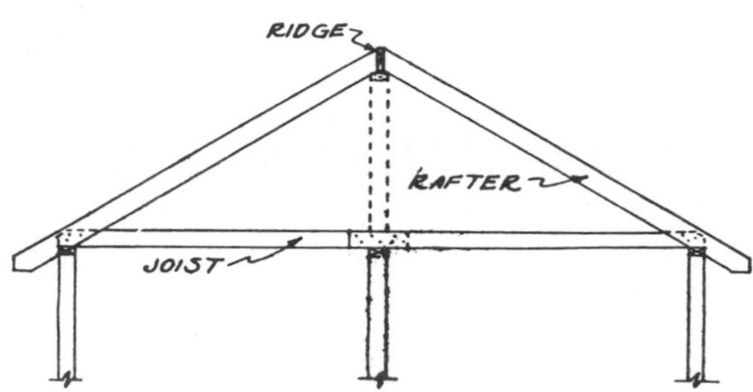

Fig. 5. Simple gable roof tied by ceiling joists.

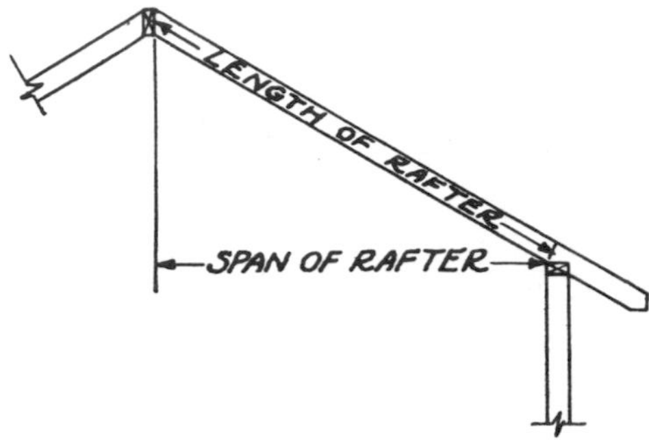

Fig. 4. Rafter span of gable roof between wall plate and ridge.

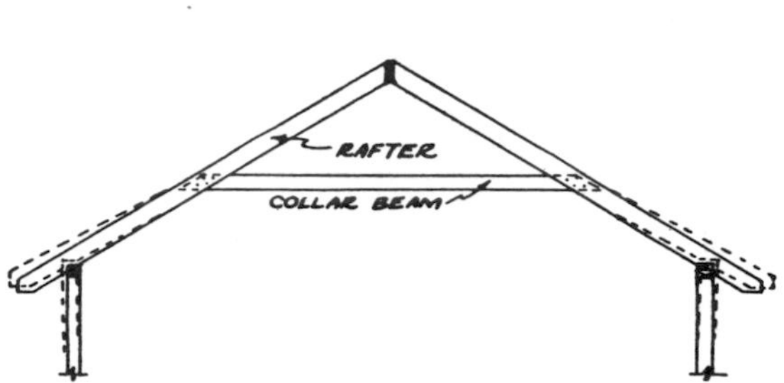

Fig. 6. Gable roof tied against outward thrust by collar beam.

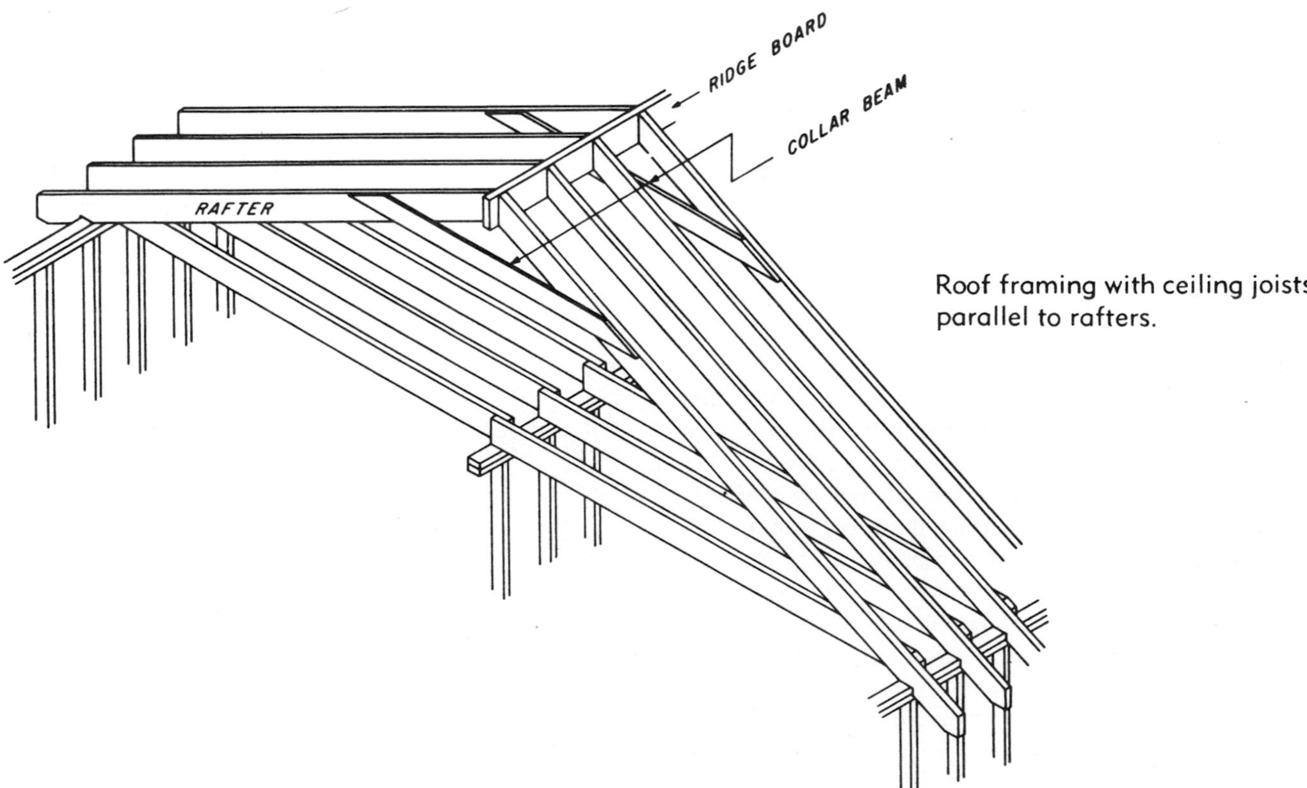

Fig. 7. Simplest type of gable roof construction, having ceiling joists running parallel to rafters and collar beam.

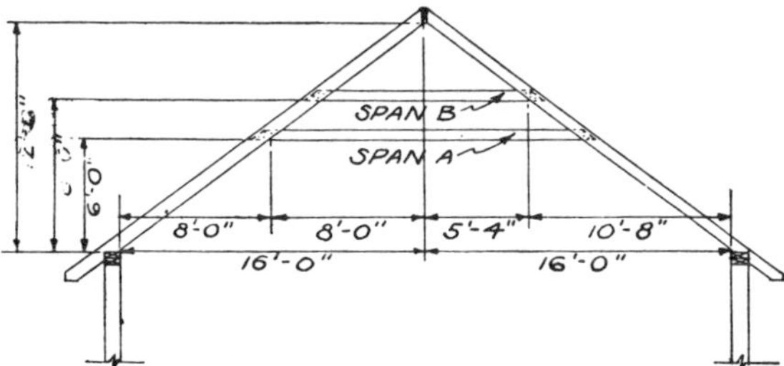

Fig. 8. Spans of rafters with collar beams: 8′ at span A, 5′ 4″ at span B, 16′ if no collar beams are installed.

Fig. 9. Knee walls with sole plates nailed to rough flooring support rafters at approximate mid-span. Collar beams across every third set of rafters add to structural strength of roof.

flat roof framing there is no angle of slope or pitch hence such construction is often used in contemporary designs of houses having carports, patios and courts. The flat roof does not provide an attic space, hence adequate insulation must be used between the joists or roof rafters in certain sections of the nation. The flat roof may also be designated as a single plane roof.

The Shed Roof. Another single plane roof is given a distinct angle of slope, or pitch, and is known as the shed or lean-to roof. This type of roof is framed with common rafters resting on wall plates, **Figure 2.** This style of roof is derived from a relatively small structure built for storage or shelter. In house construction the shed roof may be used over ground level rooms, garages and porches in connection with multi-level dwellings. In framing the rafters, the plumb cut is placed at the top of the rafter attached to the larger structure and the bird's mouth cut and tail cut at the lower end of the rafter. The end studs are notched on to the end rafter to form the framing for an exterior wall. If the ceiling is built horizontally under the shed roof a small attic space may be provided producing additional insulation for the room or space below.

The Gable Roof. The most popular and widely used roof of a house is formed by two roof planes meeting at the top most part or the ridge. A gable roof is characterized by a ridge line running the length of the house, the two planes of the roof sloping down at equal angles. The gable ends of the structure are triangles.

In the four outline drawings shown in **Figures 3, 4, 5 and 6** are shown the relation of shed roof framing to gable roof framing. **Figure 3** shows the rafter span of a shed roof measured from wall to wall, **Figure 4** the rafter span between the wall plate and ridge without a collar tie. A simple gable roof having rafters meeting at the ridge member and resting on the plate members of the exterior walls, is shown in **Figure 5.** The horizontal members are the ceiling joists which tie the rafters and side walls together. The ceiling joists may rest on an interior wall partition and a vertical support, indicated by dotted lines, may extend from the partition plate to the ridge. **Figure 6** shows the use of a horizontal collar beam instead of a vertical support. The dotted lines indicate the possible spread or bulge of the gable rafters if not tied by a collar beam or given a vertical support on a partition.

Methods of Framing The Gable Roof. The most simple type of construction of the popular gable roof is shown in **Figure 7.** The two exterior walls are framed with studs and plates. In the center is a load bearing partition wall and plate with the horizontal ceiling joists spanning from exterior plate to partition plate parallel to the rafters. The rise of the rafters is 6″ per foot of run or 1/4 pitch. The rafters rest on the plates at the outside walls and press against the ridge board at the apex. The collar beams, placed on every fourth set of rafters counteract the outward thrust of the roof load, **Figure 8.** Knee walls in the attic level of a Cape Cod house, **Figure 9,** hold rafters with the sole plate resting on the rough finished flooring. Note that the framing is not resting on any part of the chimney to allow the masonry to settle without affecting the structure.

In framing a gable roof for an artistic and protective projection or overhang at the ends and sides of the house to form eaves, rafters are extended beyond the wall plates, **Figure 10.** A ladder type of framing supports the overhanging rafter to the main roof framing by short lookout rafters resting on a plate, **Figure 11,** supported by cripple studs resting on the end wall plate. The ceiling joists are placed parallel to the rafters tieing the outside walls together. Short lookout pieces attached to the rafter and exterior sheathing form an eave, provide backing for nailing soffit panels.

Gable roofs, in conventional framing, are sometimes framed with the ceiling joists running at right angles to the rafters, **Figure 12.** The joist nearest to the side wall framing is anchored to the plate and rafter by metal straps or corner anchors attached to short ceiling joists placed parallel to the rafters and butting against the right angled ceiling joist. The subflooring across the

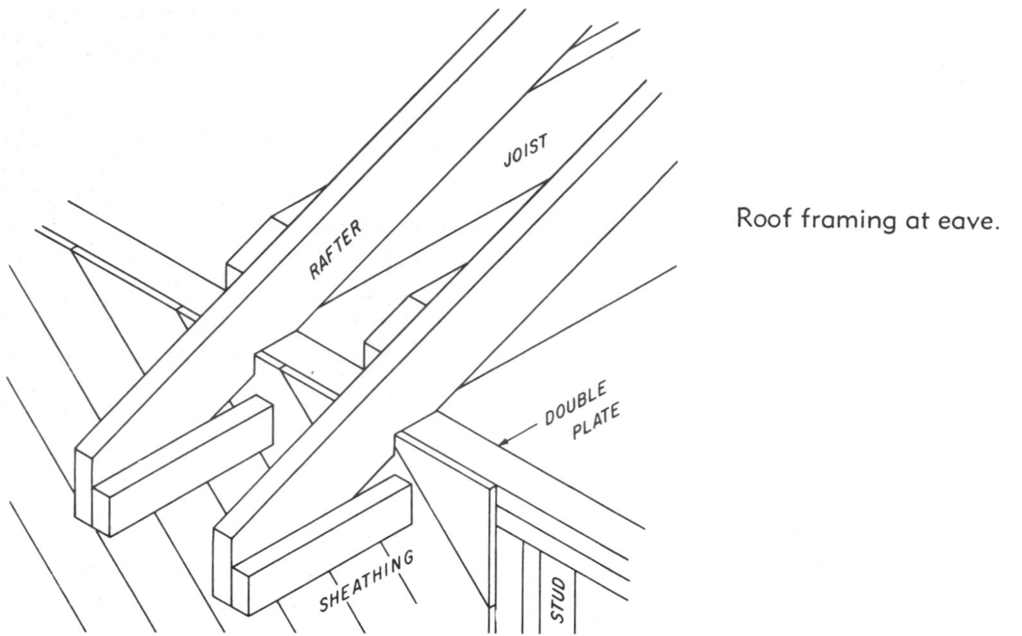

Fig. 10. Overhang at eave is formed by extending rafters beyond wall plate. Blocking supports soffit panels.

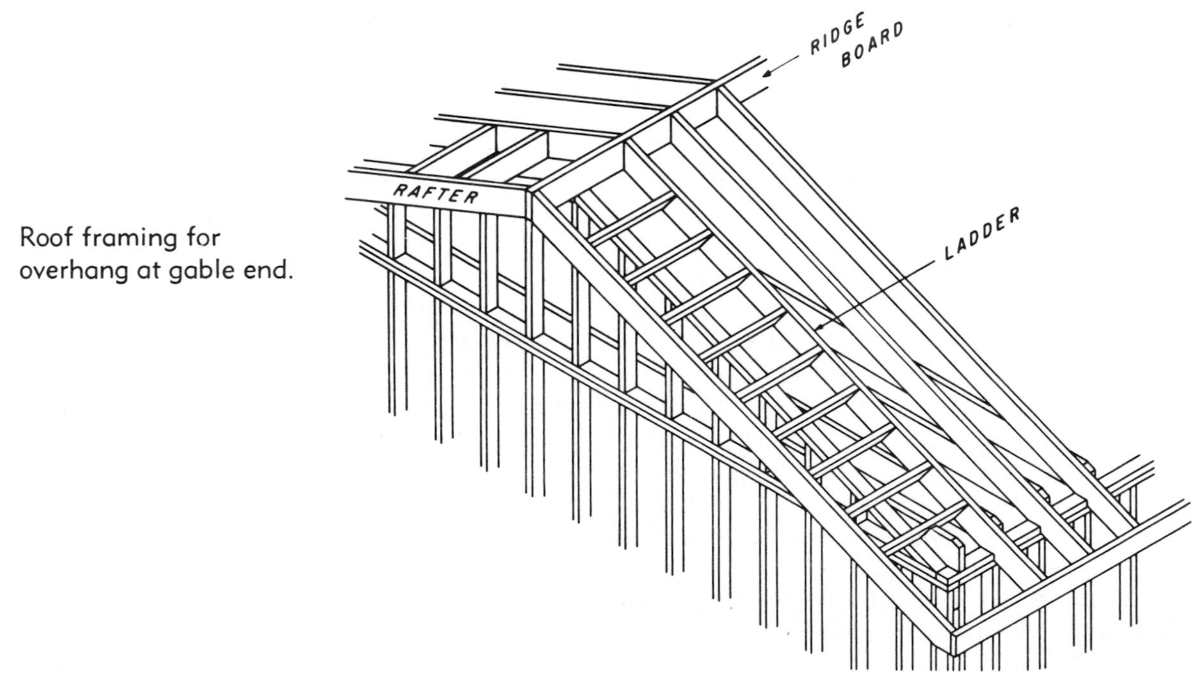

Fig. 11. Overhang at gable end is formed by ladder arrangement of short lookouts bearing on plate to support end rafter.

rafters ties the rafters against the outward thrust of roof load.

Gable roofs are also used in combinations of different pitches and at different levels and angles to express good architectural style in the one-level, the two story, the multi-level, the contemporary, the ranch, etc., type of house construction. Intersecting valleys complicate construction somewhat.

Gable Roof Intersections. In the construction of gable roofs joined at right angles, the intersecting roofs form a valley, **Figure 13.** The framing is the same for both roofs, except at the intersection which requires a valley rafter. The shortened common rafters, extending downward from the ridge boards, are known as jack rafters, and are supported by a valley rafter as indicated. The valley rafter runs from the ridge down to the outside wall plate. Note the use of bevel cuts at the ends of the jack rafters. If the gable roofs are of different heights a supporting header for the lower structure is used between the common rafters of the higher roof, **Figure 14.**

The Dormer in a Roof. In order to provide light and air in an upstairs room a window in the roof is framed as shown in **Figure 14.** This drawing shows a gable type dormer framed much like the main roof. Note the use of double trimmer rafters, double headers above and below, the common rafters of the dormer and main roof, and the jack rafters meeting the valley rafters at the intersections of the dormer roof and the main roof.

When a large dormer is desired in an upstairs room, particularly to provide greater floor space up to the outside wall a shed type dormer is framed, **Figure 15.** This style of dormer not alone provides useable floor space, but useable head room which frames off an extra room in an upper floor or in an attic formed by one half or greater pitched roof. The framing of this dormer is quite similar to the framing of the shed roof shown in **Figure 2.** Note the use of double trimmer rafters to support the cripple studs at the sides, the window opening framed in the wall above the outside wall studs, and the placing of the ceiling joists which run from the shed rafters of this dormer to another dormer or to rafters on the adjacent roof slope.

The Hip Roof. Another type of roof, often used in house construction, is the hip roof. In a hip roof the ridge does not run the full length of the house; hip rafters extend up diagonally from each corner to meet the ends of the ridge as shown in **Figure 16.**

The common rafters, as in a gable roof, extend from the ridge to the plate and, for an overhang for an eave, even farther. The hip jack rafters are the shorter members meeting the long hip rafters at acute angles. French Provincial style houses have hip roofs with a high pitch, **Figure 17.** In a hip roof over a square house, the hip rafters meet at a point and the roof resembles a pyramid.

A hip roof forms four roof planes while a gable roof

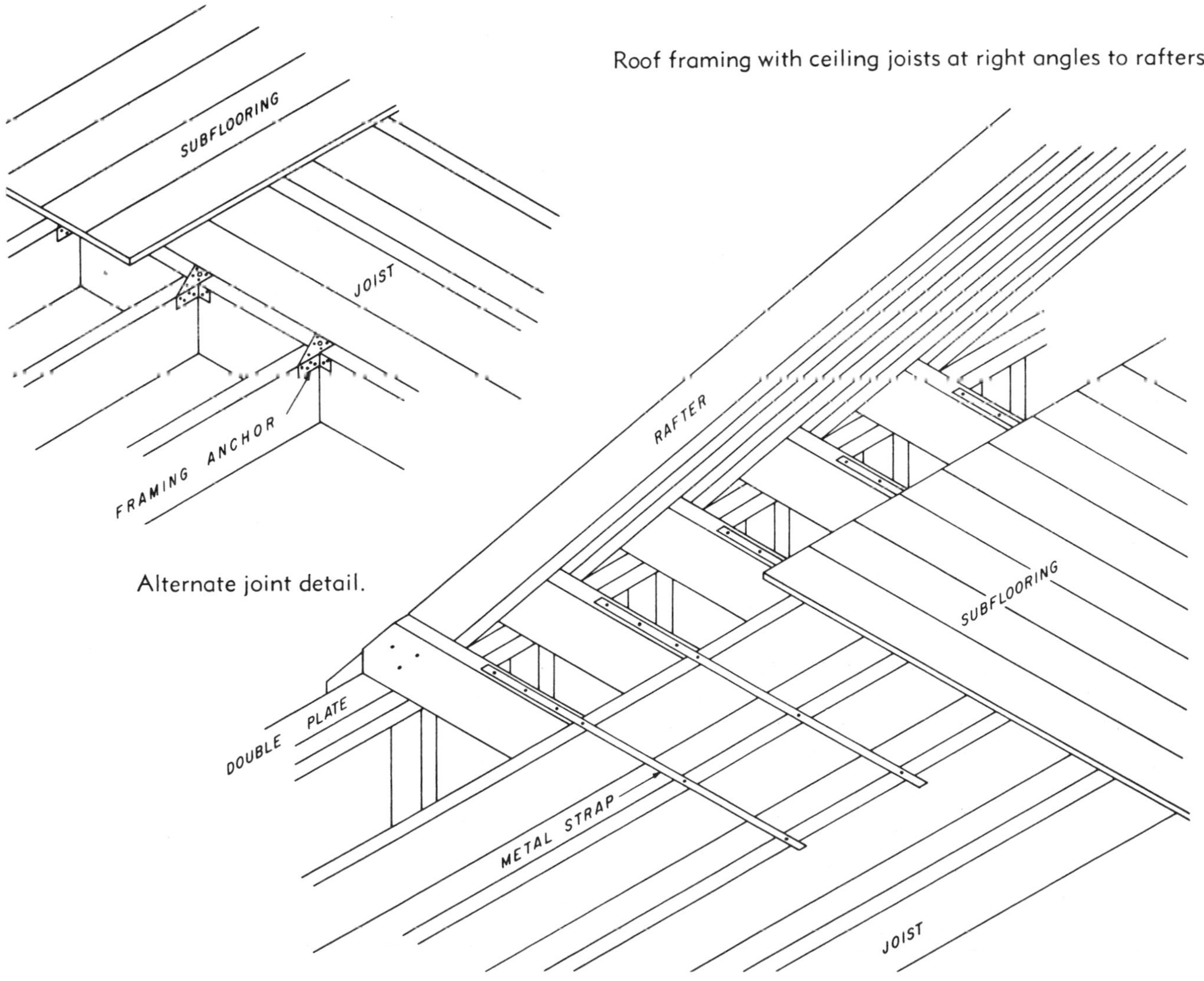

Fig. 12. When joists run at right angles to rafters, short joists and metal straps can be used to tie rafters to subflooring.

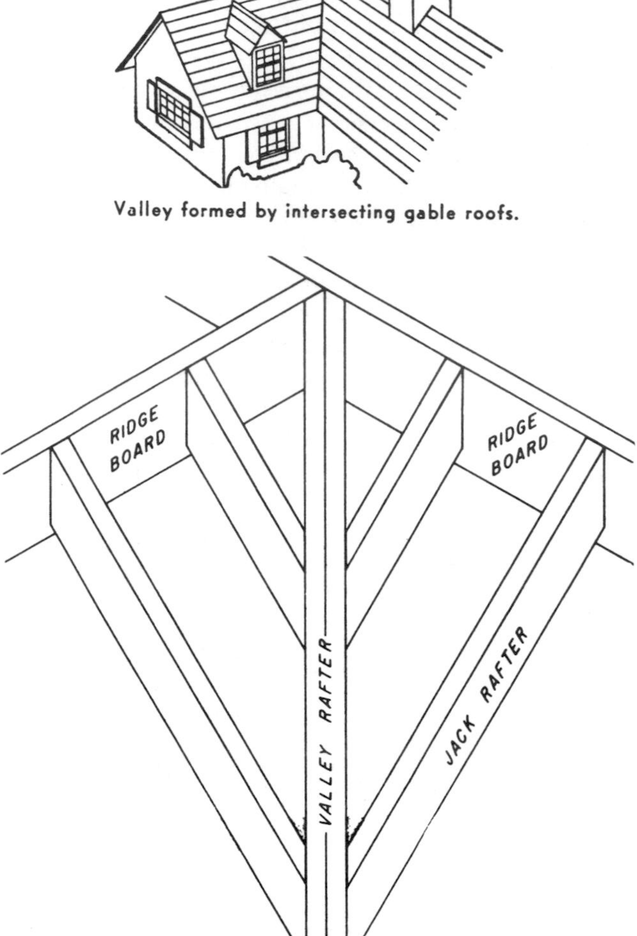

Valley formed by intersecting gable roofs.

Fig. 13. Roof valley (top) is formed by nailing jack rafters between intersecting ridge boards and diagonal valley rafters.

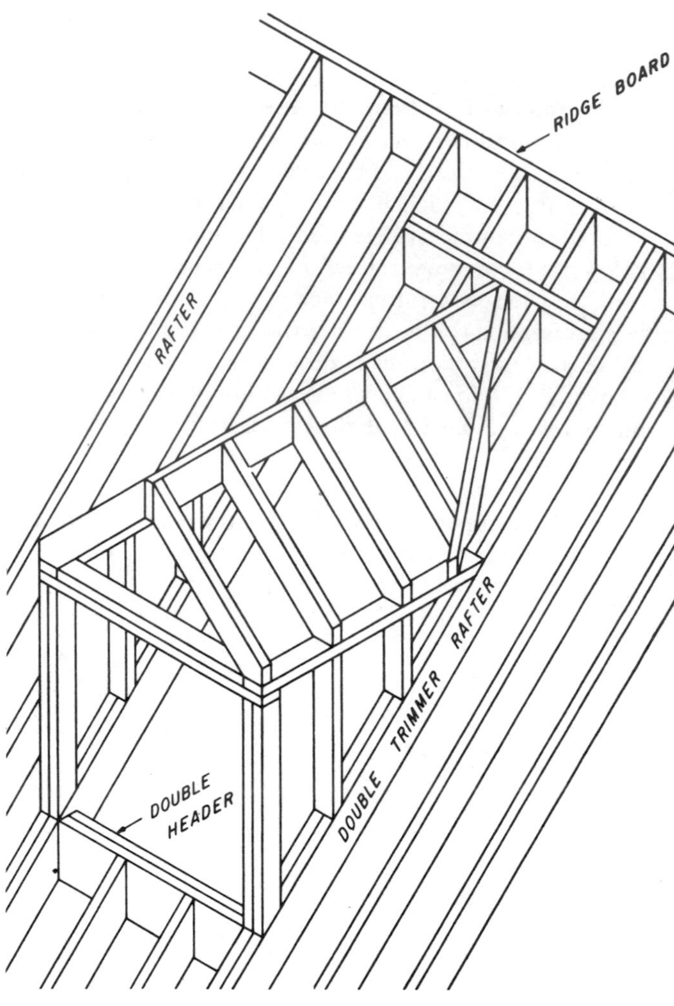

Fig. 14. Dormer on gable roof is framed over opening formed by doubled trimmer rafters and headers at top and bottom.

forms two roof planes. The hip roof is often used to form a pleasingly low roof on a ranch house or on a rambling contemporary covering a relatively large area of rooms. The low pitched hip roof affords a wide variety of room arrangements joined at pleasing angles.

Hip roofs may intersect one another at the same level or at different levels or may be used in connection with an intersecting gable roof of the same level, **Figure 18.** Note the valley rafter joining at the ridges, the hip rafters joining at the end of the ridge and the hip jack rafters and the valley jack rafters with bevelled cuts.

A gable roof for a porch may extend from a hip roof at a relatively low level, **Figure 19.** Note the intersection of the porch roof ridge at the valley rafter in the hip roof and the cripple jack rafters running from the valley to the hip.

The Gambrel Roof. The gambrel roof has a single ridge running the full length of the house, similar to the gable roof, but each roof deck is angled outward in two planes, with the lower plane much steeper, **Figure 20.** The horizontal member to which the upper and lower set of rafters are framed is called the purlin. Usually the purlins on both sides of the ridge are connected by tie or collar beams at each rafter. Gambrel framing offers more usable space than a gable or hip roof of equal height, with no greater span of rafters. **Figure 21** shows a dwelling having a gambrel roof running from front to back and an intersecting garage with a gable roof at a right angle. Note the wide overhang of the roof to provide an open cornice projection at the outside wall. The pitch of projection is the same as that of the upper part of the gambrel roof.

The Mansard Roof. A French style roof, designed by Francois Mansard, is constructed somewhat like the gambrel roof of four planes. The Mansard roof has eight planes formed by hip rafters meeting at a ridge or

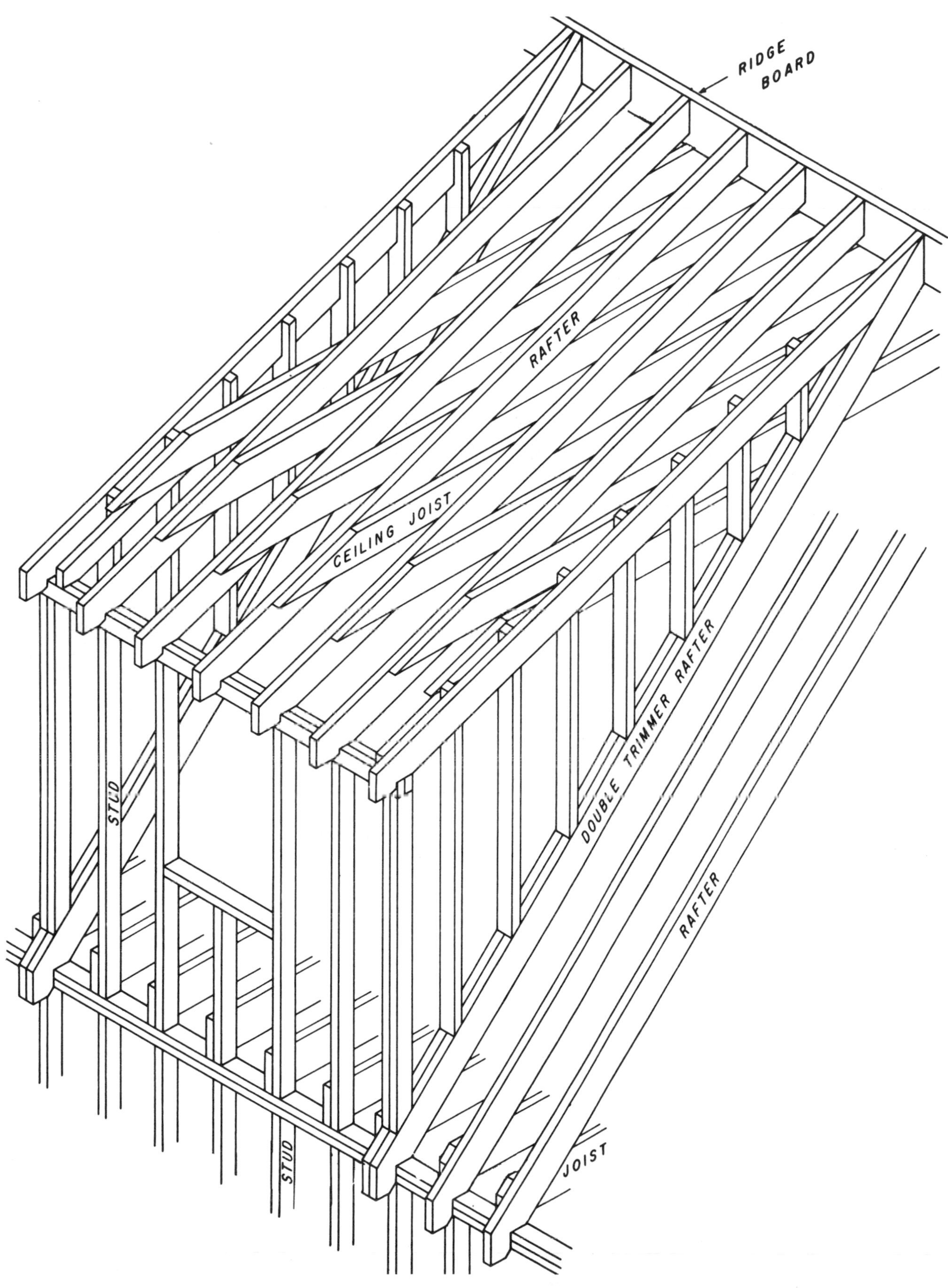

Fig. 15. Dormer framed with shed roof provides maximum floor space and headroom on the second story.

at the topmost point of the upper slopes. The Mansard roof does not have gable ends like the simple gable roof or the gambrel roof. When dormer windows are built into the steeper roof slopes, light and air are provided for the room or rooms just below the roof.

The Low Sloping Roof. The low sloping roof is almost as popular as the simple gable roof. The larger slope or plane is generally continuous over two or more floor levels, **Figure 22.** The arrows indicate the points of support for exterior and interior walls.

Additional headroom can be gained by framing a two plane, low sloping roof, over a 1 and 1/2 story house, **Figure 23.** Note the placing of knee walls, dormer walls, partition walls and ceiling joists. The smaller detail shows a method of framing the gable end to include a large window opening.

Fastening Roof Members. Roof construction should be strong in orderto withstand anticipated snow andwind loads. Members should be securely fastened to each other to provide continuity across the building and should be anchored to exterior walls. **Figures 24, 25, 26, 27, 28, 29, 30 and 31** illustrate proper methods of tying rafters in place. **Figure 24**; a cripple stud placed above the full length outside wall stud at a gable end. Note the bevel cut to house the rafter. It is also known as a notched stud. **Figure 25**; a collar beam or rafter tie holding rafters together to divert lateral pressures. Letters mark placement of 10d nails. **Figure 26**; metal straps attached to the side wall stud, shoe and rafter to resist horizontal thrust and uplift. **Figure 27**; two methods of attaching rafters to partition plate and outside wall plate. The first rafter has two bird mouth cuts and an extension to form a built up eave. The ceiling joist is nailed above the partition plate. The second rafter is cut flush with the outside plate to take side wall covering, frieze and crown molding. The ceiling joist may also be attached to this rafter. Letters mark places where nailing is done. **Figure 28**; a method of attaching rafters to an exterior wall having firestop headers placed between each floor joist. Note the use of an additional plate member placed between the rafters and nailed to the attic flooring. **Figure 29**; a method of

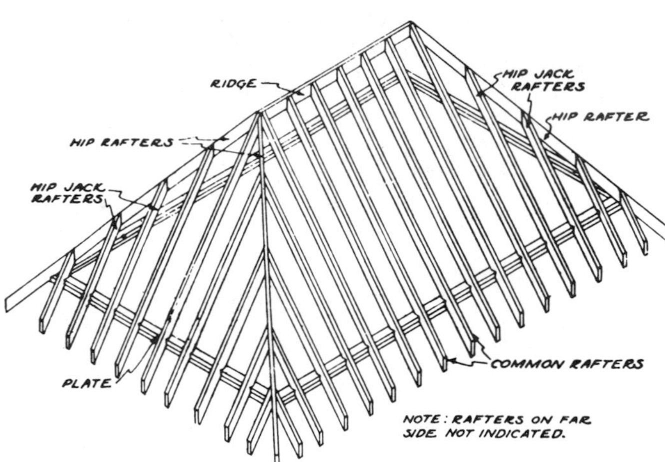

Fig. 16. Structural members used in a hip roof.

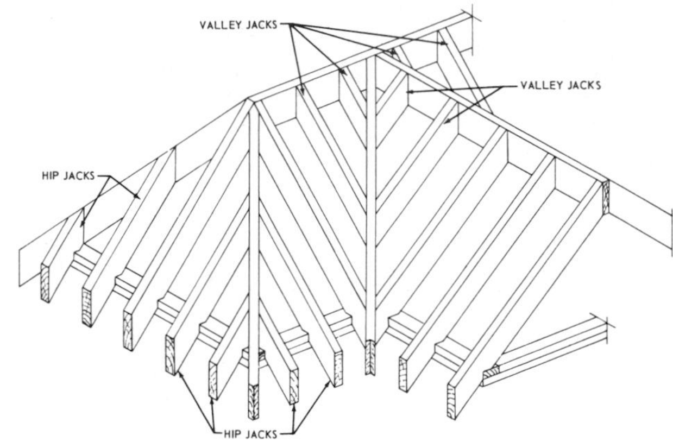

Fig. 18. A gable roof (at right of dwg.) framed at right angles to a hip roof at the same ridge level.

Fig. 17. French Provincial roof on dwelling and garage.

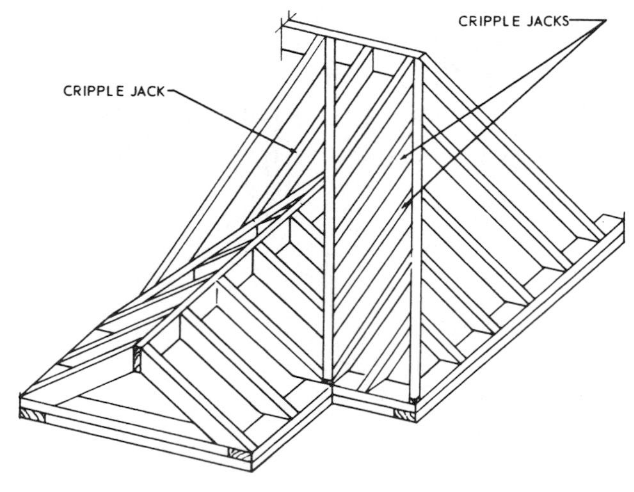

Fig. 19. Gable roof for a porch framed to a hip roof.

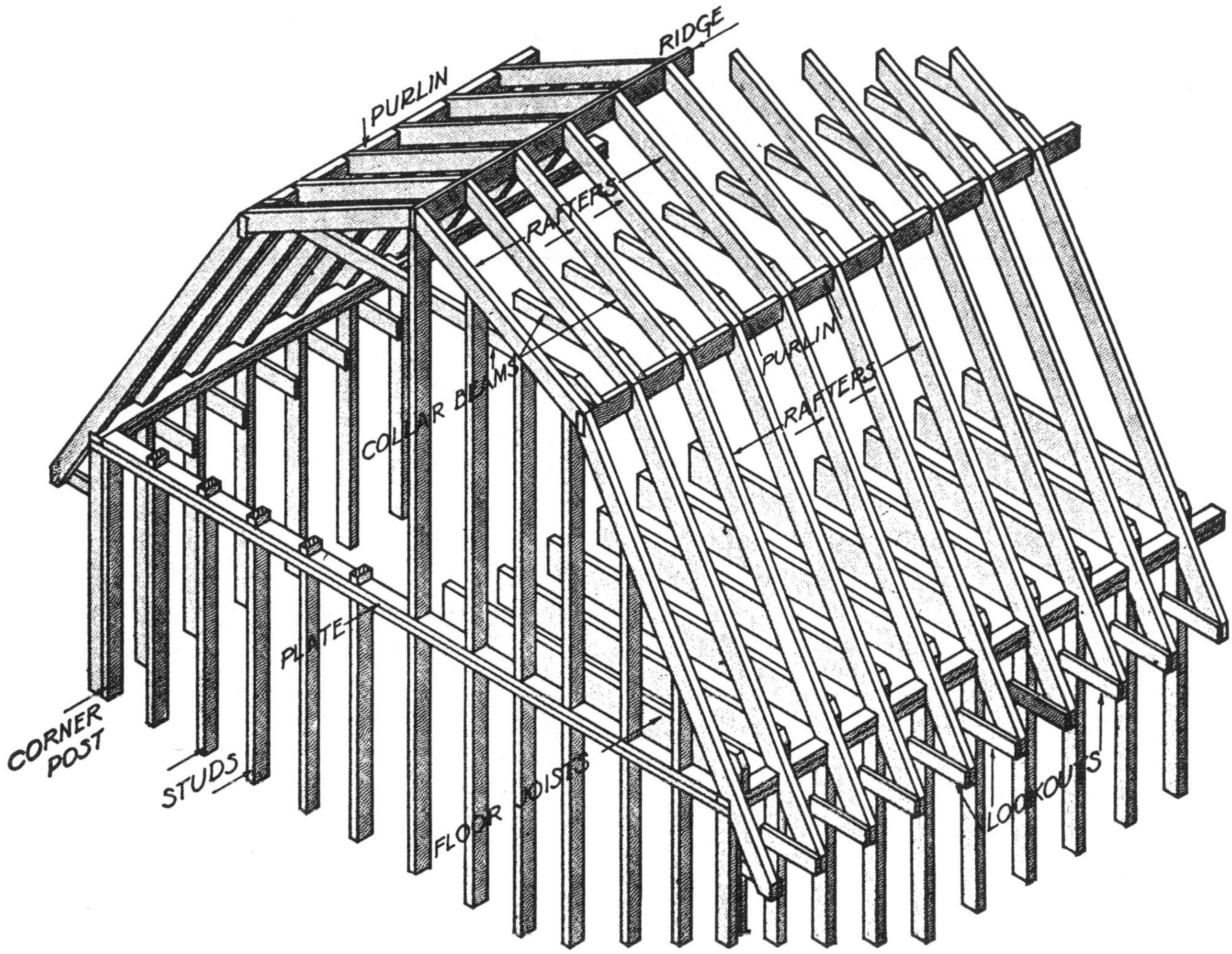

Fig. 20. Structural members used to frame a gambrel roof. This type roof offers the most useable space and headroom.

Fig. 21. Gambrel roof over dwelling is intersected by a gable roof over garage at right of photo.

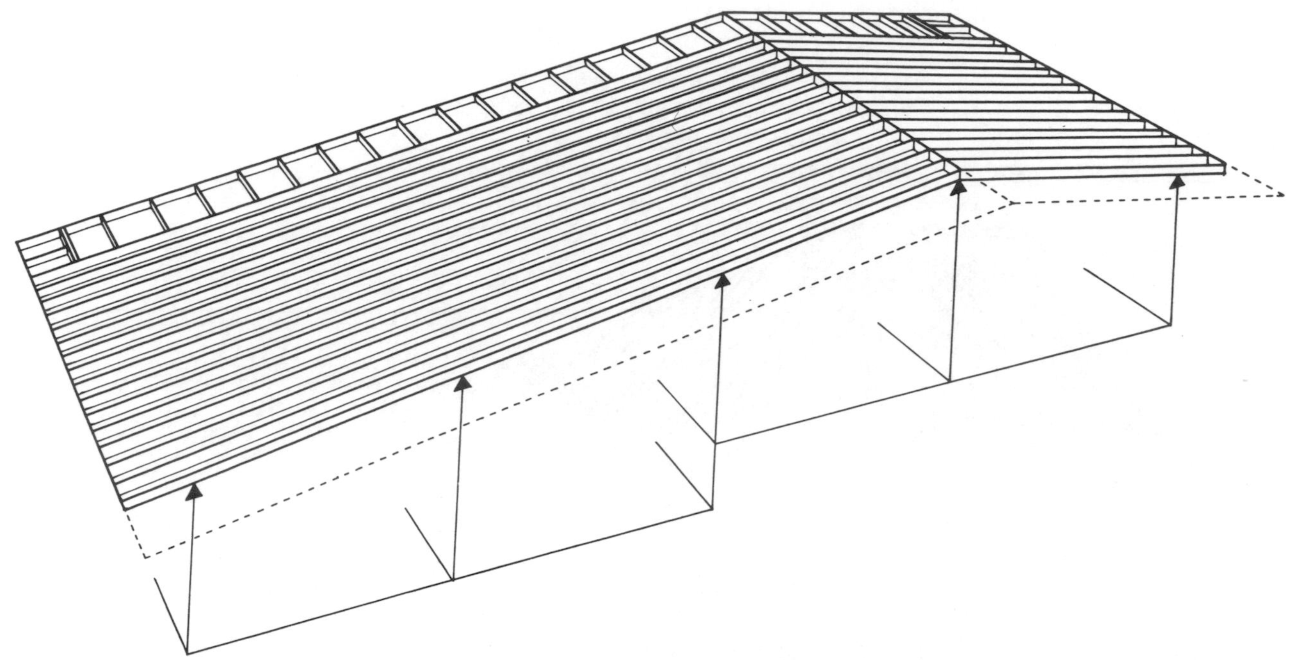

Fig. 22. Low sloping roof of two planes over a split-level arrangement of rooms.

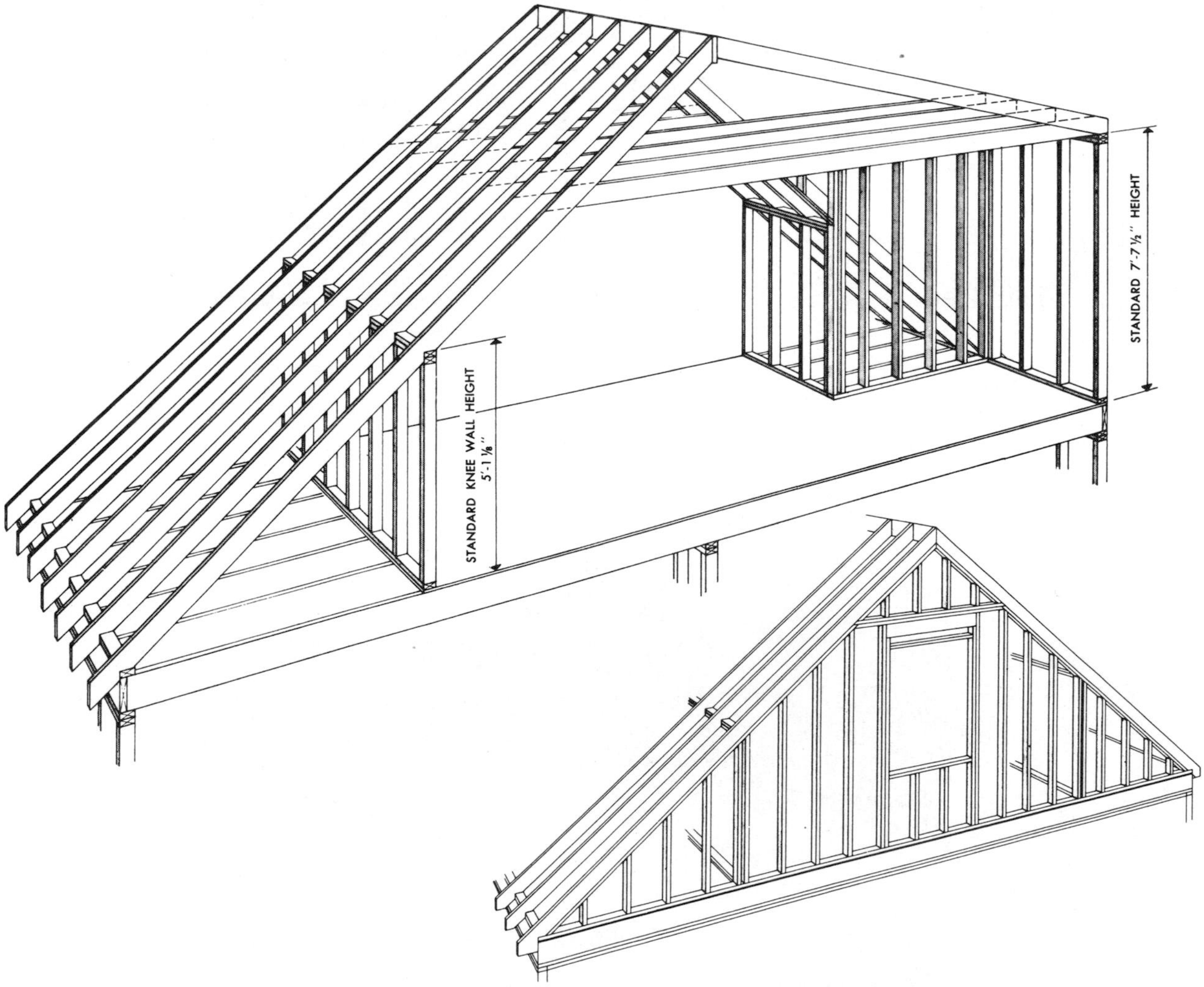

STANDARD KNEE WALL HEIGHT 5'-1 ⅛"

STANDARD 7'-7 ½" HEIGHT

Fig. 23. Framing details of two-plane roof. Lower detail shows method of framing window opening in gable end.

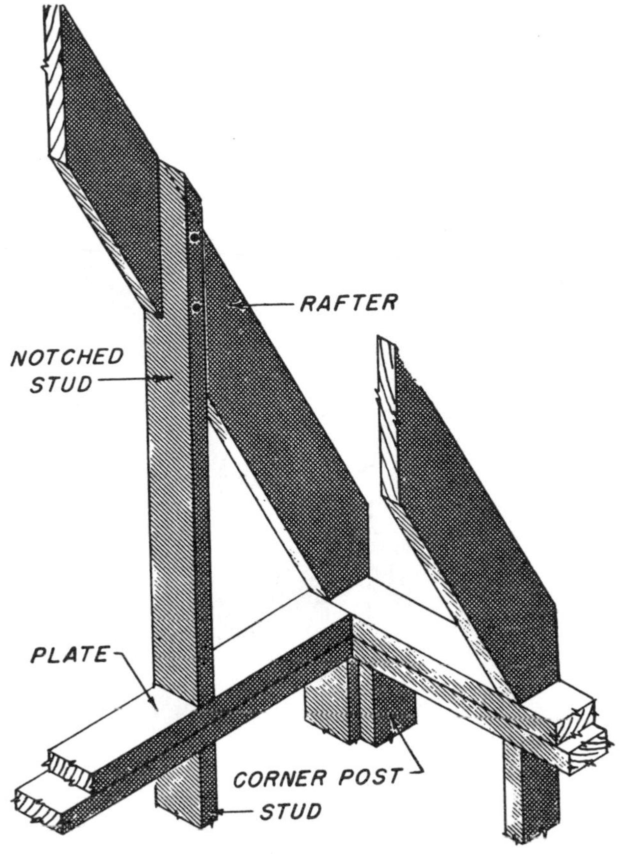

Fig. 24. Notched studs in gable end framing provide maximum support for steep rafter. Note placement directly over wall studs.

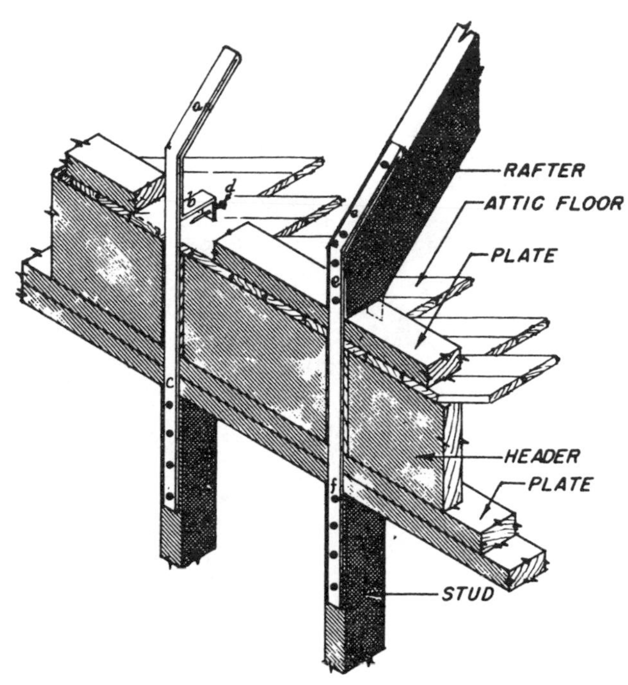

Fig. 26. Metal rafter straps tie rafters to studs to resist horizontal thrust and uplift pressure of high winds.

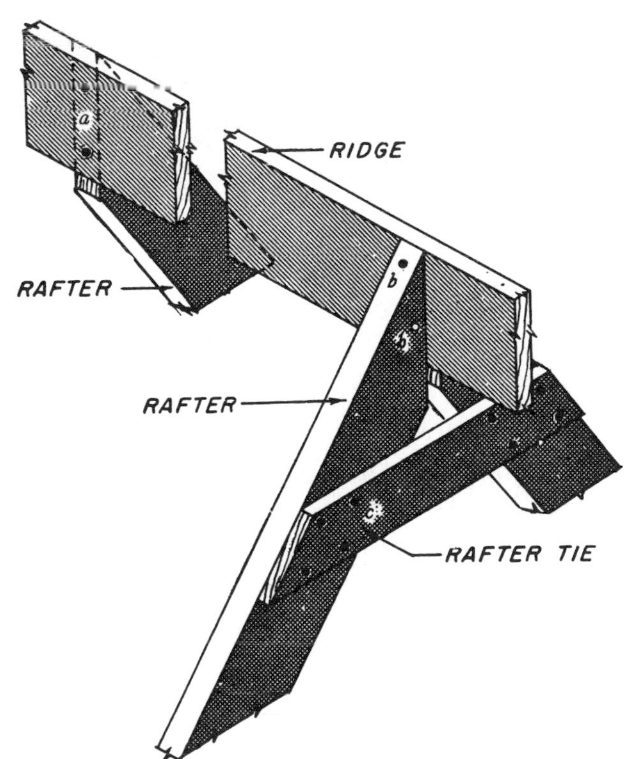

Fig. 25. Rafter tie under ridge board helps resist spreading from wind action. Letters mark placement of 10d nails.

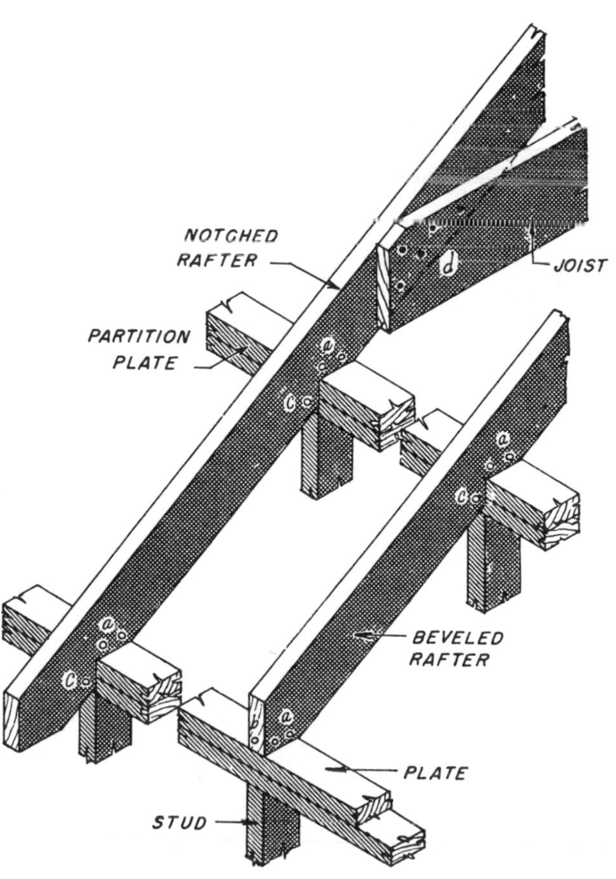

Fig. 27. Rafters notched or beveled at wall plate and notched over partition or knee wall plate with ceiling joist above.

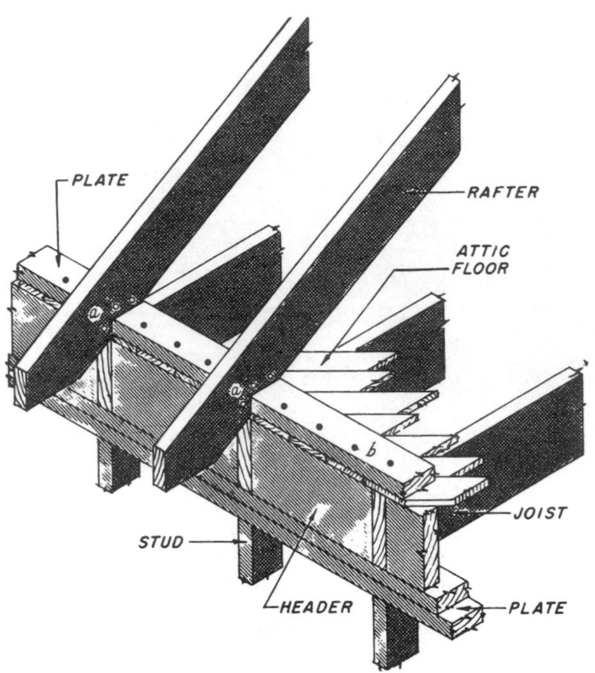

Fig. 28. Notched rafters resting on plate nailed to subflooring.

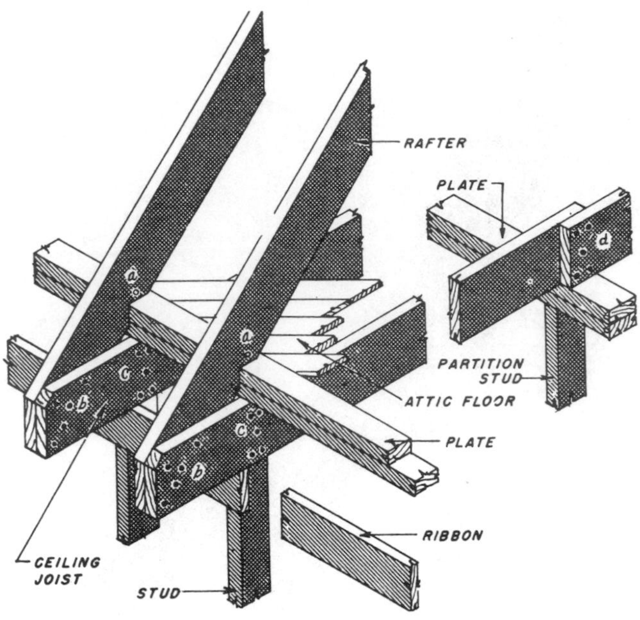

Fig. 30. Notched rafters rest on plate nailed across joists.

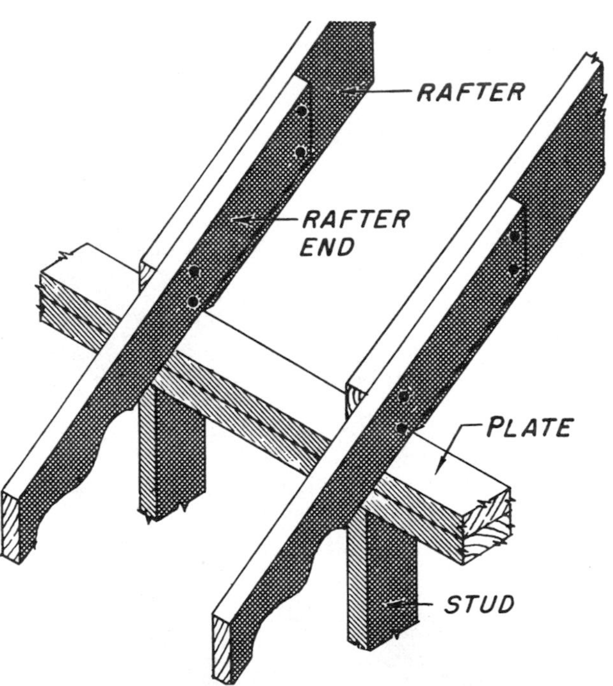

Fig. 29. False rafters are nailed to main rafters with 10d nails.

Fig. 31. False rafters nailed to 2″ × 4″ rafters in hip roof.

nailing a false rafter end to the load bearing rafter to form an elaborate open overhang of the roof. **Figure 30**; a method of framing a lookout with the extension of the ceiling joist and rafter. The rafter rests on the plate and the ceiling joist rests on a ribbon on the outside wall and the partition plate. **Figure 31**; a photograph of a corner of hip roof framing. Note rafter lookouts nailed to ends of jack rafter meeting the hip rafter. A let-in brace is placed in outside wall studding which is covered with 24″ insulating sheathing.

The Plank and Beam Roof. The use of heavy planks and beams in the erection of a roof of two planes, having a gable style in appearance and creating the cathedral ceiling, is somewhat similar to the low sloping roof design. The plank and beam roof system is essentially a skeleton framework. Transverse beams span from ridge beam to wall columns, and planks span the transverse beams parallel to the roof direction. Beam sizes vary with the span and spacing. They may be solid or glued-laminated wood, or built up of several thinner pieces,

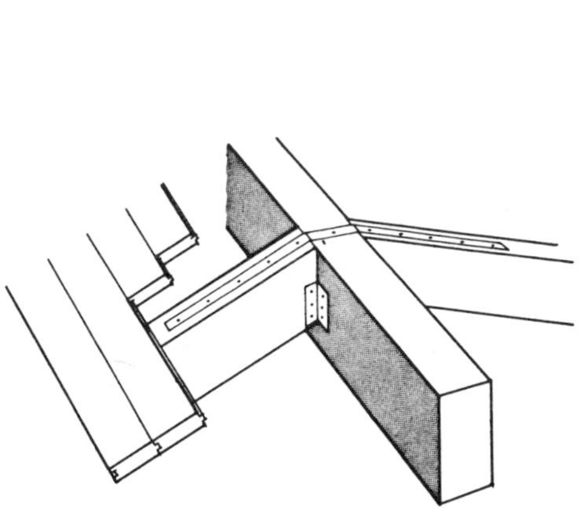

Fig. 32. Supporting roof beams on a ridge beam.

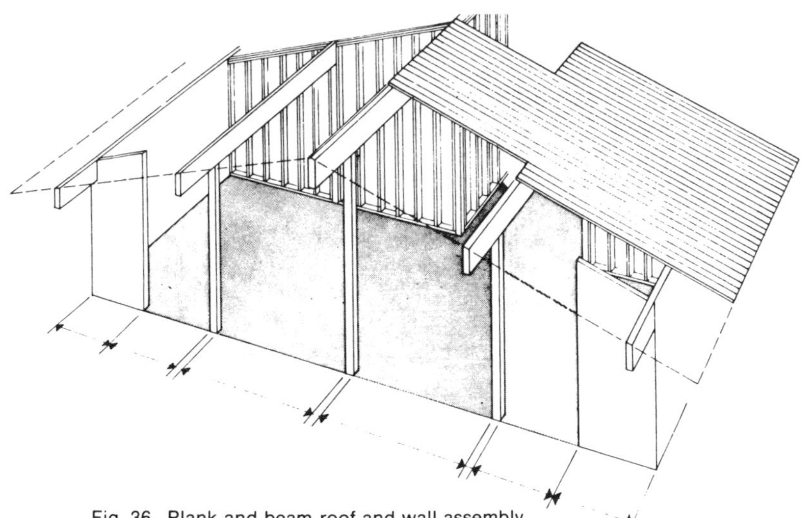

Fig. 36. Plank and beam roof and wall assembly.

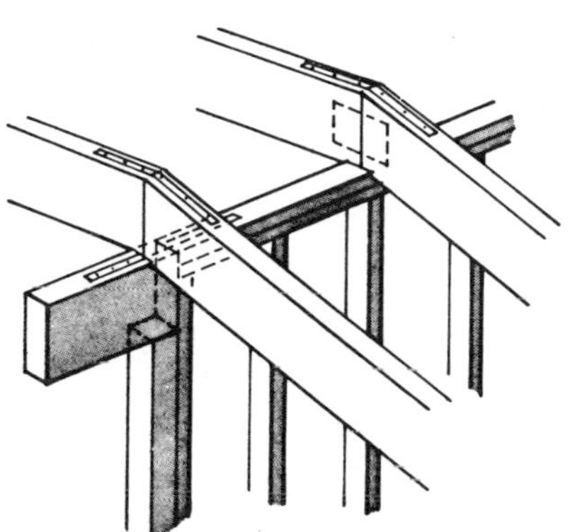

Fig. 33. Supporting roof beams on a bearing wall.

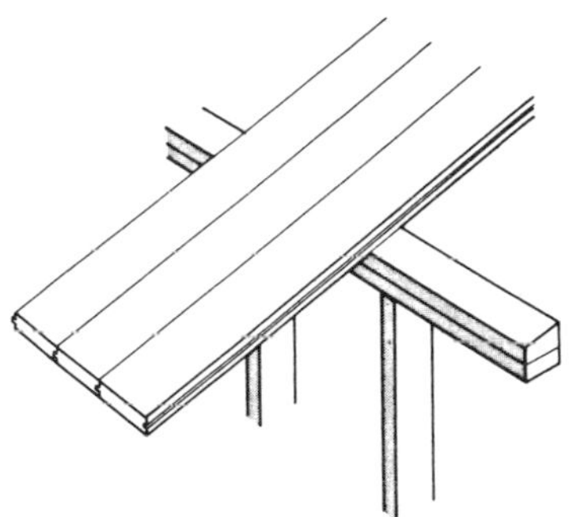

Fig. 37. Plank roof overhang at exterior wall, plate beveled.

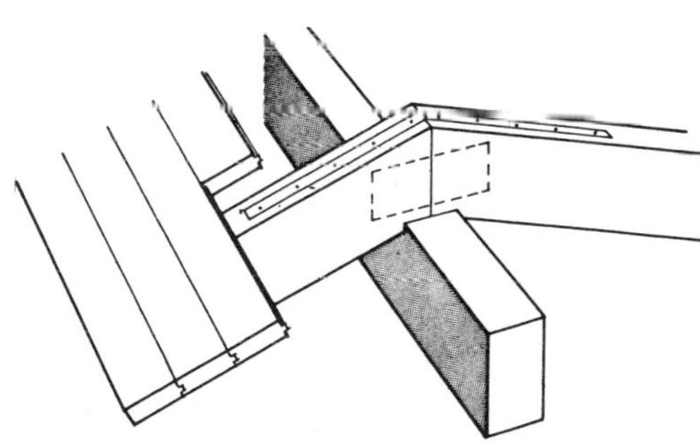

Fig. 34. Roof beams joined on top of ridge beam.

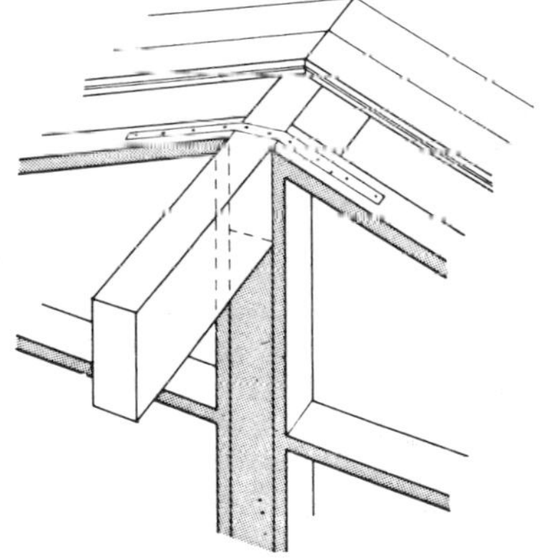

Fig. 38. Ridge beam supported by column in end wall.

Fig. 35. Roof beam supported by column in exterior wall.

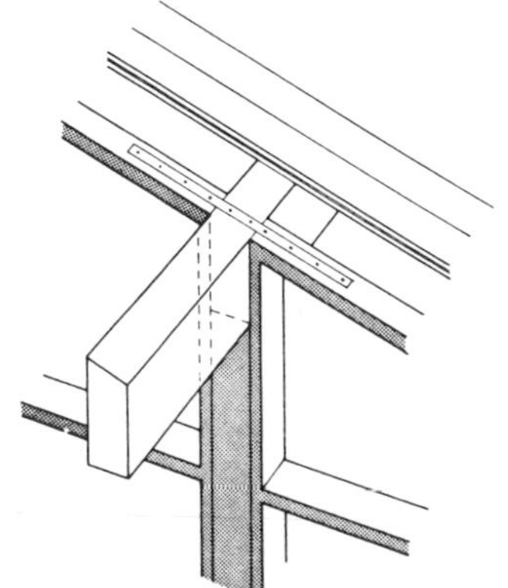

Fig. 39. Intermediate longitudinal beam on column in end wall.

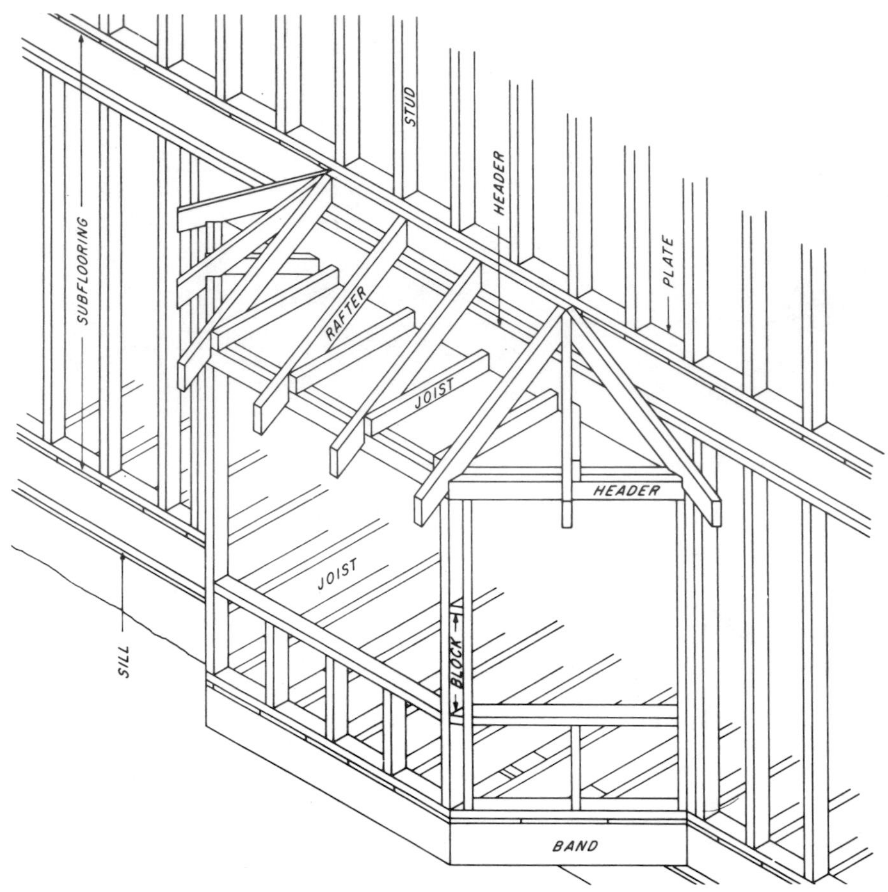

Fig. 40. Framing details for a large bay window in an exterior wall of a dwelling. Note cantilevered floor joists.

Fig. 41. Photo shows large bay window with movable sash at sides. Elevation and plan are of stock oriel window.

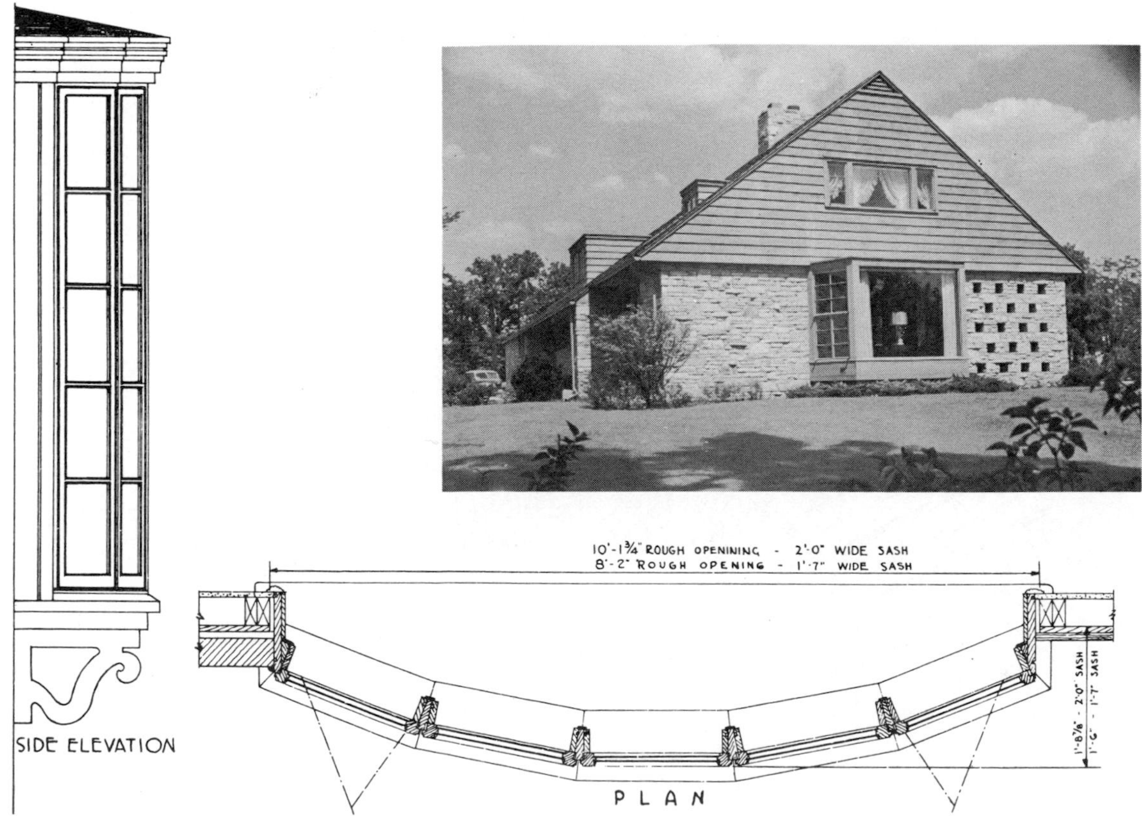

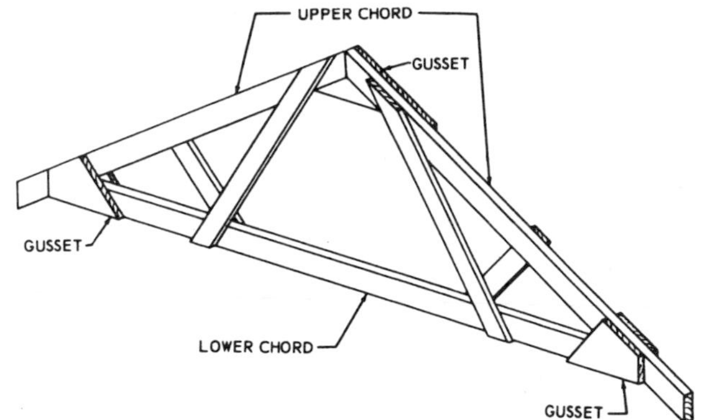

Fig. 42. A simple, lightweight truss, using three gussets to join members.

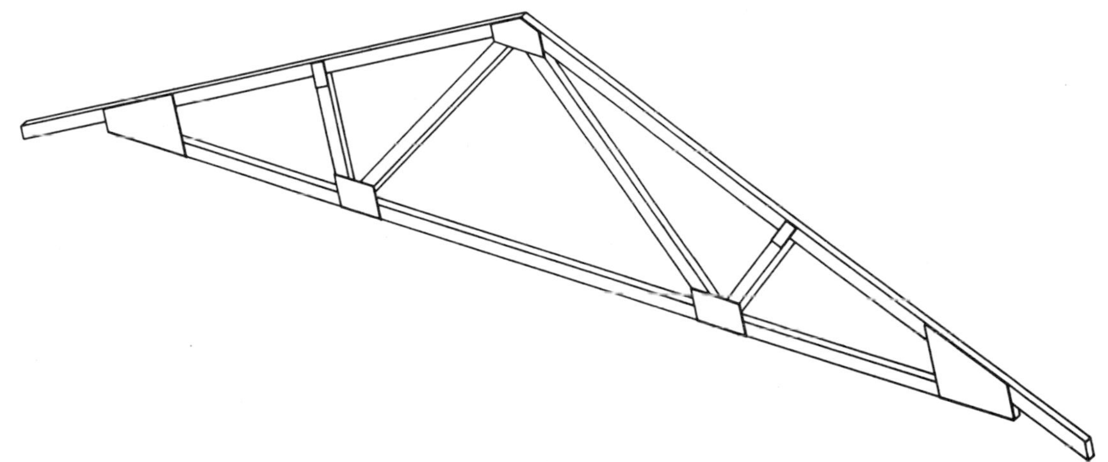

Fig. 43. A trussed rafter assembled with plywood gusset plates on one or both sides.

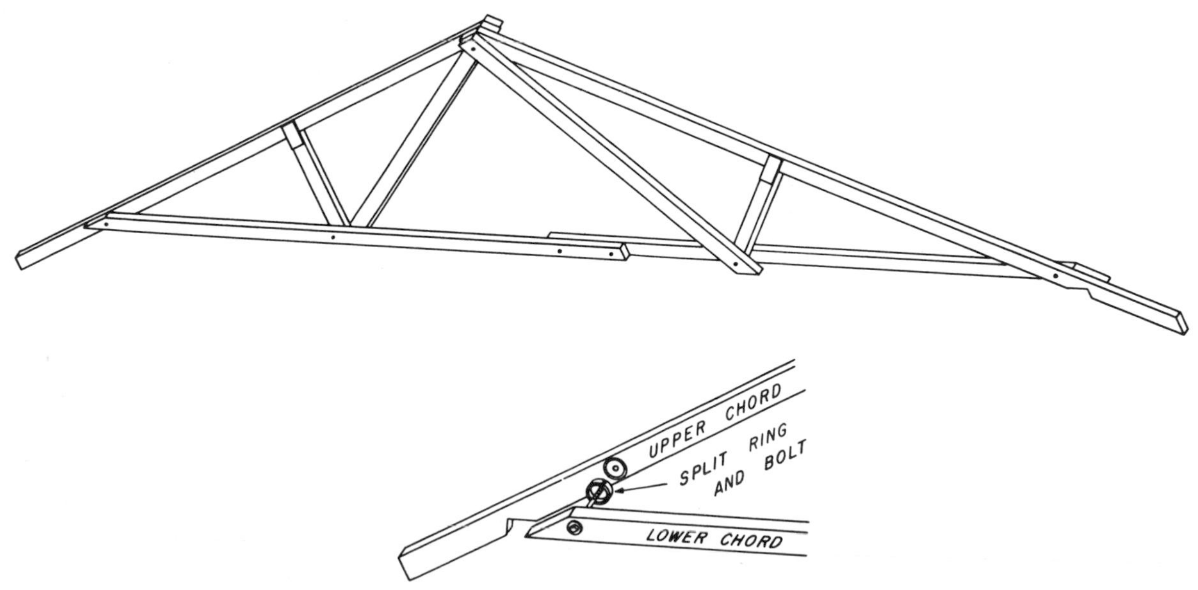

Fig. 44. A trussed rafter assembled with ring-type timber connectors and bolts.

securely nailed to each other or to spacer blocks. Strength and stiffness of plank roofs are utilized most efficiently when planks are continuous over more than one span. Good design requires that all members be properly fastened together so that the house and roof will act as a unit in resisting external forces.

Plank and beam systems may be coordinated with the modular planning concept which is a standard or unit of measure in which the proportions are regulated. **Figures 32, 33, 34 and 35** illustrate methods used in supporting roof beams. **Figure 32**; a roof beam supported on a ridge beam, with metal hangers and tie straps to absorb the horizontal thrust. The tongue and grooved planks rest on the roof beam. **Figure 33**; a roof beam supported on a bearing wall with metal tie straps. The column and false beam end are fastened to the top plate and end stud of bearing wall. **Figure 34**; the roof beams supported on top of the ridge beam, with metal tie plates to absorb the horizontal thrust. **Figure 35**; the roof beam supported by a wall column and tied with a metal plate. Note the bevel of the plate supported by the stud wall.

The placing of beams along the length of the structure varies with the span. The exposed plank and beam ceilings, whether placed parallel to the width of the structure (transverse), or to the length (longitudinal), achieve a distinct and pleasing architectural effect for dwellings, creating the cathedral ceiling needing only the desired finish. Longtitudinal beam sizes vary with the span and spacing of beams. Outstanding design variation of end walls may be achieved with the extensive use of glass and with projected roof overhangs. Wall, window and door dimensions, when coordinated with the structural member dimensions permit a completely engineered system of fabrication and erection.

Figures 36, 37, 38 and 39 illustrate methods of framing a house with longitudinal beams. **Figure 36**; the plank and beam roof assembly, the end columns, the filler panels, and the room partition studding. **Figure 37**; the roof plank overhang at the exterior wall. The planks are fastened to the sloping top plate of the exterior wall. **Figure 38**; the end wall column support of the ridge beam. A metal strap ties the adjacent panels and beams. **Figure 39**; the framing of an intermediate roof beam between the outside wall and the ridge beam. Note the metal tie strap, the overhang of the planks and the bevel of the beam.

BAY CONSTRUCTION

A bay is a framed structure housing windows, ceiling and floor space which projects from the wall of a house, usually overhanging the foundation wall. It is usually semi-octagonal in plan, of normal floor to ceiling height, with a separate roof at the outside wall.

A bow window is similar to a bay window but is semicircular in plan and not generally supported by a foundation wall but by brackets attached to the exterior wall. A bow window may be built in an exterior wall at an upper floor level as well as the first floor level.

An oriel window is often smaller than a bay or bow window, is supported by moldings or corbels on the exterior wall and may be semi-octagonal or semi-square in plan. Oriel windows are generally placed at upper floor levels.

Bay, bow and oriel windows project from the vertical exterior walls of a house, while dormer and eyebrow windows extend from the exterior slopes of roofs. There are many different shapes, styles and methods of in-

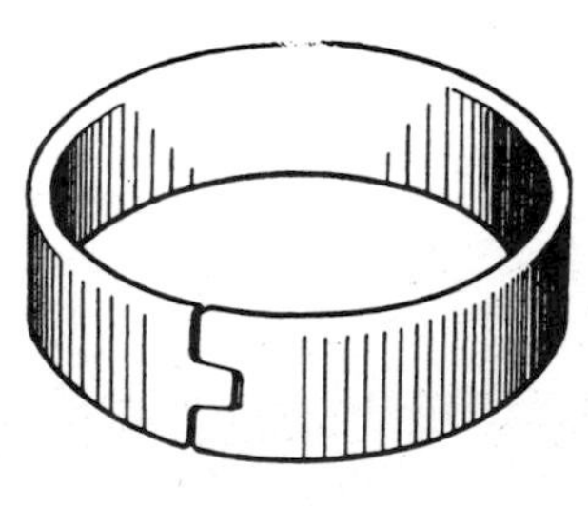

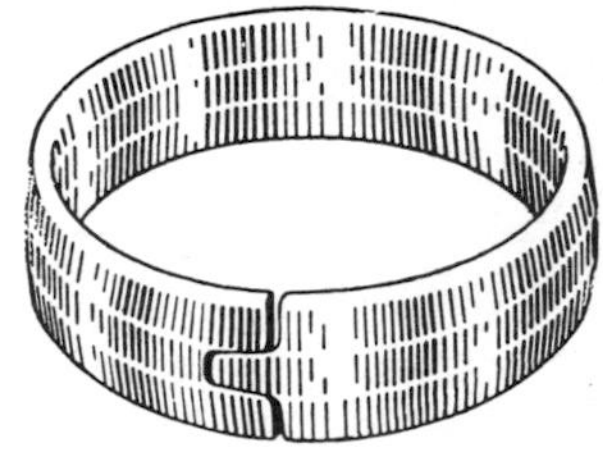

Fig. 45. Split-ring connector is set into circular grooves cut in meeting truss members, then bolted.

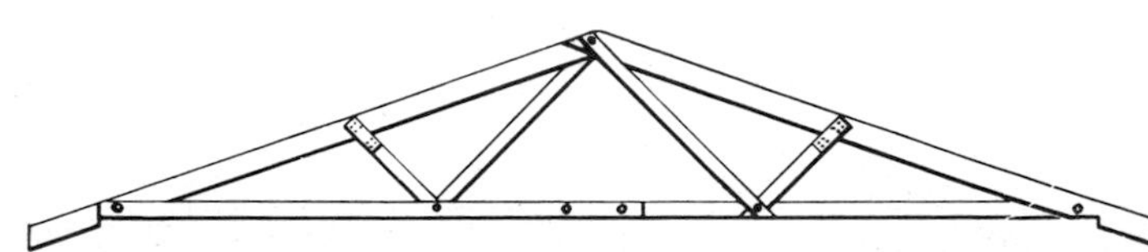

Fig. 46. "W" or "Fink" truss assembled with split-ring connectors at diagonals and lower chord.

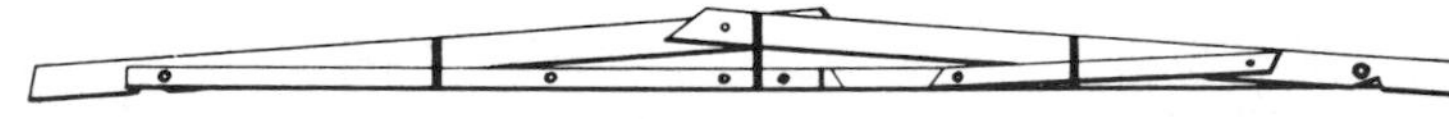

Fig. 47. Same truss with connectors in upper chord removed and members folded flat for shipment.

stalling windows in walls or windows in roofs.

A large bay window can be framed in the outside wall by extending the floor joists beyond the foundation wall as support, **Figure 40.** The roof of the bay is framed with a combination of hip and gable roof rafters. The sill header or band carries the weight of the framing to the floor joists and sill plate. Bays of this size are sometimes supported by a foundation wall of short depth, or one which extends the depth of the main foundation, forming a recess in the basement floor area.

A large bay window with movable sash at sides and a large picture window in front increases the room size and admits more light than a flush window, **Figure 41.** The window in the gable is smaller, flush with the exterior wall but of the same style. The elevation and plan shows a factory stock oriel window.

ROOF TRUSSES

When roofs are framed with trusses, the members form a rigid frame-work, the single ridge loses its structural function as the unifying member of the roof. The principal parts of a truss are the upper chord consisting of the rafters and the lower chord, corresponding to a ceiling joist and various diagonal or vertical bracing and connecting members are known as web members. **Figure 42** shows a typical, lightweight gable roof truss with gussets of wood. This truss uses but three gusset plates and is the simplest light weight truss to make on the job or in the factory. The truss in **Figure 43** has gusset plates of plywood at the seven joints, as indicated. Timber connectors, split ring and bolt, join all joints but two in the truss illustrated in **Figure 44.**

Types of Trusses. The trusses illustrated in **Figures 42, 43 and 44** as the standard "Fink" truss are often referred to as the "W" truss type. Other trusses for the gable roof are known as the "Kingpost" or "K" truss and the "Triangular Howe" or "H" type. The "Flat Howe" or "F" truss is used in flat roof framing. There is also a "Monopitch" or "M" truss for shed roof construction. The standard "Hip" or "HP" truss is used in framing a hip roof and the standard "Valley" or "V" truss for roof valleys.

With the growth of pre-cutting techniques and the purchase of prefabricated components from factories and lumber yards, ease of transportation has made two types of manufactured trusses important, that is, the folding truss and the single-plane truss.

Many builders found they could frame roofs economically by cutting conventional rafter-ridge assemblies on jigs at the building site. Many builders bought trusses from prefabricators or lumber yards (or pre-cut and pre-assembled trusses in their own work yards). The transportation of pre-assembled trusses presents a problem.

The overlapping member truss was given continued effectiveness by the use of the split-ring timber connector, **Figure 45.** A "Fink" or "W" truss, assembled with these connectors, **Figure 46,** can be folded for transporting, when the peak bolt is removed, **Figure 47.**

Fig. 48. Workmen assemble a king post type of roof truss at job site, using split-ring connectors and bolts. Truss members are precut and marked before delivery to site.

The "Fink" truss can be prepared in three ways for shipment; (a) Remove the peak bolt and fold over on itself, for example, a 24′ span truss with a 4/12 slope will then occupy 13-1/2 cu. ft. of area; (b) Folded and shipped in halves taking but 9 cu. ft. of area; and (c) Knocked down, completely removing all bolts and rings, taking but 4 cu. ft. of shipping area.

The Kingpost Truss. Figure 48 shows a photograph of workman, at right, tightening the bolt at the peak of a double kingpost truss. The other workman is nailing gusset at the diagonal brace. The notches under end bolts rest on outside wall plate.

A Special 1-1/2 Story Truss. Figure 49 shows workmen setting up a truss for a 1-1/2 story house. The pitch of the truss is 10″ in 12″. One collar tie is in place. Workmen walk on bottom chord while inserting ring and bolt connectors at right. The sectional drawing of the 1-1/2 story house, **Figure 50,** shows timber connectors at peak, collar, knee walls and plates.

The Steel Truss. An overlapping truss made of steel and field assembled to support a gable roof of low

Fig. 49. Story-and-a-half truss is erected on ceiling joists of floor below, which serve as the lower truss chords.

pitch is illustrated in **Figure 51.** Note the attic floor insulation blankets and the plywood roof decking attached to longitudinal wood purlins supported by the steel trusses.

Assembling Pre-Cut Trusses. The flat, single plane truss can be put together, for ready transportation, in several ways. The Kingpost truss, **Figure 52,** is held by gusset plates on both sides, nailed and glued at the meeting points of the butted members, lower drawing. The gable end truss, upper drawing, has 1/2″ plywood gussets glued and nailed on one side to take gable end wall covering.

Another method used in assembling the pre-cut flat truss is in the use of a connector in some form of steel plate. In one of these, the H-Brace, **Figure 53,** a thin

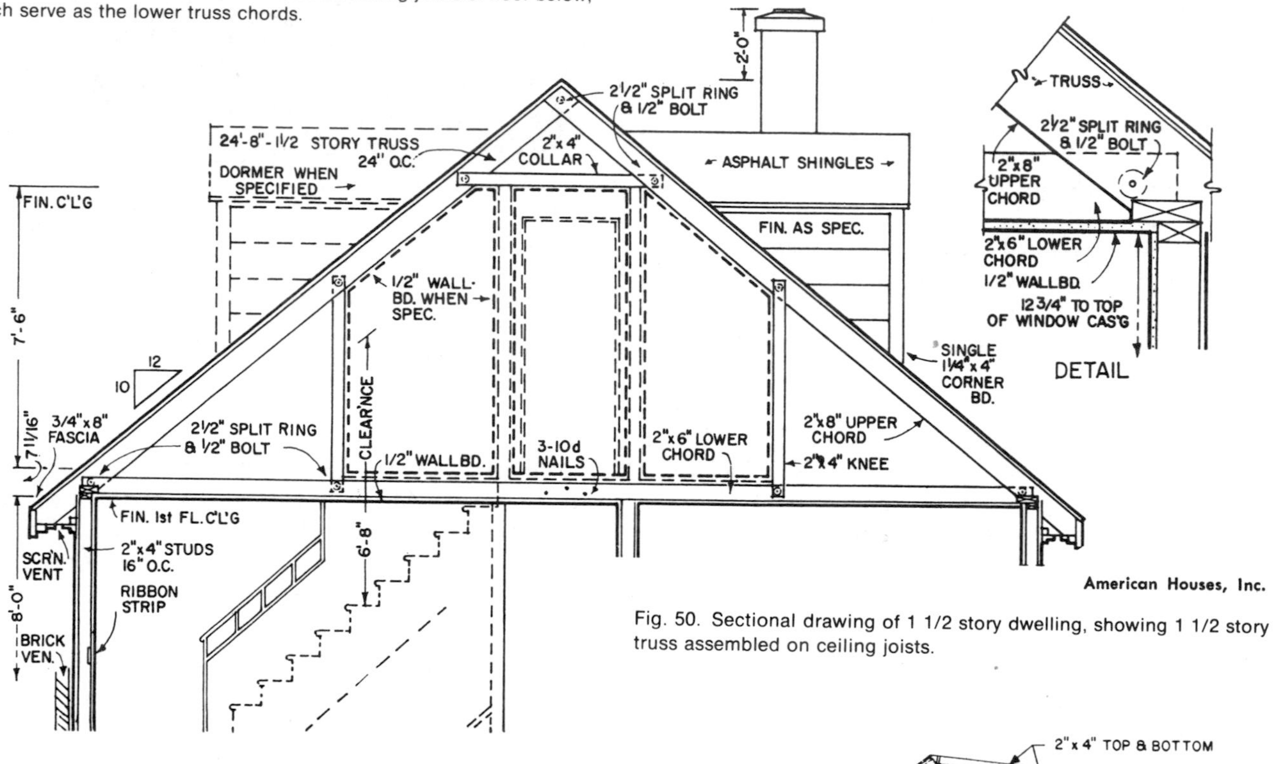

American Houses, Inc.

Fig. 50. Sectional drawing of 1 1/2 story dwelling, showing 1 1/2 story truss assembled on ceiling joists.

Fig. 51. Plywood roof decking is nailed on wood purlins fastened across low pitched steel trusses in prefab house.

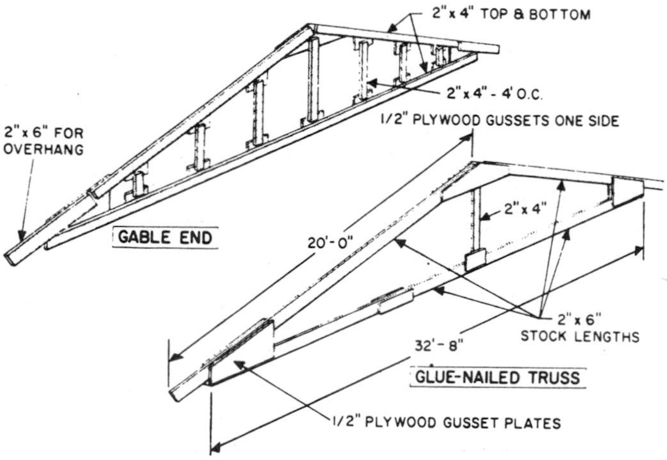

Fig. 52. Glue-nailed king post truss joined with plywood gussets, and matching gable end truss with gable studs.

piece of steel is folded and perforated in such a way that the related truss chords can be fitted and nailed to the form. Another type, **Figure 54,** is the flat steel plate punched so that triangular teeth are bent out. These steel plates, of various sizes and shapes to fit the parts of the truss to be held, are forced, usually by a power press, into each side of the wood members, **Figure 55.**

Framing Roofs With Flat-Plane Trusses. Two men can erect lightweight flat-plane trusses, **Figure 56.** The trusses are placed on the bearing walls and tilted up into position, then spaced as specified. Note the placing of gussets on both sides of the joints and the sixteen trusses stacked on the rough floor. This is a 4′ × 8′ panel, door and window system of pre-fabrication construction built on the modular system.

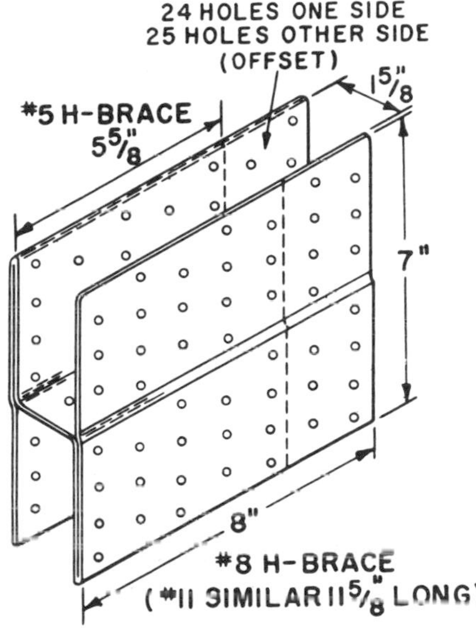

Fig. 53. Perforated metal H-brace joins truss members securely. It is fitted over meeting sections and nails driven through holes.

Fig. 54. Flat steel truss plates are punched so that teeth protrude to fasten truss members together.

Fig. 55. Portable power press forces all teeth in a truss plate into the wood with one powerful squeeze.

Fig. 56. Trusses are placed upside down on wall plates, then tilted into position, usually by only two men.

The gable and Kingpost trusses are shaped like interior trusses, but have additional vertical members for nailing on the siding, **Figure 57.** Note the plywood gussets placed on but one face of the gable end truss.

Trusses are braced with a temporary spacer strip, **Figure 58.** These trusses are cut to fit into the H-shaped sheet metal connector previously described. All framing members are on one plane.

A pre-fabricated, split-level house, having a gable roof and a hip roof is framed as shown in **Figure 59.** The

Fig. 57 Gable truss is set in position and nailed to plate on end wall. Note plywood gussets only on inside of truss.

Fig. 58. Trusses are braced with a temporary wood strip to hold them plumb and at proper spacing.

standard Kingpost flat plane trusses are set in place to form the gable roof, and workmen are setting the gable lookout rafters in place. The flat plane hip trusses of the lower structure have been set in place and fitted with hip rafters. Metal grip plate gussets are used throughout.

ROOFING

After the final structural members, the rafters or beams, are set in place the entire framed area is covered with sheathing, decking or panels and protected by the finish roofing.

Conventional Roof Sheathing. Roof sheathing depends to some degree on the kind of roofing to be applied. Wood shingles are applied to spaced sheathing boards rather than a solid deck, **Figure 60.** Wood shingles are delivered quickly to a roof by an escalator belt from a truck, **Figure 61.** Note that the entire framing is first covered with spaced sheathing. Solid wood sheathing must be of well seasoned tongue and groove or shiplap sheathing, not over 6″ in nominal width, **Figure 62.**

A good practice for T & G sheathing is to drive one 8d nail through the edge of the board and the other through the face. In butting two boards at a joint over a rafter, avoid making two such joints immediately adjacent on the same rafter. Wood shingles or shakes can be laid on either solid or spaced sheathing. In snow free areas, spaced sheathing is practical, using 1″ × 4″s or wider, spaced on centers equal to the weather exposure at which the shakes are to be laid, but not over 10″. Where wind driven snow is encountered, a roof deck of solid sheathing, normally shiplap, is recommended, unless the roof pitch is 8-in-12 or steeper. The solid sheathing should be covered completely with unsaturated building paper, such as rosin sized paper or deadening felt.

Sheathing Rafters With Panels. Single sheet panels of plywood or built up stressed skin plywood roof decking are now widely used to cover rafters, **Figure 63.** One workman moves large sheets of plywood, right, from a simple lean-to rack or ladder. At the left is a workman nailing plywood panels to an adjacent roof. Pneumatic nailers and staplers, **Figure 64,** save time, but fasteners themselves may cost more than box nails.

Plywood roof sheathing should be 1/2″ thick or over, either three or five ply, the maximum center spacing between supporting members being 24″. Many builders toenail headers or nailing strips between rafters, for nailing of each edge of butted plywood, but special H-clips are available for aligning unsupported edges of panels.

The stressed skin panel is a plywood component hav-

Fig. 59. Hip roof framed with flat plane hip trusses and joined by hip rafters at four corners.

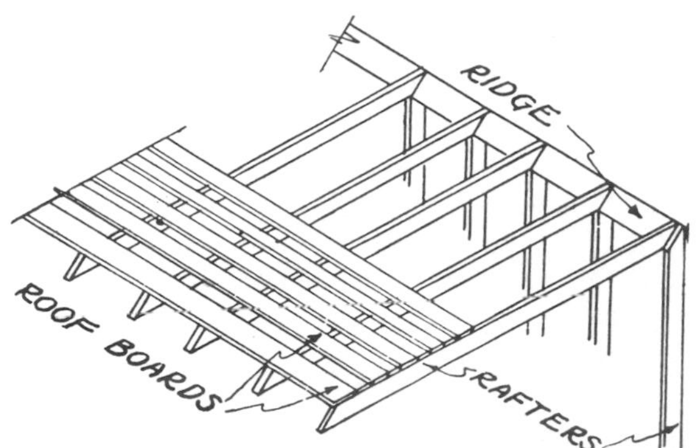

Fig. 60. Spaced sheathing boards are used in warm areas to ventilate wood shingles laid over them and keep them dry.

Fig. 61. Conveyor belt delivers bundles of shingles direct from truck to rooftop, saving much handling of material.

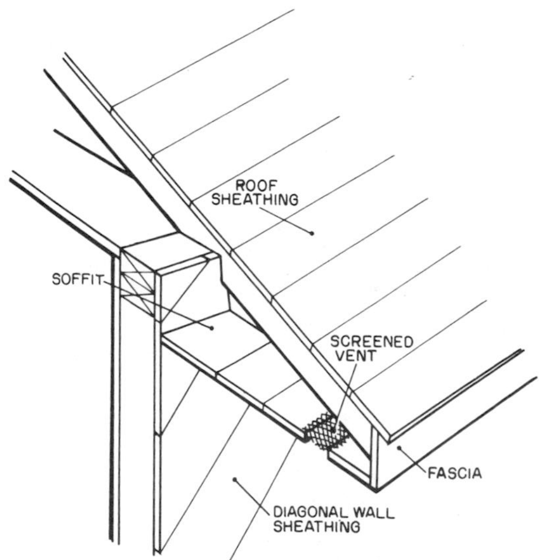

Fig. 62. If solid wood sheating is used, it should be of well seasoned T&G or shiplap boards, not more than 6" wide.

ing top and bottom skins glued to longitudinal framing members or other core materials. The glue joint between the covering skins and the framing members makes it possible for all panel elements to act as a unit in resisting loads. Insulating material can be inserted in two sided panels during manufacture or in single-skin panels on the job site. In the illustration, **Figure 31** of Chapter 3 is shown a detail of a stressed skin floor panel. Stressed skin plywood panels can be used for floor, wall or roof construction. Because of their size and weight, stressed skin panels are hoisted into position, **Figure 65.**

Combination Roof-Ceiling Decking. A heavy beam ceiling construction of a roof eliminates the need for roof sheathing and insulation. In the construction of a cathedral ceiling heavy beams are used instead of light weight rafters. Heavy beams permit the use of a combination ceiling-roof deck approximately 2′ × 8′ in size and thicknesses from 5/8″ to 1-7/8″, **Figure 66.** The finished underside of roof-ceiling decking provides ceiling and roof deck over exposed roof beams, **Figure 67.** The use of decking eliminates the need for roof sheathing and insulation, thus simplifying ceiling installation and finish and providing a perfect base for nailing shingles.

Finish Roofing—Applying Wooden Shakes. The minimum recommended pitch for a wood hand split shake is 4-in-12. Maximum weather exposure is 13″ for 32″ shakes, 10″ for 24″ shakes, and 8-1/2″ for 18″ shakes. A superior three-ply roof is achieved by reducing these exposures to 10″, 7-1/2″ and 5-1/2″ respectively.

Fig. 63. Simple lean-to rack speeds job of passing plywood roof decking to rooftop. One man slides panels off easily.

Individual shakes should be spaced apart about 1/4'' to 3/8'', for expansion. These joints should be offset in adjacent courses. Rust resistant nails, long enough for adequate penetration into sheathing boards, should be used, two for each shake at least one inch from each edge and about two inches above the butt line of the following course.

The doubled starting course projects at least an inch below the bottom roof board to make a drip edge; a smaller projection is made at the rake. Double coursing allows wider exposures. The gable edge is protected from water dripping by the insertion of beveled siding, **Figure 68.** Note the double course of shingles at the bottom. "Economy" grade shingle or single backer is often used for the under course.

Asphalt Roofing Shingles. At least 80% of American homes are covered with asphalt roofing shingles. They are used for both new construction and in remodeling where the roof deck is pitched at 4'' or more per horizontal foot. Made in a variety of styles, the most popular is the square, butt strip shingle, elongated in shape with three tabs, two tabs, or even one tab without cutouts. Much less popular today are the hex shingle, so called because it resembles a hexagon when applied and the individual shingle with interlocking or staple-down tabs.

Fig. 64. Pneumatic nailer makes fast work of fastening plywood decking, especially when compressor is set up for several roofs.

Fig. 65. Prefabricated stressed-skin panel is lowered into position on roof beams, closing in house quickly.

Fig. 66. Tongue-and-groove edges of wood fiber panels are tapped together with block and hammer for snug fit.

Fig. 67. Beams are beveled to angle of roof pitch and are left exposed beneath roof decking.

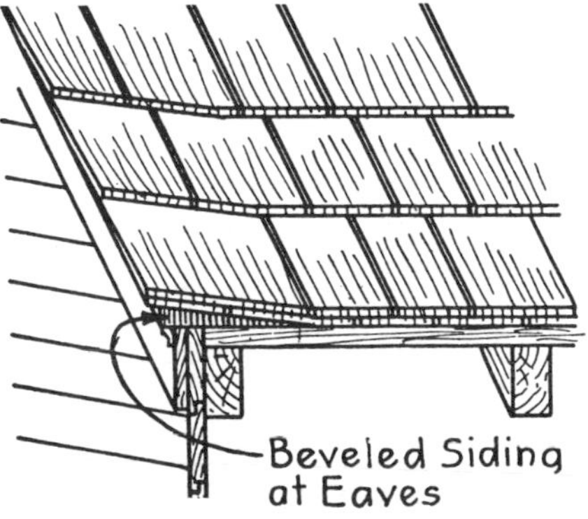

Fig. 68. Wood shakes are laid over length of bevel siding at gable edge to prevent dripping. Note double starter course.

Asphalt Roofing Mfr's. Assn.

Fig. 69. Asphalt roofing styles. Upper left: Double or triple lamination of some tabs produces rough, textured appearance. Lower left: Heavily embossed shingles that simulate wood shakes. Upper right: Many different color tones give the appearance of wood shingles. Lower right: Staggered widths and random butt lines.

Typical asphalt square butt strip shingles are made with a base mat of cellulose or glass or asbestos fibers saturated and coated with asphalt and surfaced with ceramic-coated opaque mineral granules. The asphalt provides waterproofing qualities while the mineral granules provide protection from fire and the sun's drying rays as well as giving the shingles their color.

The demand for more appealing roof materials has created a new generation of premium quality asphalt shingles in new designs, bolder textures and richer colors. The new shingles, some of which are shown in **Figure 69** may be double or triple laminated (upper left), heavily embossed to resemble wood (lower left), available in many tones resembling stained wood (upper right), and staggered butt lines with tabs or irregular thickness (lower right).

Many of the new shingles are designed to provide up to 25 years of service with only minimal maintenance. They can be used on any of the roof shapes previously described and shown in **Figure 70**.

A drip edge of corrosion-resistant sheet metal is laid along the eaves and along the rake of roof slope, **Figure 71**. An eaves flashing strip is added where there is a

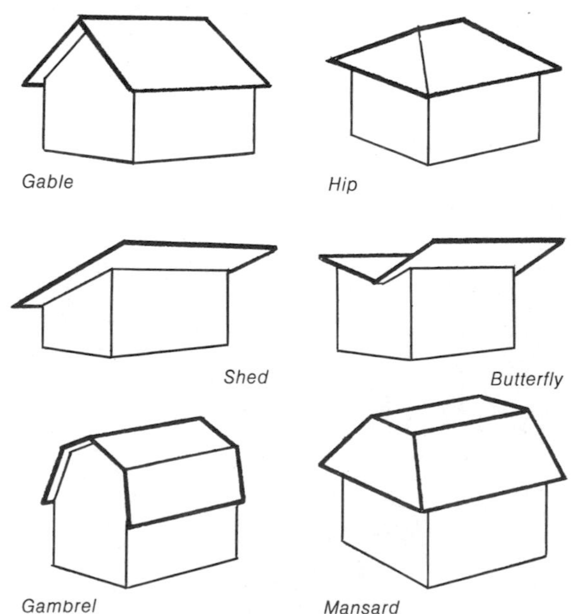

Fig. 70. Six different roof styles to which asphalt roofing shingles can be applied.

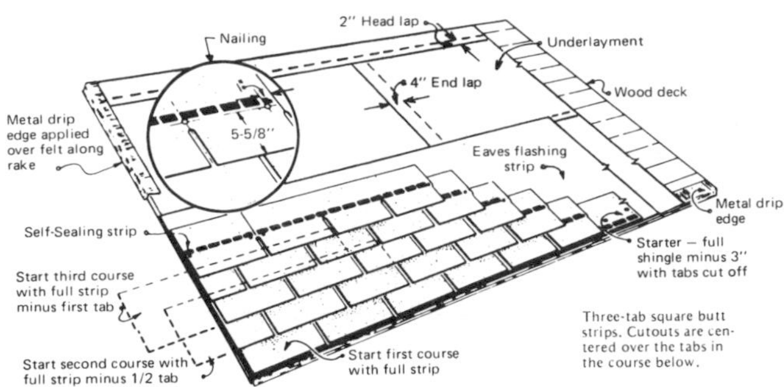

Asphalt Roofing Mfrs. Assn.

Fig. 71. Approved method of laying asphalt shingles with metal drip edge, eaves flashing strip, and full starter strip less 3" with tabs cut off.

possibility of ice forming and causing a back up of water.

A starter course of a 9′′ strip of mineral surfaced roll roofing or a row of inverted shingles is applied even with the lower edge of the eaves flashing strip. The first course is started with a full length strip; the second course is started with a cut strip, to break the joints on thirds, halves or at random, **Figure 72.** Each full strip is nailed with four nails, one nail back 1′′ from each end, one nail 5/8′′ above each cutout centerline, **Figure 73.** In windy locations a spot of quick drying asphalt cement is applied under the center of the exposed portion of each tab. However, this may be necessary only under extreme conditions since practically all asphalt shingles now have a line of adhesive dots or dashes from one side to the other above the tabs. The sun's heat makes them adhere.

Special treatments of asphalt roofing permit its use in low roof pitches of 2-in-12 or 3-in-12. One technique is a dab of cement under each tab. Another is the use of self sealing strips, in which an adhesive or mastic is applied to the shingle strip at the factory. For low slope applications the Asphalt Roofing Industry Bureau recommends two complete layers of 15 pound asphalt saturated felt; and a continuous layer of plastic asphalt cement between each layer underlayment in the eaves flashing area, **Figure 74.**

Re-Roofing With Asphalt Shingles. Old wooden shingles or shakes or old asphalt shingles can be covered with new asphalt shingles without removing the old shingles, **Figure 75.** Loose wood shingles are tightened and feathering strips are placed along each butt line to provide a smooth surface for new asphalt shingles, **Figure 76.** Often plywood sheathing is laid directly over the old roofing, after loose shingles have been removed.

Roll Roofing. Roll roofing is used on flat or low-sloped roofs. One form is the simple black-asphalt-impregnated felt; another is mineral coated, and another has a pattern edge to simulate shingle design.

An attractive treatment of low-pitched roofs is to cover with several layers of saturated felt, then with a poured layer of asphalt or pitch, then with a layer of marble chips or broken rock or gravel, usually light

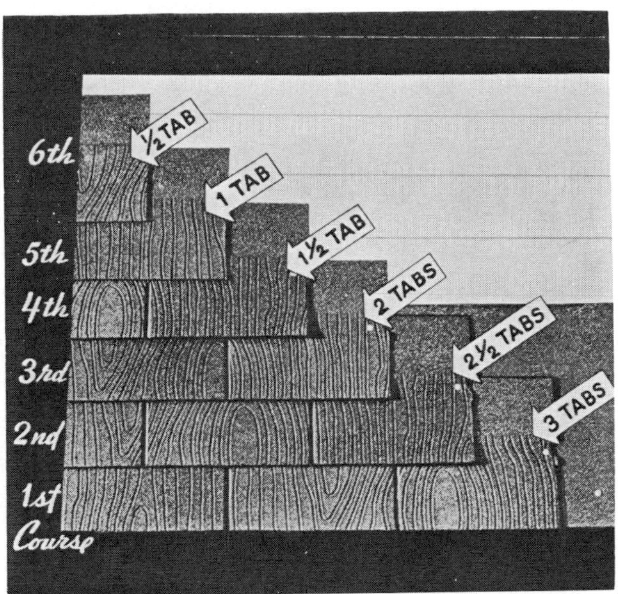

Fig. 72. Method of cutting successive courses of shingles along gable edge to stagger the joints.

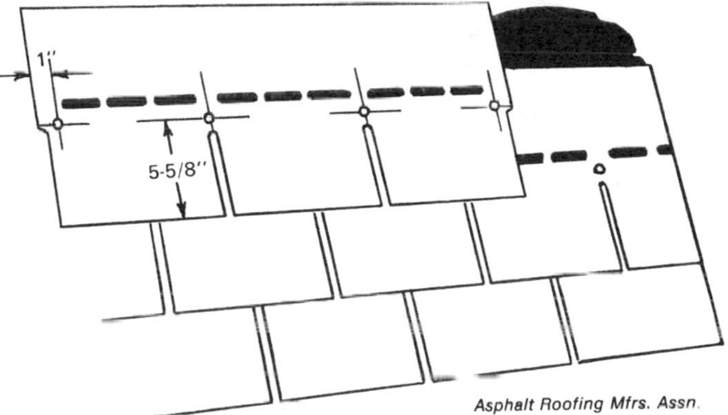

Asphalt Roofing Mfrs. Assn.

Fig. 73. Three-tab shingle is nailed with 4 nails. One nail 5/8" above each cutout and one nail back 1" from each side.

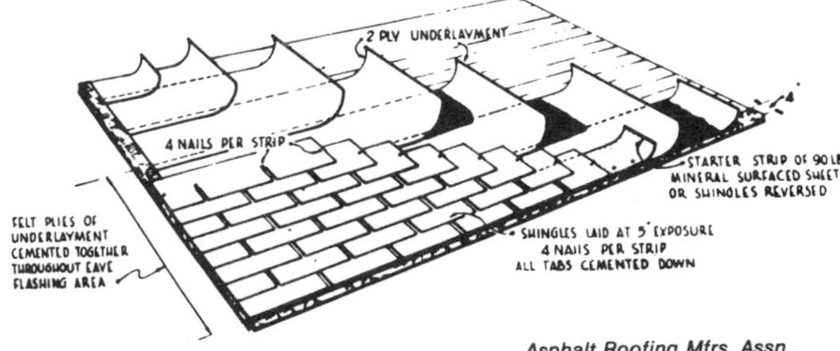

Asphalt Roofing Mfrs. Assn.

Fig. 74. Shingles are not recommended on low-slope roofs, but if used, must be laid over two layers of 15 pound asphalt felt.

colored to reflect radiation, **Figure 77.**

Corrugated aluminum strips are widely used for roofing of farm and utilitarian buildings. Aluminum roll

Fig. 75. Old shingles need not be removed when reroofing a building. New shingles can be laid over the old roof.

Fig. 76. Thick wood shingles should be tightened and feathering strips laid between them to make a smooth surface for new shingles.

Fig. 77. Built-up, low-slope roof has several layers of mopped-on felt topped with marble chips or crushed rock.

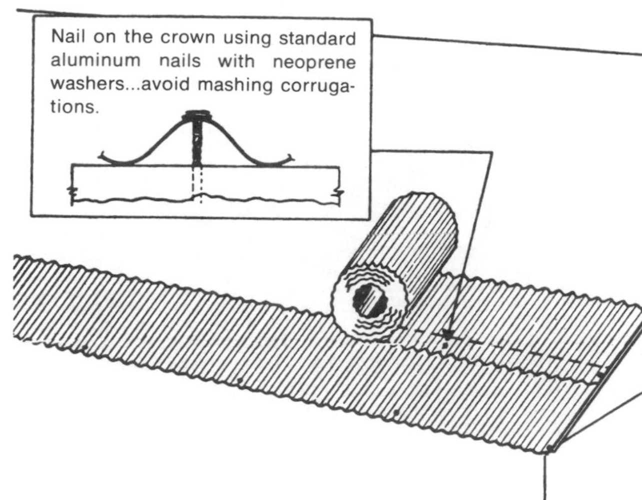

Fig. 78. Corrugated aluminum roll roofing is nailed with aluminum nails sealed with neoprene washers (insert).

Fig. 79. Terne roof has vertical interlocking seams.

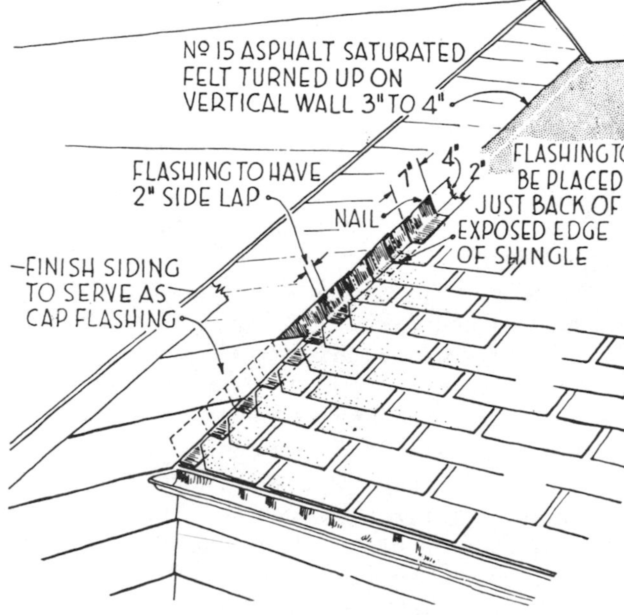

Fig. 80. Flashing makes joints watertight at intersecting surfaces.

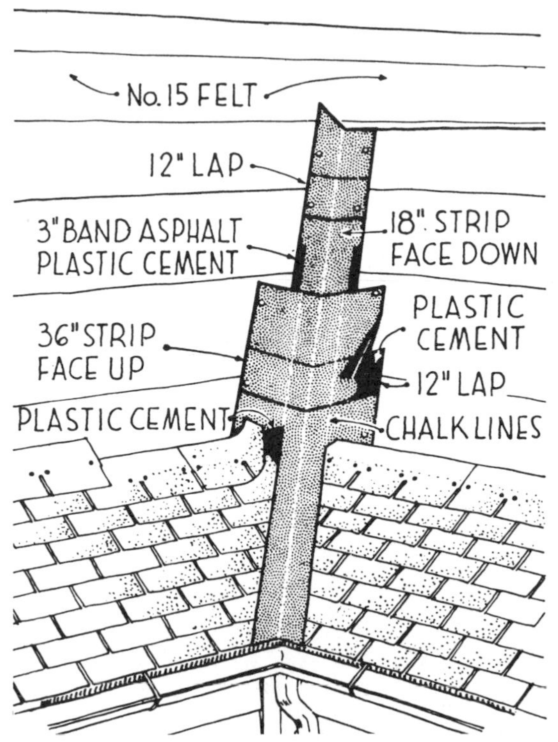

Fig. 81. Roof valley can be flashed with roll roofing.

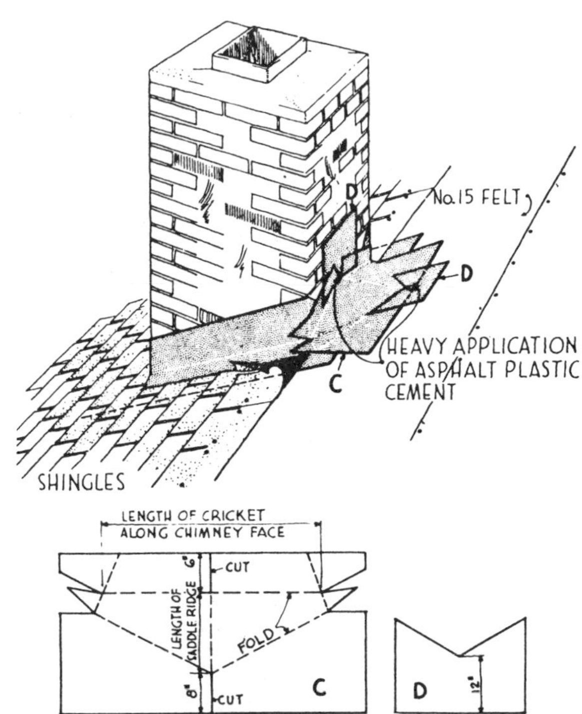

Fig. 83. Flashing over cricket built up at rear of chimney.

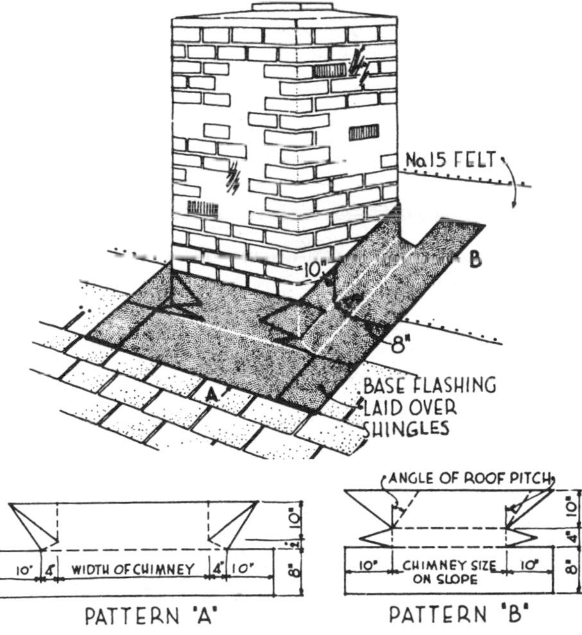

Fig. 82. Chimney base flashing patterns cut and applied.

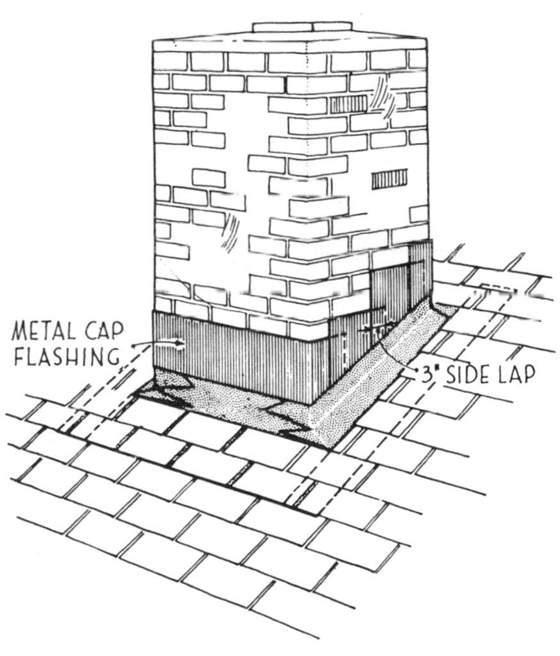

Fig. 84. Metal cap flashing applied to overlap base flashing.

roofing is nailed with aluminum nails with neoprene washers, **Figure 78.**

Tile, Slate and Terne Roofing. A tile roof is more expensive than the more common roofing materials, but is durable and desirable for interesting effects.

Slate roofs are aristocratic and durable, but are comparatively rare in new residential construction since the advent of newer, cheaper materials.

Terne roofs are used sometimes in luxury houses. Terne is a durable alloy of lead and tin over steel or iron. It is applied with interlocking seams to seal out the weather, **Figure 79.**

Plastic Roofing. Plastic and various synthetic materials are coming into use for roofing, as the base for light-reflecting chip roof, and as a coating for plywood panels, such as neoprene elastomer sprayed on 1/2″ thick plywood decking in a NAHB Research House. Caulking grade of neoprene was used in beveled edges of the panels, and a layer of hypalon plastic was rolled on after the panels were in place.

Flashing. Joints are made watertight with flashing when a roof surface intersects another roof surface, an adjoining wall, **Figure 80,** or projections through the roof, such as chimneys, ventilators or soil stacks.

Where sloping roof surfaces join it is necessary to install metal or roll roofing flashing in the valley, then continue with the roofing, **Figure 81.** Note the use of 36″ wide roll roofing at the base of the valley. Although applying roll roofing the roof valleys as in **Figure 81,** many roofers prefer to weave asphalt roofing shingles across the valley as in **Figure 87.**

Installation of flashing around a chimney is very important. Flashing with roll roofing and sheet metal to properly waterproof the intersections of roofing and masonry should be cut to fit and applied as illustrated in **Figures 82, 83, 84, 85, and 86.**

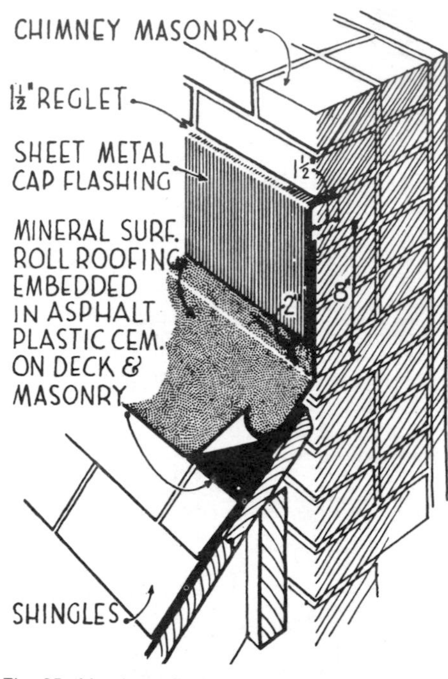

Fig. 85. Metal cap flashing is set in mortar joint.

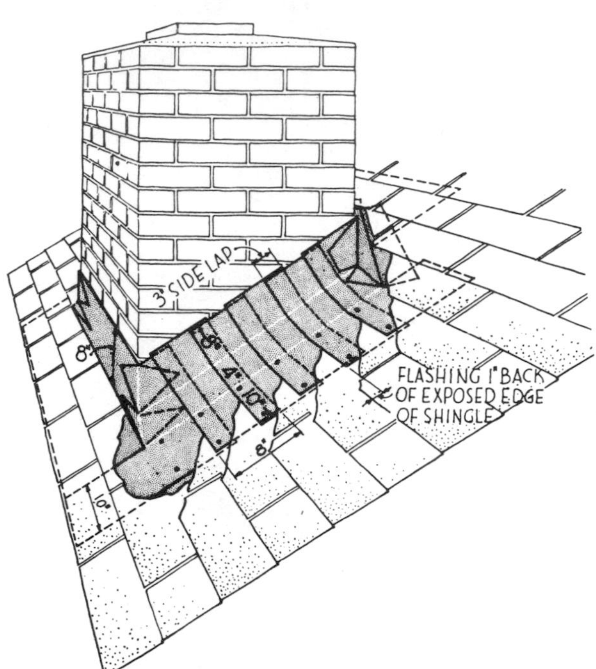

Fig. 86. Alternate base flashing known as step flashing.

Asphalt Roofing Mfrs. Assn.

Fig. 87. Textured style asphalt roofing shingles are woven across the valley already covered with roll roofing.

Chapter 7

CORNICES/PORCHES/PATIOS/ATRIUMS

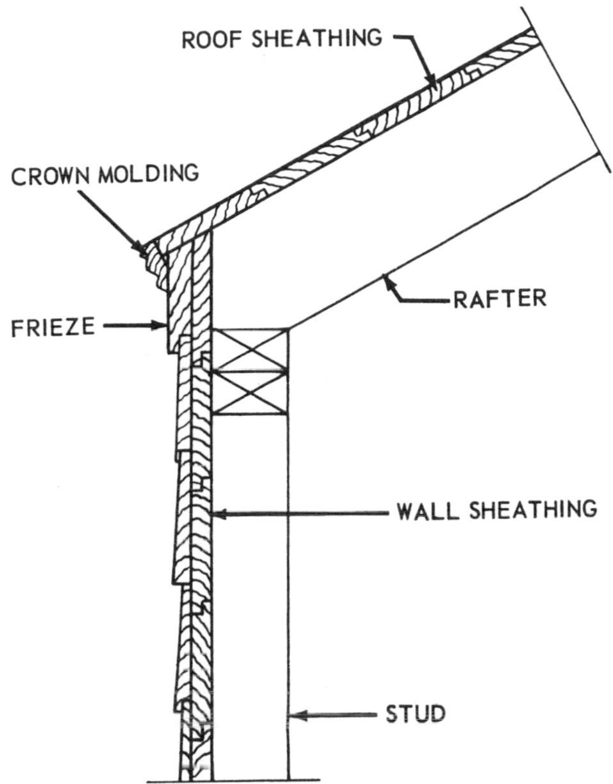

Fig. 1. The most simple type of cornice construction.

The rafter end edges of a roof are known as eaves. A hip roof has rafter end edges all the way around, and all four edges of a hip roof are therefore eaves. The rafter end or sidewall edges of a gable roof are eaves; the gable end or end wall edges are called rakes. The exterior finish at the eaves is called the cornice. In modern styles the cornice has lost most of the decorative molding and design it formerly displayed.

A Simple Cornice. The most simple type of eave and cornice trim in conventional construction consists of a frieze board attached to the exterior wall sheathing, and crown molding attached to the frieze and roof decking, **Figure 1.** The finish roofing material extends out over the crown molding to form a drip course for water run off either to the ground or to an attached gutter.

The Open Cornice. In open cornices, the rafters extending beyond the wall trim or exterior wall covering are exposed along the underside of the eave, **Figure 2.** False rafters are sometimes used in open cornices to permit the use of better material in the exposed section. These false rafters are often carved and shaped before being attached to the roof rafters resting on the plate. Tongue and grooved "ceiling boards" are also often used for the visible area of the overhang in an open cornice.

Rake or Gable Cornice. The plain or simple cornice

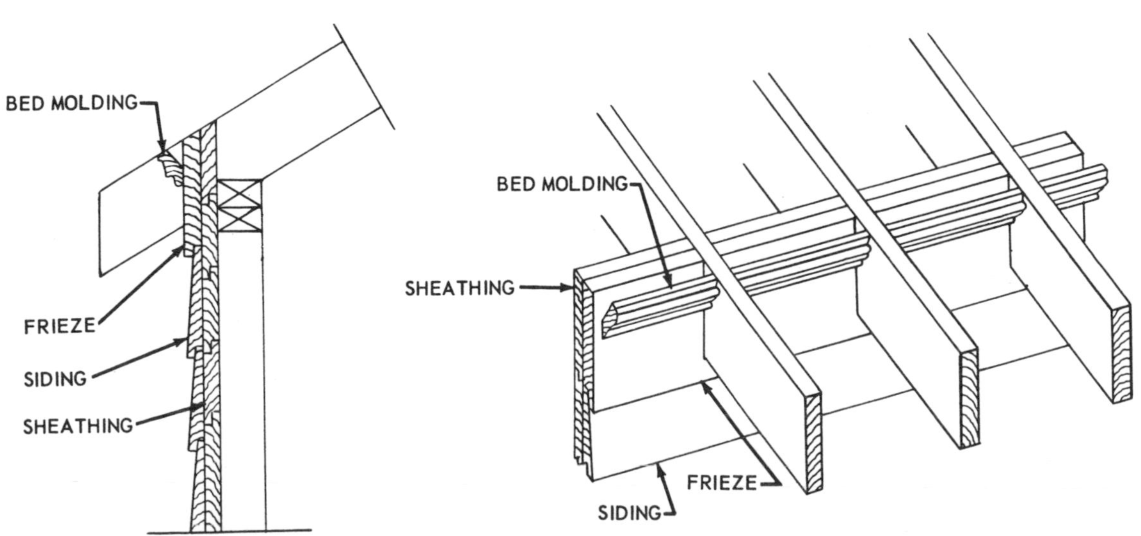

Fig. 2. Sectional and perspective drawings of an open cornice with bed molding nailed between rafters.

construction at the eaves of a house, **Figure 1,** is carried to the rake incline or gable ends of a house as in **Figure 3.** The roof decking or sheathing is carried beyond the edge of the rake molding.

The Closed Cornice. In a simple closed or box cornice, **Figure 4,** the rafters extend 18″ to 24″ beyond the sheathing and are built up by a ledger strip and lookout. The finish applied to the lookout is known as the plancier, often spelled plancher, plancia or planch and referred to as the soffit. The plancier or soffit may be built of plywood or of tongue and grooved cedar stock, **Figure 5.**

The Cornice Return. If a closed or box cornice is carried around the eaves to the rake cornice of a gable roof some of the trim is often continued horizontally a short distance, **Figure 6,** and the cornice return and rake meet in a traditionally corner blending of trim.

Note the wide overhang of the cornices along the lengths of the roof of the house pictured in **Figure 7.** The gable end treatment of the cornice is a one piece frieze along the rake. The horizontal return is a one piece strip butting at the gutter. Note the continuous louvered vent in the soffit, the hung gutter and the downspout. The decorative cupola provides ventilation for the attached garage and the main roof. Aluminum beveled siding is applied to the exterior walls and vertical aluminum siding is used in the overhanging gable ends.

A Closed Cornice on a Brick Veneered House. An exterior wall covering of brick is often placed on a house framed of wood, **Figure 8.** Note how the brick veneer is worked up to the blocking in the cornice. Gypsum board is used on the interior and exterior of the framed wall. The roof is framed with light weight trusses, which bear on the plate.

Figure 9 shows a photograph of this house during construction. Note the wide cornice and porch roof construction. The open end of the cornice shows the plywood roof decking and the soffit. Note the groove in the

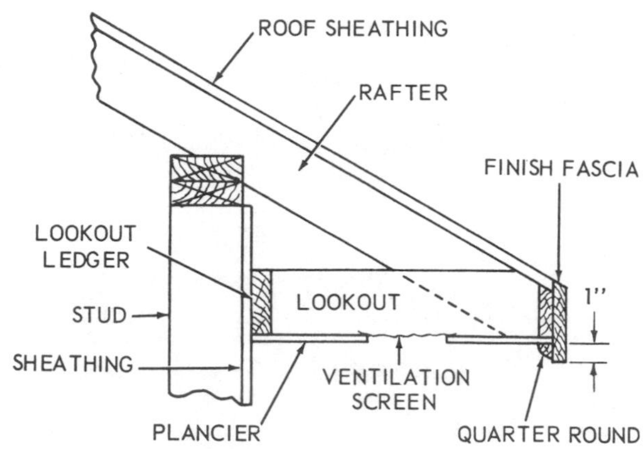

Fig. 4. Simple box cornice with lookout and plancier or soffit.

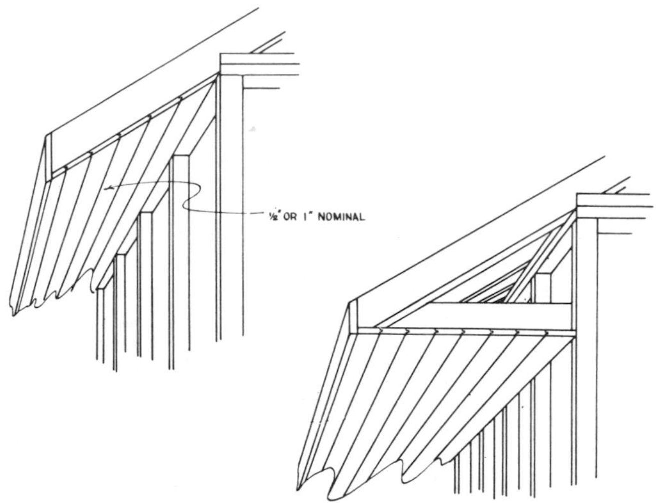

Fig. 5. Closed and box cornice covered with T&G boards.

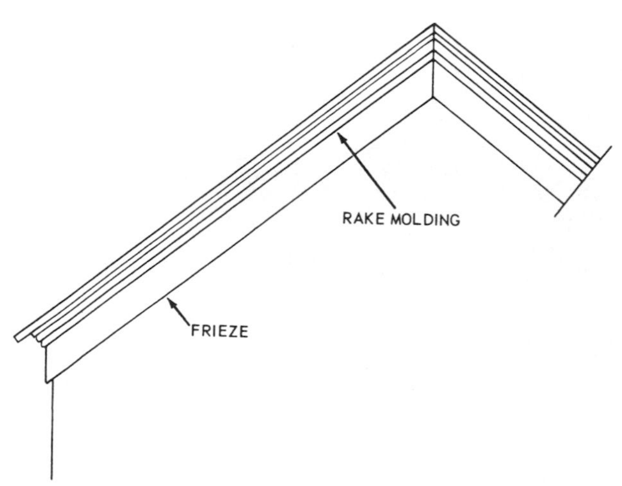

Fig. 3. Cornice trim carried up rake of gable roof.

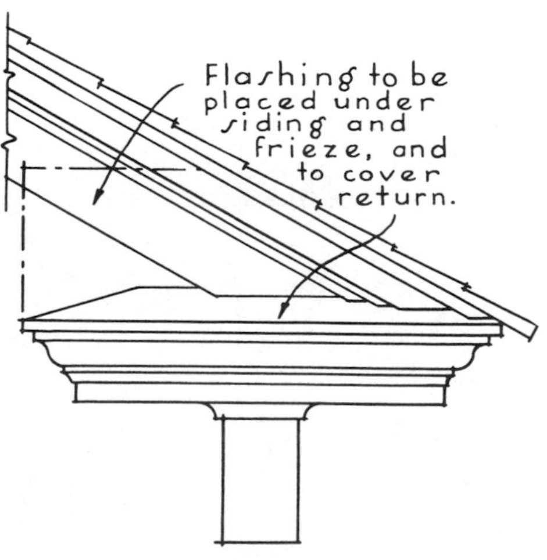

Fig. 6. Cornice return and rake meet at traditional corner.

Fig. 7. Wide cornice is accented by rake trim and return. Soffit is perforated for ventilation of attic space.

fascia trim to receive the plywood soffit and the frieze attached to the side wall to support it. Metal straps nailed to the gypsum sheathing at the studding serve as ties for the brick veneer exterior wall covering.

Construction Details of Several Types of Cornices. Figures 10, 11, 12 and 13 show construction details of various cornices. A small molding strip serves as a frieze at the top of a beveled siding wall, **Figure 10.** A plaster finish is applied to the gypsum soffit, which is sheltered from weather.

Outriggers or lookouts nailed to a 2″ × 2″ ledger strip, nailed to the studs, then nailed to the rafter end, form a base for the plywood soffit, **Figure 11.**

Irregularities or waviness can be eliminated from the fascia board by undercutting the rafter ends, **Figure 12.** The fascia is butted at the bottom against the edge of the soffit panels, which can be cut straighter than a series of rafter ends.

To improve the appearance of an open cornice, tongue-and-groove clear ceiling boards are used instead of butted sheathing boards on the overhanging rafter ends, **Figure 13.**

A Parapet for a Flat Roof. Flat roofed buildings often have a parapet or a vertical extension of the outside walls of the structure, **Figure 14.** The water collected on this type of roof is usually run to a leader or downspout through the parapet, usually in the rear of the

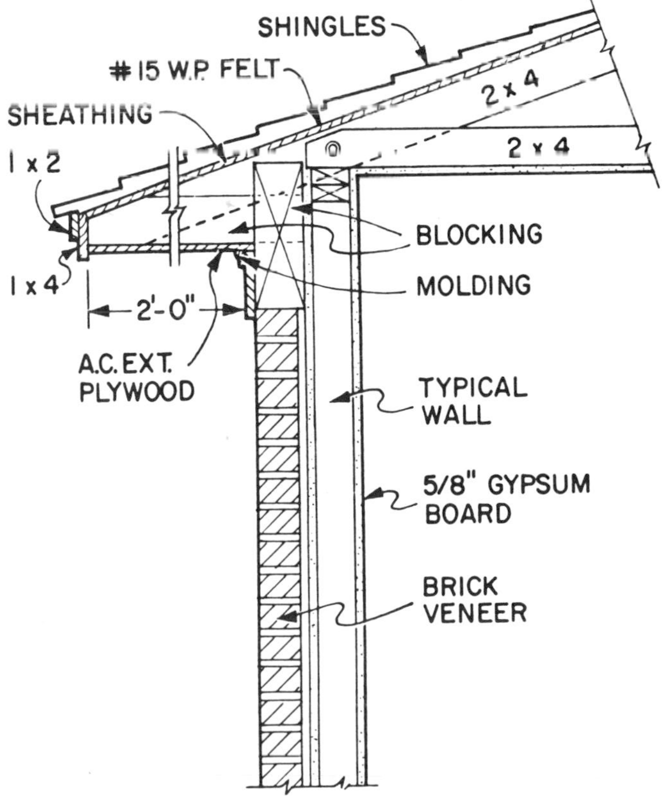

Fig. 8. Blocking frames cornice to brick veneer exterior wall. Frieze and molding cover gaps between vertical blocking.

Fig. 9. End view of house under construction shows soffit-supported by groove cut in fascia and by top edge of frieze.

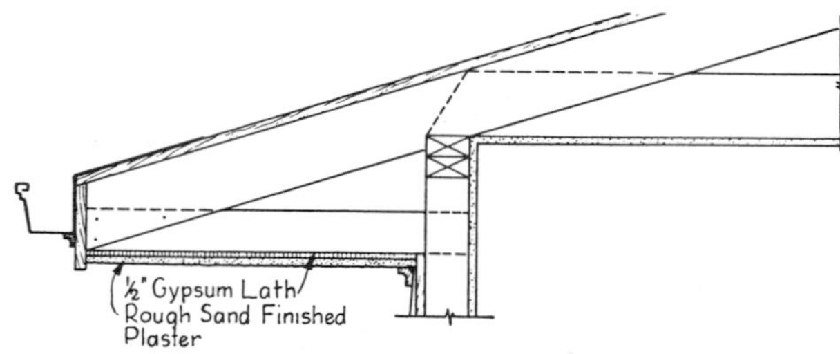

Fig. 10. Molding strip serves as frieze above lap siding.

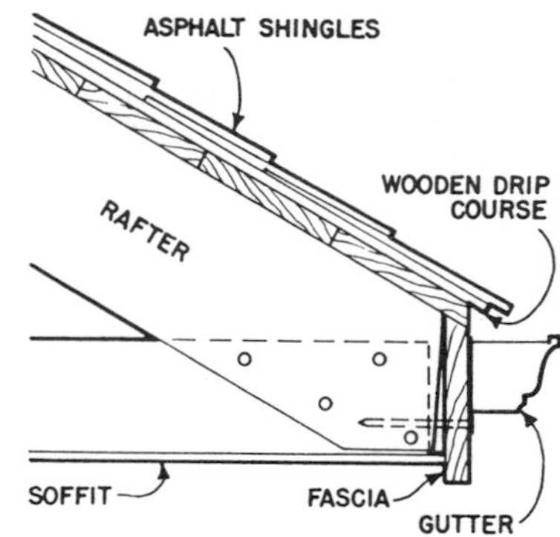

Fig. 12. Soffit extends beyond rafter to straighten fascia.

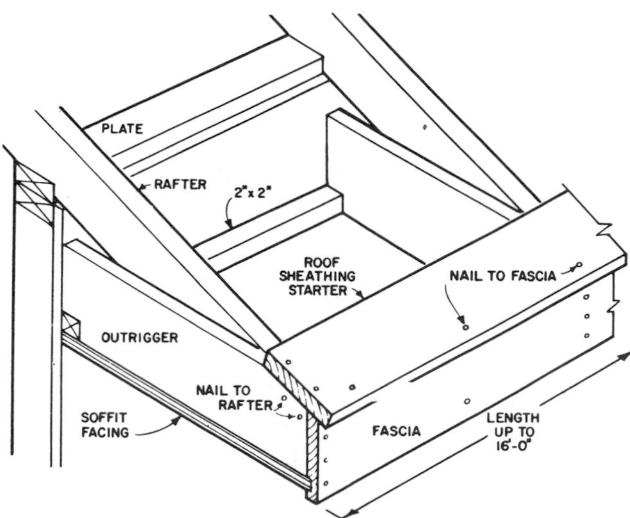

Fig. 11. Outrigger is nailed to rafter, bears on 2" x 2" strip.

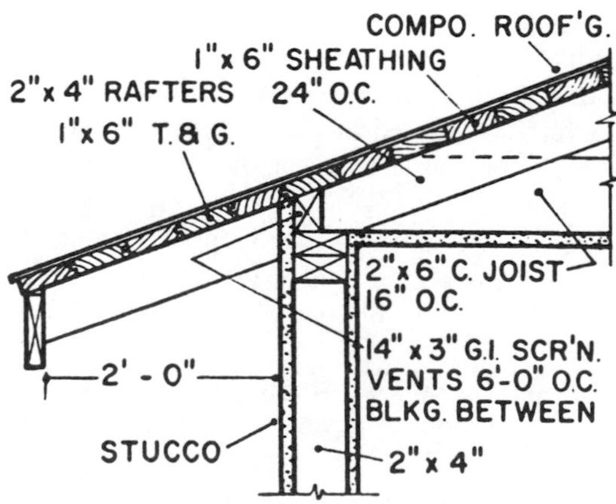

Fig. 13. T&G ceiling boards deck area above exposed rafters.

building. Note the placing of built up roofing on the roof decking and the mineral wool insulation laid between the rafters.

Ventilating the Attic. Since the importance of ventilating the attic for climate and moisture control has become widely understood in recent years, soffits have been vented and protected by screening. Headers between rafters at the plate line are omitted, to permit the circulation of air from the soffit up to the louvers in gable peaks. Often an attic fan helps to air-wash the attic. A highly efficient way of venting the attic is to install a powered roof fan that draws air through vents in the soffits on both sides of the attic. The vents may be 12″ to 14″ long and 6″ to 8″ wide and are covered by flat aluminum louvers. Where it is not possible to make such large vents, triangular or rectangular aluminum louvers can be installed in the gable

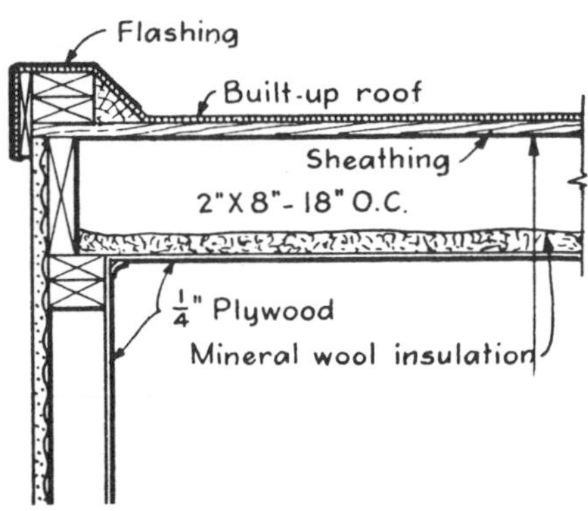

Fig. 14. Shallow parapet for a flat roof is built up of a triangular cant strip and doubled 2″ x 4″s.

Fig. 15. Continuous ventilating strip in a wide, boxed cornice will be covered with screen to keep birds and insects out of attic space. Note thick insulation between ceiling joists.

Fig. 16. Holes bored in soffit panel pictured above permit air to circulate freely through attic space.

Fig. 17. Holes in soffit on both sides of ridge beam vent heat in attic space as cool air flows in at eaves.

Fig. 18. Miniature circular aluminum vents are pressed into pre-drilled holes in soffit. When combined with roof, ridge, or gable vents they provide the cross ventilation needed to keep attics cool and minimize condensation. Diameters (L to R) include 1½″, 2″, 3″, and 4″. The 4″ size is best for attic ventilation. Screening on the inside keeps out insect pests.

Fig. 19. A wide cornice overhang protects the front entrance door and garage. Note ventilating cupola on garage roof.

ends. Another method is to insert circular screened aluminum louvers which are 2″ to 4″ in diameter into holes bored in the soffit, **Figure 18.** The attic floor itself is usually insulated to prevent loss of heat in winter and to prevent summer heat from penetrating the house. A simple method of venting an attic is to cut a continuous opening in the soffit, **Figure 15.** The opening strip will be screened to keep out birds and insects. The unfinished section at the left shows how space above the thick attic insulation blanket will permit free movement of air.

Decorative perforations in the soffits also provide a means of ventilating the attic, **Figure 16.** The vertical louvers at the left of the wide cornice ventilate the garage, which is at a lower level. A prefabricated house of post and beam framing should have vents in the overhanging panels used to close in the roof, **Figure 17.**

Porches and Patios

Porches are covered entrances to a house, usually having a separate roof. The styles and construction of porches vary a great deal, from the simple, front or rear door entrances with a simple stoop to the elaborate porch shelter extending across the entire house.

The Colonial house may have several columns supporting a two story high roof running across the front of the dwelling.

The Early American house may have a porch entrance protected by a roof blending with a roof at the same level over another part of the house.

The Contemporary house may have a terraced walk flanked by planters leading to the porch stoop, front door and entry hall.

A wide cornice overhang protects the front entrance to the foyer as well as the garage entrance, **Figure 19.** Note the double doors and side lights at the entrance and the pleasing combination of brick veneer and vertical siding on the exterior walls.

The traditional porch is often featured by an elaborate balcony and overhang requiring special cornice treatment. The front stoop and porch in **Figure 20** has a protecting canopy over the front door. The large side porch is finished with full screening, shed roof, gutter and leader. A balcony, **Figure 21** is added to a prefabricated house by attaching lookouts to every other ceiling joist. Note the protective metal railing around the balcony. A wide roof overhang may be used to shelter a handsome balcony outside a window wall, **Figure 22.** Details of such a balcony attached to the glass win-

Fig. 20. Protecting canopy over front stoop shelters the home owner while he unlocks entry door.

Fig. 21. Balcony can be built onto house by extending second floor joists or adding lookouts to them.

Fig. 22. A wide gable overhang supported by cantilevered beams shelters this handsome balcony outside a window wall.

dow wall of a post and beam framed dwelling are shown in **Figure 23.**

Patios are a modern development of the trend to open-air living first expressed in the porch. Sheltering patios from the sun has given increased functions to overhangs, cornices and roof extensions.

Patio planning takes careful thought. One must consider the view, the sun, the access to the house and the exposure to other buildings. Drainage and the slope of the land are also important. If certain areas are to be screened out, this can be done by using a block screen wall.

Patios are made of cast-in-place concrete or of precast concrete products such as patio stones. Patios are often made square or rectangular, but any shape can be built. A curved free-form patio is sometimes more attractive, particularly when it complements the contour of the lawn and is accented with proper plantings of flowers and shrubbery.

The construction of a patio is similar to that of any flat slab. The area must be cleared of all sod and topsoil. If fill is needed, it should be placed in 6″ layers and compacted thoroughly. The patio slab should be 4″ thick. Isolation joints should be placed to separate the slab from the foundation wall, walk, driveways or any other rigid section. Control joints are cut so that the patio slab is subdivided into panels no larger than 15′ square. The control joints should be cut one-fifth the thickness of the slab.

A patio may be paved with individual concrete slabs, **Figure 24.** A 6″ wide strip separates the slabs, which are 6′ × 6′, 3′ × 6′ and 3′ × 3′.

The concrete can be given any finish from rough to smooth. The wood or light metal float gives the surface uniform gritty finish. If a smooth finish is preferred the slab is finished with a steel trowel. Do not over-trowel, however. Too much troweling will produce a surface that is too smooth and one that is likely to be slippery when wet.

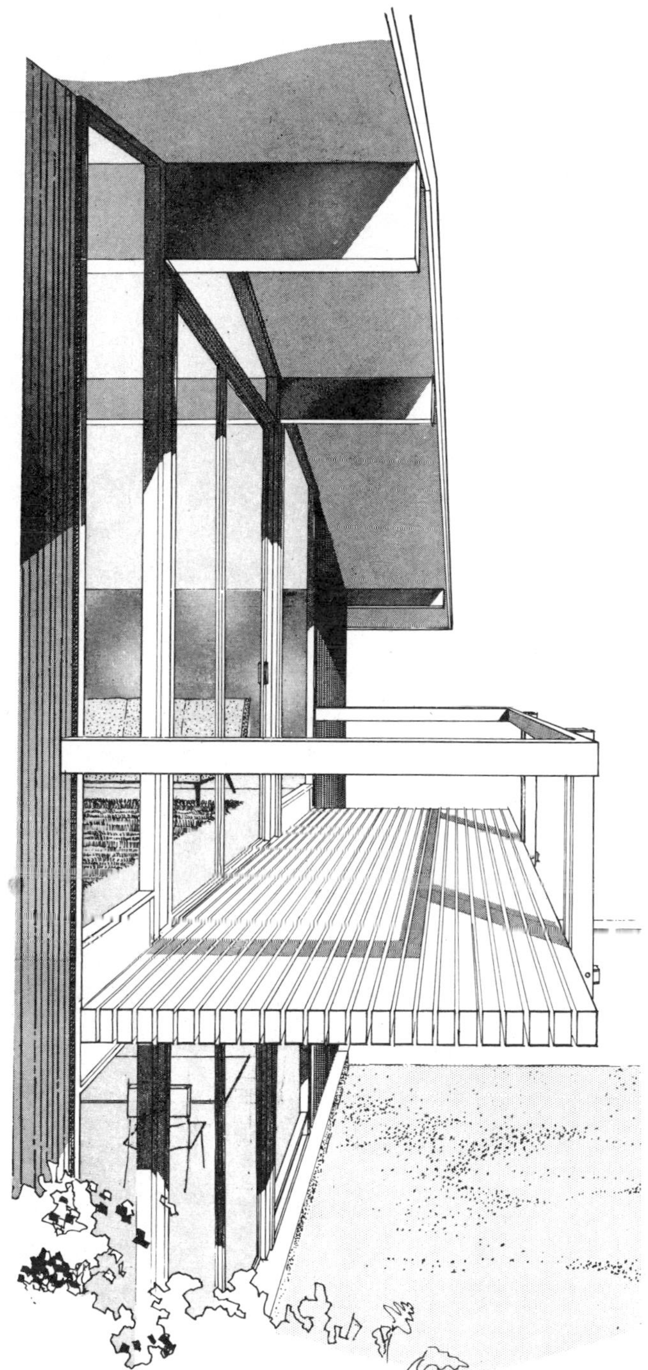

Fig. 23. Contemporary balcony in post-and-beam house has spaced wood decking laid across cantilevered floor beams.

Fig. 24. Interesting patio is made of individual concrete slabs of varying sizes with grass growing between them.

Fig. 25. Screen wall of concrete block adds privacy to patio while permitting cool breezes to pass through sculptured faces.

Fig. 26. Screen wall of wood encloses end of patio covered with slatted sun break nailed across open rafters.

Fig. 27. Outdoor barbeque shares chimney with fireplace on opposite side of living room or family room wall.

Many firms make colored concrete slabs especially for use on patios. Other firms sell units made on a regular block machine. The dimensions are usually 2″ × 8″ × 16″. These precast units are laid over a bed of sand or similar granular material. The units can be arranged to form a number of attractive patterns.

Screen Walls. Screen wall units are becoming widely available and their use in house and patio construction is becoming extensive. Screen walls offer many opportunities to beautify homes and gardens.

A screen block wall must be laid on an adequate foundation that will resist frost action, **Figure 25.** In new construction a continuous foundation will probably be the most economical. When an existing lawn is established, a grade beam foundation might be well worth considering because of the limited excavation.

A screen wall of wood provides privacy for a patio having a slatted sun break with open rafters supported by posts in planter and bench, **Fig. 26.**

Fireplaces are sometimes built in a three-way chimney stack with fireplaces in living room and family room, and an outdoor barbecue, which is a feature of the patio, **Figure 27.**

Sometimes a patio is incorporated within the outlines of the house, and is called a lanai, in Hawaiian style, **Figures 28 and 29.** Double louver doors open on to the lanai vestibule. A barbecue is built into the brick wall at the chimney stack.

Fig. 28. Full view of lanai emphasizing privacy resulting from walls on three sides of outdoor area.

Fig. 29. Looking out through double louvered doors to Hawaiian style lanai incorporated in outline of house.

Fig. 30. Glass-walled rooms on four sides look out onto open atrium located in or near center of floor plan.

Fig. 31. Addition of family room wing creates an L-shaped plan which screens the patio on two sides.

The Atrium

The trend toward outdoor living along with greater demand for privacy led to the revival of the Roman court, or atrium.

The atrium in present day construction is essentially an interior court, open to the sky, and built as an integral part of the floor plan or room arrangements of a dwelling, **Figure 30.** It is often surrounded by a study, dining room, breakfast room and living room. Sliding doors of glass provide an entrance into the inner court. The atrium is completely enclosed by window walls. Planters are placed in the atrium.

Remodeling for Outdoor Living. A new family room added to a house often creates an L-shaped or U-shaped plan, which provides screening on two or three sides for a patio. The new room features a fireplace and sliding glass doors for convenient access to the patio, **Figure 31.** The patio adds a great deal of charm and beauty to this remodeled structure.

Chapter 8

WINDOWS

A window in the wall of a house may have any or all of three functions: to transmit light; to permit air to enter or leave the house for ventilation; and to allow outward vision. Of the three, outward vision has become more important as designers of living space stress indoor-outdoor relations to make the house seem larger by seemingly taking space from the outside.

Components of a Window. The two main parts of a window are the sash, that is the glass holding element; and the frame, in which the sash is held in the structure. The sash or the frame, or both, can be made of either wood or metal. Complete window units, comprising fitted sash and frame, with hardware installed at the factory are available in both wood and metal, **Figure 1.**

Frame and sash elements are identified by the same terms whether of wood or metal. The top of the frame (usually drawn in section as seen from the side) is called the head. The side of the frame is called the jamb. The lower part is the sill.

The top of the frame, or head, consists of a horizontal member, called the yoke or head jamb, an outside head casing (moulding), and an interior head casing.

In metal frames the head and casings will form a single unit, rolled and welded and made of steel, or extruded if made of aluminum. The metal frame is sometimes fitted to a shaped wooden buck. The construction of a head for a frame building differs from that used in a brick building. A frame house requires a drip cap over the head jamb while the outside edge of a brick wall acts as its own drip cap.

The jambs, like the head, have an inside and an exterior casing or moulding. The inside moulding conceals the crack where the plaster or plasterboard meets the window frame; the outside moulding covers the crack where the sheathing* meets the window frame and butts against the siding. The bottoms of the exterior casings or mouldings rest on the window sill while the ends of the interior casings rest on the stool, a part described below.

The sill assembly of a wooden window frame consists of three milled parts: the sill proper, on which the bottom sash rests and which projects outward to shed water; the stool which projects horizontally inside the house as a ledge; and the apron which is a moulding that supports the underside of the stool and covers any gap between the bottom of the stool and the plasterboard beneath it.

The sash is made up of horizontal rails at the top and bottom, vertical stiles at the sides, and glazing. If there is more than one pane of glass, or light, in the sash, the dividing members into which the panes are fitted are known as muntins or muntin bars, **Figure 2.** These terms apply to both wood and metal sash.

Small triangular points of zinc are used to hold the glass in wooden sash and small clips are used in metal sash. Glazing compound is used to seal the glass in place, **Figure 3.**

When two or more sash elements are placed side by side in a single frame, vertical members termed mullions brace the sash on both sides as a jamb. Mullions also join windows in horizontal rows.

Classification of Windows

Windows may be classified according to how they function; as stationary elements, as sliding elements, or as hinged elements. Stationary elements are most likely to be found as attic windows, skylights, fan lights over doors, picture windows; clerestory, gable end, or stair landing windows. Glass block is one form of stationary window. Ventilators, or louvers, though unglazed, may be considered a form of window.

Fig. 1. A floor-to-ceiling multiple sash window being assembled in a mill shop prior to delivery to site.

Sliding elements are the vertical acting double-hung window and the horizontal sliding window.

Hinged or rotating elements have grown tremendously in importance and include casement, awning and hopper type windows.

Double hung Windows. One of the oldest and most popular types of window, the double-hung, is usually made of wood. In the better grades the wood is ponderosa pine, **Figure 4.** There are also metal double-hung windows but these are usually commercial and are made of steel or aluminum. Residential replacement double-hung windows of aluminum are available to replace old and badly worn wood windows but these are only a small part of the window market and are more expensive than wooden types. Like all metal windows they tend to sweat in cold weather unless they have a non-metallic insert between their inner and outer frames.

Wood double-hung windows have two sashes, each of which rides vertically in its own track. The lower sash is held in its track by the interior strip of trim and by a wood strip behind it called a parting bead. The upper sash is held in its track by the parting bead and by another wooden strip behind it called the blind stop.

In older homes the sash used to be counterbalanced with heavy sash cords, rollers, and sash weights which moved up and down in concealed boxes inside the jambs. This method has long been obsolete and no double hung windows are made today with this type of counterbalancing.

Suspension Methods for Double hung Windows. Modern double-hung windows use tempered coil springs as counterbalances which are often incorporated as part of metal track liners and weatherstripping. In some cases the tracks are vinyl plastic, zinc or aluminum. The counterbalancing springs may be concealed at the top or head of the window with cords or metal tapes connected to the sash, **Figure 5.**

A very common form of suspension is the spiral spring balance in a slotted or closed aluminum tube which also forms part of the weather stripping and tracks, **Figures 6 and 7.**

Removable Sash in Double hung Windows. Floating metal tracks with powerful corrosion resistant springs behing them are used in some double-hung windows to keep sash in place instead of counterbalancing springs. A screw in a well can be turned with a screwdriver to regulate the pressure against the side of the sash, **Figures 8 and 9.** By pushing the window against the floating track the opposite side of the sash can be freed from its track and the sash removed for easy cleaning on both sides.

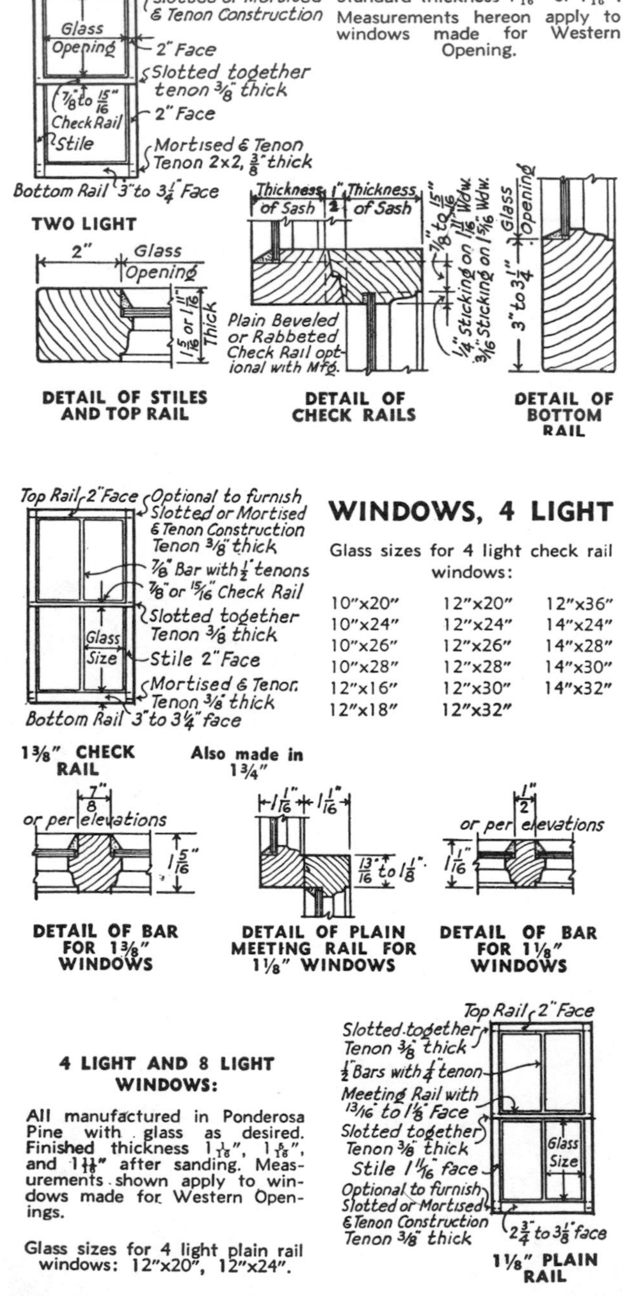

Fig. 2. Details of double hung windows, showing stiles and rails and check rail (top), and muntin bars used to divide sash into four light and eight light windows.

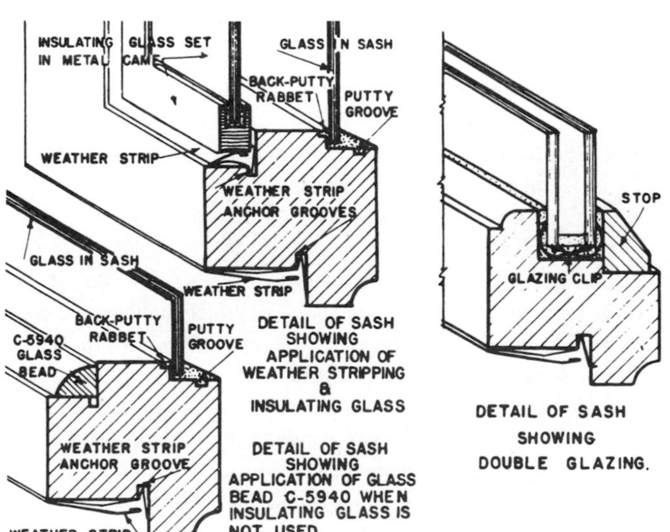

Fig. 3. Puttying details for single and double glazed windows.

National Woodwork Mfr's. Assn.

Fig. 4. Doublehung windows made of ponderosa pine.

Fig. 5. Workman fitting sash spring tension guides.

Fig. 6. Double hung window with weatherstripping and balances. Spring balances are concealed in slotted tube which forms part of the metal weatherstripping.

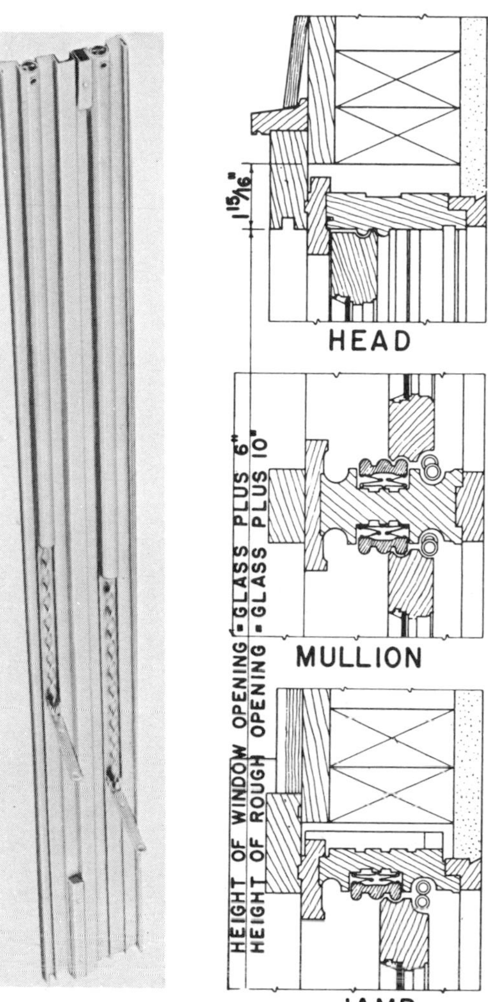

Fig. 7. Spiral spring balance for a double hung window (left). Head, mullion and jamb sections (right) show springs in place.

Fig. 8. Floating jamb is held snug against sash by steel springs which can be compressed to remove sash (see photo 9).

Fig. 9. Pressing sash against floating jamb releases opposite side, permitting easy removal for cleaning.

Fig. 10. Removing a double hung screen panel from a window frame. Panel is connected to spring balance at upper corners.

Fig. 11. This double hung sash can be pivoted inward to a convenient position for cleaning outside of glass.

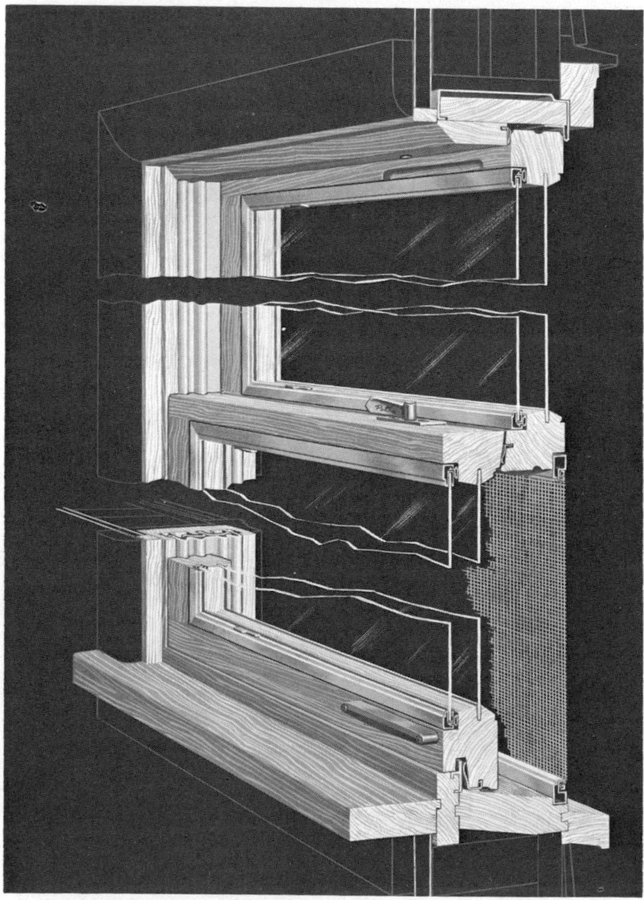

Fig. 12. Wood double hung window pivots inward on spring-loaded vinyl jamb liners. Half screen is removable from inside to allow windows to rotate. Note weatherstripping.

Some windows have removable sceen panels, **Figure 10.** Others are pivoted so they can be turned inward for easy cleaning, **Figures 11 and 12.**

Side Sliding Windows. There are excellent side sliding windows made of wood or metal which have two movable sashes, each of which slides in its own track like a pair of bypassing closet doors. However, there are wood and metal types which have one sash fixed while the other can be moved, **Figure 13, 14.** Side sliders are among the simplest of windows and require no counterbalancing springs.

Both the wood and metal types generally have sashes that can be removed from the frame by pressing the sash upward against a spring-loaded groove surface so that the bottom rail of the sash clears the track, **Figure 15,** permitting it to be pulled inward toward the room and removed from the frame. The metal types, **Figure 16,** operate in a similar manner.

Casement Windows. Probably the oldest type of window knows, this hinged type swings outward from it frame on a vertical axis. The frame and sash may be made of wood (usually ponderosa pine), aluminum extrusions or steel. Casement vents (sash) may come in pairs or singles. There are usually two extended hinges so that when the sash stands at right angles to the frame it is possible to extend one's arm between the frame and the sash and wash the outside of the glass panes, **Figure 17.**

The wood types usually come with a coat of prime (white) paint and are completely framed and glassed, ready for installation. Some models are available in double or single glass. One manufacturer supplies his casements (and other windows) with double glass and a permanent sheath of extruded white vinyl plastic over treated wood, **Figure 18.**

The screens that are available for wood casements (as well as aluminum and steel) are attached to the inside of the frame. Each sash is moved by a screw gear mechanism turned by a rotating handle. The mechanism is located under the bottom edge of the screen (when it is in place) near the lower hinge and operates a lever that moves the sash, **Figure 19.** The sash is held shut by a handle with a cam lock on the frame. The completely assembled window may be installed in exterior walls of either wood or masonry, **Figures 20, 21.**

Aluminum and Steel Casements. In general, metal casements follow the style of the wooden types as far as hinges, interior screens, and rotary operating mechanisms. The aluminum types may have a baked white enamel finish, an anodized finish, or no finish at all. Unfinished aluminum tends to become pitted in salt air or in the sulfur laden air so common in most American cities. Aluminum, in time, forms its own protective oxide but this is rather ugly and dity-looking. Anodizing forms a smooth greenish gray, permanent finish which is weather resistant. The small extra sum for the anodized or baked white enamel finish is therefore worthwhile.

Large Steel Casements. This is rather special type which has long been popular with apartment house builders. These casements are very large and have many fixed lights or panes and two movable sashes. Such windows, not infrequently have a hinged hopper type sash which opens inward toward the room and directs incoming air upward toward the ceiling. They are very serviceable windows and usually require little care other than painting.

Carelessness in handling during construction often results in their being bent, whch is the main reason why they sometimes do no close tightly and are drafty in cold weather. Lack of adequate weatherstripping is another reason for their tendency to leak air. However, various companies produce neoprene or rubber weatherstrips which can be cemented around the edges of the movable sash so that they can be sealed up tightly when locked, **Figure 22.**

Awning Windows. Like casements, awning windows open outward but are hinged at the top or have a sliding lever mechanism at the sides which permits them to simultaneously open outward and slide downward several inches. Long favored by architects and builders for hospitals, schools, and other institutional buildings, the awning window has become very popular for one-family residences. Available in aluminum

Fig. 13. Right hand sash of this wood sliding window is stationary. Left sash is movable.

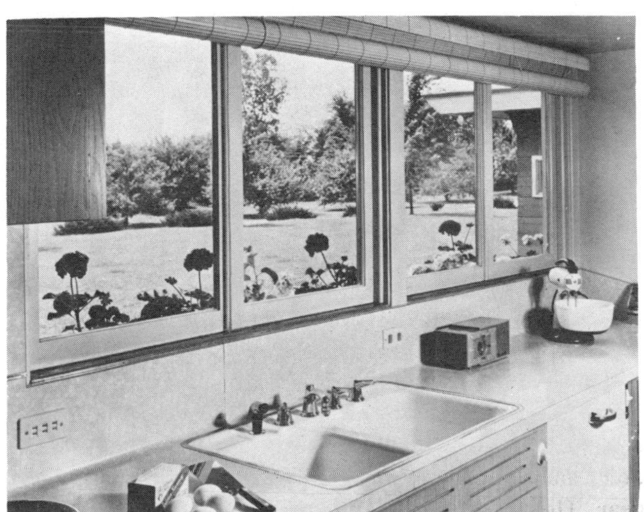

National Woodwork Mfr's. Assn.

Fig. 14. Side sliding windows above a kitchen sink. Outer (left) sash moves smoothly and easily.

Fig. 15. By-pass sliding window is removed by pressing upward and pulling the bottom inward toward the room.

Fig. 16. Horizontal sliding window of aluminum. Both sash move and are removed by pressing upward against spring-loaded track.

or wood, the awning window may be a simple "push-out" hinged at the top and held in position by a jointed metal stay, **Figure 23,** or it may be one of 9 to 12 or more individual windows in a window wall, **Figure 24,** each operated by a rotary screw gear mechanism. The individual glass panes, each mounted in metal or wood, may be ganged one over the other and simultaneously operated by a single rotary screw gear. The glass panes may be as narrow as 8 or 9 inches, **Figure 25,** but on an average tend to be 12 to 14 inches (or more) in height, **Figure 26.**

National Woodwork Mfr's. Assn.

Fig. 17. Double casements with fixed sash between them. Extended hinge permits washing of outer surfaces when sash stands at right angles to wall.

Andersen Corp.

Fig. 18. Double insulated glass and white vinyl cladding over treated wood are features of this casement window.

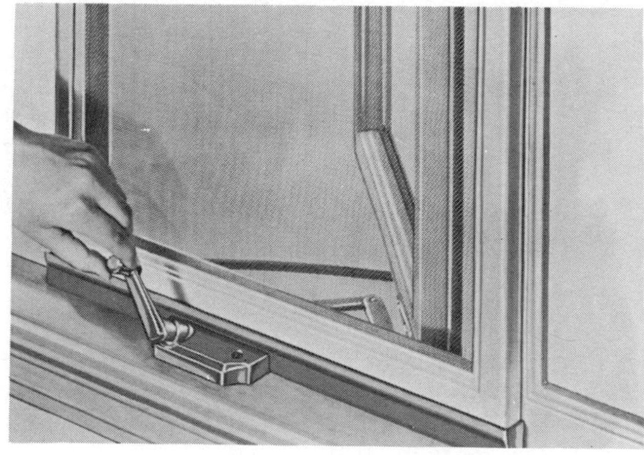

Fig. 19. A casement window operated by a screw type gear.

Fig. 20. Bow window made up of double casements flanking fixed windows in the center

Fig. 21. Double casements over a kitchen sink. Note locking handles on inner edges and operating handles on sills to left and right.

Fig. 22. Large steel casement with movable vents on each side of fixed panes.

Fig. 23. Simple pushout awning window held open by sliding arms and metal stay.

National Woodwork Mfrs. Assn.

Fig. 24. Window walls made up of multiple awning windows with individual geared operators.

Since these windows open outward, both the storm glass and screen panels usually clip on the inside of the window frame. The framing of both storm and screen panels is aluminum whether the windows themselves are wood or metal. The screens are locked into a rim channel with a vinyl or aluminum spline. The glass storm panels are held in a groove by a U-shaped vinyl weathertight channel that wraps around the edge of the glass.

Hopper Windows. These are the reverse of awning windows since they open inward toward the room and are hinged at the bottom. In the main they are used for certain types of basement windows, or as large vents beneath fixed glass panes, **Figure 27.** They have window locks or catches and simple jointed or straight stays at the side to keep them in position.

Basement Windows. Actually, this type of window is simply a variant of hopper and awning types, **Figure 28,** some being hinged at the top, others at the bottom. They may be made of steel or wood and the frame is usually set right into the concrete or masonry of the foundation. Some, especially metal ones, have separate storm and screen panels that clip onto the frame. There are also special types with removable glass panels, **Figure 29,** and others which are short jalousies, **Figure 30.**

Glass Jalousies. Jalousies seem to have originated in the Caribbean Islands where they functioned mainly as wooden shutters. In modern form they have extruded aluminum frames with plate glass louvers held at the ends by aluminum clips attached to the sides of the frame, **Figure 31.** The glass louvers overlap each other by about ⅛″ to ½″ and have elastic vinyl weatherstripping at the sides and at the top. Each louver is connected at one end to a steel or aluminum vertical bar which, in turn, is operated by a rotary worm gear mechanism so that the louvers turn simultaneously from 0° to 90°.

The modern type of jalousie with proper weatherstripping can close tightly enough to be used as a regular house window by and large it is favored as a porch or patio enclosure, **Figure 32.** Complete glass walls can be created by mullioning several windows together. Sufficient height for 8-foot ceilings can easily be achieved by joining two 4-foot windows vertically with self-tapping aluminum screws, **Figure 33.**

As is true of other outward-opening windows, aluminum framed glass (storm) and screen panels that clip on to the frame (described under casement windows) are available for jalousies.

Storm Windows. A storm window is an extra sash that fits closely against the outside of the window frame to create an insulating air space outside the main or house window. Storm

National Woodwork Mfrs. Assn.

Fig. 25. Narrow awning windows above and below fixed sash provide ample ventilation.

National Woodwork Mfrs. Assn.

Fig. 26. Large awning windows flanking fixed picture windows.

windows reduce the heat loss from the house. In modern homes it also minimizes loss of cooled air so expensively created by air conditioners.

Older types of storm sash have wood frames and are hung on clips or hooks at the top in the fall and removed in the spring when wood framed insect screens are substituted. Such storm windows are found only on older homes built before World War II, **Figure 34.** These single-unit storm sash are hung at the top and swing out for ventilation. A modification the full length single unit storm sash is one hinged at the middle as shown in **Figure 35.** This style of sash permits ventilation with more exact control than in the single unit type.

The modern type of storm sash is the permanent combination window, usually of aluminum, made up of a frame and interchangeable glass and screen components called "inserts", **Figure 36.** Usually the lower of he double hung elements is stored in a channel inside the upper part of the aluminum frame when not in use. The screen is raised and the lower glass insert lowered in winter. The screen is lowered and the bottom glass insert raised in summer. The glass elements can be tilted out or removed for cleaning.

To permit easy movement of the inserts, the track or channels in which the inserts move are often made of tough vinyl plastic. The corners of the main frame and of the aluminum frame around the storm (glass) and screen panels are often reinforced by die cast zinc corners which are inserted in the hollow aluminum frames and fastened with screws.

Fixed and Movable Window Combinations. Picture windows are often used not only to provide large amounts of light but to present a pleasing exterior scene easily visible from the interior of the room. A picture window may be simply a fixed sheet of double strength glass, ¼-inch plate glass, or double sheets of insulated glass. Large fixed picture windows are often used as the center piece of a bay window with smaller double-hung or casement windows at the sides for ventilation, **Figure 37 and 38.**

Bow windows are another type which combine fixed and movable glass panes. These are very large windows that "bow" out in a gentle curve and may be as high as 6 feet and

National Woodwork Mfrs. Assn.

Fig. 27. Hopper windows beneath fixed glass panes direct breeze upward and can stay open during rain.

Fig. 29. In-swinging basement awning window is easily removed from hangers mounted on upper corners of sash.

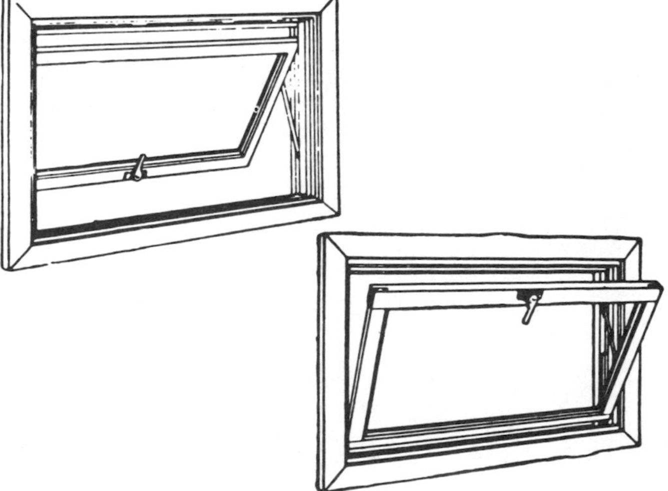

Fig. 28. Awning (top) and hopper windows (lower) are often used as basement windows.

Fig. 30. Jalousie window has aluminum frame sized to fit into course of concrete block for basement installation.

Fig. 31. Double jalousie window in metal frame. Note drip caps and metal clips holding glass louvers.

Fig. 33. Jalousies joined one above the other and mullioned together make porch enclosures with excellent ventilation.

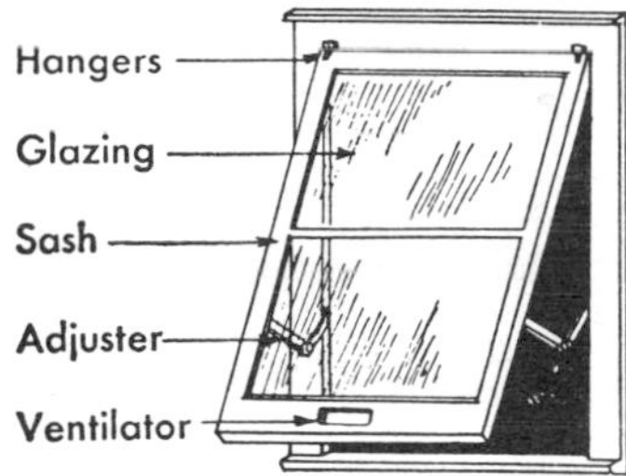

Fig. 34. Single unit storm sash is hung on hangers at top, swings outward as an awning window for ventilation. Such units are now found only on older houses.

Fig. 32. Jalousies at right are ideal for porch and patio enclosures when mullioned together.

Fig. 35. Jointed storm sash is also hung at top, but lower half swings outward for better control of ventilation.

Fig. 36. Self storing aluminum combination storm sash and screen.

Fig. 37. Colonial style bay window with many small lights and flanking casements.

as wide as 12 feet or more. They are made up of panes of glass from 12 to 18 inches square. At least 3 or 4 of the panes are operable and swing out like an awning window for ventilation, **Figure 39.**

Large fixed window walls are used in certain types of modernistic architecture. Sometimes a whole end wall may consist of glass panes framed in wood with or without movable or operable panes, **Figure 40.** In **Figure 41** the top row opens like awning windows, the middle row is fixed and the bottom row is made up of hopper windows opening inward.

Fixed Special Windows. Various types of small, fixed windows are used to provide added light in different parts of the house. In some cases, in addition to providing light, they are also decorative and may or may not be operable such as those shown in **Figure 42.** These are special types that provide light or ventilation or both for hallways, stairwells, attics and other places that are not living quarters. Many Georgian or colonial style front entrances have fixed window panes on one or both sides of a wooden door to allow natural light to enter a dark hallway.

Areas of glass blocks, either clear or translucent, are placed in exterior walls to illuminate stair landings. When both light and privacy are desired the glass blocks are translucent, **Figure 43.**

Clerestory windows are used where natural light and privacy are desired. Sometimes called "ribbon windows", they are often used in bedrooms and are located high on the wall near the ceiling and may run the entire length of the wall. Such windows are long and narrow, often no more than 12 inches high, and fill the space between studs, **Figures 44 and 45.** Skylights are another form of fixed window that add light to rooms, **Figures 46 and 47.**

Windows As A Ventilation Factor. Double hung windows are often the cheapest and most available and have been widely used in most building projects. However, from the point of view of ventilation doublehung windows have their drawbacks. When fully opened with both sashes fully up or fully

Fig. 38. A bay window with fixed glass in center is flanked by operating casement windows for ventilation.

Fig. 39. Large bow window with four movable lights, one in each corner.

Fig. 40. End wall of contemporary house is built almost entirely of glass.

Fig. 41. Playroom window wall combines hopper windows with fixed glass in same size frames. Wide roof overhang shelters windows.

down this type of window offers no more than 50% of the window area. In northern or cold climate areas, this may not be a disadvantage since doublehungs seal up more tightly than most others even when worn and thus keep cold air out better. On the other hand, homes in cool climate areas that have hot summers suffer from inadequate ventilation because their doublehung windows cannot be opened to the full size of the window opening. However, most room air conditioners are made for double hung windows and homes with other types of windows require more expensive through-the-wall installations of room air conditioners.

Side sliding windows that have one fixed and one movable sash have basically the same advantages and disadvantages as doublehung types since they can only open to 50% of their total window area, **Figure 48.** Sliding widows in which both sashes can be removed offer 100% opening for ventilation although this is rarely done.

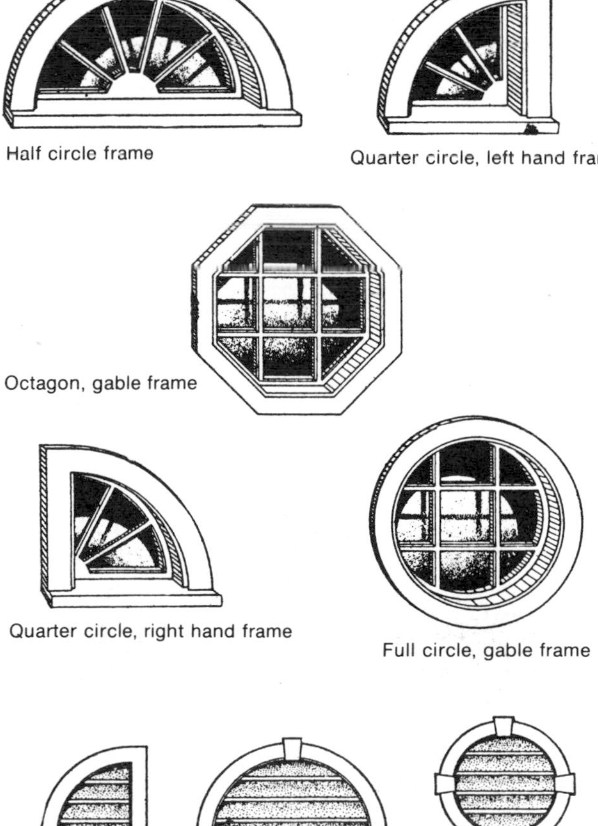

Fig. 42. Types of special purpose windows for hallways and stairwells, and ventilators for gable walls.

Fig. 43. Translucent glass block provides light with privacy, has insulating qualities because of air sealed inside.

Casements are ideal for complete ventilation since they open out to 100% of the window area. This makes them desirable in warm weather climates but they are less desirable where the climate is cold because they admit a great deal of air even when slightly open.

Awning windows rate high when it comes to ventilation since they can be opened to about 90% of the window opening and they can be made to close quite tightly in cold weather. With properly fitted storm sash they can cope with any cold weather.

Hopper windows do not usually open to more than about 45° although some have stays at the side which permit slightly more opening. In general, they provide no more ventilation

Fig. 44. Clerestory windows above flat roof admit light to bedroom wing beneath shed roof at rear of house.

Fig. 45. View from porch shows large fixed glass picture window flanked by panes of fluted opaque glass in awning sash. Picture window on far side of living room is topped with panes of fixed glass set in clerestory framing at ceiling level.

Fig. 46. New dormer-shaped plastic skylights are less conspicuous than older dome types for residential applications.

Fig. 47. Skylight installed in cathedral ceiling brings soft, natural light to center of kitchen, supplementing windows.

than doublehungs and are difficult to stormsash.

Jalousies provide almost as much ventilation as casements when the louvers are opened out so that they are horizontal. They can be opened to almost 95% of the window opening. The can be kept slightly open in any rain. When well made and properly weatherstripped they can keep cold air out in winter. However, tightly fitting interior storm sash are a necessity in cold weather. Screen panels held by turn buttons replace the interior storm sash in summer.

Window Installation

Most windows are delivered to the site complete with glass, sash, frame, and hardware installed at the factory, **Figure 49.** The least expensive types may come with raw wood surfaces but the better wood types usually have at least a prime coat of white paint. The Perma-Shield windows come with a film of white vinyl plastic extruded directly over the wooden parts as a permanent weatherproof covering.

The complete units are installed in the rough openings cut through the sheathing on siding, **Figure 50.** Since the rough openings are often as much as two inches larger than the frame of the window, it is necessary to square the frame within the openings. The window frame is squared with a level and held in place with wedges and wooden blocks, **Figure 51.** It is usually necessary to remove the upper and lower sash before squaring the window in order to nail the sides of the frame through the blocks to the studs on both sides. The studs on either side of the window are double, the 2×4's that support the bottom of the window frame are also double. The header above the window frame between the doubled studs may be double 2×4's laid flat or double 2×6's which function as a lintel to support the structure above the top of the window, **Figure 52.**

Once the frame is nailed in place and the sash restored, the mitered exterior casing trim or mouldings is trued up with a

Fig. 48. Aluminum side slider window fitted with insulated glass. Right side is fixed and only left hand sash moves allowing only 50% of total area for ventilation.

Fig. 49. Awning type sash being installed in openings on both sides of a fixed glass picture window.

Fig. 51. Casing trim is trued with level before nailing.

Fig. 50. Held square by temporary corner braces, sash frame is set in place with shims to hold it plumb on opening.

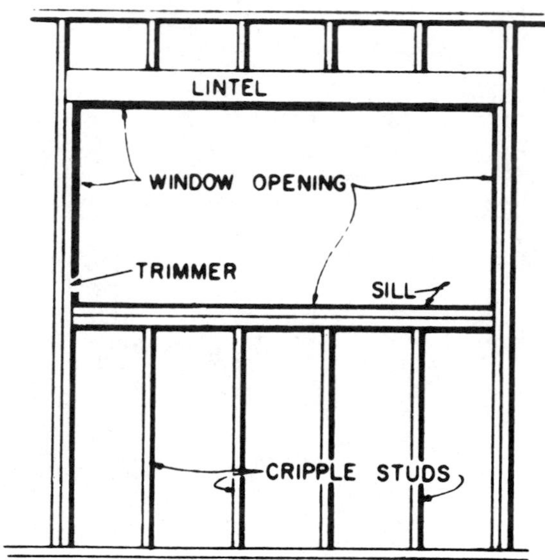

Fig. 52. Framing fo window with double sill at bottom, double sides and 2×6 lintel at top.

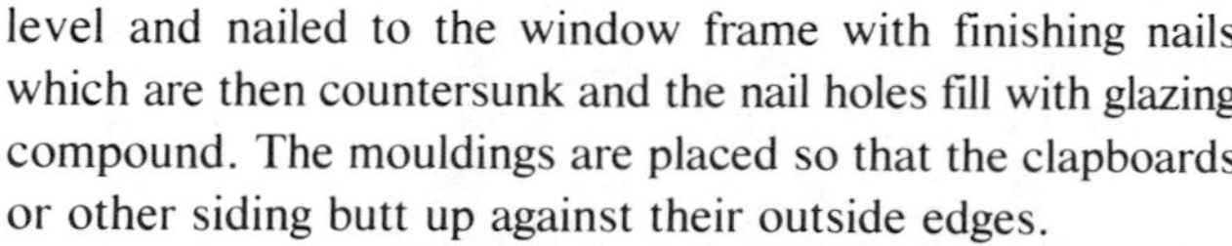

level and nailed to the window frame with finishing nails which are then countersunk and the nail holes fill with glazing compound. The mouldings are placed so that the clapboards or other siding butt up against their outside edges.

Summary of Developments in Window Details. One of the most significant developments in wood windows is their treatment with chemical perservatives to make them water, fungus and termite repellent. The chemical (pentachlorophenol) impregnates the wood and guards it against rotting, warping, shrinking and swelling without interfering with the ability of the wood to take paint or other finish.

Another important development is the removable double-hung and side-sliding sash previously described in this chapter.

Screen and metal double hung elements are similarly removed. The basic technique is to push the sash to one side against a flexible pressure jamb liner permitting the other side of the sash to be pulled free. In side sliding sash the unit is lifted against a pressure plate in the head jamb and the bottom is pulled inward toward the room. Basement awning windows frequently come with a hanger system that permits easy removal.

Removable Muntins. In addition to sash, screens and storm windows being removable and adjustable, muntins are removable, too, for easy cleaning of glass, **Figure 53.** The traditional patterns or diamond shaped muntins are 'plugged in' by inserting doweled ends into small metal grommets located in the sash.

Fig. 53. Removable muntin bars of wood or plastic are snapped into small metal grommets in sash, making it easier to clean one large pane rather than six or eight small panes.

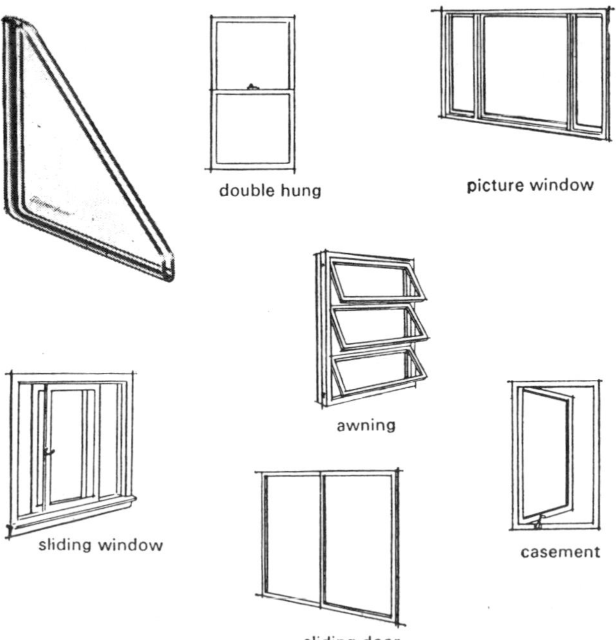

Fig. 54. Section of doubled insulating glass (above left) and six types of windows in which it may be installed.

Insulating double glass panes with completely sealed edges, **Figure 54,** are no longer new and are now applied to a wide variety of windows. Even more effective is the application of an added storm window panel behind the double glass. The added panel increases the insulating value of the window by 35% over the double pane insulating glass. The extra glass panel is easily placed or removed, **Figure 55.**

Window Planning Principles

The following window planning principles are recommended by the University of Illinois Small Homes Council for private homes.

1. If one large opening is provided instead of several small ones, a more desirable distribution of light is secured. Dark areas between openings are eliminated.

2. Windows in more than one wall give more effective lighting than windows in just one wall.

3. A horizontal window gives a wider spread of light than a vertical window of the same area.

4. The higher the window is placed in the wall, the deeper is the penetration of light into the room.

5. Large glass areas can be used to extend indoor space outward.

6. Horizontal window division members are undesirable when they interfere with view.

7. Provide glass areas in excess of 20 percent of the floor area of the room.

8. Place principle window areas toward the south.

9. Screen only those parts of the window that open for ventilation; full screens absorb as much as 50% of available daylight; half screens absorb only 15% of the daylight.

10. Locate ventilation openings to take advantage of the prevailing breeze and to direct ventilation to occupants' level Window types that deflect air movement downward are not restricted to low placement.

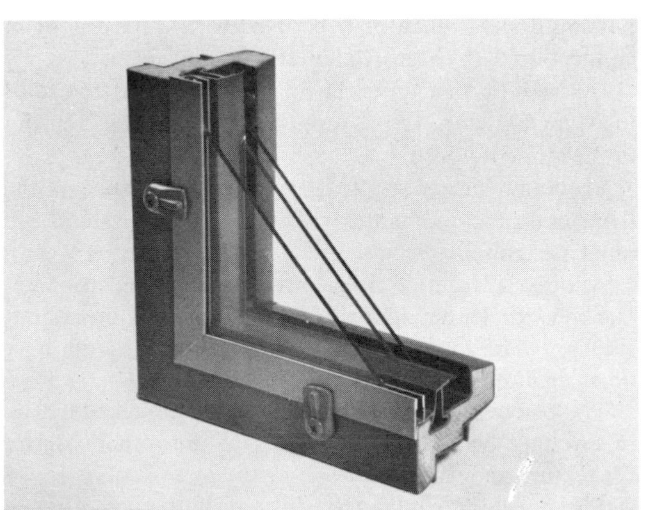

Andersen Corp.

Fig. 55. Perma-shield casement with added storm panel over double insulated glass. The triple glass increases the insulating value of these windows by more than 35% over double insulated glass.

Chapter 9

DOORS

Doors, in the modern house, are as functional and varied as windows. Basically, a door is a barrier, by which an entry is closed or opened. While most doors operate by swinging on hinges, sliding doors, including folding doors and pocket doors, have grown in popularity with the need to make every bit of living space count. Various kinds of doors form a significant part of the trend to prefabrication and stock components.

Most doors are made of wood, but metal is coming into greater use as frames for sliding glass units, and combination storm and screen doors. Plastics are used in several ways such as flexible coverings for folding doors and as translucent full size panels of bath tub and shower enclosures. Glass is used for large sliding doors leading to gardens, patios, lanai or atria and as lights in French doors and exterior and interior doors. Various types of solar glass can control sunlight and reduce heat gain in summer, heat loss in winter.

Components of a Door. The two main parts of a door are the sliding or swinging member and the frame to which it is attached for closing, opening, sliding or folding. Complete door units of every type are now factory built and shipped ready for fast installation.

Classification of Doors. Doors may be classified as to function and size when used in house construction. The largest door made for a house is the garage door. Sliding glass patio door units are next in size, having two by-passing doors, which may be used with a fixed door in connection with extensive fenestration in a wall.

By-passing door units, bi-folding door units and folding door units are large, too, and their use is increasing in the modern house.

Probably the most attractive door of a house is the front entrance door unit, with side lights, pilasters, cap and base trim.

Another attractive door often used in the front entrance is the Dutch door or double door, with or without sidelights. In some front door units the sidelights may be opened to permit ventilation.

The conventional panel door, long a popular favorite in dwelling construction, appeals to the owner's desire for texture and variety.

The development of the flush or flat door, used in connection with panel doors, adds charm to the house.

Other doors, such as the louver doors; the traditional cafe doors, louvered or paneled; mirror bi-folding door units; wardrobe closet units; china case doors and sliding cabinet doors all have their function in today's house of attractive appointments. Not to be omitted are the popular, interchangeable screen-storm door; the French door; the jalousie door and shower enclosure doors.

How Doors Function

Doors, as well as windows, may be classified as to the method of functioning or operating in addition to size and placement in house construction.

Folding doors hang from an overhead track and are fastened to one jamb and close against the opposite one. If the door opening is wide enough, two doors are used which meet in the middle and close against each other. Bifold doors consist of two panels of equal size hinged together. Each pair of panels is pivoted at the top and bottom next to the jambs and the outer panels slide in an upper and sometimes also a lower track.

Bi-passing doors consist of two door members hung on parallel overhead tracks which permit the doors to be opened by passing left or right. The doors are closed by moving one door to extreme right and the other to the extreme left.

Swinging doors, such as cafe doors, swing on double acting or two way hinges, permitting the doors to swing in either direction. The action of gravity returns the doors to a normal closed position.

Hand of Doors. Interior or exterior entrance doors, or doors hinged on either left or right hand side have a direction of swing known as the hand of door. The hand of a lock and door is determined from the outside of an entrance door or from the corridor or hall side of a room door. As you stand to insert the key in the lock, if the hinge is on the left and the door opens away from you, it is a simple "left hand" door. If the door opens toward you, with the hinge on the left and the knob or handle on the right, it is a "left hand-reverse bevel" door. If the hinge is on the right and the door opens inward (away from you), it is a "right hand" door. If the hinge is on the right and the door opens outward, toward you, it is a "right hand-reverse bevel" door. Reference should be made to the chapter on Hardware for complete information and diagrams.

Garage Doors

A garage door is the largest moving part of a house. When a garage is built as an integral part of a house, being accessible from the front, the door often covers nearly one-fourth of the front wall elevation. If accessible from the side or rear of a house the garage door covers a like area and its construction and styling are equally important to the exterior design.

Garage doors, in the modern house, are most often made of wooden panels or sections, hinged along the horizontal rails and opening upward and inward on vertical-horizontal tracks.

Garage doors are made in a wide range of sizes from 8′ to 18′, **Figure 1.** A one-car-wide door is available in 8′, 9′ or 10′ widths. The two-car-wide door, center sketch, may be 15′ or 16′ wide up to 18′ wide as shown in the lower sketch. Two 8′ or 9′ doors may be used for an extra wide garage.

4-SECTION-HIGH PANEL DOORS

1-CAR WIDE (8′, 9′ and 10′)

2-CAR WIDE (15′ and 16′)

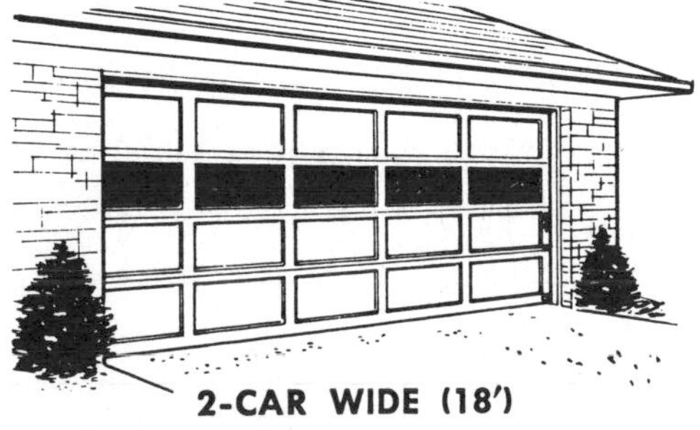

2-CAR WIDE (18′)

Fig. 1. Garage doors range from 8′ to 18′ in width, usually four panel sections high. Two single doors may be used for two cars.

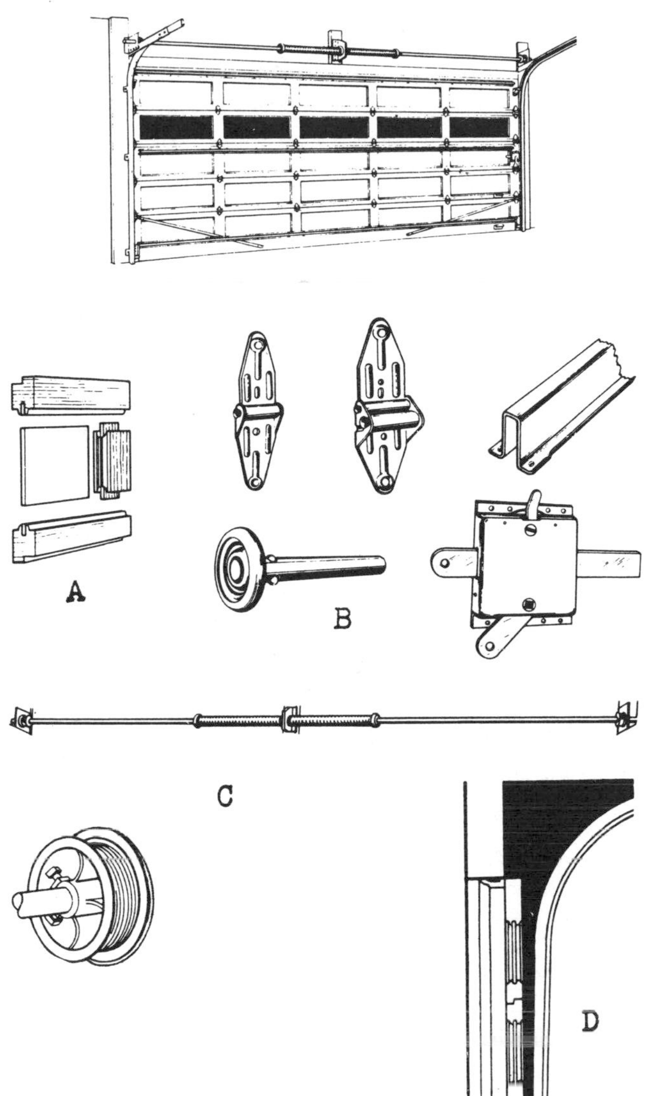

Fig. 2. Garage door and its parts: A. Rails, panel and stile. B. Hinges, roller, stiffener and lock. C. Torsion shaft, springs and drum. D. Section of track which carries door upward and overhead. Tension type balance spring may be used instead of torsion spring.

The Construction of a Garage Door. The door may be made of 1-3/8″ thick Douglas Fir with hardboard panels. The rails and stiles have continuous keep mortise and tenon joints into which the panels are glued and steel doweled.

The panels are hinged together at every stile with sturdy embossed hinges. The two-car wide door is reinforced and stiffened, with steel struts and braces to minimize sagging in the open position. Ball bearing rollers attached to the door fit securely in steel channel tracks mounted at the jambs and header of the door frame. The vertical-horizontal tracks are rigidly supported and braced, **Figure 2.** The smaller sketches show the features of construction of this large five-panel door. A. The rails, panel and stile. B. The hinges, steel strut, ball bearing roller and the lock. C. The shaft, torsion springs and drum. D. The track or vertical channel housing the rollers.

Fig. 3. Flush panels may be used plain or trimmed with molding as at left.

A wide variety of designs may be created from combinations of raised panels with routed panels, **Figure 3.** A flush or flat panel, both inside and out, may be used or decorative molding applied to door faces to enhance their styling and add character.

While torsion springs are used on many garage doors, extension coil springs as counterbalances are more popular. **Figure 4** shows the ballbearing rollers in the horizontal part of the track of a newly installed garage door.

Sectional garage doors come in a variety of styles and materials. They may be flush with hardboard "skins" glued to wood frames or the skin may be 1/8″ hardboard with a rough sawn surface, **Figure 5,** or the panels may be made of steel, aluminum, or white fiberglass framed in aluminum. **Figure 6** shows a steel sectional overhead garage door with windows and optional shutters. The horizontal lines of the sectional steel door in **Figure 7** match the lines of the siding. Factory primed in white on both sides, its galvanized, corrosion-resistant steel assures unusual durability under all weather conditions.

The light weight aluminum sectional door in **Figure 8,** has a 2″ tempered aluminum framework, overlapping horizontal joints to keep out weather, heavy duty tubular vinyl astragal, white pebble grained panels, and plastic foam that provides a weatherseal between panels and end stiles.

Complete parts of a fiberglass overhead garage door just before assembly are shown in **Figure 9.** Dent resistant panels are made of thermo-setting acrylic modified polyester reinforced with glass fiber threads, nylon rollers assure smooth operation. Flexible vinyl weatherstripping keeps out drafts. The white translucent fiberglass allows a soft filtered light to enter the garage even when the door is closed.

Automatic Garage Door Operator. An overhead garage door may be opened by an automatic, radio controlled operator attached to the door and track, **Figure 10 A.** A small battery-powered transmitter,

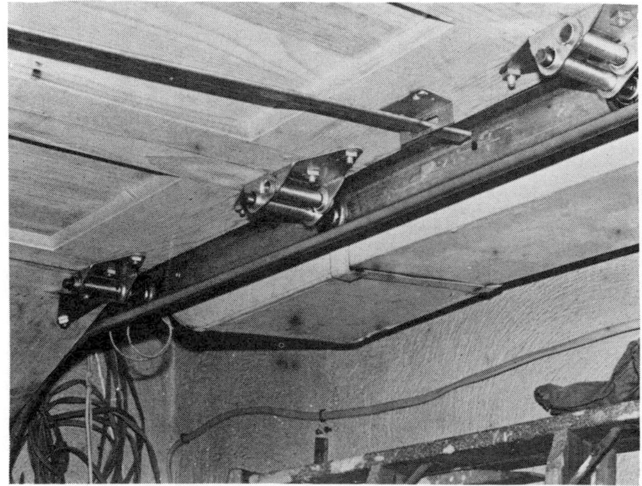

Fig. 4. Wood panel garage door rolls smoothly on ballbearing rollers in horizontal part of galvanized steel track. Coil springs are outside track.

Clopay Corp.

Fig. 5. Four-section overhead garage doors with rough sawn hardboard over wood frames.

Stanley Vemco Co.

Fig. 6. Steel sectional overhead garage door features nylon rollers, bottom weatherstrip, windows and optional shutters.

Clopay Co.

Fig. 7. Factory primed on both sides, this galvanized corrosion-resistant steel door has unusual durability under all weather conditions.

kept in the glove compartment, or attached to the car dash, controls the automatic door opener. The control unit can be transferred from one car to another. When the door is opened, a light is turned on. It remains lit until the door is closed by pushing a button mounted inside or outside the garage.

The system consists of a transistorized portable hand transmitter, a receiver and door operator mounted inside the garage, and proper wiring. The operator is powered by a 1/5 horse power electric motor operating on ordinary household electrical circuits. The door op-

Clopay Co.

Fig. 8. Light weight aluminum sectional door has 2" tempered aluminum framework, pebble grained panels and foam weatherseal between panels.

Stanley Vemco Co.

Fig. 9. Complete parts of a fiberglass overhead garage door just before installation.

DOORS

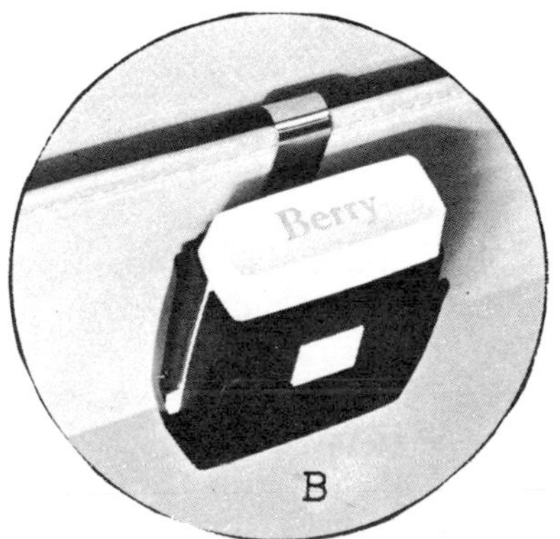

Fig. 10. Automatic garage door is opened by electric motor drive (A) controlled by cigarette-pack size transmitter (B).

erator is 121″ in length with a draw bar pull of 90 pounds. The radio transmitter range is from 40 to 60 feet inside the car and can be activated only at the radio wave length to which the door operator is tuned. The operator will lift a 7′ high sectional door at the rate of 10″ to 12″ per second. The control button and radio transmitter is no larger than a package of cigarettes, **Figure 10 B.** In some systems a second touch of the button closes the door and locks it while the driver remains in the car. The automatic garage door operator provides valuable personal protection as well as modern convenience.

An improved type of transmitter and receiver has switches which enable the owner to select his own frequency from 1024 possible settings. Thus if a nearby garage door opener has the same radio frequency code that opens his garage door, the owner of the new type of transmitter and receiver merely has to flip a few miniature switches to reset his own code, **Figure 11.**

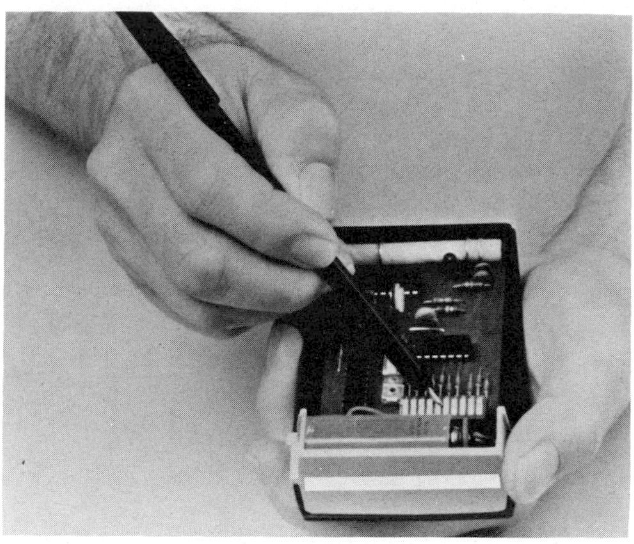

Stanley Vemco Co.

Fig. 11. Garage door opener radio transmitter control has 10 miniature switches which enables owner to change frequencies if any nearby garage door opener has code that matches his own.

Front Entrances

The main entrance to a dwelling is most always at the front of the house facing the street or roadway. It is somewhat more ornate than a side or rear entrance since it provides a first impression of the house to callers.

In the conventional construction of a front entrance door, especially where there is little protection from the weather, the parts of the frame of the door itself correspond to the parts of a window frame. At the top is the head jamb, with outside casing often surmounted by a drip cap, at the side are the jambs, also with their casings, and at the bottom is the sill, sometimes fitted with a narrower piece of trim, called the threshold, to fit underneath the door.

The door stop is either nailed to the side jamb or inserted into a plowed-out groove. It may also be rabbeted into the jamb. The door frame must be adapted to the construction of the building, whether frame, or brick veneer, or masonry, or steel framed. For example, steel angles or lintels must be used over the head jamb to support the flat masonry arch, or the bricks must be arranged to distribute the weight of the wall, **Figure 12.** Identical door assemblies can be installed in an outside wall of wood, or in a brick wall, **Figure 13.** Casing and trim fit over framing members and edges of jamb **Figure 14.** Door stop can be nailed on jamb or inserted in a plowed-out groove. The jamb may also be rabbeted to form a stop on either side, **Figure 15.** The wedges between the jamb and rough frame hold the door frame in vertical position.

An Attractive Doorway. The entrance doorway is one of the most attractive features of any house in either traditional or contemporary style, **Figure 16.** This is a formal Georgian style entrance, with broken pediment, fluted pilasters, panel door, antique knocker and lock assembly. **Figure 17** shows a Colonial entrance of a brick house. The windows have small panes of glass, with shutters to harmonize with the Colonial style.

A Modern Doorway Having a Flush Door. A modern, exterior flush door often has simple trim, with glass or plastic panel, on one or both sides, **Figure 18. Figure 19** shows a formal double entrance door in Georgian style with fluted pilasters, broken pediment, panel doors and small glass panes to light the hallway.

Various Colonial Front Entrances. The panel door, patterned after the Colonial or Early American period of architecture, is surrounded by highly decorated mouldings, carved pilasters and raised panels, **Figure 20.** The decorative cornices, either curved, figured or slanted, require careful flashing, especially when built in a masonry wall, **Figure 21.** Flashing may be installed vertically (**Figure 21,** left), or diagonally from the cornice (**Figure 21,** right).

INEXPENSIVE TYPE OF FLAT OR "JACK" ARCH. BRICKS NOT RUBBED TO WEDGE SHAPE HORIZONTAL JOINTS AT RIGHT ANGLES TO RADIUS, OF BRICK. BRICK RUBBED AT SOFFIT AND TOP ONLY

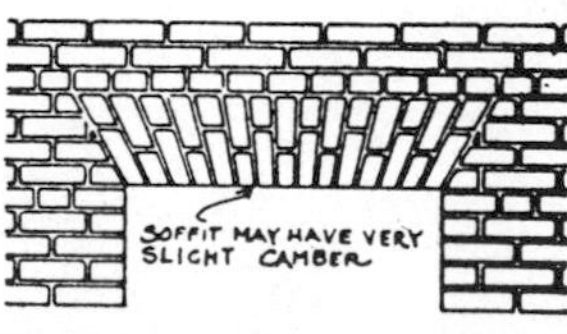

BETTER TYPE OF FLAT OR "JACK" ARCH. BRICK RUBBED TO WEDGE SHAPE AND RUBBED TO FORM HORIZONTAL JOINTS TOP AND BOTTOM OF EACH BRICK

Fig. 12. Typical flat arches of brick to support masonry above openings in exterior walls for doors and windows.

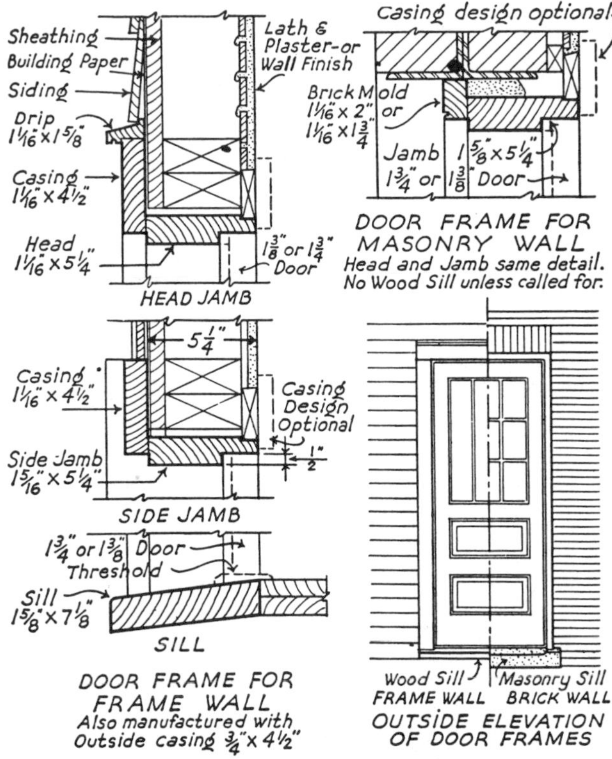

Fig. 13. Sections of door frame showing how an identical door may be installed in an exterior wall of wood or brick.

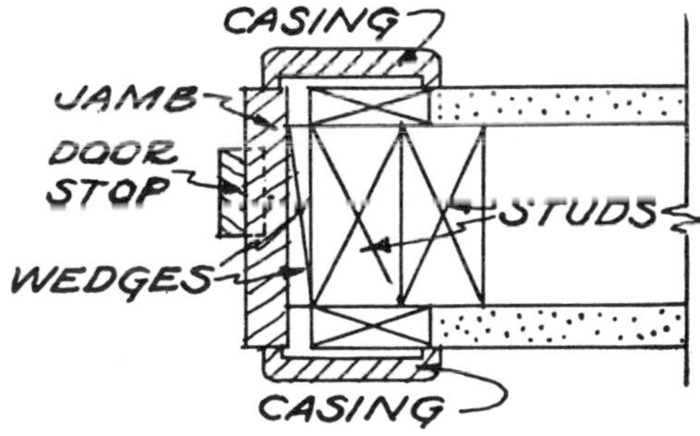

Fig. 14. Jamb section shows casing trim and alternate methods of setting stop by nailing it on jamb or inserting in groove.

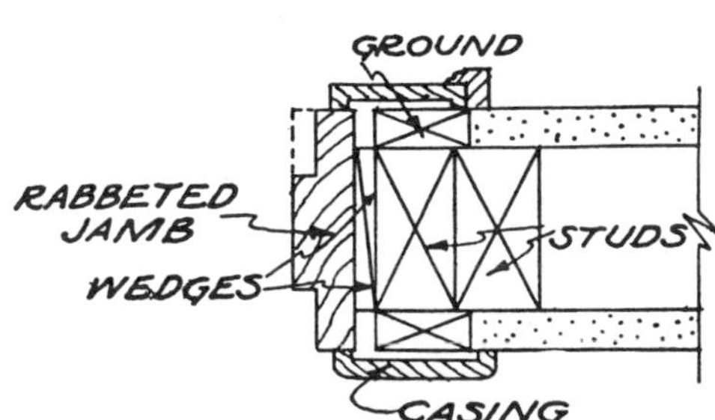

Fig. 15. Jamb with stop rabbetted on either side. Wedges between jamb and rough frame set door plumb.

Fig. 16. Formal Georgian style front entrance, with broken pediment, fluted pilasters, panel door, knocker and lock.

Fig. 17. Colonial style front entrance, with panel door and glazed transom above to bring soft light into entry hall.

Fig. 18. Modern entrance with flush door and fixed glass side panel fitted with drapes for privacy.

National Woodwork Mfrs. Assn.

Fig. 19. Double entrance door in a brick wall has broken pediment and glass panes to light hall behind it.

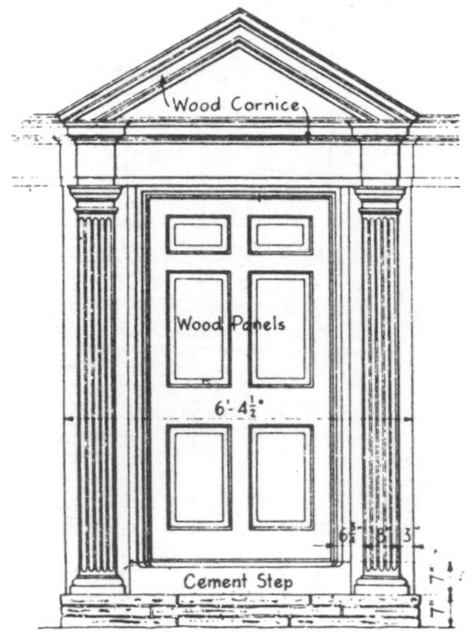

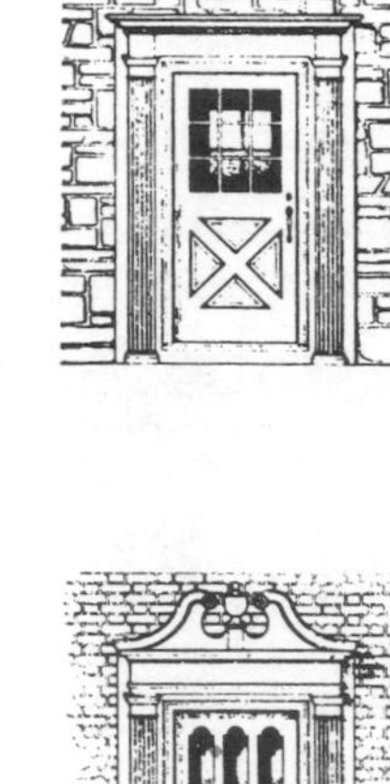

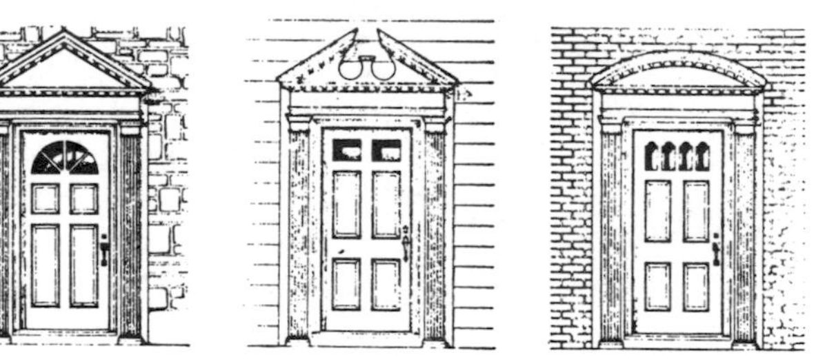

Fig. 20. Suggested designs for Colonial front entrances.

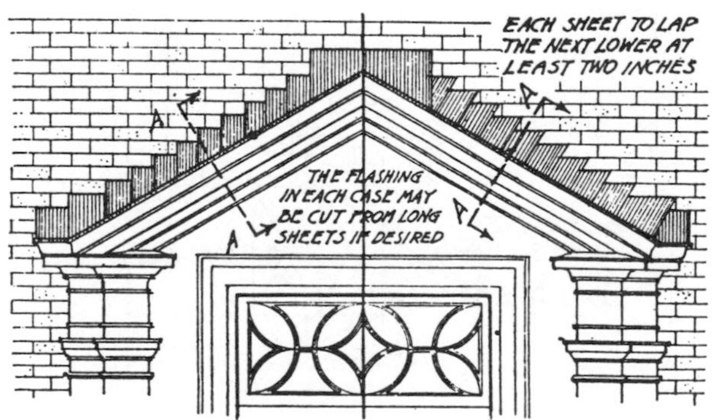

Fig. 21. Vertical and diagonal methods of flashing doorway cornice against a brick wall to seal out weather.

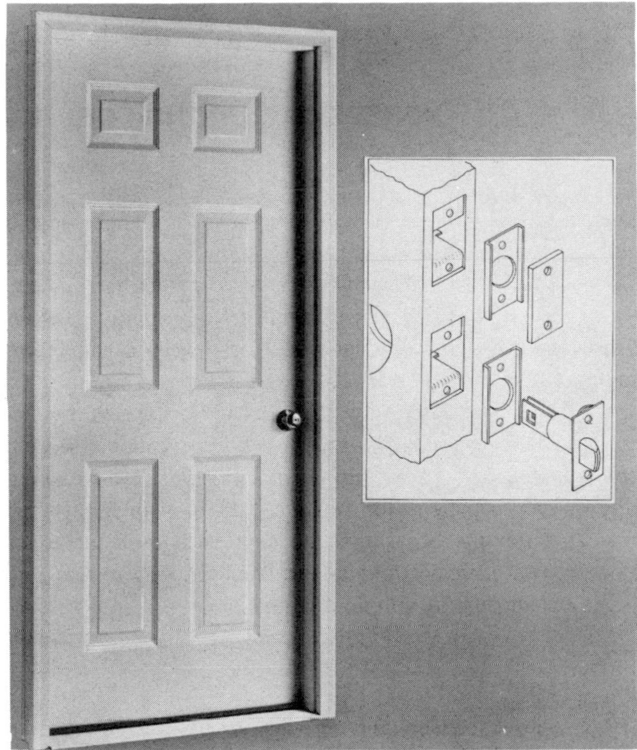

General Products Co.
Fig. 22. Steel clad door is pre-hung in wooden frames or come with steel frames knocked down. Comes pre-drilled and cut to receive lock sets and latch plates.

Insulated Steel Doors

One of the interesting developments in home building and remodeling is the use of insulated steel doors. Some are flush, others are stamped or embossed to look like wood panel doors and many are available with lights. All are filled with polyurethane or other plastic foam which provides at least twice as much insulation as a wooden door plus storm door. Another advantage of steel doors is that they do not split, warp, or rot. They are thoroughly rustproofed and usually come with a smooth prime coat or a sprayed white finish.

Some, like the Benchmark door, **Figure 22,** are prehung in wooden frames or come with steel frames knocked down. Unlike many steel doors, this one comes predrilled and cut to receive latch plates, lock sets and dead bolts. Magnetic weatherstripping encased in plastic is used on the top and strike sides and double compression plastic tubing on the hinge side. Safety measures include tempered glass in models with lights. Insulated glass is offered by some manufacturers as an option.

Like the wood doors which they so closely resemble, many are offered as double doors with numerous small lights, **Figure 23.** Some insulated steel doors have the appearance of deeply carved wood, **Figure 24.**

Many insulated steel entry doors are decorated with moldings and plaques, **Figure 25,** which gives them an attractive appearance. Typical details of the structure of steel clad insulated doors may be seen in **Figure 26.**

Stanley Co.
Fig. 23. Double steel insulated entry doors look like wood and have numberous small lights.

Lake Shore Industries
Fig. 24. Insulated steel door heavily embossed to resemble deeply carved wood entry door.

The Popular Panel Door

The conventional panel door, usually made of Ponderosa pine, for interior or exterior use, offers a pleasing texture and variety of design, **Figure 27.** The arrangement of parts and size of panels in the main frame of a typical door are shown in **Figure 28.**

How Panel Doors Are Made. Panel doors are factory made and assembled. Door panels fit into stiles and rails milled in cove and bead or ovolo designs, **Figure 29.** Panels may be raised to flat. Flat panels are usually of three-ply plywood stock. The stiles and rails are joined by dowelling or blind mortise-and-tenon joints, or both.

The dimensions of typical interior and exterior doors, showing variations in panels, stiles and rails, and glazed openings are shown in **Figures 30, 31, and 32.**

French doors, often called casement doors, are used as interior doors or interior doors opening on to porches or terraces, **Figure 33.** The doors are glazed with small lights from top to bottom rail and hinged to meet at the middle of the doorway. Exterior French doors are now often replaced by gliding or sliding glass doors framed in metal or wood. The dimensions, standard sizes and designs of typical casement doors are shown in **Figure 34.**

Kinkead Industries

Fig. 25. Moldings and plaques were used on these insulated steel double doors to make look like conventional wood entry doors.

Flush Doors

Interior or exterior doors are often built with a solid or flush surface of pleasing texture but of somewhat less artistic design than the paneled doors. Flush doors are still used extensively for interiors in institutions and many homes. The one shown in **Figure 35** is an entry door that is quite striking because of its long glass light.

How Flush Doors Are Made. Some flush doors are built up with a smooth, one-piece, zinc coated and bonderized steel facing. These are often used in the doorway between the house and attached garage to satisfy local fire code requirements. Some flush doors of wood are built up of a solid core of panel sections and core strips, **Figure 36.** This solid core door has a cross banding of wood veneer and a facing or surface veneer of figured Philippine Mahogany, Swedish hardboard or birch wood.

Solid flush doors are made of narrow blocks often of random lengths, of the same wood, glued together, usually covered with two veneers, the under cross-banding with the grain running horizontally, and the outer veneer with the grain running vertically, in the same general direction as the core blocks. The edges are of the same veneer as the surfaces.

When Ponderosa pine is used for flush doors, the wood

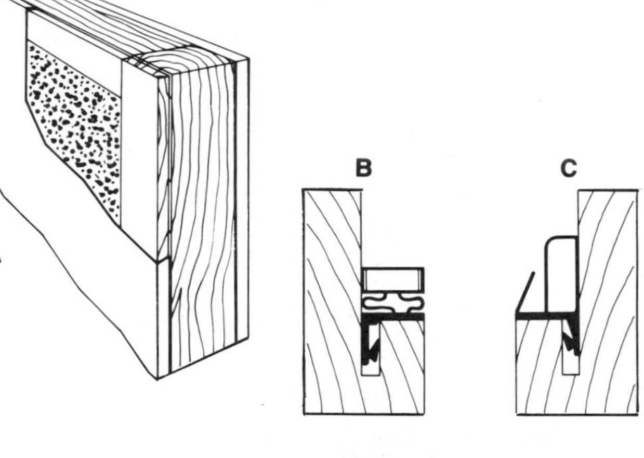

Kinkead Industries

Fig. 26. Insulated steel door details. A-Polystyrene foam core fills wood frame and provides R-18 insulation. Wood frame acts as thermal break between front and back steel covers. B-Continuous magnetic weather-stripping around head and strike jamb with compression on hinge jamb provides positive seal against drafts. C-Aluminum threshold is adjustable at top to mate with bottom sweep and can be raised or lowered without disassembly.

Fig. 27. Interior panel type door of ponderosa pine.

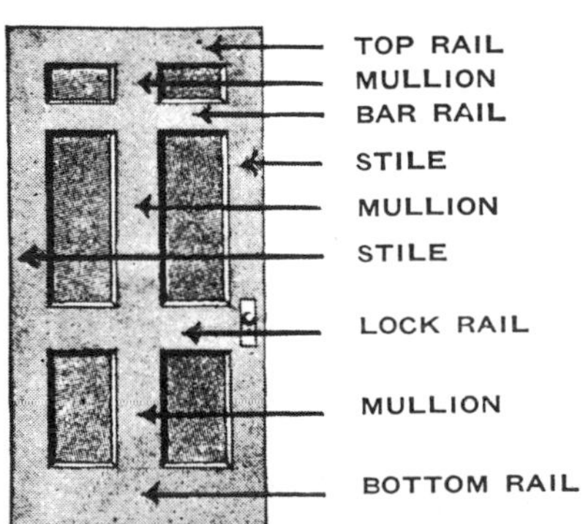

Fig. 28. Parts of a stock interior panel door.

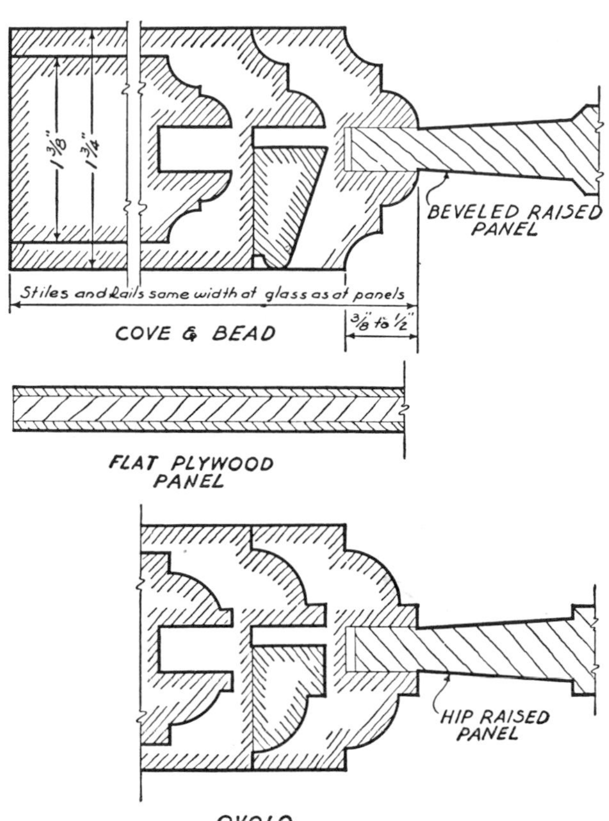

Fig. 29. Sticking and panel details of panel doors.

INTERIOR DOORS

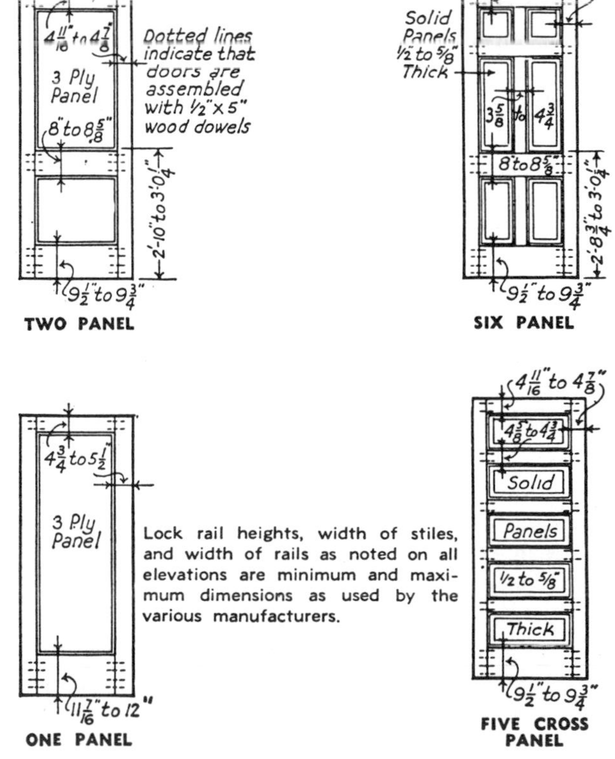

Fig. 30. Specifications of interior doors of various designs.

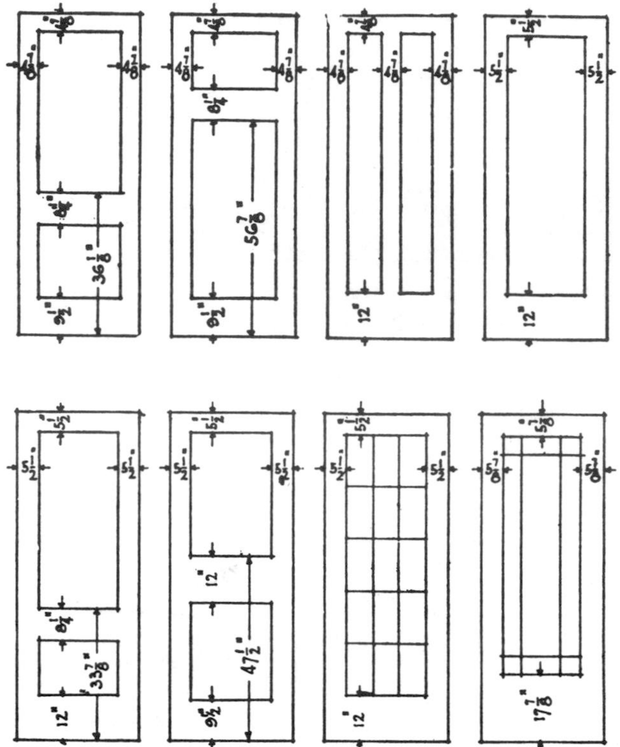

Fig. 31. Typical interior doors showing stiles and rails.

EXTERIOR DOORS

Manufactured in Ponderosa Pine with pine panels as shown on elevations. Moulded B&C, C&B or Ovolo sticking. Standard thickness of doors 1 3/8" or 1 3/4".

STANDARD SIZES

2' 6"x6'6"	2' 10"x6' 10"
2' 8"x6'8"	2' 8"x7' 0"
3' 0"x6'8"	3' 0"x7' 0"

GLASS DIVISIONS

All glass openings in exterior doors can be divided into smaller lights as desired. Usual divisions are:

3 lights wide
4 lights (2 wide—2 high)
6 lights (3 wide—2 high)
9 lights (3 wide—3 high)

Fig. 32. Specifications of exterior glass and panel doors.

must be kiln-dried and the cores made of low-intensity blocks not more than 2-1/2" wide. Stile and rail cores are constructed with blocks running in the direction of the longitudinal dimension of the stile or rail, although some newer styles of flush doors with pine cores run all the core blocks vertically, surrounded with continuous edge strips under the edge veneer, **Figure 31.** Some exterior flush doors are made with a wide edge strip to allow for cutting the width and height of the doors in hanging.

The veneer cross-banding is generally not less than 1/16" thick, nor more than 1/8" thick, and the face veneer is from 1/16" to 1/4" thick before sanding, except where V-grooving is required, when it must be 1/4" thick.

Hollow core doors are about one-third as heavy, in weight, as a solid door. The inner portion may be made of a square grid of thin interlocked wooden strips, **Figure 38.**

Stiles and rails vary in width, but all hollow core doors must have solid lock blocks, in addition to side stiles, long enough to afford freedom in placing lock hardware and hinges.

Door Units For Rough Openings. Most paneled or flush doors are now factory built, with accompanying trim, to be installed in exterior or interior openings. Rough openings vary from 1'-7-3/4" × 6'-10" to 3'1-3/4" × 6'-10" to take units having door sizes from 1'-6" × 6'-8" to 3'-0" × 6'-8".

Fig. 33. A pair of typical ten light French doors.

CASEMENT DOORS

TEN OR FIFTEEN LIGHT CASEMENT DESIGN

Manufactured in Ponderosa Pine with glass as desired. Moulded B&C, C&B, O.G., or Ovolo sticking. Standard thickness of doors 1 $\frac{5}{16}$" and 1 $\frac{11}{16}$".

Interior Casement Doors are also made in any hardwood with veneered stiles and rails and solid division bars.

STANDARD SIZES

4'-0" opening, 2'-0"x6'-8" or 2'-0"x7'-0"
4'-8" opening, 2'-4"x6'-8" or 2'-4"x7'-0"
5'-0" opening, 2'-6"x6'-8" or 2'-6"x7'-0"
5'-4" opening, 2'-8"x6'-8" or 2'-8"x7'-0"

CASEMENT DESIGNS

Casement doors can also be divided into:

8 lights (2 wide—4 high) and
12 lights (3 wide—4 high).

Pairs of casement doors in openings less than 5'-0" wide have 3 $\frac{9}{16}$" stiles as shown while pairs in openings 5'-0" wide and wider have 4¼" stiles.

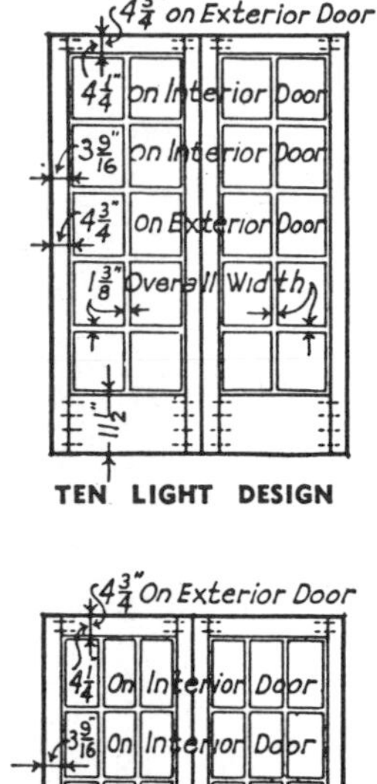

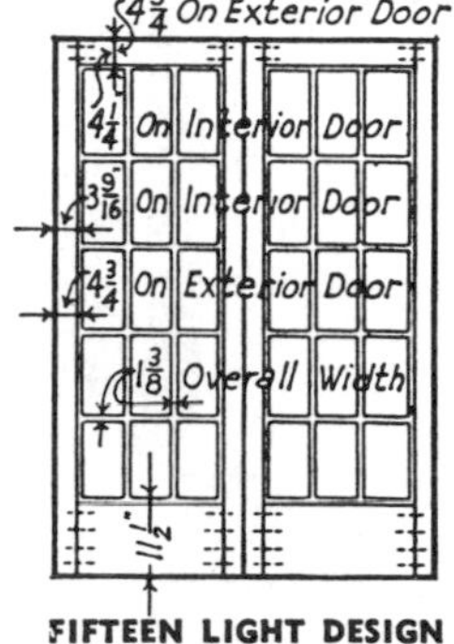

Fig. 34. Specifications of French or casement doors.

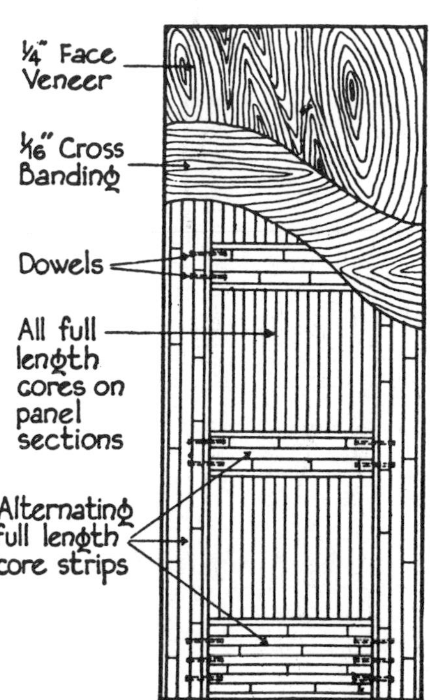

Fig. 36. Solid core flush door having cross-banding and face veneer on each side. Glued blocks are doweled at stiles and rails.

Mohawk Door Co.

Fig. 35. Flush entry doors with a very long narrow light is striking in appearance and provides light for hallway behind it.

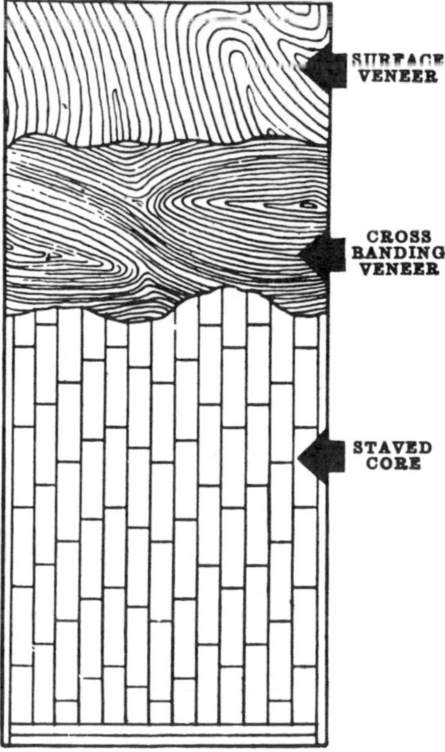

Fig. 37. Flush door with core blocks running in same direction.

Wall thicknesses vary from 3-5/8″ to 4-5/8″ or 5-1/4″ depending upon the type of material used in the wall.

A door unit consists of a door, an especially designed frame made up of a wooden sub-frame on the hinge side to be set and plumbed in the rough opening, and steel enameled jambs, stop and header. The hinge jamb is slipped over the plumbed stud of wood, the header trim and the strike jamb are locked together with countersunk turn screws. The metal jambs are aligned to the proper width, top and bottom, and the jambs are then nailed at the bottom.

The door is mounted on the hinges, which are factory applied on the door and hinge jamb, and the hinge pins dropped in place. The door is factory bored for the lock set, which is selected by the owner or builder.

Wall Framing for Doors

The framing of a wall for a door opening varies with inside or outside walls and whether the wall is a weight bearing component or not. In conventional exterior wall construction, the cripple studs (a), **Figure 39,** are supported by a double 2″ × 6″ lintel, properly nailed to full length studding (b) and supported by addi-

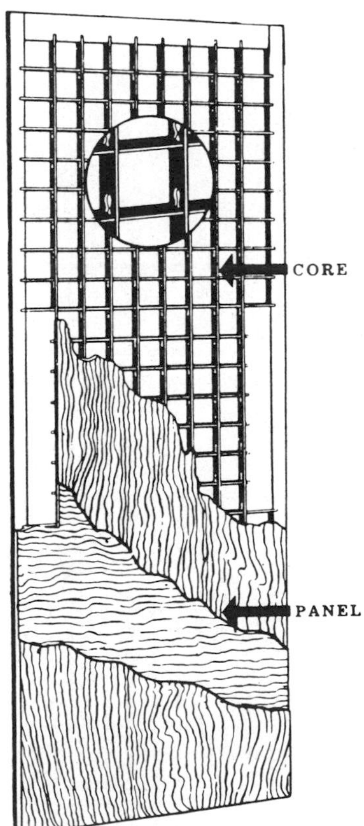

Fig. 38. Hollow core flush door with interlocking grid of wood strips glued into slots in stiles, rails and lock blocks. Framed openings can be had in hollow core doors.

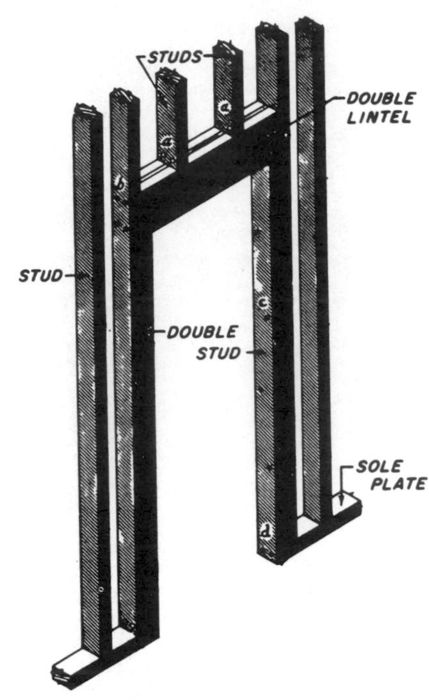

Fig. 39. Framing of door opening with double lintel and studs.

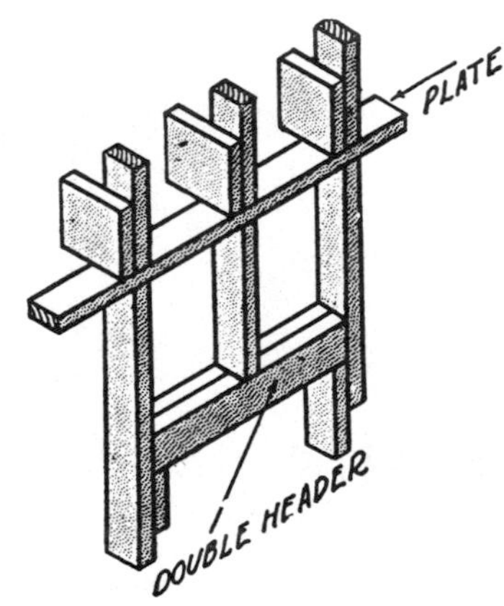

Fig. 40. Double 2″ × 4″ lintel over opening in bearing partition.

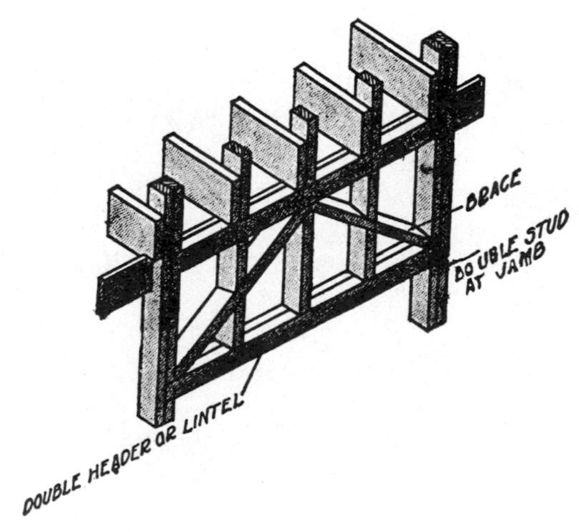

Fig. 41. Trussed header distributes load to jamb studs.

Fig. 42. An aluminum combination door built as part of a factory prehung exterior door unit.

tional studding (c). Ten penny nails are used in this rough framing.

In interior walls, the rough opening in a bearing partition is often framed by using two, 2″ × 4″ members, for the head or lintel, **Figure 40,** or trussing arrangement, **Figure 41.** The weight is supported by a truss brace let into the framing. Often a steel flitch plate is sandwiched between the double headers when the expanse is great. For a large glass sliding door, an angle iron may be used to support the lintel.

Combination Doors

Combination doors, of various and pleasing styles, have a sturdy frame which holds a screen insert in summer or a glazed or paneled insert, as a storm door, in winter. Combination doors are used in connection with regular door units being hinged on the frame of the heavier and more ornate exterior doors.

Sometimes full storm doors are used in place of combination doors, and are available in standard sizes to fit openings. These doors are removable and are hung on pin butt hinges. A full screen door is lighter in weight than a combination or storm door, and is designed to admit the maximum amount of air while excluding insects. One disadvantage is removing, handling and storing a full sized door from season to season.

An Aluminum Combination Door. The modern combination door is made of aluminum, built as an integral part of a standard factory built exterior door unit. The frames have 1″ × 2-1/4″ stiles, top and bottom rails mitered at the corners and blocked inside with

Fig. 43. Colonial style combination doors. Cross buck is fixed, with interchangeable screen-storm panel at top. Louvered panel and screen in lower door can be replaced by storm sash.

corner gussets permanently joined with stainless steel fasteners, **Figure 42.**

This combination door has two glazed storm sash panels and two screen insert panels. The glass panels are glazed with tempered safety gloss glass set in a vinyl glazing channel. The screen panels have a gray, fiberglass screen cloth held in place by removable splines. An adjustable door sweep is attached to the bottom rail, its plastic edge keeps out the weather and compensates for eneven sills. The door is furnished with hinges, push button latch set, a pneumatic door closer and chain stop. Door sizes vary from 2′-8″ × 6′-8″ to 3′-0″ × 6′-8″.

Colonial Style Combination Doors. Colonial styles of combination doors, **Figure 43,** are made of 1-1/8″ thick vertical grain hemlock wood. These doors have doweled joint construction throughout. The stiles, top rail and lock rail are 4-1/2″ wide and the bottom rail is 9-3/8″ wide. The lower section of the upper door has 9/16″ solid raised panels. The glass panel insert is glazed with double strength glass bedded in putty. The screen panel insert is an aluminum screen cloth held in place with flush moulding.

The door at bottom has a scalloped edge glass insert at the top and a louvered bottom insert panel. Both panels are raised slightly above the outside face of the door for a pleasing appearance and shadow line. The stiles, top rail and lock rail are 3-5/8″ wide, and the bottom rail is 7″ wide. The screen inserts have aluminum frames with fiber glass screen cloth. The storm panels have double strength glass set in aluminum frames. Both screen and storm panels are held in place, in the door, by screws.

Fig. 44. Dutch doors hung in an exterior opening. Doors open inward.

Dutch Doors

A door divided horizontally, having independently swinging units, is known as a Dutch door or a double door. Double doors or Dutch doors have two 2′-8″ × 3′-4″ units each hung on hinges. Dutch door units are now factory made and assembled in a frame, suitable for wood framed, masonry or brick veneer walls, with exterior trim casing, 1-3/4″ doors, sill, threshold, weatherstrip and hinges, **Figure 44.** Notching, boring, mortising and fitting are done in the factory. The lock set, handle, escutcheon and interior trim may be ordered with the door set.

Doors in House Design

Doors, in the same manner as windows, add much to the appearance and usefulness of a dwelling. In recent years door areas have increased, together with different types and styles.

Sliding Glass Doors in House Design. Fenestration in the dwelling of today is accomplished by a pleasing arrangement and design of windows and glass doors, making the outdoor scene a part of the indoor decor, **Figures 45 and 46.** Sometimes floor to ceiling, by-passing, sliding or gliding door units are used in a wall, and may be combined with panels of fixed glass. These units are factory built consisting of frames of wood or aluminum housing large door panels of glass. The units have two or three sliding doors. The size of the units vary in height for a rough opening, from 6′-9-7/8″ to 6′-10-3/8″ to a width of 6′-1-1/6″ for two doors to a width of 11′-10-3/4″ for three doors.

Standard glazing is available in 3/16″ plate glass. Glazing is also available in tempered glass, 5/8″ thick, which is heat treated for safety. If tempered glass is broken, pieces fall in harmless, rounded granules. Another method of glazing is in the use of insulating

National Woodwork Mfrs. Assn.

Fig. 45. Sliding glass doors create window wall with a view and fill the room with natural light.

National Woodwork Mfrs. Assn.

Fig. 46. Wide expanse of glass provided by sliding glass doors helps make the outdoors a part of the living room.

glass, 5/8″ thick, which has two sheets of glass bonded together forming a dead-air space between the sheets. Glazing in doors is also available in gray glass, 5/8″ thick, which has a pleasing neutral color thus reducing eye strain caused by harsh sun glare. Gray glass also cuts down heat transmission from the sun.

The door panels operate on ball bearing rollers. Doors are equipped with a heavy duty security latch with a walnut wood handle on the inside and a metal handle on the outside with a cylinder key lock. A large screen is available made of aluminum cloth held in an aluminum frame. The frames are weatherstripped with woven pile and flexible vinyl.

Fig. 47. Bi-fold doors, hinged at center and suspended on track, swing flat against jamb instead of out into room.

Other Types of Doors

Bifold doors may be made of wood, wood and hardboard, or steel. Steel bifold closet doors are a great favorite of builders because they are easy to install, do not warp, have trouble-free hard, baked enamel finishes, and come in a wide variety of styles and sizes. Bifolds, whether wood or steel, may be flush, fully louvered or louvered at the top only, ornamented with moldings in simple designs or have full length mirrors, **Figures 47 and 48.**

Closet doors of this type consist of two panels hinged in the middle. Sizes of panels range from 8″ wide and 6′ 8″ high to 18″ wide by 8′ high. Each pair of panels has 4 pins. The top and bottom pins nearest the jamb ride in pivot bearings locked into the aluminum upper and lower tracks. The top pivot pin is spring loaded to make insertion into the bearing easier. The lower pivot pin is threaded to permit leveling of the door in relation to the tracks and the opposite 2-panel door that rides on the same tracks. The second panel has top and bottom guide pins which slide in the tracks on nylon sliders or the pins themselves are nylon tipped for easy movement and quiet.

Some bifold doors do not have bottom tracks but they are not as stable and quiet as those which do although they are usually less costly. Generally top and bottom tracks are identical aluminum extrusions. At least one manufacturer (Leigh Mfg. Co.) makes one track higher than the other for use on the floor to permit the panels to swing over a thick rug. If there is no rug problem both tracks can be used interchangeably.

Louvered cafe doors are used in entrances between kitchen and dining alcove for artistic effect. These doors are generally made of Ponderosa pine, 1-1/8″ thick,

Leigh Products

Fig. 48. Steel louvered bifold closet door has upper and lower tracks and is finished in white baked enamel.

National Woodwork Mfrs. Assn.

Fig. 49. Ponderosa pine cafe doors swing back and forth on double acting spring hinges.

National Woodwork Mfrs. Assn.

Fig. 50. Louvered bi-passing wood doors hang from an upper track and have a nylon guide on the floor to align them.

National Woodwork Mfrs. Assn.

Fig. 51. Bipassing panel type closet doors of ponderosa pine.

Leigh Products

Fig. 52. Unusual Snap-Fit bipassing closet door makes use of any paneling from ⅛" to ¼". Snap-on metal moldings form the sides and top and bottom of the panel used. Panels are covered with wallpaper with same pattern as cloth of bed.

and hung in pairs, **Figure 49.** A pair of doors will fit openings from 2'-6" to 3'-0". The height of the doors is generally 4'. Each door is hung on large (3") double acting, spring hinges as shown in the insert.

By-Passing Sliding Closet Doors. There are almost no doors that are specifically made for this purpose. They may be made of wood or steel. In wood they may be flush, louvered or paneled and are usually 1-3/8" in thickness, 24" to 36" in width, and 6' 8" to 7' 11" in height. Steel doors are flush, louvered, or given the appearance of paneling by means of moldings. Their dimensions are the same as those for wood except that they are considerably thinner.

The doors are suspended from a double aluminum track which is screwed to the header above the closet opening. Many tracks have their own fascias. Adjustable hangers with metal or plastic wheels are fastened to the backs of the door near the top. The wheels, often ballbearing with nylon tires, ride in the tracks. Each door has 2 sets of wheels and rides in its own track. In most cases the doors overlap by about 1". Many bi-passing doors do not have a bottom track, instead they may have a nylon guide, **Figures 50 and 51,** which aligns both doors. Where there is a bottom track there are metal guides which fit into slots in this track.

An interesting development in this kind of door is the Snap-Fit type which makes use of any paneling from ⅛" to ¼" thick and has no actual frame of its own but functions as a bi-passing door, **Figure 52.** Snap-on metal stiles for the sides of the paneling and metal trim for the top and bottom form the frame for the paneling and hold the roller and guide brackets for the upper and lower tracks. The panels may be hardboard, ¼" plywood, or wood grain paneling and may be papered or painted to match the walls.

Kinkead Industries

Fig. 53. Plastic tub enclosure framed in aluminum moves on upper and lower tracks and opens at either end.

Fig. 54. Opaque glass shower enclosure in metal frame.

Shower or Tub Enclosures. Rigid plastic, sometimes corrugated, and glass are used as bathtub and shower enclosures, usually in metal frames, **Figures 53 and 54.** Frames are fastened to walls at each end of the tub and calked to prevent leakage.

Jalousie Doors. Jalousie doors, similar to jalousie windows, are housed in a frame and hung on hinges in standard openings. The doors have adjustable glass louvers which can be locked in any position. A removable screen of aluminum is placed on the inside of the door. Clear or opaque, 7/32″ glass, is used in the louvers set in an aluminum frame.

Fig. 55. Folding door separates utility room from dining area.

Fig. 56. Folding door of wood slats closes wide opening between family room and living room, providing privacy.

Partition Screens. Folding or accordion doors, are most often made of flexible plastic, over metal expansible frames. These have come into wide use where hinged doors would be unhandy, and where there is no space for a sliding or a pocket door, such as between a dining nook and a utility room, **Figure 55.**

Note the use of wood slat folding doors to separate a living room from a family room, **Figure 65.** They are also used frequently as closet doors, **Figure 57.**

Folding doors or screens are also made of wood, firm plastic, and even bamboo strips, **Figure 58.**

Basement Door. A new style to an old standby is created in the use of a direct access door to the modern basement of the house, **Figure 59.** The modern steel basement door opens onto the rear area of a parcel of property. This type of door can readily be used when the finished grade line is near the top of the foundation wall of the basement, making the basement of a dwelling more accessible for recreation, hobbies and storage.

The stairway is built into the stairwell, **Figure 60,** which provides ready access to the rear yard where recreational facilities may be developed.

A vertical door, in the basement wall at the stairway, gives added protection in the basement against heat loss and condensation, may be fitted with a lock for additional security.

Eastern Roper

Fig. 57. Vinyl covered folding door makes attractive closet door.

Fig. 58. Bamboo strips form a curtain door which turns a corner to close off kitchen, hallway and dining room from each other.

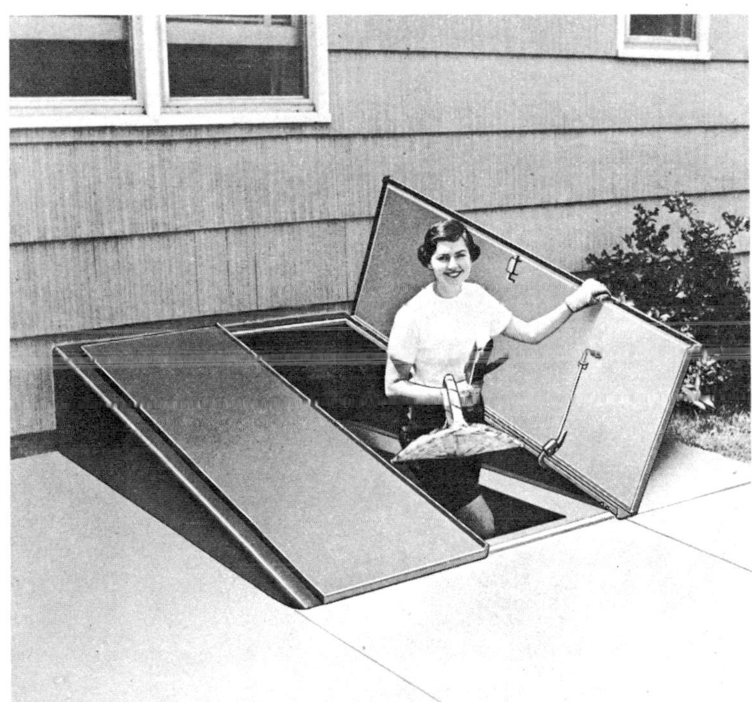

Fig. 59. Steel door over concrete or masonry stairwell provides easy access between basement and rear yard or patio.

Fig. 60. Steel stringers are fitted with wood treads to make sturdy stairway.

Chapter 10

EXTERIOR WALL COVERING

The exterior wall covering is the final protective layer of material attached to the frame or the sheathed or masonry walls of a building. Various materials are used for the exterior wall covering; such as boards in various forms, plywood, shingles, brick, stone, stucco, concrete, aluminum, wood compositions, asphalt, asbestos, plastics, glass, fiberglass, hardboard and steel.

Most exterior surfaces are applied at the building site, including masonry veneer on prefabricated houses; however, some factory built houses have the exterior walls covered and the entire completed wall panel is shipped and hoisted in place.

Wood and Wood Products on Exterior Walls

Wood siding, trim and other exterior woodwork should be kept at least six inches above the ground to avoid decay and attack by termites.

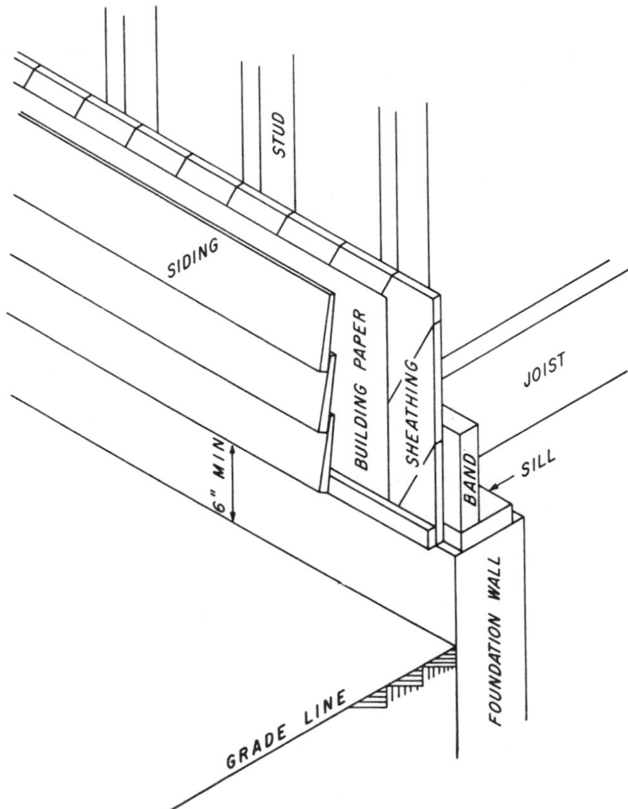

Fig. 1. Plain beveled siding applied over building felt stapled to sheathed exterior wall. Note clearance above grade.

Applying Beveled Siding. In conventional construction, beveled siding is nailed over wood sheathing, **Figure 1.** Note the clearance of at least 6″ above the grade line. This style of siding is also known as clapboard or weatherboard, with a bevel on the face from 1/8″ at the top of the board to 3/8″ at the bottom. The siding is applied horizontally with an overlap of about 1″. The siding should be nailed every 24″ if applied over wood sheathing. If other types of sheathing are used, nails should be driven through the sheathing into the studs at each bearing. The length of the nail will vary with the thickness of the siding and types of sheathing.

The corner treatment of beveled siding depends upon the overall house design and may involve corner boards, mitered corners, metal corners, or alternately lapped corners, **Figure 2.** Mitered corners on the exterior are difficult to make weather tight.

Siding and exterior trim should be installed with corrosion resistant nails, usually galvanized steel or aluminum. Nails should be driven just above the lap to permit possible movement due to the change in moisture conditions. **Figure 3,** shows a method of nailing rabetted bevel siding, using a 6d. nail.

Beveled siding is furnished by the mill in several types. Smooth beveled siding is furnished in clear red cedar, 1/2″ × 7-1/2″ or 3/4″ × 9-1/2″ wide by 6′ through 16′ lengths, with most lengths being 14′ or longer. Rabbeted siding is milled 3/4″ × 7-1/2″, with a 1/2″ rabbet for a 7″ exposure. Random lengths run from 3′ to 13′. Rustic sawn siding has a rough-sawn-texture face, 5/8″ × 9-1/2″ in actual size. Random lengths run from 3′ to 16′. Narrow Cape Cod beveled siding is milled in 1/2″ × 4-1/2″ size with random lengths running from 3′ to 16′ in size. The Williamsburg beaded edge siding is milled 3/4″ × 9-1/2″ in size with random lengths running from 6′ to 16′ in size.

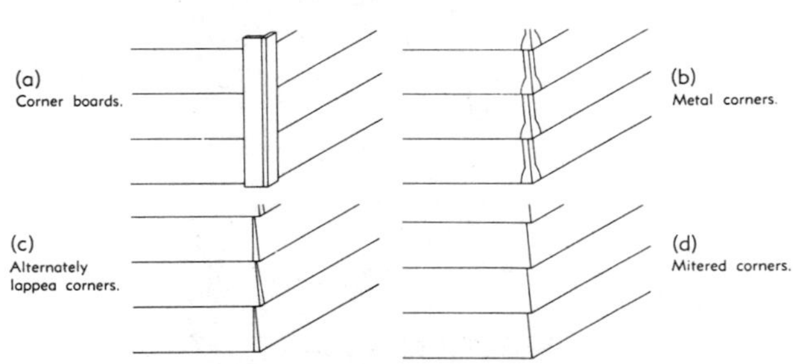

Fig. 2. Four methods of finishing outside corners of beveled siding to close meeting ends against the weather.

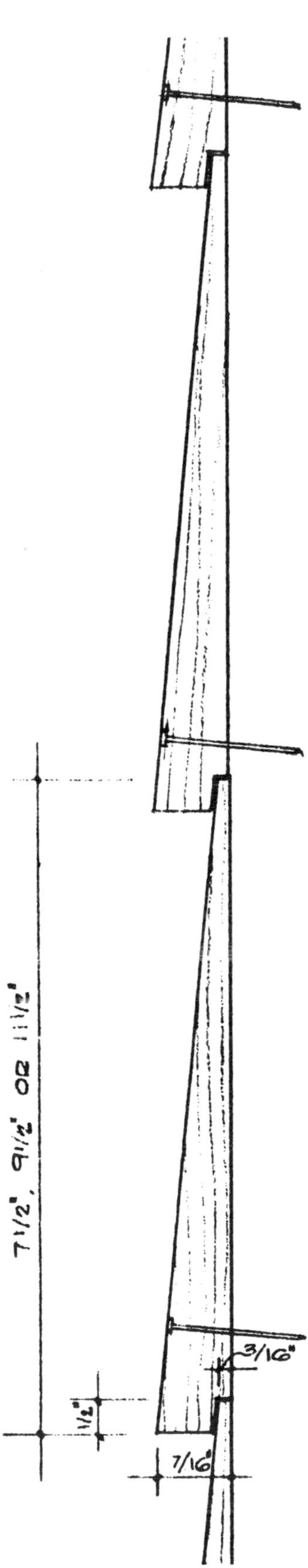

Fig. 3. Nails are driven just above the lap in plain or rabbeted bevel siding to allow for expansion and contraction.

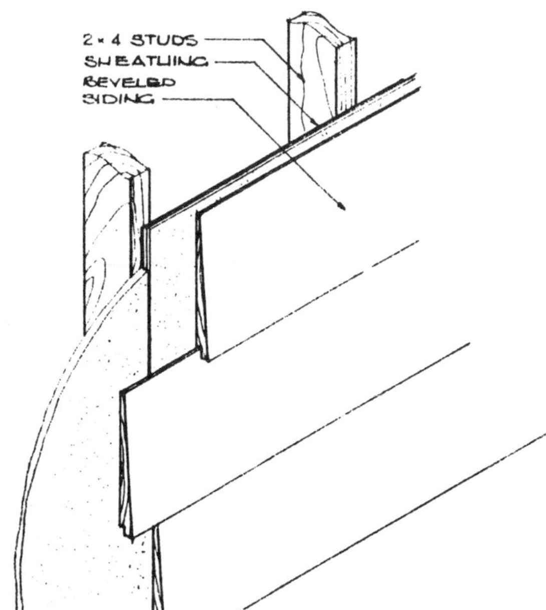

Fig. 4. Bevel siding should be applied over sheathing covered with asphalt saturated building felt. Joints are staggered.

Most siding is given a priming coat of alkyd resin primer at the mill. **Figure 4** shows how horizontal beveled siding is applied when the rabbeted joint is used.

Applying Paint Prefinished Beveled Siding With Siding Fasteners.

Conventional styles of beveled siding, milled and factory prefinished with a complete paint system, is a product of modern technology. This siding is made of kiln dried, vertical grain redwood, to which is applied a primer of alkyd resin and a topcoat of acrylic latex emul-

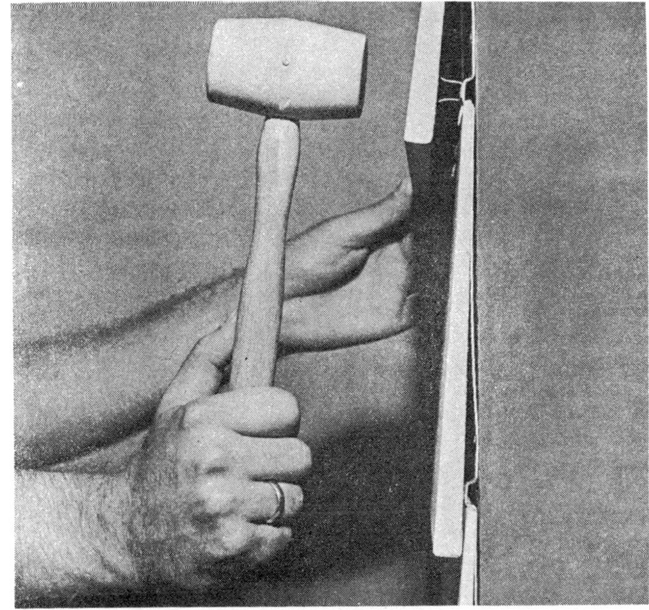

Fig. 5. Factory finished siding system being driven onto spaced prongs of special fastener strips nailed to sheathing.

sion paint. The back face of the siding is sealed with a specially compounded water repellent.

The siding is held in place on the exterior sheathing and sheathing paper by a concealed fastening system, **Figure 5,** which eliminates puncturing of the paint film on the exposed face and provides automatic venting of moisture. The siding is furnished in 3/4″ × 10″ and 5/8″ × 10″ sizes. A white rubber mallet is used to drive the siding on the prongs, **Figure 5.**

The fastener strip is attached to each edge of the exterior wall, **Figure 6,** using 4d or 5d nails. Strips are aligned with a chalk line running from corner to corner, and nailed at each stud to receive the prefinished siding.

The corners are finished with metal corners, **Figure 7.**

Applying Clapboard Siding. Applied horizontally as an exterior wall covering, clapboard siding is heavier than beveled siding, measuring 1/2″ thick and 12″ wide, **Figure 8.** Present day clapboard siding is a manufactured wood fiber board product that looks like wood, works like wood but is denser than western red cedar. This wood fiber product has no internal grain to warp or twist and is free of surface grain. All pieces are straight, flat, square and free from knots and splinters. This wood fiber board can be obtained with factory applied vinyl color coat of paint similar to the prefinished bevel siding.

Nailing can be done along the top edge or bottom edge of the board with 2-3/8″ aluminum sinker nails. Each course should be held by an adhesive caulking applied by a gun along the top edge of each board. The inside and outside corners are trimmed with metal corners nailed in place.

Applying Drop Siding. Drop siding, **Figure 9,** is seldom used in modern dwelling construction but the variety of milled shapes still appeals to some over the conventional beveled pattern. One pattern, **Figure 10** is milled of nominal 1″ stock, 6″ wide, with tongue and groove, and blind nailed by a 6d finish nail. If a wider stock is used, 8″ or 10″, the siding should be

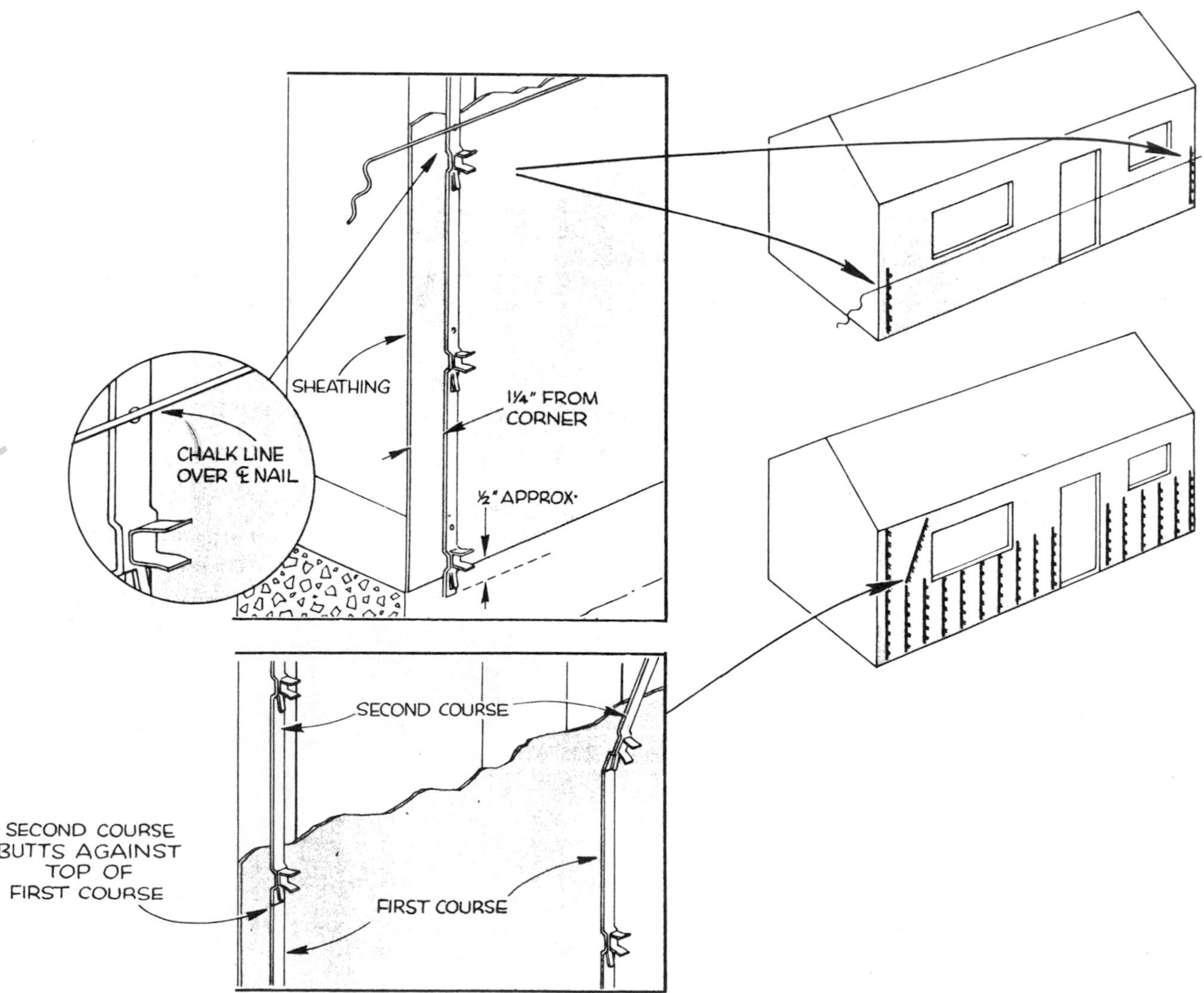

Fig. 6. Pronged fastening strips are leveled and nailed to ends of wall. Inner strips are aligned with end strips.

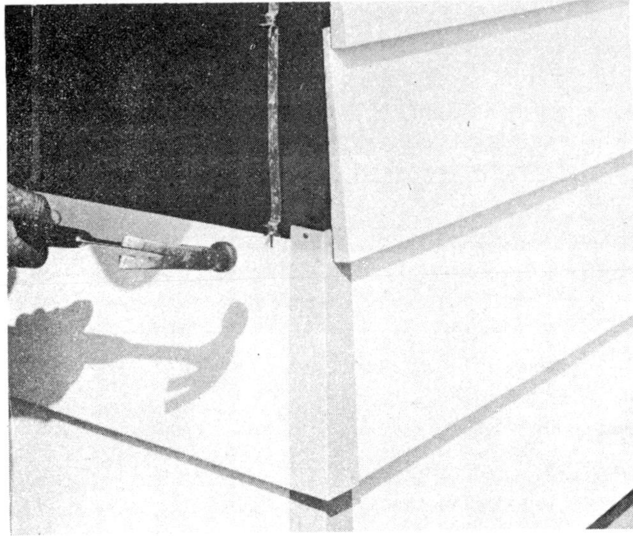

Fig. 7. Metal corners cover meeting ends of siding.

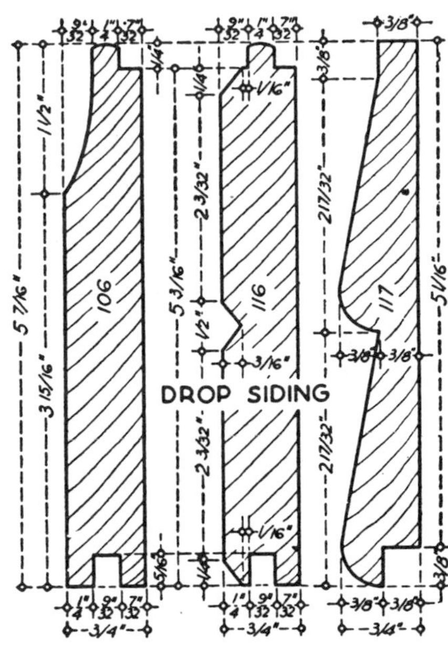

Fig. 9. Drop siding is milled in a variety of shapes which add pleasing texture and shadow lines to exterior walls.

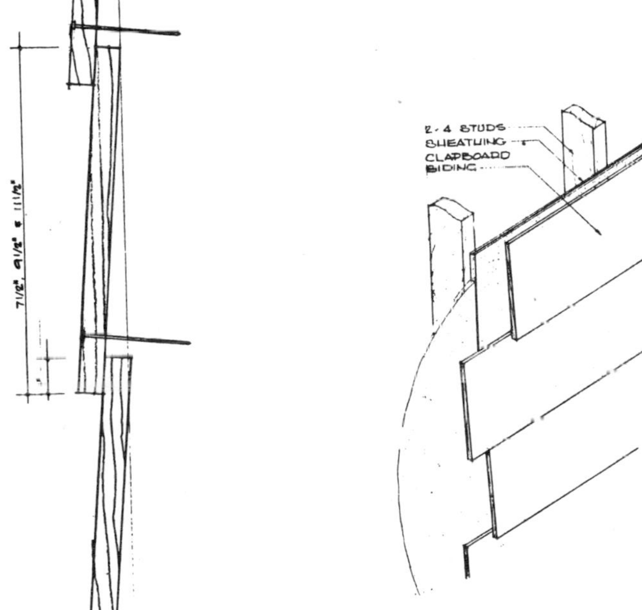

Fig. 8. Clapboard siding is uniform in thickness, usually ½", is applied horizontally, overlapped 1" and nailed as shown.

Fig. 10. Nails can be completely concealed in tongue-and-groove siding, using blind nailing method shown above.

face nailed with two 8d nails.

Applying Horizontal Wood Siding. Horizontal siding of wood, of any type or style, should be applied with care in order to protect the finish and surface and enhance the appearance of the dwelling.

Suggestions for nailing three types of beveled siding are illustrated in **Figure 11.** A water repellent, such as paint or primer, should be applied to siding, **Figure 12.** Ends of siding, cut to length, should be treated with a water repellent, **Figure 13.** The placing of caulking and flashing around window or door frames, **Figure 14,** seals out weather.

A screened ventilator should be installed in siding when the foundation of the house has a crawl space, **Figure 15.** The vent areas should equal at least 1/150th of the floor space of the house. A vapor barrier installed inside the insulation prevents moisture from inside the house from getting to the sheathing and siding, **Figure 16.** Flashing is placed behind the siding and over a concrete porch or patio slab, **Figure 17,** or under the siding and starter strip where a porch roof is attached to the sheathing of an exterior wall, **Figure 18.**

An Exterior Finish of Anzac Siding. One of the most aesthetic of the horizontal wood siding materials is the Redwood Anzac beveled siding derived from a New Zealand pattern. Anzac siding has a relatively thick

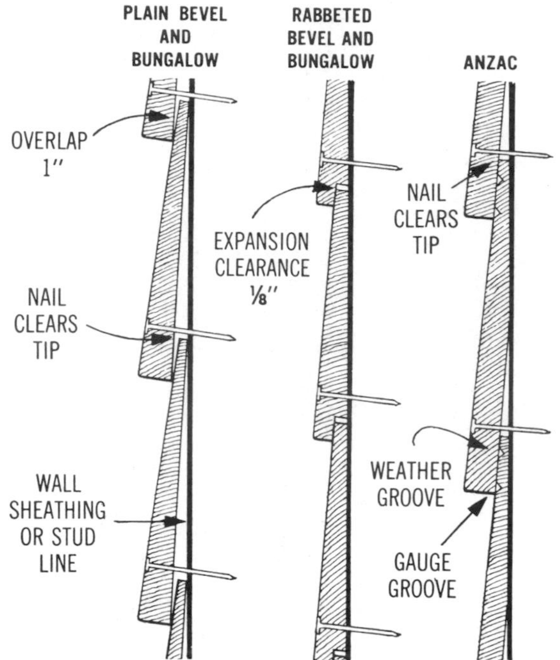

Fig. 11. Three types of beveled siding and method of nailing.

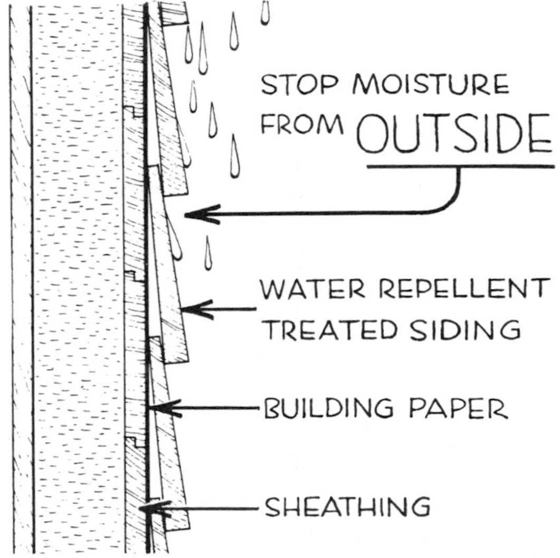

Fig. 12. Water repellent helps siding shed moisture.

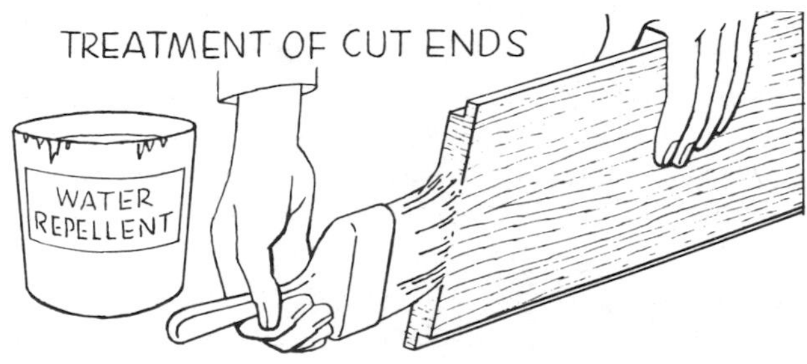

Fig. 13. Cut ends of siding should be sealed with repellent.

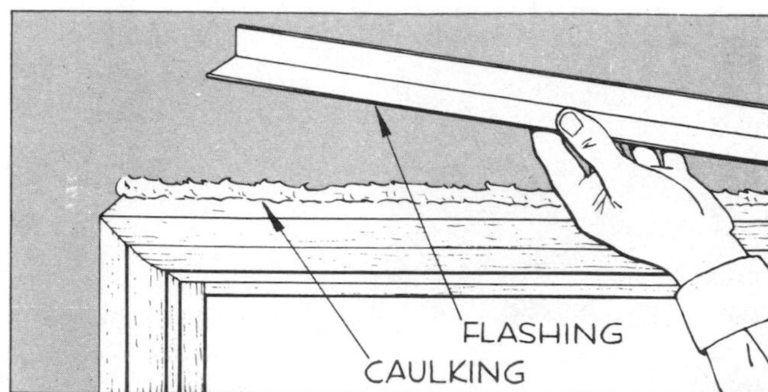

Fig. 14. Metal flashing and caulking prevents water from seeping behind wood frames of doors and windows.

butt edge, providing a heavy shadow line under each course, **Figure 19.** On the face of the pattern are two grooves. The deep, upper groove acts as a water barrier, preventing the movement of wind-driven rain up under each course. The lower groove is a guide line which aids in laying the siding properly and quickly. The reverse side is machined so that each piece will lie flat against the exterior wall sheathing.

Applying Vertical Siding. Contemporary or traditional styles of architecture can well be expressed in the use of board and batten or vertical siding on house exteriors. Santa Rosa siding, consisting of nominal 1/2″ vertical boards laid under 1″ standard boards, **Figure 20,** gives the appearance of a reverse board-and-batten wall covering.

Traditionally, a batten is a relatively thin, narrow strip of wood stock used to seal or reinforce a joint between wider pieces of wood stock, **Figure 21.** True board-and-batten siding is made of nominal 1″ × 4″ to 12″ nominal cedar stock nailed to exterior wall sheathing and protected by nominal 1″ × 2″ cedar stock nailed over joints with 10d nails. When square edge boards are used with battens, the boards are spaced about 1/2″ apart and nailed only at centers. The batten is attached by one nail driven through the center so that it passes between the boards. This permits movement with change in moisture conditions. When other than wood sheathing is used, blocking should be placed between wall studs to permit nailing of the vertical siding.

Fig. 15. A screened ventilator should be installed in the siding below floor level when house is built on crawl space.

Applying Horizontal and Vertical Siding. Occasionally horizontal and vertical siding may be used on the same exterior wall, especially to break a wide expanse of one style of siding or at a floor level, **Figure 22.** Building paper, drip cap, and flashing are used where tongue and groove V joint vertical siding is placed above a plain pattern of horizontal bevel siding. The same care must be taken in building up exterior walls where brick, metal or wood products are used in portions of the walls.

Other Styles of Solid Wood Vertical Siding. Solid wood siding provides a relatively high degree of insulation. In insulating value a redwood board of 1″ thickness is equal to 6″ of brick; 9-1/2″ of cement block; and, 15″ of cement or stucco, thus keeping heating and air conditioning expenses at a minumum. **Figure 23** shows a pattern of vertical siding known as inland rustic siding. **Figure 24** illustrates shiplap and rustic tongue and groove solid wood siding of redwood, cedar or cypress wood. **Figure 25** shows the use of tongue and groove pecky cypress of random widths placed vertically on a one floor plan slab dwelling. Note the carport and roof decks supported by steel columns. A prefabricated house with an exterior of rough-sawn cedar boards and battens is illustrated in **Figure 26.**

Siding Made of Wood Products

In addition to the wood fiber board siding, a number of present day siding materials are made of plywood. One such siding material is a medium-density resin-impregnated fiber, permanently bonded to the plywood in factory hot-plate presses. Pre-cut siding, of overlaid fir plywood is available in stock, 12″, 16″ and 24″ widths standard panels 4′ to 8′ in length. Pre-cut fir plywood panels may be used as horizontal exterior wall covering, **Figure 27.** Standard panels are also available in 4′ × 8′ in size, and may be applied as exterior siding, with battens nailed on the surface to create a board-and-batten effect, **Figure 28.**

Textured Plywood Sidings. Several styles and types of plywood having a minimum thickness of 3/8″, 48″ wide and furnished in lengths of 8′, 9′ and 10′ are now manufactured for direct studding application in exterior wall covering in house construction, **Figure 29.**

Overlaid plywood siding provides a surface designed as a base for superior paint finishes. The plywood is for exterior use being bonded with 100% waterproof phenolic resin glue and cellulose fibers pressed to a thickness of .014″. The overlay may be bonded to one or both sides of the panel.

Some plywood panels, 5/16″ thick, are manufactured for use over outside wall sheathing.

Figure 30 shows a heavy, 5/8″ thick, plywood panel for exterior use, scored with parallel grooves,

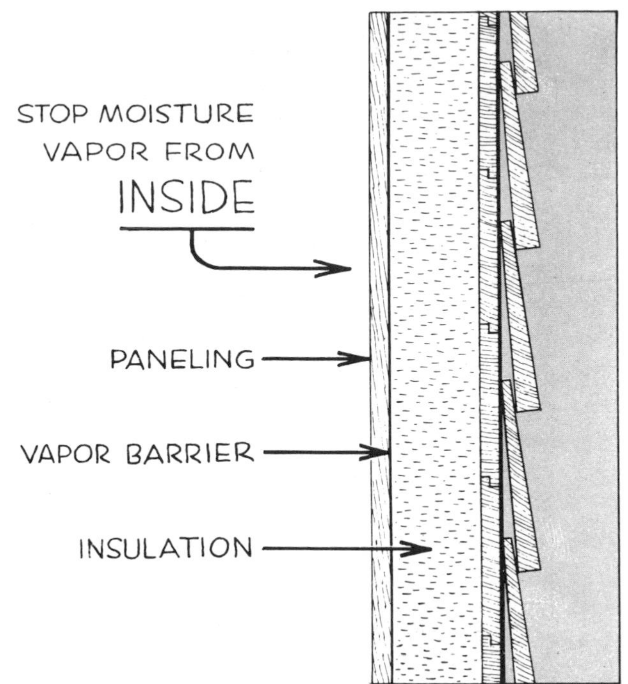

Fig. 16. Vapor barrier prevents moisture from passing through inside wall and condensing in insulation or behind siding.

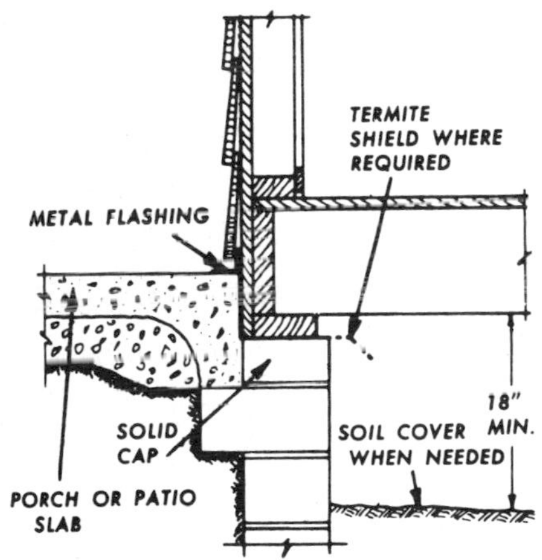

Fig. 17. Metal flashing protects sheathing and framing from water which might seep behind concrete stoop or patio.

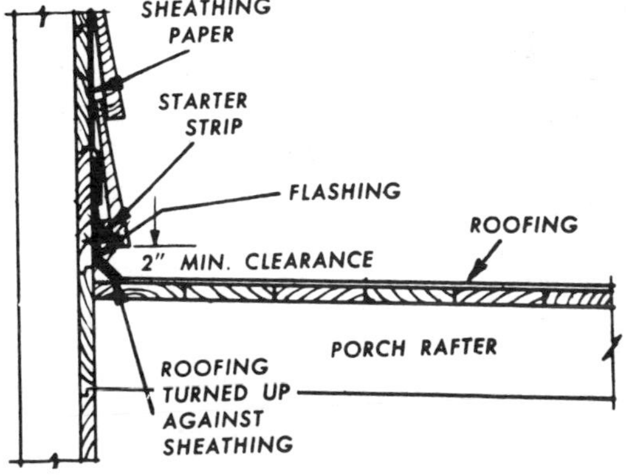

Fig. 18. Flashing seals joint at wall and low roof.

Fig. 19. Redwood Anzac beveled siding is made with a thick butt edge which creates heavy horizontal shadow lines.

Fig. 20. Santa Rosa redwood, bleached to a driftwood gray, is applied vertically in reverse board-and-batten pattern.

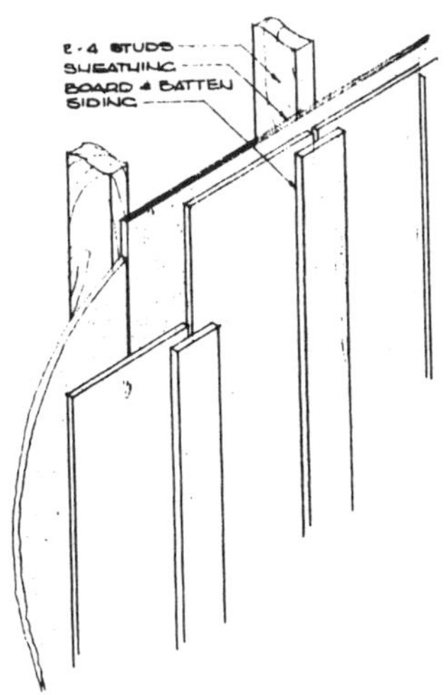

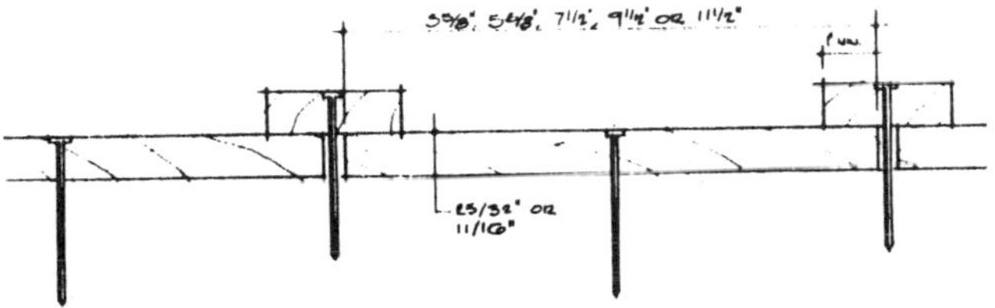

Fig. 21. Board-and-batten siding is applied vertically with narrow battens overlapping boards and nailed as shown.

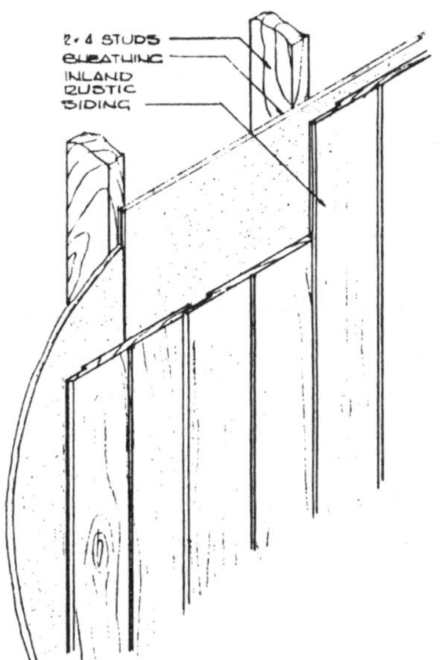

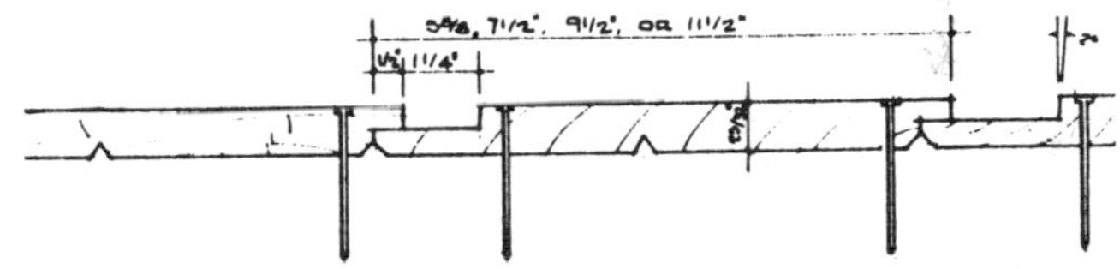

Fig. 23. Inland rustic siding may be applied with V-groove exposed (top) or reversed with lap and channel exposed (bottom).

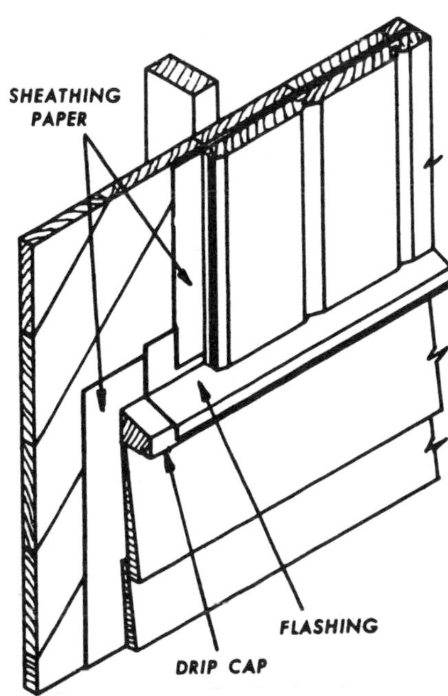

Fig. 22. Drip cap and flashing are used where vertical and horizontal siding join, or where siding is applied over brick.

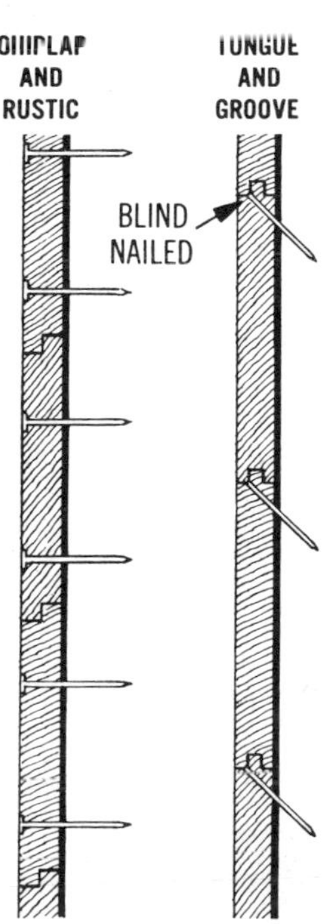

Fig. 24. Weathertight joints in face-nailed shiplap siding (left) and blind-nailed T & G siding (right) are shown in plan view.

Fig. 25. Tongue-and-grooved pecky cypress in random widths adds interest and texture to exterior walls.

Fig. 26. Rough-sawn cedar board-and-batten siding contrasts horizontal louvers in vent installed in gable end.

Fig. 27. Workman applies pre-cut bonded plywood siding.

Fig. 28. Dummy battens are evenly spaced between battens covering joints in plywood siding for board-and-batten effect.

either 2″ or 4″ on center, being set in place on an exterior wall.

Striated fir plywood panels are set and nailed in place vertically, **Figure 31.** Note the pre-cut opening to fit around a window frame.

Plywood Sidings Having a Polyvinyl Fluoride Film. In addition to wood fiber board and textured plywood for walls, modern technology has produced a siding material available in panels and patterns which are covered with durable polyvinyl fluoride film, in colors, and in many styles, such as 3/8″ × 48″ flat plywood panels, 3/8″ × 12″ lap siding, and a 3/8″ × 48″ vertical batten plywood panel with shiplap joints, **Figure 32.**

Wood Shakes and Shingles. Wood shakes and shingles are applied as siding in much the same manner as they are applied on roofs, except that there can be greater exposure when applied to vertical surfaces. A

Fig. 29. Five types of textured plywood siding.

Textured reverse board and batten
Wide grooves cut into wire-brushed, coarse-sanded, or natural textured surface. Long edges are shiplapped for continuous application and weather seal without battens.

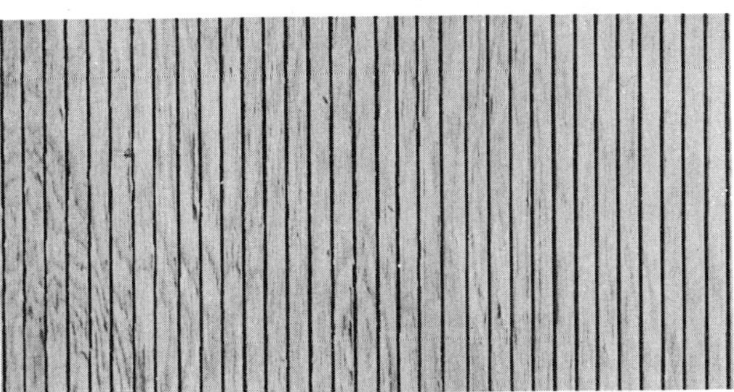

Pin-striped surface
Fine grooves cut into surface to provide distinctive striped effect. Grooving reduces surface checking, adds durability.

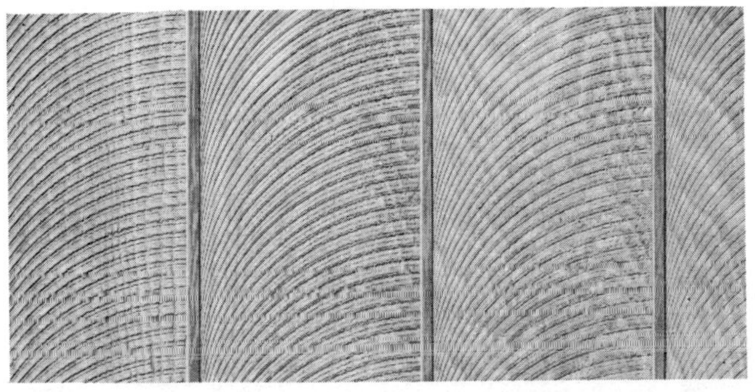

Textured plank
Rough-sawn sections create a vertical pattern with grooves spaced 8" apart. Long edges are shiplapped for continuous application. Especially suitable for exterior stains.

Striated
Closely spaced grooves of random width conceal joints between meeting panels as well as nailbeds. Finished with exterior paint or pigmented stain.

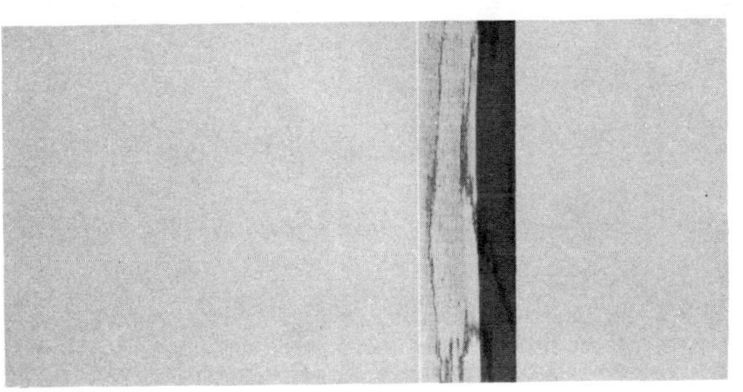

Overlaid reverse board and batten
Wide grooves cut into medium density overlaid surface, which facilitates a quality paint job. Grooves provide deep shadow lines. Long edges are shiplapped for continuous application.

Fig. 30. Grooved panel for use as exterior siding is bonded by waterproof glue. Edges are shiplapped.

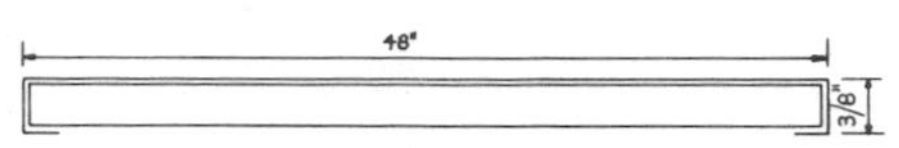

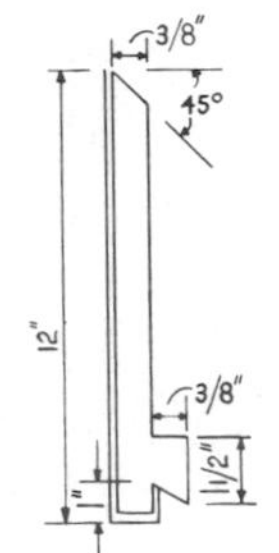

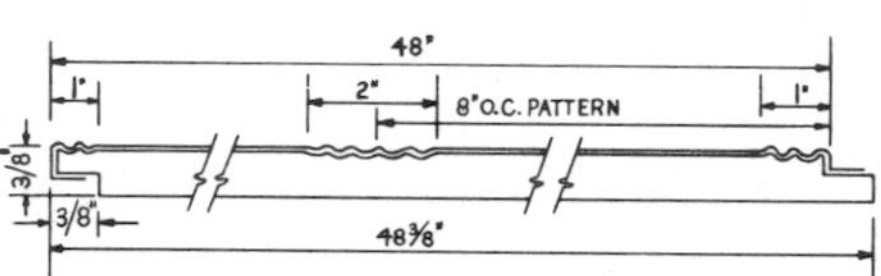

Fig. 32. Sections of panel siding, interlocking lap siding and pattern siding surfaced with Tedlar polyvinyl fluoride film.

Fig. 31. A pin-striped plywood panel is set in place after being cut to fit around window frame.

Fig. 33. House is covered quickly with barn shake shingles bonded to fiber backer board 46-3/4" long.

heavy waterproof building paper is placed over the sheathing; the starting course is doubled, and fitted snugly against the water table, which serves to carry dripping water away from the foundation. The corners may butt against corner boards or may be overlapped or laced at alternate courses.

Hand split, resawn, cedar shakes may be used on roofs and sidewalls of new or existing buildings. Most

Fig. 34. Applying new 18" red cedar shingles over deteriorated clapboard siding. Bottom course is doubled, then a straightedge is tacked along exposure line for easy positioning. Although not necessary, aluminum-foil-surfaced paper is stapled over old siding.

shakes are available in natural cedar color measuring 1/2'' to 3/4'' in thickness to 24'' in length, packed 18 courses to the carton and five cartons equal 100 square feet of coverage. If used on roofs a 10'' exposure is given and if used on side walls an 11-1/2'' exposure is given.

Barn shake panels are cedar shakes bonded electronically to a fiber backer board, then trimmed to an overall length of 46-3/4'' and a height of 16'' to 16-3/4'', with a staggered edge. Eleven panels, in a carton, will cover 50 square feet with a 14'' exposure, **Figure 33.**

Existing walls can be recovered with 18'' red cedar shingles or shakes, **Figure 34.** The bottom course is doubled then a straight edge is laid along the line for easy positioning of upper courses. Aluminum foil is tacked between the old and new siding, although not necessary for good application. The bottom three courses on the gable end wall have been painted.

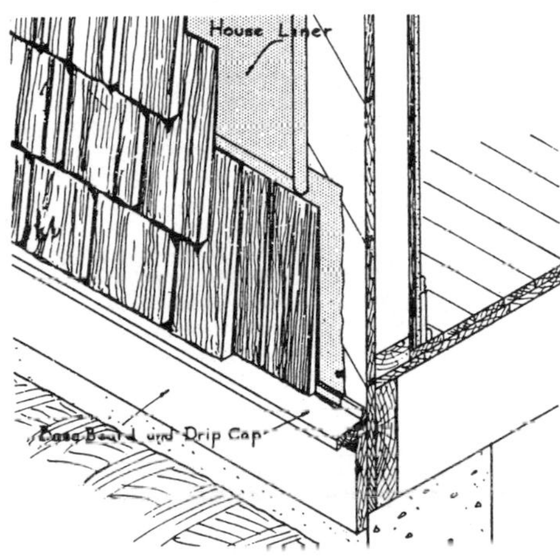

Fig. 35. Hand-split shakes applied over building paper. Starting course is doubled, fitted snug against drip cap.

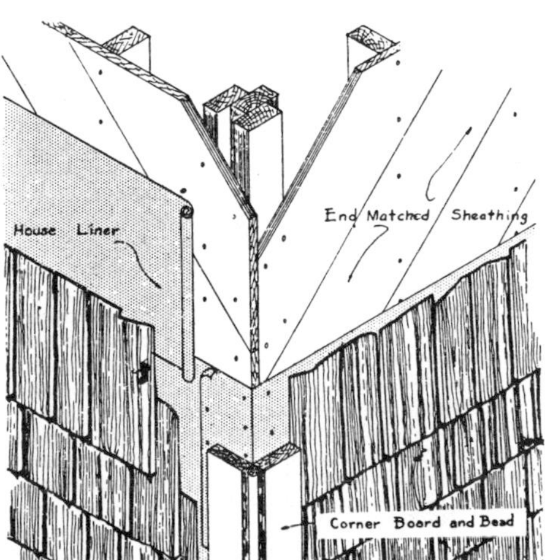

Fig. 36. Shingles are nailed under the lap with two 6d zinc-coated nails and fitted snug against corner boards.

Hand split shakes are applied over a heavy building paper, **Figure 35.** The starting course is double, with the first course fitted snugly against the drip cap.

Shingles may be fitted against a corner board and bead, **Figure 36,** with each shingle nailed under the lap with not less than two 6d zinc coated nails.

Stucco and Concrete Walls

In favorable climates concrete block is used in dwellings, in both foundation and superstructure walls, **Figure 37.** The porous surface is usually coated with mortar to present a stuccoed appearance. In wet climates, waterproofing of the foundations with asphalt coatings or pitch and felt is necessary, as well as installation of drain tile surrounding the footings.

Stucco is applied as a finish coat to both existing houses and new houses. It can be finished to give a number of interesting textures to conform with either traditional architectural styling. Pigments can be added to the stucco mix to produce many different colors. If deeper colors are desired, the stucco can be painted in the same manner as block.

Stucco can be applied over block, masonry, or frame construction. It is usually applied directly to block without metal reinforcement, **Figure 38.** Reinforcement is necessary when stucco is applied over wood or glazed masonry and in unusual concrete masonry conditions such as on painted block walls. Metal reinforcement should be made of corrosion-resistant or galvanized metal, or treated with rust inhibitor. The openings should not exceed 4'' square in area and the weight of the reinforcement should be at least 1.8 pounds per square yard. Metal reinforcement should be held 1/4'' from the surface of the wall with special nails or other type of anchor.

Drip caps and flashing should be used with stucco over concrete block, **Figure 39.** Stucco can be applied over studs without exterior sheathing by nailing wire mesh over heavy building felt, **Figure 40.**

Application of Stucco. Stucco is applied in three coats. If the masonry wall is extremely dry, it should be sprinkled lightly and uniformly with a water spray to control suction by the wall. The first coat is 3/8'' thick and is called the "scratch coat". The scratch coat is raked or combed to produce a rough texture for the bonding of the second coat. The wire reinforcement is embedded in the scratch coat. The second coat also 3/8'' thick, is called the brown coat. It is usually applied 8 to 24 hours after the scratch coat is put on, although it can be applied as early as four hours. The brown coat is finished with a wood or light metal float. The surface must be true and even, since the third coat is but 1/8'' thick and will not cover large imperfections and uneven areas. The third or finish coat should not be applied until the brown coat is 48 hours old.

Mixes to Use. The same stucco is used for the scratch and brown coats. It consists of 1 part of Portland cement, up to 1/4'' part of type S hydrated lime, and 3 to 5 parts of masonry sand. Enough water is added to give the mix the desired workability. The mix for the finish coat usually consists of 1 part of white or gray Portland cement, up to 1/4 part of type S hydrated lime, and 2 to 3 parts of masonry sand, to which mineral color pigments may be added.

A factory prepared finish coat mix is recommended. Some companies blend and sell a packaged finish coat mix containing all necessary ingredients including sand. These mixes are available in many colors. Light colored or white sand is recommended for the finish coat mix.

When colored stucco is desired, mineral pigments are added to the mix. Follow the pigment manufacturer's instructions closely. The mineral pigments must be carefully measured and mixed to ensure uniformly colored batches.

Curing the Stucco Coats. The scratch coat must be cured to strengthen it and to reduce the possibility of shrinkage cracking. If the scratch coat stands longer than 8 to 12 hours before application of the brown coat, it should be cured. Curing is done by fogging lightly and uniformly with a fine water spray. Fogging is done at frequent intervals for 48 hours or until the brown coat is applied. The brown coat is cured in the same manner for 48 hours after it is applied. The finish coat need be cured only when there is excessive drying. Be careful not to apply the water too early or with too much pressure.

"Artificial Stone". Artificial or simulated stone is a concrete product widely used as a veneer, most often to resurface older buildings. Sometimes the stone is molded at the site, then cemented to the scratch coat previously applied to the prepared wall surface; or the molds are applied directly to the wall. Color is applied to the material to simulate various types of stone.

Sometimes cast blocks, about the size of bricks, are broken into uniform sizes, but with a rough exterior surface to provide an attractive textured exterior veneer, planter, or fireplace material, **Figures 41 and 42.** Roman rough is the name of one type of broken block.

Asbestos-Cement Siding

Asbestos-cement products are made of portland cement and asbestos fibers. Asbestos cement siding and roofing materials, pipe, sandwich panels, corrugated sheets, and perforated sheets are used widely in house construction. Asbestos-cement siding is available in a wide range of attractive colors attained by using special paints and

Fig. 37. Concrete block walls are laid up on poured footings.

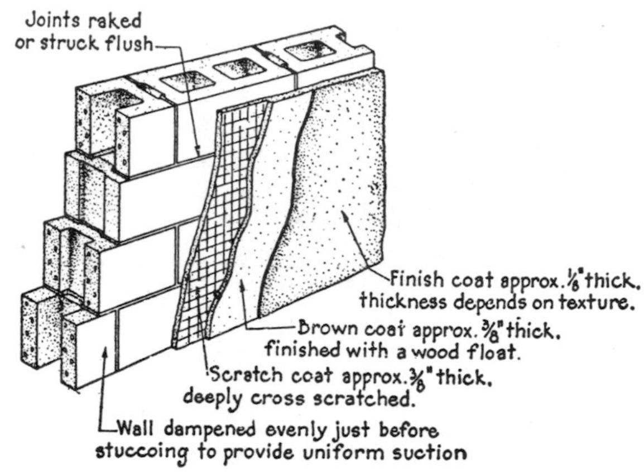

Fig. 38. Stucco is applied in three coats over concrete block.

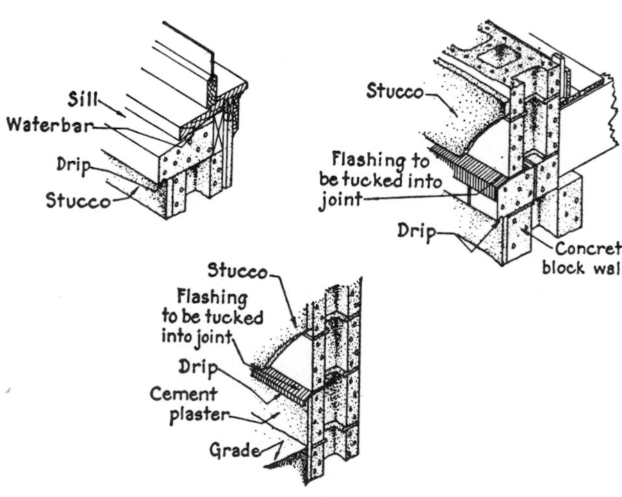

Fig. 39. Drips caps and flashing close edges of stucco.

Fig. 40. Stucco can be applied over studs by stretching wire mesh tightly over heavy building felt.

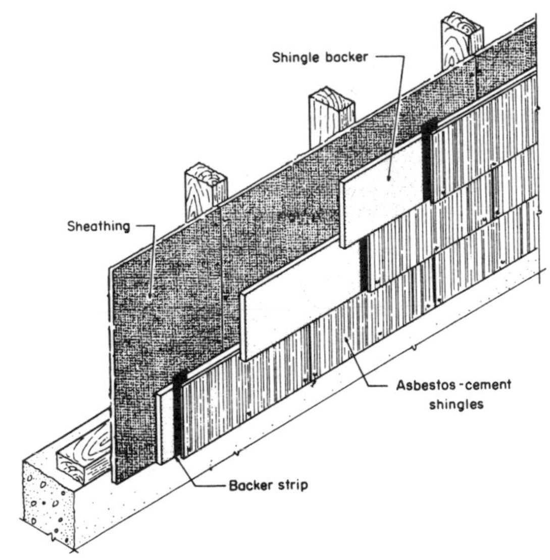

Fig. 43. Thin asbestos-cement shingles are nailed over backer boards for heavier shadow lines. Shingles are predrilled for nails.

Fig. 41. Rough edges of broken block add texture to masonry walls. Note glass block window mortared into masonry.

Fig. 42. Upper third of walls is light colored broken block laid up over dark brick lower wall. Planter is light block.

plastic coverings. They have a long life and excellent color-fastness.

Asbestos-cement is fireproof, durable and easy to work. It is vermin, rodent and termite proof. It does not corrode, requires little maintenance, and is a non-conductor of electricity. The shrinkage of asbestos-cement sheets is remarkably small and the joints between individual pieces stay tight.

Applying Asbestos-Cement Siding. Asbestos-cement shingles should have a solid backing. On new construction the shingles are nailed directly over wood sheathing or insulation sheathing. Backer strips are used to keep out water. To accent the lines of the siding and to give a deep shadow line, 1/8″ to 1/4″ strips, are placed under the bottom of each shingle, **Figure 43.** These strips eliminate the need to use special fasteners when applying asbestos-cement over the softer backing boards.

Some siding units are pre-drilled while others can be nailed without drilling. When applying asbestos-cement over existing siding, it is necessary to use furring strips to obtain an even surface.

Board and batten construction is ideal with this type of surfacing. Flat sheets are fastened to the wall and vertical batten strips are nailed over them. The sheets are available from 2′ × 4′ and from 4′ × 12′ in size. However, asbestos cement shingles have declined in popularity and are no longer used as frequently as in the past.

Simulated Stone or Brick Siding

Applying Fiberglass Simulated Masonry Panels. A fiberglass exterior wall covering is available in 4′ × 8′ panels and is applied with 1-1/2″ galvanized nails in the same manner as plywood panels are fastened to sheathing or studding.

Fig. 44. Panels of fiberglass simulating brick and fieldstone may be used as an interior or exterior wall covering.

Fiber glass panels are available in random rubble, field stone, colonial brick, split face stone, etc., patterns, **Figure 44.** Fiber glass masonry has the color, texture and feel of real masonry, will not warp, crack, fade or peel. Cutting is done with a metal cutting saw. The nail heads and joints are sealed with a compound similar to fiber glass, a hardener, mortar paint and dusting powder.

Stone as an Exterior

Stone continues to be popular due to its rich texture and its symbolism of solidity and wealth. It is commonly applied as a veneer over standard framing, in combination with other surfaces, **Figure 45.**

Ashlar stone, cut with rectangular corners, is the most common shape. Rubble stone, having unshaped corners, is seldom used in ordinary house construction, but may be used in a feature wall for dramatic effect or to create a rustic appearance, **Figure 46.** Stone is common in planters. Often a fireplace wall is decorated with an indoor planter. Sometimes a stone accent wall is carried through an exterior glass wall, so that it is difficult to tell inside from outside.

Brick as an Exterior

Brick maintains its popularity as a dignified and confidence-inspring exterior surface, although most present day "brick" houses are really a single-width

Fig. 45. Ashlar stone, cut into rectangular shapes, is commonly applied as a veneer over standard wood-framed walls.

Fig. 46. Rubble stone, of natural random shapes, often is used in a feature wall to create a rustic appearance.

veneer laid outside a standard stud-and-sheathing frame, **Figure 47.** This brick veneer or facing lends itself to interesting combinations with other surface materials, especially wood, **Figure 50.** Many new houses are enhanced by a veneer of old brick salvaged from demolished structures.

Brick veneer provides an economical means of rehabilitating an older structure or facing a new one at less cost than an all masonry wall. Most of the details on framing round openings, flashing, and anchoring shown in **Figure 49** apply equally to new construction and remodeling. In areas subject to heavy rainfall and high winds, continuous flashing should be installed at the base of the wall, just above the grade line. Weep holes should be provided to permit moisture that might penetrate the facing to drain to the outside.

SCR brick, developed by the Structural Clay Research Foundation, offers new opportunities for building with brick. Its nominal 6″ width (the actual dimensions are 2-3/16″ × 5-1/2″ × 11-1/2″) permits its acceptance under nationally recognized codes as the main load bearing wall for one-story, single family dwellings where the wall height does not exceed 9′ to the eaves or 15′ to the peak of the gables. It requires 450 units to cover 100 square feet of wall area. Bed and head mortar joints are 1/2″ thick.

The unit is vertically cored with 10 holes. Nominal 2″ × 2″ furring is used with the brick, attached by clips. The furring permits easy installation of electrical facilities and application of blanket insulation.

Cavity Wall Brick Construction. Cavity wall brick construction is less common than it used to be, one drawback being the difficulty of adequate insulation. One solution to this problem is to have the cavity wall composed of an exterior course of brick and tile, and the cavity filled with loose blown mineral wool insula-

Fig. 47. Laying brick veneer on exterior walls. Note the vertical siding of wood between the two front windows.

Fig. 48. Brick veneer in combination with board-and-batten siding.

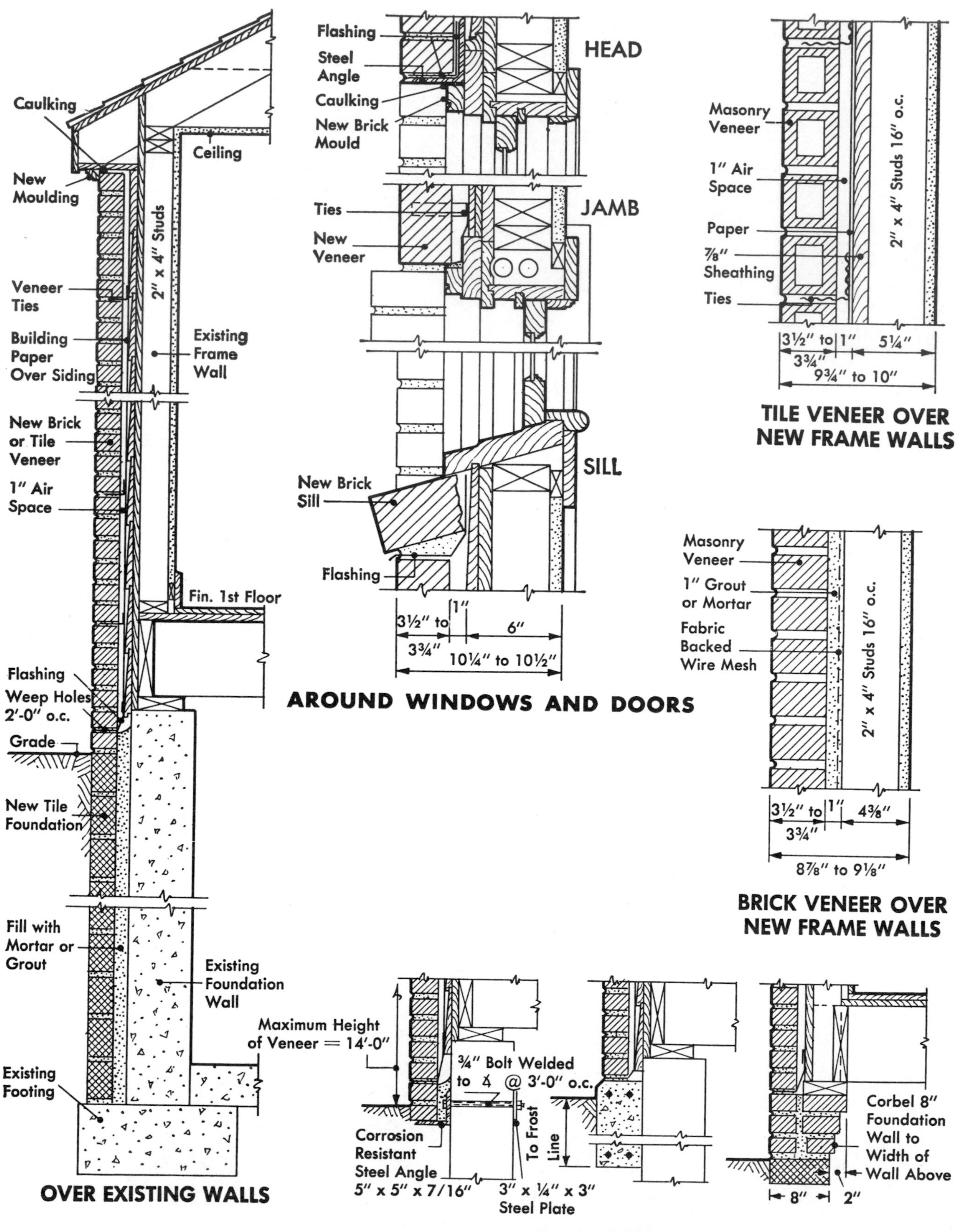

Fig. 49. Typical construction details for laying up brick veneer over framed walls and ground openings for doors and windows.

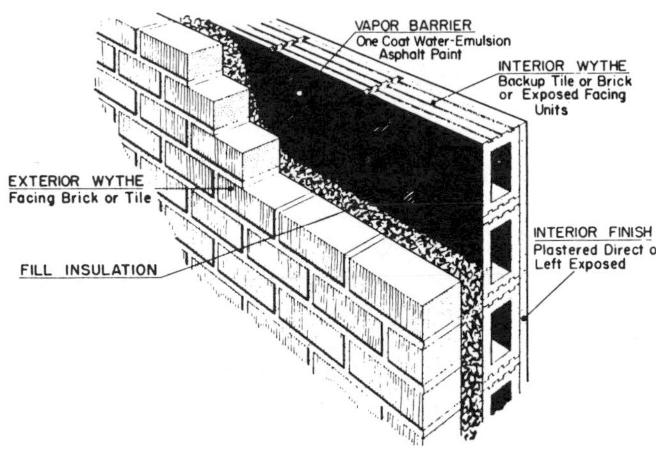

Fig. 50. Masonry cavity wall, with water-resistant fill insulation between outer brick wythe and inner tile wythe.

tion, **Figure 50.**

Tile walls. Hollow tile may be used for partitions and loadbearing exterior walls, **Figure 51.** Face brick is laid directly against the tile. Plastering is possible directly on the inner wall, without furring.

Panels of Thin Brick. Panels of thin brick are nailed to the sheathing, to the existing siding, or to concrete block. Mortar may be applied for appearance. The nails are driven into the holes in the "brick" units, or to clamps, **Figures 52 and 53.**

Applying Vermiculite Brick. Modern technology has provided the building industry with a special formula featuring a mica product of light weight vermiculite. It is a non-inflammable brick like product applied with mastic adhesive. Unlike fiber glass panels, this product is made in single brick units, 3/8″ × 3″ × 12″ in size. After the bricks are set in place the mortar line is brushed with an adhesive to form mortar. The brick can be sawed and filed, **Figure 54.**

Metal as an Exterior

Steel and aluminum are used as exterior wall covering of sheathed studded walls in patterns resembling siding of wood, **Figure 55.** The shaped, steel base is given a zinc coating, a bonding coating and a finish baked enamel coating. The shaped aluminum base is coated with a double baked finish. Both steel and aluminum siding are available in white and colors.

Aluminum Siding Backed With Polystyrene. Aluminum siding backed with insulating polystyrene is used as an exterior wall covering on the present day house. Aluminum siding with baked color enamel is bonded over a 3/8″ backing of polystyrene foam insulation to provide a low cost, water resistant siding, **Figure 56.** Aluminum siding can be easily installed over older bevel siding of wood, **Figure 57.**

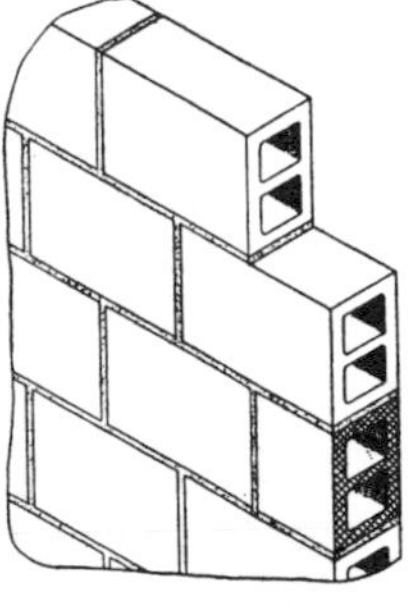

5″ TILE WALL

5x8x12 tile for 5″ partition walls with 1/2″ mortar joints.

MATERIALS PER SQUARE

135 5x8x12 tile
4 cu. ft. mortar.

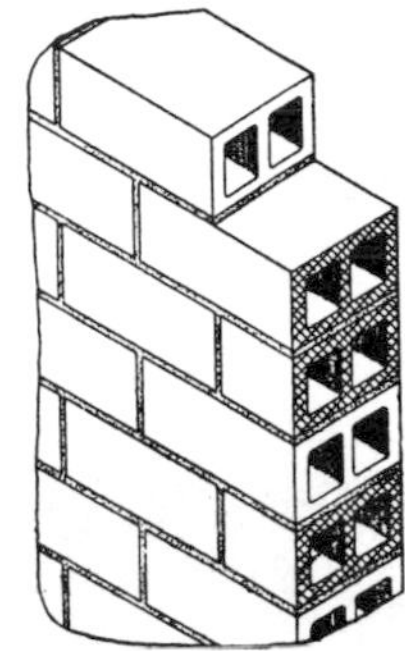

8″ TILE WALL

5x8x12 Load Bearing tile for 8″ walls with 1/2″ mortar joints.

MATERIALS PER SQUARE

210 5x8x12 tile
8.5 cu. ft. mortar.

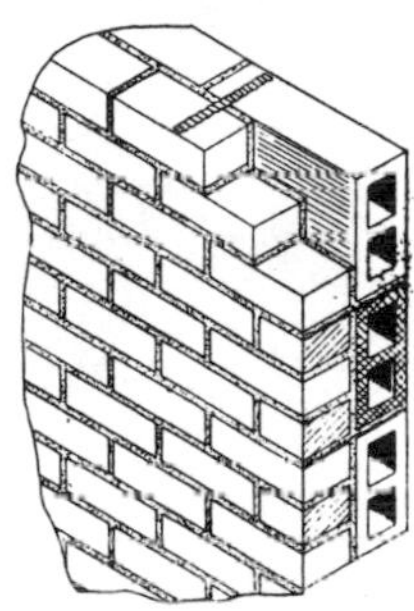

8″ BRICK & TILE WALL

Face brick backed with 4″ x 8″ x 12″ tile on edge with metal ties every third brick course.

MATERIALS PER SQUARE

616 Face Brick
137 4x8x12 tile
50 wall ties
14 cu. ft. mortar

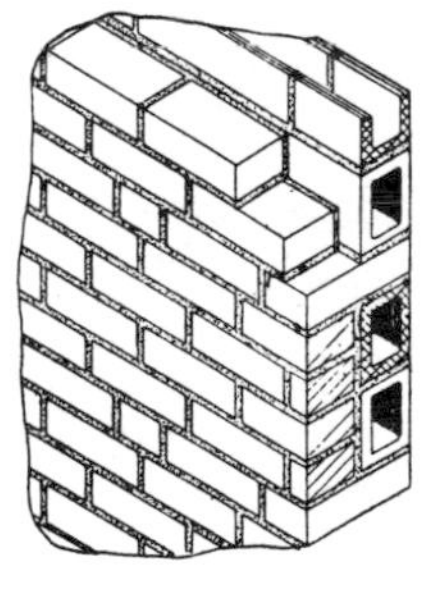

8″ BRICK & TILE WALL

Face brick backed with 4″ x 5″ x 12″ tile on edge with Flemish Header every 5th course.

MATERIALS PER SQUARE

660 Face Brick
80 Common Brick
164 4x5x12 tile
14 cu. ft. mortar

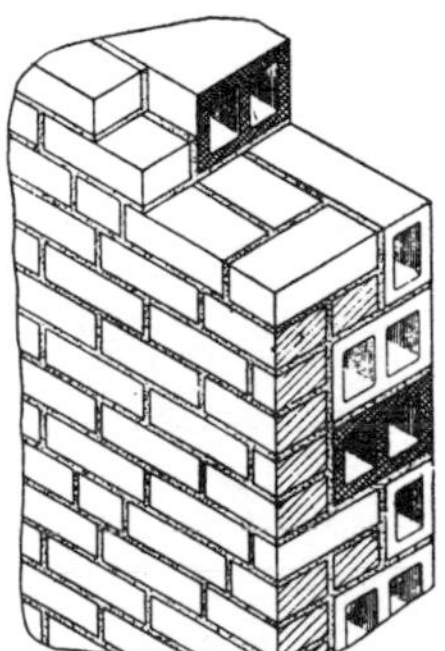

12″ BRICK & TILE WALL

Face brick with 5x8x12 backup tile and 4x5x12 tile to back up Flemish header every 6th course.

MATERIALS PER SQUARE

650 Face Brick
172 Common Brick
138 5x8x12 tile
69 4x5x12 tile
21 cu. ft. mortar

Fig. 51. Five types of hollow tile and brick walls.

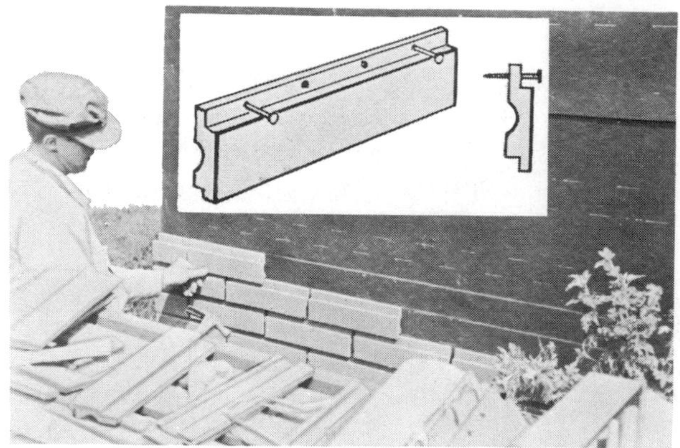

Fig. 52. Thin pre-drilled brick panels are nailed to sheathing and mortar pressed into spaces to simulate real brick.

Fig. 53. Metal hangers hold these brick panels in place. Workman applies mortar as brick is nailed on.

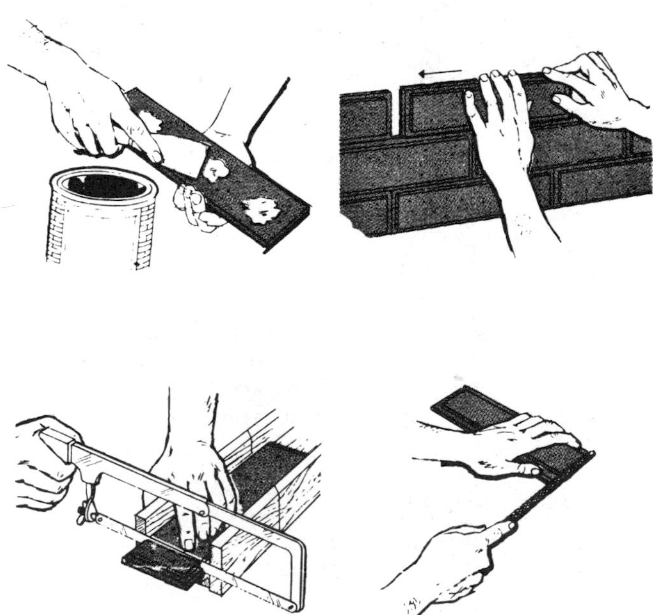

Fig. 54. Lightweight vermiculite brick is set in mastic. It may be cut with a hacksaw and filed for precise fitting.

Fig. 55. Prefinished aluminum siding may be had embossed to resemble wood grain in white or colors.

Aluminum Siding on a Prefabricated House. Aluminum siding, having a baked enamel or acrylic finish, applied over framed exterior wall sections or panels is often featured in factory built houses, **Figure 58.** Horizontal and vertical panels are combined on the exterior walls of this one-story slab-on-grade dwelling.

Panels of Steel on a Prefabricated House. Enameled panels of steel are also used as exterior wall covering.

Solid Vinyl Siding. Extruded solid vinyl is available in single courses of 8″ exposure, or double 4″ exposures per course, **Figure 59.** Vinyl is worked with metal cutting tools and is applied in much the same manner as aluminum siding. **Figures 59, 60, 61, 62, 63, 64, 65.** Vinyl siding should not be confused with plastic coated, boardback siding. The color is molded in the vinyl, and matching trim and corner boards can be had.

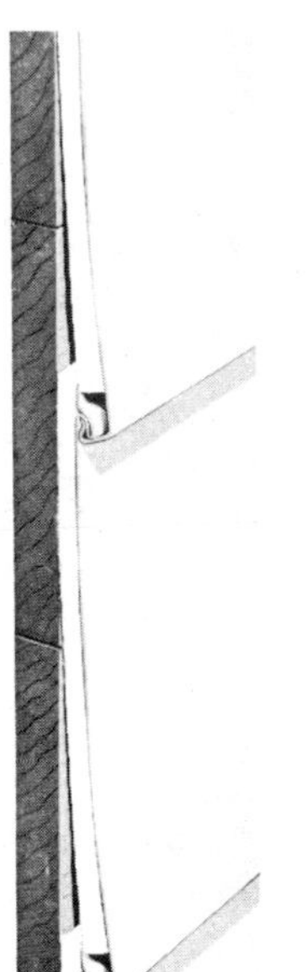

Fig. 56. Aluminum horizontal siding is backed with polystyrene foam for high insulation value and low maintenance surface.

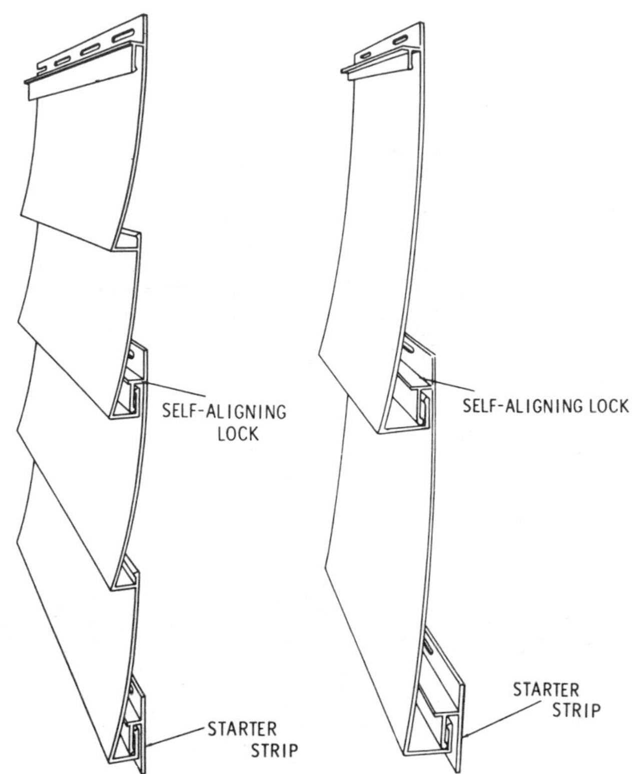

Fig. 59. Extruded solid vinyl siding in 8″ or double 4″ exposure has color molded in, never needs painting.

Fig. 57. Interlocking joints in aluminum siding make it especially easy to apply over old siding.

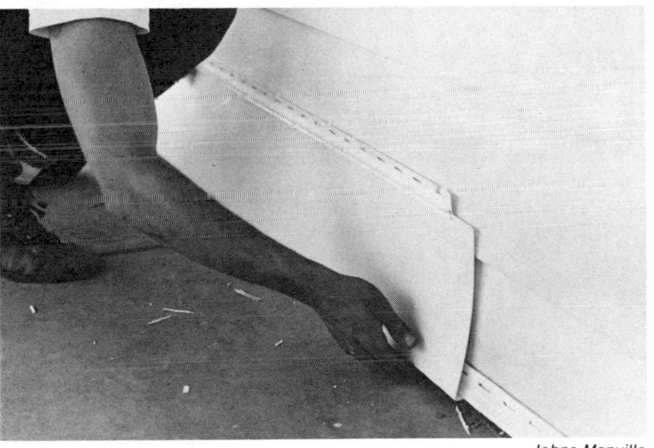

Johns-Manville

Fig. 60. First course of solid vinyl siding is hooked under previously nailed starter strip.

Fig. 58. Baked enamel or acrylic finish on these panels of vertical and horizontal siding eliminate painting for many years.

Johns-Manville

Fig. 61. Vinyl corner strip is nailed on before siding application begins.

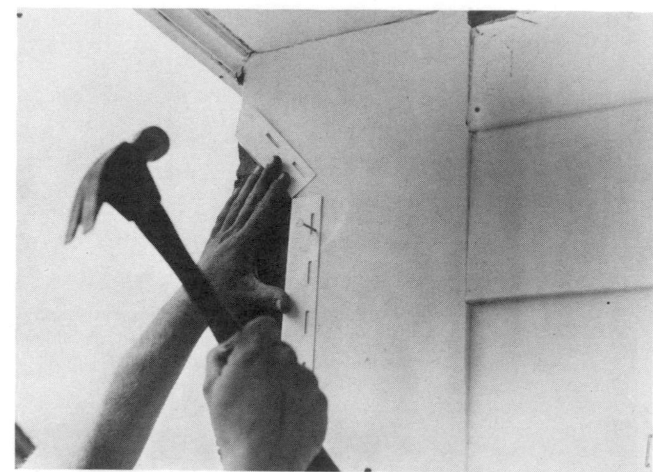

Johns-Manville

Fig. 63. Nailing corner strip over the old siding.

Johns-Manville

Fig. 62. Window trim is cut with a hacksaw.

Bird & Son

Fig. 64. Water repellent building paper is stapled to sheathing as vinyl siding goes up.

Fig. 65. Vinyl trim strip, a siding accessory, is installed over pediment to channel rainwater to sides of entry door.

Chapter 11

INTERIOR WALL COVERING/INTERIOR TRIM

Interior wall coverings are the protective and decorative materials applied to inside walls and ceilings. Formerly plaster under paint or wallpaper, and wood either painted or with a natural finish, were virtually all that was used as interior wall covering. Today the builder has a wide choice of materials such as wood-derived composition board, plywood, masonry, glass, tile, imitation brick, and numerous plastic-coated wallpapers, vinyl wall coverings, and vinyl covered hardboard.

Interior trim is the woodwork or other finish stock placed around the windows, doors and on walls. Interior trim consists of such items as door and window trim, turned spindles used on planters and room dividers, wainscot molding, decorative sculptured moldings, and dentil, corner, shoe, and baseboard moldings.

Gypsum Drywall Wallboard. Wet plastering with its requirement of three coats of plaster over rough wood lath, perforated gypsum lath, or wallboard has largely been abandoned in favor of gypsum drywall which can be applied much more quickly and at less labor cost. The same can be said about metal lath and plaster which is now used mostly in commercial buildings. However, expanded metal lath and plaster are used to form the curved surfaces of archways between rooms.

Gypsum wallboard is made of a glass fiber reinforced core of gypsum rock enclosed between two layers of tough paper. The durable cream face paper is folded around the long edges to reinforce and protect the core. Edges of regular gypsum wallboard are made in square, tapered and beveled form, **Figure 1.** Gypsum wallboard should not be applied to nailing members spaced over 16″ o.c., **Figure 2.** Gypsum wallboard is available in 1/2″, 3/8″ and 5/8″ thicknesses, in 4′ widths and in 8′ to 16′ lengths.

One occasional defect of wallboard construction is that nails may pop. This is most often caused by improper moisture content of the studs. Finishing of a drywall interior partition includes inside corner reinforcing bead (left center) outside corner bead (right center), feathered and taped horizontal joints between panels, and smoothed nail marks, **Figure 3.** Wallboard drywall can be painted, or finished with wallpaper. The joints are taped and feathered and sanded before paint is applied, **Figure 4.** The perforated paper tape embedded in the joint cement prevents cracks at joints.

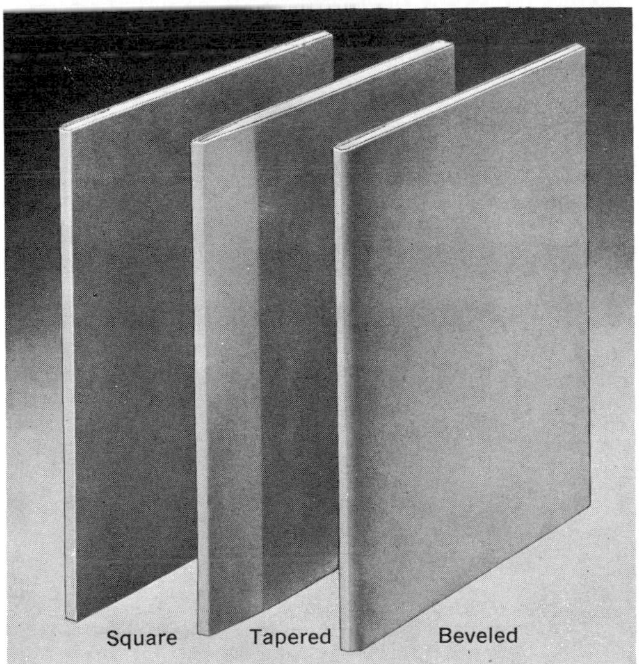

Fig. 1. Gypsum wallboard or drywall has core of fiber-reinforced gypsum covered with tough paper. Tapered edge is most common.

Fig. 2. Nailing plasterboard to studs on 16″ centers.

Fig. 3. Drywall finishing includes reinforcement of inside and outside corners with metal bead, taping and cementing of butting edges and application of cement over nailheads.

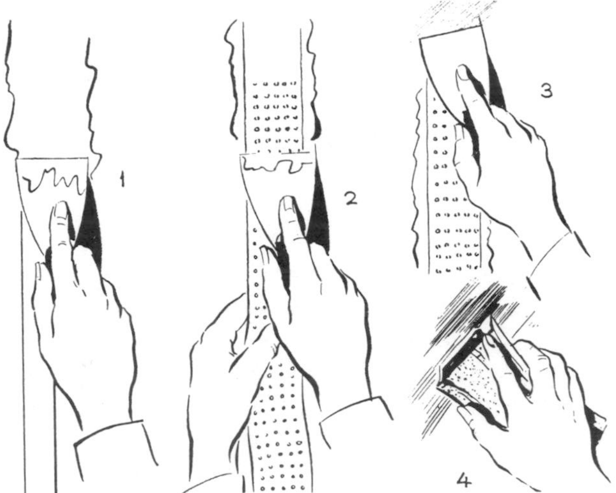

Fig. 4. Four steps in covering joint with tape and cement.

Pre-decorated Gypsum Paneling

Wood grained gypsum paneling is regular gypsum wallboard prefinished with reproductions of true wood patterns. The fireproof gypsum core of the panels is reinforced with glass fibers for extra strength. The deep linen textured vinyl surface provides permanent scuff, crack and chip resistance. The panels are available in 3/8″ or 1/2″ thicknesses, 4′ widths and 8′, 9′ or 10′ lengths. The panels are nailed on wall framing with matching color headed nails. Wood batten strips may be used to cover joints if desired.

In addition to wood grain paneling the gypsum panels with vinyl surfaces are available in colors in standard cloth textures, burlap, travertine and solid colors. The panels may also be attached with drywall adhesive to studding, gypsum sheathing or existing wall covering.

Fig. 5. Random width pine boards with shiplap edges are blind-nailed to nailing strips across kneewall studs in attic.

Interior Walls of Solid Wood

Solid wood planks or boards may be used as a wall covering, generally to impart the informal look in living room, play room, or finished basement or attic. **Figure 5** shows random width pine boards being nailed to horizontal furring to cover a knee-wall in an attic. Paneling is usually full wall height.

Board paneling tends to be simple, with less tendency toward milled patterns than formerly. Boards are easily

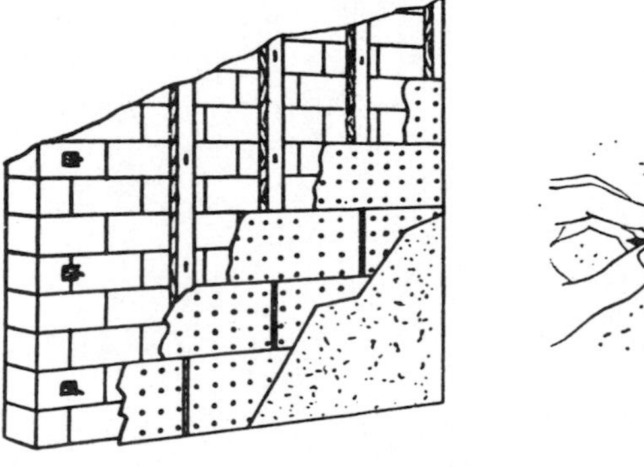

Fig. 6. Anchor nails adhered to masonry wall to secure furring strips. Nails are inserted in holes in strips, then clinched.

attached, in remodeling, to old plaster walls, with furring strips, and to masonry walls, with furring strips and clips.

When board paneling is applied over masonry of brick or concrete block, **Figure 6,** anchor nails are fixed to masonry wall by adhesive. Holes are drilled in furring strip to receive shank of nails and ends are bent over to secure strip. Insulation board may be nailed on furring strips and the vertical or horizontal wood paneling nailed or glued to the insulation board.

Fig. 7. Solid planks of Philippine mahogany cover fireplace wall. Note recessed shelves for bric-a-brac.

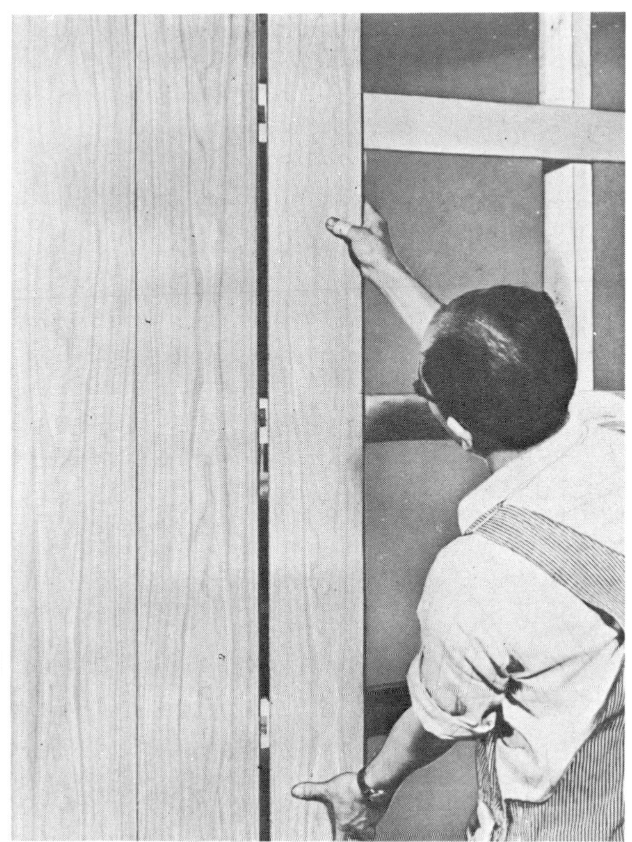

Fig. 8. Tongue-and-groove planking is installed with clips nailed to furring strips, concealing method of fastening.

Fig. 9. Texture 1-11 groved exterior plywood is used on interior wall and continued out to exterior wall with good effect.

Solid boards of Philippine mahogany can be used effectively to cover a fireplace wall, **Figure 7.**

Figure 8 shows solid tongue and grooved wood plank panel installed with clips nailed to furring strips.

Interior Walls of Plywood

Plywood is made from logs from the lower portion of a tree. The short logs, generally 8′ or 9′ in length and 2′ to 5′ in diameter are placed in a lathe and rotated against a knife which peels the wood into sheets of various thicknesses up to 3/16″. After being dried and cut, the sheets are glued evenly, placed cross-wise in layers, then pressed and sanded. All this, of course, is done in a wood manufacturing plant. Various types of wood are used, with the desired wood on one or both faces of the plywood veneered board. There are many grades of plywood both of waterproof exterior type and moisture resistant type.

Plywood, of both soft and hard woods, is often used for interior finishes. The Texture 1-11 design, primarily an exterior plywood, is used with success on interiors, **Figure 9.** In this room texture plywood is used on the far wall, picture window glass is used on the wall to the right and masonry is used on the chimney wall to the left. The ceiling interior is finished in plank and beam construction. In **Figure 10** panels of swirl-textured plywood are used on the wall at the left.

Plywood Interiors of Hardwood. Hardwood plywood offers a rich range of color in smooth textures depending on the grain and color for a rich, formal effect, **Figure 11.** Panels can be bonded with glue, nailed, or clipped to the surface beneath. Permanent and durable factory prefinishes are a useful development in hardwood plywood. Some panels are available in 16-1/4″ widths instead of 4′ × 8′ in size, and are grooved for overlapping to achieve plank effect. Grooving is a method of achieving texture; the effect of separate random planks is obtained by grooving 4′ × 8′ panels in 9-7, 12-4 and 8-8 spacing to permit finish nailing to studs in the groove lines, **Figure 12.**

Fine effects are achieved by careful matching of grain within each panel and from panel to panel, **Figure 13.**

Interior Walls of Hardboard

Composition boards or hardboard were at first simply a means of utilizing sawdust and sawmill waste, and were given only the most utilitarian uses. A growing variety of surface effects now make pressed boards, or composition boards, attractive and pleasing as finish materials. One such product is a hardboard made of selected wood fibers and manufactured in easy-to-use panel form. Small wood chips are "exploded", cleaned and refined, leaving the desirable cellulose fibers and wood's natural binding agent, lignin. These two materials are reunited under heat and pressure to form a hardboard with neither knots nor grain, which offers many advantages natural wood does not possess.

Fig. 10. Textured plywood squares are set at right angles.

Fig. 11. A formal wall of prefinished hardwood plywood.

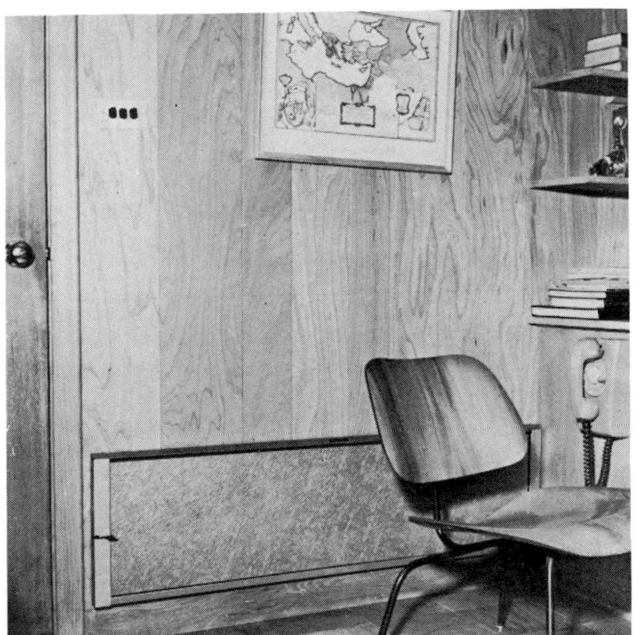

Fig. 12. Grooved plywood paneling resembles random plank.

Fig. 13. Matched grain in birch plywood shows care taken in selection and placement of panels.

Some types are striated or combed to resemble plywood as shown in **Figures 14 and 15.** Newer patterns resemble leather, tile and travertine marble.

The most popular and most readily available hardboard panels are those that simulate a woodgrain surface. These have a layer of paper with the printed pattern of the grain in the natural color of the wood. The paper is glued to the smooth surface of the hardboard and a layer of clear plastic or vinyl film is bonded over the printed woodgrain. The plastic film not only protects the printed surface but also assures easy cleaning with a damp cloth. The printed woodgrains simulate a wide variety of American hardwoods although walnut in light and dark shades is predominant. Exotic tropical hardwoods are also simulated including teak and Brazilian rosewood, **Figure 16.** Woodgrain colors are available in a wide variety of shades and may be very light, **Figure 17,** or have a warm medium glow, **Figure 18.** While the majority of panels have smooth surfaces, some feature what looks like rough-sawn planking with authentically reproduced knots, splits and cracks, **Figure 19.**

Applying Hardboard Panels to Walls. Hardboard panels are applied to open studs or solid backing with adhesives or colored ring groove nails or both. The contact cement widely used in the past has given way to a more convenient panel cement which is a thick paste supplied in cartridges used in a caulking gun. The plastic nozzle on the cartridge is cut at an angle to permit the gun to extrude a bead about ⅜″ or ½″ thick. Unlike contact cement, it is only necessary to apply

Fig. 14. Embossed and grooved ¼″ hardboard panel resembles wire-brushed wood grain in texture and appearance.

Fig. 15. Random striations scored in hardboard surface make it almost impossible to see butt joints between adjacent panels.

panel cement to one surface, either the open studs or the backs of the panels when they are applied to solid backing. When applied to the studs the bead may be a continuous zigzag, **Figure 20,** or in 3″ lengths, **Figure 21.** In many communities local building codes require solid backing for ¼″ paneling and fire codes sometimes require that paneling be applied only over gypsum board. Where solid backing is required, a continuous bead is applied to the back of the panel all around its perimeter ½″ from the edge. Diagonal beads from corner to opposite corner in the shape of an X are then applied to the panel back. The panel can then be pressed against the wall and adjusted to fit adjacent panels and the ceiling if necessary. The panel cement has quick grab and easily holds ¼″ 4′ × 8′ panels, however, it remains flexible long enough to permit adjustment.

Nails can be used along with the cement along the top and bottom of the panel where they will be covered with moldings. Colored nails that blend well with the color of the panel and are virtually invisible need not be covered with moldings. The colored ring groove nails that are usually supplied with the panels can be used to fasten panels without cement if desired. Nails are spaced 4″ apart on all edges and 8″ apart along the studs, **Figure 20.**

Perforated Hardboard or Pegboard Uses. Perforated hardboard or pegboard is widely used in kitchen, dens and garages for hanging utensils, pictures or tools. Hardboard panels perforated with holes 1″ o.c., are available in 6 panel types, 8 standard sizes with 1/8″

Permaneer Corp.

Fig. 16. Brazilian rosewood woodgrain design reverse printed on the underside of a protective sheet of vinyl laminated to a smooth-surfaced particleboard.

Masonite Corp.

Fig. 18. Random grooved woodgrain hardboard paneling that simulates hickory wood and has chocolate brown grooves, grains, and knots.

Masonite Corp.

Fig. 17. Hardboard paneling with an exotic lauan woodgrain effect in a very light champagne color.

Masonite Corp

Fig. 19. Prefinished hardboard paneling with random width grooves that give the appearance of rustic rough sawn barnwood with carefully reproduced knots, splits and cracks.

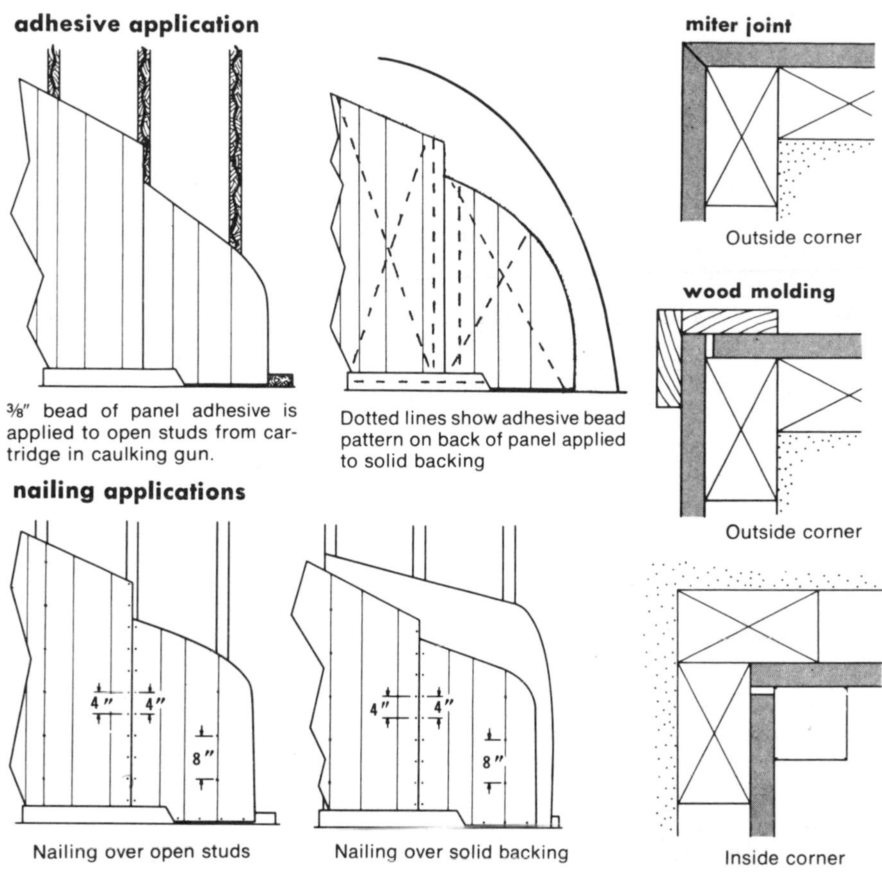

Fig. 20. Applying hardboard paneling to studs or gypsum board with adhesive or nails. Note treatment of corners.

or 1/4″ thicknesses. Holes are 3/16″ on 1/8″ board and 9/32″ on 1/4″ board. Panels should be fastened to wood framing or furring strips spaced not more than 16″ o.c. Panels should be fastened with nails or screws. There are over 60 different peg-board metal hanging fixtures, **Figure 22.**

Covering Ceilings

Ceilings may be covered with all materials previously described for use on interior walls. **Figure 23** shows how panels of simulated marble hardboard are attached to existing ceilings and to open joists or newly framed ceilings.

Suspended Ceilings. Ceiling panels and lighting panels of light-weight fiberglass materials are available in a variety of colors in 2′ × 2′ and 2′ × 4′ sizes. The large, light weight panels are easy to handle; flexible to permit bending for installation in tight, confined areas, and are easily cut to fit around ceiling obstructions.

Installing a Suspended Ceiling. A suspended ceiling may be installed by first making a diagram of the room layout, **Figure 24.** Place the wall angles around the perimeter of the room, bringing the angles to true level and desired height, (step 1.) **Figure 25.** Position the main runners, with hanger wires, on 4′ centers at right angles to the joists (step 2). Locate the 4′ cross tees between the main runners and lock in place (step 3). The standard 2′ × 4′ or 2′ × 2′ panel may not fit evenly along the edges. In this case, space the difference evenly on both sides of the room or put the odd panels all on one side. Insert the panels in the grid formed by the suspended tees (step 4).

Other Methods of Installing Ceilings. In addition to the suspended grid system of installing ceiling material, the clip strip method is used when the material is not suspended but held by metal strips nailed direct to ceiling joists, **Figure 26.** Workman is applying 12″ cellulose squares, or tiles, to furring strips spaced one foot apart, nailed to joists through flange of blanket insulation. The tiles are held in place by clips and nails.

Ceiling tile may be applied by adhesives on a plaster, gypsum board or an above grade concrete in sound condition. In some instances application of the tile is done from the center of the ceiling to the walls, allowing parts of tiles to form a border.

Ceiling tile may also be applied by a "piggy-back"

1.

2.

3.

Fig. 21. Fastening hardboard paneling to open studs. 1) Adhesive is applied to studs in 3″ beads about 6″ apart. 2) Panel is tacked with nails at the top. 3) Firm uniform pressure is applied, to permit adhesive beads to spread evenly between studs and panel.

stapling system over gypsum wallboard, **Figure 30.** The locking of the tile to the wallboard is accomplished by two staples riding piggy back, into the surface. A stapling gun is used to drive home the two staples. This method can be used on gypsum wallboard or gypsum lath surfaces that are level and square.

Another method of attaching celing material and tile to joists and existing gypsum ceilings is Armstrong's Integrid system. This uses rigid metal furring channels. The metal furring channels require only about 15 nails compared to about 250 wood furring strips for an average 12′ × 12′ room. The metal channels do not absorb water or warp as sometimes happens to wood strips with the result that ceiling tile is forced out of line. The channels are rigid enough to bridge ceiling or joist irregularities and are thus self-leveling.

In installation wall angles are installed around the four walls about two inches below the joists or existing ceiling. The light steel channels are nailed at right angles to joists 48″ apart. Cross tees are snapped on to the flanges of the furring channels and moved along the channel until the edge of the tee slipps into a slot in the tile which may be 12″ square or 12″ × 48″. The tiles are square at the side and shiplapped at the ends so that the textured surface shows no dividing lines and the channels and cross tees cannot be seen, **Figures 27, 28, 29.**

Masonry, Ceramic and Glass

Masonry, ceramic materials and glass are also used more extensively as interior coverings, with the growth of informal decoration and its reliance on varied textures. Bare brick walls, often of old brick, **Figure 31,** and cut stone, **Figure 32,** are favored as a relief from smooth walls. Imitation stone in the form of prefinished hardboard paneling, **Figure 33,** and fiberglass panels made with powdered limestone provide texture wall surfaces and the appearance of stone at low cost, **Figure 34.** Concrete block, sometimes with the addition of molded sections, is another attractive form of wall texture, **Figures 35, 36, and 37.**

Ceramic Tile Wall Covering. Ceramic tile is a good example of how old materials take on new life when given wider use. In a rich variety of new colors and sizes the tile is used not only on kitchen and bathroom walls but also as counter surfaces, built-in range surrounds, and work surfaces, **Figure 38, 39, 40, 41.**

Fiberglass Simulated Panels. The fiber glass simulated masonry panels described in Chapter 10, Exterior Wall Covering, are used as Interior Wall Covering, too, and are applied in the same manner.

Glass Block as Wall Covering. Interior wall covering of glass block may be built up in panels as shown in **Figure 42.** Panels of glass block must be built up with care and properly supported to limit movement and settlement.

Imitation Masonry Plastics. Various imitations of

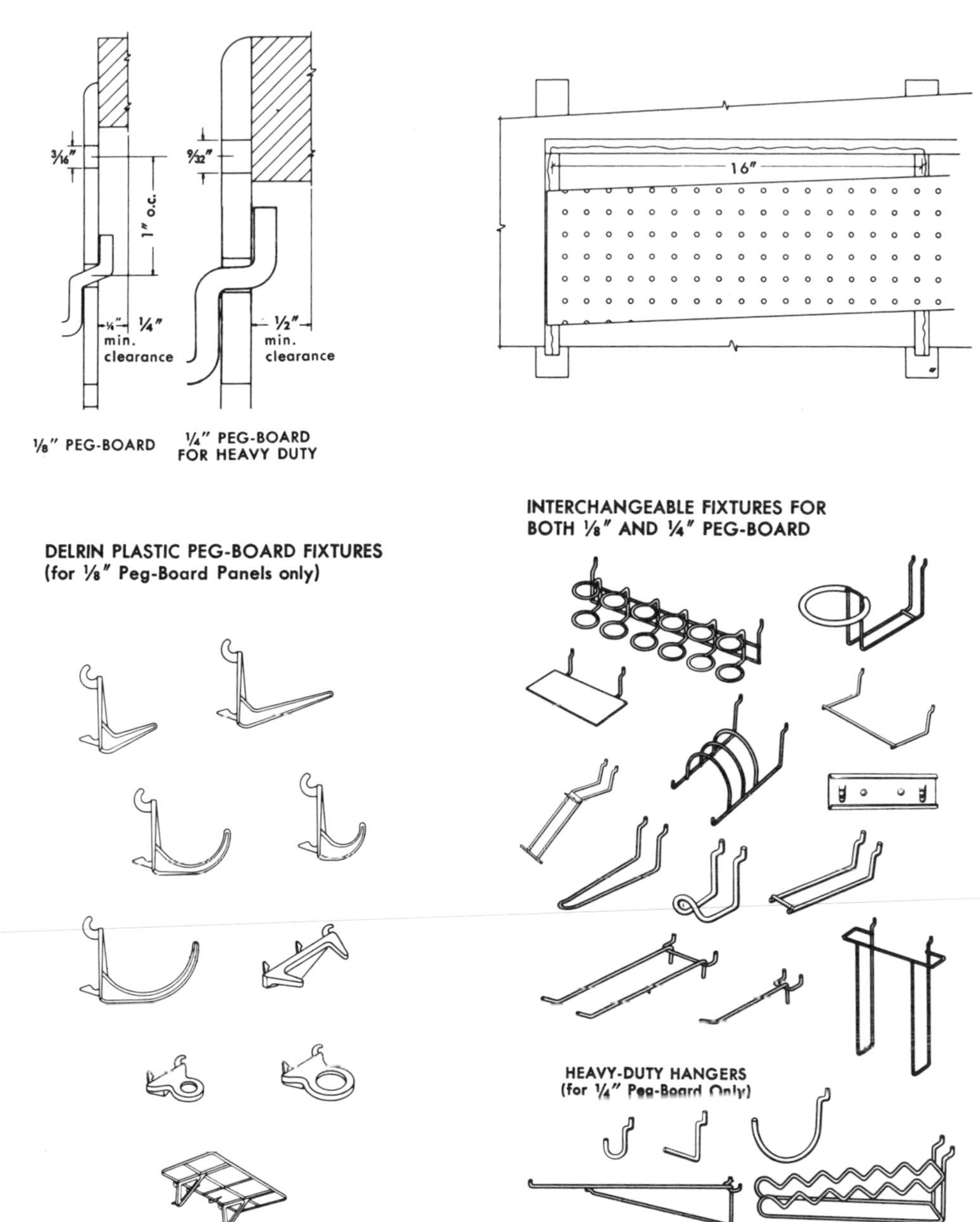

Fig. 22. A variety of plastic and metal hangers are available to make perforated hardboard functional in kitchen, closets, shop.

brick pattern and texture are used as interior wall covering. Hollow pressed flexible plastic, or thin cementitious veneers can be nailed to unfinished surfaces for brick effect, **Figure 43.**

Masonry Blocks in Ashlar or Brick. Broken cement blocks in ashlar or brick size are common fireplace materials, and often appear as indoor planters, **Figures 44 and 45.**

Metal as Interior Wall Covering

Stainless steel or copper-colored tiles are used mainly in kitchens, **Figure 46.** Perforated or expanded metal panels are used for accents; a room divider may feature wrought metal railings, as shown in **Figure 47.**

Modern Wallpaper

Wallpaper is second only to paint as the most acceptable and popular type of wall covering. Changes in wallpaper have been little short of revolutionary. So many wallpapers are actually plastics, especially vinyl, that the industry perfers to call its product "wall coverings." Older types of papers had to be trimmed at the sides and applied with a wheat paste adhesive, a rather messy job. Today's product is pre-trimmed pre-pasted, and strippable. Buyers are supplied with a waxed waterproof cardboard box that can be filled with lukewarm water into which a strip of wallpaper is dipped for a few minutes, **Figure 48.** The strip can then be applied to the wall, adjusted for proper fit and smoothed with a brush to eliminate air bubbles. The adhesive holds perfectly yet the wall cover-

Over Existing Ceiling

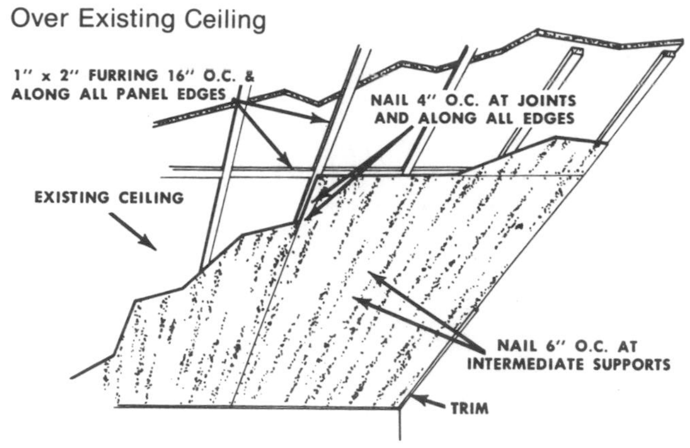

Over Open Joists

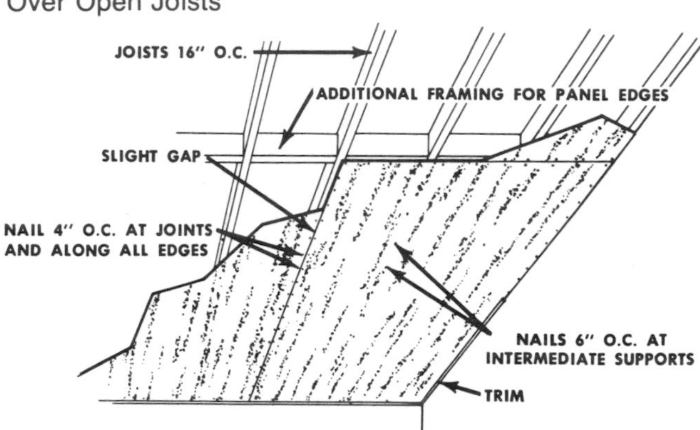

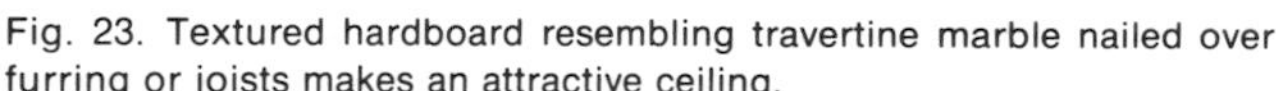

Fig. 23. Textured hardboard resembling travertine marble nailed over furring or joists makes an attractive ceiling.

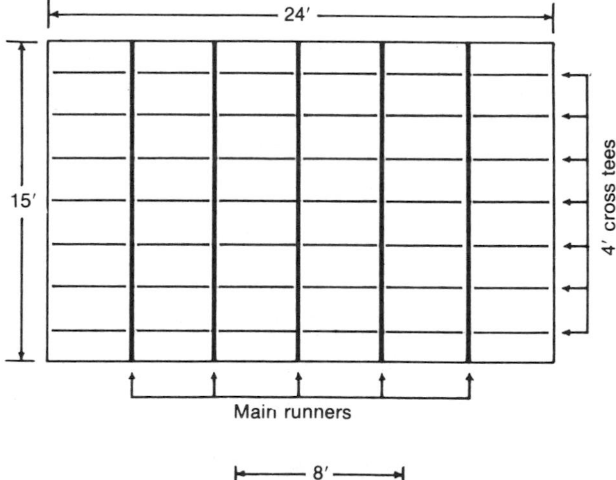

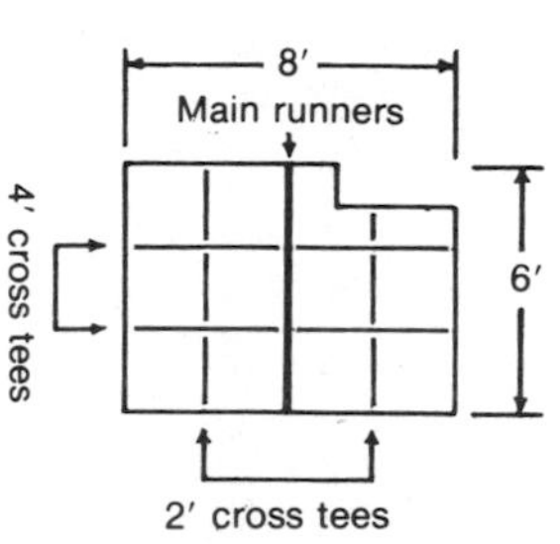

Fig. 24. Making a diagram of the room and grid layout is the first step in installing suspended ceiling panels.

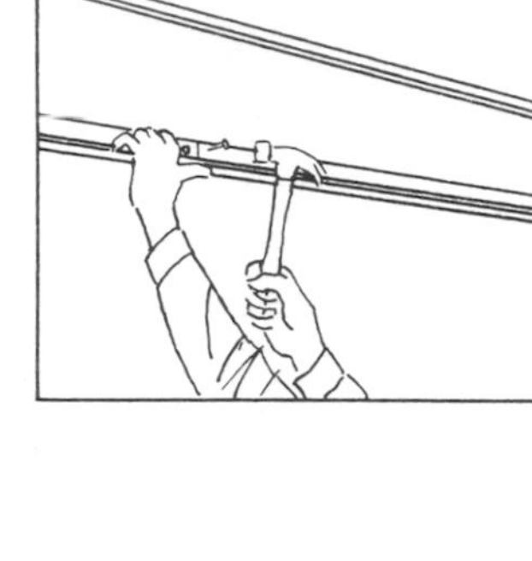

1. Attach Wall Angle to wall around room at new ceiling height.

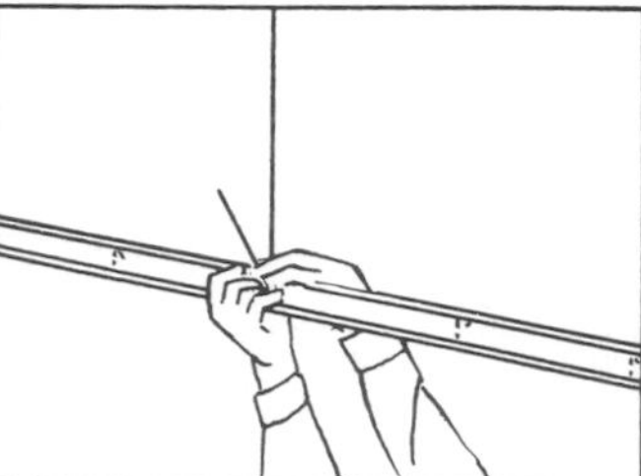

2. Position Main Tees with hanger wires.

3. Install Cross Tees between Main Tees and lock with Tension Pins.

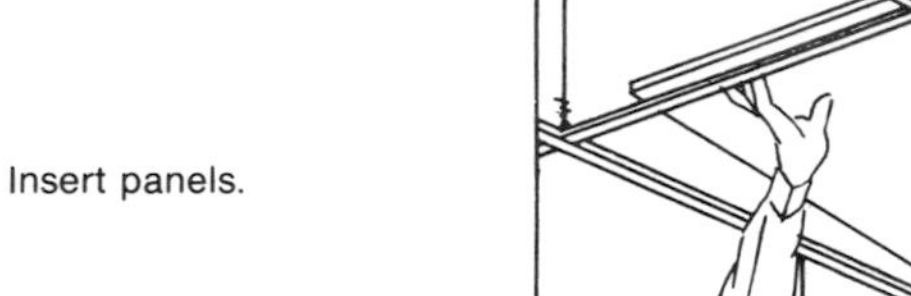

4. Insert panels.

Fig. 25. Only four basic steps are required to install the grid and panels in a suspended ceiling.

Fig. 26. Ceiling tiles may be attached to clips fastened to furring strips nailed across joists.

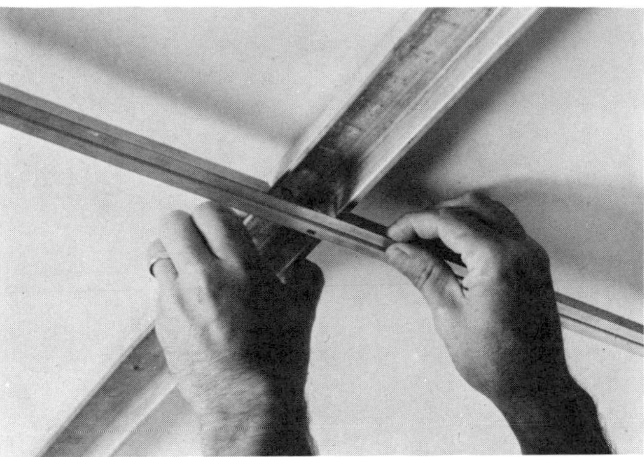

Armstrong Cork Co.

Fig. 27. Metal furring channels eliminate wood furring strips and speed installation of ceiling tile. Cross tee which holds tile is shown being snapped onto furring channel.

Armstrong Cork Co.

Fig 28. Cross tee in Intogrid system is moved along furring channel until its flange can be inserted in slot in side of 12" × 48" tile.

Armstrong Cork Co.

Fig. 29. Textured surface of tile shows no dividing lines and supporting furring channels and cross tees cannot be seen.

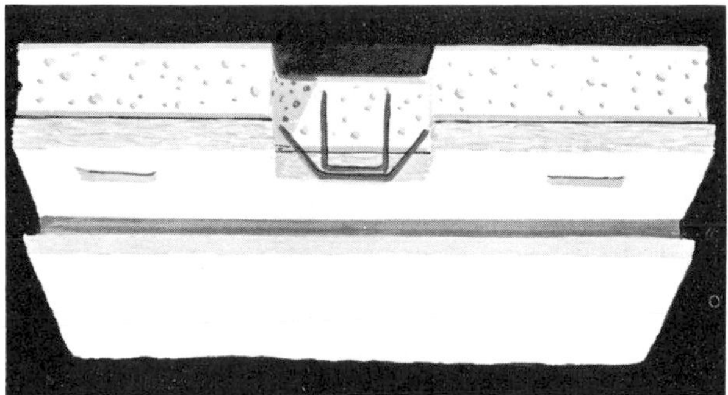

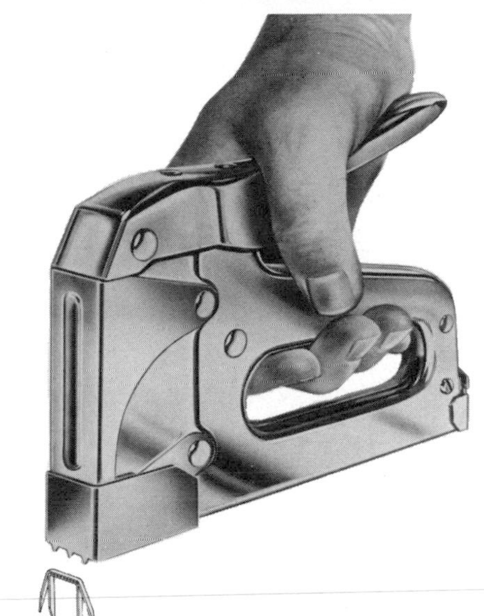

Fig. 30. "Piggy-back" method of stapling ceiling tile to gypsum ceiling panels. Stapling gun with special attachment drives two staples, one over the other (dwg. at top). The first staple spreads the second staple for a secure grip in the gypsum.

Fig. 31. Old brick, including brick painted different colors, is popular for fireplaces and accent walls.

Fig. 32. Cut stone in Ashlar pattern is used to cover fireplace wall and is carried beyond glass wall to exterior patio wall.

Masonite Corp.

Fig. 33. Authentic to the smallest detail, this prefinished hardboard paneling duplicates the look of Indiana limestone in random Ashlar pattern. Panels are fastened with nails and can be inverted to avoid repetition of pattern.

ing can be stripped from the wall easily years later without steaming, soaking or chemicals. In addition, most modern wall coverings can be wiped clean with a damp cloth or even scrubbed except those that are flocked or interwoven with metal foils.

Because these improved coverings are so easy to apply or remove, hanging them has become a largely do-it-yourself job.

Marlite Co.

Fig. 34. Imitation stone panels made from crushed limestone and reinforced with fiberglass can be attached with nails. Mortar is applied with ordinary caulking gun.

Interior Trim

Interior trim must be complementary to interior walls, ceilings, floors, windows, doors, stairs, etc., built into various rooms of a dwelling. Interior trim must complete the materials, styles and textures expressed in the larger interior areas.

Interior Trim of Wood. Conventional trim, **Figure 49,** is made of Ponderosa Pine, Western Red Cedar or Oak stock to match the materials used in doors, windows, closets, stairs and Interior Wall Coverings. In addition to these standard moldings, there are cove, cove and inside corner, outside corner, dentil and combination base and show moldings, **Figure 50.**

Most moldings are prefinished or furnished with a primer coat to match interiors. There are many applications for moldings on walls in interior work, **Figure 51.**

Portland Cement Assn.

Fig. 35. Concrete blocks set in simple brick pattern form one wall of this living room.

Portland Cement Assn.

Fig. 36. Two different types of blocks form an attractive wall design.

Portland Cement Assn.

Fig. 37. Molded concrete blocks break monotony of rectangular blocks.

Tile Council of America

Fig. 38. Glazed tile on kitchen counters and walls between upper and lower cabinets provides durable easily cleaned surfaces.

Tile Council of America
Fig. 39. Glazed decorative tile that goes up to the ceiling resists cooking oils and grease and is easily washed.

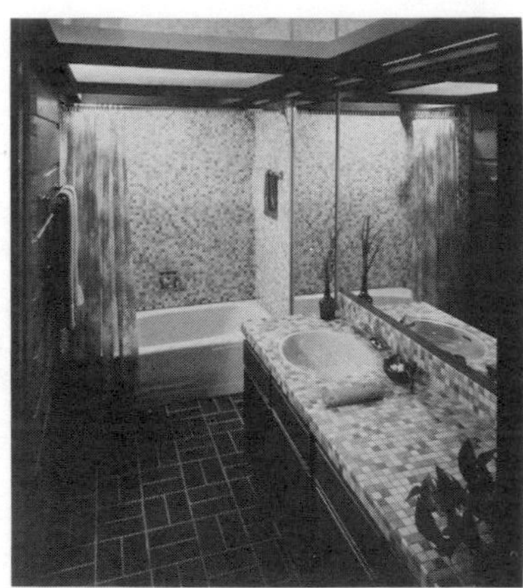

Tile Council of America
Fig. 40. Classic mosaic tile on vanity top and around bath and quarry tile on the floor lend a luxurious note to this bathroom.

Tile Council of America
Fig. 41. Sunken bath tub and seat finished in crackle glazed tile is the epitome of luxury.

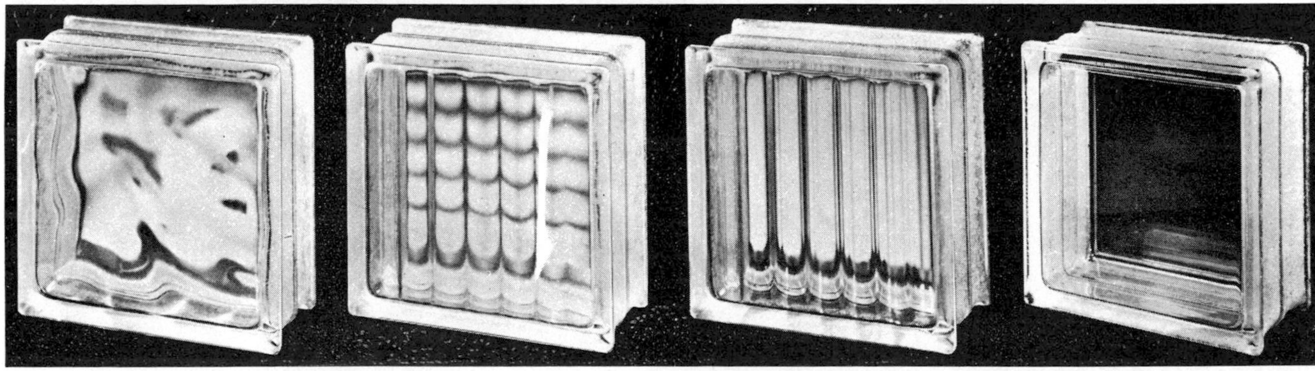

Fig. 42. Panels of glass block laid up in exterior walls admit soft natural light while obscuring view from outside.

Other Decorative Moldings. Other decorative moldings used as interior trim are turned spindles, 2″ × 2″ × 48″ and 3″ × 3″ × 60″ in size for use as planter spindles and room dividers, **Figure 52,** and decorative wainscot molding sets of Ponderosa Pine. These are available in 1-1/8″ × 9/16″ size with overall pattern sizes running from 21-1/2″ × 25-1/8″ to 25-1/8″ to 70-3/8″ in size.

Sculptured Moldings of Fiberglass-Reinforced Plastic. A new flexible molding giving a richness to furniture, doors, room dividers, walls, cabinets, mirrors, frames, screens, lamps, etc., is made of flexible glass reinforced plastic. This molding can be applied with glue, nails, staples or screws, and is worked like wood. It is furnished in 1-5/8″ × 5/8″, 1″ × 1/4″, 1″ × 5/16″, 7/8″ × 1/2″, 3/8″ × 3/8″ and 1-3/16″ × 5/16″, in many patterns, **Figure 53.**

Molding of Aluminum for Pegboard Hardboard of

Fig. 43. Panels of brick-patterned plastic are adhered to wall.

Fig. 44. Accent wall of broken block is lighted to show texture.

Fig. 45. Island range counter and planter built of block.

Fig. 46. Stainless steel, aluminum and copper tiles vary reflections according to direction of satin finish grain.

Fig. 47. Expanded metal lath is fabricated into screen panels for room dividers and stair rails.

Fig. 48. Dipping a roll of pre-pasted wallpaper in a box of water before applying it to the wall.

Hardboard Paneling. Extruded aluminum moldings to form outside corners, inside corners, joints between panels, and trim around windows are available to match patterns and colors of pegboard or hardboard paneling, **Figure 54.**

Applying Metal Moldings. Metal outside and inside corner moldings are applied after a panel on the adjacent wall has been fitted and installed. Edging moldings are applied before fitting and installing the last panel. Fasten metal moldings through flanges with lath nails. When metal moldings are used the 4″ o.c. nailing along the edge of the panel is not required. Panels should not be pushed clear to the bottom of the metal molding, but a slight space should be kept for expansion, **Figure 55.**

Hardboard Moldings for Hardboard Panels. Wood grained, prefinished moldings of hardboard are available for hardboard panel installation, **Figure 56,** solving the problem of staining wood molding to match factory finishes.

Moldings for Interior Walls of Plywood. Joints and corners in interior walls of plywood, may be concealed and trimmed with molding made of plywood, and standard solid wood moldings **Figure 57.**

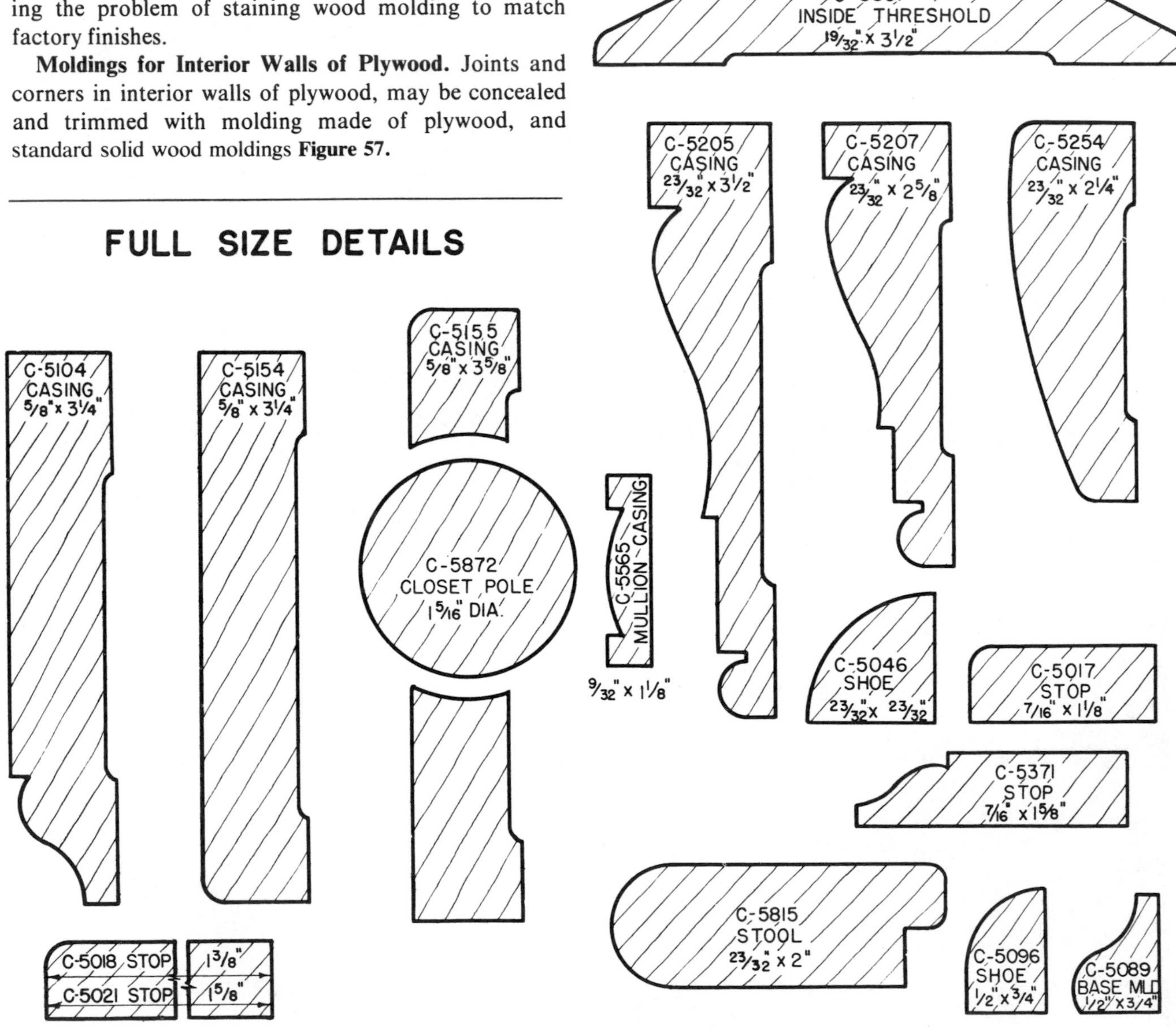

Fig. 49. Interior moldings: casings, base moldings, stools, stops, and thresholds commonly used in homebuilding.

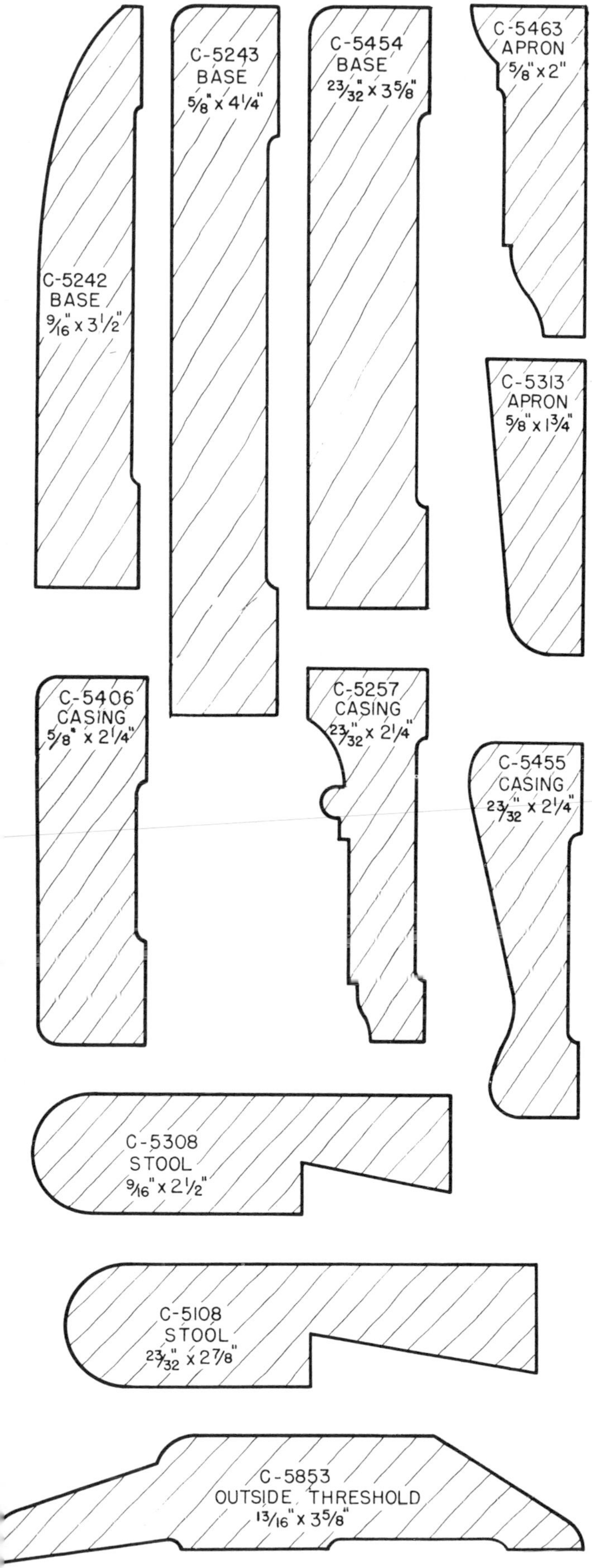

C-5242
BASE
9/16" x 3 1/2"
C-5243
BASE
5/8" x 4 1/4"
C-5454
BASE
23/32" x 3 5/8"
C-5463
APRON
5/8" x 2"
C-5313
APRON
5/8" x 1 3/4"
C-5406
CASING
5/8" x 2 1/4"
C-5257
CASING
23/32" x 2 1/4"
C-5455
CASING
23/32" x 2 1/4"
C-5308
STOOL
9/16" x 2 1/2"
C-5108
STOOL
23/32" x 2 7/8"
C-5853
OUTSIDE THRESHOLD
13/16" x 3 5/8"

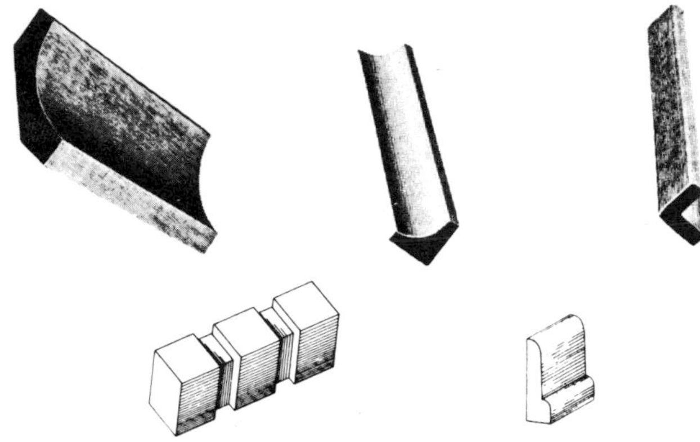

Fig. 50. Molding from left to right, top: cove, cove and inside corner, outside corner; bottom: dentil, combination base and shoe.

COVE
STOP
CASING
CHAIR RAIL
STOP
CASING
INSIDE CORNER
SHOE
APRON
STOOL
BASE
OUTSIDE CORNER

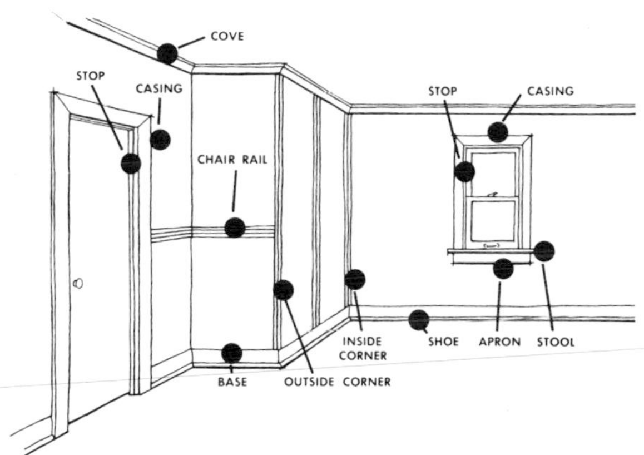

Fig. 51. Typical molding applications on interior walls.

ACTUAL SIZE

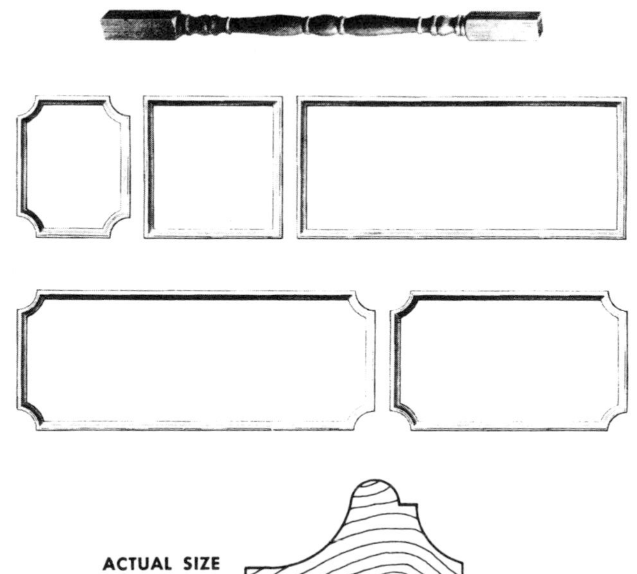

Fig. 52. Turned spindle and wainscot molding sets.

Fig. 53. Decorative sculptured molding of fiberglass-reinforced plastic are worked like wood.

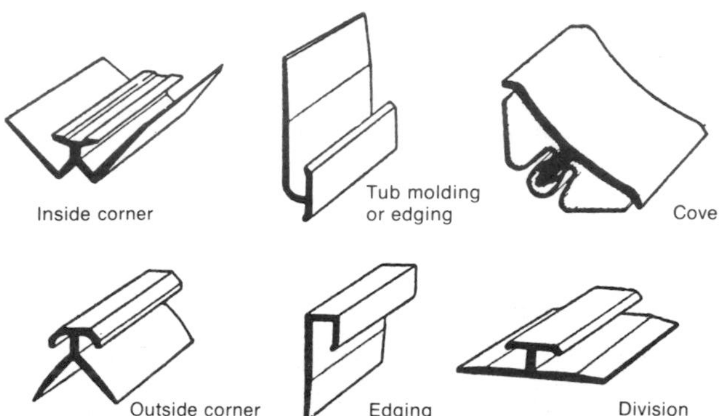

Fig. 54. Extruded metal moldings for wall paneling.

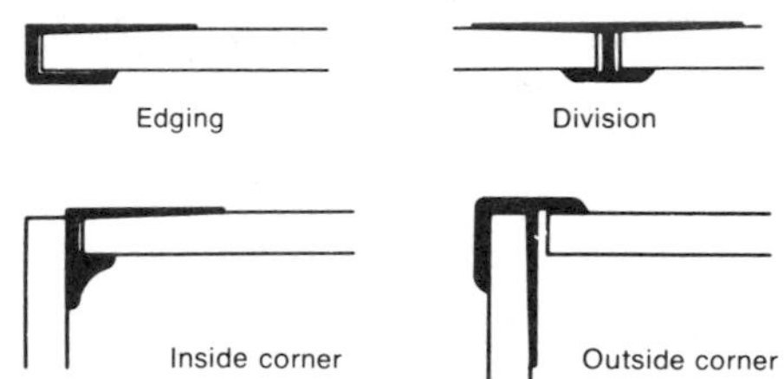

Fig. 55. How moldings are used.

Fig. 56. Prefinished wood and hardboard moldings solve the problem of matching molding to prefinished panels.

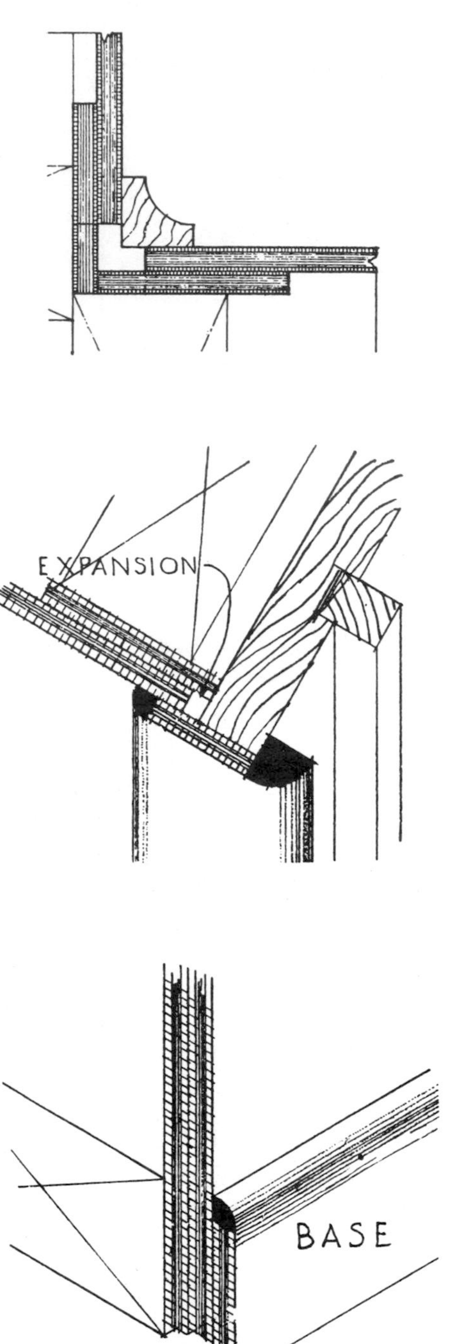

Fig. 57. Plywood trim fitted with stock molding to conceal edges may be used to finish walls paneled with plywood.

Chapter 12

INSULATION

Economy is another basic objective achieved through insulating walls, floors and ceilings of comfort conditioned houses. In cold climates, there is a material saving in fuel. Smaller heating plants can be used. In warm regions, insulation can effectively reduce the size of required air conditioning systems, as well as their cost of operation. The savings in equipment, fuel, and power more than pay for the added cost of insulation, in a relatively short time.

Insulating the Wood Frame House

Of all house building methods and materials commonly employed, wood frame construction lends itself most readily to the installation of effective insulating materials. Most structural and finish materials offer some resistance to the passage of heat; some to a higher degree than others. Millions of tiny air cells, entrapped within its cellular structure, make wood a natural barrier to heat and cold. One inch of wood provides the same resistance to heat flow as six inches of common brick or fifteen inches of concrete or stone.

Air is an excellent insulator, whether trapped in the small cells of a material, or in an enclosed air space. Air spaces created between structural elements, as those between wood studs in an exterior wall, are very effective heat barriers, and are always considered when determining the thermal resistance of building sections.

Modern wood frame construction combines wood framing and air spaces in a structural assembly having high thermal efficiency. More important, the spaces between framing members make it easy to install insulating materials, without increasing the thickness of the wall, roof, or floor section.

Other types of uninsulated walls are normally thicker than those of wood frame construction. Adding insulation to these walls increases their dimensions even further, thus reducing the interior living area.

Wood frame construction provides a structurally sound wall, thinner than other types and capable of being insulated to a higher degree. No other method of wall construction, in general use, provides as much protection against heat flow as an insulated wood frame wall.

Insulating Against Heat Loss by Conduction and Convection. Heat may be transmitted or conducted through the particles making up a substance, or heat may be transferred or convected by circulation around a substance. Modern insulation against the conduction or convection of heat is made of blankets, loose fill, batts, or rigid slabs. These materials contain trapped air and are in themselves very poor conductors of heat.

In house construction, various insulating materials are used to prevent or impede the transfer of heat or sound. Water vapor is controlled by a plastic or asphaltic vapor barrier, usually placed on the warm side of insulation.

Controlling Heat by Insulation. Prospective house buyers of today may choose from an unlimited variety of styles and sizes in new houses. They also have an opportunity to select houses that will provide more com fort for their families at a reasonable cost, whether in cold Northern climates or in the warm Southern regions.

Fig. 1. Energy Research and Development Administration Brookhaven National Laboratory test house which sves 50% of heating and cooling costs.

Fig. 2. Types of insulation: batt (upper left), blanket (center), pouring wool (right), rigid perimeter insulation (lower left).

Fig. 3. Foil or asphalt wrapping prevents passage of vapor.

Since the cost of comfort is frequently the largest item of operating expense, economy of heating and air-conditioning systems is of considerable importance and should exert a major influence upon the house buyer's decision.

Builders are aware of this important criterion and have modernized construction practices by increasing the use of thermal insulation in all types of houses and in all sections of the country.

Increased comfort is a basic objective in providing properly insulated houses. During cold weather, houses must be heated to maintain comfortable indoor temperatures. Insulation plays an important part in attainment of uniformly comfortable conditions. Ceiling, floor and wall surfaces, in insulated houses, will be warmer. In hot weather, insulation is most effective in maintaining cooler indoor temperatures.

Insulation and the Energy Problem

The rapidly increasing cost of gas, oil and electricity has changed the thinking of builders, manufacturers and even homeowners about home insulation. The changes are not so much about types and forms as they are about quantities and efficiency. The efficiency of insulation today is measured not so much in thickness as in "R" values. R stands for the resistance of heat through the insulation. The longer the insulation delays the passage of heat the higher the R value. A mineral wool batt of one manufacturer marked R-19 may be 6 inches thick while a similar batt with the same R value may be 8-3/4 inches thick, yet both are of equal efficiency.

Greater quantities of insulation are now recommended for walls, floors, and ceilings by insulation manufacturers, ERDA (Energy Research and Development Administration) and its Brookhaven National Laboratory, and many builders. Thus the test house built to specifications of the Brookhaven Lab near Stony Brook, L.I., N.Y., **Figure 1,** has 12 inches (R-38) of insulation in the attic floor, 6 inches (R-19) under its floors, and 6 inches in its walls. The 6 inches in the walls was made possible by having 2×6 studs rather than the usual 2×4's which can only take 3½ to 4 inches (R-11). However, many builders are coming to accept the idea of 2×6 studs even though it increases the cost of the house. For the home buyer

Fig. 4. Mineral wool insulation is made in various thicknesses for application in walls, floors and ceilings.

the additional insulation helps repay the extra cost in fuel savings. The Brookhaven test house achieved savings in heating and cooling costs of some 50%.

However, insulation alone did not accomplish this saving by itself. Other contributing factors included satisfactory attic ventilation (See Chapter 4), insulated windows (Chapter 8), and insulated entry doors (Chapter 9).

Other recommendations are that ducts be placed **below** rather than in the attic since this results in significant reductions of heat loss and heat gain. Caulking should be used to seal the sill plates against drafts and moisture. Electrical wiring should be placed in notches at the bottom of 2×6 studs and lie on the sole plate so that there will be no compression of insulation. Wiring in the attic should also be attached to trusses high enough not to interfere with a complete uninterrupted insulation blanket.

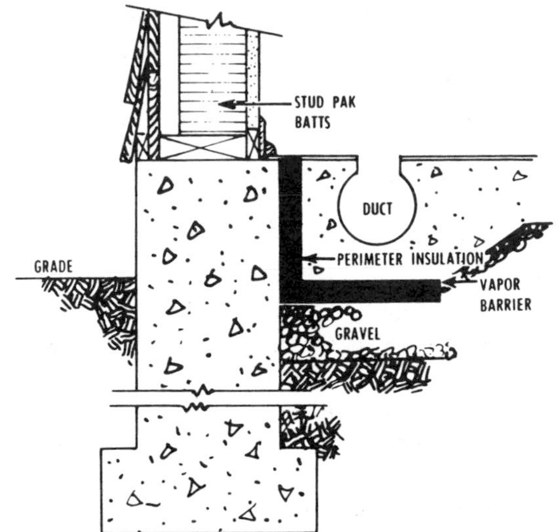

Fig. 5. Rigid perimeter insulation is set in place before slab is poured. It reduces heat loss around edges of slab.

Controlling Heat Loss by Conduction and Convection

Blanket Insulation. Blanket insulation is available in long rolls, usually 15′′ wide, to fit between studs and joists, and of various thicknesses. Most blanket insulation comes in a paper or foil wrapper, with a flange on each side to permit stapling to the edges or the sides of the framing members, **Figure 2.** The wrapper may be on one side only, as is common with the six inch thickness; or there may be a foil or asphalt-impregnated paper on one side, to serve as a vapor barrier, and plain or perforated paper on the other, **Figure 3.**

Blanket insulation is most commonly made of mineral wool, which includes rock wool and glass fiber. Cellulose fibers, a wood conversion product, are also available in blanket form. The rolls of blanket are resilient, springing out to their full thickness when released.

Batt Insulation. Batt insulation or batts have the same construction and general appearance as blankets, but come in short lengths, either 4′ or 8′.

Poured Insulation. Poured insulation is available in paper sacks for the do-it-yourself trade. It may be mineral wool, cellulose, or vermiculite pellets (a form of exploded mica).

Poured Insulation. Like all other forms of insulation, the poured type is also rated in terms of R values. The user will usually find a table printed on each bag which gives the number of square feet each bag should cover to achieve a specific R value. To know how many bags to use in an attic, its length is multiplied by its width, which gives you the number of square feet. The square feet are multiplied by .90 to subtract space occupied by joists 16″ o.c. If the joists are 24″ o.c. the multiplication is by .94.

Mineral wool and cellulose poured insulation should be fluffed up to approximately three times its volume in the bag

Fig. 6. Rigid polystyrene foam board is widely used for perimeter insulation. Scoring makes it easy to break to right width.

W.R. Grace Co.

Fig. 7. Home owner adds vermiculite insulation to inadequate depth of attic insulation to bring it up to R-19 or 6 inches.

INSULATION

where it is greatly compressed. The contents of the bag should be poured into a box and stirred vigorously with a stick or hand before being poured.

As a safety precaution, operators pouring mineral wool by hand should wear gloves, long sleeved shirts, goggles, and a mask covering nose and mouth to prevent irritation to the skin, nose, mouth, and eyes.

Vermiculite is not a fluffy substance and is easy to pour, **Figure 7,** but its R value has to be determined by its thickness with a ruler or tape measure.

Since poured insulation has no vapor barrier of its own, the operator should first determine whether there is any under the attic floor before any pouring be done. If there is none, the operator should first lay down 2-mil sheets of polyethelene plastic on the gypsum sheets under the joists, always on the warm side of the insulation.

Contractors may use pneumatic machines and a hose to blow insulation wool into attics and walls. This process is described below under "Applying Insulation."

Rigid Insulation. Rigid insulation is used for the perimeter of slabs. Since there is a significant heat loss from the edge of the concrete slab to the foundation, a barrier of rigid insulation is placed inside the foundation before the slab is poured. The material may be compressed mineral wool with an asphalt stiffener adhesive, or an expanded polystyrene plastic, **Figures 5 and 6.**

Applying Insulation

Applying Loose Fill Insulation. Blowing machines for loose fill insulation are usually installed in trucks, which also carry the sacks of blowing wool. The insulation fibers are blown through a special hose that comes in 50′ lengths. The machines are either hand-fed or automatic; feeding is controlled by a buzzer or bell signal line pressed by the hose man.

Mineral wool fibers and cellulose fibers or flakes can be blown pneumatically into stud spaces in walls, and between ceiling joists, **Figures 8, 9 and 10.** Blowing is the only way to insulate walls of existing houses, short of complete removal of the inside walls for the application of blanket insulation.

In new construction, blankets are most commonly used for both ceilings and walls, sometimes wall spaces receive blankets, and the ceiling is blown; least often both walls and ceilings are blown. In new construction, too, a plastic film vapor barrier is placed on the inside of the studs, beneath the drywall or plaster; the outside sheathing is drilled in each stud space, to receive the pneumatic hose nozzle.

Insulating wool can be blown into the space between a flat roof deck and the ceiling of the top floor of an apartment building, **Figure 11.** In many remodeling jobs, brick is removed and insulation blown into wall of an existing house, **Figure 12.**

Applying Blanket Insulation. Blanket insulation is applied to the interior of a masonry wall by attaching furring strips to the masonry and attaching blankets with a stapling hammer **Figure 13.**

In new frame houses, the most common practice is to staple blanket insulation between the studs. The edges of the studs, **Figure 14,** provide the best protection and the greatest effectiveness to the vapor barrier function of the wrapper, but some drywall applicators find it hard to get a smooth drywall joint over the edge stapled flanges. Side stapling requires careful stapling at no greater than 6″ intervals to prevent gaps at the edge of the blanket, **Figure 15.**

Fig. 8. Blowing loose wool into holes cut in stucco wall. Skilled mechanic can patch holes to escape detection.

Mineral Wool Assn.

Fig. 9. Blowing wool requires professional application. Air pressure from machinery in a truck parked outside the house forces insulation through a hose. A skilled workman makes sure coverage is even.

Fig. 10. To avoid breaking brick wall, first course of roofing and deck was removed to blow mineral wool into wall.

Fig. 11. Blowing wool insulation into the space between a flat roof and the ceiling of the top floor.

Fig. 12. Even if wool is blown into the top of a brick wall, a brick should be removed beneath each window to fill that space.

Fig. 13. Furring strips are attached to a masonry wall and blanket insulation is stapled to the furring strips.

Stapling flanges to the sides of studs and joists is essential when the vapor barrier has an aluminum foil facing. This allows for a ¾″ air space between the foil and the gypsum board or paneling, which permits the foil to reflect heat back into a heated room. Note the use of an aluminum foil vapor barrier between raters in **Figure 16.**

It is essential that insulation be closely fitted in narrow spaces, **Figure 17,** and diagonal bracing, **Figure 18.** Loose insulation torn from batts or blankets should be stuffed into cracks around windows, doors, outlet boxes, service entrances, **Figure 19,** and around ceiling lights, **Figure 20.** A cardboard or metal circle, about 6″ in diameter, should be placed above each light housing that protrudes above the ceiling to keep insulation away from the top of the housing.

Applying Batt Insulation. The ceiling requires the most effective insulation to prevent heat loss as heat rises and Batts should fit tightly in the space between joists without compression. Batts are especially useful under floors because of their shorter lengths, which makes them easier to handle where they must be cut to fit tightly against bridging, **Figure 21.** Where they are used under floors over cellars or crawl spaces with earth floors, they must be applied with their vapor barriers upward and supported with slats, **Figure 22,** or chicken wire secured with staples and lath, **Figure 23.** Thickness of batts here should be R-19 or R-22.

Where walls have "cats" or midway bracing and the studs are 2×4's, the batts should be 4-ft. lengths, 3½" to 4" thick (R-11), with flanges stapled to the studs, **Figure 24.**

In existing houses home owners are urged to add to their attic insulation to bring it up to at least R-19 and even to R-38 (12" thick.) Since R numbers are additive, two layers of R-19 batts can be used to achieve a total of R-38. The first layer should have a vapor barrier placed face down. The second layer should have no vapor barrier or if this unobtainable its barrier should be slashed, **Figure 25.**

Where loose fill insulation is already on the attic floor that already has a vapor barrier underneath it, additional R-19 batts without vapor barrier can be added, **Figure 26.**

Applying Batt Insulation. The ceiling requires the most effective insulation to prevent and reduce the transmission of heat. Six inch batts between the attic rafters or floor joists (according to whether the attic space is finished or not), or 6" of blown or poured

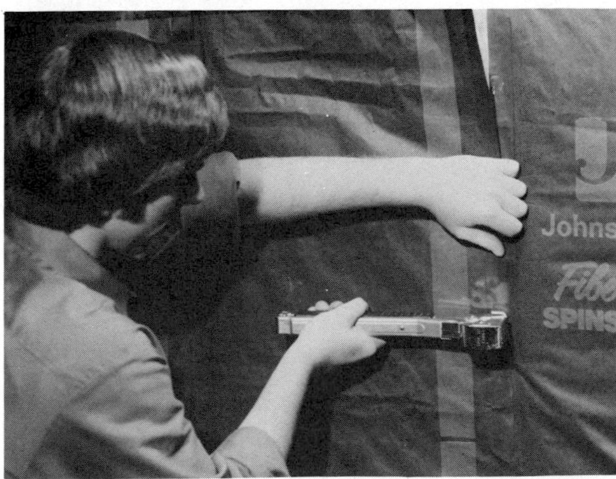

Johns-Manville Corp.

Fig. 14. Above, flanges of batt insulation are stapled to edges of studs, rafters and collar beams to insulate attic. Below, fiberglass insulation being stapled tightly through its flanges every 6" against front edge of stud to form vapor barrier.

Fig. 15. Stapling insulation flanges to face of studs requires more care to avoid gaps which let vapor pass.

Johns-Manville Corp.

Fig. 16. Mineral wool blankets faced with aluminum foil are side stapled to rafters above. Batts are used on end wall here and flanges stapled to sides.

Fig. 17. Blanket is cut lengthwise to fit narrow space between studs and edges of vapor barrier are stapled to studs.

Fig. 20. Loose insulation is tucked around light fixture to insure full protection against fire and draft.

Fig. 18. Full thick blanket insulation is carefully cut to fit above and below diagonal bracing in corner of wall.

Fig. 21. Batts are cut to fit snug against bridging and electric cable to avoid gaps in continuity of insulation.

Fig. 19. Strips cut from batts, or loose wool are stuffed into spaces around electric service entrance and piping.

Fig. 22. Insulation beneath main floor is supported by slats.

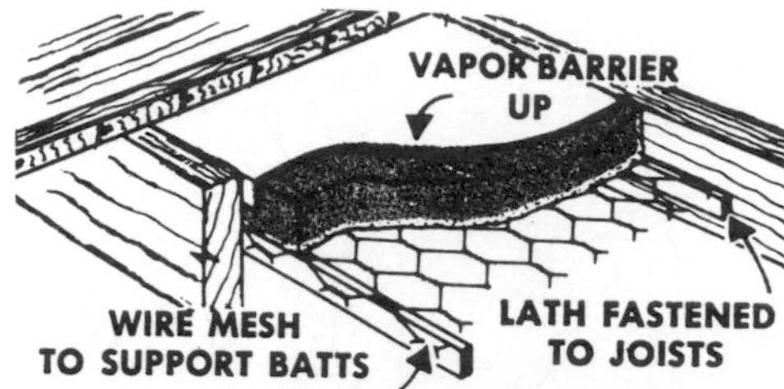

Fig. 23. Above, insulation under main floor is supported by wire mesh. Below, six-inch batts fit snugly between ceiling joists. Flanges overlap when stapled, forming continuous vapor barrier.

insulation are the standard installations.

Applying Rigid Insulation. In addition to the use of rigid insulation in a slab foundation and slab floor; rigid and blanket insulation are used in stressed skin panels made up as plywood components in roof panels as shown in **Figure 27.**

Air conditioning ducts are available in flat sheets of fiber glass, scored for bending. These ducts are faced with aluminum foil, **Figure 28.** The shaping is held by adhesive and tape. The lengths are joined together and held with adhesive and caulked.

Insulation and Comfort

By retarding heat transfer, insulation has a significant influence on indoor comfort conditions and the ease with which they can be controlled. Most adults are comfortable at temperatures of 70 to 72 degrees F. in mild weather, but prefer higher temperatures during severe cold periods. The passage of radiant heat from the human body to wall, floor, and ceiling surfaces, of markedly lower temperature, makes it necessary for room temperatures to be raised to compensate for the loss of body heat.

Although there are no existing standards which establish surface temperatures of enclosing walls, floors, and ceilings, it is generally agreed that these surface temperatures, when more than 10 degrees F. below the average room temperature, can cause discomfort. In extremely cold weather, a difference of about 6 degrees F. between room air and wall surfaces, and 4 degrees F. for ceilings under unheated attics, will provide reasonable comfort.

Adequate insulation installed in ceilings, floors and side walls will materially reduce these temperature differences in cold weather. In warm climates, insulation will retard the inflow of heat, or heat gain, particularly through roof and wall surfaces exposed to direct rays of the sun.

Minimizing the differences in temperature between enclosing surfaces and room air not only increases the comfort of the occupants, but also results in significant savings in fuel by reducing heat losses from the home.

Where Heat is Lost

Homeowners must buy enough heat in fuel to offset the escape of heat from the house, plus the efficiency loss of the heating system. The amount of heat lost from the average house varies with its type of construction, style, amount of insulation, and many other factors.

Johns-Manville Corp.

Fig. 24. the 2×4 studs get R-11 (3½") of mineral wool in this wall which is the most that can be used with 2×4 studs.

Mineral Wool Assn.

Fig. 25. Home owner installs a second layer of mineral wool batts in an attic floor to achieve a thermal resistance value of R-38. Resistance values of 30 and 38 are used more and more frequently. The additional cost is justified by increased fuel prices.

The principal areas and approximate amounts of heat loss from an uninsulated two-story house, with a basement are:

Walls	33%
Window glass and doors	30%
Ceilings and roof	22%
Air leakage (infiltration)	15%
Floors	Negligible
Total heat loss	100%

Heat gain, during the summer months, will roughly equal the heat loss percentage for each part of the house.

Houses of different architectural style will vary in their heat loss characteristics. A single-story house, for example, would contain less wall area and a proportionately greater ceiling area than a two-story house. Greater heat loss through floors would be expected in houses erected over concrete slabs or unheated crawl spaces.

Windows and doors are generally sources of greater heat loss than either walls or ceilings. Discomfort caused by these colder surfaces should be offset, as far as possible, by reducing the heat transfer through other sections of the house. Weather stripping around windows and doors will reduce air infiltration more than 50%. In colder sections of the country, heat losses through glass surfaces can be reduced 50% or more by installing double glazed windows, or by adding storm sash and storm doors.

The transfer of heat through uninsulated ceilings, walls or floors can be reduced almost any desired amount by incorporating insulation. Maximum quantities of insulation in these areas can cut heat losses more than 90%.

Controlling Water Vapor By Insulation. Modern construction results in a tight-house, one in which there is no easy movement of free air through sheathing, side walls, doors and windows. Properly installed vapor barriers will protect ceilings, walls, and floors from condensation of moisture originating within the home. Such barriers should be applied carefully on the warm side of insulation to provide a complete envelope, thus preventing moisture vapor or water vapor from entering enclosed wall spaces.

Weather stripping seals gaps, and panel sheathing of cellulose board or plywood is less permeable than rough diagonal board sheathing. One result is that moisture, loosed into the atmosphere within a house, from cooking, washing, breathing and other processes, does not freely move out of the house. Unless prevented, moisture or water vapor will move slowly into and through the walls. Reference should be made to the chapter titled, 'Painting and Finishing'.

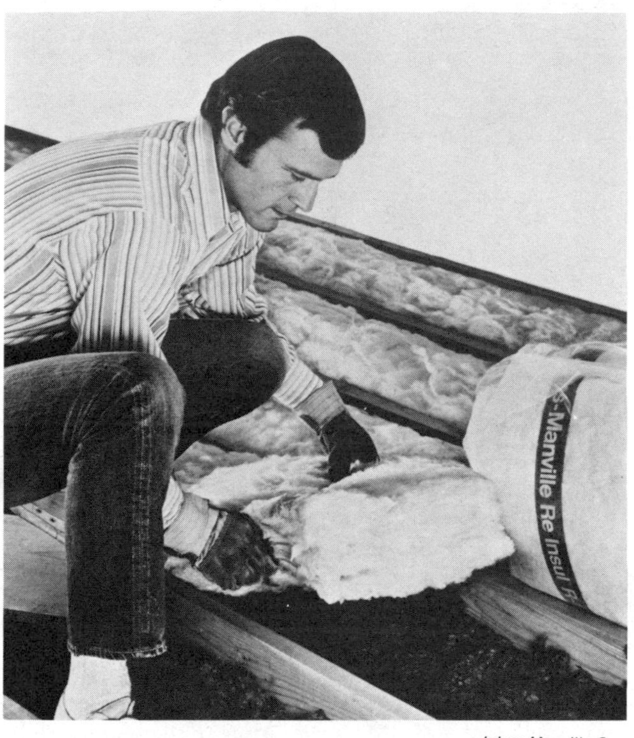

Johns-Manville Corp.

Fig. 26. Home owner adds fiber glass batts without vapor barrier to inadequate rock wool between attic joists to bring his insulation up to R-19 value.

Fig. 28. Faced with aluminum foil, these fiberglass heating and air conditioning ducts are made in flat sheets, scored for bending. Shape is held by adhesive and tape. Lengths are taped together, intersecting joints adhered and caulked.

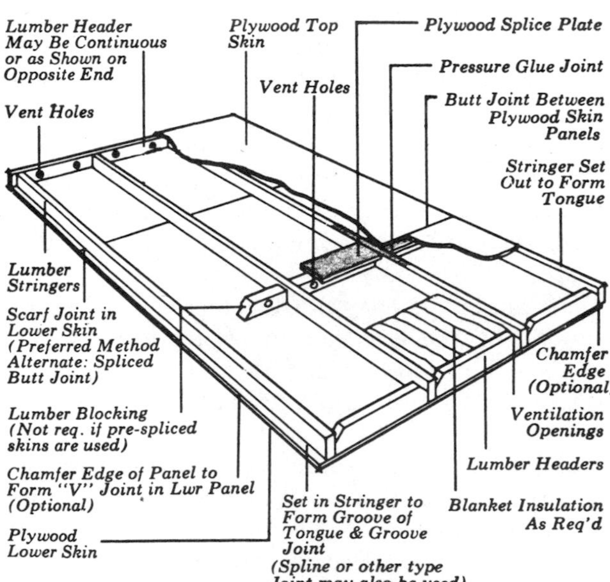

Fig. 27. Plywood stressed-skin panel has insulation inserted between stringers before top skin is nailed on.

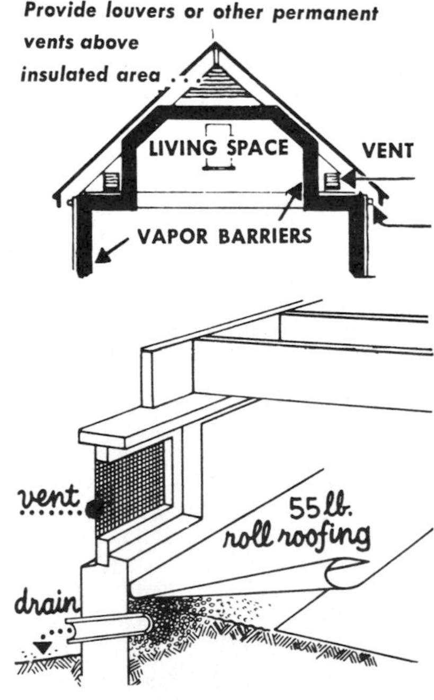

Fig. 29. Ventilation should be provided around attic space (top) and in crawl space. Note vent, drain and ground vapor barrier.

There is less harm from this in warmer months, when the moisture remains in vapor form. But in colder periods, the water vapor making its way outward will meet a cold layer near the outside, usually at the sheathing. There it will condense, and eventually damage the sheathing and cause paint peeling of the siding and efflorescence on brick. Moisture may also saturate the insulation, reducing its effectiveness.

Insulation Controlling Water Vapor

Moisture must be prevented from taking this escape route from the interior of the house. That is why most blanket insulation, whether of mineral wool, cellulose, or aluminum of copper foil layers, has its inner wrapper of an impervious material, such as asphalt impregnated paper or aluminum foil. Where no vapor barrier is provided on insulation, sheets of polyethylene film stapled to studs after installing batts will keep walls dry.

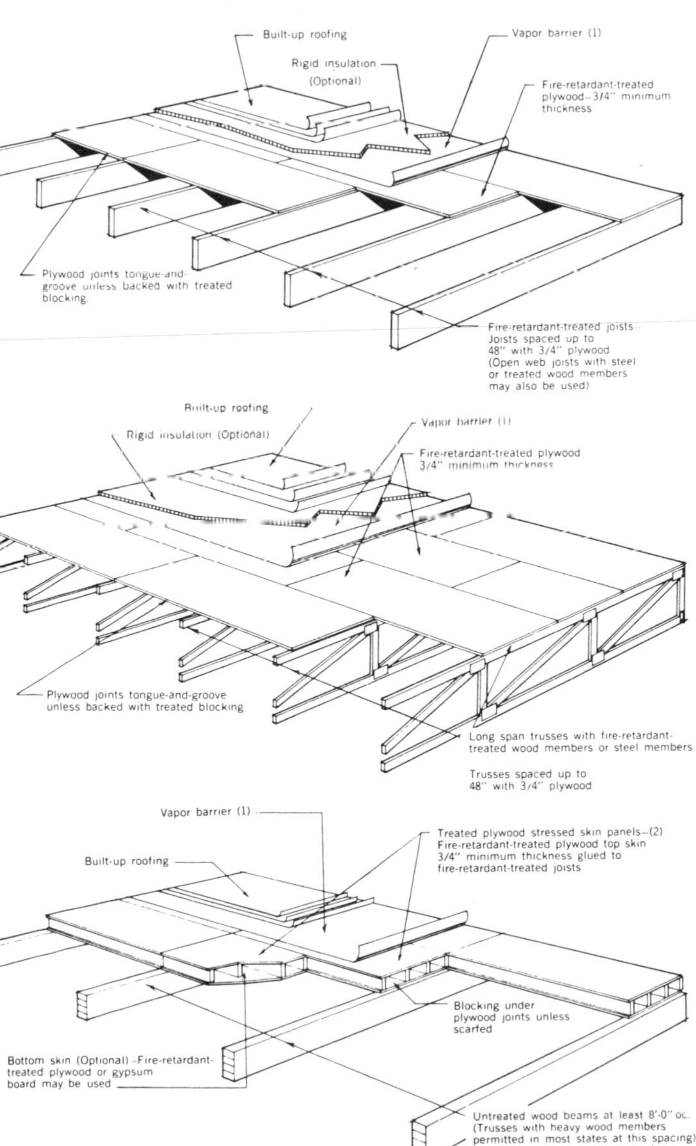

Fig. 30. A vapor barrier must be installed beneath built-up roofing to prevent moisture from separating the layers.

Applying Water Vapor Insulation

The common expression, "a house must breathe," has done a great deal of harm when wrongly applied. Some builders think it means spaces must be left at top and bottom of the insulation in each stud space. This not only reduces the insulation value but also destroys the vapor barrier. If anything, a "house breathing" means that the outer surface of the stud space should have access to the outside atmosphere, so that moisture that makes its way inside the walls can escape.

The simple rule is: the water vapor barrier insulation must face the warm side during the colder weeks of the year. Attic insulation is placed with the impervious wrapper side down, toward the ceiling of the rooms below. Underfloor and crawl space insulation is placed with the water vapor barrier up, near the subflooring.

Applying Vapor Barrier Under Roofing. Figure 30 shows the use of vapor barriers in conjunction with the framing of roofs using fire retarding plywood and framing stock in joist, truss and stressed skin panel roof framing systems. The vapor barrier is stapled to roof decking before applying insulation and built up roofing.

Insulation Controlling Sound Transmission

Sound is the sensation perceived by the sense of hearing. Noise is a noticeably loud, harsh or discordant sound. Noise can be transmitted from source to listener entirely through the air or often through the air and the building structure. In the latter case, the air borne sound striking a surface or by mechanical impacts directly on and to the structural materials. Noise control procedures interrupt the transmission path at some point either by reflecting or absorbing the energy. One common method of insulating against sound is to weave blanket insulation between staggered studs, **Figure 31.**

In controlling sound transmission in walls, a slit stud wall with a 1-1/2″ blanket insulation hung from the top plate offers greater resistance to noise transmission than standard stud walls, **Figure 32,** and does not increase the wall thickness.

Staggered steel studs with 3/8″ gypsum lath and plaster compares favorably with staggered wood studs and insulation, **Figure 33,** in reducing noise transmission.

Even better reduction in the transmission of noise can be achieved by using 2×6 top and sole plates, staggering 2×4 studs and weaving 4″ mineral wool blankets between the studs or placing 4″ batts between them, **Figure 34.**

Sound Controlling Insulation and House Planning

The arrangement of rooms in a house and the location of the structure on the site are important factors in controlling sound. At one time a large front porch was an important factor in house design. Now, generally speaking, the front has been reduced to an attractive entrance and, sometimes, a stoop, due to the ever increasing noises from the roadway or street.

A windowless wall or garage, or both, may act as adequate sound barriers against street noises or play yards of neighbors.

Present day house design often places the outdoor-indoor living areas to the rear of the house, away from street noises.

The separation of quiet rooms such as bedrooms or study, from noisy rooms such as play areas, living room, kitchen, utility room or workshop is a matter of room design and layout. The quiet rooms of a house might well be separated from the noisy rooms by closets, halls or a fireplace. Wall, floor and ceiling construction using adequate insulation, **Figure 35,** is necessary to control sound.

Controlling Home Equipment Noises. Sound control or noise control may also be brought about by the reduction of vibration noise produced by a furnace, an air conditioner, kitchen or laundry appliances. The use of felt, cork or rubber mounts for appliances or utilities is always advisable to reduce vibration noise. The need for repair, adjustment and lubrication of household appliances and utilities is a must if the home owner desires to reduce unwanted and unpleasant noises.

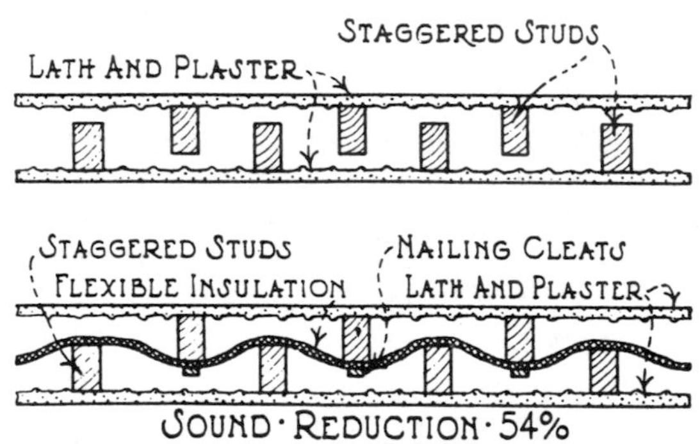

Fig. 31. Blanket insulation between staggered studs.

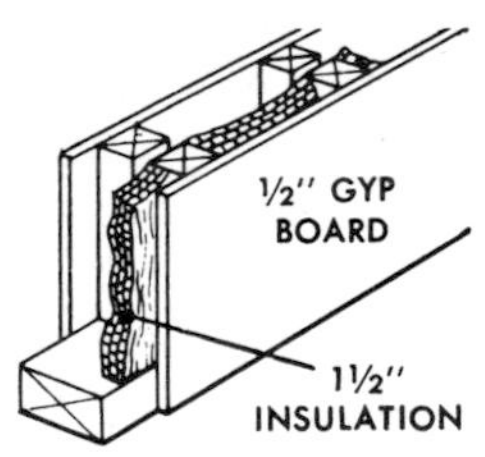

Fig. 32. Slit studs retain standard wall thickness.

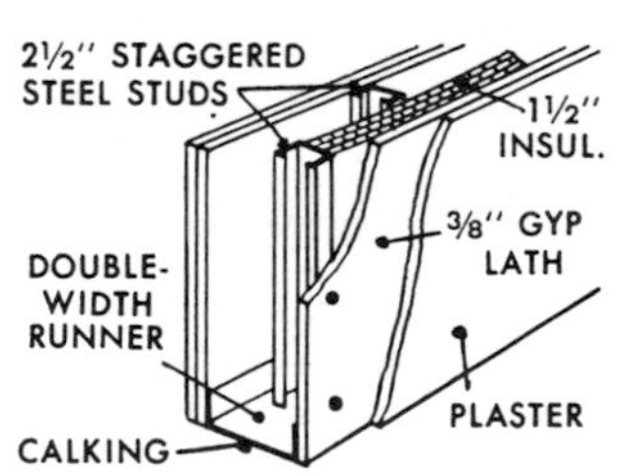

Fig. 33. Staggered steel studs set on double width runner.

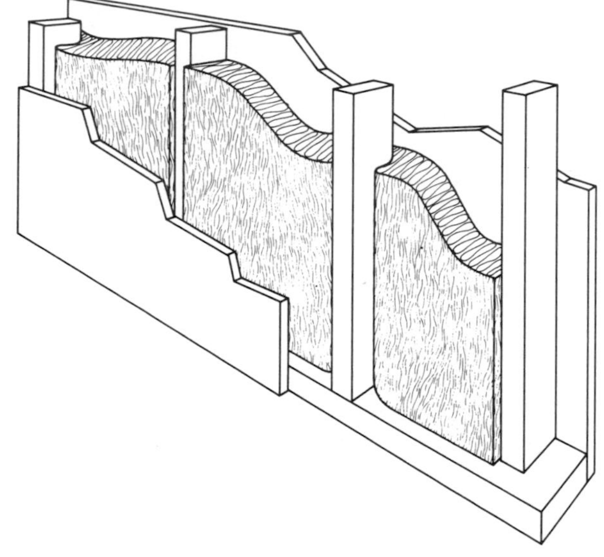

Johns-Manville Corp.

Fig. 34. Staggered 2" × 4" wood studs: single layer 1/2" gypsum board each side: one thickness 4" fiber glass Sound Control Batts on 2×6 ceiling and sole plates will produce Sound Transmission Class 52 which will prevent very loud voices from being heard.

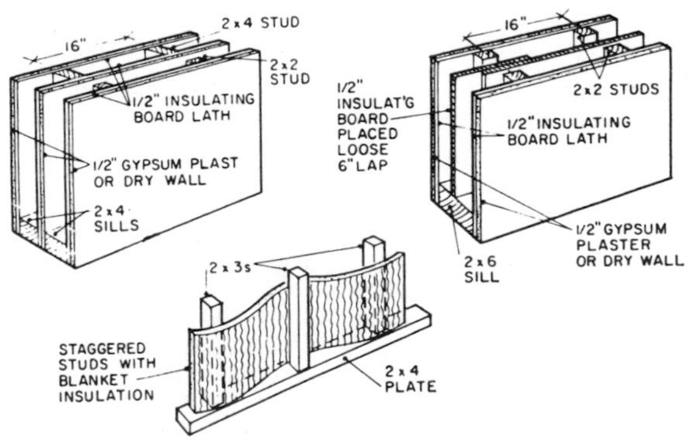

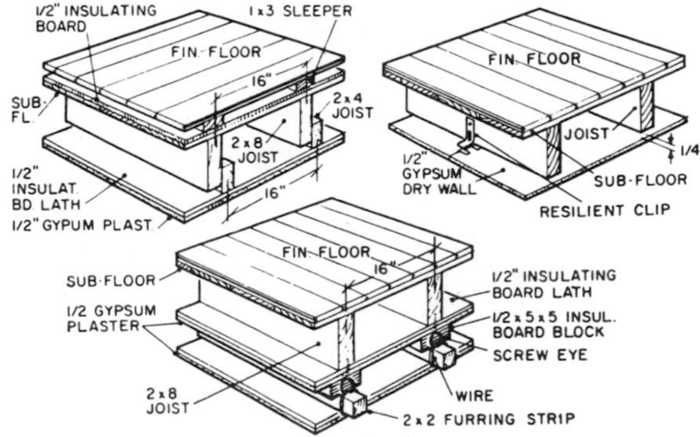

Fig. 35. Methods of reducing sound transmission in walls, floors and ceilings of single family homes and apartments.

Chapter 13

STAIR CONSTRUCTION

A stair is a series of steps or flight of steps used to walk from one floor level to another. A staircase is a completed stairway consisting of stringers, treads, risers, hand rail and often balusters, newel post and other stair parts.

A staircase may be made of wood, stone, concrete or metal or a combination of these materials.

A stairwell is a shaft, through one or more floors of a house, in which the stairway or staircase is built or placed.

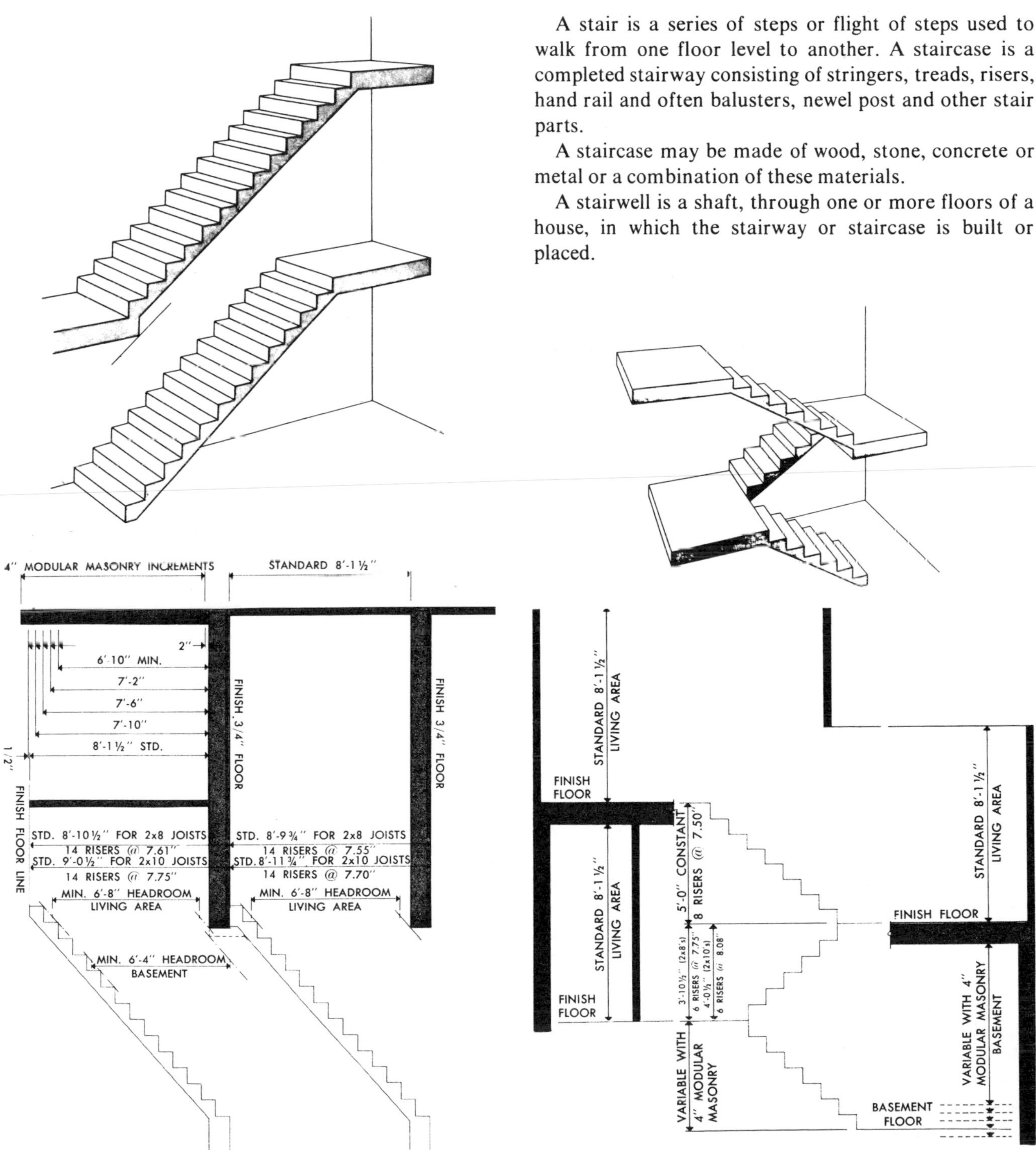

Fig. 1. Standards for main stairs and basement stairs.

Fig. 2. Standards for split-level stairs.

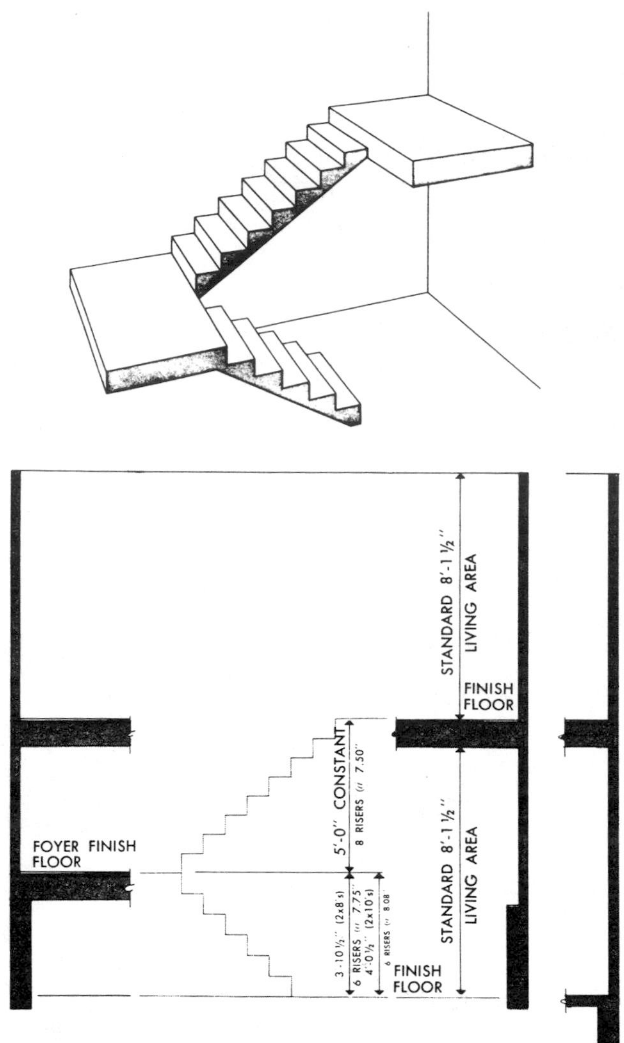

Fig. 3. Standards for bi-level stairs.

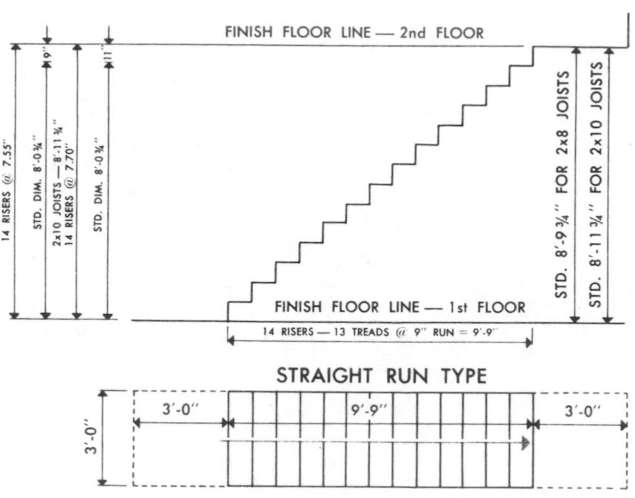

Fig. 4. Dimensions and design of straight run stairs.

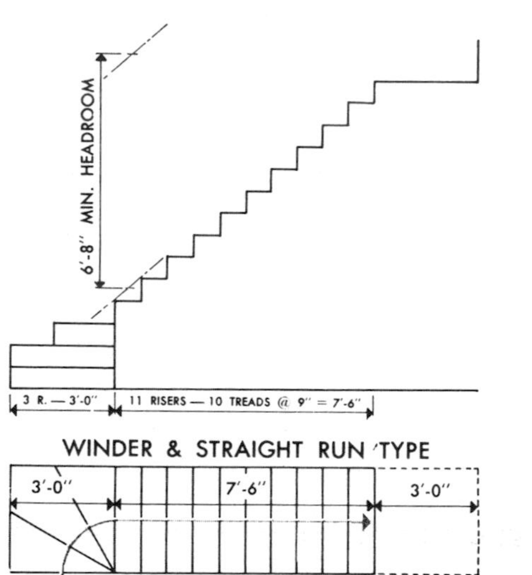

Fig. 5. Straight run stairs with winders.

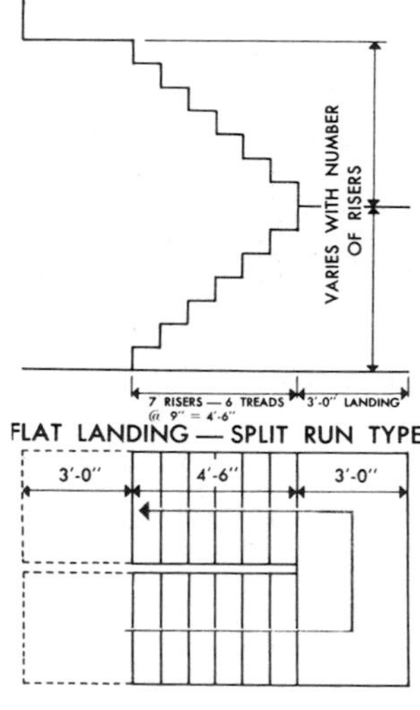

Fig. 6. Split run stairs with a flat landing.

Standards for Stairs

Good stairways require careful planning to eliminate dangers created by poor design. Stairs must be of sound construction and of good materials; both requirements can be advantageously controlled in shop-fabricated component units. Basic stair standards require the minimum number of different units to satisfy the greatest flexible use of variables ordinarily encountered in the basic house types.

Floor-to-floor dimensions, between any two major living levels, are always constant when standard wall heights are used, varying only with the joist size in floors. One standard dimension for 2″ × 8″ upper floor joists, and a 2″ larger one for 2″ × 10″ joists are provided in **Figure 1.**

Main stair standards provide for 14 equal risers and 13 equal treads, except where winders or landings are required. Basement stairs are coordinated in design height with 4″ modular masonry increments, as indicated. Basement stair standards also allow for 2″ × 8″ and 2″ × 10″ first floor joists.

Curved, circular and special angular stairway runs are not included in these particular stairway standards. Minimum clear headroom for main stairs is 6′ - 8″ and, for basement stairs, 6′ - 4″.

Standards for Split-Level and Bi-Level Stairs. Split-level houses usually have three major living levels, **Figure 2.** Bi-level houses, **Figure 3,** have only two. Stairway designs, however, for the split-foyer entrance of the bi-level are comparable to the middle level of split-level houses. Floor-to-floor relationships are identical for standard stairs, when standard wall heights are used for the major living levels. All minimum and maximum dimensional requirements are the same for split stairs and main stairs. Although floor-level relationships may vary with design conditions, the stair flight between middle and upper levels remains a constant dimension with either 2″ × 8″ or 2″ × 10″ joists. Two standard riser dimensions are provided between the middle and lower levels to adjust for the two joist depths. A basement stair may be used for the optional fourth level. The standards for the number of risers and the riser height, required for this stair, are coordinated with 4″ modular masonry increments. The other floor levels and stairs are also coordinated with modular masonry dimensions.

Standards for Stairs with Winders or Flat Landing and Split Run. Main stairways may be of straight run design, **Figure 4,** or may vary the run direction by the addition of winders or landings at the stair turn, **Figures 5 and 6.** Regardless of the run variables, riser height should be the same in each flight. The maximum stair or open-landing width is 3′. For stairs that turn with winders, two risers are the maximum number across the turn. The run per rise on winders, is 18″ from the converging end, shall not be less than 9″. Handrails are needed on at least one side of each flight of stairs which exceeds three risers.

Standards for Widths of Stairs. Main stairs should have a minimum width of 2′ - 8″ clear of the handrail. A basement stairs should have a minimum 2′ - 6″ clearance, **Figure 7.**

Standards for Stair Landing. The minimum dimensions for a stair landing is 2′ - 6″ when a door is placed at the top of any stair run that swings toward the stair, **Figure 8.**

Standards for Open Riser Stairs. The minimum stair run per rise should be 9″ in basement stair construction. With closed riser construction, the minimum width of the tread should be 10-1/8″; with an open riser construction, **Figure 9,** the minimum width of the tread should be 9-1/2″. The maximum rise should be 8-1/4″.

Standards for Closed Riser Stairs. In closed riser stair construction the main stair tread width should be a minimum of 9″. With open or closed riser, the minimum tread width should be 10-1/8″. The maximum rise should be 8-1/4″, **Figure 10.**

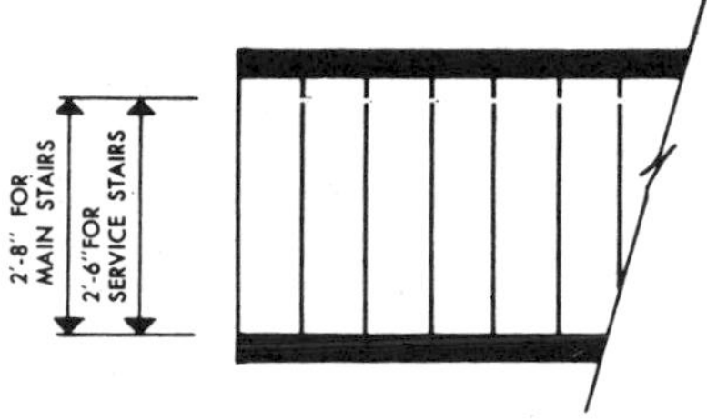

Fig. 7. Minimum widths for main stairs and service stairs.

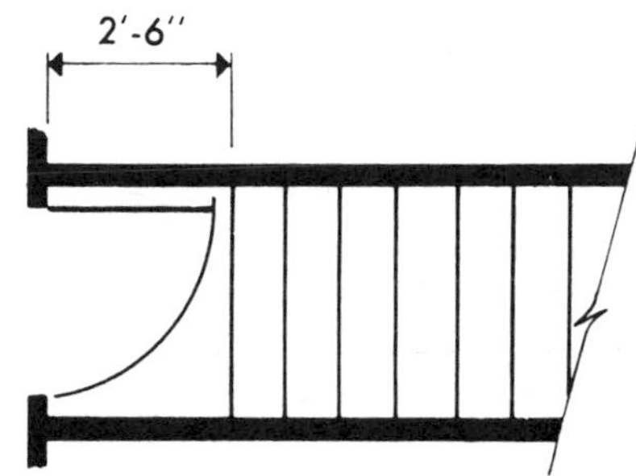

Fig. 8. Stair landing with door at top of run.

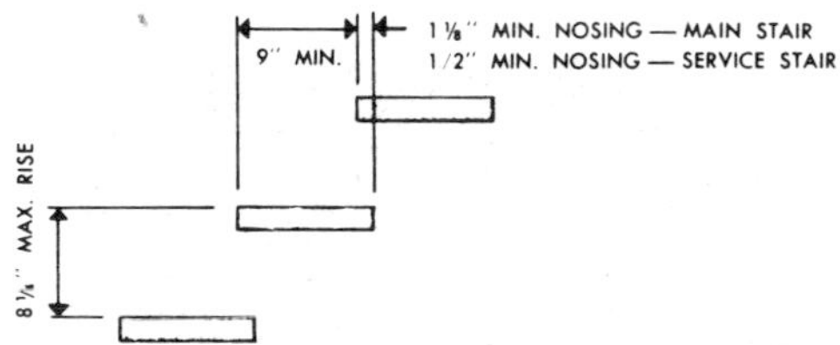

Fig. 9. Dimensions for open riser construction.

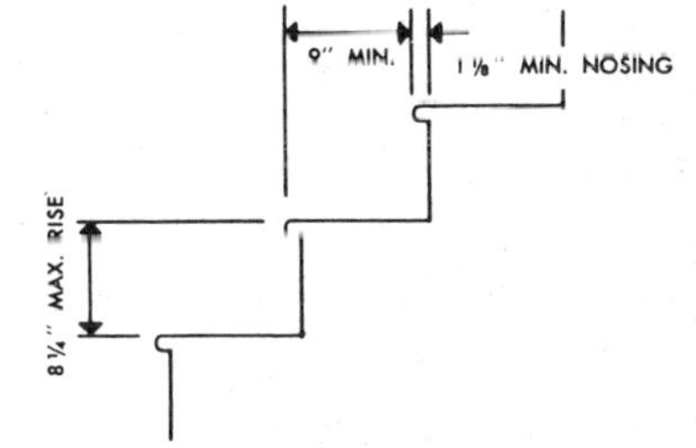

Fig. 10. Dimensions for closed riser construction.

Safety Factors for Stairs, Ladders, Ramps and Inclines

Statistics show that 40% of home accidents are falls, about a third of which take place on stairways and steps. Certain desirable dimensions for stair risers and treads have been worked out under the Safety Engineering Department of the National Workmen's Compensation

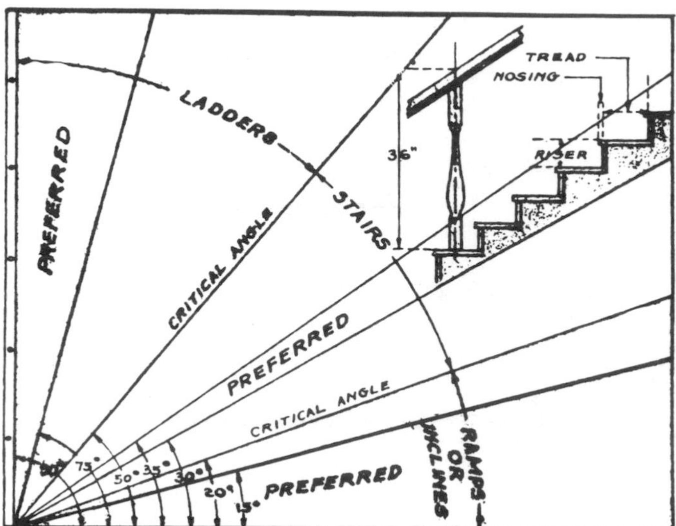

Fig. 11. Table of preferred and critical angles of stairs ladders, ramps, and inclines or ramps.

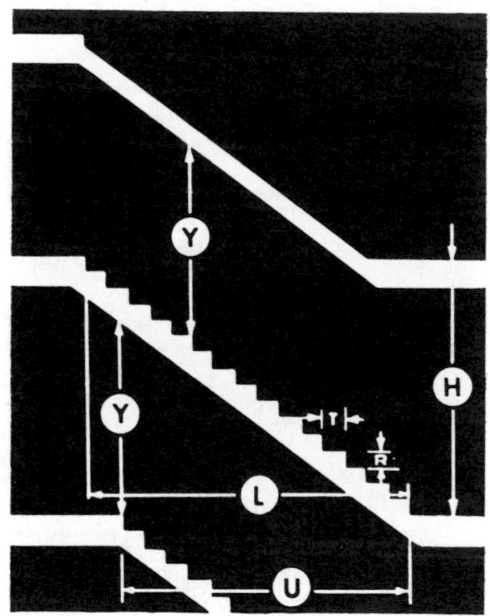

Height Floor to Floor H	Number of Risers	Height of Risers R	Width of Treads T	Total Run L	Minimum Head Room Y	Well Opening U
	12	8"	9"	8'-3"	6'-6"	8'-1"
8'-0"	13	7⅜"+	9½"	9'-6"	6'-6"	9'-2½"
	13	7⅜"+	10"	10'-0"	6'-6"	9'-8½"
	13	7⅞"–	9"	9'-0"	6'-6"	8'-3"
8'-6"	14	7 5/16"–	9½"	10'-3½"	6'-6"	9'-4"
	14	7 5/16"–	10"	10'-10"	6'-6"	9'-10"
	14	7 11/16"+	9"	9'-9"	6'-6"	8'-5"
9'-0"	15	7 3/16"+	9½"	11'-1"	6'-6"	9'-6½"
	15	7 3/16"+	10"	11'-8"	6'-6"	9'-11½"

Fig. 12. Rise and run: H indicates distance floor to floor; Y indicates minimum headroom; L indicates total run (horizontal distance between first and last riser); U indicates well openings (usually based on 6'-6" minimum headroom.

Bureau. Satisfactory values are 6-3/4" to 7" height for the riser, and 10-1/2" to 10-3/4" for the width of the tread.

Slight variations may be made, but the sum of tread and riser should equal about 17-1/2", exclusive of the nosing, and the angle of rise of the staircase should be between 30 and 36 degrees. A good principle is: **the sum of two risers and the tread should be between 24" and 25".** Safety standards for stairs, ladders, ramps, and inclines are indicated in **Figure 11.**

The general safety standards for stair construction recommended by the National Workmen's Compensation Bureau are as follows:

1. Stairs should be free from winders.
2. The dimensions of landings should equal or be greater than the width of stairways between handrails (or handrail and wall).
3. Landings should be level, and free from intermediate steps between the main up flight and the main down flight.
4. All treads should be equal and all risers should be equal in any one flight.
5. The sum of one tread and one riser, exclusive of the nosing, should be not more than 18" nor less than 17".
6. The nosing should not exceed 1¾".
7. All stairs should be equipped with permanent and substantial handrails 36" in height from the center of the tread.
8. All handrails should have rounded corners and a surface that is smooth and free from splinters.
9. The angle of stairs with the horizontal should not be more than 50 degrees nor less than 20 degrees.
10. Stair treads, if used, should be splipproof and firmly secured, with no protruding bolts, screws, or nails.

Conventional Stairwell and Stair Construction

Most stairs, in modern dwelling construction, are straight for economy and simplicity. The components of staircases, adequate for most one floor plan, split-level, or bi-level houses, are stock items.

Basement stairs are often open at both ends and do not have risers. These simple, stringer-tread stairs, are often prefabricated in the builder's shop or can easily be cut at the building site. In the building of a two story house, with a half-way stair landing, the half-flight stairs and the six or seven step stairs of split-level houses, are often shop fabricated for ease in transporting.

The main components of a staircase are the stringers, the treads, and the risers. The design of the stringers is

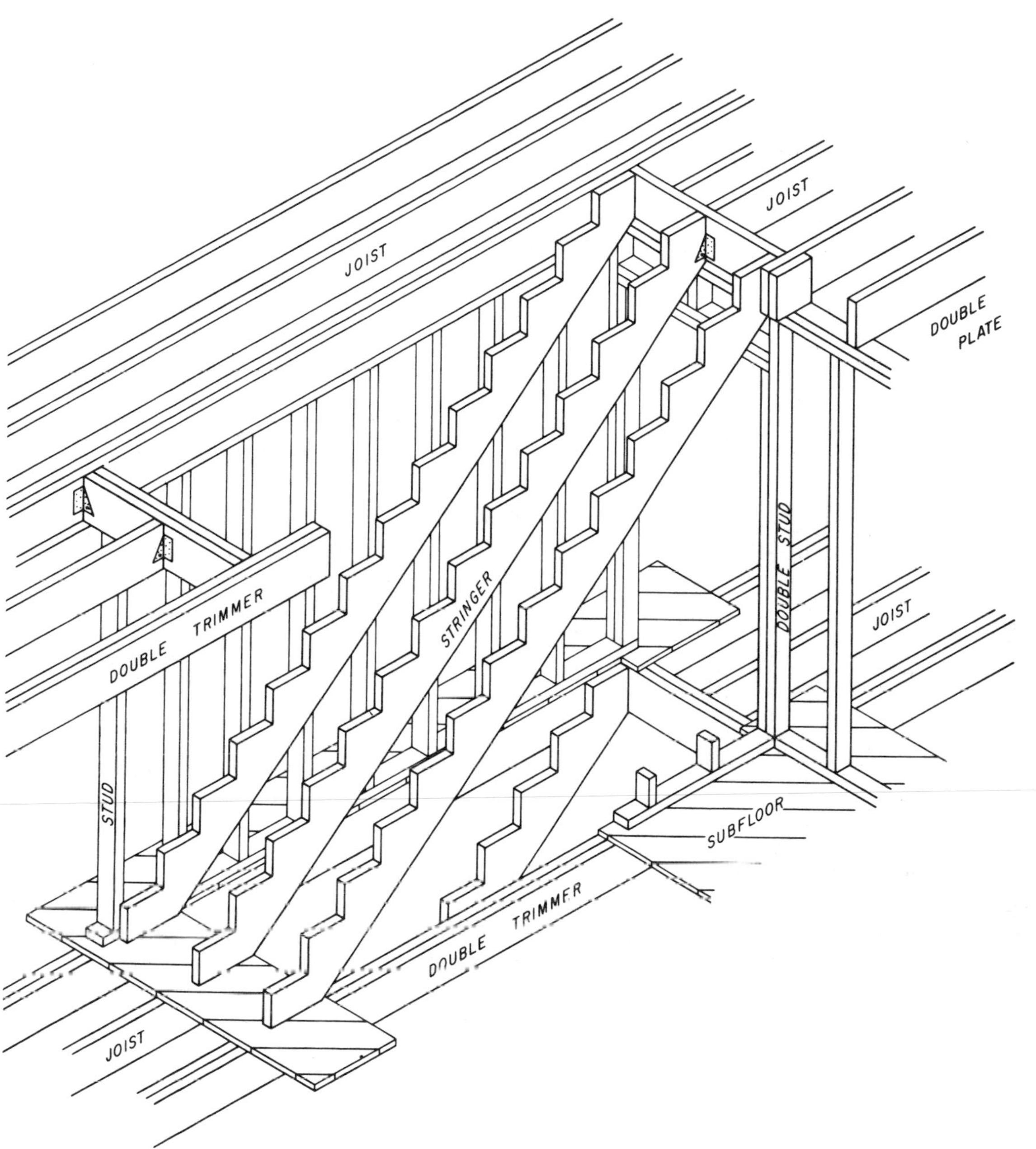

Fig. 13. Framing details for a stairwell running from the basement to the first and second floor levels.

governed by the needed rise and run of the stair. The rise is the vertical distance between the two levels to be connected. The run is the horizontal distance between the bottom riser and the perpendicular drop from the top riser. The steeper the rise, the taller each riser will be compared with the depth of the treads, **Figure 12.**

Various Terms Used in Stair Work. Stair stringers are also called carriages or horses, with or without an exact difference in meaning. Generally, the term "stringer" means a heavy plank, 2″ × 10″ or 2″ × 12″ in size, cut at right angles to receive the treads and risers. This may also be called an "open stringer" A "carriage" often refers to an uncut plank to which triangular blocks are glued and nailed to receive the treads and risers. This is seldom done in modern construction.

A Stair Well. A stair well is the framed opening in the floor or floors of the rooms in which the stairway is built or the staircase is placed. **Figure 13** shows a set of three stair stringers placed in the stair well which extends from a first floor level to a second floor level. In the stair well opening which extends from the ground floor level to the basement floor level, but one stair stringer is in place. One other stringer is yet to be placed against the double trimmer in the foreground.

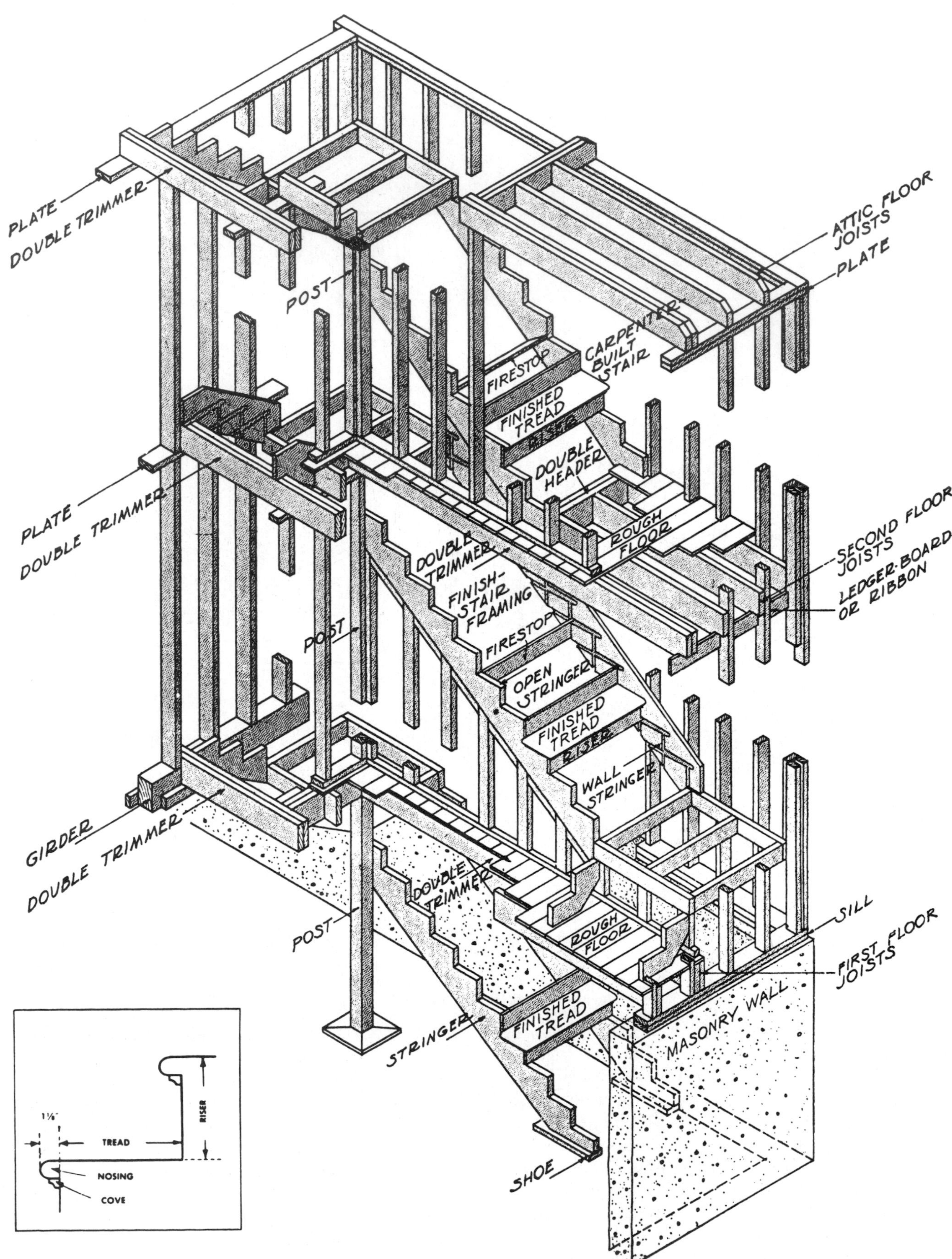

Fig. 14. Stairway details, showing double trimmers around openings and two types of commonly used stringers.

A Four Level Stair Well and Stairway. When a rough basement staircase runs along a wall, the outside stringer is likely to be "open" or exposed, **Figure 14.** The wall stringer is often set next to an uncut sloping member fastened to the wall. For finished work, the stringers are not cut out, but are dadoed to receive the treads and risers in incised grooves as shown in the wall stringer running from the first to the second floor levels.

This series of stairs extends from the basement floor level to the attic floor level, forming a four level stair well, accurately and sturdily built with horizontal and vertical framing members. Wedges are applied to fix the members tightly, then treads and risers are glued together with blocks to prevent loosening and cracking, **Figure 15.** The use of wedges and glue blocks in a grooved stringer is termed "housing."

The Balustrade in Stair Construction. An open stair, that is, one using open balusters and handrail, or a free standing rail, usually has a housed outside stringer for firmness and appearance, **Figure 16.**

Fig. 15. Wedges driven into housing of wall stringers hold treads and risers without nails. Risers may be let into the tread above and grooved for the tongue of the tread below.

Fig. 17. Free-floating handrail supported by wall brackets.

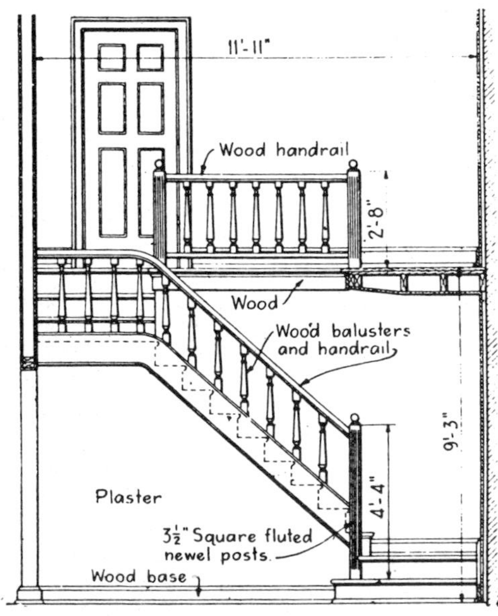

Fig. 16. Elevation of open stairs with protecting handrail. Outside stringer is housed; handrail is free-standing.

Fig. 18. Cased free side of stairway with floating handrail.

Fig. 19. Rail and balcony treatment in split-level house.

Hand rails are necessary both for safety and to protect the wall finish, when the rail is fixed to the wall side. Today the trend is to free-floating handrails, supported at intervals by blocks or brackets, **Figure 17.**

The outside of a stair is often cased in, where formerly one would find a graceful balustrade and landing rail, **Figure 18.**

Various Types of Stairways in Today's House. Due to the fact that the split-level house offers varied choices in balconies over the living or family room, the stair rail and the upper balcony rail offer a variety of treatments, **Figure 19.** Metal railings, **Figure 20,** of wrought iron or of shaped aluminum, are widely used for informal effect.

The classical handrail of mill-shaped wood is supported by vertical balusters. The balustrade starts at the newel post, which is housed on the bottom tread. This tread extends in a semi-circle, quarter-circle, or scroll to receive the newel in formal balustrades. A simple square newel is often let into the edge of the bottom tread, **Figure 21.** Colonial stairs often include a gooseneck as part of the handrail. Like the handrails, goosenecks are available as prefabricated stair parts and may be made of birch or oak. They are used as decorative sections to join handrails to newel posts at landings and upper floors, **Figure 21.**

If the starting newel is placed on an extended tread, the rail must curve out to meet it. The curve or crook is called an easement. A volute is the final graceful curve which rests on the newel post. Sometimes the volute is supported not alone by the newel post but by a ring of slender balusters, **Figure 22.** Handrails may also be started with a turnout easement and newel cap. Like volutes, turnouts may curve to the right or to the left and are another prefabricated part milled to the style of the handrail, **Figure 23.**

If the stair case includes a landing, there may be a landing newel post that repeats the design of the starting newel. The landing, or angle newel, often descends below the landing in a decorative "drop." If the balustrade ends at a wall, a half newel is used to receive the rail, or a simple round rosette serves the purpose, **Figure 24.**

Treads and Risers in Stair Construction. The flat, horizontal steps, or treads, bearing the direct weight of the person using the steps, are usually made of clear oak wood, 1-1/16″ to 1-1/4″ thick × 10-1/2″ wide, either solid or glued up. The forward edge, called the nosing, is rounded, beveled or chamfered. Nosing often is eliminated when carpeting is installed in new homes to allow a better fit over square tread corners. In good construc-

Fig. 20. Wrought iron railing is set on finished base.

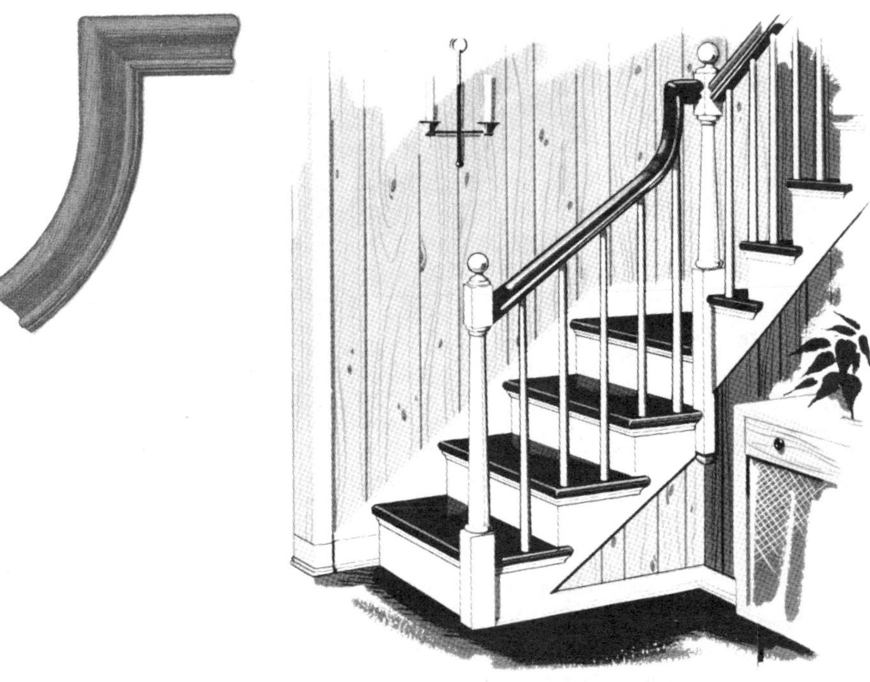

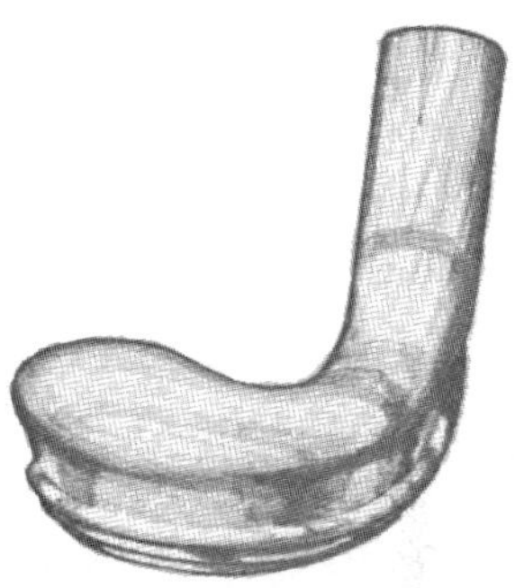

National Woodwork Mfrs. Assn.

Fig. 21. Square newel post let into the edge of a bottom tread. Note gooseneck in inset and as part of handrail at upper newel post.

National Woodwork Mfrs. Assn.

Fig. 22. Easement and volute support by small newel post and ring of balusters on extended bottom tread. Volute is a prefabricated handrail part (inset) and may curve to right or left.

National Woodwork Mfrs. Assn.

Fig. 23. Simple turnout and easement with newel cap starts handrail of this old American stairway. Turnouts may curve left or right. Note turnout in inset.

tion the riser is tongue and grooved to the tread above and ploughed for the tongue of the tread below (**See Figure 15.**) The treads are often drilled to receive the balusters, 1-1/4″ × 1-1/4″ × 31″ to 39″ in length, turned, of oak or beech stock. A sturdy balustrade can be built by dovetailing the balusters to the treads. A piece of side nosing, mitered to fit the front nosing of the tread, will conceal the dovetailing, **Figure 25.**

A Wrought Iron Handrail. Another style or type of free floating handrail is made of wrought iron, supported only by mounting brackets and screws, **Figure 26.** This handrail is available in 12′ lengths, with adjustable mounting brackets and molded end caps, which are easily attached after the rail is cut to desired length.

A Straight Run Suspended Stair

Suspended stairs having oak treads securely anchored to aluminum tubes which fit inside a standard stair well floor opening can be bought ready to install for an exclusive decorative effect in the modern house, **Figure 27.**

No side wall nor end wall support is necessary.

The finished-floor to finished-floor opening of the stairwell should be 3′-0-1/4″ in width × 9′11-1/2″ in length for two models or 3′-6-1/2″ in width × 9′-11-1/2″ in length for two other models.

The suspended stairs are furnished for a maximum rise of 8-1/4″ per tread, for a finished-floor to finished-floor dimension of 8′-11-1/4″ for two particular models or for a finished-floor to finished-floor dimension of 9′-7-1/2″ for two other models.

The rise is fully adjustable for lesser dimensions by cutting the tubes to the desired length. The finished opening in the floor should be 3′-0-1/4″ × 9′-11-1/2″ for two models and 3′-6-1/2″ × 9′-11-1/2″ for two other models. The finished opening size or stair well opening may be reduced to a minimum length of 7′-7-1/2″.

The stair treads are made of 1-1/2″ × 11-1/2″ × 36″ to 42″ unfinished laminated oak stock, pre-drilled with set screw attachment hardware installed. The aluminum tubes are 3/4″ in diameter, pre-cut with a peel-off protective coating.

The oak support rails, 2-1/8″ × 5-1/2″ × 9′-11-1/2″ in size, are pre-drilled and ready for nailing directly to the joists inside the finished floor opening.

A Spiraling Suspended Stair

A stair of exclusive decoration and graceful beauty is created in the smooth spiraling combination of oak treads and gold finished steel ring assembly which anchors aluminum tubing, **Figure 28.** The entire stair assembly is suspended from a structural steel ring making it independent of walls for side support.

The resulting design freedom at both floor levels gives complete flexibility in using the stair. The spiral direction may be either clockwise, or counter-clockwise. The

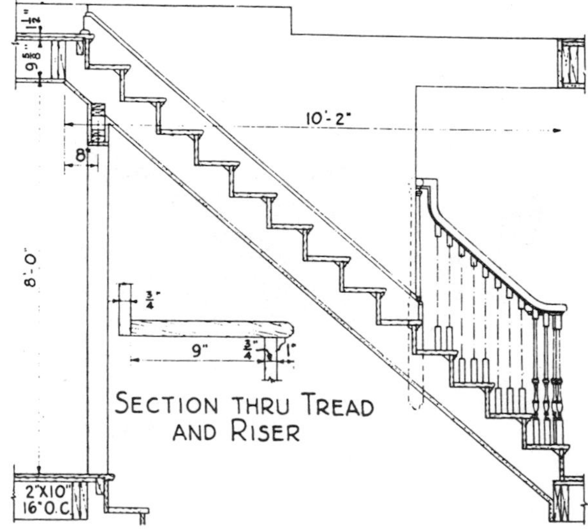

Fig. 24. Balustrade drops from half newel at wall.

Fig. 25. Balusters are dovetailed to edges of treads and joints concealed by side nosing mitered to front nosing.

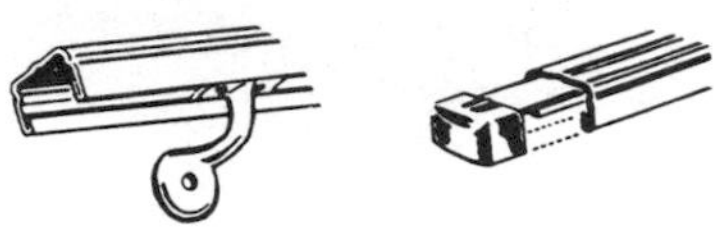

Fig. 26. Wrought iron handrail has adjustable mounting brackets and matching caps to close open ends.

Fig. 27. Prefabricated straight run suspended stair is installed with minimum effort in contemporary house.

Fig. 28. Graceful spiral is created by oak treads suspended from vertical aluminum tubing. Stair can be left or right hand.

oak treads have a minimum width of 6″, a maximum width of 20″, and a center dimension of 13″ for comfortable stepping room. The treads are supported by gold anodized aluminum tubes which are securely anchored to the steel ring.

The inner tubes are 32-1/4″ from the outer tubes providing ample passage up and down. The tubes are spaced so that no handrail is needed. The stair units are furnished for a maximum rise of 8-1/4″, which is proper for a finished-floor to finished-floor dimension of 9′-7-1/2″ for one model and 8′-11-1/4″ for another model. The rise is fully adjustable for lesser dimensions. To determine the actual rise, divide the actual finished-floor to finished-floor dimension by 14 for one model and by 13 for another model.

The stair well opening for the spiraling stairs should be 7′-0-1/4″ square. Headers should be doubled. The sub and finished floors should extend into one-fourth of the stair well opening to provide an upper landing for the stair. The stairs should be installed after opening finish has been applied.

A Staircase with Winders

Winders are stair treads of wedge shape used in a staircase without a landing platform. The narrow portion of the winder is housed at the angle newel, **Figure 29.** Sometimes the use of winders becomes undesirable due to the demands of floor space for a straight run of stairs. There is little space to place the foot near the newel, and stumbling is more likely than on a straight stair or on a spiral stair in which, although the treads are not rectangular, they are at least of uniform shape.

The most dangerous winders are those coming to a point at the newel. Winders tapering to a small tread at the newel are safer.

A Framed Spiraling Staircase

A spiraling staircase of wood, extending from one finished-floor level to another finished floor-level is shown in **Figure 30.**

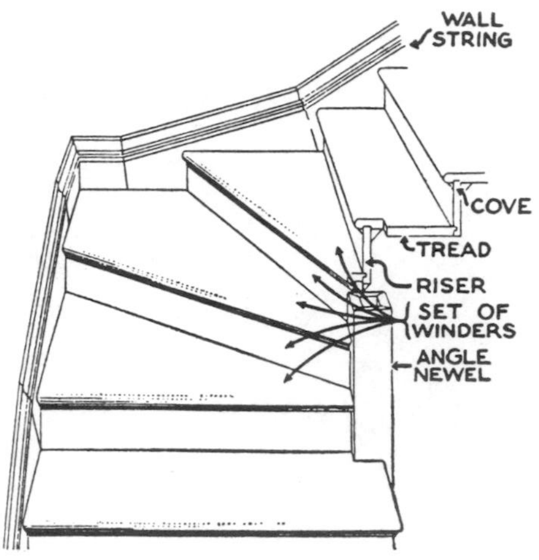

Fig. 29. Winder should be framed so that treads are sufficiently wide at newel post instead of tapering to a point.

Fig. 30. Spiral staircase is attractive feature in entry hall.

The method of framing a spiraling staircase requires the utmost care after a full size layout of the shape of the treads; the length of the risers; all angle cuts of treads and risers; and the shape of the winder stringers have been graphically detailed.

The drawing or layout can be made on the floor of the lower finish floor where the stairway begins or full scale paper patterns can be made for a particular staircase.

A factory built spiraling staircase would be less expensive to purchase and install than a staircase cut and built up on the building site.

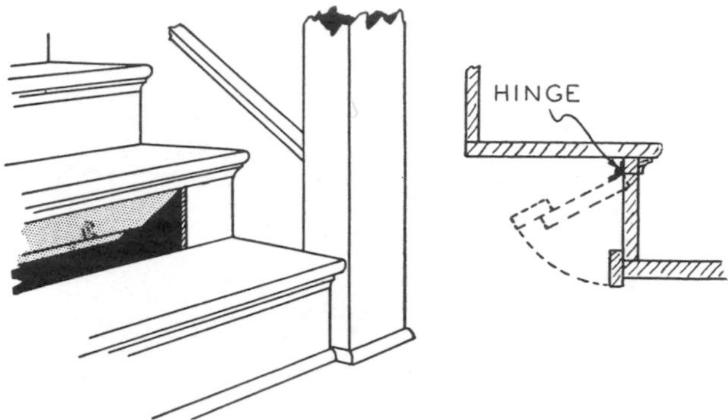

Fig. 31. Riser mounted on spring hinge serves as door to chute running to laundry room in basement.

An Inconspicuous Laundry Chute

A novel and practical laundry chute may be built into a stairway, **Figure 31.** A second or third riser is cut and hinged to provide ready access to the basement laundry. A spring hinge closes the riser automatically. If the chute opening is longer than 12″ the tread above should be braced with a horizontal metal support to stiffen it.

A Stoop, Stairway, and Steps of Concrete

An entrance platform, or stoop, with or without steps, is required for every door. Large platforms or steps should be fixed firmly to or supported at the foundation wall by anchors, piers or corbels built with the foundation wall.

The outer edge of the slab should be supported on a foundation wall or grade beam and piers if the soil under the slab is fill or susceptible to frost action. All fill must be carefully placed in 6″ layers and each layer of soil compacted thoroughly. The entrance platform, or stoop, **Figure 32,** should be reinforced with steel bars if the span is greater than 3-1/2′.

Stairways of Concrete. Stair approaches of concrete are built for porches, stoops, patios, pools and sunken gardens exposed to the weather. All stairs must be reinforced with 1/2″ steel bars, laid 6-1/2″ o.c., **Figure 33.** Form work for a one flight stair beam must be solidly built. Commercial forms are sold by a number of companies. Forms may also be rented. All forms should be oiled to facilitate removal. The concrete must be vibrated or tamped as it is poured to prevent honeycombing and to obtain a smooth, dense surface. The concrete should be spaded along the faces of the forms with a flat tool. In places inaccessible to a mechanical vibrator the forms should be tapped lightly with a heavy hammer to consolidate the concrete.

Fig. 32. Form boards are erected over well compacted fill for poured concrete stoop. Note steel reinforcing bars.

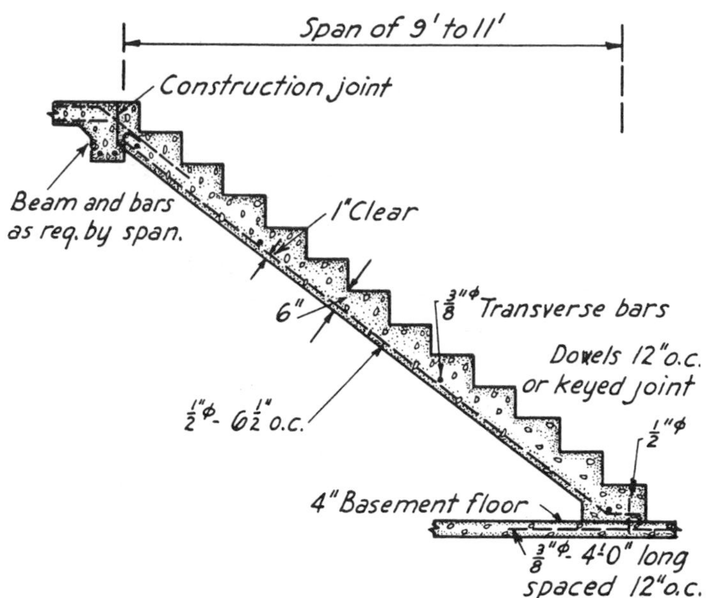

Fig. 33. Concrete stairway, supported by cross beam at top, is reinforced with 1/2" steel bars spaced 6-1/2" o.c.

A two flight concrete stairway is designed as simple beams, supporting a live load of 40 pounds per square foot, **Figure 34.** The span of the lower flight is assumed at 10′; for convenience the same amount of reinforcement is used in the shorter upper flight span.

Steps of Concrete. Treads of reinforced concrete slabs serve as steps for the several stairs leading from one outdoor level to another in **Figure 35.** The treads are supported by reinforced stringers of concrete. A hand rail of metal, as provided at the entrance of the swimming pool, should run along the flight of each series of steps.

This split-level contemporary dwelling of wood and concrete emphasizes modern outdoor-indoor living. The pool and each level is well protected with screen walls of masonry.

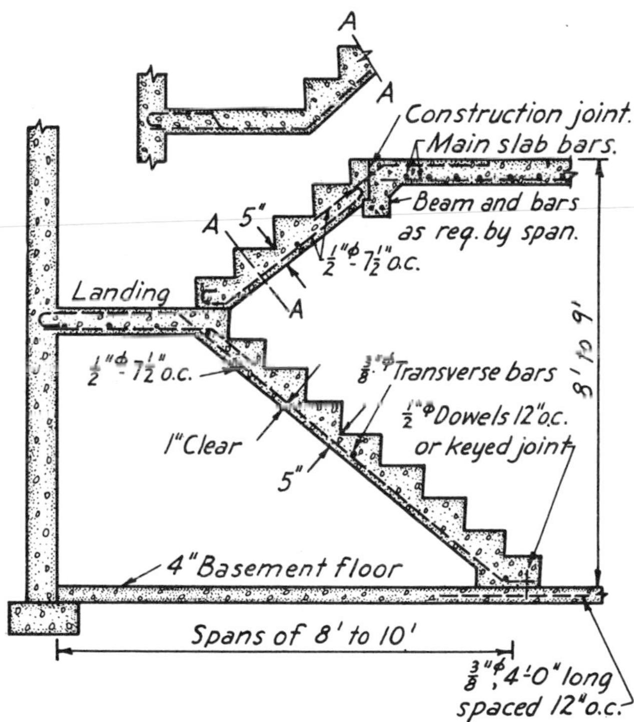

Fig. 34. A two-flight stairway of concrete supported by a landing and reinforced with steel bars. Dowels anchor base.

Disappearing or Folding Stairway

The folding or extension attic stairway is widely used for easy access to an attic, finished living quarters or to closet and storage space. The use of the folding stairway saves the floor space of the floor below the device, which allows for freer planning. One typical folding stairway, **Figure 36,** is designed so that the bottom two treads fold under; the next section folds upward to fit within the area of the plywood panel, which fits the ceiling opening. Another type of folding stairway mounting, **Figure 37,** when not in use, has the stairway ladder thrust up to clear the hinged panel; when lowered, the doweled frame swings upward and over, bringing the top of the ladder to the edge of the opening.

Many houses are now built with the disappearing staircase already installed as standard equipment, **Figure 38.** Often a framed opening, with trim, is built into the ceiling to house the stairway.

Spring balances, or counterweights, **Figure 39,** are used to compensate for the weight of the stairway and coil springs are used to give necessary lateral balance.

Fig. 35. Precast reinforced concrete treads rest on concrete stringers, adding to exterior beauty of contemporary house.

Fig. 36. Stairway to attic folds in several places for flat storage within dimensions of panel which closes opening.

A Basement Stairway Leading Directly to the Rear Yard

When the finished grade line at the rear of a house is near the top of the foundation wall, a prefabricated basement stairway fits readily into a stairwell to provide easy access from the basement to the outside. **Figure 40.** Wood treads are supported by steel stringers fastened to the concrete or masonry walls. A sloping access door is installed on top of the stairwell.

Fig. 37. Simple type of folding stairway is pushed up into opening and stored beneath rafters as shown.

A Steel Staircase Component

Staircase components of steel are available as well as staircases components of wood and are often used as basement stairs or stairs to an attic where there is a 7′ or 8′ ceiling height, **Figure 41.**

A Stair Glide

A stair glide or home stairway elevator is a device of great convenience in the home when a member of the household is disabled or should not exert himself in using stairs.

The stair glide may also be used as a handy dumbwaiter which eliminates exhausting and dangerous trips up or down stairs. Food trays, medicines, vacuum sweepers, laundry and luggage can be placed on the seat and moved up or down.

The stair glide, **Figure 42,** is easy to install. It runs on an aluminum alloy track, 12″ wide, which is placed against the wall or the wall stringer trim. The track is held at three places by rubber shims at the bottom, center, and landing of the stairway and rests against the nosing of every step. There are no screws nor attaching devices driven in the stairway or wall. The glide is driven by a one-third horse power electric motor operating a small steel cable and drum. The motor is controlled by a push button on the seat of the glide and by "call-send" buttons at each end of the track. When in use the motor will not interfere with radio or television reception and its source of power is the regular 110 volt house current.

The stair glide is ideally suited for nearly all stairways, but if a second stairway is encountered, a second stair glide must be used since the device will not go across a stair landing and up another flight of steps.

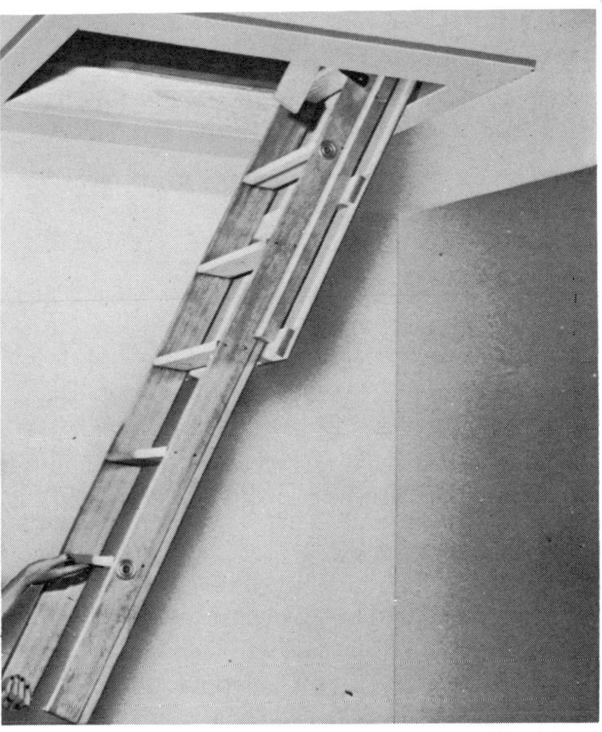

Fig. 38. Bottom section of this stairway slides upward.

Fig. 39. Counterbalances or springs compensate for weight of disappearing stairways, making operation easy.

Fig. 40. Steel stringers are notched for wood treads to speed installation of access door to rear yard or patio.

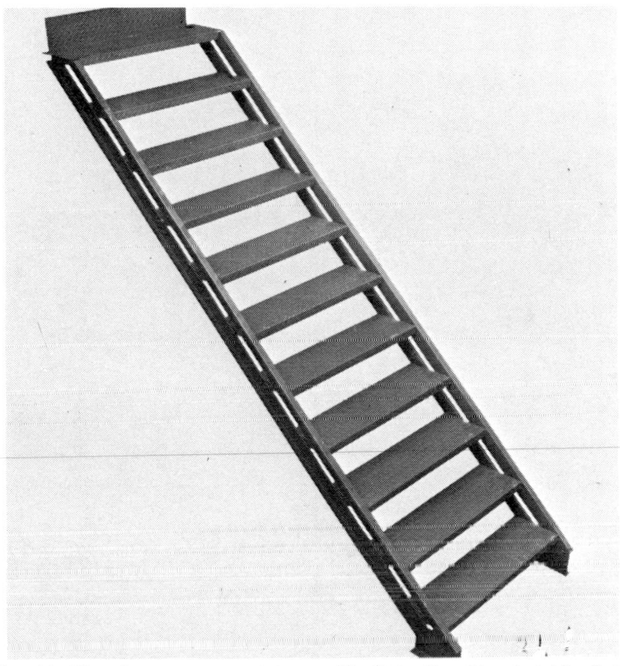

Fig. 41. Manufactured steel staircase fits 7′ to 8′ ceiling heights. Self leveling shoes adjust treads automatically.

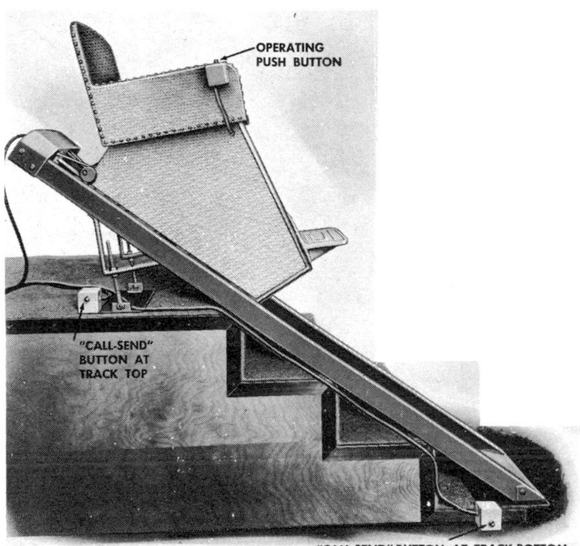

Fig. 42. Stairway elevator, powered by electric motor, rides on tracks to carry disabled person from floor to floor.

Chapter 14

HARDWARE

Hardware used in the construction of a house may be classified as operative hardware and as holding hardware as follows:

Operative Hardware, such as hinges, locks, catches, latches, stops, closet rods, bi-folding door hardware, bi-passing door hardware, shelf hardware, snap in window grilles, window roto release operators, mantle beams, sliding cabinet door track are used to operate, support, or hold an operating door, window, shelf, etc., in place and are known as operative hardware.

Holding Hardware, such as nails of many types, screws of many types, bolts, glue, mastic, adhesive tapes, glass fiber insulation, glass, building paper, roofing, are used to hold materials together or weather out and are known as holding hardware.

The Operative Hardware of a Door. One of the most important items used in house construction is the door. The large two-car garage door at the front, side or rear entrance of a house employs numerous important items of operative hardware which have been described in the chapter on DOORS.

The operative hardware for an entrance door is more elaborate than other doors. An important item about a

RIM LOCK
OUTSIDE
MORTISE LOCK
OUTSIDE
LEFT HAND
RIGHT HAND
MORTISE LOCK
OUTSIDE
RIM LOCK
OUTSIDE
LEFT HAND REVERSE BEVEL
RIGHT HAND REVERSE BEVEL
RIGHT HAND
LEFT HAND
OUTSIDE
OPENING OUT
LEFT HAND
RIGHT HAND
OUTSIDE
OPENING IN
CASEMENT WINDOWS - SPECIFY WHETHER IN- OR OUT-OPENING
ALSO APPLIES TO CREMONE BOLTS
NOTE - THESE CONVENTIONS ARE EXACTLY OPPOSITE TO THOSE FOLLOWED BY THE METAL WINDOW INSTITUTE
LEFT HAND DOOR & LOCK
RIGHT HAND DOOR & LOCK
CUPBOARDS, CABINETS, BOOKCASES
NOTE - REFRIGERATOR DOORS FOLLOW THESE CONVENTIONS
THICKNESS VARIES
3°35'
2"
STANDARD BEVEL

Fig. 1. "Hand" of doors, casements, cupboards and cabinets. Bevel permits door to swing open without binding against jamb.

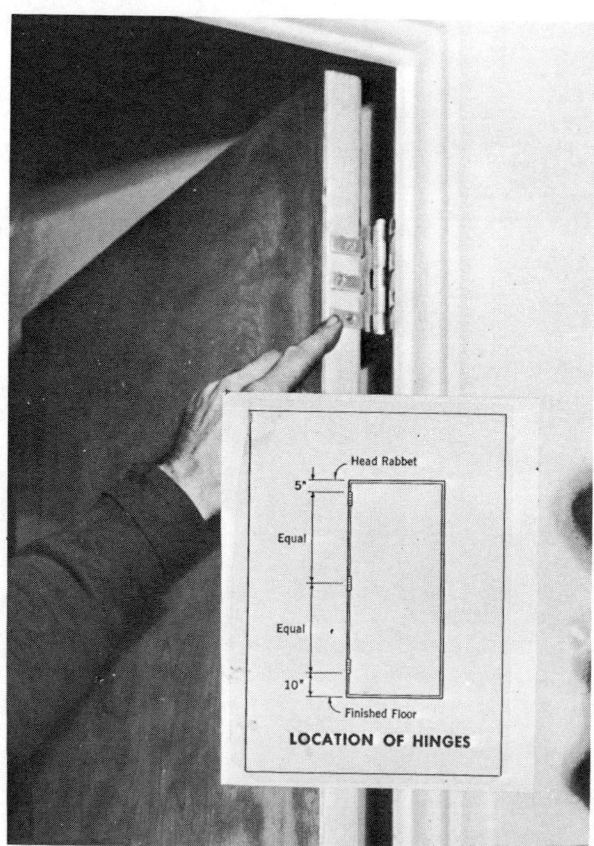

Fig. 2. Non-mortise hinge is screwed flush against edge of door and jamb. Slotted leaves fold together to single 1/8" thickness. Insert shows approved placement of hinges on door.

door is the way it swings in relation to the person opening or closing it. This is termed the "hand" of the door, **Figure 1.** The hand of a lock and door is determined from the outside of an entrance door or from the corridor or hall side of a room door. As you stand to insert the key in the lock, if the hinge is on the left and the door opens away from you, the door is a simple, "left hand," door. If the door opens toward you, with the hinge on the left and the knob or handle on the right, the door is a "left hand reverse bevel."

If the hinge is on the right and the door opens inward (away from you), the door is a "right hand" door.

If the hinge is on the right and the door opens outward, toward you, the door is a "right hand reverse bevel" door.

Since casement windows and doors with cremone bolts are generally opened and closed from the inside, the hand of such doors and windows may be considered in relation to the person standing inside. Thus, standing inside, one finds the casement with the hinge on the left called the "left hand" and the other the "right hand."

Since one opens cupboards, cabinets, and bookcases from the outside, the "hand" is the same as the exterior doors: hinge on left: "left hand."

One may find that many locks are marked "reversible," meaning that they are interchangeably right or left hand; in these instances no reference to hand or bevel of lock is necessary. The operation of these locks is alike on both sides; or they can be inverted in order to reverse the locking functions.

Hinges for Doors. Factory made exterior and interior door sets are now available as a building component completely furnished with hinges, lock sets, glazed if required, and with storm doors. There are numerous types and styles of hinges, such as the non-mortise hinge, **Figure 2.** A non-mortise hinge is screwed flush to the edge of the door and to the jamb, giving 1/8″ clearance, the thickness of the shoulder of the hinge; the hinge fingers alternate, giving only a single thickness of metal when closed together.

Hinges are specified by number (in pairs); type of bearing (ball or oilite); type of tips (button, ball, cone, steeple, or non-removable); type of screws; type of hinge; height; width (if for full mortise hinges); weight and finish.

Template hinges are made to standard templates or patterns, usually in accordance with government standards, to insure exact matching of the hinge and its screw holes to metal doors and jambs fabricated by other manufacturers.

A minimum of three hinges is recommended for all doors, and four are advisable if the door is over 7′-6″ high. The bottom of the lowest hinge is set 10″ above the finish floor; and the top of the hinge is 5″ from the head rabbet. The space between is evenly divided by the other hinge or hinges.

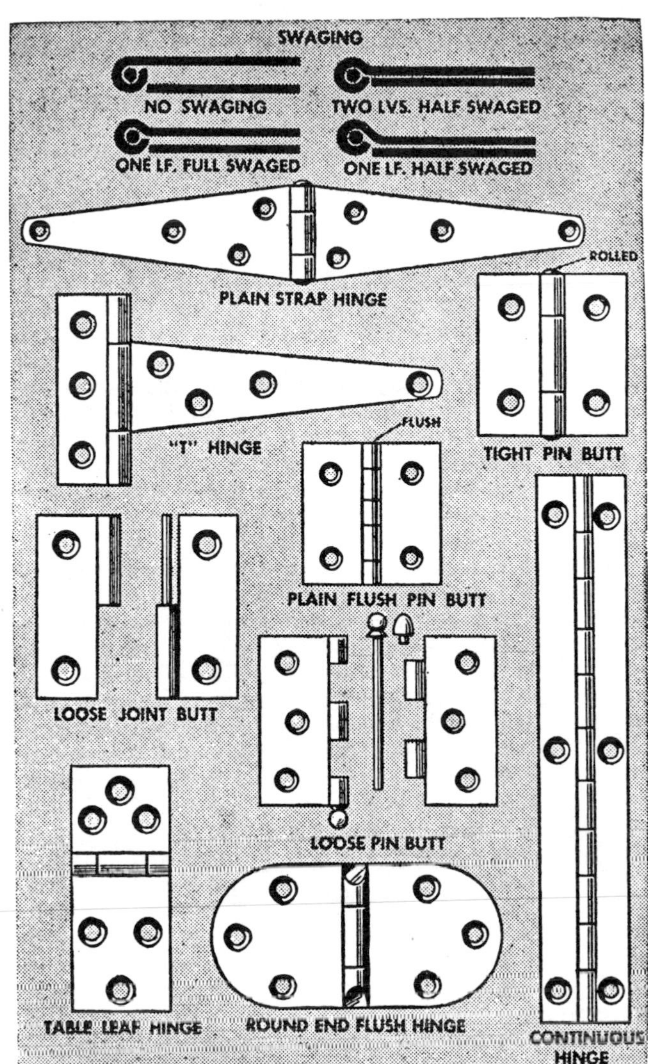

Fig. 3. Most common types of hinges used in house construction.

Hinges and Locks

Various types of hinges used in building construction are shown in **Figure 3.** The strap hinge, originally strictly utilitarian for rough surfaces, such as barn and cellar doors, is sometimes used in kitchen cabinets for its rustic appearance. The T-hinge combines one strap leaf and a butt. There is a great variety of special hinges, many of them patented, for cabinets; some constructed so as not to show at all when the doors are closed, some for full display, such as H-shaped hinges and various other rustic types, often in rough forged finishes, **Figures 4, 5, 6, and 7.** Piano or continuous hinges have long pins and knuckles and narrow leaves, and are often used for combination storm doors.

There are various devices used to keep hinge pins from rising during use, without preventing easy removal when desired. The pin may be kept down by making it rotate

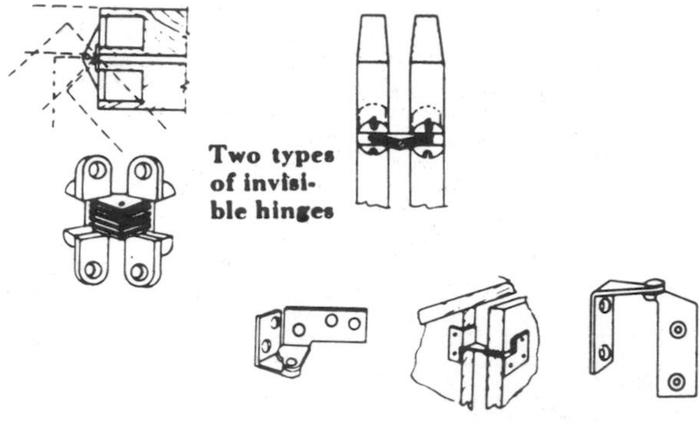

Fig. 4. Invisible and semi-visible hinges used in cabinets.

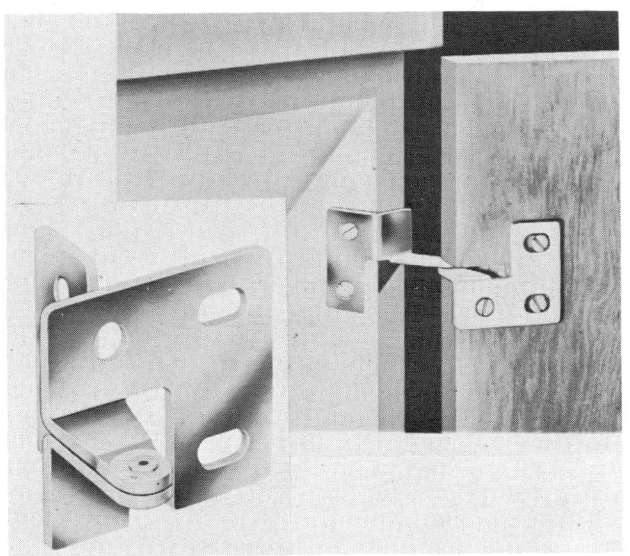

Fig. 5. Semi-visible hinge requires narrow slot in edge of door. Oval screw holes permit precise adjustment.

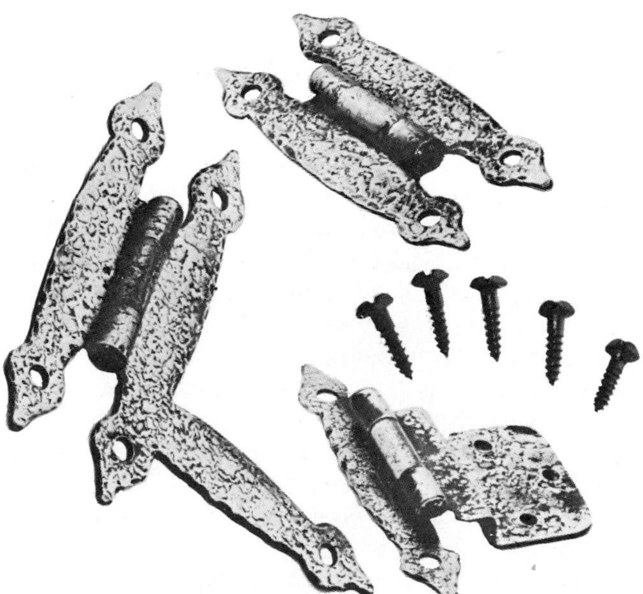

Fig. 6. Forged iron hinges for flush doors (left and top) and offset doors (right) simulate traditional hardware.

both ways, rather than in only one direction; or a shoulder may be built into the pin to keep it in place in ordinary use, **Figure 8.** Some hinges are built so as to lift the door enough to clear the carpet as it opens, **Figure 9.**

Various types of latches, handles, and pulls come under the head of operative hardware, from the simple hand grip to various shapes of knobs and pulls. Flush pulls, either round or rectangular, are used with sliding and by-pass doors, **Figure 10.**

The Butt Hinge. The most common door hinge is the full-mortise butt hinge, in which one leaf is mortised into the inside of the door jamb, the other into the edge of the door, so that the leaves meet flush. The screw holes of the hinge are beveled so that the flat-headed screws are flush with or below the face of the leaf. The full-surface butt hinge has the jamb leaf attached to the face of the jamb, and the door leaf to the face of the door stile.

A half-surface butt hinge has one leaf mortised into the door jamb, the other applied to the surface of the door. A half-mortise butt hinge has one leaf fixed to the casing of the frame, the other leaf mortised into the edge of the door, **Figures 11 and 12.**

Spring Hinges. Spring hinges have springs within cylinders placed at the juncture of the leaves called the knuckle. Single acting butt hinges may use a hanging strip to keep the spring cylinder clear of the jamb, or the cylinder may be kept clear by angling both leaves, **Figure 13.** Double-acting spring butt hinges permit the door to swing both ways, **Figure 14.**

Spring pivot hinges may be placed in the head jamb, in the top of the door, in the bottom rail of the door, **Figure 15,** or below the flooring.

Door Checks

Door checks are made in various types. The mechanism may be attached to the door, **Figure 16,** or to the soffit. It may be semi-concealed or fully concealed, with the mechanism in the head jamb, the top door rail, in the floor, or the bottom door rail, **Figure 17.**

There are numerous types of door stops, holders, bumpers, stays, and shock absorbers, **Figure 18,** to prevent doors from damaging walls when fully opened.

Locks and Latches

As operative hardware, locks and latches may be classified in several ways: according to manner of insertion in the door (rim, mortised, bored-in, cut-out, **Figure 19**); kind of locking mechanism (bit key or cylindrical); extent of locking function (simple sliding latch, locking latch, latch and dead lock, or various combinations).

Fig. 7. Forged iron hinges and matching door and drawer pulls enhance rustic appearance of wood cabinets.

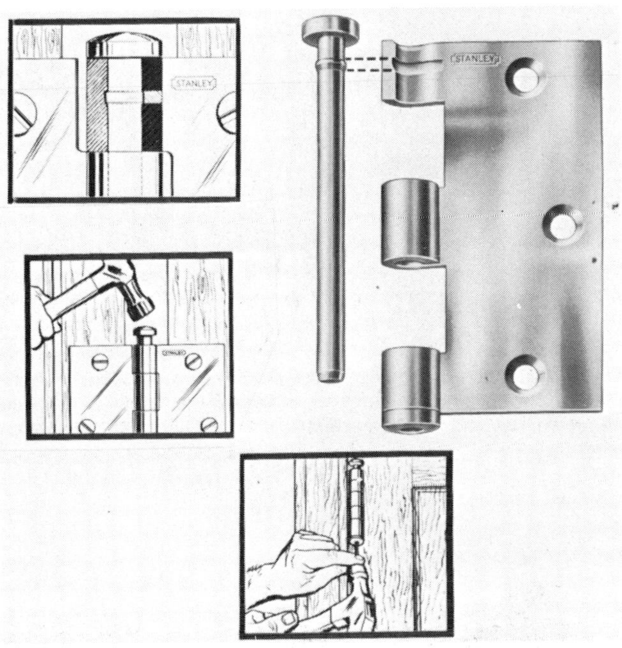

Fig. 8. Non-rising hinge pin revolves freely, but a shoulder snaps firmly into groove in top hinge knuckle.

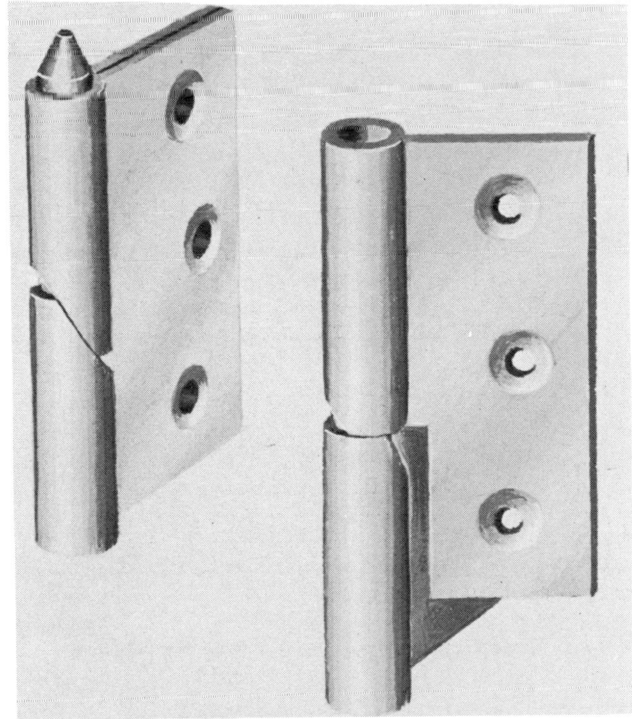

Fig. 9. Door-lift hinge raises door as it opens to clear carpeting, lowers door as it is closed.

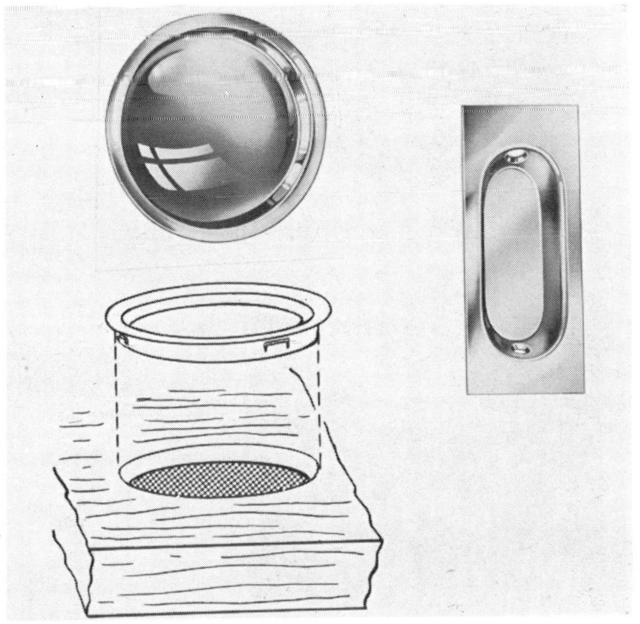

Fig. 10. Round and rectangular flush pulls for sliding doors. Round pull snaps into hole, rectangular pull is held by screws.

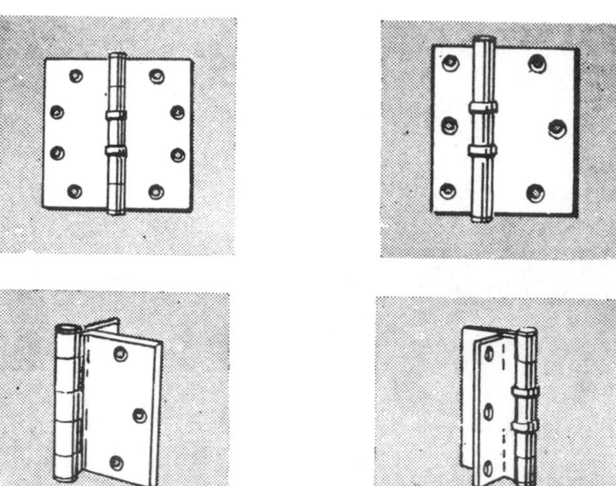

Fig. 11. Full mortise butt hinge (top left) is mortised into door and jamb. Full surface (top right) has jamb leaf on jamb face, door leaf to face of stile. Half surface (lower left) is applied to surface of door, but mortised into jamb. Half mortise (lower right) is mortised into door, on surface of jamb.

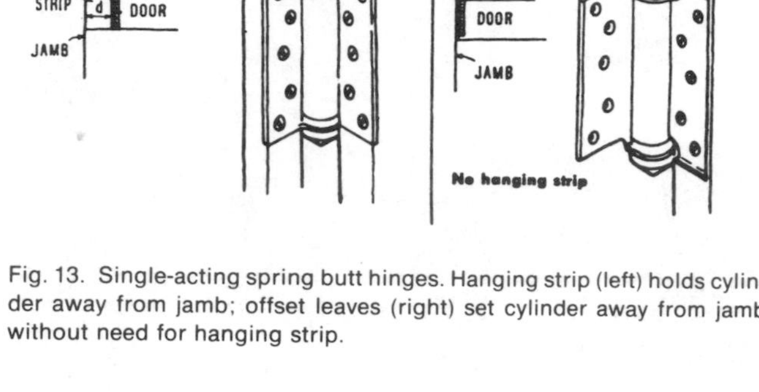

Fig. 13. Single-acting spring butt hinges. Hanging strip (left) holds cylinder away from jamb; offset leaves (right) set cylinder away from jamb without need for hanging strip.

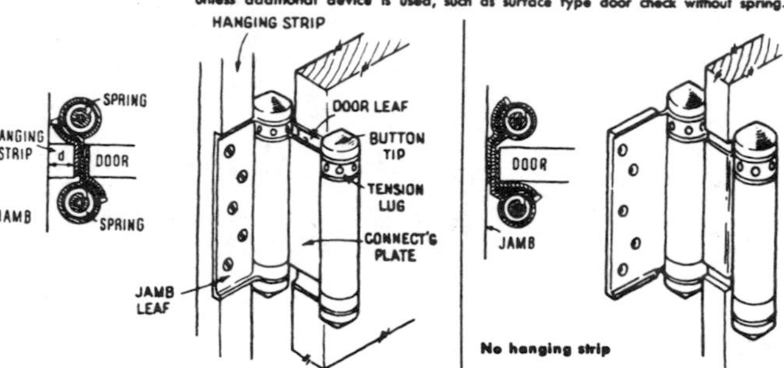

Fig. 14. Double-acting spring butt hinges, mounted on handing strip or with curved leaves similar to single acting type.

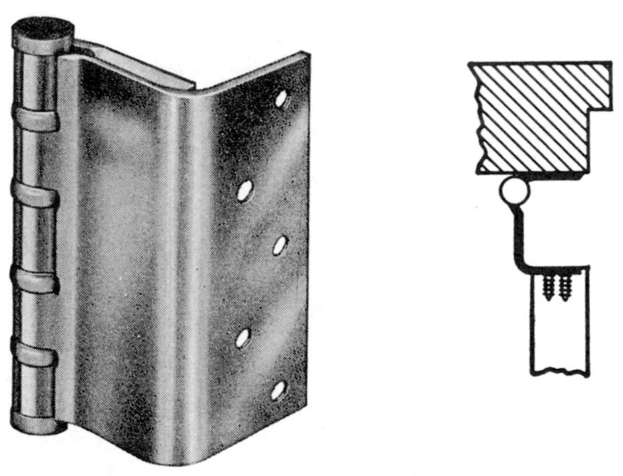

Half Mortise Swing-Clear Hinge

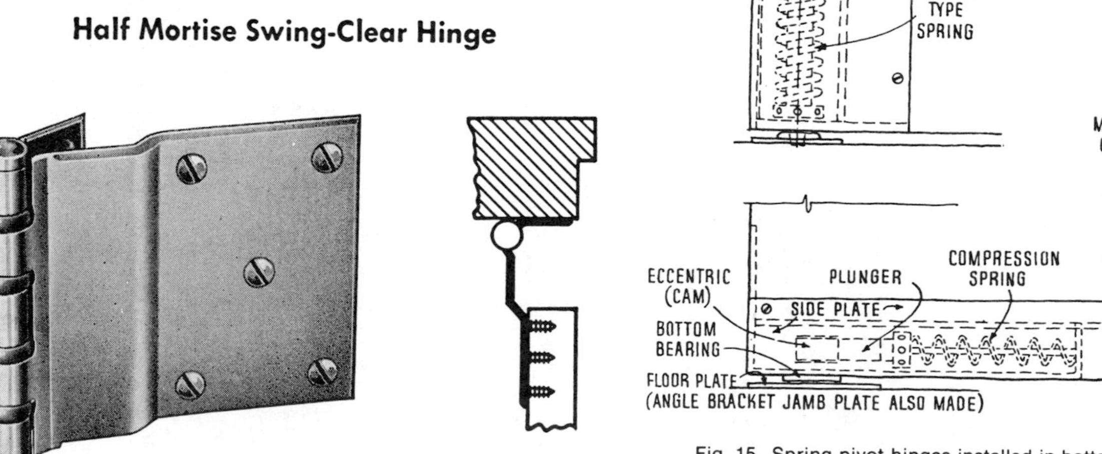

Full Surface Swing-Clear Hinge

Fig. 12. Swing-clear hinges mount on edge or face of door.

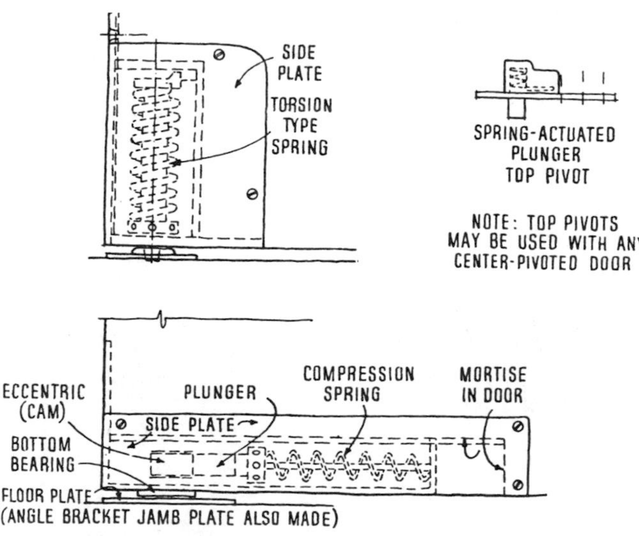

Fig. 15. Spring pivot hinges installed in bottom rail of door.

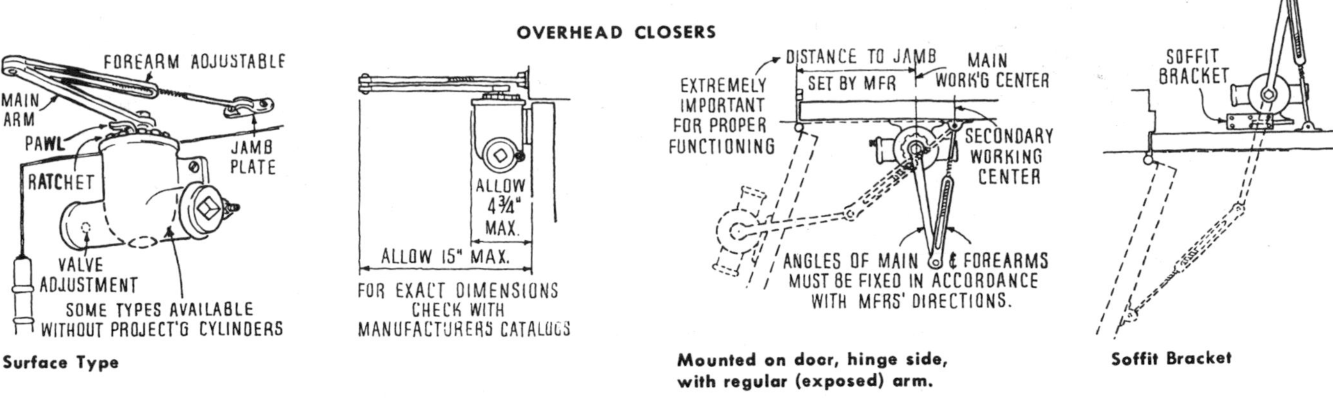

Fig. 16. Door checks, mounted on door and top jamb, close door at a controlled rate.

Fig. 17. Cutaway showing mechanism of overhead door check.

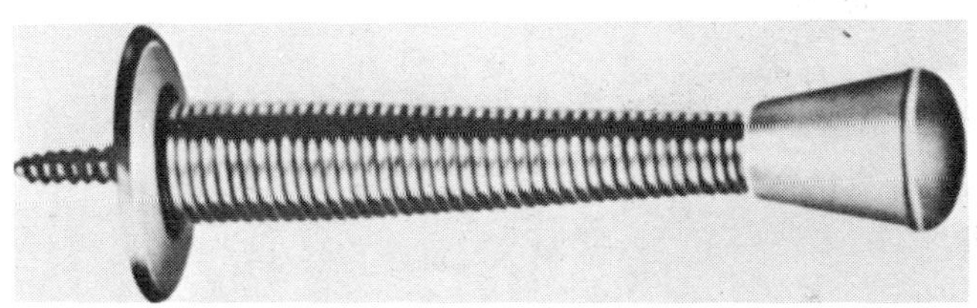

Fig. 18. Flexible, rubber-tipped door stop is screwed into base molding to prevent door from damaging wall when opened.

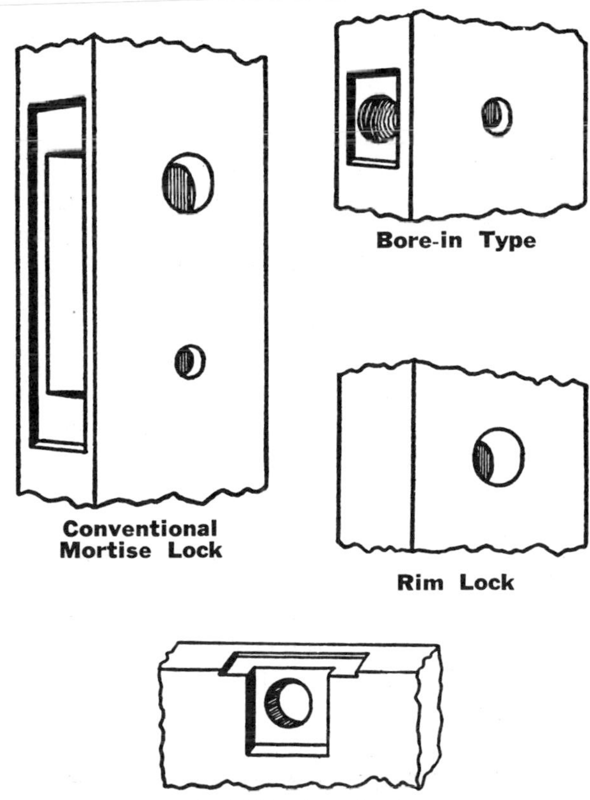

Fig. 19. Lock installations: mortise, bore-in, rim, half mortise.

The Mortised Lock. The mortised lock once was standard for most doors, in preference to the rim lock, which is less elegant, though easier to install, **Figure 20.** Mortising requires an extensive cut into the edge of the door, and exact measurement to coincide with the placement of the spindle and key holes in the stile.

With the development of the cylindrical lock mechanism, which usually places the key in the knob, mortised locks became easier to install. The much easier bored in lock, which requires merely a round bore from the edge, plus a meeting hole in the stile for the knob, containing a cylindrical lock has generally supplanted the mortise in new construction, **Figure 21, 22 and 23.** A jig is usually used to align the bores in edge and stile, speeding installation.

The Simple Latch. The simple latch slides against the strike and holds the door closed, but can be opened from either side. With button mechanism and split spindle and hub, the latch may be opened from the inside but requires a key for opening from the outside when the button is in closed position. The button is commonly on the door edge in a mortised lock, on the knob in a cylindrical, bored-in, key-in-the-knob lock, **Figure 24.**

The Dead Lock. The dead lock must be specifically turned to close or open, by key or a catch. A dead lock is often added to a lock-latch mechanism, so that the same key from the outside can open both latch and dead lock. For maximum security dead locks are installed separ-

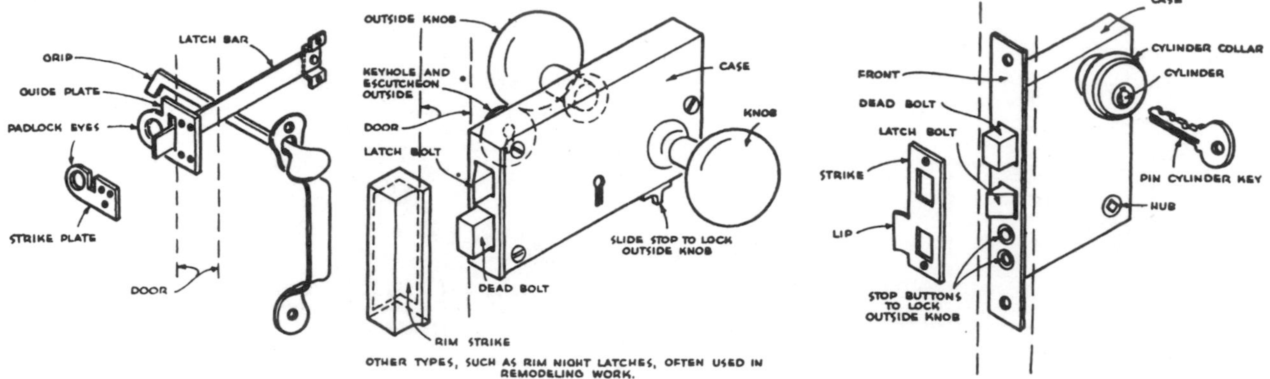

Fig. 20. Thumb latch (left) is simple to install, difficult to adjust, and may be padlocked. Rim lock and latch (center) has case and strike mounted on face of door and jamb. Mortise lock and latch (right) requires deep cut in wood doors.

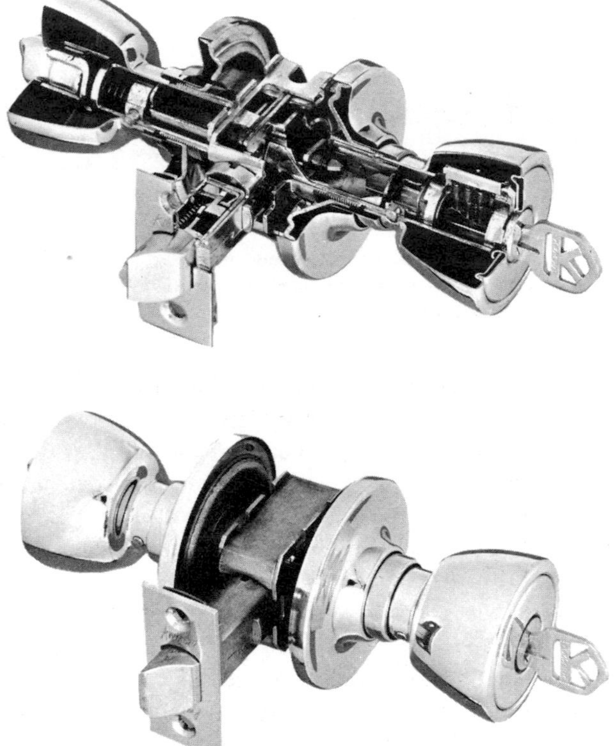

Fig. 21. Key-in-knob lockset for exterior doors has deadlocking latchbolt. Outer knob has four pin cylinder lock, inner knob has turn-button lock control.

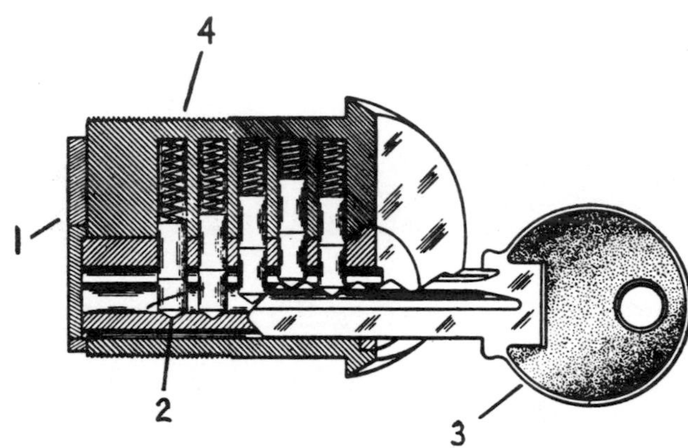

Fig. 22. Cylinder lock: (1) cylinder; (2) pins; (3) key; (4) springs. Cut edge of key aligns pins to turn cylinder.

ately from the latch mechanism, **Figure 25**, with specially slotted screws which can be tightened easily, but are difficult to remove.

A Bit Key. On a bit key, only the end, the slotted wing, serves to actuate the lock. Wheel wards permit the key to pass obstructions to get into place, and the slots cut in the end of the wing or bit, engage the tumblers. Cylinder lock keys, which are functional along their whole length up to the grasp, allow much greater complication and keying function. The proper cylindrical lock key fits into a barrel and permits tumbler pins to fall into the key slots in such a way that the entire barrel can rotate, thus turning a cam which in turn governs latch or dead lock, or both. The wrong key will prevent the pins from settling, so that they prevent the barrel from turning in its shell.

Door Bolts. Door bolts are actuated directly by hand, rather than by keys, although locks may be added to prevent bolts from moving. The simplest bolts are attached to the surface, like the barrel bolt or the chain bolt. Flush bolts, mortise bolts, panic bolts actuated by a horizontal lever, cremone and espagnolette are types of bolts, **Figure 26.**

Closing Devices. A variety of closing mechanisms is manufactured for cupboards and cabinets and screen doors: latches with turn-buttons to open, finger-lift latches, hook-spring catches that are lifted to disengage, friction catches, rollers on spring grips, and magnetic catches, **Figure 27.**

Other Operative Hardware for Doors

Simple Deadlock. A two pin tumbler deadlock provides additional security for 1-1/8″ to 2-1/4″ doors. The

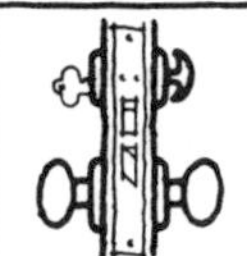

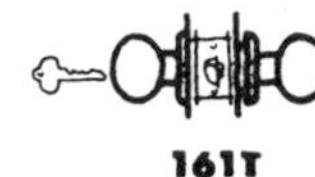

Fig. 23. Latch and deadlock. Numbers shown are U.S. Federal specification numbers from FF-H-106a.

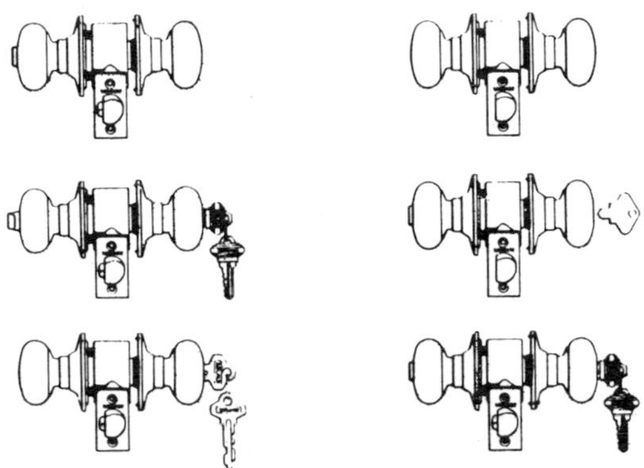

Fig. 24. Types of locks, by function: Knob latch (upper left) for passage and closet doors; both knobs always free, latch bolt will retract with either knob. Button lock (upper right) for basement and sun-deck doors; pushing inside button locks outside knob, turning inside knob releases button. Emergency key lock (middle left) for bathroom and bedroom doors: Works like button lock, but can be opened from outside with key. Exterior door lock (middle right); turning inside button locks outside knob; to release, button must be turned to vertical position or key used from outside. All-purpose lock (lower left); pushing inside button locks outside knob; turning inside knob or using key releases button; closing door does not release button. Utility lock (lower right) may be locked or unlocked by key in outside knob. Inside knob always free for ready exit. Latch bolt automatically deadlocks when door is closed, sometimes called "key control locks."

dead bolt is operated by key from the outside, by a knob from the inside. The case is streamlined as well as the bolt, generally of brass, **Figure 28.**

Night Latch With Chain Guard. A double duty security plus a built-in convenience in that the key can release both the latch and chain or the chain guard stays locked if the latch bolt is retracted, **Figure 29.**

Through-the-Door Viewers. The through the door viewer permits the occupant to examine a caller without being seen. For use on any door without a window. The wide angle viewer permits one way vision only, from the occupant of the house to the outside. The viewer is made of rust proof aluminum, finished in anodized bright brass. Installed by drilling a 1/2″ hole through the door at viewing height. The lens half of the tube is inserted from the outside, the eye-piece half is screwed in from the inside, **Figure 30,** to tighten the unit securely in place.

Door Knocker and Built-In Viewer. A Colonial designed door knocker can be had with an added safety feature of a through-the-door viewer. The viewer is inconspicious from the outside blending with the door knocker. Knocker may be finished in bright brass, colonial black or satin-chrome. It is particularly attractive on any exterior or interior door on a hall, **Figure 31.**

Slip Over Door Escutcheons. Die-cast trim escutcheons that slip over the door knob and fasten to the door with screws convert a plain door into an Early American or Contemporary style. Eschutcheons are finished in old iron, antique brass or white and gold, **Figure 32.**

Entrance Pulls. Door pulls are mounted on the exterior by two through bolts, inside trim grommets are used to receive the bolts. Doors should not be predrilled, pulls can be used right or left hand, or in pairs on

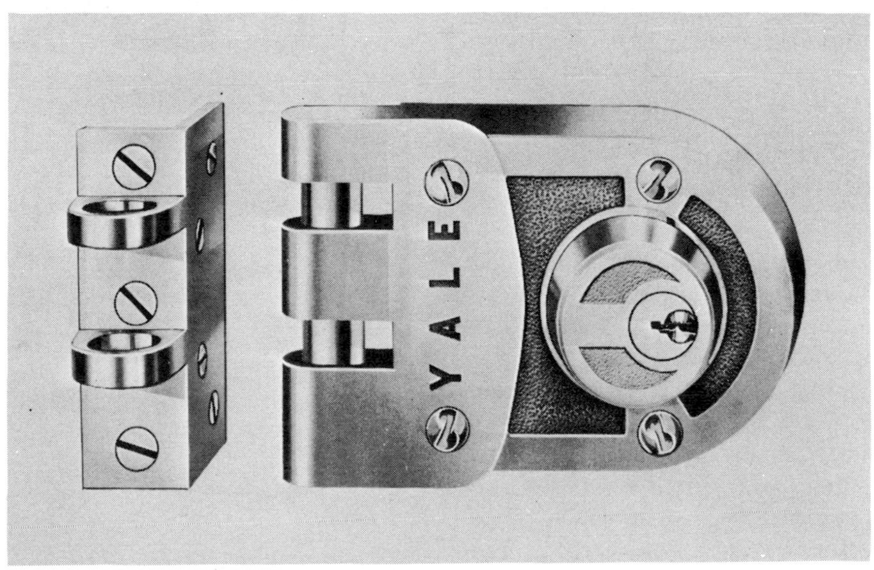

Fig. 25. Double cylinder auxiliary vertical bolt rim deadlock must be operated by a key from inside as well as outside.

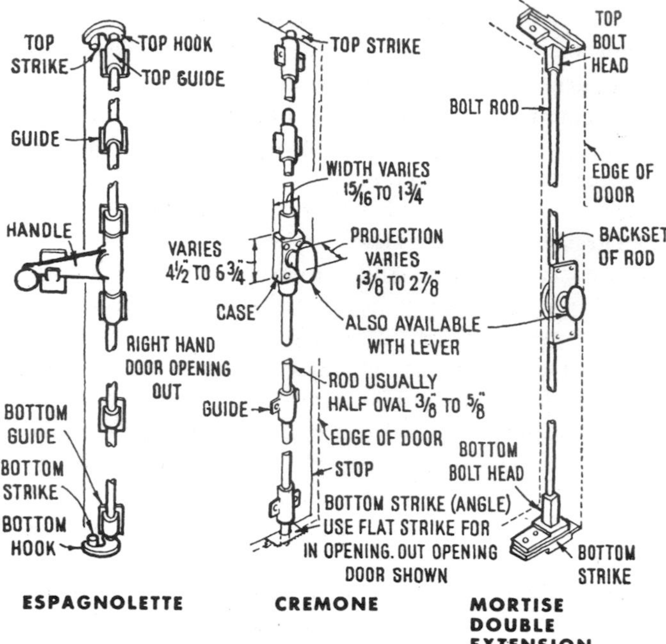

Fig. 26. Bolts running full length of door engage strikes at top and bottom, usually are operated only from inside.

double doors, **Figure 33.**

A Sound Control Accessory for Doors. A sound control device for interior doors, 2′-8″ or 3′-0″ wide by 6′-8″ high, is available in sets. Each set, **Figure 34,** consists of an automatic door bottom strip made of a durable, flexible vinyl strip locked in a sturdy aluminum frame, and a set of header and side stops of contoured pine with self adjusting soft vinyl lip that seals out sound around door's edge.

Weather Strip for Doors. A sturdy, extruded aluminum weather strip with tough, durable vinyl is available for doors of wood or metal. Note placing of weather strip, **Figure 35,** at the head of the door, along the side of the door and at the bottom of the door.

Sliding, Bi-Passing, Bi-Folding Door Hardware

With the growth in popularity of the sliding, bi-passing, bi-folding door for closets, folding partitions, or glass sliding exterior doors, there has been a rich development of sliding and roller mechanisms. The main elements of these assemblies are the overhead track and the rollers, **Figures 36, 37, 38, 39.**

Most so-called sliding doors are really suspended rolling doors. Most sliding doors are double, with two segments by-passing each other. Most double glass doors are bi-pass doors, and require double tracks. The pocket door, which opens into a frame set inside a partition, needs only single track, **Figure 40.**

Many light weight bi-pass and pocket doors dispense with a floor track, and depend for stability when closed on separators, guides, or meeting catch mechanisms where leaves meet or where the door meets the frame, **Figures 37, 41.**

Refinements are continually being added to hanging hardware mechanism to permit height adjustment installation, elimination of side sway, and to insure quiet operation. Nylon rollers are widely used. The frames, often of complicated shape, are extruded when made of aluminum.

Sliding Mirror Doors. Sliding mirror doors are available, **Figure 42,** in two bi-pass door units with track to fit standard 4′, 5′ or 6′ wide by 6′-8″ high finished openings. The trim is fitted around the opening before installing the track, which is screwed to the head of the door opening. Some doors also have floor tracks.

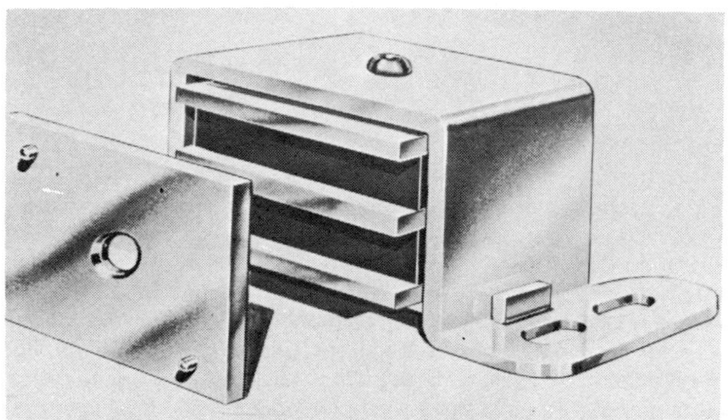

Fig. 27. Magnetic catch for cabinet door has 30 lb. holding power and floating action to prevent misalignment.

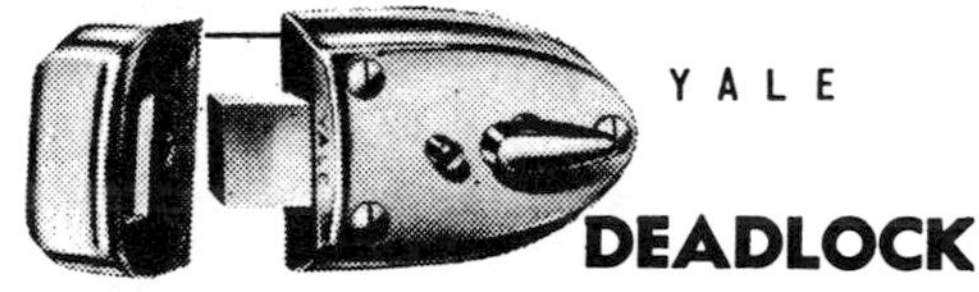

Fig. 28. Simple deadlock has knob inside, key lock outside.

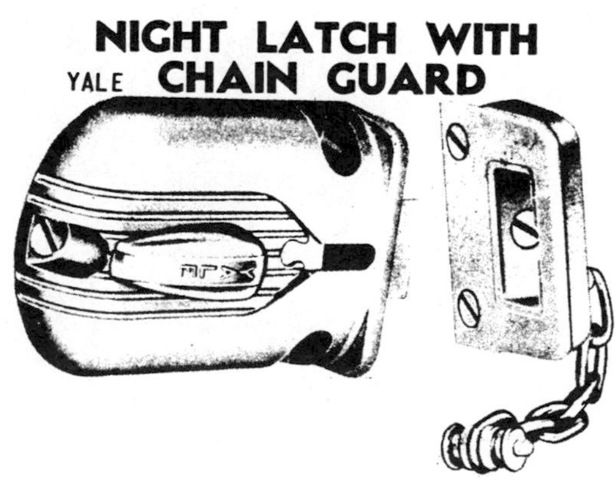

Fig. 29. Night latch has chain guard in addition to bolt.

Fig. 30. Through-the-door viewer has wide angle viewing lens.

Fig. 32. Slip-on escutcheons.

Fig. 31. Viewer in knocker.

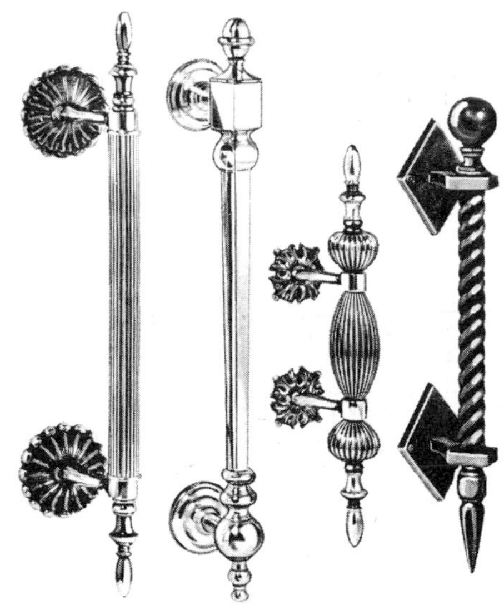

Fig. 33. Elaborate entrance pulls used singly or in pairs.

Sliding Door Hardware. Sliding door hardware consisting of adjustable hangers and door guides are installed as shown in **Figure 43.** Doors can be adjusted by using hammer and block to tap adjustment arm of hanger to right or left. Door guides are fastened to the floor.

Operative hardware for windows is described and illustrated in the chapter titled **WINDOWS.** Operative hardware for closets and built-in equipment is described in the chapter titled **CLOSETS, SHELVES and BUILT-IN EQUIPMENT.** Additional description of operative hardware for doors is given in the chapter on **DOORS.**

Fig. 34. Vinyl strips reduce sound transmission.

Holding Hardware

Holding hardware, sometimes termed rough hardware, consists of nails, screws and bolts, connector plates, staples, joist and beam hangers, framing anchors, metal bridging, flat brackets, gussets, and numerous other devices used to holding parts in place.

Most nails are made of steel wire, being termed wire nails, and are cylindrical in form and range from 4d to 60d in size as shown on the chart, **Figure 44.**

Nails are made in various shapes and sizes from the small head brad to the broad headed roofing nail. Sharp-pointed nails have a greater tendency to split wood. For this reason cut nails are used for oak flooring. Cut nails, the older type of nail, are cut from flat iron plate, tapering from the head to the blunt point. Steel cut nails are used to nail wood furring strips to masonry, **Figures 45, 46 and 47.**

To determine the length in inches of common nails or brads for a given "penny" (d) size, up to and including 10d, divide the penny size by 4, and add 1/2″. For

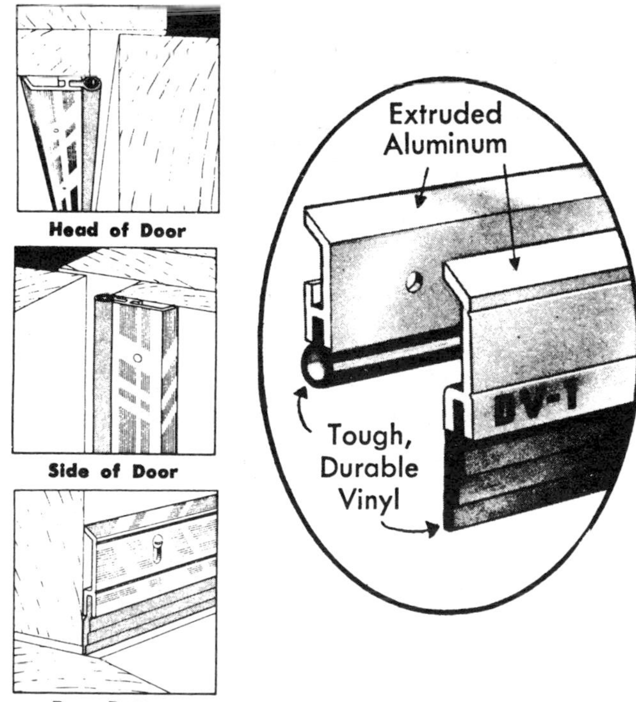

Fig. 35. Weatherstrip set seals head, side and bottom of doors.

WHEEL SUPPORTED, CHANNEL TRACK

TRACK: rolled steel, formed steel, or extruded aluminum. BEARING: plain, bushed, Oilite bushed, steel balls or steel rollers. WHEELS: steel, brass, fibre, rubber or plastic.

WHEEL SUPPORTED, I-BEAM TRACK, TOP MOUNTED, ADJUSTABLE

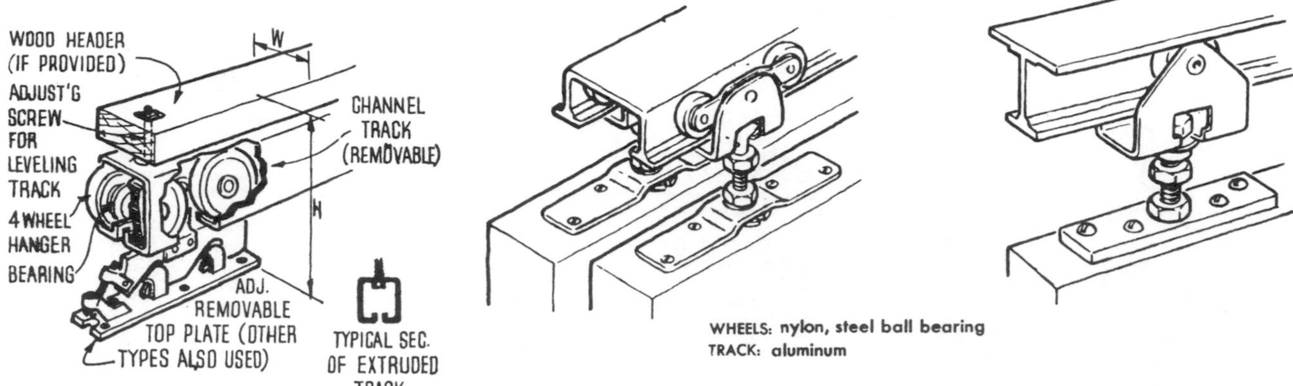

WHEELS: nylon, steel ball bearing
TRACK: aluminum

Fig. 36. Types of by-pass door tracks and carriers.

TYPICAL EXTRUDED SECTIONS FOR BY-PASSING DOORS

TYPICAL DOOR FASTENING DEVICES

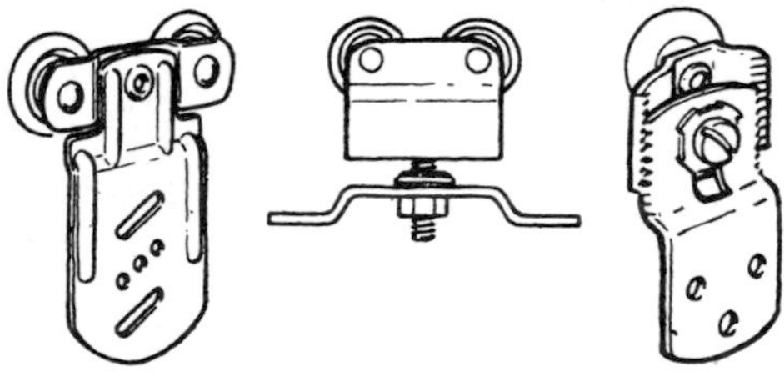

TYPICAL FLOOR GUIDES & TRACK

THRESHOLD TYPE

FLUSH TYPE

TRACK FOR FOLDING & ACCORDION DOORS

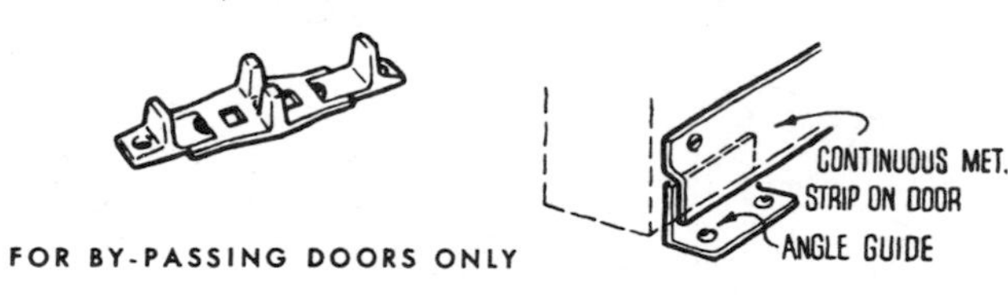

FOR BY-PASSING DOORS ONLY

Fig. 37. Sliding door hardware: extruded track, hangers fastened to top of doors, floor track and guides for door bottom.

Fig. 38. Cutaway shows extruded double track screwed to header and plaster applied to thin lip on face edge of track.

Fig. 39. Nylon rollers ride in double track grooves. Note elongated screw holes for precise adjustment of carriers.

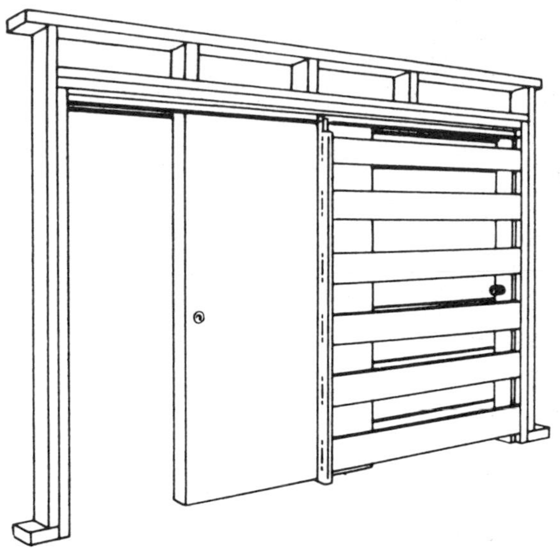

Fig. 40. Pocket door vanishes into frame built in partition.

example, a 5d nail measures 1-3/4″ (5/4 plus 1/2 equals 1-3/4).

Anchor, helical and spiral shank nails are available for greater holding power for roofing, siding, and drywall. Galvanized-steel nails and nails made of aluminum, copper and stainless steel are used where corrosion resistance is important.

Wood and Metal Screws. Wood screws vary in shape of head and thread, **Figure 48.** The flat head screw is the most commonly used in building; the round and oval head are used where appearance is a consideration.

Wood screws lengths vary from 1/4″ to 6″ with a 1/8″ increase between 1/4″ and 1″; a 1/4″ increase between 1″ and 3″; and a 1/2″ increase between 3″ and 5″.

Wood screws are used properly as shown in the three steps, **Figure 49.** After locating the place for the screw, a pilot hole the diameter of the screw body is bored in the upper part of the stock. Then a starter hole equal to the root diameter of the screw is bored through the upper stock to the lower piece of stock. Finally, the hole bored in the upper stock is countersunk and the screw is driven home as shown in the section.

Screws for assembling metal parts, **Figure 50,** are made of steel and brass with flat, round, oval and fillister heads, with straight or Phillips type slots.

Lag screws, **Figure 51,** are used when ordinary wood screws would be too short or too light and spikes would not be strong enough. Combined with expansion anchors lag screws are used to frame timbers to masonry.

Anchor Bolts and Toggle Bolts. Anchor bolts and toggle bolts are used in anchoring stock to masonry and plasterboard, **Figure 52.**

Stove Bolts and Expansion Bolts. Stove bolts are less precisely made than machine bolts. They are made with flat or round slotted heads and may have threads extending over the full length of the body. They are generally used with square nuts and applied, metal to metal, wood to wood, or wood to metal. If flatheaded they are countersunk; if roundheaded, they are drawn flush to the surface. An expansion bolt is used in conjunction with an expansion shield, **Figure 53,** to provide anchorage in substances in which a threaded fastener alone is useless. The shield, or expansion anchor inserted in a predrilled hole, expands when the bolt is driven into it and becomes wedged firmly in the hole, providing a secure base for the grip of the fastener.

Corrugated Fastener. The corrugated fastener is one of the many means by which joints and splices are fastened in small boards. It is used particularly in the miter joint. Corrugated fasteners are made of sheet

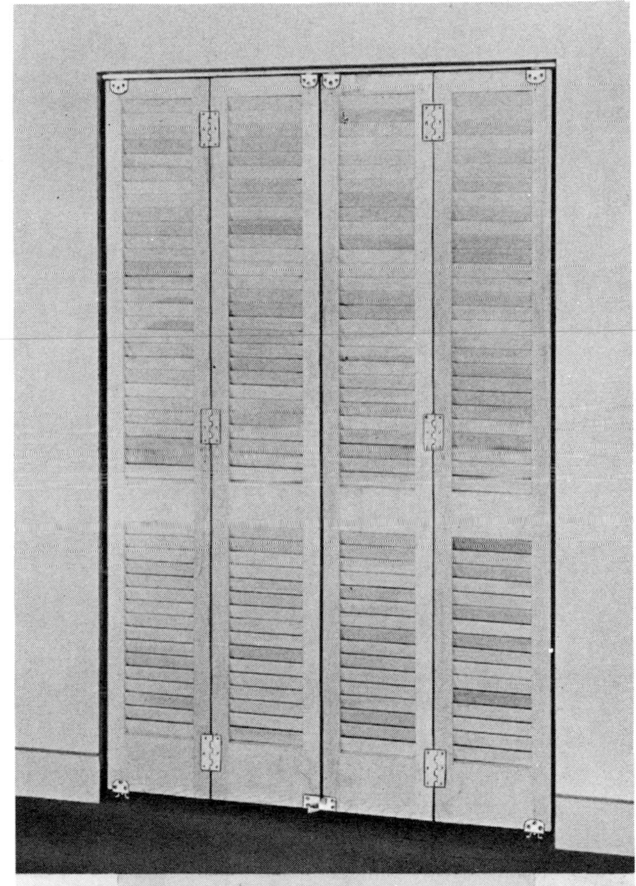

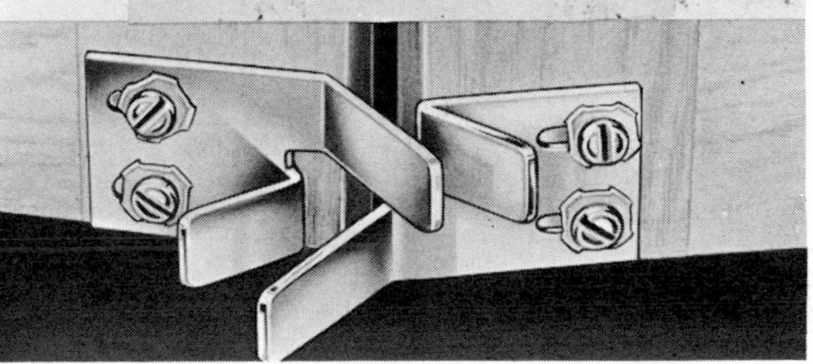

Fig. 41. Door aligners, fastened to bottom of meeting bi-fold doors, eliminate bottom guides and slide doors into perfect alignment. Doors are shown as seen from inside closet (top).

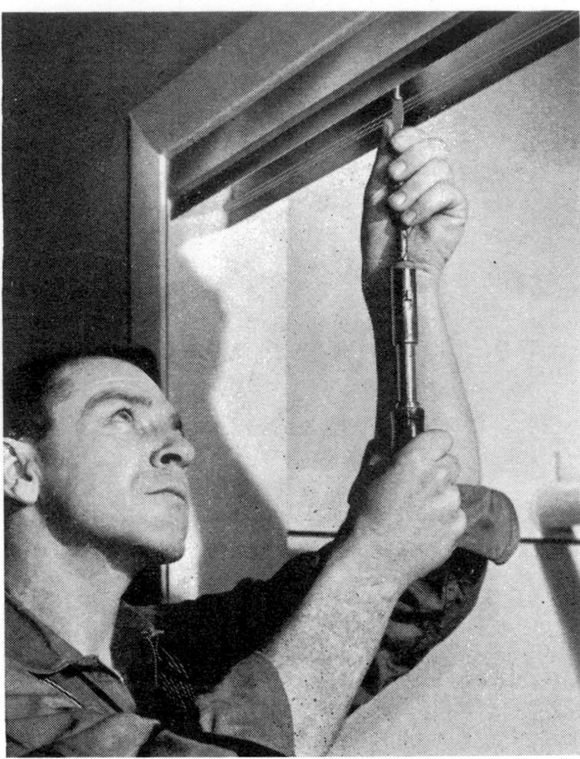

Fig. 42. Workman installing track across wide opening for sliding mirror doors on closet.

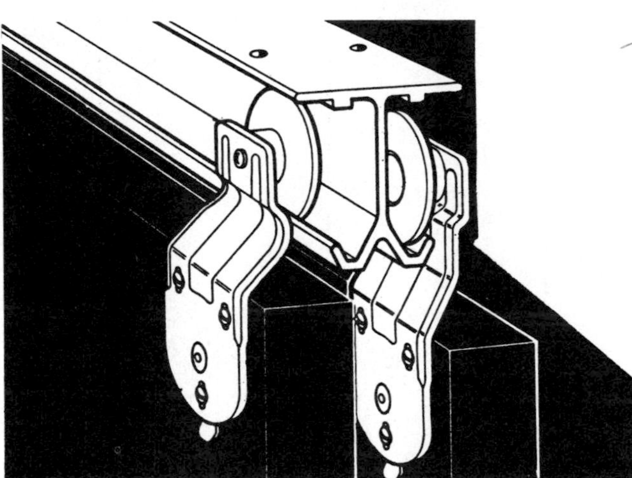

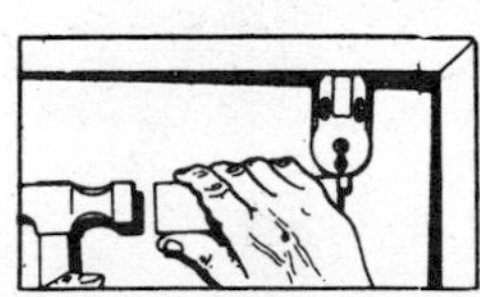

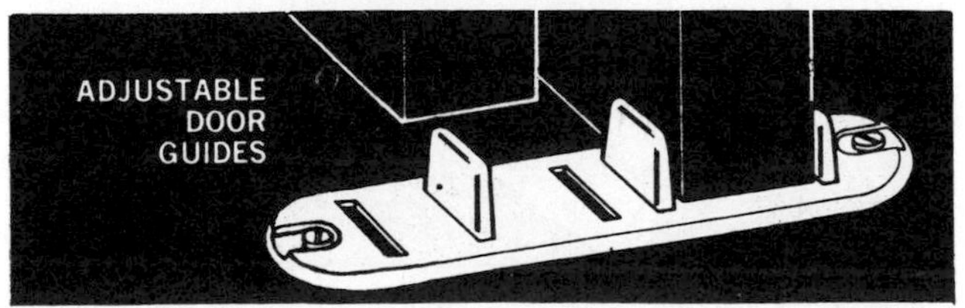

Fig. 43. Tapping adjustment arm with hammer and block (insert) raises or lowers door carriers. Bottom guides adjust laterally.

metal of 18 to 22 gage with alternate ridges and grooves; the ridges vary from 3/16 to 5/16 inch, center to center. One end is cut square; the other end is sharpened with beveled edges. There are two types of corrugated fasteners: One with ridges running parallel, **Figure 54,** the other with ridges running at a slight angle to one another. The latter type has a tendency to pull the material together since the ridges and grooves are closer at the top than at the bottom.

The width of the fasteners varies from 5/8 to 1-1/8 inches, while the length varies from 1/4 to 3/4 inches. The fasteners also are made with different numbers of ridges, ranging from three to six ridges per fastener.

Corrugated fasteners are used in a number of ways; to fasten parallel boards together, to make any type of joint, and as a substitute for nails where nails might split the timber.

Timber Connectors. Timber connectors are metal devices for increasing the joint strength in timber structures. Efficient connections for either timber-to-timber joints or timber-to-steel joints are provided by the several types of timber connectors. The appropriate type for a specific structure is determined primarily by the kind of joint to be made and the load to be carried. The connectors eliminate much complicated framing of joints. They simplify the design of heavy construction, eliminate lateral movement, give greater efficiency of material and reduce the quantity of timber, hardware, time and labor.

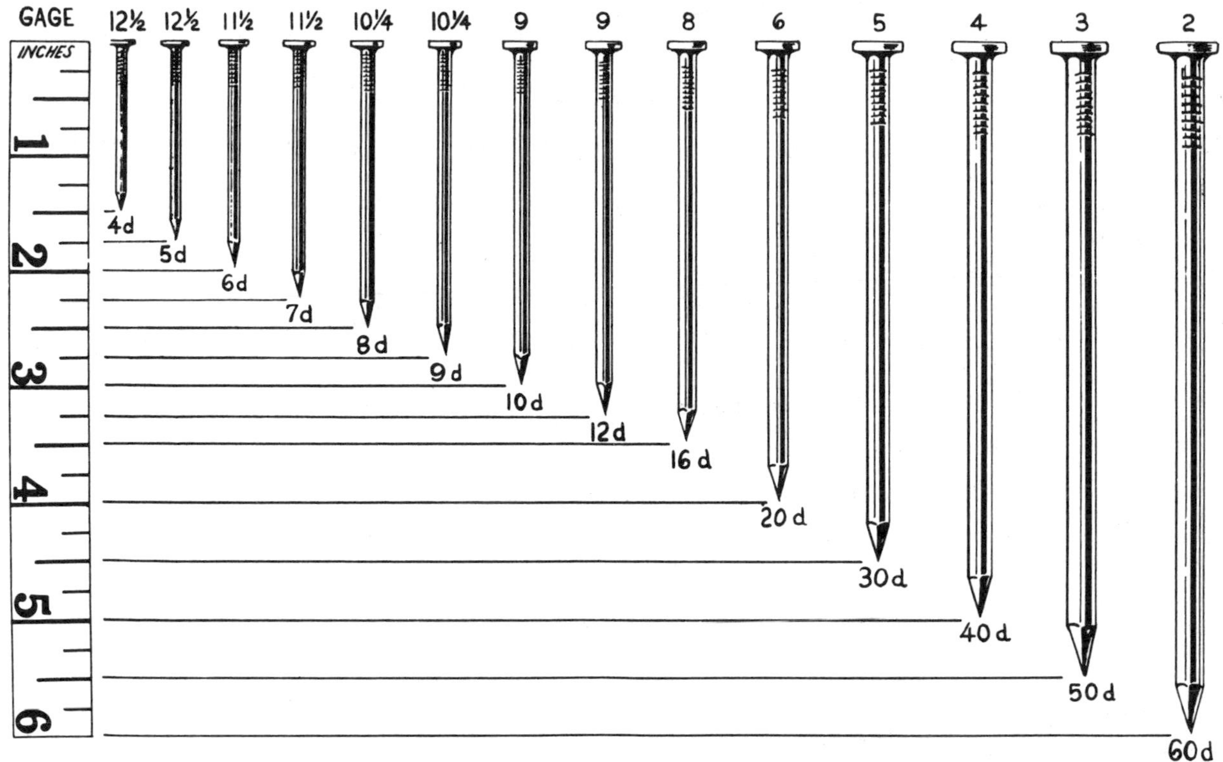

Fig. 44. Length and gauge of common wire nails from 4d (1-3/4") to 60d (6"). Gauge is equivalent to wire size.

HARDWARE

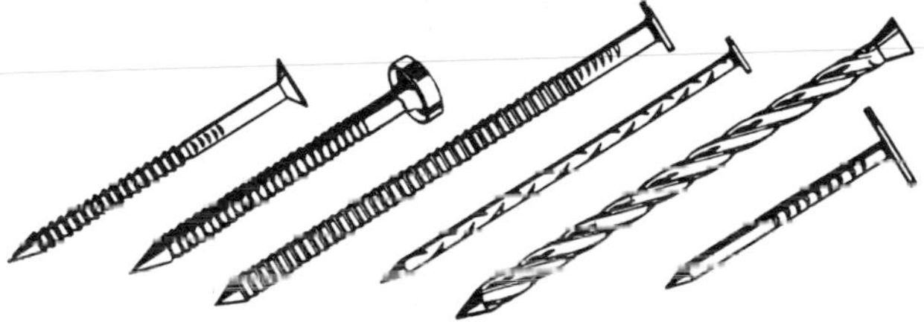

Fig. 45. Annular or ringed shank nails and spiral shank nails resist "popping."

Split rings are made of low-carbon steel and are available in 2-1/2 and 4 inch diameters. They are used between two timber faces for heavy construction. They fit into grooves which are cut half the depth of the ring, **Figure 55,** into each of the timber faces. The grooves are made with a special bit used in an electric, air or hand drill, **Figure 56.** Trusses assembled with split ring connectors and bolts have great strength and rigidity.

The tongue-and-groove split in the ring permits simultaneous ring bearing against the cone wall and outer wall of the groove into which the ring is placed. The inside bevel and mill edge facilitate installation.

Toothed rings are corrugated and toothed, and are made from 16-gauge plate low carbon steel, **Figure 57.** They are used between two timber frames for comparatively light construction and are embedded into the contact faces by pressure, **Figure 58.**

Typical Residential Truss. Plywood gusset plates often are used to build up roof trusses as shown in **Figure**

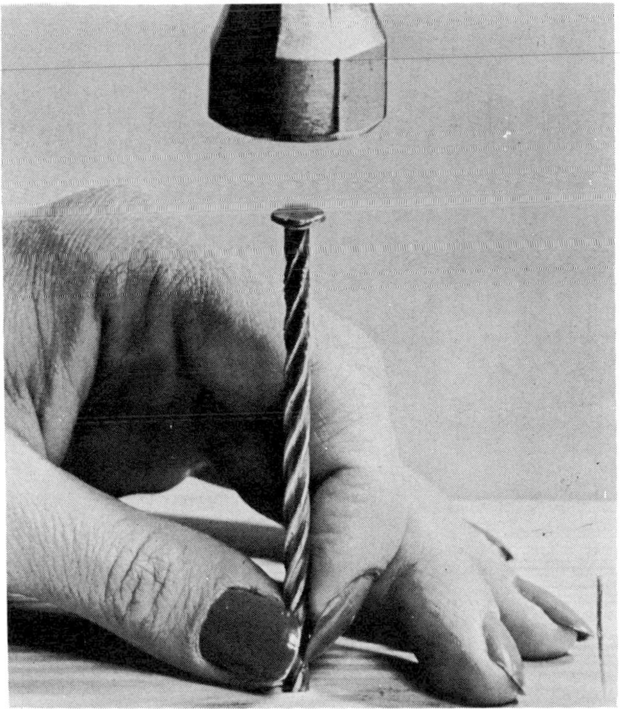

Fig. 46. Spiral shank nail has greater holding power than other type nails of the same size.

59. This method of joining timbers permits members to be assembled without having to be overlapped at the joints. Greater stiffness and strength are obtained when the gussets are pressure glued or nailed and glued.

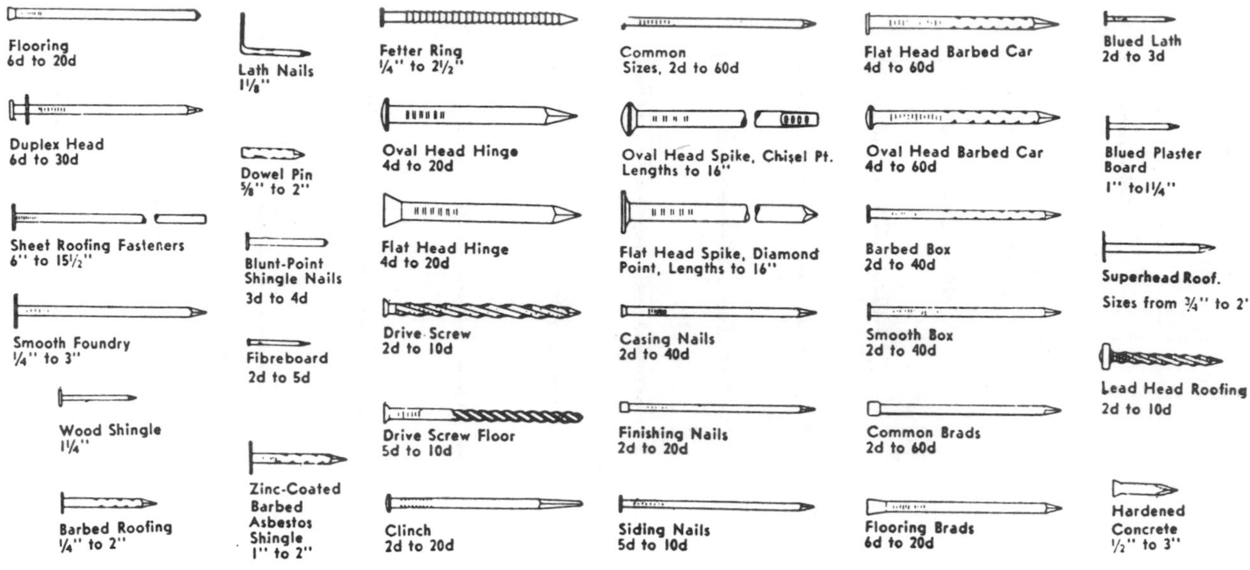

Fig. 47. Thirty four types of nails used in building.

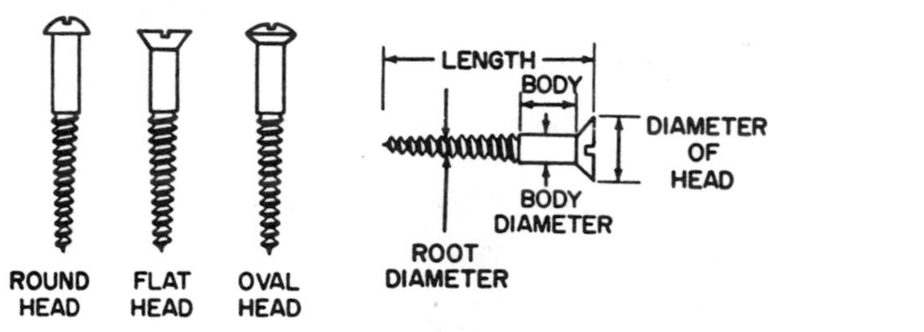

Fig. 48. Types of wood screws and nomenclature.

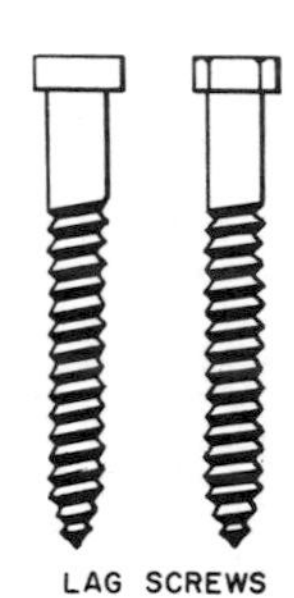

Fig. 51. Square and hexagonal head lag screws.

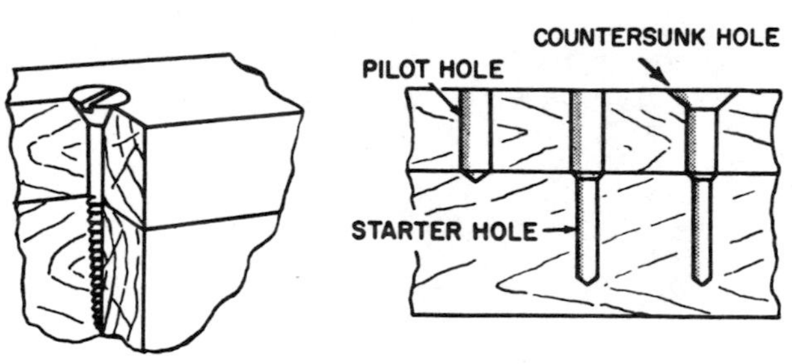

Fig. 49. Screw is driven into predrilled holes to join stock.

Fig. 50. Screws used for assembling metal parts.

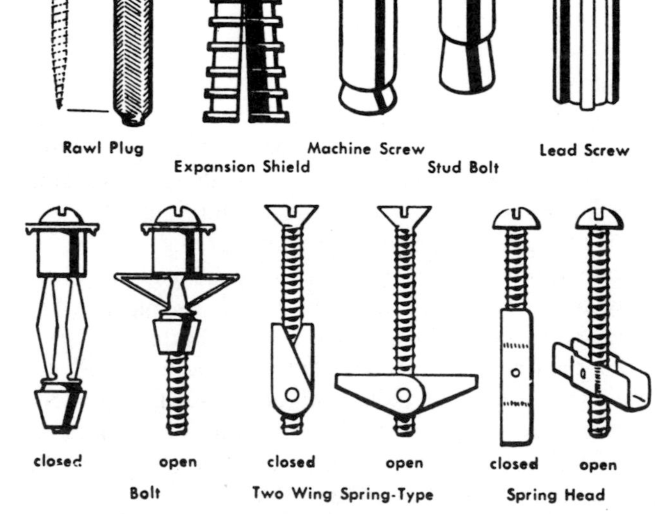

Fig. 52. Bolt anchors expand inside a hole; toggle bolts pass through a hole, then spread to bear against rear of wall.

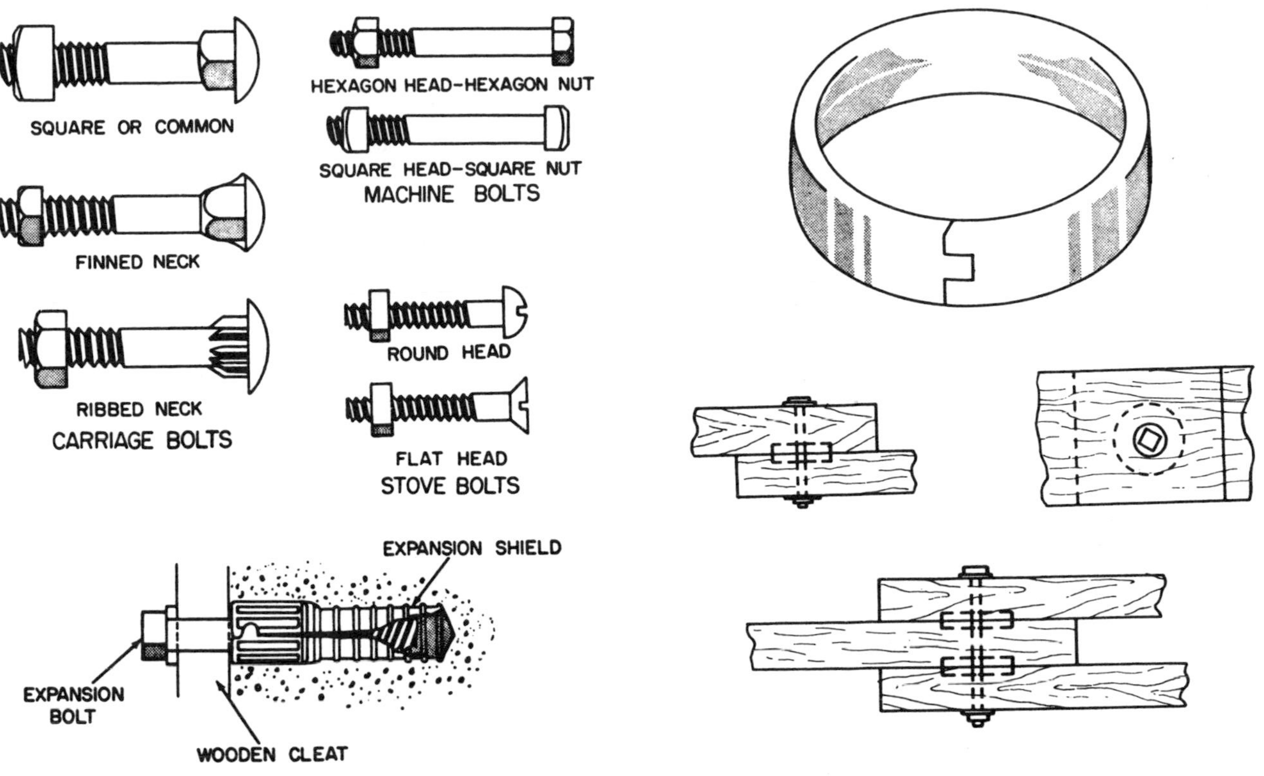

Fig. 53. Types of bolts, including expansion bolt.

Fig. 55. A split-ring timber connector and its installation between two or three structural framing members.

HARDWARE

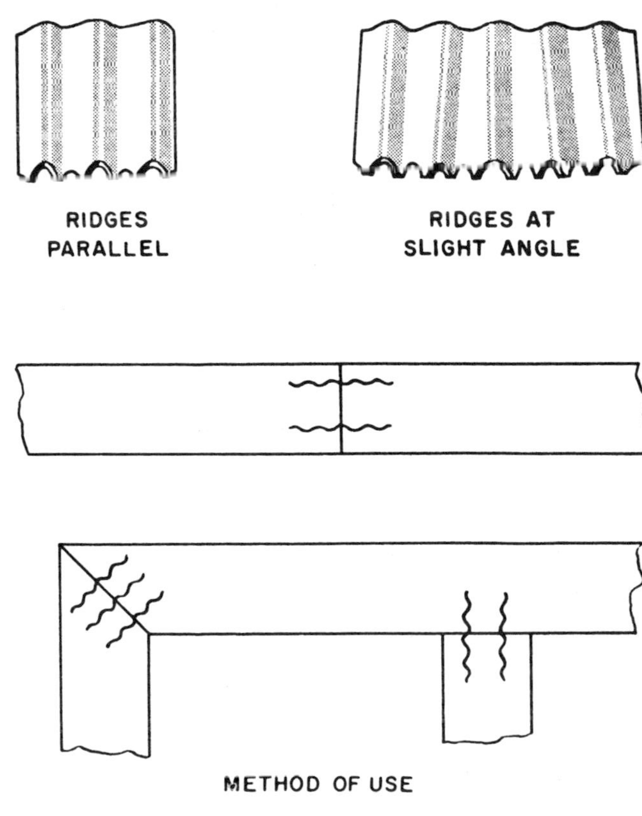

Fig. 54. Corrugated fasteners join butting edges of stock.

Fig. 56. Half-depth ring grooves are cut in meeting members for connectors, using electric drill and special cutter.

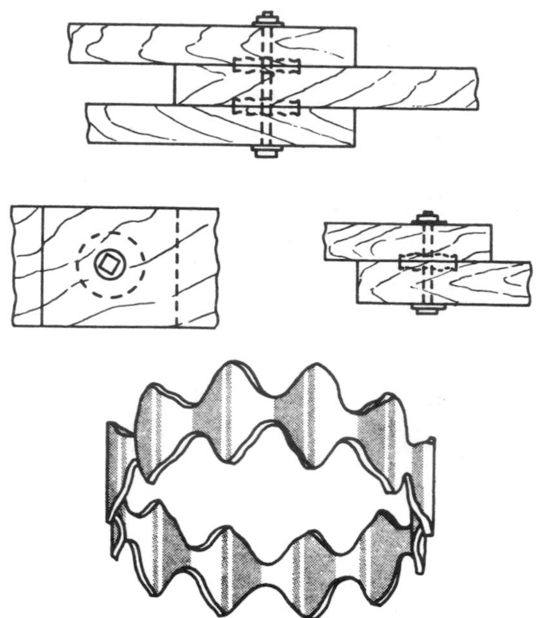

Fig. 57. Toothed ring connectors are used much the same as split rings, but for comparatively lighter construction.

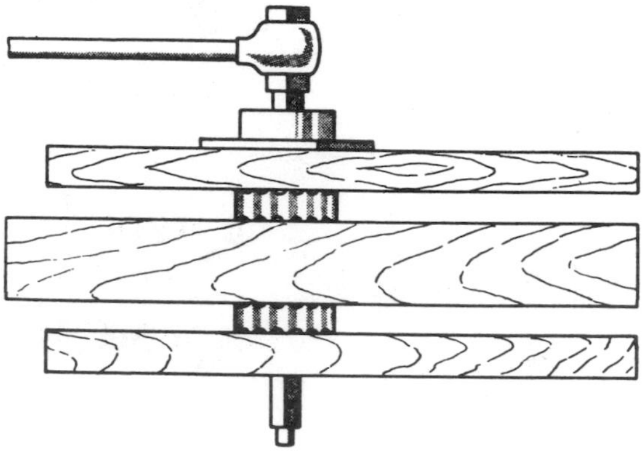

Fig. 58. Toothed ring is forced into lumber by pressure of heavy bolt holding members together.

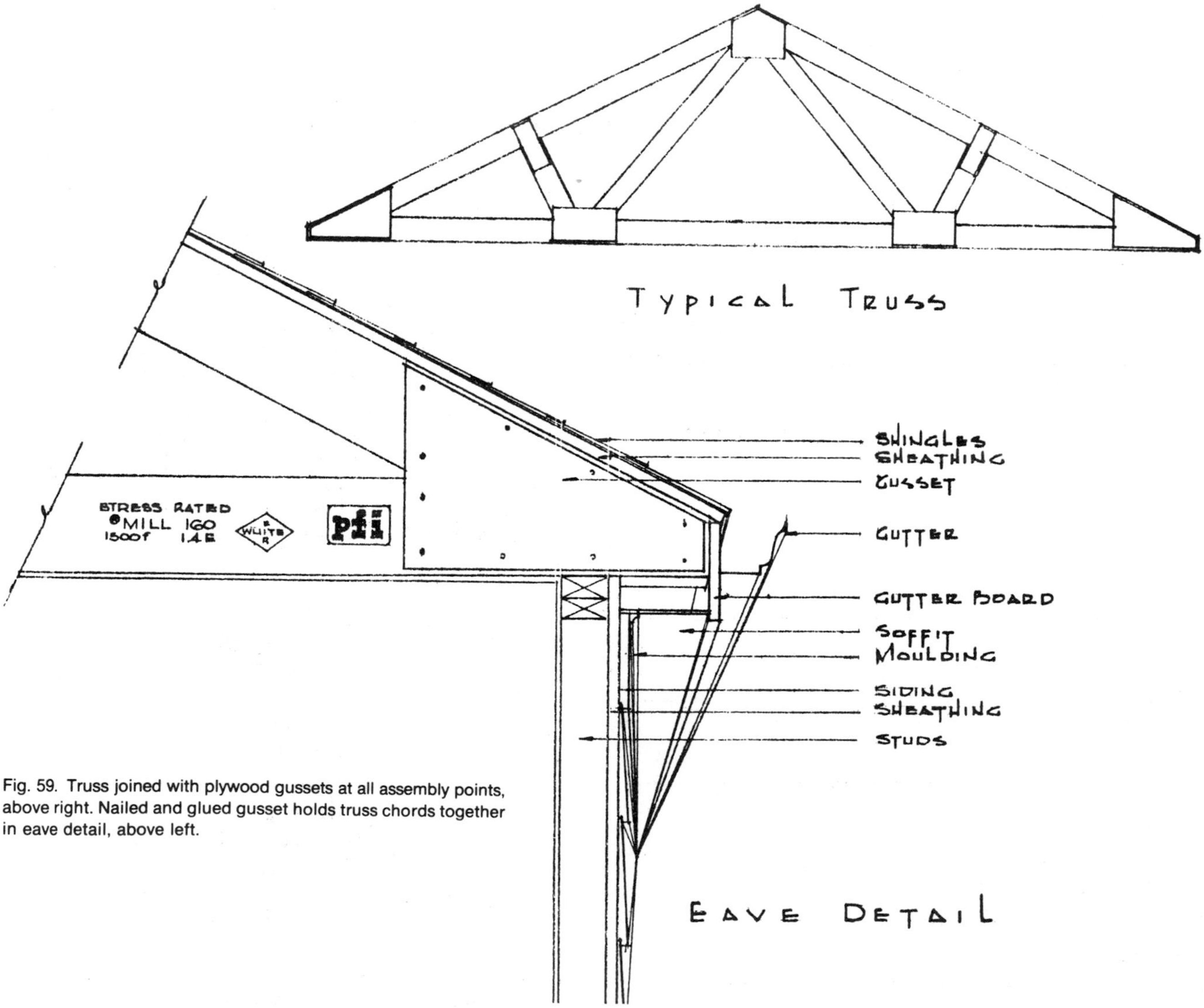

TYPICAL TRUSS

STRESS RATED
MILL 160
1500f 1.4E
WHITE
pfi

SHINGLES
SHEATHING
GUSSET
GUTTER
GUTTER BOARD
SOFFIT
MOULDING
SIDING
SHEATHING
STUDS

EAVE DETAIL

Fig. 59. Truss joined with plywood gussets at all assembly points, above right. Nailed and glued gusset holds truss chords together in eave detail, above left.

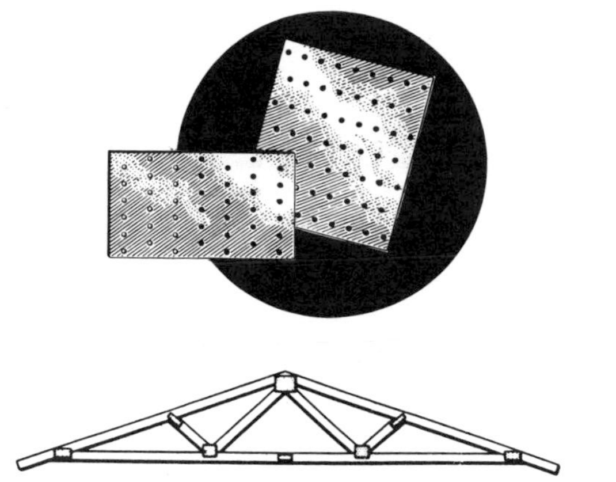

Fig. 60. Teeth formed in metal connector plates by punching holes are pressed into truss members to connect them.

Truss plates of metal are also used with use of single gauge zinc coated steel plate, **Figure 60.** Wood members are pre-cut, assembled and truss plates applied at joints by pressure. Teeth formed in the plates penetrate the lumber, eliminating nails.

Universal Building Bracket. The universal building bracket is furnished in 21 gage galvanized steel for use on standard lumber sizes. Brackets are bent as required to form holding brackets of different shapes for joining framing stock, **Figure 61.** Bracket size is 4-7/8″ square and may be bent to support the joint then nailed in place. A variety of pre-formed beam hangers and connectors for framing members is also available to speed assembly of structures and eliminate toenailing, **Figure 62.**

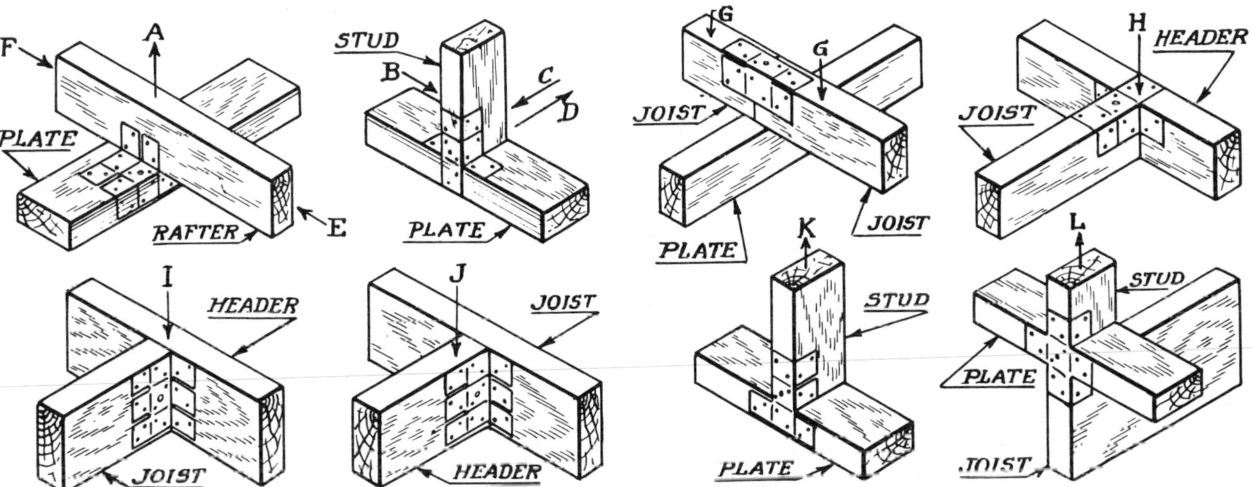

Fig. 61. Universal building bracket can be bent to required shape for many framing joints illustrated.

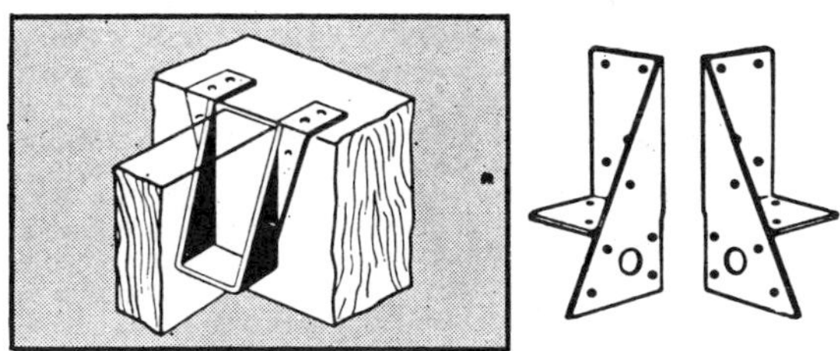

Fig. 62. Metal joist hangers and brackets speed framing.

Chapter 15

CLOSETS/SHELVES/BUILT-IN EQUIPMENT

Storage space is as important for a well functioning home as living space. Inadequate closet and shelf space is a vexation, and poorly arranged or inaccessible storage space is an expensive waste.

Closet appliances, such as shelving, racks, and hanging devices are continually being improved for greater efficiency. Careful study of clothing and storeable possessions result in specialized application of space, as in separate closets or areas for men's, women's, and children's belongings. Storage walls and other forms of built-ins now make every inch of floor space count, often reducing builders' costs by substituting them for room partitions.

Three developments have pushed the growth of built-in storage equipment in recent years: first, the factory production of component parts of houses; second, the refinement of modular theory and practice, so that furniture fits readily into houses designed on modular principals; third, the use of trusses for roof construction, which eliminates the need for interior bearing walls, permitting the easy installation of ceiling height storage walls, **Figure 1.**

Factory Built Cabinets, Closets and Storage Units. An important phase of prefabrication is the standardized kitchen cabinet, bedroom closet, and carport storage unit, shipped from factory to the new building in much

Fig. 1. Storage wall (right) of factory-made wardrobes, fits under furring strips nailed to H-brace trusses. Similar storage wall is used beyond partition at left. Ceiling height storage walls such as these can be used anywhere beneath trusses.

the same manner as hardware, comfort conditioning equipment, intercommunication units, appliances, etc., instead of being cut, assembled, and fine-finished on the job site. Designers often specify commercial brands of storage units, where they formerly detailed the custom construction of cabinets.

Since modules of three or four inches are the basis of much modern design of windows, doors, and wall panels, it is reasonable that storage units should be chosen to fit in with the basic scheme of the multiple size unit. An exception to this technique is found in another development of home prefabrication, that is, when the entire wall panel, as a unit, including window and door frames and storage units, based not necessarily on a common multiple, but on the over-all shape of the entire wall, is shipped in a single piece, **Figure 2.**

Since trussed rafters are planned to rest only on the outside walls, it is often good practice in both truss-framed and post-and-beam dwellings to finish all floors and ceilings to some extent before setting in any partitions except the bath. Full height storage walls or units are then brought in to separate some of the rooms and are supplemented with glass or plastic partitions.

The demand to save space calls for shallow and wide closets, rather than deep ones requiring walk in space, **Figure 3.** Where hinged closet doors are used, they are often fitted to contribute to the storage function with tie-racks, hat racks,shoe racks,and a mirror, **Figures 4 and 5.** Shallow clothes closets are likely to have sliding doors, either by-pass or hinged-flap, to lay the full width of the closet within reach.

Closet hardware continues to benefit from the inventor's ingenuity. Expandable shelves, easily attached clothes hooks and poles, special appliances for hats, shoes, etc., and telescoping and extending coat racks all add to convenience and efficiency, **Figure 6.**

Allocation of space for hanging garments and for shelves and drawers is now related to the size of the garments. Men's and women's clothing take up different amounts of space, and hanger poles need to be different heights from the floor, **Figure 7.** A woman's coat requires 46″ from pole to hem, while a man's coat requires about 54″. A woman's full length evening gown may need 60″ to hang freely. If overcoats are stored elsewhere, a man's shirt, trousers, and suit jackets will require only between 31″ and 33″.

Fig. 2. Kitchen cabinet unit, complete with wall section, is set in place as a partition. Counter top, sink and plumbing are installed after cabinet is secured.

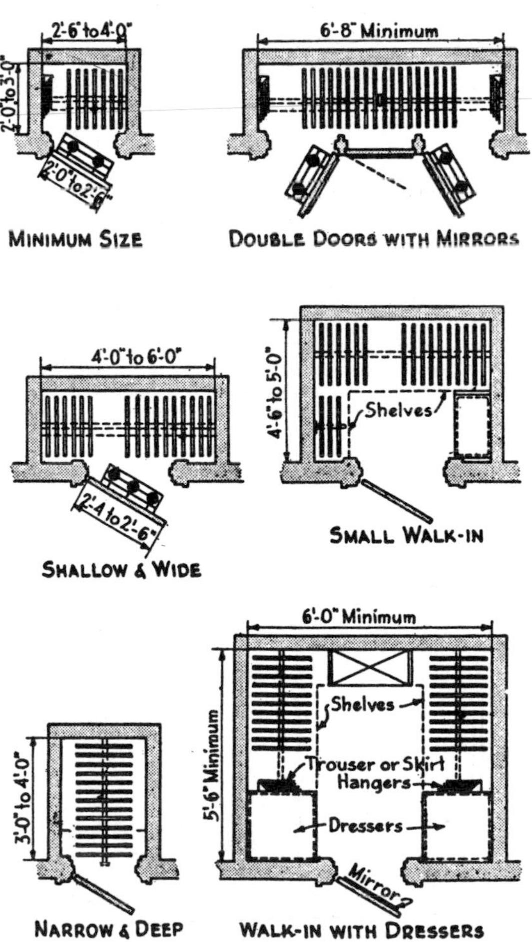

Fig. 3. Closet plans to fit various available space.

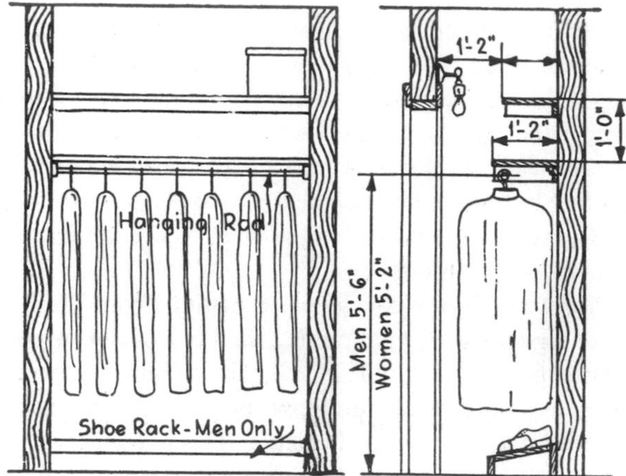

Fig. 4. Shallow hanging closet with clothes pole and shoe rack.

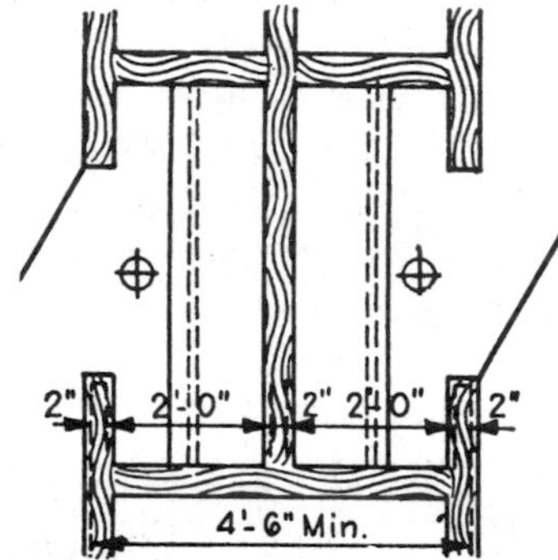

Fig. 5. Shallow closets framed back to back in adjacent rooms.

Fig. 6. Closet hardware in early Research House of National Association of Home Builders can be rearranged to suit needs.

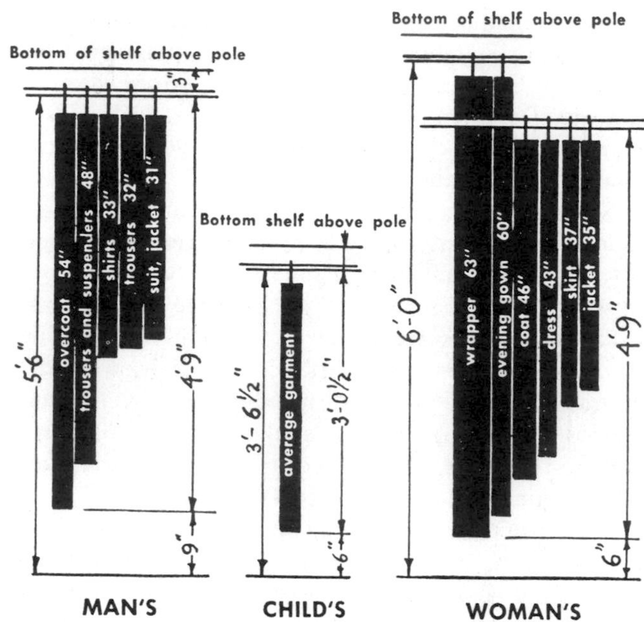

Fig. 7. Typical lengths of various articles of clothing.

Children's closets need less hanging space and more shelf and drawer space, for games, sporting equipment, "treasures," collections, etc. The pole should be placed within easy reach.

Built-In Equipment. Built-in equipment, whether appliances, cabinets, shelves, room dividers, or work surfaces, are likely to be found in any room of the house. Most often they are simply set in place as they come from the supplier. There is often a combination of factory prefabricated components and custom on-site setting. Finished sliding doors and hardware are often supplied for a conventionally built closet, or a large storage unit, or a combination of stock units are set into a plastered or furred down framework, **Figures 8 and 9.**

The Kitchen

Today's kitchen cabinets are furnished with easily accessible storage space both above and below the working counters. The doors of cabinets are ruggedly constructed of tough, wear resistant core faced with oak, birch, or maple with a full ⅜" dust proof seal, **Figure 10,** or plywood with hardwood veneer front and back.

The drawer fronts are constructed of solid ¾" hardwood stock. The hand grips may be cut into at the bottom edge of each drawer, eliminating pulls. The track attached to the top of each drawer, or those at its sides, glide on nylon rollers in a smooth quiet operation. This sink is of porcelain or stainless steel set in a counter top of Formica furnished in white and other desired colors or patterns, **Figure 11.**

Because of the daily use of numerous articles, the kitchen has special need of much easily accessible

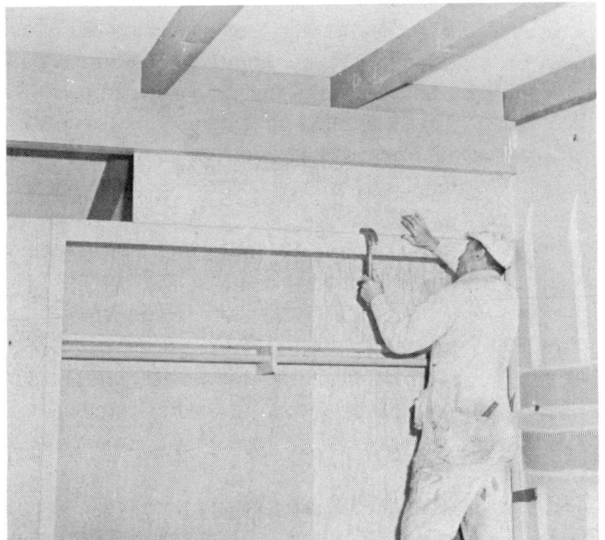

Fig. 8. A filler insert of plywood is nailed above wardrobe used as room partition beneath sloping beamed ceiling.

Fig. 9. Combination of factory-made cabinets fits into bedroom wall under fixed-glass in gable end.

storage space. Kitchen cabinets used to be free standing pieces of furniture with specialized compartments, such as flour bins and cutlery drawers. Today, practically all kitchen storage is a permanent fixture. Cabinets are of two general kinds: base units and wall units. Kitchen cabinets are now factory made. A large percentage of prefabricated cabinets, in recent years, have been made of clear ponderosa pine body and hardwood plywood veneer doors and surfaces, **Figures 12 and 13.**

Plastic Laminate Surfaces. Since the development of plastic laminate surfaces, base cabinets are made without a top, and the counter or work surface is of plastic laminate supplied separately. A single length work surface may cover several cabinet units, **Figures 14 and 15.**

Ceramic tile work surfaces are also applied over factory made as well as custom made cabinets, **Figure 16.**

Special cabinets with special shapes and functions are often used, some stock, some custom made: corner units, some with revolving shelves to make the inner recessed space easily available; tilting bins, vegetable storage bins, tray cabinets, sliding shelves for automatic mixers; pantries, pull-out garbage pails, sliding towel racks, pull-out bread boards, and other types, **Figure 17 and 18.**

A Family Center. The arrangements of rooms and work centers are striking and exceptionally unique features of this family center. The prefabricated house, or manufactured component could well be expressed in these three rooms.

Section one may be known as the "Pavilion" or eating area. It gives the diners a feeling of privacy in the large, multi-purpose room which measures 9′-6″ × 29′. Just a step away from the table is the cooking center with its spacious double oven electric range. It is easy to install a free standing model blending with the custom made exhaust hood.

Section two in the foreground of the floor plan layout, **Figure 19,** may be known as the "Periphery Kitchen." It has its own sink with a built in food waste disposer, an under counter dishwasher and extra counter and cupboard space. Note how the two kitchen areas share the same frost free refrigerator, one that requires no defrosting, not even in the freezer section.

Section three at the far end of the floor plan layout may be known as the "Laundry Center-Plus." This section has an automatic washer and dryer in its own enclosure, with storage shelves. In this section there is space for a sewing center and a play area for children. There is also room enough for a food freezer holding 588 pounds of frozen food.

Ranges, Oven, Refrigerators and Freezers. A significant development has been the built-in separate range unit and the in the wall oven, **Figure 20, 21, and 22.** Much difficulty arose from the variety of sizes of the various oven units, many of which require venting to the outside.

Attention to arrangement of appliances and counters for efficiency is important in original design, since built-ins are permanent, **Figure 23, 24, and 25.** The arrangement should provide enough counter space to meet the minimums for the various functions of food preparation and storage next to each appliance. Wherever possible the arrangement of counter space and appliances should be compact enough to minimize the amount of walking from the refrigerator (storage) to sink (preparation) to range or oven (cooking and serving). Even such detail as having counter space to the left of a refrigerator that opens to the left is important in saving steps, **Figure 23.**

With the popularity of in-counter range units, both gas and electric, there developed a demand for built-in range hoods, usually equipped with a ventilating exhaust fan, and often blended with the color of the appliance,

Fig. 10. A factory-made kitchen cabinet set finished with a new, durable synthetic resin and plastic-laminate counter.

Fig. 11. An island working counter and cabinets with sink and range installed along two walls.

Fig. 12. Rounded shelves are factory built as an accessory.

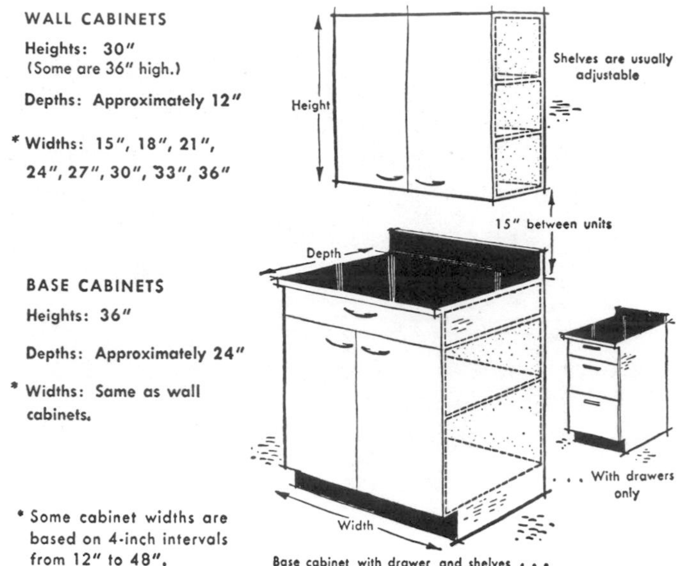

Fig. 13. Standard dimensions of wall and base cabinets.

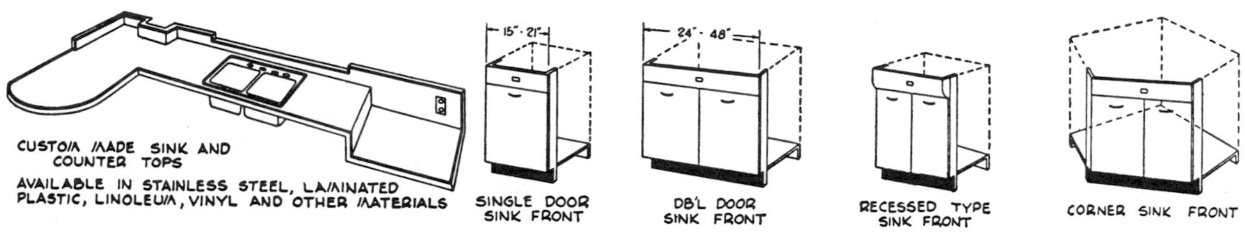

Fig. 14. Sink and counter tops are custom made to fit over various combinations of cabinets.

Fig. 15. Laminated plastic is used for countertops and for sliding doors of wall-hung cabinets.

Figure 26.

Refrigerators and freezers may also be built into the cabinet structure.

Component Manufacture. The trend toward component manufacture resulted in large kitchen units combining both appliances and cabinet space. Wired at the factory for such popular appliances as garbage disposal units and dishwashers, as well as range units and ovens, the one piece kitchen center is available in various sizes, to fit various kitchen walls, and in various colors, **Figure 27.**

Work surfaces and counter tops are variously of stainless steel, linoleum, ceramic tile, as well as laminated plastic. Walls may be finished with ceramic, or metal tile, paper, vinyl plastic, waterproofed wallpaper, or traditional kitchen paint. Sliding door panels of cabinets may be of pressed hardboard, particle board, painted perforated board or glass.

Popular styles in wood cabinets lean toward doors and fronts with raised panels, dark rich stain and clear visibility of grain. Maintenance free clear finishes, self closing hinges, drawers with side-mounted roller bearing suspension, and solid hardwood construction are among the desirable features of the better grade cabinets, **Figure 28.**

Fig. 16. Ceramic tile work surface and backsplash on wall behind counter are durable, easy to keep clean.

Fig. 17. Custom-built lazy Susan in corner cabinet makes maximum use of difficult-to-reach storage space.

National Kitchen Assn.

Fig. 18. Special kitchen cabinets and convenience devices include pantry with shelves on doors (left), vegetable bins, pull-out garbage pail, and pull-out towel rack (center), and pull-out bread board and bread storage bin (right).

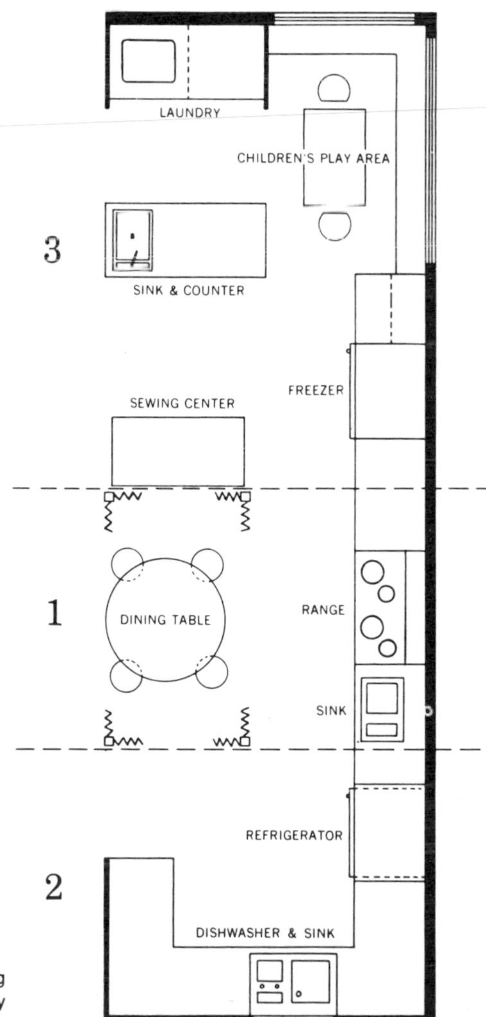

Fig. 19. Family center room arrangement provides cooking and dining area (1), food preparation and clean-up area (2), and laundry, hobby and play area (3) in 9′-6″ × 29′ space.

Fig. 20. Range and wall oven are built into cabinets.

Fig. 21. Because of depth required, cabinet for wall oven is flush with base cabinets rather than wall-hung cabinets.

Fig. 22. Frame installed in cabinet supports wall oven.

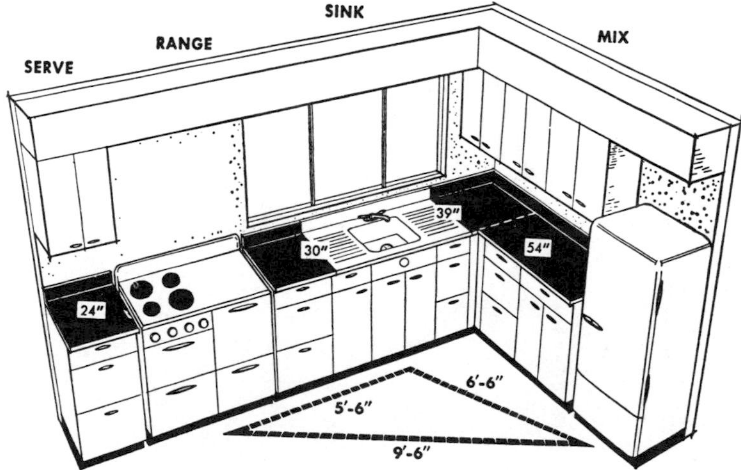

The kitchen above has well-defined centers arranged in normal sequence. Enough counter is provided to meet minimums needed for the various functions of food preparation as well as for storage requirements. The sum of the distances between appliances is 21'-6".

Fig. 23. Stock cabinet (left) has space for standard size oven. Efficient kitchen layout (above) has triangular traffic pattern to save steps. Each appliance has adjacent work space.

National Kitchen Cabinet Assn.

Fig. 24. L-shaped kitchen with counter space next to refrigerator sink and oven and range so food preparation can be moved in normal sequence from refrigerator to sink to range or oven.

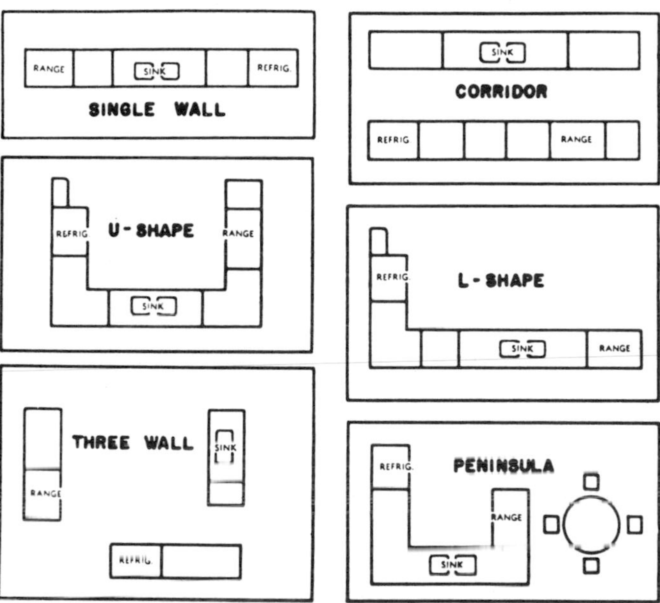

Fig. 25. Efficient kitchen layouts fit wide range of room shapes.

Fig. 26. Traditional design hood and ventilating fan carries cooking odors and vapor away from built-in range.

Fig. 27. Manufactured work center has plumbing and wiring for clothes washing, sink, garbage disposal unit and dishwasher installed and connected at factory.

Excell Wood Products Co.

Fig. 28. Raised panels, dark rich colors, clear visibility of grain, drawers with roller bearing suspension, and maintenance free finishes are attractive features of modern kitchen cabinets.

The Dining Room

When the dining area is given a special room, its storage area is usually devoted to flat drawers for table linen, place mats in shallow drawers, and deeper drawers for cutlery. The old china closet built into a corner of the room remains popular both as a display piece and as a way of showing attractive plates and pottery, **Figures 29 and 30.** Wall hung cabinets with glass doors are available for the same purpose.

Often the dining area is an extension of the kitchen, separated by a counter; sometimes it is a part of the family room. It can thus share storage space with the family room. A common practice is to place cabinets under the dividing counter or pass through, with doors on both sides. Similarly, hanging cabinets, suspended on posts or from the ceiling, may have sliding doors on both kitchen and dining area sides, **Figure 31.** A modified "L" shaped floor plan, **Figure 32,** places the kitchen, family eating area and family recreation area all together.

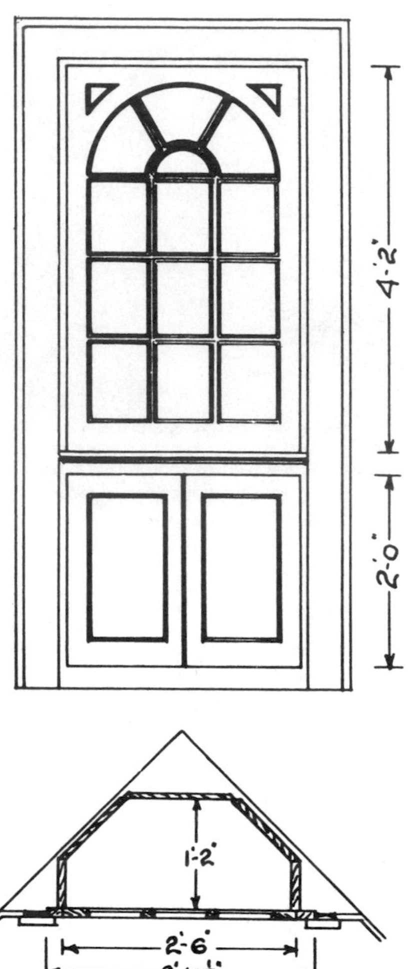

Fig. 30. Built-in china closet showing simple construction and method of joining wall material to edges beneath trim.

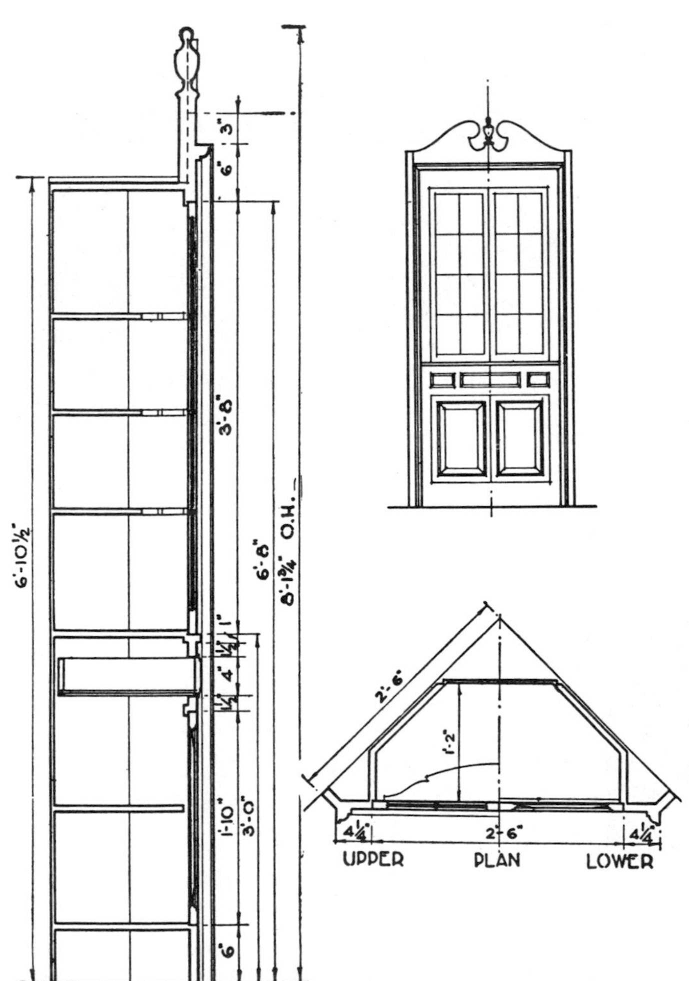

Fig. 29. Elevation, plan and section of stock china closet.

Fig. 31. Pass-through cabinets between kitchen and dining-living room have sliding doors which open from both sides.

Fig. 32. Open L-shaped floor plan brings three areas together.

Fig. 33. Built-in television and music equipment in custom dividing wall. Note cloth screen over speaker cabinet beneath horse ornament, and long sliding panel doors.

Fig. 34. Basement remodeling job features plywood storage wall divided into wardrobe, space for sewing machine, music equipment and record storage, and compact home ofice.

Living Room and Family Room

Since the requirements for the living room are decorative as well as utilitarian, great craftsmanship is expended on shelves, built-ins, and free standing storage units. As the family room has grown separate from the living room in function and appearance, the types of storage likewise have changed; the living room has grown more formal, the family room has taken on the casual quality.

Living room storage space is of several kinds: shelves for books, bric a brac, silver plate; built-in desks, radio and musical components, and television, **Figures 33, 34 and 35.**

Room dividers are used to break up space, to offer some decorative storage space for small items and space for plants, **Figures 36 and 37.** Closed storage, with drawer space, and shelves with swinging or sliding doors, **Figure 38,** is built into full height storage partitions.

A Pre-Cut Storage Unit. A storage unit can be made of precision assembled panels that fit together perfectly in such a way that the panel frames lock rigidly into position before being glued and nailed, **Figure 39.** All mitering, notching and slotting is factory done. All doors are reversible, choose either the Modern Flush on one side or the Colonial Panelled on the other, whichever fits the

Fig. 35. Similar storage wall of knotty pine includes television, music equipment and built-in fish tanks.

Fig. 36. One side of a two-way room dividing unit.

Fig. 37. Other side of unit. Some shelves open to both sides.

room decor. The unique hinges are self positioning, no notching of frame or door required, the doors can be installed to swing right or left.

All framing members, shelves and drawers are made of clear Ponderosa pine, door panels are of pine plywood, side panels of Fir plywood, backs and drawer bottoms of hardboard. All parts are smoothly sanded, ready for paint or stain. All necessary glue, nails, drawer guides, wood knobs, shelves, hinges and door catches are included in this pre-cut storage unit which can be built up on the job.

The large wardrobe unit includes a metal pull out clothes hanger bar, and the desk units include drop chains and a special writing surface insert panel to be used when the flush door face is used to the outside.

Custom Made Room Dividers. Room dividers are often installed after a house is completed. Usually they are relatively light structures with open shelves and a few closed compartments with sliding or hinged doors, a shelf door that swings down and serves as a light desk or space for ornaments. Some are home made, others are custom built by professional carpenters, **Figure 40.**

The Bedroom

Some of the greatest advances in storage and cabinet activity have been made in bedroom closets. One of the most widespread closet treatments is the full ceiling height wardrobe with two sets of sliding doors: the tall lower pair, opening on a full height clothes hanging

Fig. 38. Living room storage unit backs up kitchen cabinets.

Fig. 39. A manufactured storage wall, which can be used as a room divider.

Fig. 40. Light room divider with adjustable shelves, doors for closed compartments and a door that swings down to form a desk.

space and banks of drawers and shelves, and an upper pair enclosing seldom needed items, such as luggage, **Figure 41.**

Tract builders design shallow closets as part of the plan, and equip them with by pass doors on standard roller hardware, **Figure 42.** Contemporary style buildings, with sloped ceilings or trusses or post and beam construction, often use ceiling height storage walls as room dividers, **Figure 43.** Such closets often permit flexible arrangement of shelves and drawer space, **Figures 44, 45, 46.**

Closet Doors. Regular, filled and veneered panel doors are most commonly used for closet and wardrobe doors. The doors stay straight and perfectly aligned. The excellent screw holding strength enables standard hardware to be fastened securely and permanently. Tight, void free edges can be left exposed, filled and painted or stained to blend with door faces, or they can be banded with lumber or veneer strips.

Phenolic overlaid doors have a smooth, paintable surface that is highly resistant to moisture and abrasion. Fire retardant panels for safety are also recommended for this type of application. Finishes can be duplicated on both faces and backs of doors for positive protection against warping.

Shallow closets designed to save space in smaller rooms may be equipped with folding or accordion doors which do not project into the room yet to give almost full access to the closet. Where door openings are narrower than the width of

Fig. 42. Valance conceals door track on bedroom closet.

Fig. 41. End of wardrobe room divider is fitted with spacious linen closet and swing-out laundry hamper.

Fig. 43. Wardrobe room divider being set in place against wall paneled to blend with unit. Track has already been installed at top and bottom for sliding by-pass doors.

Fig. 44. Adjustable shelves and drawers make unit flexible.

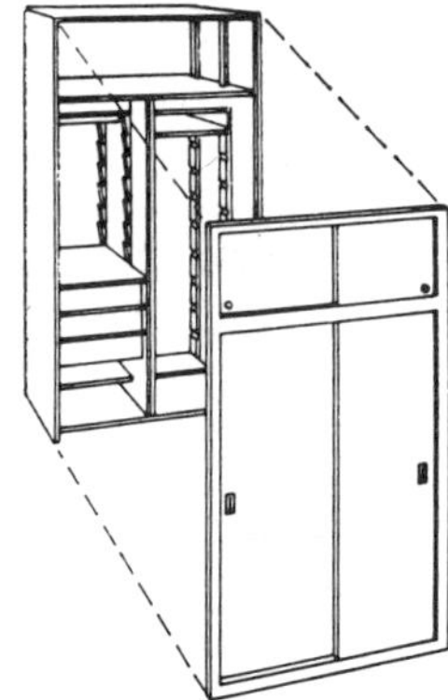

Fig. 45. Frame for sliding doors at top and bottom is set in place after fitting out interior of wardrobe.

Fig. 46. Cedar wardrobe has bi-fold doors at top and bottom.

the closet, shelves may be built into the sides. The door hangs from an overhead track and moves on rollers. It is covered with vinyl and has an internal expansible metal framework, **Figure 47.**

Children's Bedrooms. Dormitory rooms and children's bedrooms present special problems in storage. Often the bed itself is built into a bunk arrangement, with catch all drawers below the mattress, and open and closed shelves for play equipment, **Figure 48.** Built-in and double bunks are popular, to conserve floor space, **Figure 49.**

The Trundle Bed. Another space saver for two children is the trundle bed on rollers that fits under a normal height bed during the day, and is pulled out for sleeping, a throwback to colonial days. Combination dressing tables and wardrobes are handsome novelties; some other built-in ideas are "lazy susan" swivel clothes racks and window seat alcoves with drawers, **Figure 50, 51 and 52.**

Attic Built-ins. Expansion attics offer ideal places for built-in storage. Most storage built-ins are installed long after the house is completed and the work is done by the home owner himself or by a remodeling contractor. They are usually built of plywood and have a paint finish. The one shown in **Figure 53** is a good example of multi-purpose storage that provides drawers for cloths and shelves for linen, books, etc.

Fig. 47. Shallow closet with built-in shelves has vinyl folding door which saves floor space because it does not project into room.

Fig. 48. Child's bedroom with built-in drawers and shelves.

Fig. 50. Closet swivels 180 degrees on center pivot to expose entire length of curved clothes pole. Unit closes flush.

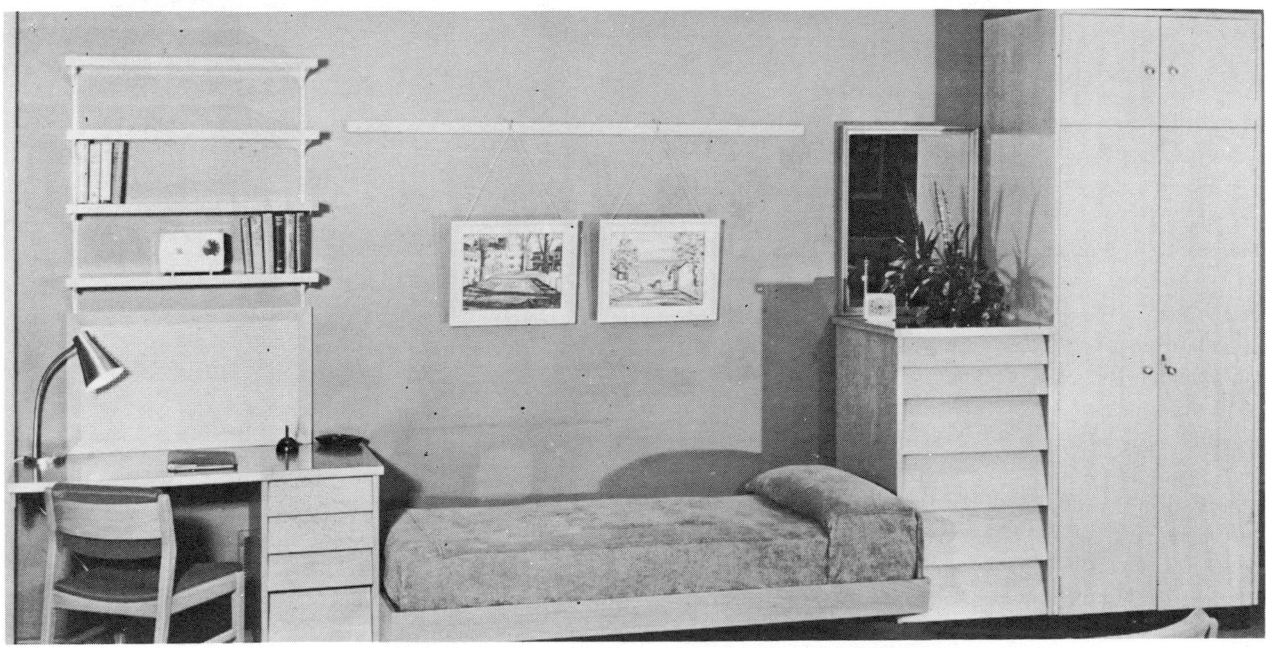

Fig. 49. Bunk built into storage units conserves floor space, makes cleaning easier.

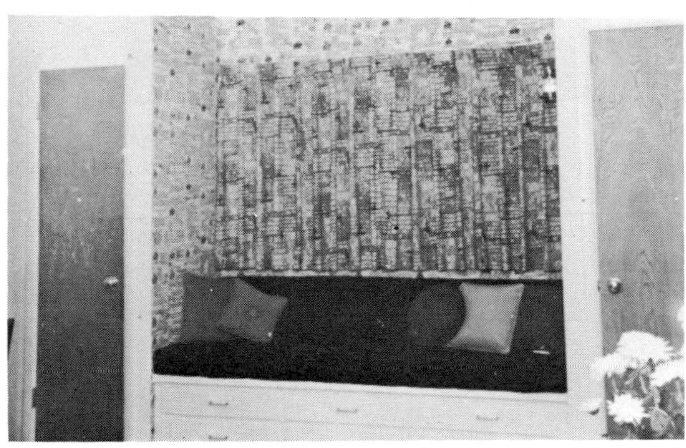

Fig. 51. Cushioned window seat has drawers built in.

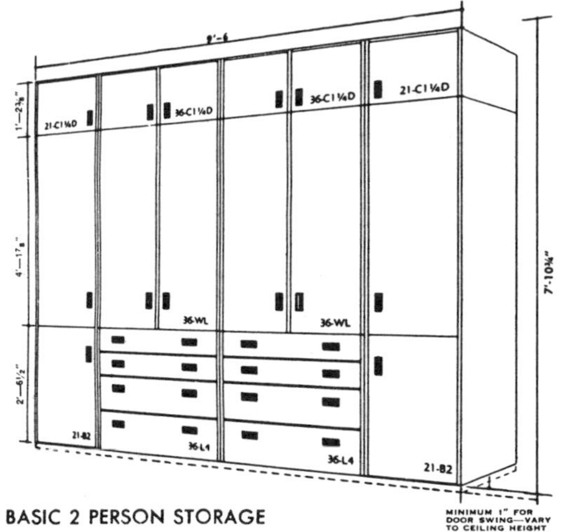

Fig. 52. Twin storage closet assembled with stock units.

Fig. 53. Built-in storage closet in an attic bedroom that provides for clothing, linen, and books.

The Bathroom

The main development in bathroom storage has been the integration of lavatory, vanity table, and storage shelving into large built in units. The counters of the units may be ceramic tile or laminated plastic; the sliding doors may be of hardwood plywood, plastic faced hardboard or even pegbboard, **Figures 54, 55, and 56.**

In simpler bathrooms, a smaller combination may be composed of lavatory, small counter space serving as a vanity, and a shelf or two beneath. The manufactured in-the-wall medicine cabinet remains standard, **Figure 58.** Other items of bathroom equipment have come into wide use, such as vanishing toilet tissue holders,tumbler and tooth brush containers, collapsible drying racks, etc.

The one story ranch type house on a slab has forced builders to make provisions for laundry space. Even where there are basements, automatic washer and dryer equipment is often placed in the bathroom or adjacent to it in the bedroom wing, in special housings containing work and storage shelves. Folding and sliding doors or screens are excellent equipment for such locations, **Figure 59.**

Privacy is achieved by compartmentalization of the various bathroom fixtures and by shower and tub enclosures, usually in aluminum or chrome plated frames with translucent glass or plastic panels, **Figure 57.**

Fig. 54. Factory-made lavatory-medicine chest unit is shipped separately and joined to adjacent unit at site.

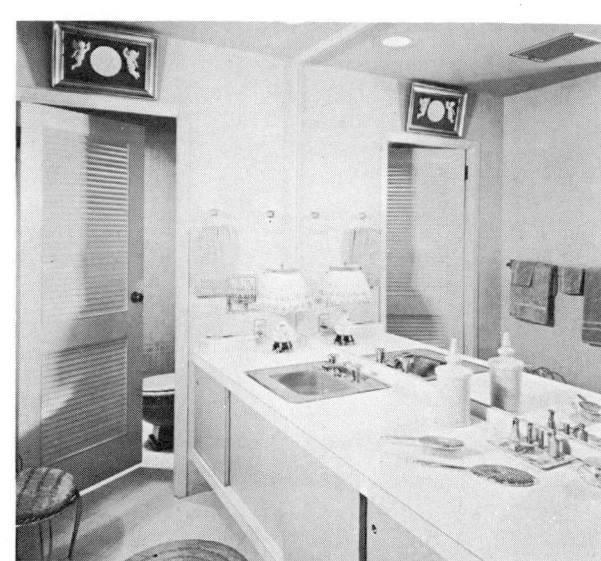

Fig. 56. Louvered door screens water closet in separate compartment, permitting multiple use of bathroom with privacy.

Fig. 55. Twin lavatories, set in ceramic tile counter top, have ample storage beneath enclosed by sliding doors.

Better Homes & Gardens

Fig. 57. Sliding pocket door converts master bath into two separate privacy areas. Textured panels on walls are 16″ × 8′. Left photo shows vanity area with prefab vanity and laminated plastic top. Note waterproofed shutter used as shower doors.

Hall, Hobby Shop, Etc.

With emphasis on space saving and efficiency in modern home construction, the house heating and cooling and hot water heating equipment is often located in a utility room or closet in the hall- often adjacent to a similar alcove for the laundry. Since provision must be made for easy movement of return air, the closet doors are louvered, perforated, or fitted with grilles, or they may be cut off a couple of inches from the floor, **Figures 60 and 61,** to permit air flow under them.

Garages are often equipped with workshop storage space, as well as units for gardening and athletic equipment.

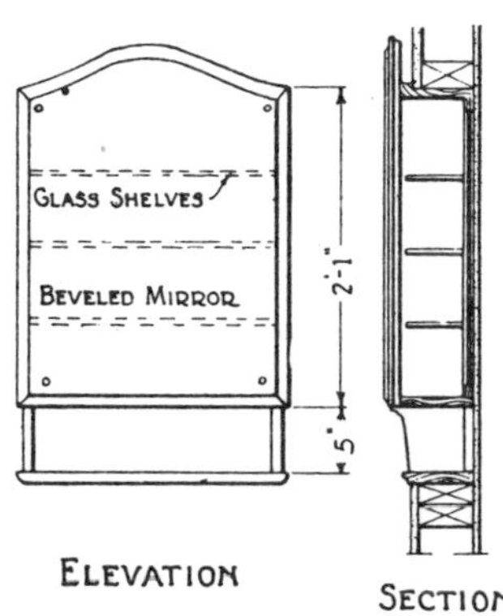

Fig. 58. Medicine cabinet is built into bathroom wall.

Fig. 60. Louvered doors have larger openings in lower panels for return air to heating-cooling equipment in hallway utility room.

Fig. 59. Laundry alcove in bedroom wing and adjacent to bathroom, is concealed by folding screen of bamboo.

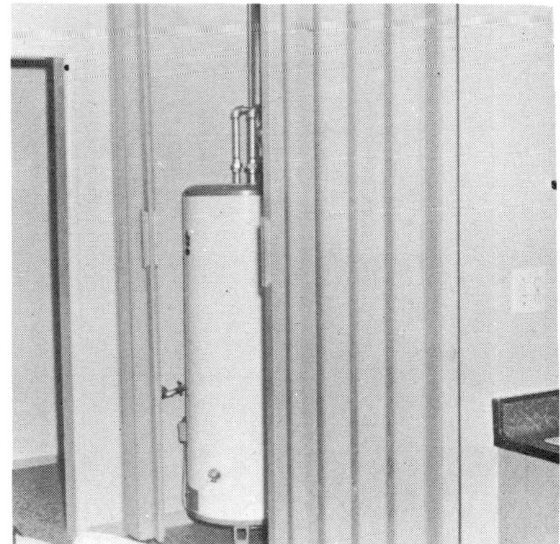

Fig. 61. Utility room, closed off with wood-slat folding doors, is close to kitchen for efficient placement of plumbing.

Chapter 16

FINISH FLOORING

Finish flooring is the final covering of material placed over the rough floor or subfloor of rooms in a dwelling. Wood is still the preferred flooring for the formal display areas of the house, such as the living room and dining room, but other rooms may be covered with linoleum, tile of various composition, cement, brick, ceramic tile, or natural stone.

Adhesives and Underlayments. The development of various adhesives and underlayments has greatly increased the variety of choice in flooring. Wood, linoleum, various types of tile, and carpeting have all benefited from these advances, which permit, for example, a diversity of tiles laid below grade instead of asphalt, now no longer used in residential construction. There is also a greater choice of subflooring, including plywood, hardboard, particleboard, and screeds on concrete.

Colors and Patterns. Colors and patterns in flooring are available in a wide variety of selections. Wood strips and squares are now available, factory-finished, in light or dark tones; cork patterns are popular in various kinds of tiles, and there are strip bands and metal inserts for linoleum and tile. Ceramic and vinyl tile are available in various sizes and shapes that can be arranged into decorative color patterns by skilled applicators.

Wood as Finish Flooring

The traditional subfloor is of boards, about 1″ × 4″, laid diagonally on the joists. It is of utility grade lumber, and face nailed. Above the subfloor is placed a layer of building paper or asphalt impregnated felt, overlapped, **Figure 1.** The paper prevents air and dust movement from the basement or the crawl space, and tends to cushion out squeaking caused by wood rubbing together.

An isometric drawing of this method of laying strip flooring is shown in **Figure 2.** Note the placing of the baseboard and shoe molding at the ground strip of wood which secures baseboard and shoe molding and acts as a gauge for the plaster finish at the partition.

The wood flooring, when in strip form, is generally laid parallel to the longer side of the room, although sometimes the direction of the flooring is governed by that of the flooring of an adjacent room, which may be of

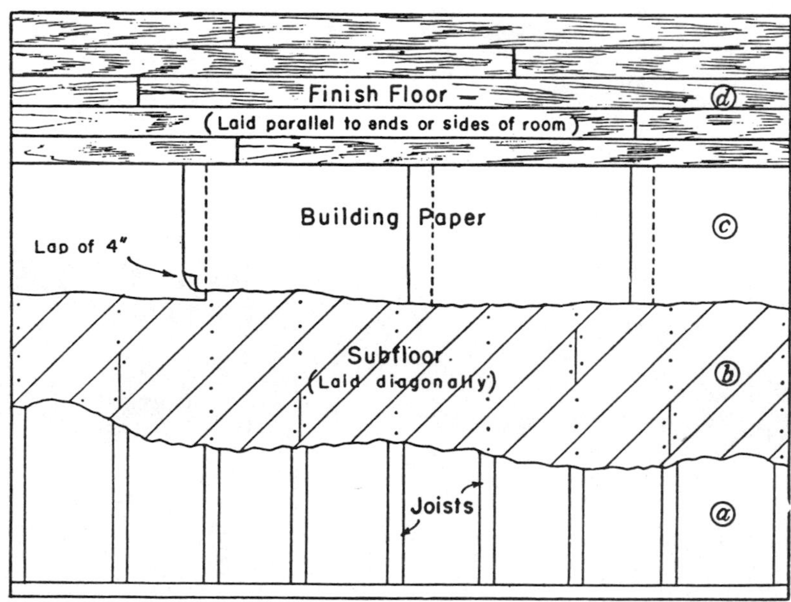

Fig. 1. Cutaway plan view of a floor. Floor joists (a) support subfloor (b), laid diagonally and face-nailed to joist. The subfloor, leveled and cleaned, receives the building paper (c), an asphalt-saturated felt or Kraft paper. Flooring strips (d) are started parallel with walls and run along longest length.

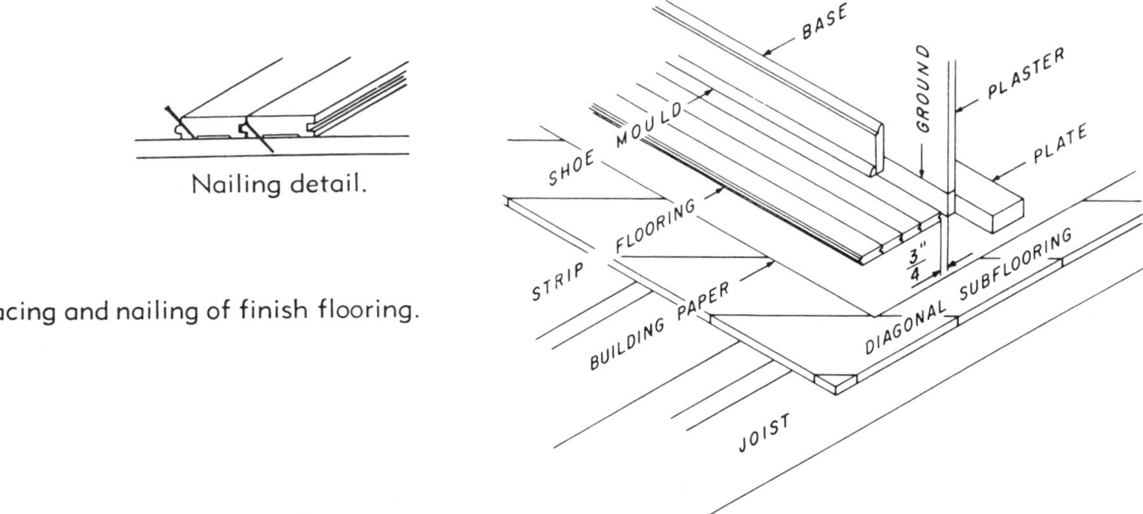

Fig. 2. First strip of flooring is nailed ¾" from wall to allow for expansion. Strips are blind-nailed through tongues.

Fig. 3. In an open floor plan such as this, strip flooring may run in one direction in all rooms, regardless of room size.

Fig. 4. Plywood subflooring is laid with staggered joints to provide smooth surface for resilient tiles set in adhesive.

Fig. 5. Layer of ¼" hardboard panels is laid over diagonal subflooring to provide smooth underlayment for tile.

greater size. Generally, the flooring is run without direction change from one room to the next, **Figure 3.**

Plywood subflooring is laid more quickly than diagonal boards, and requires less fitting. Joints of adjacent plywood panels are staggered, **Figure 4.** Sometimes 1/4″ panels of hardboard are placed over the plywood or board subfloor as a smooth base for tile, **Figure 5.** Whatever the subfloor, its surface must be carefully cleaned and leveled before receiving the layer of building paper or adhesive.

Hardwood Strip Flooring. The most common hardwood flooring boards are 25/32″ thick and 2-1/4″ or 3-1/4″ wide, but tongue and groove strip flooring comes in a variety of widths and thicknesses, **Figure 6.** Oak is the most common strip flooring material, but beech, birch, hard maple, and pecan are other hardwoods used for finish flooring. Yellow pine is the most common softwood flooring.

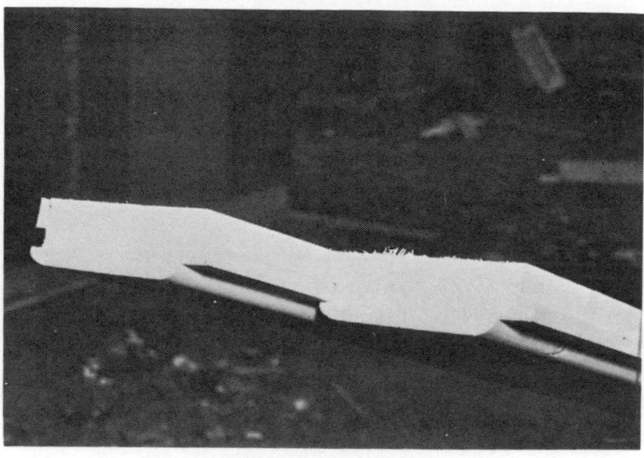

Fig. 6. Strip flooring, showing tongue-and-groove match along edges. These pieces are cut square and not end matched.

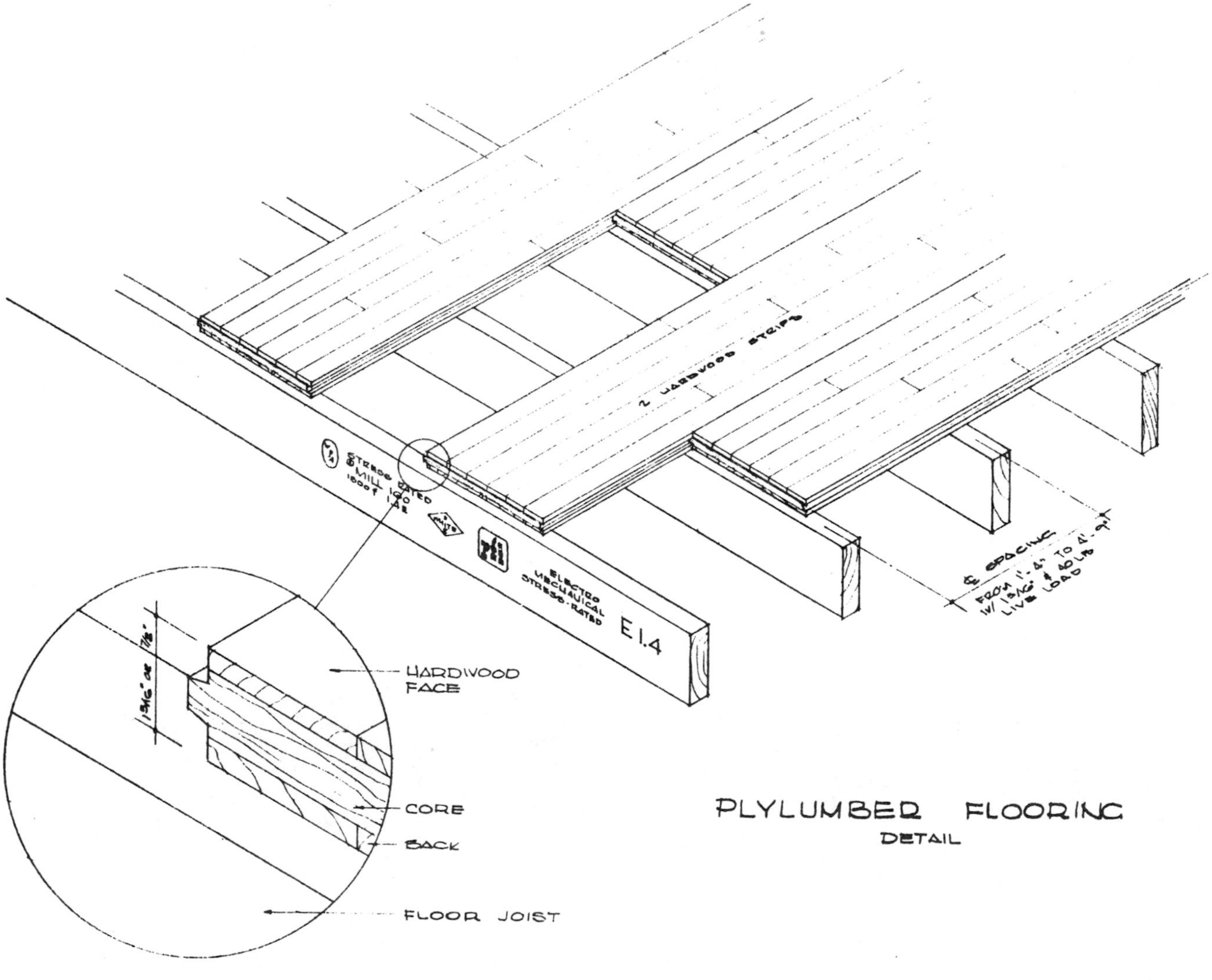

Fig. 7. Hardwood-faced plylumber flooring laid directly over joists with end joints staggered for stronger application.

Plywood Sheet Flooring. Finish flooring is also available in plywood sheets having a hardwood face, a tag core, tag end matched, and a backing which rests directly on the floor joists. The surface is made up of 2″ wide hardwood strips, staggered over the core to form a prefinished wood flooring, **Figure 7.**

Flooring Patterns. Most hardwood flooring is available in tongue and groove, and end matched (a tongue in one end, a groove in the other) for a perfect fit at every joining edge. The lower face of the flooring piece is often hollowed along the length to insure a perfect fit to the subfloor, **Figures 8 and 9.**

Outline drawings of various flooring patterns are shown in **Figure 8,** with exact dimensions of each face. Three thicknesses are shown: 25/32″, 1-1/16″, and 1-5/16″, each in five widths.

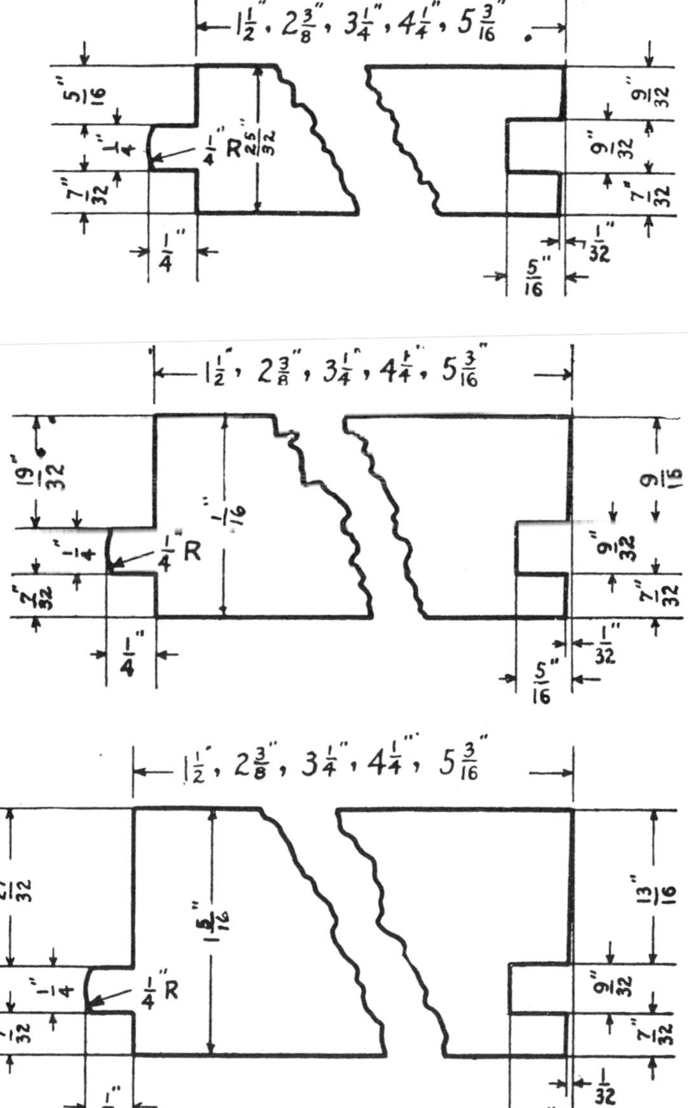

Fig. 8. Three thicknesses and patterns of standard tongue-and-groove flooring, which is also made in five widths.

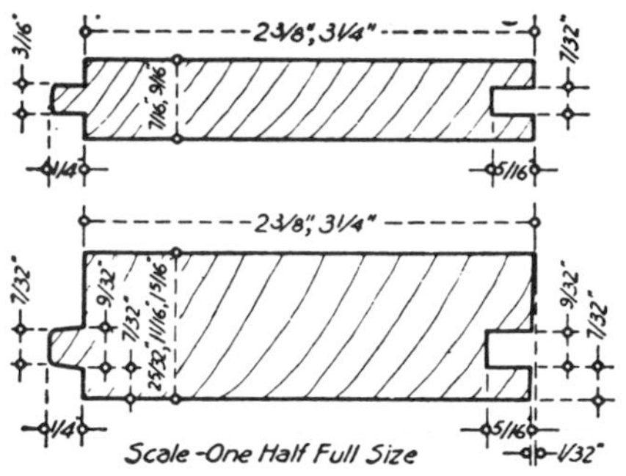

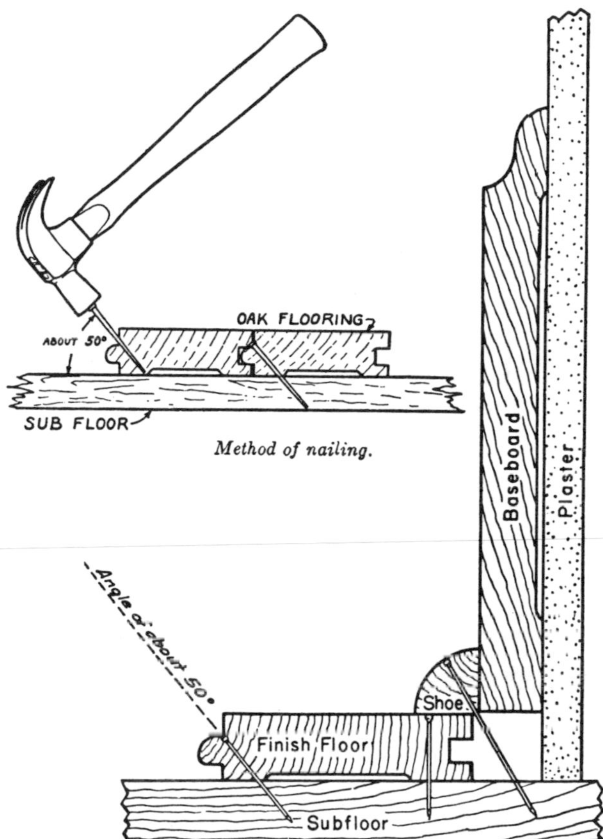

Method of nailing.

Fig. 9. Shoe molding is nailed to subfloor or baseboard to allow flooring to move beneath it without pulling it loose (lower dwg.). Blind nailing is done through tongue of flooring at 50 degree angle (center). Edge-grain flooring (top) has least shrinkage.

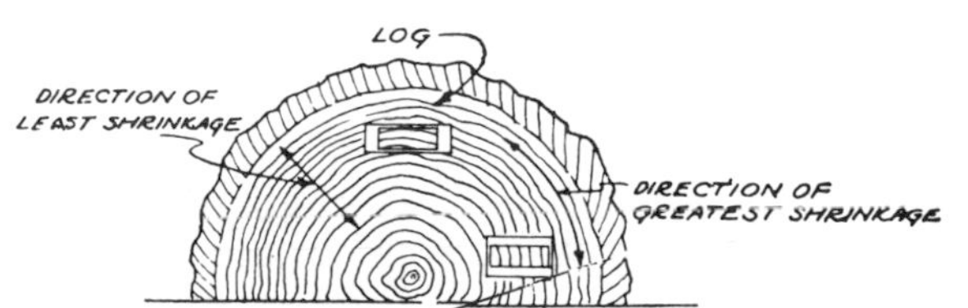

Fig. 10. Flat grain flooring (top of log) shrinks most.

FINISH FLOORING

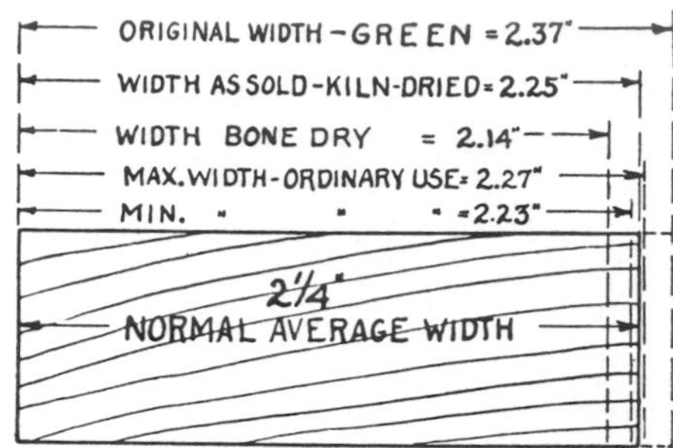

Fig. 11. Shrinkage of flat-grain floor under varying conditions of moisture content and use.

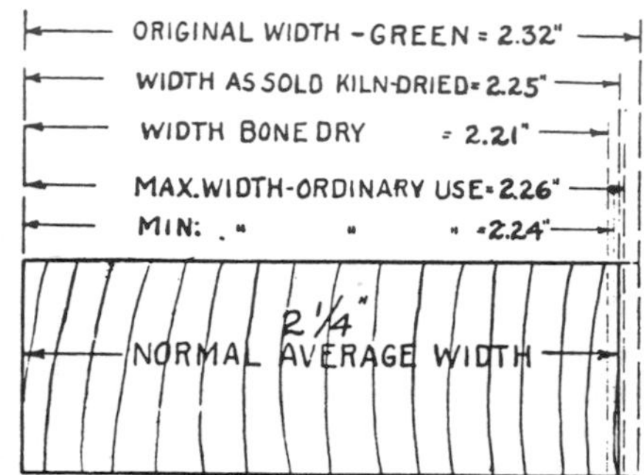

Fig. 12. Comparatively minor shrinkage of edge-grain flooring under identical conditions shown for flat-grain flooring.

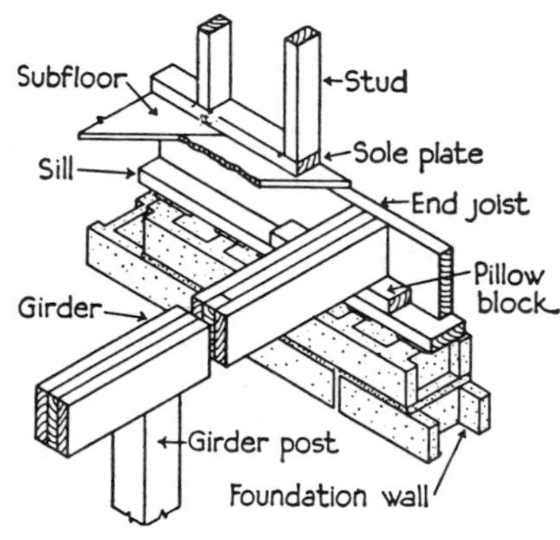

Fig. 13. Typical framing and subflooring at first floor level.

Method of Nailing Strip Flooring. The first strip of flooring is face nailed to the subfloor at the groove end, which is set toward the wall. The shoe molding is not nailed to the finish floor board, but to the subfloor, **Figure 9,** or to the baseboard, so that any contraction of the finish floor would not pull the molding away from the wall. These two ways permit the floor to expand or contract under the shoe molding without creating an unsightly gap.

How Wood Is Selected for Flooring. The selection of wood for flooring begins with the consideration of the portion of the log from which the wood is obtained. The shrinkage of wood from end to end is negligible, but in cross section it shrinks about one half as much at right angles to the annual rings as it does parallel to them. Edge grain or vertical grain flooring, if not properly seasoned, will shrink only one half the amount of similarly unseasoned flat grain flooring. The cross sections show in an exaggerated manner the results of seasoning in different parts of the log, **Figure 10.** The flat grain piece of 2-1/4″ flooring, **Figure 11,** shrinks from an original green width of 2.37″ to a bone dry width of 2.14″. The edge grain strip, **Figure 12,** may shrink from 2.32″ when green to 2.21″ when bone dry.

Framing at Different Floor Levels. Framing and subflooring methods at first and second floor levels are shown in **Figures 13 and 14.**

Dropping of Subfloor for Tile Surfacing. The subflooring, in traditional manner, is prepared for ceramic tile finish as shown in **Figure 15.** Note the framing for the soil pipe. More common in recent times is plywood subflooring under vinyl or vinyl asbestos tiles.

Types of Nails Used in Flooring. Cut nails with blunt points are often used for flooring strips to prevent splitting the tongue of the strips, **Figure 16.** Screw nails with small heads have come into wide use for flooring. Nailing machines are handy for flooring because they set the exact angle for effective nailing, **Figure 17.**

Plank Flooring. Plank floors are reproductions of the style of floors used in early American houses. Solid stock, generally of oak or maple, in random widths of 4, 6, or 8 inches, is used, with a slight bevel at the top edges.

To provide for expansion, a spacer blade is used to produce a hairline space as the pieces are toenailed or blind nailed. The planks must not touch any vertical surface at any point, that is, baseboard, plinth block,

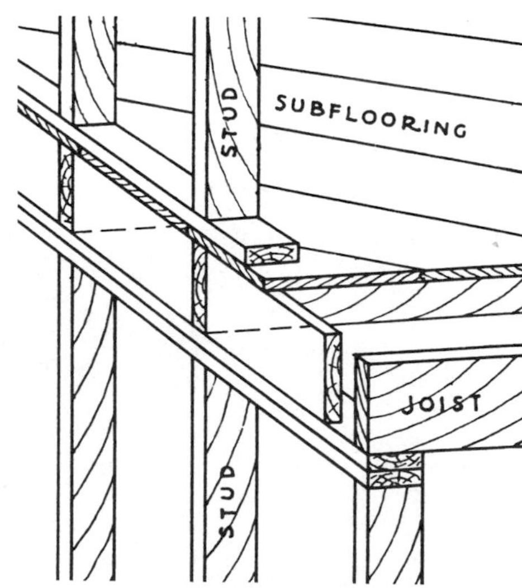

Fig. 14. Framing and subflooring at second floor level.

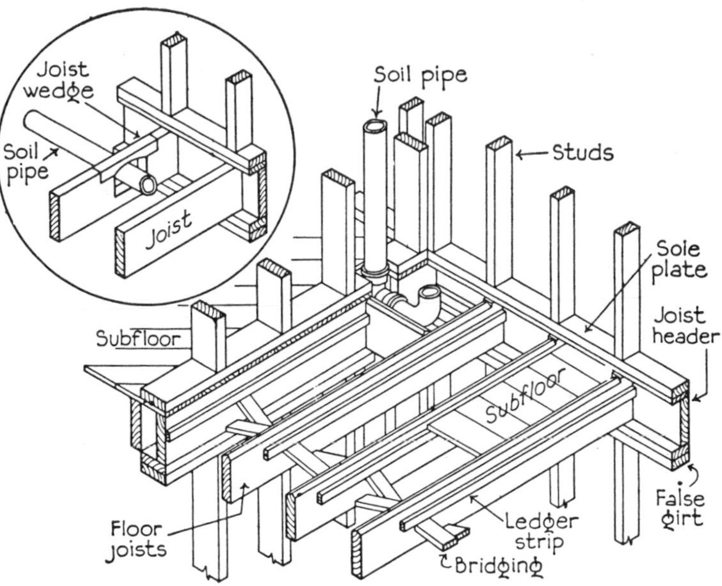

Fig. 15. Placing subflooring below level of joists in second floor bath in preparation for tiling. Note framing for soil pipe.

Fig. 16. Types of flooring nails for different flooring widths.

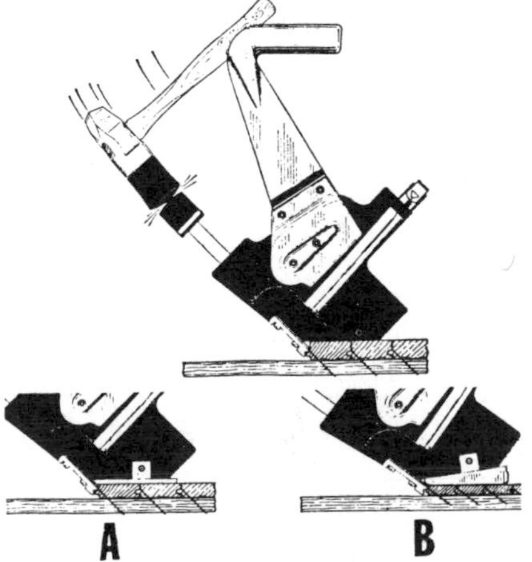

Fig. 17. Nailing machine automatically sets nail at correct angle. Adapter plates raise machine on thinner flooring.

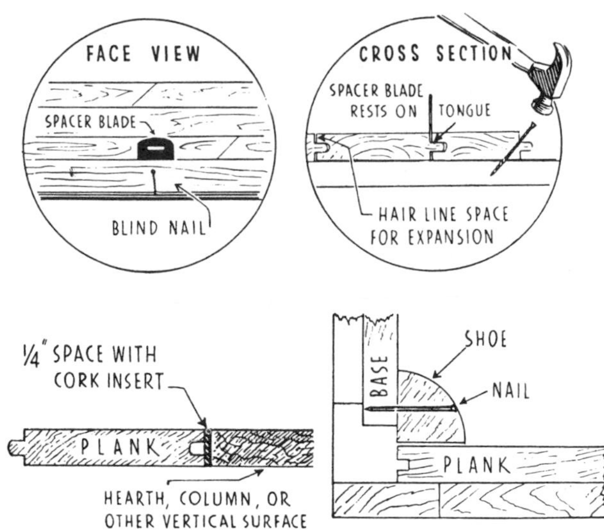

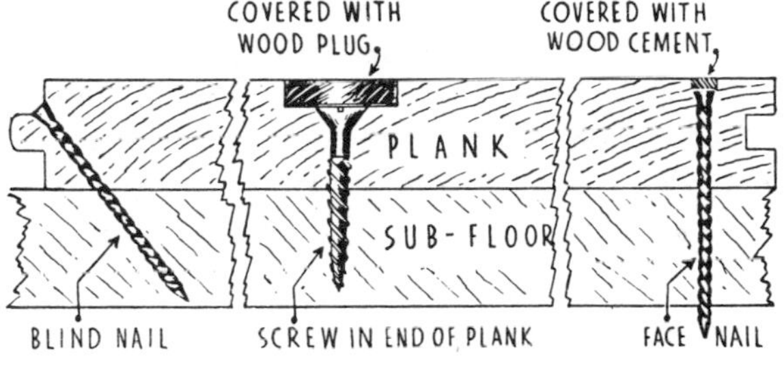

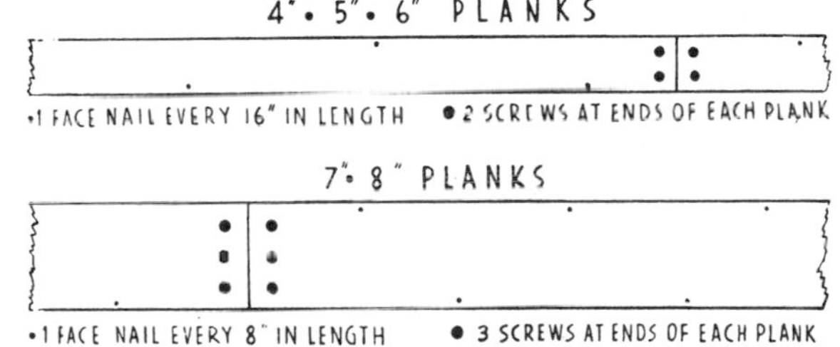

4"• 5"• 6" PLANKS

•1 FACE NAIL EVERY 16" IN LENGTH ● 2 SCREWS AT ENDS OF EACH PLANK

7"• 8" PLANKS

•1 FACE NAIL EVERY 8" IN LENGTH ● 3 SCREWS AT ENDS OF EACH PLANK

Fig. 18. Use of spacer blade, cork and air space to allow for expansion (top four dwgs.). Section in center shows how plug covers screw head. Nails are used in addition to screws (bottom).

Fig. 19. Prefinished dark fireside plank flooring.

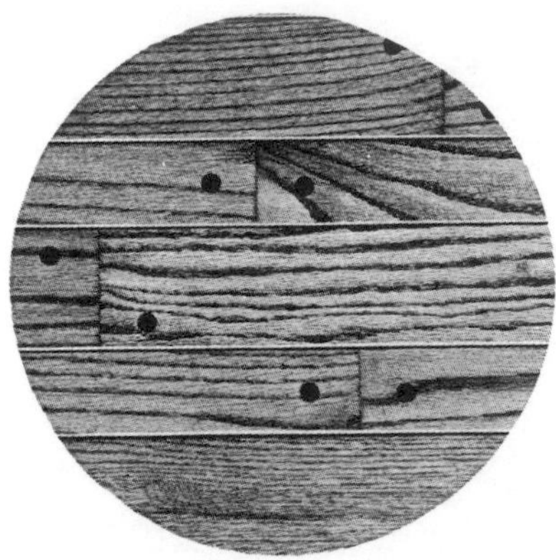

Fig. 20. Prefinished ranch plank flooring with plugs.

Fig. 21. Random width planking with factory applied plugs.

column, hearth, or threshold. To give the effect of antique style pegs, screws are used for face fastening, then covered with a plug.

Nails used for face fastening should be countersunk and the indentures filled with wood putty. Narrow planks should be nailed every 16″ and the wider planks face nailed every 8″, **Figure 18.** In modern factory finished planking, the peg or dowel effect is incorporated in the planking.

A 25/32″ thick, oak flooring, **Figure 19** with alternate 2-1/4″ and 3-1/4″ widths is available. This dark fireside plank flooring has wide beveled edges and a factory applied finish. Its appearance installed is similar to that of an expensive random plank laid floor.

This plank flooring is made up of #1 common and better grade oak stock. It is tongue and grooved and end matched. Bundles are 8′ long, containing an equal number of two widths. Pieces are as short as 1-1/4′ and average 3′ in length.

A prefinished ranch plank flooring, **Figure 20** is available for Contemporary or Colonial architecture. It is composed of alternate strips of 2-1/4″ and 3-1/4″ factory finished stock. The walnut pegs are installed at all end joints, glued, sanded and finished at the factory. This flooring is 25/32″ thick, comes in random lengths, 1-1/4 feet and up, averaging 3 feet in length. It is made of clear select and #1 common stock. Bundles are 8′ long and end capped for protection.

Figure 21 shows how the peg or dowel effect of factory finished plank flooring harmonizes with furniture.

Block Flooring. Block flooring takes several forms. One form is the solid glued block of narrow slats. No attempt is generally made to match the slats exactly in color and grain design, since the variety is of interest, **Figures 22 and 23.**

Some styles of blocks are tongue and grooved to permit expansion or shrinkage as a unit. These blocks are available in an equal number of right tongue and left tongue blocks, so that tongues are always exposed for blind nailing as the floor is laid. The tongues have drilled nail holes to prevent splitting.

Fig. 22. Solid glued oak block flooring gives luxury look.

Fig. 23. Hardwood block flooring is accented by area rug.

The more common and generally cheaper wood block form is the hardwood veneer block, in which the fine grained oak is laminated under pressure to a utility body. The veneer is usually a single piece, and variety is obtained by alternating face directions. Mastic is the usual adhesive for veneer block, **Figure 24.**

Fig. 24. Cross ply structure of blocks compensates for expansion, so blocks can be laid close to wall. Blocks shown are being laid in mastic spread over concrete slab floor.

Fig. 25. Finish flooring of prefinished hardwood strip blocks.

Prefinished Hardwood Strip Blocks. Strip blocks are 25/32'' thick and 9'' square. Each is made up of 4 strips of 2-1/4'' tongued and grooved, end matched flooring. Each block has a continuous tongue along one side and end, and a continuous groove along the other side and end. Tongues and grooves are designed for installation with mastic, not by nailing. Each block is sanded, finished and waxed at the factory.

Metal splines clamp each block together on the back and prevent the block from cupping and the joints between the strips from coming apart. The edges of each block are slightly bevelled to accentuate the block effect. The block is not recommended for radiant heated floors. An expansion space of at least 1'' must be left at every wall and obstruction, **Figure 25.**

Prefinished Laminated Oak Blocks. Laminated blocks are three ply, built up under heat and pressure with moisture resistant glue. All three plies are oak, hence the stress is equalized. The blocks are 1/2'' thick and 12" square. They are grooved on opposite sides and tongued on opposite sides so that each block can be laid with the grain running in the opposite direction to those it adjoins. Tongues and grooves are designed for installation with mastic, not nailing. Each block is sanded, finished and waxed at the factory, providing a surface that will not scratch, chip or peel, **Figure 26.**

Applying Strip Flooring to Concrete and Wood Floors. The development of waterproofing compounds has made it possible to use an inexpensive technique of applying strip flooring to concrete slabs. Mastic is spread over the waterproofed slab, then 2'' × 4'' screeds of various lengths are laid in the mastic in a scattered pattern; the oak flooring strips are then nailed across the screeds, **Figure 27.** Strip flooring applied over wood subfloors should have building paper between the subfloor and finish floor. Strips are blind nailed through the tongues, **Figures 1 and 28.**

Fig. 26. Finish flooring of prefinished laminated oak blocks.

Fig. 27. Screeds of 2″ × 4″ are laid in mastic on a concrete slab floor, then strip flooring is nailed to the screeds.

Fig. 28. Blind nailing oak strip flooring through the tongues of the strips and into subfloor. Note black building paper under strip floor to minimize squeaks.

Fig. 29. Random width, wide beveled plank flooring is factory finished in a dark tone with tough, long lasting materials.

Fig. 30. Laying hardboard underlayment in preparation for resilient tile Sheets are fastened with ring groove nails.

Fig. 31. Hardboard underlayment sheets for resilient tile are laid with offset joints. Blade of linoleum knife is used to space boards for expansion.

Armstrong Cork Co.

Fig. 32. Butcherblock design vinyl tile is self adhering and has a clear vinyl wear surface.

Finishes on Wood Floors. Finishes on wood floors are subject to severe wear. Much wood flooring comes from the factory with the finish baked on, then waxed and polished, **Figure 29.** Maintenance is more important than the type of floor used. There are many varnishes and plastic coatings available, some of which turn darker sooner than others.

The essential point in maintaining any floor finish that includes a surface coating is to re-coat before the old coating has worn down to the wood, permitting it to discolor or darken.

Good floor paints or varnishes require less frequent renewal than wax finishes under similar conditions of wear, but each renewal is more expensive. Floor waxes are applied over a sealer to prevent discoloring the floor, and are easily applied with motor driven floor polishers and buffers.

GAF Co.

Fig. 35. Spanish tile motif has embossed aggregate chips around marble insets. Factory apply adhesive on back makes installation easy.

Armstrong Cork Co.

Fig. 33. Spanish Court pattern has the look of Mediterranean tile. Clear vinyl wear layer eliminates need for waxing.

Flintkote Co.

Fig. 36. No-wax, high gloss tile resists stains, spills, and scuffs. The 12" × 12" tile can be laid on, above or below grade.

GAF Co.

Fig. 34. Non-directional chip pattern makes this tile look like one sheet. Clear urethane wear layer retains gloss and keeps out dirt.

Flintkote Co.

Fig. 37. Brick style is very popular and high resistance to dirt and stain makes this vinyl asbestos tile ideal for kitchen use.

Resilient Flooring

Linoleum, asphalt tile, cork tile, rubber tile, vinyl asbestos tile, pure vinyl tile and sheet all fall in the category of resilent flooring. In the past, builders tended to specify vinyl asbestos tile for kitchens because it provided a durable, easily cleaned surface that required less care than wood. To a great extent this practice is still being followed, especially in middle priced and less costly homes. In more expensive homes pure vinyl tile and sheet are used in kitchens because of the greater variety of attactive patterns and resistance to scuffing. Vinyl asbestos tile has become the great do-it-yourself flooring of home owners and its ease of application is being heavily promoted to the home owning audience, particularly to women. In fact, once the old floor is properly prepared, anybody can install this product, especially if the self-sticking type is used.

Linoleum is rarely used today and is more likely to be found on commercial desks than on residential floors.

Asphalt tile was formerly the only kind of resilient flooring that could be applied to concrete floors below grade. It was very durable when properly maintained with water emulsion wax and was resistant to the alkaline salts in concrete. However, its colors were dull and it was difficult to produce it in attractive patterns. The new backings on vinyl asbestos, vinyl tile and vinyl sheet, which are resistant to alkaline salts and therefore permit their use on below grade concrete, have pretty

Fig. 39. Vinyl sheet floring has foam cushion under it and no seams to catch dirt in this handsome kitchen.

Fig. 38. Sheet vinyl flooring has no seams and comes in 6′ and 12′ widths.

Armstrong Cork Co.

Fig. 40. Non-wax, high gloss sheet vinyl flooring in this kitchen just needs damp mop to be cleaned.

Armstrong Cork Co.

Fig. 41. Easy to install, this highly flexible cushioned vinyl sheet is trimmed with a sharp knife and straight edge.

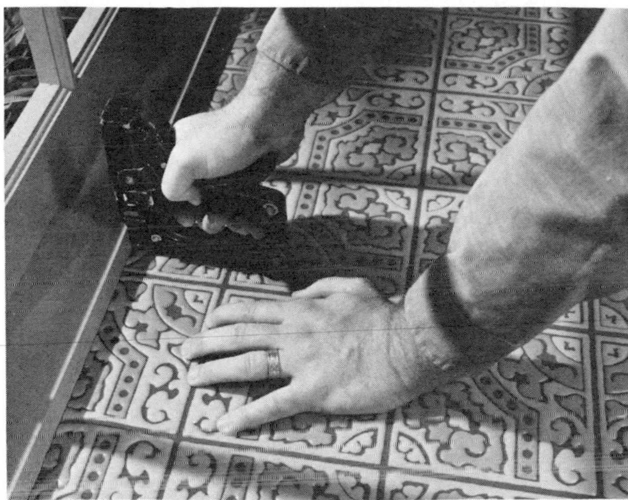

Armstrong Cork Co.

Fig. 42. Vinyl sheet flooring is fastened at edges with staples which are later covered by shoe molding.

much eliminated asphalt tile from residential use. Rubber tile is only used on certain commercial jobs.

Cork tile, even with a special vinyl surface is no longer seen in residential use. Popular taste has shifted to the more colorful effects that can be secured with vinyl, and it is therefore vinyl asbestos, and improved versions of vinyl tile and vinyl sheet which now dominate the field.

Preparing Old Floors for Resilient Tile and Sheet. Old wood floors must be clean, free of oil and grease, and level. High spots must be planed down or sanded and low spots filled. If the old boards are badly cupped they must be filled with floor leveling compound of portland cement and latex. If this is not practical the floor must be covered with sheets of plywood, particle board, or hardboard. The ¼″ plywood must be smooth sanded on one side or its grain will "telegraph" through the resilient flooring. The particle board must be of the filled and smooth sanded underlayment type. The most commonly used underlayment is a special type of hardboard, **Figure 30,** which is nailed over the old floor with ring groove nails. To allow for the expansion of the hardboard underlayment, a space is allowed between the hardboard sheets no greater than the thickness of a nickel or the blade of a linoleum knife, **Figure 31.** The hardboard sheets must be laid so that all joints are offset.

New tile or sheet can be laid over old resilent tile provided it is level and smooth, free of dust, oil and grease, and is not embossed. Embossed designs tend to telegraph their shape through to the new flooring.

Improved Vinyl Tile. New types have a clear vinyl wear layer which is more resistant to dirt, grease and stains than vinyl asbestos. Some have no asbestos filler in the body but have a vinyl composition instead, and most have a factory applied adhesive on the back protected by special paper which is peeled off before application. The new vinyl types may also have a ¼″ of white vinyl foam on the back, which makes

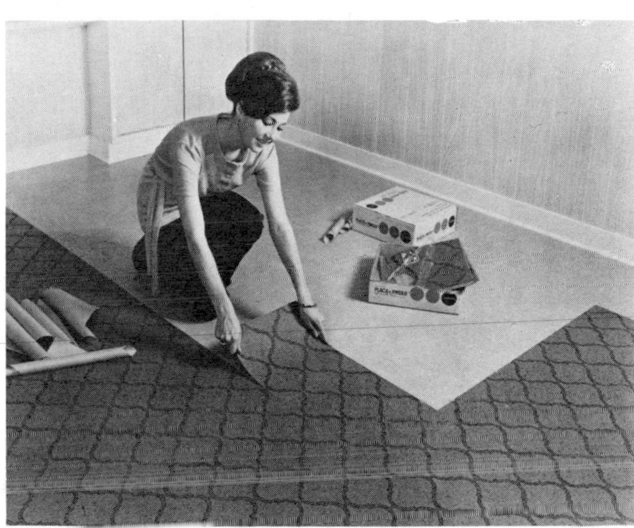

Fig. 43. Factory applied adhesive on backs of tile makes installation easy. Just peel paper off back and press in place.

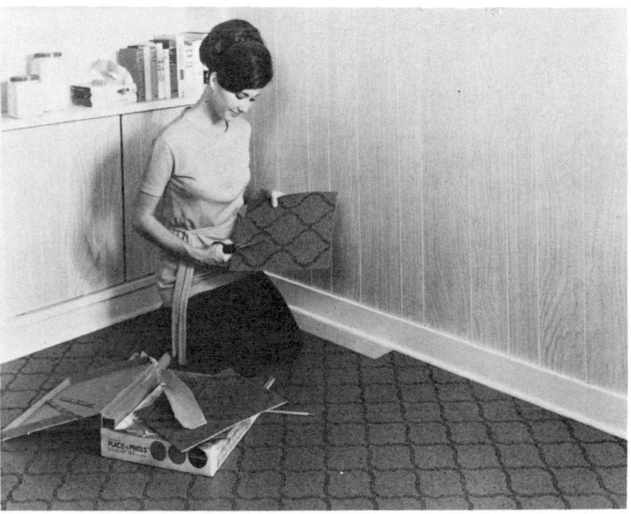

Fig. 44. Vinyl tile is easy to cut with scissors when fitting pieces of border tile against wall.

Fig. 45. Quarry tile enhances rustic furnishings, is serviceable surface for floor and counters in kitchen.

Fig. 47. Ceramic tile counter top and backsplash complements used brick wall and knotty pine cabinets in kitchen.

Fig. 46. Bathroom is tiled with different patterns on floor, wall and counter. Note glass block window with hopper insert.

Fig. 48. Small ceramic tiles on lavatory counter and floor contrast with large tiles on wall around sunken tub.

Fig. 49. Modern bathroom features tiles on face as well as top of counter, and checkerboard pattern on floor.

Fig. 50. Rectangles of stone are used on floor of upper dining level, on steps, and on floor of adjoining informal room, blending nicely with brick wall for a warm effect.

Fig. 52. Wide grouting provides a dramatic outline for each irregularly shaped flagstone in this traditional kitchen. Flagstones are set in bed of portland cement.

Fig. 51. Vari-colored slate makes interesting entry hall.

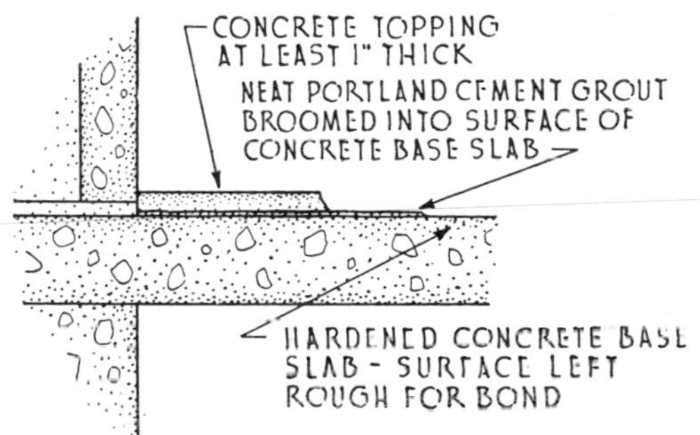

Fig. 53. Finish topping of concrete applied over slab.

them more comfortable underfoot. The butcherblock and Spanish court designs in **Figure 32 and 33** are typical of types that are self adhering and have a vinyl wear surface.

Others have a vinyl asbestos body but have a urethane wear surface that assures resistance to dirt, grease, and most stains. They also have factory applied adhesive on the back, **Figures 34, 35.** Not all of the new types are guaranteed against alkaline salts which result from wet concrete. Some manufacturers warn that their product can be applied to concrete only if there is no evidence of moisture. Still other vinyl asbestos types have high gloss wear resistant top layers that minimize any need for waxing but can be applied above, on, or below

Fig. 54. Brick laid without mortar on well-packed sand bed has interesting texture in low afternoon sunlight.

Fig. 55. Brick floor is carried indoors from the patio.

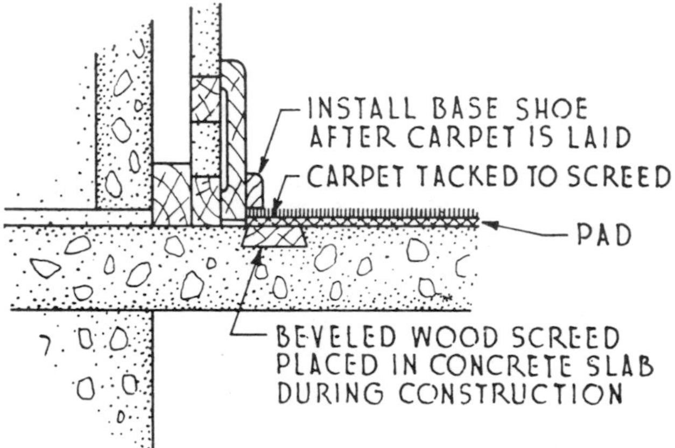

Fig. 56. Wall-to-wall carpeting is fastened to wood screed set in slab floor when concrete was placed. Lacking screeds, a toothed tackless strip may be anchored or adhered to the slab floor to secure the edges of carpeting.

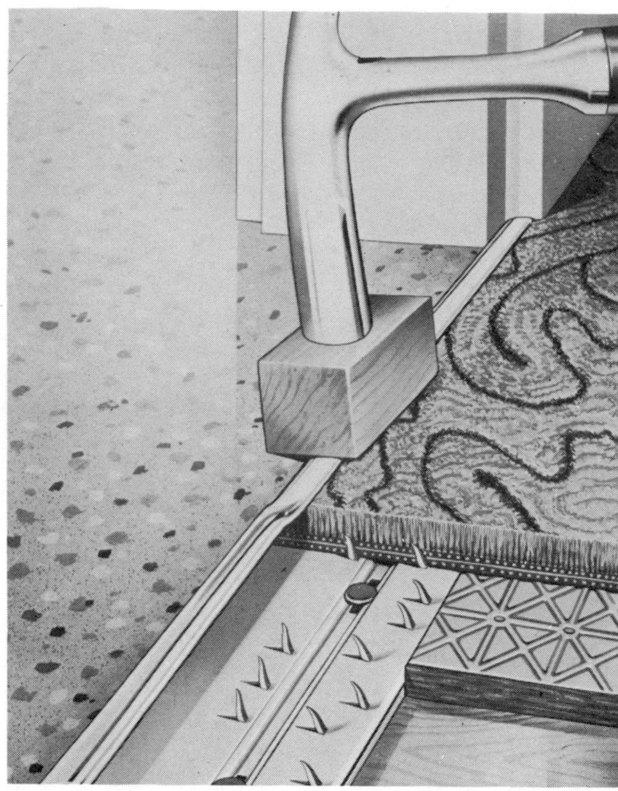

Fig. 57. Teeth in tackless edging strip fastened to floor grip carpeting securely at edges. Edging strip is closed with block.

ground level and can be used over concrete, wood, steel, and most old floor coverings without cement, **Figures 36 and 37.**

Cushioned Vinyl Sheet. Designed for application by home owners, cushioned sheet is remarkably flexible. It can be bent and folded like paper without cracking. Available in 6-foot and 12-foot widths, it offers the home owner the advantage of a seamless floor, a high gloss wear surface that eliminates waxing and is comfortable underfoot. The absence of seams makes it ideal for kitchens because there are no cracks to catch dirt, **Figures 38, 39, 40.**

Installation of Cushioned Sheet. Installation is made easy by the extreme flexibility of the sheet which permits easy trimming at the walls with a sharp knife and a straight edge, **Figure 41.** Staples are then used to fasten the edges, **Figure 42,** which are then concealed by the shoe molding. Where staples cannot be used, a narrow ribbon of special glue is spread under the edges. No adhesive is used elswhere other than the edges.

Tile Installation. Tile is even easier to apply than sheet. The room is divided into four quarters by snapping a chalk line from the midpoint of one wall to the center of the opposite wall. The same is done with the remaining two walls. Tiling then begins from the point where the two lines cross with each quarter of the room being filled in sequence until only the narrow space against the walls remains to be filled. At this point the tiles are cut with scissors to fit the empty spaces. No adhesive is used because the tiles already have adhesive on their backs, **Figures 43 and 44.**

Ceramic Tile. Ceramic tile has come into wide popularity with the trend toward the rustic quality of the ranch type house. Quarry tile, an unglazed tile, is popular for its earthy color, and is used on kitchen floors, counters, family rooms and entrance hallways, **Figure 45.** Glazed tile is used in bathrooms and on work surfaces, **Figures 46 and 47. Figures 48 and 49** illustrate two distinguished bath designs.

Flagstone Pavement. Another effect of the informal trend in modern indoor-outdoor living is the carrying of outdoor flagstone pavement into the house, **Figure 50.**

Slate is admired for its rich varied coloring and its irregular rectangular shapes, **Figure 51.** Entrance ways and family rooms opening onto outdoor terraces are popular places for stone floors.

The deluxe kitchen, **Figure 52,** is an authentic center having authentic Early American hardwood cabinets which harmonize with Italian and French Provincial decor. The cabinets are finished in cherry, have magnetic catches and rounded shelf edges. The finish flooring is of irregular flagstone properly separated by wide grouting.

Finish Flooring of Concrete. Concrete is handled not merely as a subflooring but sometimes as a finish flooring. Concrete topping can be troweled smooth and waxed and polished, and permanent color can be incorporated in the concrete mix, **Figure 53.** It can be scored or marked off into squares, diamonds or other geometrical patterns.

Marble, cut into thin squares of 9″ × 12″ in size, or squares of 9″ × 9″ or 12″ × 12″ are set in mastic for special luxurious floor effects.

Finish Flooring of Brick. Brick is another outdoor material that sometimes makes its way into the house to obtain an indoor-outdoor continuity, **Figure 54.** Finished flooring of brick is continued into the interior from a patio floor in **Figure 55.**

Carpeting

Carpeting is sometimes used as a substitute for finished flooring, being laid over slab, or rough subflooring, on a fiber or foam rubber underlayment. A combination of plywood subflooring, rubberized underlayment, and carpeting is available in square units for new houses, **Figures 56, 57, 58 and 59.** FHA's decision to include carpeting in the mortgage encouraged builders to use more of this material in new homes.

Four types of flooring, wall to wall carpeting, quarter inch oak, vinyl, and resilient flooring tile, are laid over plywood subflooring to create the attractive combination illustrated in **Figure 60.**

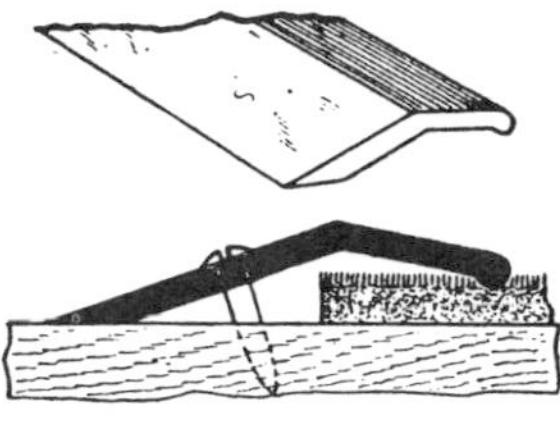

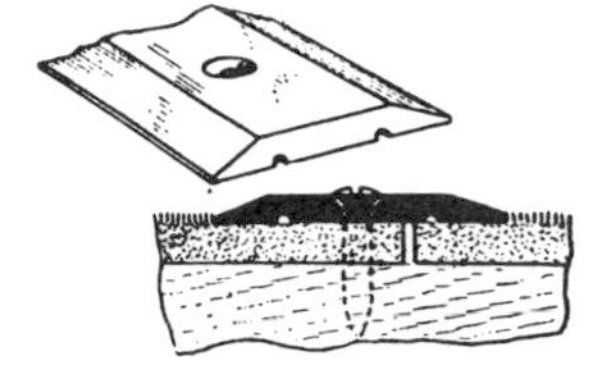

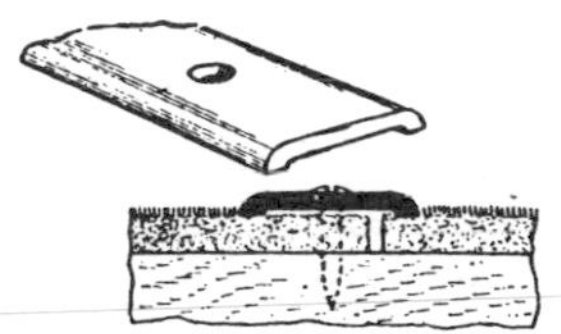

Fig. 59. Metal edgings for use with carpeting.

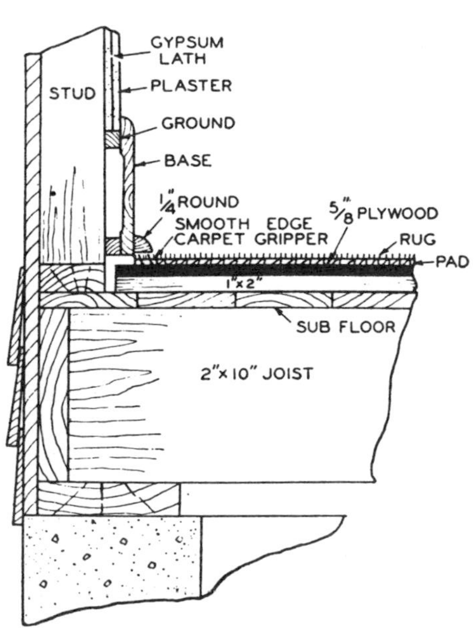

Fig. 58. Carpet and pad installed on wood subfloor.

Fig. 60. Four types of flooring laid over plywood subflooring: wall-to-wall carpeting (upper left), oak strip flooring (upper right), resilient tile (center), vinyl (foreground).

FINISH FLOORING

Chapter 17

CHIMNEYS AND FIREPLACES

A chimney may house one or more flues from a fireplace, stove, furnace or heating device, through which heated air, gases, or smoke escapes into the open air.

Since, in modern homes, fireplaces are only occasionally needed for the heat they give, their function is mostly decorative or atmospheric, as a focal point of interest. For this reason they are often built into impressive masonry walls. Then too, the fireplace wall is often the location for wood paneling, of plywood or solid planking.

The mantel, which used to be universal in older fireplace design, is often omitted from the masonry fireplace wall, but planters, niches, or projections are sometimes built into it, **Figure 1.**

Fig. 1. Fireplace wall has built-in planter, continuous ledge.

Fireplaces of Ceramic Tile. Ceramic tile is often used for very colorful effect in the construction of a fireplace. Often the fireplace projects into the room or serves as a divider between two areas, such as a dining room and a living room. Steel columns are sometimes used to support the upper section of the fireplace. The flue section of the fireplace is often cantilevered out over the base, **Figures 2 and 3.**

Storage space for firewood is often built into the masonry of the fireplace as well as shelves for books and ornaments, **Figure 4.**

Fireplace Flanked By Glass Walls. When a fireplace is located in the outside wall, its masonry and that of the chimney form an interesting unifying element between indoors and outdoors, especially when the fireplace is flanked by glass walls, **Figure 5.**

Framing Around a Fireplace. Chimney stacks are built into exterior walls of wood. A chimney stack of brick is shown in **Figure 6,** one of field stone in **Figure 7.** The brick of the chimney stack rests on a foundation built out from the foundation wall of the framed house, **Figure 8.** This foundation wall must be independent of the foundation walls for the exterior of the house. **Figure 9** shows construction details for an old style mantel piece. Factory made mantels are of period design having a shelf length of about 72″, a shelf depth of about 10″ and placed at a height of 51″ above the hearth.

Several Flues In a Chimney Stack. An exterior wall chimney stack may incorporate an outside barbecue in the main stack, **Figure 10.** Sometimes a single stack section may serve several purposes; a fireplace in the living room; another fireplace in the family room adjoining; and the outside barbecue, **Figures 11 and 12.** There may even be another fireplace in the den below. The fireplace chimney stack often accommodates the flue of the central heating system.

Chimneys and Fireplaces. Chimneys are generally constructed of masonry supported on a suitable foundation. A chimney must be capable of producing sufficient draft for a fire, and of carrying away harmful gases from the fuel burning equipment and from other utilities. Light-weight prefabricated chimneys that require no masonry protection and no concrete foundation are now accepted for certain uses by fire underwriters and are available for use.

A fireplace is a luxury except in mild climates or in locations where other heating systems are not available. Since an ordinary fireplace has an efficiency of only about 10%, its value as a heating unit is low compared to its decorative value and to the cheerful and homelike atmosphere it creates.

The heating efficiency of a fireplace can be materially increased by the use of a factory made metal unit incorporated in the fireplace structure that allows air to be heated and circulated in a system separate from the direct heat of the fire. The chimney and the fireplace must be carefully built in order to be free of fire hazards, and it is desirable to have them in harmony with the architectural style of the house.

Chimney Construction. The chimney should be built upon a solid masonry foundation properly proportioned

Fig. 2. Modern fireplace is faced with vertical wood paneling, is accented by ceramic tile panel at left of hearth.

Fig. 3. Steel columns support open end of fireplace wall which separates living room from dining room while warming both.

Fig. 4. Sloping masonry fireplace breast has storage space for firewood at right and shelves for books or ornaments at left.

Fig. 5. Masonry fireplace is flanked by floor-to-ceiling fixed glass installed in gable end framing.

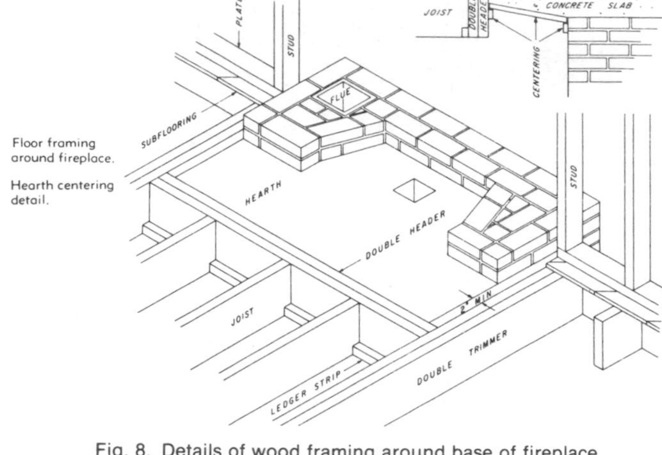

Fig. 8. Details of wood framing around base of fireplace.

Fig. 6. Brick chimney stands outside wall of wood-framed house.

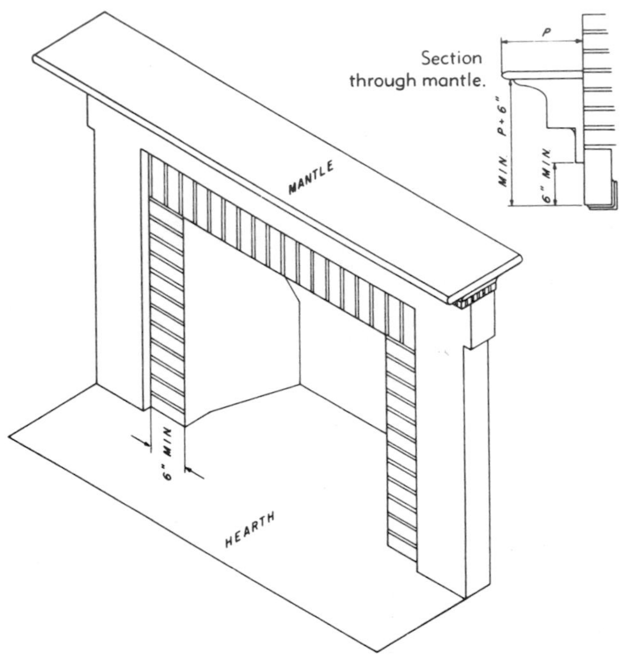

Fig. 9. Trim details for a traditional mantle.

Fig. 7. Massive fieldstone chimney has its own foundation.

to carry the loads imposed without settlement or cracking. The footing for an exterior chimney should start below the frostline.

The minimum size of a chimney will depend upon the number, arrangement, and size of flues. The flue from a heater should be of sufficient cross sectional area and height to create a good draft and to develop the rated output of the equipment in accordance with the manufacturer's recommendations. The chimney should be so located with reference to any higher buildings nearby that wind currents will not form eddies and force the air downward. When downdrafts from surrounding trees or buildings hit the top of the chimney and prevent proper draft, a hood or cover may be helpful in deflecting these currents. The total area of the side openings where a cover is used should be at least four times the area of the chimney flue.

Most building codes require that the tops of chimneys be carried high enough to avoid downdrafts caused by turbulence of wind as it sweeps around nearby obstructions or over sloping roofs. In no case should the height be less than 2′ above the ridge of the roof at the chimney site and not less than 2′ above the highest ridge within 10′ of the chimney. The height above a flat roof should be not less than 3′, more if called for by the manufacturer's requirements, **Figures 13 and 14.**

In some instances it is necessary to offset or corbel a chimney to obtain an architectural effect. The total offset, corbel or overhang of an independent chimney should not exceed three eighths the width of the chimney in the direction of the offset, **Figure 15.**

A single-flue chimney extends through the roof and flashing is applied as shown in **Figure 16.**

A three-flue chimney stack passes through the roof and flashing is built up as shown in **Figure 17.**

Fireplace Construction. When hollow tile is used in fireplace and chimney construction, the successive courses of tile should be staggered to prevent cracks in the mortar, **Figures 18 and 19.**

Fireplaces which extend through the exterior wall, **Figure 20,** and those which back up against an interior stud wall, **Figure 21,** present special problems in framing. Backing up two fireplaces in a brick party wall presents another problem in getting the proper spacing

Fig. 11. Barbecue section of multiple stock has wood bin beneath grille. Fireplace is in living room on opposite side.

Fig. 10. Inside-outside fireplace has five flues, including flue from furnace in basement.

Fig. 12. Chimney for main fireplace in living room on second level is cantilevered out over outside barbeque on grade.

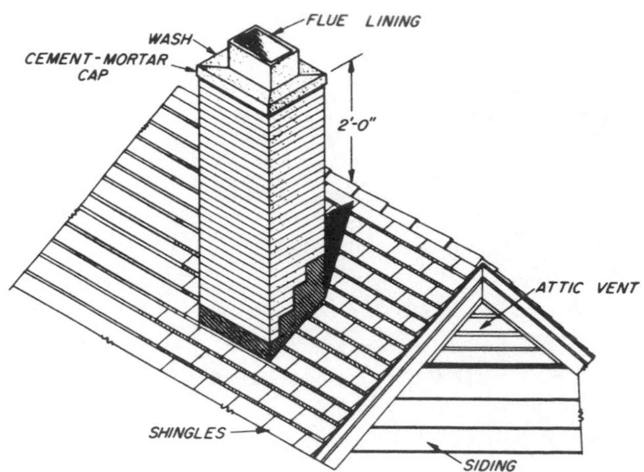

Fig. 13. A chimney must extend at least 2′ above the ridge of the roof to avoid downdrafts and a smoky fire.

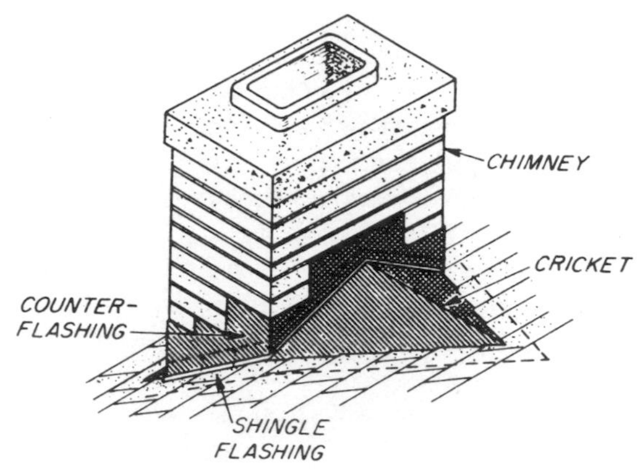

Fig. 16. Counter flashing, shingle flashing, and cricket built behind single flue chimney to divert rainwater.

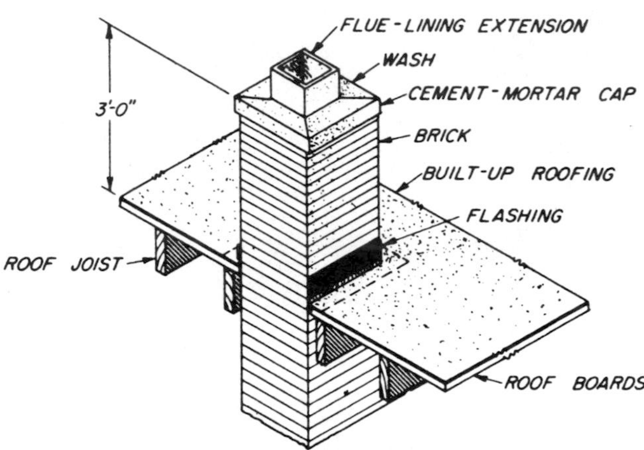

Fig. 14. A chimney must extend at least 3′ above the surface of a flat roof to avoid downdrafts.

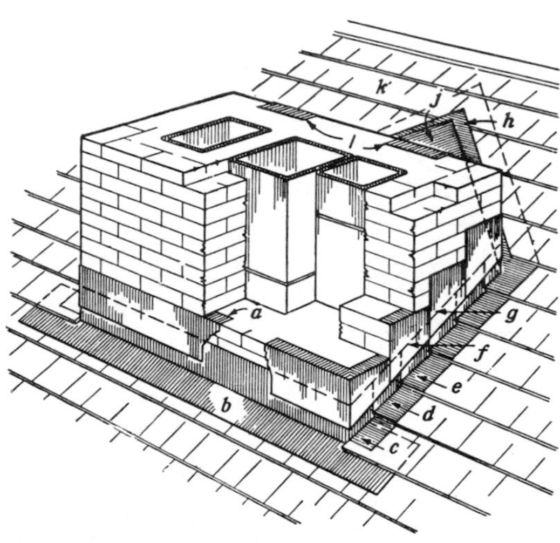

Fig. 17. Flashing is bedded onto mortar joints of three-flue chimney as it passes through the roof of a dwelling.

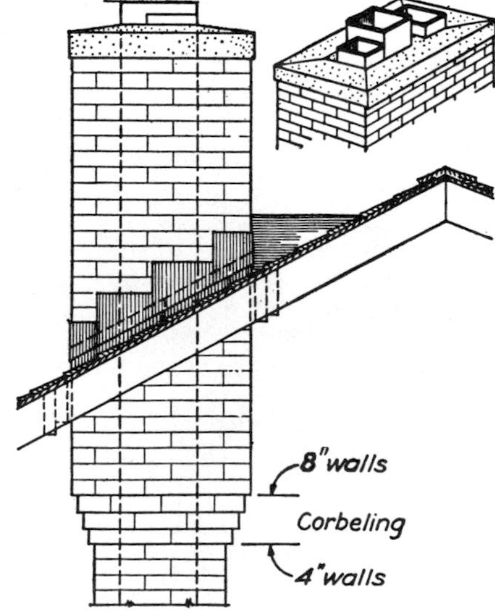

Fig. 15. Corbeling narrow chimney for architectural effect.

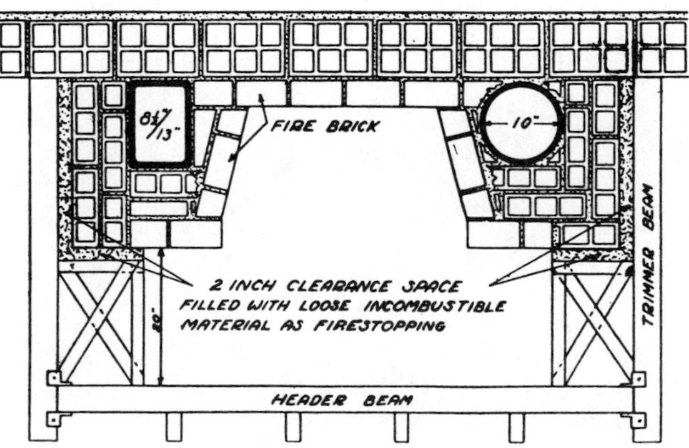

Fig. 18. The use of hollow tile lightens fireplace construction.

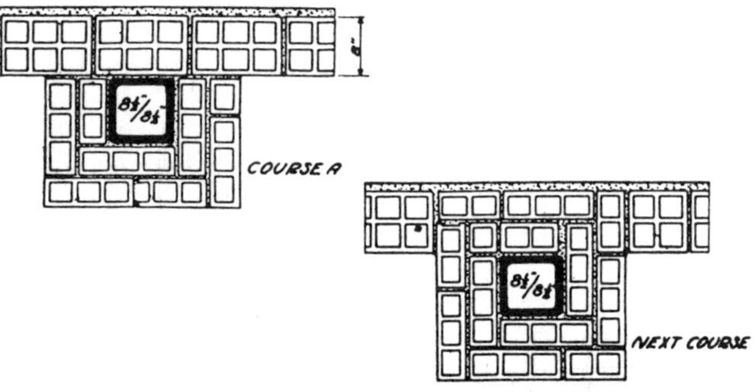

Fig. 19. Plan views showing staggering of joints of hollow tile.

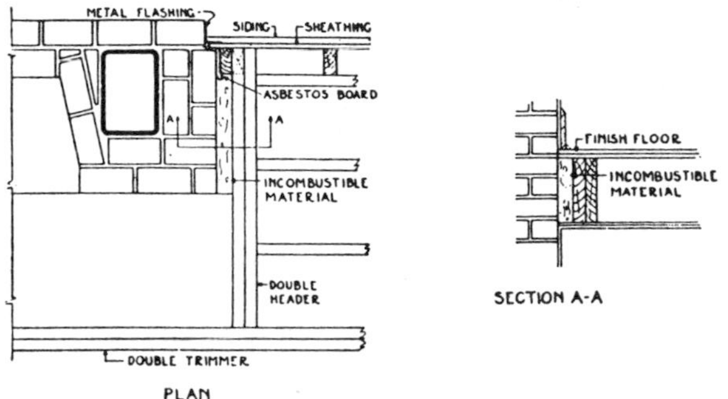

Fig. 20. Incombustible material is placed between fireplace and wood framing. Elevation of floor is shown in section A-A.

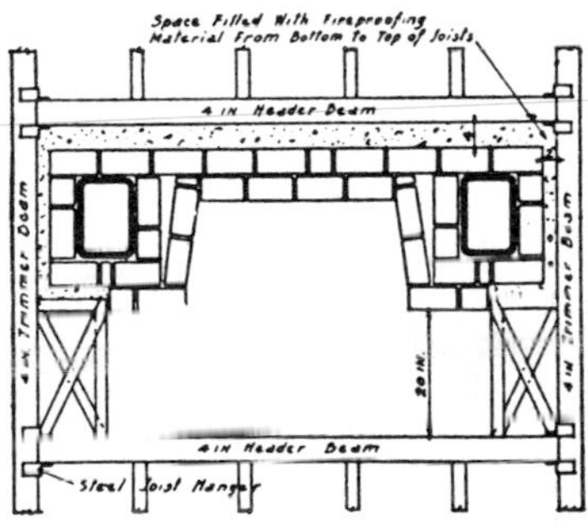

Fig. 21. Plan view of floor framing around a single fireplace built against a stud wall in the interior of a house.

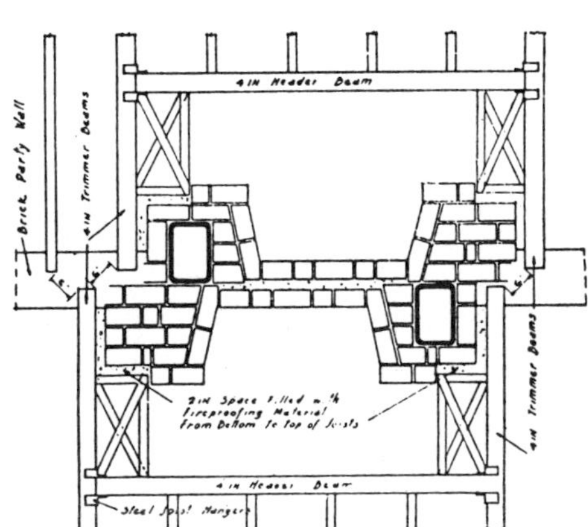

Fig. 22. Plan view of floor framing and method of building two fireplaces back to back, in a brick party wall to insure proper spacing between ends of floor joists.

between the ends of the floor joists, **Figure 22.**

The slab below a fireplace should be constructed at a level which permits the finishing hearth of brick, tile or other material to be laid at finish floor level, or at whatever height is desired for a raised hearth. At damper level, the enclosure narrows to form the smoke chamber. It is important that the slope of its two sides be identical, the flue taking off from the center.

The recommended vertical position of the damper is 4″ to 8″ above the breast wall of the fireplace, which is often supported by a stiff steel angle. The damper must afford a smooth metal throat for the passage of smoke and fumes. A vertical front flange permits the damper to rest snugly against the masonry of the fore wall. The opening must be narrow, from front to rear, to continue the plane back slope and leave room for the smoke shelf. The valve plate must be removable and must operate for the full width of the fireplace, to form an effective barrier along the smoke shelf.

Figure 27 shows the plan section of the back to back fireplaces shown in **Figures 11 and 25.** Note the position

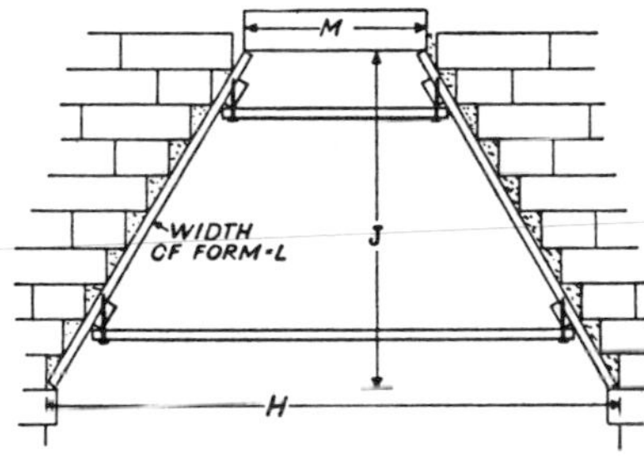

Fig. 23. Form boards help to slope brick and mortar creating smooth surface for unobstructed discharge of smoke.

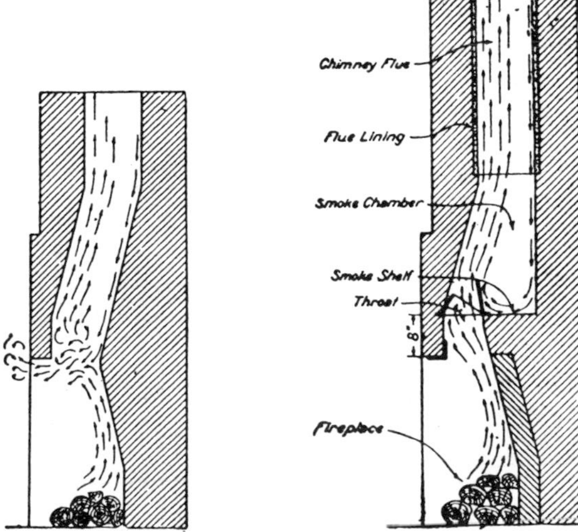

Fig. 24. Smoke shelf (right) is essential to avoid downdraft shown at left, which discharges smoke into room. Damper helps arrest downdraft and deflects it into rising current of warm air. Note 8″ throat beneath damper.

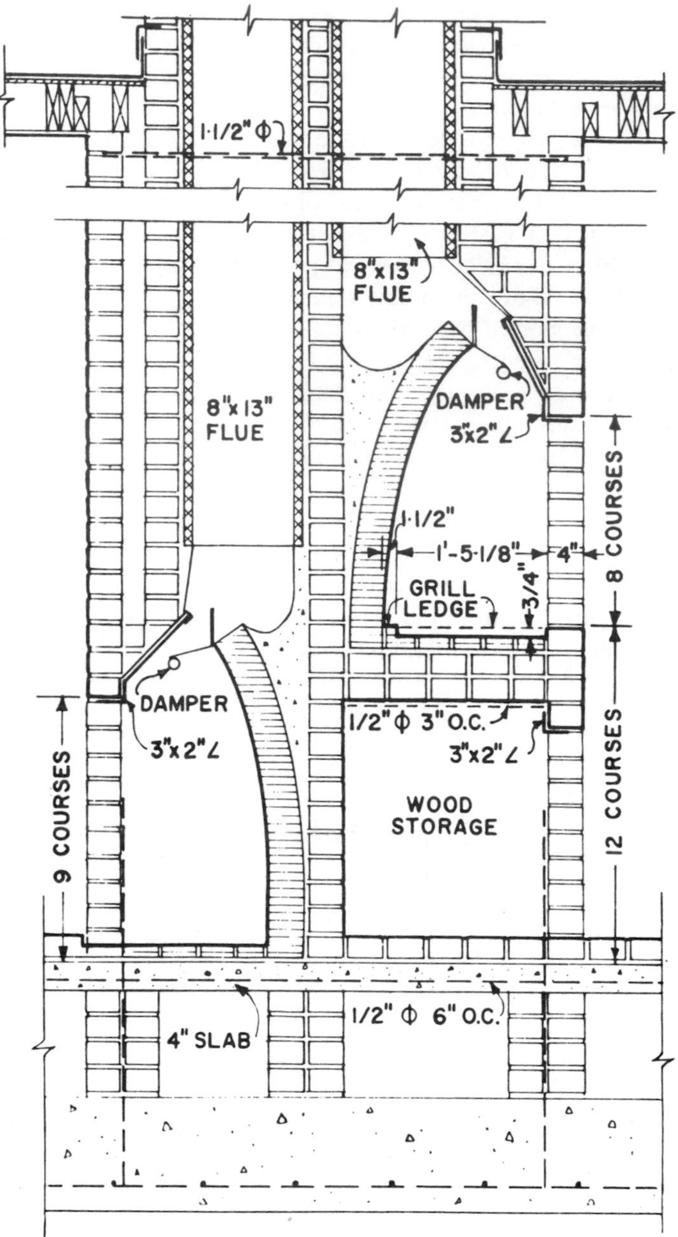

Fig. 25. Section of multiple stack fireplace, similar to that shown in Figure 11. Note reinforced concrete footing.

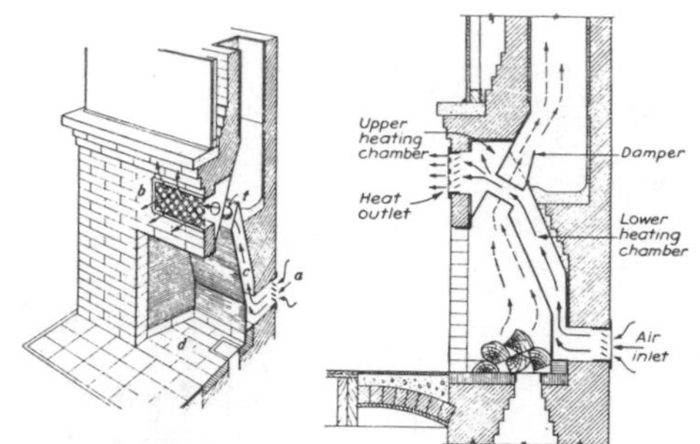

Fig. 26. Air-circulating, heater type fireplace.

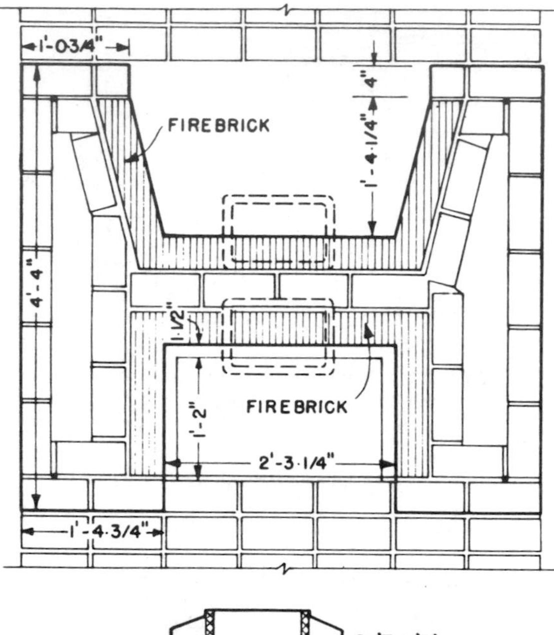

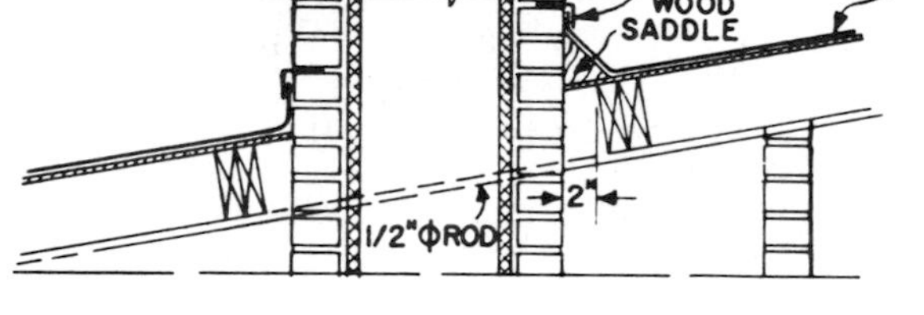

Fig. 27. Plan of fireplace at left, showing arrangement of flues and flashing where chimney passes through roof.

of the two flues. The lower sectional elevation of the chimney shows the flues as they project through the roof.

Flues and Chimney Linings. The most efficient chimney is one built perfectly straight with a round or nearly round flue and a smooth interior surface, **Figure 28.** A fire clay lining is required to prevent disintegration of mortar and bricks through the action of flue gases. Linings are especially important when the chimney has an offset, because loose bricks or mortar may fall and become lodged in the angle, **Figure 29.** Without a lining, the rough brick and mortar sets up turbulence which cuts down the effective diameterof the flue.

All standard masonry chimneys should be built from the ground up; none of the weight should be carried by any part of the building.

Every fireplace, stove, or ventilating register should have its own flue outlet for adequate draft. A division wall of at least 4″ of brick should separate each flue from others in the same chimney. The plan views of two chimneys, **Figure 29** right, show good bonds for the bricks and flue linings. The linings should be tightly

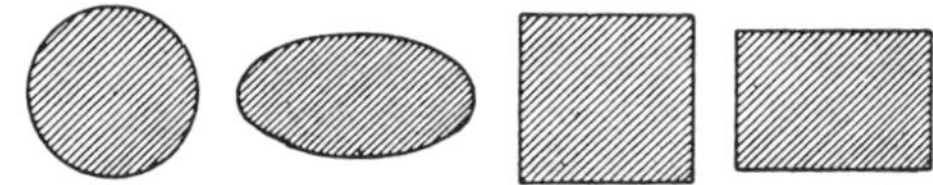

Fig. 28. Four types of flue liner, rectangular is most common.

cemented together, to prevent suction of smoke through the cracks in the masonry and down one flue, while smoke ascends in the other. Unequal projection of flues above the stack is a safeguard against smoke pouring out of one flue and down the other. The proximity of a tree or high building may be a hazard to the free discharge of smoke.

A chimney should stand independent of the framing around it, and a foundation wall should form one of the walls of the ash pit. Good practice calls for a dead air space between a plastered wall and the chimney, and the use of incombustible materials at the header, **Figure 30.**

Figures 31 and 32 show how roof framing is carried around chimneys and not tied into the masonry, so that settling of the chimney will not place strain on the house structure.

Flashing and Capping

Flashing and counter flashing of chimneys are usually made of metal. The term flashing is applied to the pieces of tin, copper, lead or other metal that are nailed onto the roof along with the shingles, and bent up against the chimney wall. Counter flashing is the term for the pieces of metal set into the brickwork and bent downward over the flashing to form a water tight joint, **Figure 33.**

Cap and saddle flashing, **Figure 34** should be provided around chimneys, with a lap of at least 3″. If the flashing is not built into the wall it should be secured by 1″ wide lead plugs on an 8″ center. **Figure 34B** shows poor construction, because it does not allow for the shrinkage or settling of materials.

Waterproofing Tops of Chimneys. Capping protects and waterproofs brick or masonry tops of chimneys. Moisture may permeate the porous materials and joints, and on freezing it will crack the chimney tops and crumble mortar from the joints. This deterioration affects not only the chimney, but also spreads to structural members of the house.

With the development of masonry waterproofing compounds, silicone sealers, and synthetic rubber sealing compounds, deterioration of masonry caps is less of a problem than it used to be. Chimney caps are often made of lead, in one piece, covering the entire chimney top and extending down the sides, turning under the overhang where it is held in place by lead cleats built into the masonry. The corners should be lapped on the inside and dressed snugly against the chimney. For the flues, a square or round cut is made in the cap, allowing sufficient excess material to turn down inside the flue lining, flashing this joint. Lead is not affected by the ordinary corrosive action of flue gases. Prefabricated metal chimneys often include a cap to prevent entrance of snow and rain, **Figure 35.**

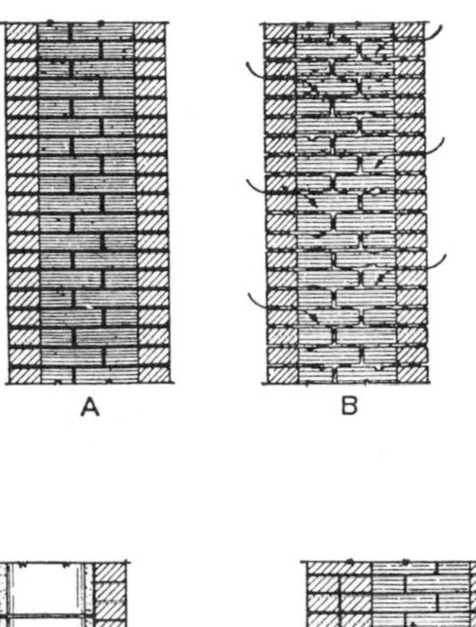

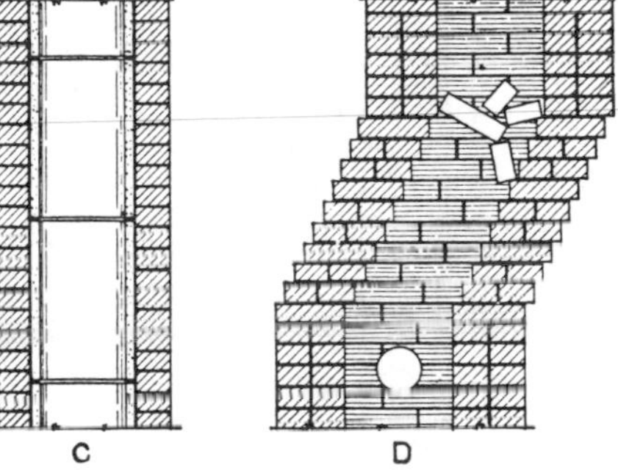

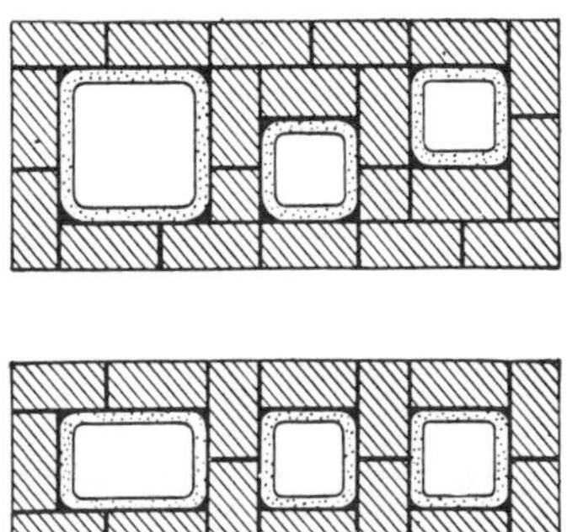

Fig. 29. Unlined chimney (A) develops leaks due to disintegrating effect of heat, gases and weather on mortar (B). Chimney (C) is fitted with flue lining to prevent escape of gases. Unlined chimney (D) has an offset, in which loose bricks may fall and become lodged. Plan views of chimneys at right illustrate methods of laying brick for effective bonding around wide and narrow flues.

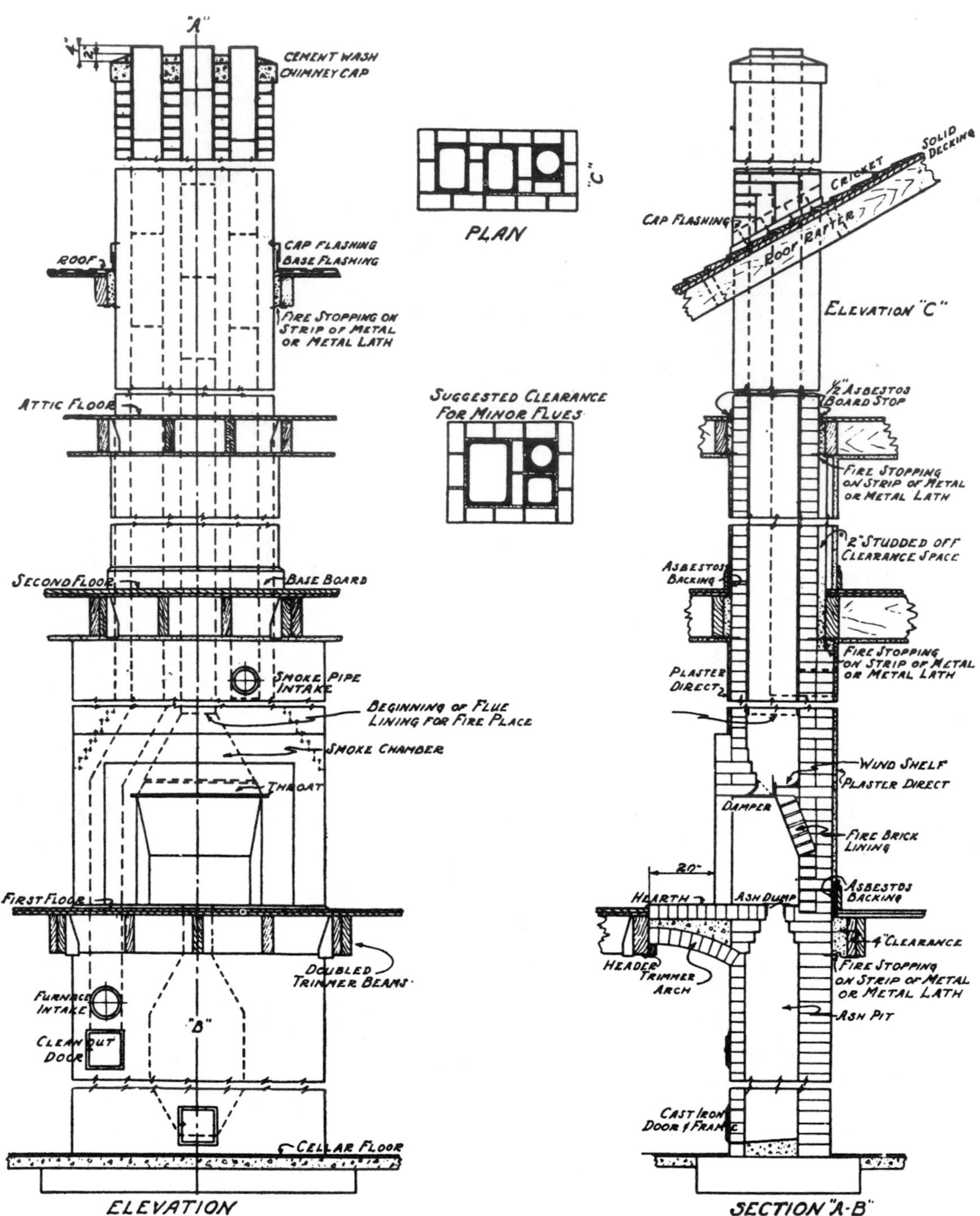

Fig. 30. Elevation and section of an interior chimney in a wood-framed house, showing how flues are placed for the furnace, fireplace and kitchen stove. Chimney is built on separate footing.

Prefabricated Fireplaces

An important modern development has been the acceptance of the prefabricated fireplace. This device may be of two kinds: (1) The single package, including firebox and flue, freestanding or attached to a wall without modification of the house design, **Figures 36, 37, 38, 39 and 40;** (2) the manufactured metal heat box, which needs to be enclosed in a masonry housing on the building site, **Figures 41, 42 and 43.**

A Prefabricated Fireplace as a Room Divider. Prefabricated fireplaces are not alone furnished as free standing and wall attachments but as units which can be used as an integral part of a room wall divider, **Figure 44.** In this unit the complete metal fireplace, as well as the flue and chimney are furnished.

The room divider is panelled, **Figure 45,** together with

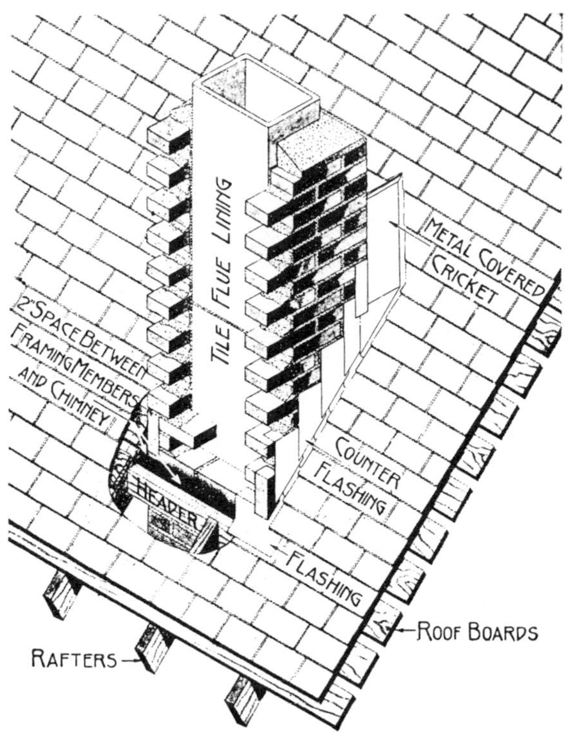

Fig. 31. Details of brickwork, tile flue lining, roof framing around chimney, flashing, and counterflashing.

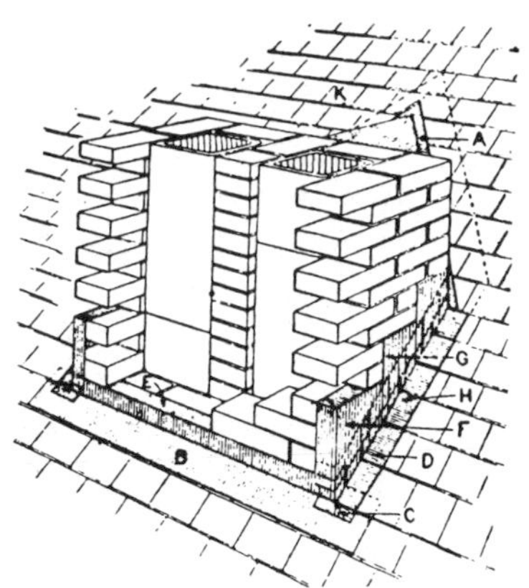

Fig. 33. Two-flue chimney details: (A) cricket and flashing, (B) apron flashing, (C) (D) (H) base flashing, (E) upper apron flashing, (F) (G) cap flashing, (K) roofing material.

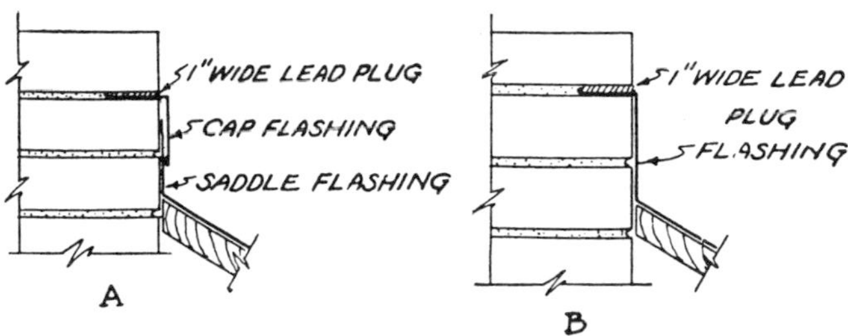

Fig. 34. Overlapping cap and saddle flashing (A). Solid flashing (B) does not allow for shrinkage or settling.

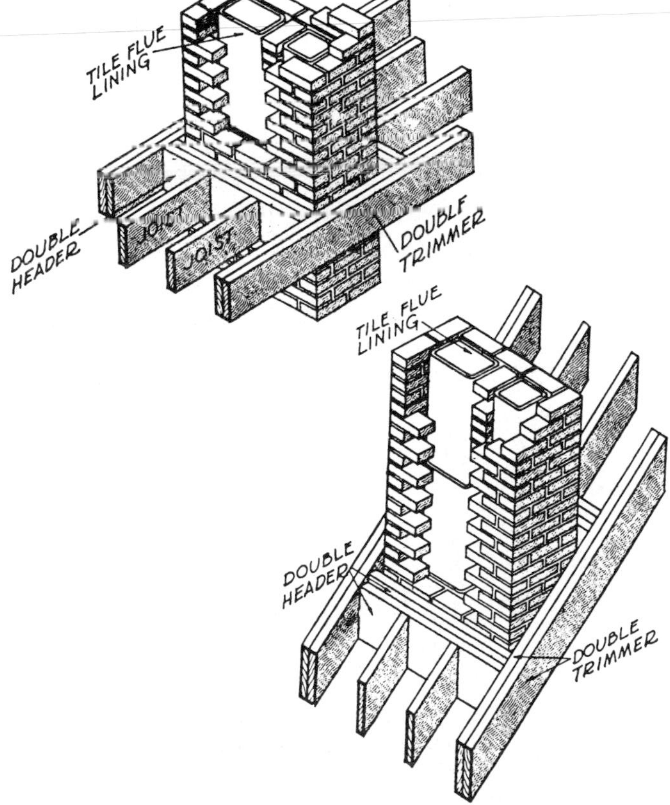

Fig. 32. Joists, rafters, trimmers and headers are framed around chimney but do not touch or bear on brickwork.

Fig. 35. Prefabricated weatherproof metal chimney and caps, used with prefabricated chimney of metal and asbestos.

Fig. 36. A free-standing informal fireplace having base and hood in one piece and flue fitted with ceiling flange.

Fig. 38. After hanging the base on the wall, the interlocking ceramic firebox sections are snapped in place.

Fig. 37. Prefabricated fireplace is hung on wall, needs no footing because of its light weight.

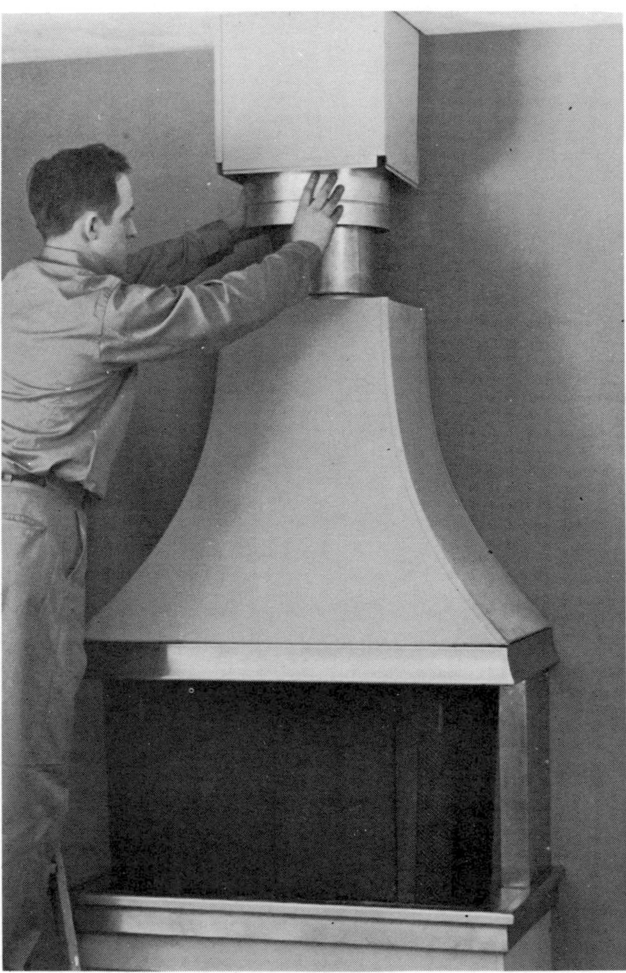

Fig. 39. Lightweight flue is attached after hood is installed.

a built in bookcase. Note the placement of a protective screen which is drawn in front of the entire fireplace whenever a fire is built.

Prefabricated Chimneys. Chimneys, too, are available in prefabricated packages, requiring simply a framed opening for the refractory clay flue, usually round, through the ceiling and roof, and a simulated brick housing above the roof, **Figures 46, 47 and 48.**

Fig. 40. Metal chimney housing and cap are flashied into roofing to enclose the prefabricated metal chimney.

Outdoor Fireplaces and Barbecues

With the integration of the patio into the design of the house, an outdoor fireplace often becomes its center of interest. Sometimes it is built against the house so that the flue can run into the main chimney stack.

The outdoor fireplace is often built away from the house, as a separate structure, with its own flue. Portable barbecue pans, some quite elaborate, often are used as a substitute for an outside fireplace, **Figure 49.**

Most outside fireplaces are built with metal grilles and grates; some with ash pits. Some have spit accessories for elaborate roasting. With the widespread use of charcoal and pressed briquettes which give off little smoke, the flue structure becomes less important than indoor fireplaces, **Figure 50.**

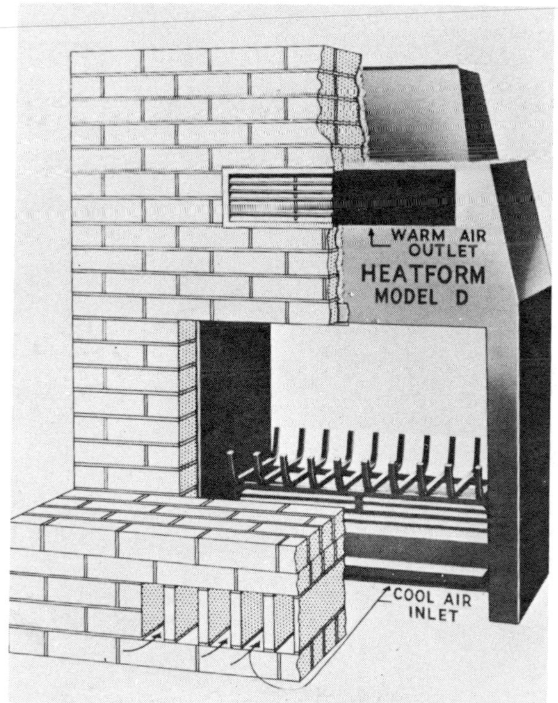

Fig. 41. Cutaway of an air circulating, heater type fireplace installed in a masonry housing with a lined masonry chimney.

Fig. 42. Completed installation of heater type fireplace shows cool air inlets at hearth level, warm air outlet beneath mantle.

Fig. 43. Custom made hood is hung on masonry veneered wall.

Where there is frost, the masonry of the outside fireplace must go below the frost line, or the fireplace may be supported on a reinforced concrete slab.

Incinerators are sometimes built in the outdoor equipment as an added feature, **Figure 51.**

Barbecue cooking has made its way indoors so that the barbecue pit, with or without an automatic turning spit, is often found in kitchens or family rooms, **Figures 52, 53 and 54.**

Fireplaces Constructed of Concrete Block

Precast blocks of concrete are sometimes used in the construction of fireplaces and chimneys. The chimney footing can be cast on solid earth, of concrete machine mixed in approximate proportions of 1 volume of Portland cement, 2-1/2 volumes of sand and 3-1/2 volumes of coarse aggregate. Not more than 6 gallons of water per sack of cement should be used.

In laying up the chimney blocks and flue linings, the mortar should be a mixture of 1 volume of Portland cement and 3 volumes of damp, loose sand, to which may be added plastic agents not to exceed 10 pounds per sack of cement. Mortar joints should not be more than 1/2″ thick, **Figures 55 and 56.**

When two flues are contained in a chimney, the joints of the adjoining sections of the flue linings should be staggered at least 7″. Where there are more than two flues, at least every third flue should be separated by masonry at least 3-3/4″ thick, bonded into the walls of the chimney. Masonry must be at least 8″ thick on the exposed sides of a chimney built into the exterior wall.

A chimney of concrete block may be unlined, but it must be built of solid masonry at least 8″ thick. The units must be laid and the mortar struck off or pointed so as to produce a smooth inside wall surface. A lined or unlined chimney is capped with a precast concrete cap having a projection of 1-1/2″ over the outside face, with a groove along the lower outer edge to provide a drip.

Fig. 44. A prefabricated metal fireplace with throat and damper can be built into a room divider.

Fig. 45. Prefabricated fireplace shown in Figure 44 installed in a room divider faced with wood paneling.

Fig. 47. Embossed metal chimney housing simulates brick.

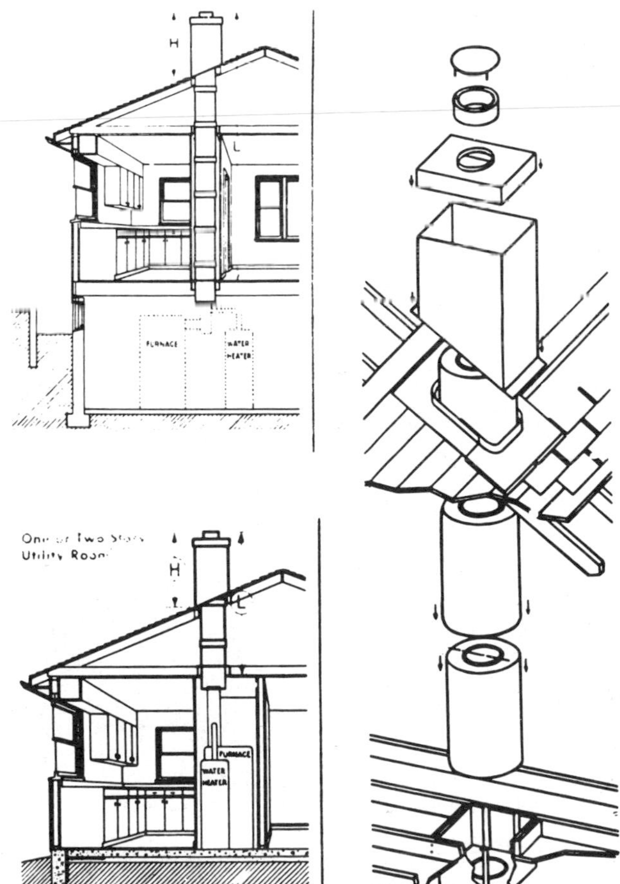

Fig. 46. Lightweight prefabricated chimney has refractory-tile-lined flue with vermiculite concrete insulating wall.

Fig. 48. Photos of chimney illustrated at left being installed. Metal base supports chimney, perforated straps are clamped around chimney and nailed to headers for lateral support.

Fig. 49. Portable barbeque, rather than built-in, used on patio.

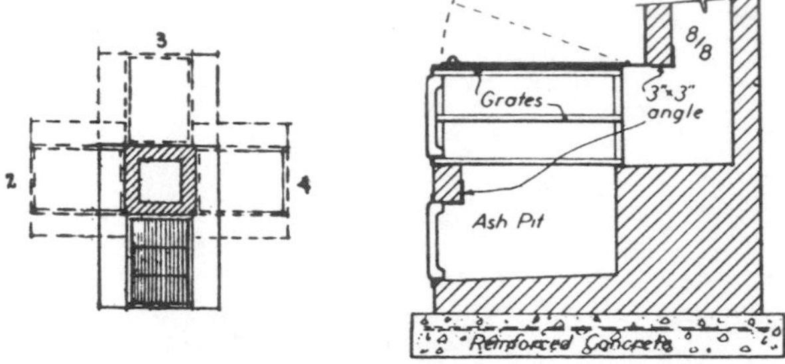

Fig. 50. Sketch of outdoor fireplace shows how four cooking areas can be grouped around a single, central flue.

Fig. 52. Flue for kitchen barbeque also serves range vent.

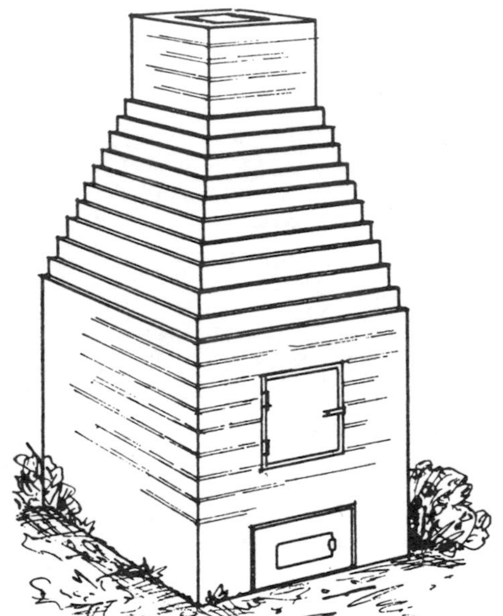

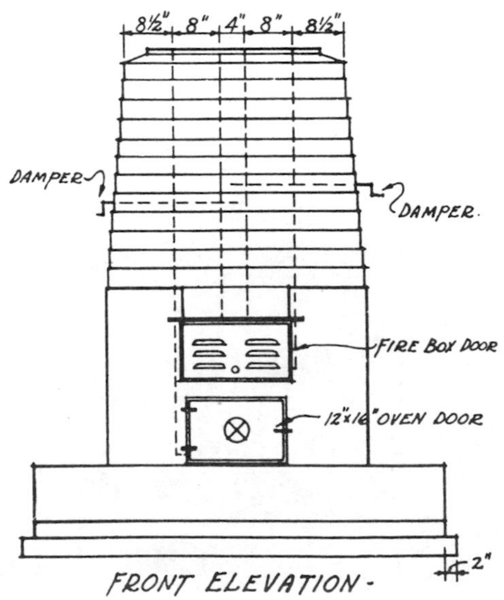

Fig. 51. Back yard incinerator should be built away from house.

Fig. 53. Handsome spit-equipped barbeque is built into a mantle. Hearth is faced with glazed tiles for easy cleaning.

Fig. 54. Metal barbeque is built into brick housing.

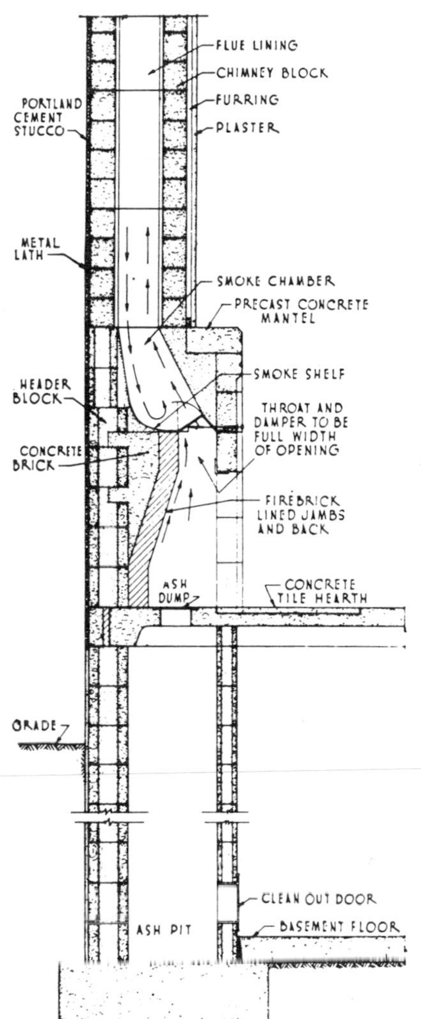

Fig. 55. Stucco applied over metal lath is an attractive finish for chimneys built of concrete block.

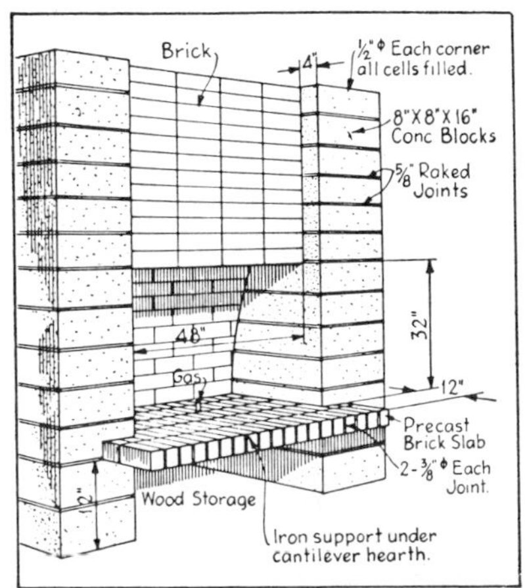

Fig. 56. Cantilevered brick hearth is reinforced with steel rods set in mortar between rows of brick.

Chapter 18

SCAFFOLDS AND HOISTS

The scaffold is a temporary structure serving as a support for workmen and building materials. Scaffolds may be built of wood, using framing lumber nailed together, or assembled with metal special purpose equipment, usually collapsible and transferable from job to job.

Some scaffolds are on rollers for mobility on the job. Others can be raised or lowered to different elevations. The well known painters' and masons' hoistable swing scaffold is not classified as a hoist because its main purpose is not to lift materials.

A hoist or elevator is a mechanical lifting device used to convey materials from one level to another. Some hoists direct force straight upward on the hydraulic jack principle, others use the winch motor and line, others use the endless chain or tractor tread technique.

Fig. 2. Collapsible sawhorse: Legs are double-hinged to fold together and under the crosspiece as shown at left.

Scaffolds

The simplest scaffold is the trestle, a platform stretched between two supports. A plank on wooden sawhorses is the most familiar example of the trestle. Sawhorses may be made of framing stock nailed together as shown in **Figure 1.** No matter how modest or elaborate the job this solidly built saw horse is used quite

Fig. 1. A solidly built and braced sawhorse (foreground) is still a popular favorite on the most modern construction jobs.

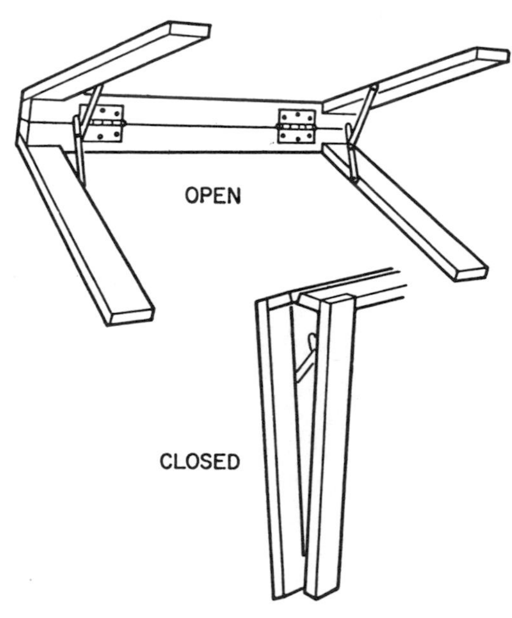

Fig. 3. A pair of butt hinges and two folding brackets are used to collapse this horse and lock it in the open position.

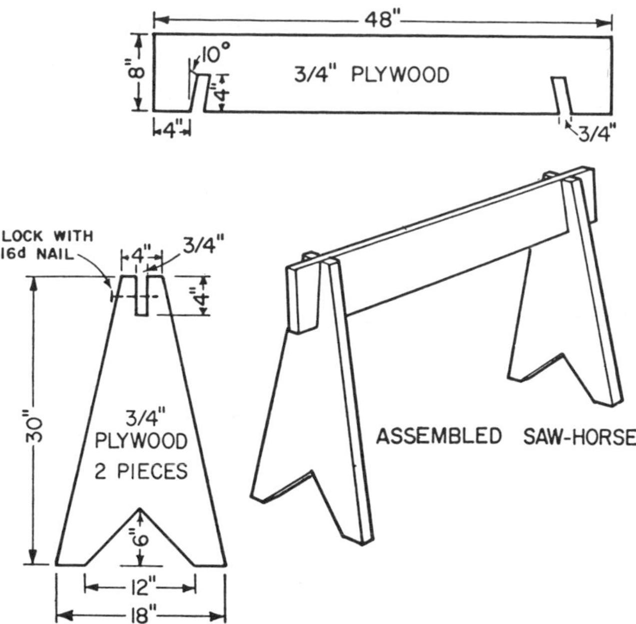

Fig. 4. A 16d nail is used as a lock pin to secure the legs to the cross-piece of this interlocking sawhorse of ¾" plywood.

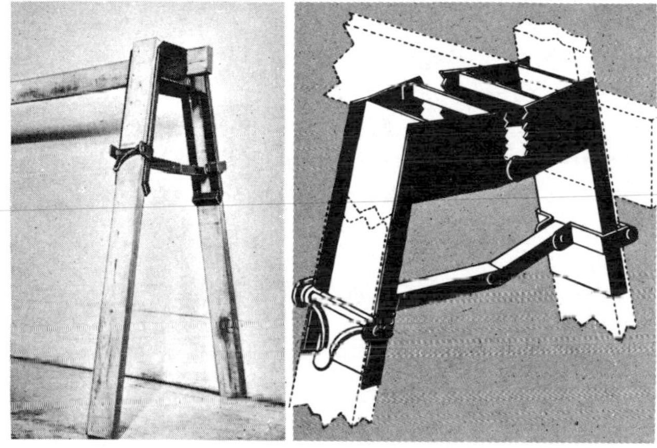

Fig. 5. Toothed edges of this lockjaw sawhorse clamp bite into the cross-piece to hold it securely on the legs.

Fig. 6. Metal tubular sawhorses used as trestle supports.

Fig. 7. Tubular metal trestle with removable cross bracing. The work platform can be moved to different levels.

Fig. 8. Allen wrench is used to tighten setcrews in special fitting which joins sections of tubular scaffolding.

Fig. 9. Three-way fitting joins scaffolding of 1" standard pipe without threads. Cap screws lock pipe securely.

frequently. For portability, sawhorses can be made collapsible and demountable using special hinges and braces, **Figures 2, 3 and 4.**

A lockjaw jack is used on construction jobs as shown in **Figure 5.** The metal jaws hold a standard 2 × 4 or 2 × 6 piece of framing stock which can be inserted between the jaws. Cam action holds the horizontal piece in place. The jacks can be used for heights from 2′ to 16′. The sawhorse principle is utilized in folding inverted "V" frames of metal, of various sizes, **Figure 6.**

Trestles. Trestles can easily be put together with two upright members and a cross platform. Two folding stepladders for bases and plank platforms, raised and lowered on the steps, are common trestles. The extensive development of light, tubular metals has brought a large variety of patented trestle systems, **Figure 7.** Various techniques are used for clamping the tubular sections or lengths of pipe, **Figures 8 and 9.**

Trestle With Materials Platform. A materials-working platform of welded tubular steel is available for masons, **Figure 10.** The materials are placed on the upper platform while the workmen stand on the lower platform. Both platforms are 25″ wide. The working level adjusts in 2″ stages from 1′-6″ to 4′-6″. For extra height the workmen can stand on the materials platform when it is turned against the wall.

A Trestle of Single Height. A heavy duty trestle for ceiling work is made of four jacks and framing stock platform, **Figure 11.** Different size jacks provide heights from 7′ to 14′.

Adjustable Trestle. A heavy duty trestle with adjustable jacks is available for heavy scaffold boards,

Fig. 10. An adjustable work scaffold or trestle.

Fig. 11. Trestle jacks support crosspiece beneath planks.

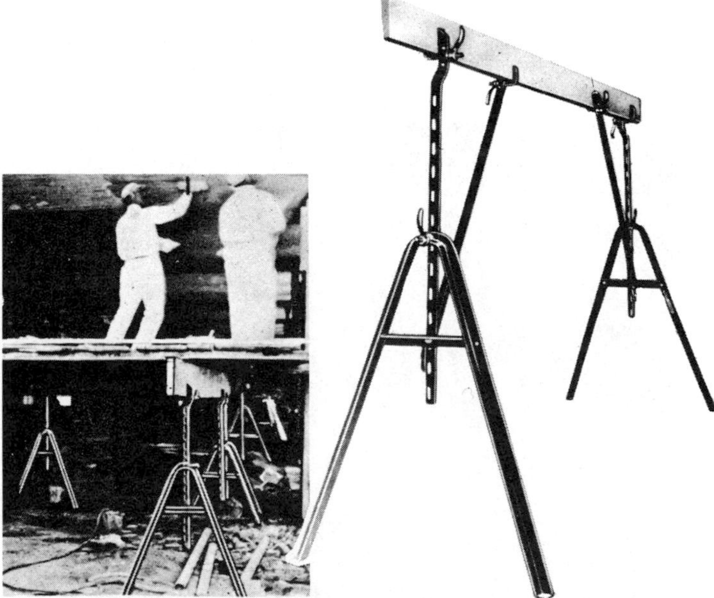

Fig. 12. Heavy duty jacks are adjustable to working heights from 7′ to 12′. Lever handles lock adjustment and crosspiece.

Figure 12. The working heights are from 7′ to 12′. Special handle nuts lock all height adjustments.

Tubular Trestle. Usually each vertical side member of a tubular trestle has horizontal members on which the platform, or platforms rest. The platforms themselves may or may not contribute to the stability of the equipment, **Figures 13, 14 and 15.**

A Stepladder Type Trestle. A trestle for scaffolds, on the stepladder principle, **Figure 16,** requires the following materials: 2 pieces of 1″ × 4″ × 6′ stock for the cross brace; 8 pieces of 1″ × 4″ × 14″ stock for the rungs; 4 pieces of 2″ × 4″ × 7′ stock for the legs; (2) 3/4″ × 4″ carriage bolts; (2) 3/4″ wing nuts; (2) large washers; (1) eye bolt; (1) hook bolt; (40) No. 12, 1-3/4″ flat head screws; 4″ of light chain. Fir or hardwood stock can be used.

A frame for supporting scaffold platforms is made of 2″ × 4″ framing stock, braced at the corners by 1″ × 6″ stock, and 3′ to 4′ square, **Figure 17.** The frames are placed on edge against the wall, about every six to eight feet and braced by sheathing boards, crossed diagonally.

A brace for supporting these frames is made by setting up two frames and marking the template as shown in **Figure 18.** Lay off right angle of cut and saw the notch as shown in the second figure. The third drawing shows the brace in place between two frames.

A Trestle of Ladder and Tire Chain. A simple scaffolding rig for occasional repairs utilizes a tire chain as a hanger. The ends of the staging plank are supported by a

Fig. 13. Platforms are notched to fit over horizontal supports. Set diagonally as shown, platforms also serve as steps.

Fig. 14. Tubular trestle rests on supports for leveling. Interrupted rungs permit wider platforms at some levels.

Fig. 15. Tubular trestle has a 10′ high base with sleeve-type extensions. Platform arm is wide enough for wheelbarrow.

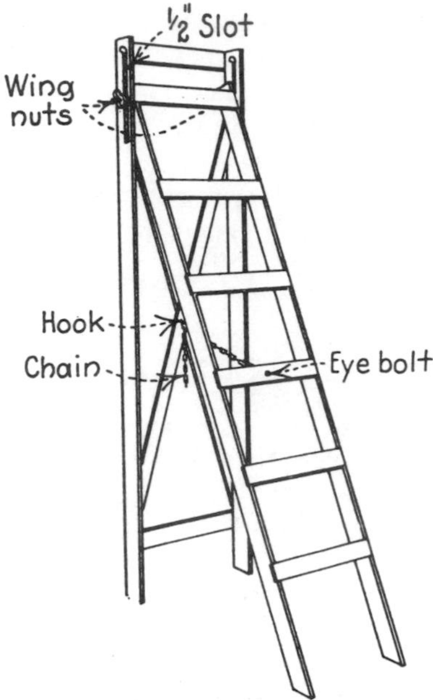

Fig. 16. Ladder-type trestle can be built of framing lumber.

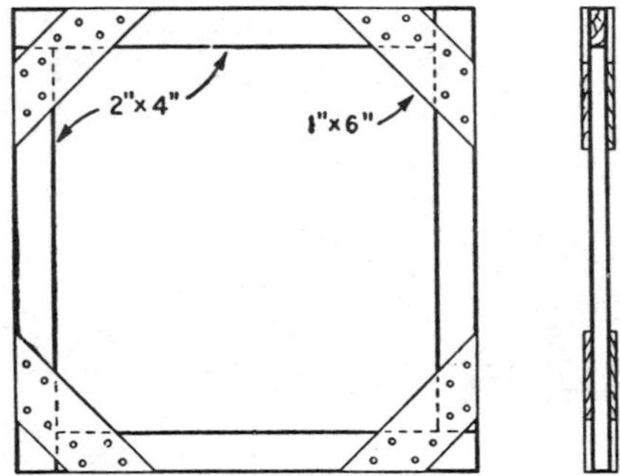

Fig. 17. A simple frame can be used to support scaffolds.

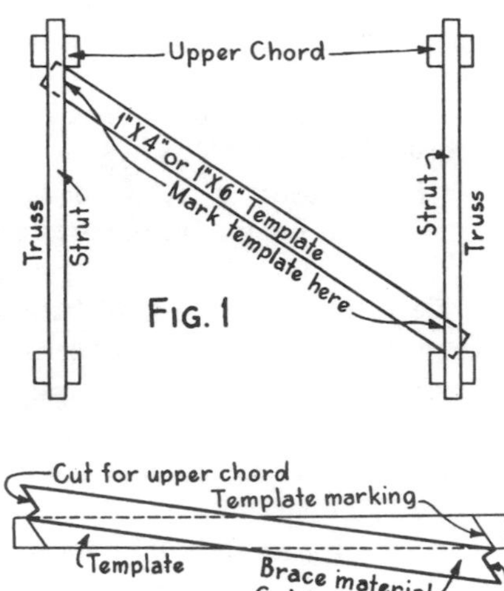

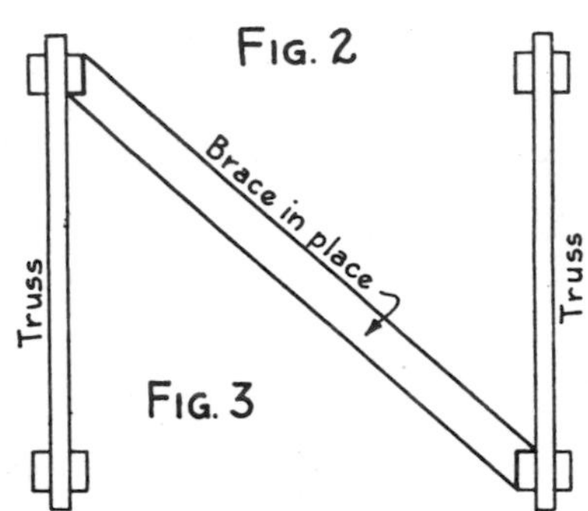

Fig. 18. Diagonal bracing for frames shown in Figure 17.

Fig. 19. Tire chain on ladder supports scaffolding.

readily adjustable triangle formed by ladder, chain, and a short piece of lumber cleated on the underside to engage the rung of the ladder, **Figure 19.**

Ladders and Stepladders. Various types of ladders are used in construction work from the old style stepladder and wooden rung straight ladder, **Figure 20,** to the light weight ladders made of magnesium, **Figure 21.** The magnesium ladder weighs one third less than a ladder made of aluminum.

The magnesium stepladder is made of 3″ wide channel rails with 3″ reinforced steps. It has heavy front and back bracing. There is also a sturdy bucket rack and hard rubber shoes for the rails.

The platform ladder has a 12″ × 18″ platform 2′ below the top of the ladder.

The extension and straight ladders have fluted nonskid rungs, on 3″ I-beam side rails. Extension ladders have interlocking sections for safety and sliding ease.

Fig. 20. Folding wood stepladder, straight ladder and extension ladder should be used for work at different heights.

Fig. 21. Lightweight magnesium stepladder and extension ladder.

Rope, pulley, safety lock and rubber shoes are available with the ladders.

Not to be overlooked is the old style ladder, **Figure 22,** made of 2″ × 6″ framing stock and the rungs or steps made of 1″ × 3″ trim stock nailed to the rails. This type of ladder is often heavy and cumbersome to use, but can be made on the job quickly, when needed.

A Safety Hook for Rung Ladders. An important use of scaffolding is for roofing work. A ladder hook is a brace at the end of a simple rung ladder, designed to fit over the ridge of the roof to prevent the ladder from sliding, **Figure 23.** A chicken run ladder, made by nailing cleats to a plank, may be used.

A safety hook of metal may be used with an ordinary rung ladder. Insert shows how the hook is attached to a ladder, and how it is hooked over the ridge of a roof, **Figure 24.**

A Roof Saddle for Materials and Tools. A roof saddle is placed over the ridge of a roof to form a platform for the materials and tools of various trades. A 4′ × 4′ roof saddle weights about 50 pounds. If plywood is used for the platform, instead of heavy planks, the saddle will be of less weight, **Figure 25.**

Ladder Step and Tool Rest. A ladder step for extension rung ladders, **Figure 26,** hooks over a rung and rests on the ladder rails. It provides a flat platform on which

Fig. 22. A straight ladder is often built on the job.

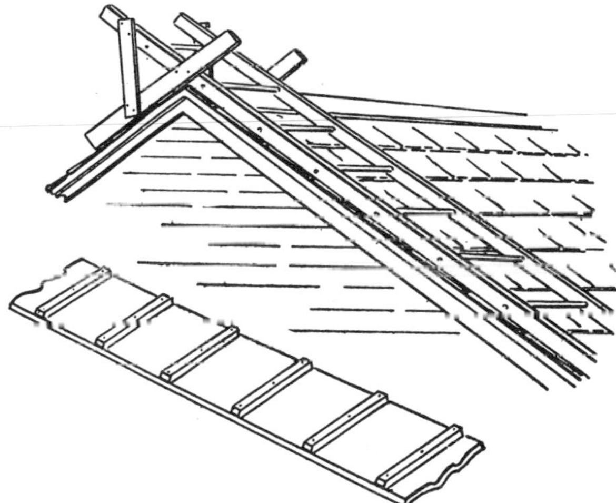

Fig. 23. Roofing ladder with rungs or cleats on plank.

Fig. 24. Metal hook on rungs hangs roofing ladder from ridge.

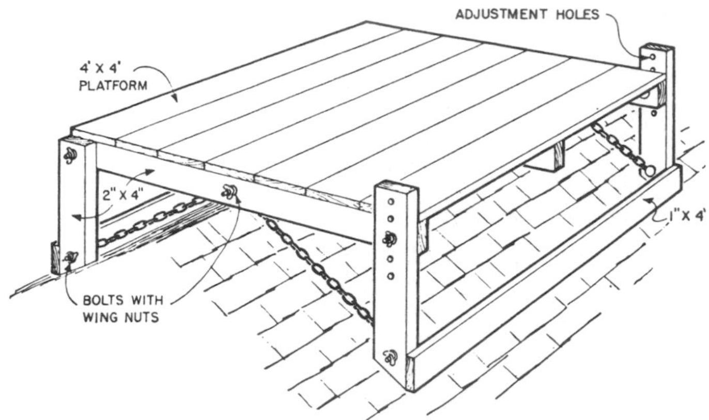

Fig. 25. Roof saddle is placed over ridge of a roof to provide a level surface for tools and materials.

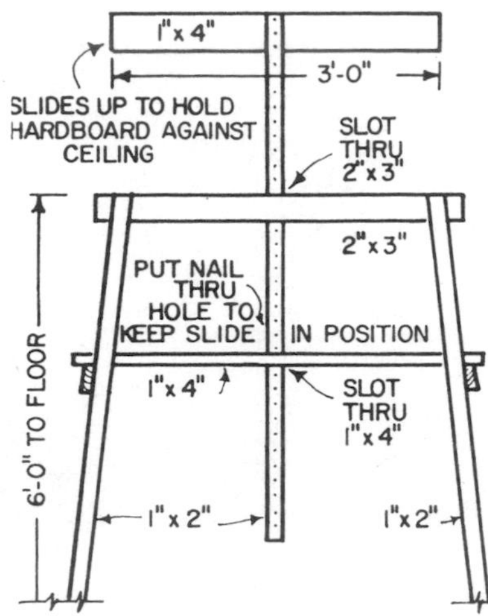

Fig. 29. Adjustable T-brace holds ceiling panels firmly in place against joists while nails are driven.

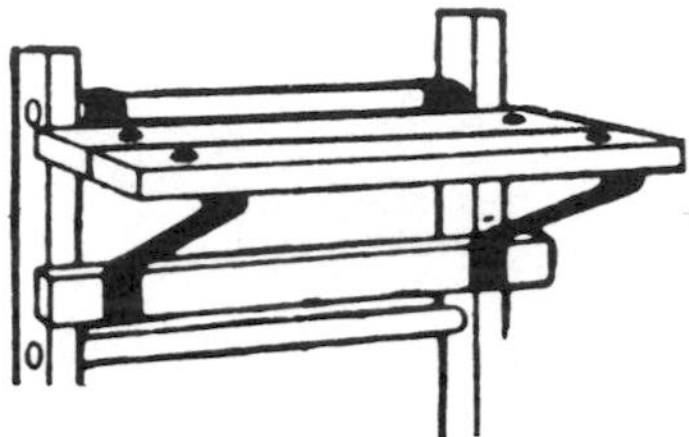

Fig. 26. Hook-on ladder step is a foot saver on long jobs.

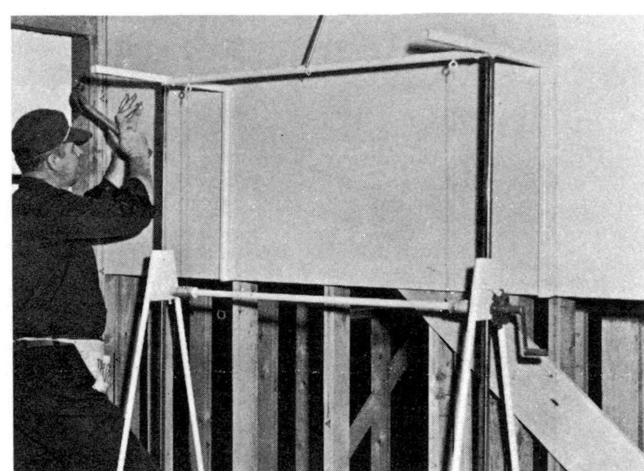

Fig. 30. Trestle holds wall panels firmly against studs, leaving carpenter's hands free for driving nails.

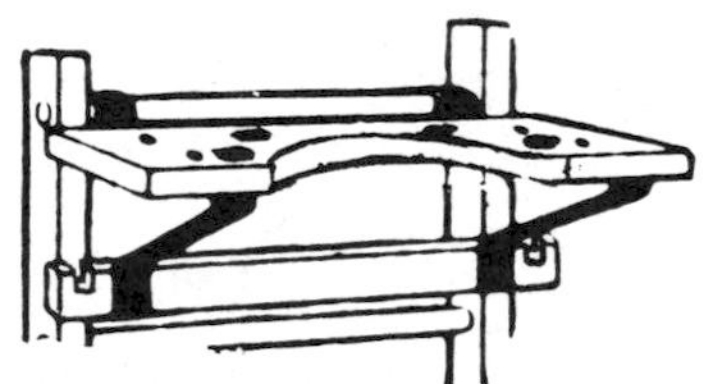

Fig. 27. Hook-on tool rest is drilled to hold tools.

Fig. 28. Work platform hooks over sill of window.

Fig. 31. Trestle shown in Figure 30 used to raise and hold ceiling panels in place while nails are driven.

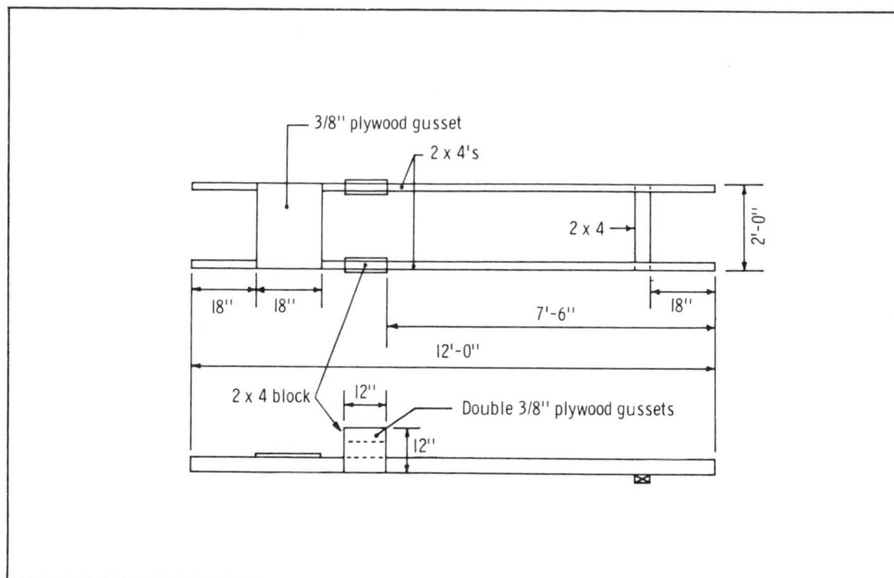

Fig. 32. A simple lean-to rack for holding plywood roof decking within easy reach of carpenters.

to stand. A tool rest, **Figure 27,** may be used in connection with the ladder step, serving as a platform for tools and supplies. The tool rest has holes drilled in it to receive various tools.

A Window Step. A strong, safe step or platform, **Figure 28,** is made of tubular steel with rubber bumpers at contact points which prevents marring surfaces. It is used by tradesmen, homeowners and industrial plant workers for repairs and replacements about windows.

A Lift or Trestle. A trestle to hold drywall and plywood against a partition or ceiling as an aid in nailing, **Figure 29,** is a very practical device. The sliding top fits through slots in the horizontal members of the trestle, then is held in position by a nail. Two trestles hold a panel against a sidewall or ceiling, leaving the carpenter's hands free for nailing.

Another type of lift with adjustable members is used to hold drywall or plywood against a partition or on the ceiling as shown in **Figures 30 and 31.**

A Lean-to Rack. A simple lean-to rack provides a trestle or lift for easy access to plywood roof sheathing by the workmen at the roof level, **Figure 32.** The drawing of the plywood rack, left, shows how easy it is to construct.

A Platform Trestle. A platform trestle for the temporary storing of plywood sheets for roof sheathing is

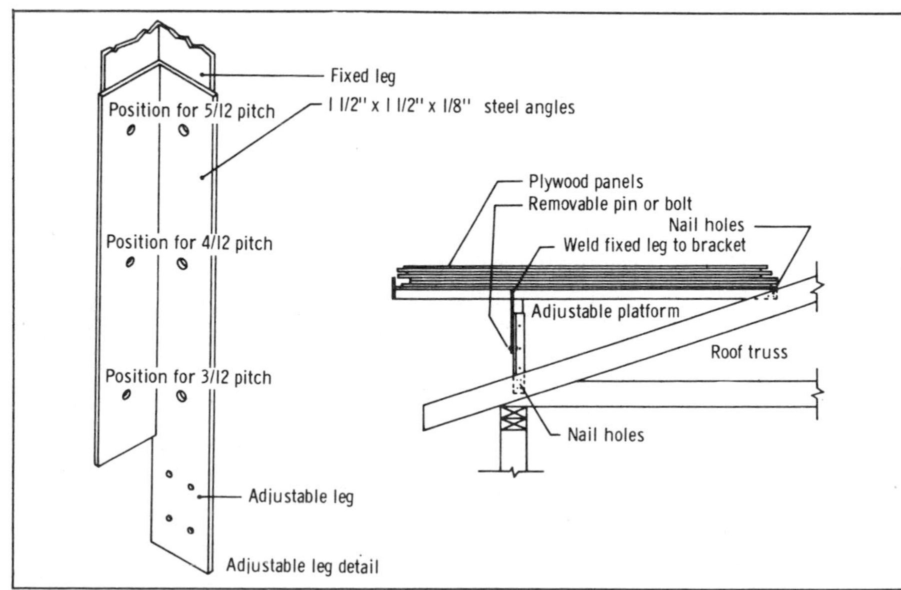

Fig. 33. Platform on adjustable legs is temporarily nailed to rafters to hold roof decking hoisted from ground.

Fig. 34. Working platform can be raised and lowered on 6″ × 6″ posts fastened to house with forked braces.

shown in **Figure 33.** The plywood sheets can be delivered direct by a hoist to the platform thus saving extra handling. Adjustable legs are nailed to trusses or rafters to hold the platform level.

Catwalk Scaffolding Fixed Against the Side of Building. An important group of scaffolding equipment uses brackets fixed to or braced against the side of the building or forms of a poured foundation. These brackets support catwalk scaffolds. Some scaffolds are supported by nails into the framing or siding, and are easily removed. One widely used type of bracket is a frame that projects from an upright piece of framing lumber, which is then braced to the building by an arm, forked to prevent sway, **Figures 34, 35, 36 and 37.**

Exterior Wall Covering Applied From a Multi-Tier Catwalk. Plastic sheets, 4′ × 8′ in size, are passed from one level to another and fastened to a concrete exterior wall with a special compound providing a floating grip that expands with temperature changes, **Figure 38.**

Workmen operate at different levels of scaffold catwalks which are fastened to tubular steel ladder supports.

Steel Scaffold Frames and Brackets. Steel scaffold frames and brackets are available to clamp on rung ladders leaning against a building, **Figure 39.** In **Figure 40** a workman is applying insulation paper over exterior sheathed walls while sitting on staging supported by steel scaffold brackets on rung ladders. The lower workman is

Fig. 35. Scaffold erected on job: Horizontal supports are nailed to framing and to vertical members resting on grade.

Fig. 36. Metal angle braces are attached to lower foundation forms to provide catwalk for crew erecting upper forms.

Fig. 37. Scaffold catwalk brackets have keyhole slots for easy mounting and dismantling. Sway braces hold brackets securely.

Fig. 40. Workmen operate from two levels of staging supported by steel scaffold brackets to apply exterior wall coverings.

Fig. 38. Workmen raise plastic wall panels from level to level of catwalks supported by ladder trestle.

Fig. 41. Work stilts clamped to legs put ceiling within easy reach of insulation installer.

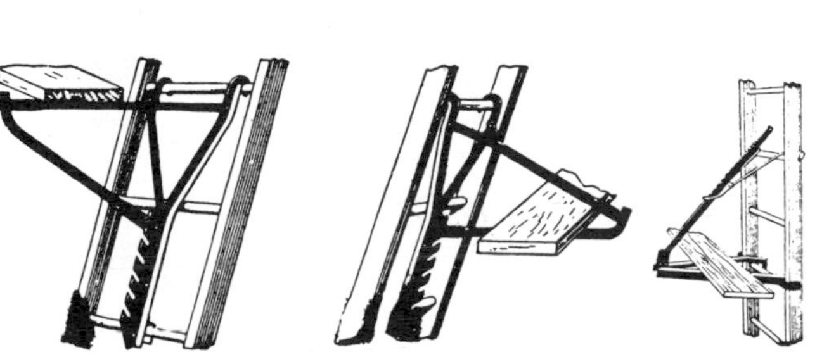

Fig. 39. Steel scaffold brackets can be used on rung ladders.

Fig. 42. Lightweight aluminum stilts have broad bases.

Fig. 43. Swinging scaffold of aluminum is raised by rope pulleys. Aluminum guard rail protects workmen.

Fig. 44. Hand-cranked ratchet raises swinging scaffold of wood.

applying metal siding over the paper while standing on the lower level of staging.

A Walking Scaffold or Stilts. Various types of "steady stilts" have been developed to give a workman enough height to work on a ceiling without losing balance and mobility, **Figures 41 and 42.**

Swinging Scaffolds or Stages. Swinging scaffolds or stages can be raised or lowered along the side of a building as the work progresses. The movement is controlled by the workmen on the scaffold, either by manual operation of a pulley, **Figure 43,** or by activating a winch, **Figure 44.** Some have a direct ratchet for

Fig. 45. Hydraulic mechanism raises mobile scaffold on casters. Note extension ladder and retractable ground supports.

Fig. 46. Self-contained hydraulic scaffold is built on trailer. It can raise load of 2,500 pounds to 16′ height.

Fig. 48. One-ton capacity erection pole has telescoping boom on heavy steel frame with swivel casters for maneuverability and outriggers for stability. Unit is manually operated, may be bolted to truck bed for greater mobility.

Fig. 47. Two-column hand operated jack hoist raises heavy air conditioner compressor to attic opening.

Fig. 49. Motorized belt conveyor of aluminum lifts materials to second floor level. Conveyor can be moved as work progresses.

Fig. 50. Belt conveyor is equipped with bucket and chute for placing concrete above grade. Height is easily adjusted.

Fig. 51. Telescoping hoist is mounted on tractor frame.

raising and a worm gear for lowering. Practically all stages are fitted with braces to support a waist-high back guard. Ropes are made of fiber or steel. Decking is often of metal. For moderate heights, such as one and two story houses, mobile scaffolds are available with motors to lift the working deck, **Figures 45 and 46.**

Hoists

A Portable and Demountable Hoist. Hoists have to be portable when moved from job to job. They are usually demountable, can be transported piece by piece, and set up again with ease, **Figure 47.** Pulley hoists, for example, require only a simple lumber frame work, which can be erected quickly.

Motorized Hoists. Motorized hoists are usually equipped with wheels, especially if the hoist is of heavy metal, **Figure 48.** The most common motorized hoist is the endless belt conveyor used to lift bricks, blocks, lumber and roofing materials to upper story levels, **Figures 49 and 50.**

The Crane Hoist. The crane hoist or fork lift is used quite often to raise materials to high levels. The crane is mounted on a truck or tractor frame, **Figure 51.**

The Boom Hoist. A heavy spar or boom hoist is used to move materials from delivery trucks to high places. **Figure 52** shows a boom moving stressed skin roofing panels in place. Prefabricated trusses, **Figure 53,** are being set in place by a long heavy boom hoist, which can swing them from a trailer on the street to their precise position on top of the building.

Fig. 52. Boom or spar hoist raises prefabricated stressed skin panels onto roof beams.

Fig. 53. Extension jib on boom swings trusses from trailer on street to precise position on top of building at rear.

Chapter 19

GARAGES AND CARPORTS

The garage is primarily a shelter for housing an automobile. Most dwellings today require a shelter for at least two cars. Some garages are built as a distinct and separate structure a short distance from the dwelling, but today's dwellings are generally built with the garage as an integral part of the house.

A carport is an open sided shelter for the parking of one or two automobiles.

Today's garage not only serves as a shelter and parking place for the automobile, but as a storage room for lawn and garden equipment; often a workshop equipped with tools, and a place where house comfort conditioning equipment may be built in to service the entire dwelling.

A Carporch. In addition to the built-in garage, the separate structure garage and the carport or sheltered parking area, there is the carporch. This is a separate structure built like a carport, but having a storage cupboard on one side which can be completely closed off to provide storage space for valuable equipment.

The carporch, **Figure 1,** is a unique combination of carport and an open storage area lined with perforated hardboard for hanging tools, and a larger enclosed closet complete with double louver doors. The overall dimensions are 15′-6″ × 20′. It is possible to screen in the carport area for a summertime entertainment area.

Fig. 1. Carporch and storage unit shelters car and garden equipment, can be screened to serve as recreation area in summer.

Fig. 2. Detached garage with gable roof, cupola, and side entry door is available as a prefabricated package.

Fig. 3. Garage at right is backdrop for plants in this secluded patio off main entrance.

The decorative full height louver panels give the supporting columns an attractive appearance.

The perforated panels are at the left of the carporch and the storage cabinet is at the right. The gable ends and vertical siding are of rustic redwood planks. The roof is finished with asphalt shingles. The entire unit is factory fabricated.

The Detached Suburban Garage. This structure, **Figure 2,** is factory built for a single car. It is 16′-7/8″ × 22′-7/8″ in size, is often referred to as a garage-and-a-half. A detached garage should be connected to the dwelling by a breezeway, porch or walk. Whenever possible, a turn around area should be part of the driveway, so that a forward approach to the street is possible when driving out.

A garage, if possible, should be so placed on a building lot to protect the dwelling from the summer sun thus helping to keep the house cooler.

This garage is made of assembled wall panels with 2 × 4 studs, 16″ o.c. double nailed to a single bottom plate and lower top plate. The siding is 3/8″ thick cedar with a single batten strip applied to the panels.

The roof trusses are placed 2′ on center and covered with 3/8″ exterior grade plywood roof sheathing. A 15 pound roofing felt is applied over the plywood and a heavy duty asphalt shingle is used.

The large door of the garage is 7′ × 9′ in size, opening overhead. A side passage door, 2′-8″ × 6′-8″ is made of 1-3/4″ Douglas fir. Ample storage space alongside the car is available, extending from the side door to the back of the garage. A window should be placed in the end or

Fig. 4. Attached garage blends with living area, adds length to two-story house.

Fig. 5. A built-in garage (right) and a carport (left) are optional features of sectional houses.

opposite side for ventilation.

Electronic garage door controls and operator are available for garages, and are described elsewhere in this book.

A roof cupola and weathervane are available for additional ventilation and decoration. A gutter and downspout may be used, especially if the garage is near a dwelling. A two car garage also is available with ample storage space. The two car garage is 20′-7/8″ × 24′-7/8″ in size with a single overhead door.

The Garage as an Integral Part of a House

Due to present day integration into the design of the house the garage, attached to a house, is no longer an unsightly side-element or after thought.

Often a garage forms a portion of a secluded entrance patio, **Figure 3.** Here the patio is hidden on two sides by the house and on the third side by a garage built to harmonize with the design of the dwelling. Plants are placed around the perimeter of the patio. Entrance to the patio is from the side of the dwelling and the rear of the garage.

Harmonizing Roof Lines. Often the roof of a porch is carried over an attached garage, **Figure 4,** to lengthen the horizontal lines of a house. In this house the height of the porch roof is the same as the height of the garage roof and the pitch of the garage roof is the same as the roof over the second story. Entrance to the house is gained from the deep garage which harmonizes with the entire building.

An Attached Garage and a Carport. The attached garage, **Figure 5,** right, is a built in feature of this sectional house. The roof extends over the main structure, the front porch and over the garage which is built as an integral part of the building.

At the left, **Figure 5,** is a sectional house with a carport protected by a roof extending from the gable end of the dwelling.

Side and Rear Garage Entrances. When the car entrance and driveway are at the side or rear of the house, the front wall of the garage has a window and trim finished to harmonize with the entire elevation, **Figure 6.** In this structure the garage is at the right hand side and is an integral part of the house. The driveway is on the right and the entrance to the garage is at the end or side.

In the placement of the garage, **Figure 7,** the garage is at the rear of the house. The driveway leads around the right hand side of the house providing a turn around area for the cars and a spacious play or recreational area.

In order to take advantage of a sloping lot a garage may be built at the foundation level and entrance gained

Fig. 6. Garage (right) is disguised by window and dummy gable.

Fig. 7. Driveway runs around right side of house to turn-around area outside rear entrance to garage.

Fig. 8. Rear entry garage on sloping lot has its roof utilized as a large porch at the right of the house.

Fig. 9. Front slope of main roof over lower level is continued over garage to create a pleasing, continuous eave line.

at the rear, **Figure 8.** This type of construction may also be used when a garage is added to an existing house. In either instance, the garage roof may be used as a porch, with railings and posts installed to support an appropriate roof framed over the porch. A field stone wall retains the grading of the soil to provide space for an adequate driveway.

Garage Entrances at the Front. When a garage is built as an integral part of the house, with the entrance at the front, additional care must be taken to have the garage in harmony with the entire dwelling. Since the garage door is the biggest door of the entire house it must blend with the features of the front elevation.

The split level house, **Figure 9,** illustrates the careful blending of colors and exterior wall covering. The garage is faced with brick veneer, as well as the front wall at the porch and a portion of the front wall of the lower level of the house. The roof line extends over half of the front of the house and is met by roof lines at a right angle but of the same pitch, over the two story portion of the house.

Similar gable and exterior wall covering is found in the one floor plan dwelling, **Figure 10,** with the garage at the left and the bedroom at the right.

Harmony and balance are obtained in the front elevation arrangement, **Figure 11,** of this wider house. Both front wings of the structure have the same scotch hip roof. The window space in the garage door does not dominate over the window spaces at the right even though the garage is built for two cars.

Garage Entrances Below Living Areas. When the garage is built under a living space of the house the garage should be finished in much the same manner as another room to prevent the loss of heat in the living areas and to protect these areas from possible fires. **Figure 12** shows two single-car garages placed under half the front elevation area. These garages, as well as the front entrance to the left, must be well insulated to prevent heat loss by radiation.

Garages built below the grade, **Figure 13,** present a problem of heat loss by radiation, and a problem of clearing the driveway of snow and water. Drainage must be adequate both in the driveway and in the garage.

Garage Built Adjacent to the Kitchen. The garage is often built next to the kitchen, **Figure 14.** This is a convenience in carrying supplies from the car to the house. It also provides handy storage for supplies and household cleaning equipment, as well as easy access to the work bench. If a freezer is used it may be kept in the garage. Living space is often saved by placing heating and cooling units in the garage instead of in a closet inside the house proper.

Distribution ducts can be placed in the attic and diffusers installed in the ceilings of the various rooms.

Sometimes a house can be cooled by placing an exhaust fan in the garage ceiling. The main garage door is closed, the door to the kitchen is opened and the fan pulls heat from the house and discharges it through the garage ceiling, into the attic and out the gable and soffit louvers.

The Garage as a Main Room. The floor plan, **Figure 15,** shows an elaborate use of a garage space occupying nearly one fourth of the floor plan of the entire structure. In this instance the garage becomes more than a storage space for the car, and must be finished in the same manner as the other rooms in the house.

Fig. 10. The garage dominates the front of this U-shaped house, while creating a private entry count between the wings.

Fig. 11. Scotch hip roof treatment and large window in wing of living area balance the two-car garage in opposite wing.

Fig. 12. Garages are adjacent to front entrance in this split-level house. Bedroom floors above garage must be insulated.

Fig. 13. When garage is below grade, sloping driveway from street level must have adequate drainage to avoid flooding.

Garage Door Designs. An attached garage often represents up to 40% of the front of the house surface. The design must add to the charm of the dwelling, **Figures 16, 17 and 18.** A garage door, too massive in size and too ornate detracts from the entire good taste of the structure. In **Figure 19,** the garage door is too big, the decoration too massive, and the front wall of the dwelling should be broken by windows.

Garage Door Construction. Garage doors once were hung and operated on a series of T-hinges and swung outward. This type of mounting and operation had serious disadvantages, one of which was keeping the doors open and out of the way while the car was driven in or out on windy days.

To overcome this situation the rigid, upward swinging overhead type of door, **Figure 20,** was devised. One great disadvantage of this door was the inability to keep it weathertight. Today's garage door is factory fabricated, in sections, mounted on a roller track system, **Figure 21,** and hand operated.

A system of springs, **Figure 22,** counterbalances the weight of the door, making for easy handling.

A more modern device is the sectional door having an automatic door operator. The operator is mounted on the garage ceiling, **Figure 23.** A small battery powered transistorized control unit is kept in the glove compartment of the automobile or attached to the visor or dash. It can be transferred from one car to another. The control unit activates the automatic garage door opener by a remote radio signal. The signal opens the door and

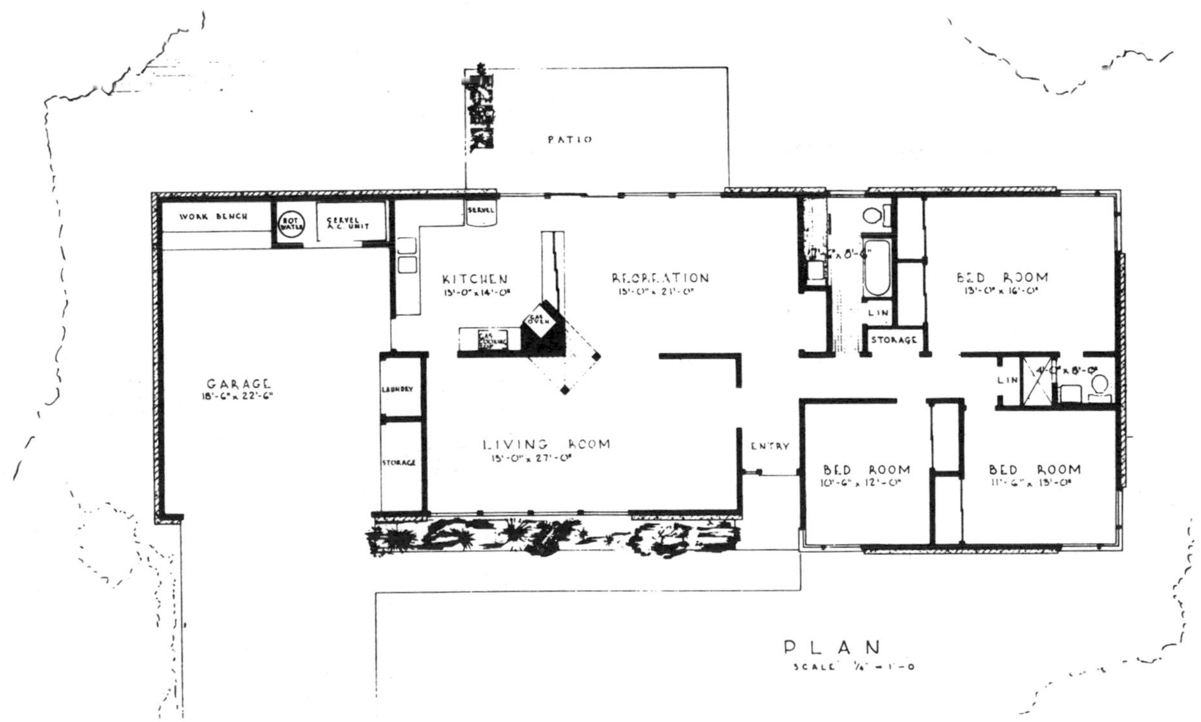

Fig. 14. Extra length garage has workshop and heating-cooling equipment located next to kitchen and side entry door.

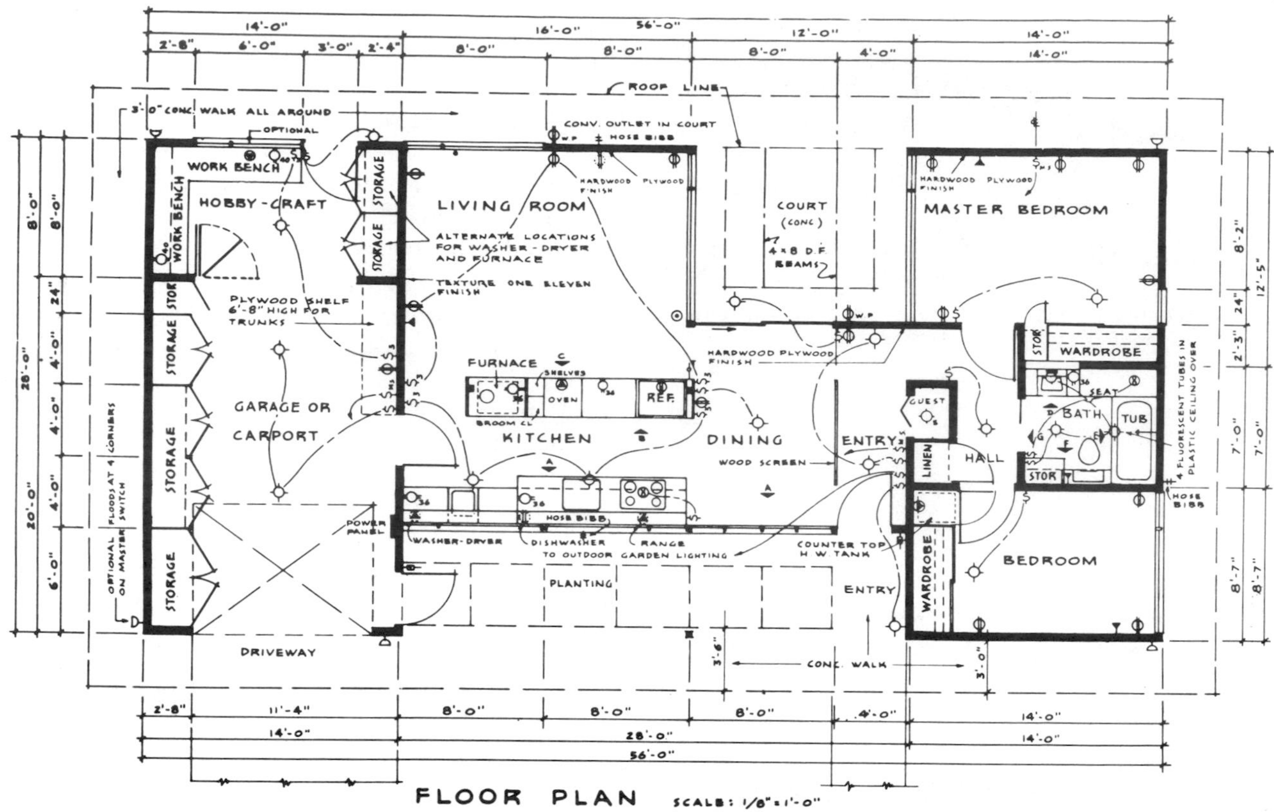

Fig. 15. Garage wing includes generous workshop and ample storage area in this house built without a basement.

turns on a light which stays on until the door is closed automatically. The radio range is 40′ to 60′ from inside the automobile. A friction safety clutch acts as a special safety feature, stopping the door if it should hit an obstruction while operating. See the chapter on doors for more information on these radio signal devices.

Fig. 16. Styling of garage door must blend with style of house.

Prefabricated Garages

Prefabricated garages are available in wood and metal. Wood garages commonly use a truss and modular wall panel system, with windows designed into the panels. **Figures 24 and 25** show a prefabricated garage of steel framework and steel or aluminum sidewalls, door and roof. The garage is placed on a concrete slab. A factory fabricated stressed skin panel system, of wood, **Figure 26,** uses 2″ ×2″ studs and various sizes of panels in constructing a 12′ × 24′ garage.

For the person desirous of framing a garage from mill fabricated stock, the framing details are shown in **Figure 27,** the floor plan and various construction details in **Figure 28,** and a pictorial detail, **Figure 29,** illustrates the complete assembly of framed members. Skill must be used in each step of framing and finishing. The overhead sectional door should be installed by a person skilled in this work.

Fig. 17. Trim applied to garage door in long, horizontal pattern emphasizes low silhouette of house.

Fig. 18. The pleasing design of this garage door blends with the house, even to the shutters on the windows.

Fig. 21. Hinges, lock and track system of sectional door.

Fig. 19. Decoration is overdone on this massive garage door. Windows in door and wall above would relieve blank effect.

Fig. 22. Tension springs counterbalance hand-operated garage door. Track is fastened to joists before finishing ceiling

Fig. 20. Rigid overhead garage door swings out and up.

Fig. 23. Automatic garage door opener opens, closes, locks and unlocks door on signal from radio transmitter in car.

Fig. 24. Steel framework of a prefabricated garage is assembled and anchored to concrete slab.

Fig. 25. Garage frame shown in Figure 24 is closed in with steel or aluminum siding, roof and doors.

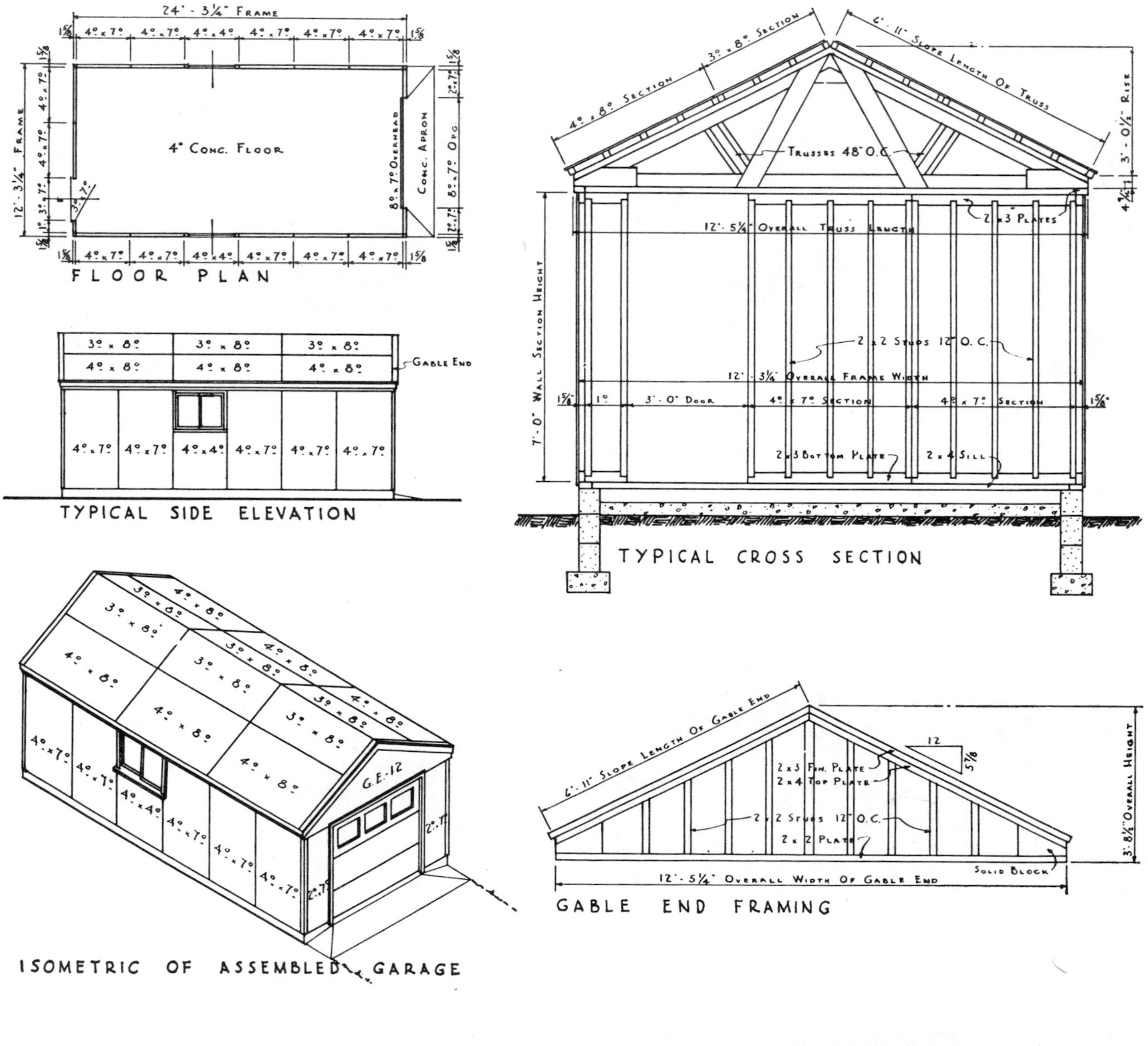

FLOOR PLAN

TYPICAL SIDE ELEVATION

TYPICAL CROSS SECTION

ISOMETRIC OF ASSEMBLED GARAGE

GABLE END FRAMING

Fig. 26. Garage constructed of stressed skin panels assembled to form walls and laid over trusses spaced 4' o.c. Modular panel sizes simplify fabrication of all but gable ends.

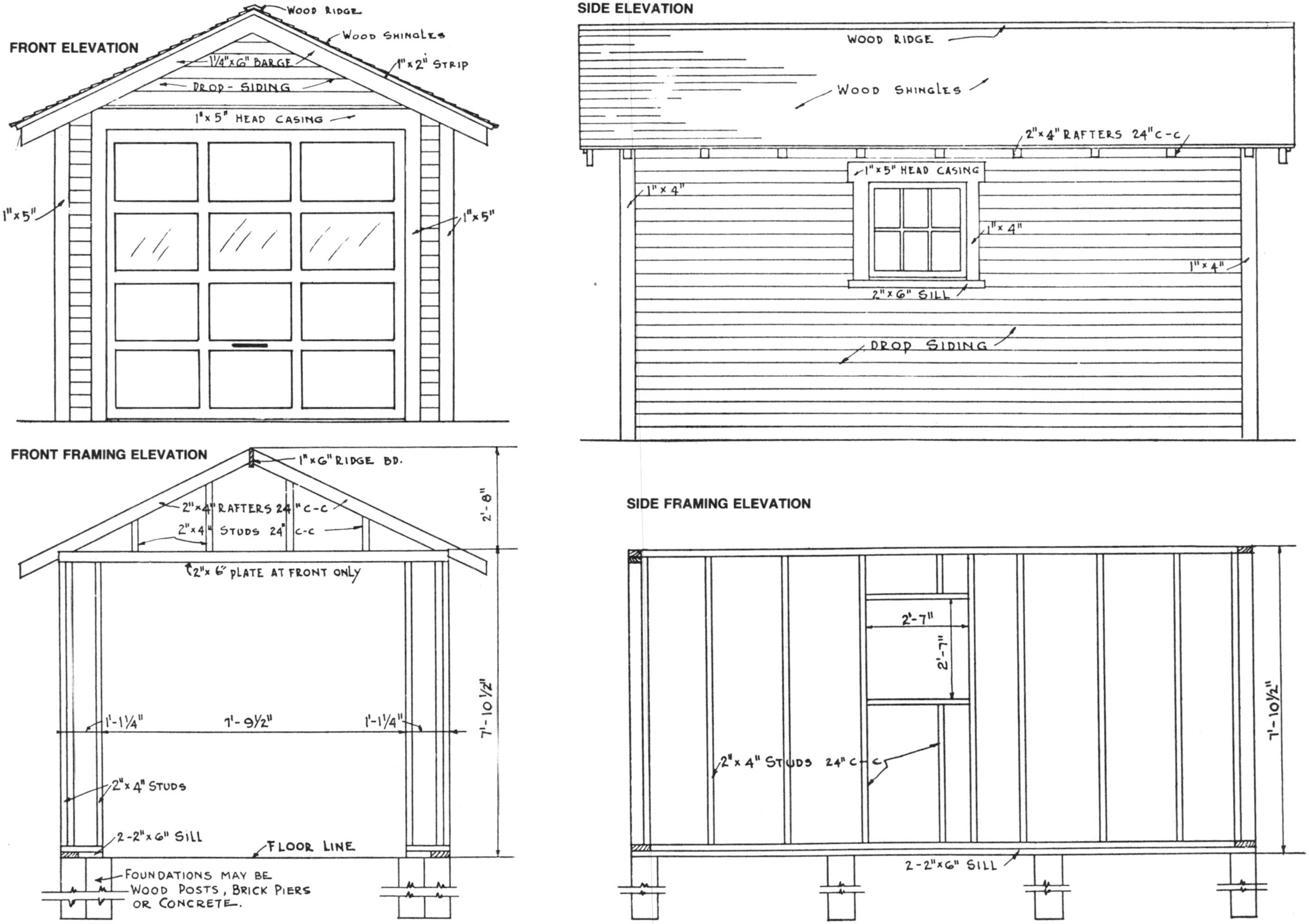

FRONT ELEVATION

WOOD RIDGE
WOOD SHINGLES
1¼"x6" BARGE
1"x2" STRIP
DROP - SIDING
1"x5" HEAD CASING
1"x5"
1"x5"

SIDE ELEVATION

WOOD RIDGE
WOOD SHINGLES
2"x4" RAFTERS 24" C-C
1"x5" HEAD CASING
1"x4"
1"x4"
1"x4"
2"x6" SILL
DROP SIDING

FRONT FRAMING ELEVATION

1"x6" RIDGE BD.
2"x4" RAFTERS 24" C-C
2"x4" STUDS 24" C-C
2'-8"
2"x6" PLATE AT FRONT ONLY
7'-10½"
1'-1¼"
7'-9½"
1'-1¼"
2"x4" STUDS
2-2"x6" SILL
FLOOR LINE
FOUNDATIONS MAY BE WOOD POSTS, BRICK PIERS OR CONCRETE.

SIDE FRAMING ELEVATION

2'-7"
2'-7"
2"x4" STUDS 24" C-C
7'-10½"
2-2"x6" SILL

Fig. 27. Details of a conventionally framed one car garage finished with drop siding and wood roof shingles.

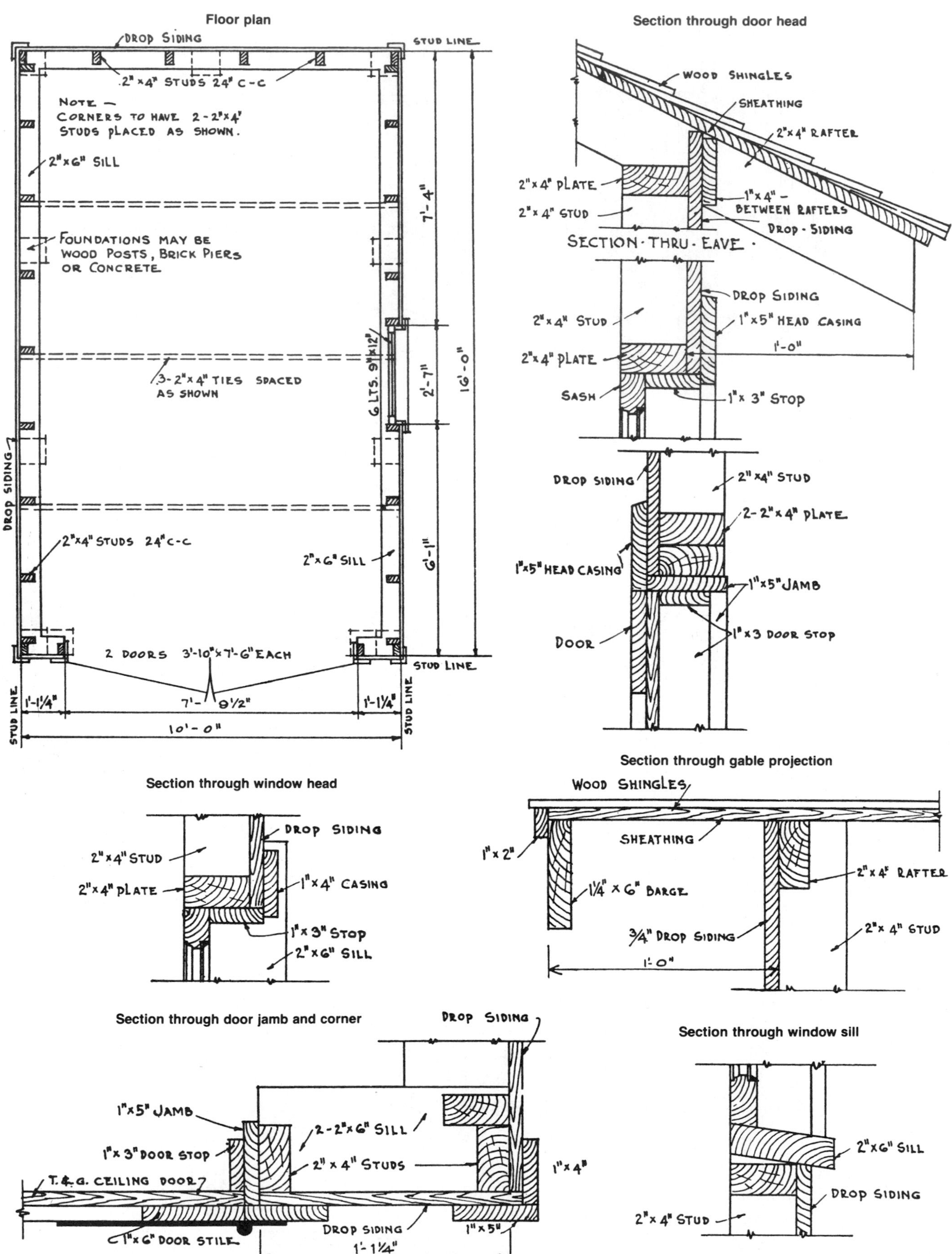

Fig. 28. Floor plan and construction details of wood-framed garage shown in Figure 27.

Garages, Breezeways and Houses

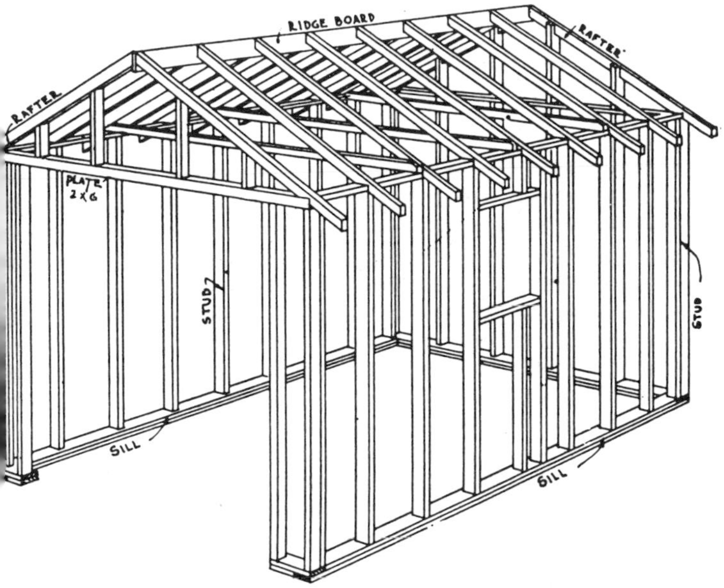

A long roof line is achieved by extending it without differentiation over the garage, breezeway and dwelling, **Figure 30.** In this instance the roofed over passageway or breezeway gives the appearance of another room with the use of jalousie windows and door.

This open area may be paved or treated as a garden or patio with access to both garage and kitchen or family room. The exterior wall covering is often continued over the breezeway. The well designed hip roof covers the entire structure with graceful charm.

A single roof line, **Figure 31,** extends over the breezeway, an enclosed storage patio and a two car garage in a contemporary style prefabricated dwelling.

Fig. 29. Framing of garage detailed in Figures 27 and 28.

Fig. 30. Breezeway between house and garage becomes an extra room when enclosed by continuous hip roof, windows and door.

Fig. 31. Unbroken roof line extends over breezeway, storage room and garage in this long, low contemporary model.

Fig. 32. A drive-through carport is a convenience when the building site is large enough for a circular driveway.

Carports

Carports are widely used as substitutes for garages since they provide economical shelter. At one time they were considered a makeshift covering with a graveled floor sufficient to provide drainage. Today the carport is designed as part of the house, its lines contributing to a central design, **Figure 32,** either as a simple extension of the house and roof,or a more subtly planned arrangement of horizontal trim.

More elaborate carpoŕts are easily converted into garages, especially when storage walls form one side of the carport. Storage has become an important function of both carport and garage. Not merely toys, bicycles, garden tools and car equipment, but overflow from inside storage finds its way into outside cabinets. As a result the carport storage is more than a redwood bin; it is now finished in design similar to that of the house, and sturdily built and supplied with locks, **Figures 33 and 34.**

Carports and breezeways are also designed to serve as a protected play space for small children, **Figure 35,** sometimes with a gate across the front.

A carport may be built with a protective roof supported by wrought iron supports fixed in a planter of brick matching the chimney stack and masonry veneer on a portion of the front of the house, **Figure 36.**

Fig. 33. Large storage space is built into rear of carport, adding feeling of solidity and continuity to open shelter.

Fig. 34. Carport storage wall appears to be a continuation of vertical siding on house.

Fig. 35. Breezeway with gate at rear serves as supervised play area for small children. Mother can watch from kitchen.

Fig. 36. Simple carport, framed around chimney on gable end and supported by low brick planter, is in harmony with house.

Chapter 20

COMFORT CONDITIONING

Comfort, as a year-round condition in a dwelling, is a more descriptive term than heating and cooling to describe the effect of controlling indoor environment. Comfort not only involves temperature control but the control of air contaminants, the movement of air and air pressure. Total comfort requires heating and cooling equipment; humidification and dehumidification equipment; air cleaning equipment; insulation, shading and solar orientation of a house.

Temperature Control

Solar Orientation. Any dwelling built in the northern sections of continental United States, should be designed, built and placed on the site with respect to the rays of the sun. Such planning would put the house at an angle, on the building site, so that large windows and main rooms would face the south to take advantage of the warming winter rays of sun which strike the earth at a low angle. The rays of the hot summer sun, being higher in the sky, can be kept out of rooms by wide, overhanging roofs; by the use of trellises and vines; and, by properly placed trees around the house.

In colder climates, large window areas should be installed with double panes of glass, otherwise gains in heat from sunlight will be offset by heat loss during the night.

Figure 1, illustrates the placement of a dwelling on a building site with reference to the position of the sun at noon, during the winter and during the summer, for a latitude of forty degrees North. This latitude is about midway through the state of New Jersey and westward through the southern portion of Pennsylvania, the cen-

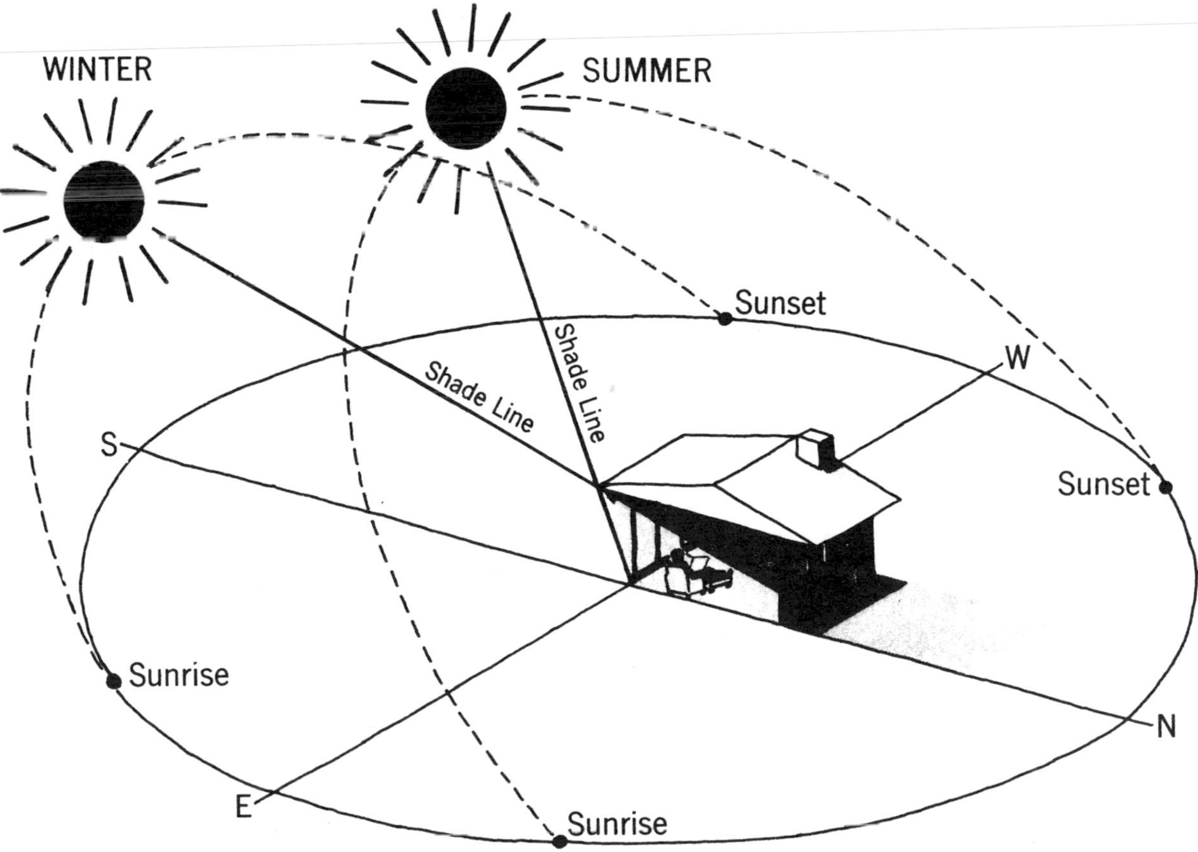

Fig. 1. Diagram showing dwelling oriented with reference to the position of the sun at noon for a lattitude of 40 degrees.

tral portions of Ohio, Indiana and Illinois; and then, through the northern portions of Missouri, Kansas, Colorado, Utah, Nevada and California.

Solar orientation may result in periods of overheating and periods of underheating. If the thermostat of an automatically controlled heating plant is in one of the rooms receiving solar radiation, the adjoining rooms may be too cold. If the thermostat is placed elsewhere, the solar heated rooms may be too hot, hence fuel will not be saved unless radiators or heating outlets are turned off during the solar heating period.

Insulation. The temperature in a house is also controlled by insulation. Comfort in a dwelling can be greatly enhanced by good, tight construction and superior skill in building. A dwelling must be insulated against the rapid transfer of heat from within, during the heating season; and a transfer of heat, from without, during the hot summer days, by the use of thermal (heat) insulation in ceilings, walls, and floors.

Thermal insulation materials are made of mineral wool, cellulose fiber, mineral pellets, aluminum foil or fiberglass. These materials are available in blanket, batt, poured, blown, foil or rigid form.

Further additional insulation in a dwelling can be accomplished by the use of weatherstripping around windows and doors and the installation of storm sash and doors at the proper places. If an older house is reconditioned it is good economy to insulate extensively, if needed, before a new heating system is installed.

Fig. 2. Old fashioned paddle-blade ceiling fan has been made popular again by antique enthusiasts, is available commercially.

Fig. 3. Fan for wall or ceiling vents heat and moisture.

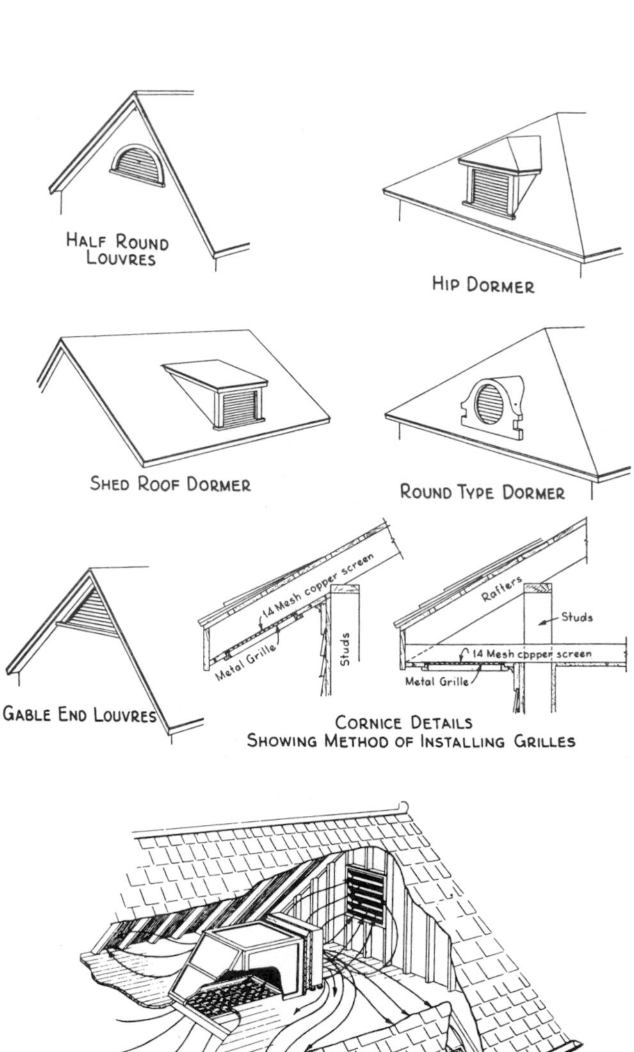

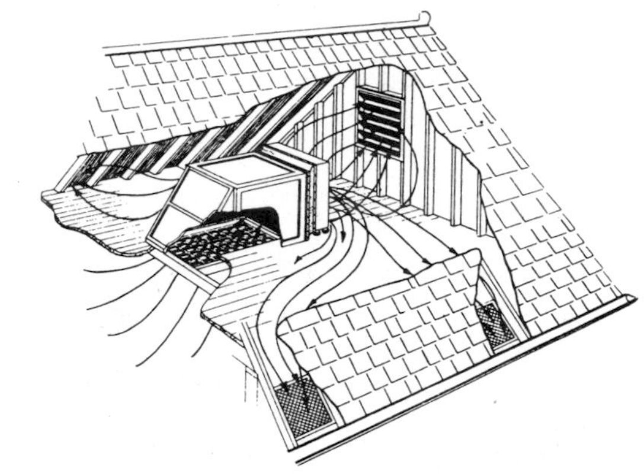

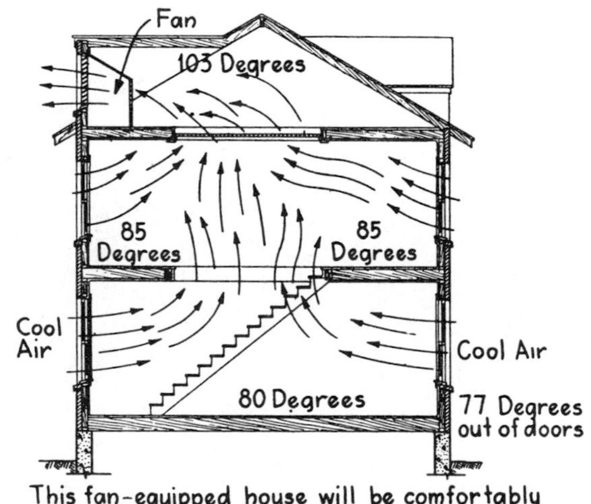

Fig. 4. Common methods of ventilating house through attic using attic fan on floor of attic in suction box.

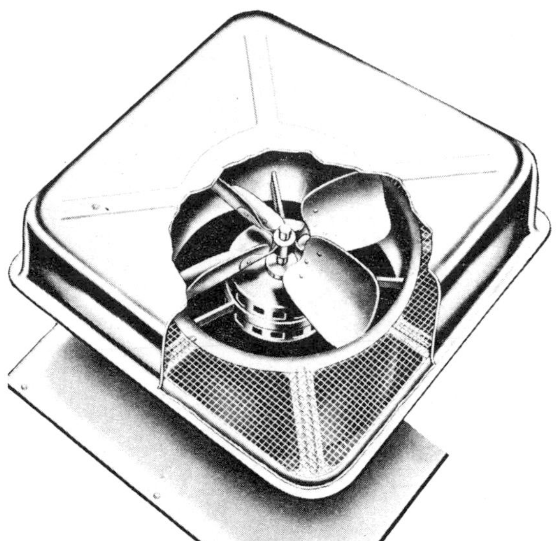

Fig. 5. Roof fan is installed in opening between rafters.

Fig. 6. The old Franklin space heater, still available today.

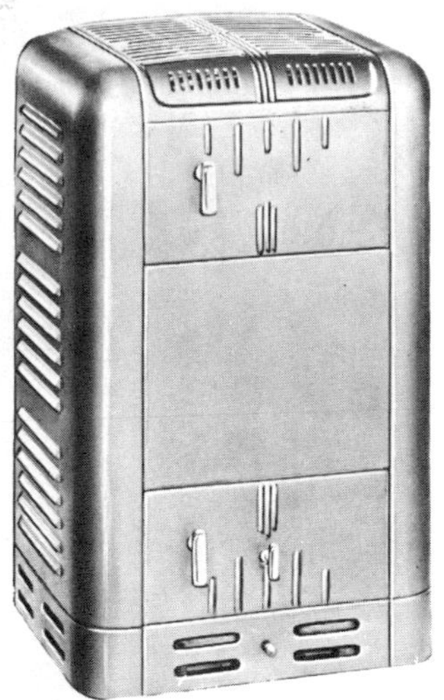

Fig. 7. Portable space heater burns gas, fuel oil or kerosene.

A light-colored roof covering will reflect the heat and keep the building cooler in summer. A dry, well-built basement, under all or part of a dwelling, will tend to make the house cooler during the warmer months and somewhat warmer during the colder months.

Detailed information and many illustrations of insulation are shown in the chapter devoted to the topic of Insulation.

Shading. The temperature of rooms is also controlled at windows and doors by window shades that can be lowered to cut off the entrance of heat from the direct rays of the sun. Where large windows or sliding doors are installed draw drapes can be hung to provide comfort and protect against the entrance of the sun's rays.

Venetian blinds are also used at windows and doors. Shutters are sometimes used to protect windows and doors. Awnings are attached to the exterior of a house to keep out the sun's rays. Porch roofs and tall trees shade windows and doors.

Glass, other than single or double strength, such as actinic, decorated, figured, florentine, frosted, mat surface, opalescent, stained, vita or blocks are used in windows to control the temperature of a room.

Sometimes shades, blinds and shutters are drawn at windows during very cold spells to keep out the cold and keep in the heat of a room to control temperature.

An attached garage, if properly placed and constructed as an integral part of the house does much to control temperatures in rooms and the pressure of winds during the colder weeks of a year.

Temperature Control by Circulating Air

Fans. Room temperature can be controlled by the forced circulation of air by fans. A portable four blade, electric fan may be used, but antique enthusiasts have revived the popularity of the old fashioned paddle blade ceiling fan, **Figure 2.** It is especially nice on porches or patios to keep air in circulation and to blow insects away.

Another type of ceiling or wall fan, **Figure 3,** circulates air and may be installed directly to rafters or studding. This fan has a protective grille and a 3″ exhaust duct which conducts the warm air directly to the outside.

A Roof Ventilator. Various types of louvers and vents for circulating air in an attic with or without a fan are built into a house, **Figure 4.**

The temperature, in an attic, often reaches 150 degrees and may be substantially reduced by circulating air by a power roof ventilator or attic fan, **Figure 5.** The fan may be mounted in a pitched, vertical or horizontal position and fits between 16″ oc studs or rafters. The motor is set to start at 100 degrees and shut off at 85 degrees temperature. This type of air circulation is recommended in air conditioned houses because reduced attic temperatures lighten air conditioning work load as much as 30%.

Temperature Control by Heating

Comfort conditioning involves not alone solar orientation, insulation, shading and air circulation by fans but the heating of the air in a house where outside temperatures drop below 72 degrees. Temperature control by heating involves the burning of fuels or electricity to create heat by several devices.

The Space Heater. Portable or installed space heaters are numerous using wood, coal, gas, oil or electricity. One of the first space heaters, and still popular, is the

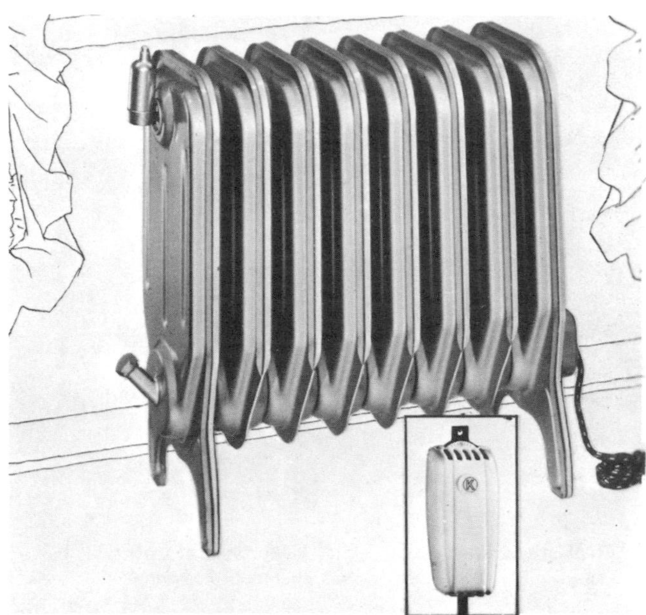

Fig. 8. Portable space heater uses electricity as fuel to heat water contained in the radiator.

Fig. 9. Heavy duty portable electric space heater has a blower mounted behind rows of resistance heating coils.

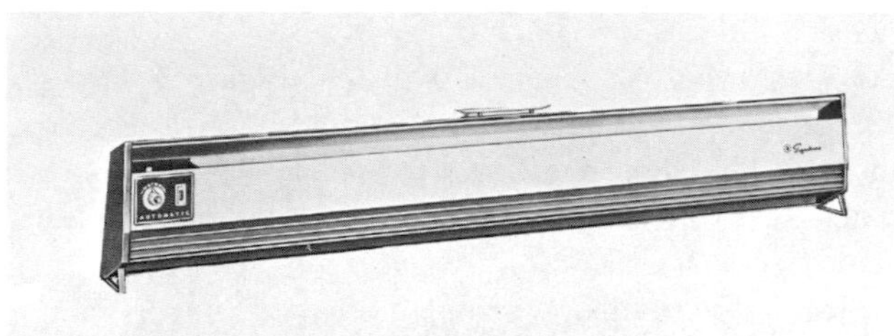

Fig. 10. Portable radiant electric heater can be placed handily along the edge of a room at the baseboard.

Fig. 11. A console style portable electric heater has blower fan enclosed.

fireplace. A chapter of this book is devoted to chimney and fireplace construction.

The Franklin Heater. One of the most widely used space heaters, running a close second to the open fireplace, is the Franklin space heater, **Figure 6.** Portable heaters of this type often are the sole source of heat in vacation homes. However, unvented heaters of this type must be used in room where a window is kept partly open to prevent burning up of oxygen and to allow escape of gases formed by combustion.

Space Heater With an Enameled Jacket. A demountable space heater, **Figure 7,** using gas, fuel oil or kerosene as fuel is sometimes used as a room heater, especially in areas not requiring constant heat over a long period of time. Like the Frankin heater, unvented heaters of this type must be used in rooms with partly open windows.

Radiator Type of Portable Space Heater. A portable space heater using electricity to heat water is shown in **Figure 8.** The water intake is at the lower left and the heating element attachment is at the lower right.

Portable Electric Heaters. Space heaters of different styles using electricity are shown in **Figures 9, 10 and 11.** Some can be attached to the 110 volt household electrical current supply, others consume higher wattage and require special wiring. Some have fans enclosed to distribute the heat gently instead of by radiation alone.

Electric Floor Insert Heater. Where space or other limitations make it impractical to install other types of heaters this space heater may be inserted into the floor adjacent to sliding doors, near large picture windows, stairwells or extra large rooms, **Figure 12.** A cut off switch shuts off the power if air circulation is blocked for any reason. An automatic disconnect switch facilitates

Fig. 12. Electric heater may be inserted in the floor beneath sliding glass doors or picture windows.

Fig. 14. Infra-red electric heater, installed in ceiling of bathroom, beams its warmth down over a wide area.

Fig. 13. Electric wall heater can be built into bathroom wall as supplementary heating. It has its own thermostat.

Fig. 15. Realistic ceramic logs for fireplace are heated by gas flame to beam warmth out into room.

easy cleaning of the heater.

Bathroom Heaters. Electric heaters are ideal for bathrooms where heat is not required in other rooms in the home. The insert or built-in wall heater, **Figure 13** provides a flow of warmth through an effective down flow principle for more uniform temperatures throughout the room area. An infra-red heater, **Figure 14** provides an instant sun like warmth on one where you want it, surrounding one with a blanket of warmth. The air is not heated, people or objects absorb the warming rays. It is easily installed on the ceiling or wall like an ordinary fluorescent fixture, or it can be built in.

Realistic Gas Logs. A space heater burning natural or LP gas, to be inserted into a fireplace is available for heating a room. One type of gas log heater has removable ceramic logs, **Figure 15,** having a high or low setting, with an automatic pilot and a safety shut off valve.

Another gas log heater has radiant 'Fiberfrax' logs, with high or low setting, an automatic pilot and a safety shut off valve. The heater is capable of developing 45,000 BTU of heat.

Room Temperature Control by Cooling

Fig. 16. Room air conditioner designed for installation in a horizontal sliding window.

Temperature control not alone involves heating the air of a house or separate rooms by space heaters but cooling the air by space coolers or room air conditioners. Room air conditioners cool one or two rooms. They are operated by electricity, most generally on the 110 volt household circuit.

The Window Air Conditioner. The window model may be installed in double hung windows, casement, or sliding windows, **Figure 16.** The unit for the double hung window may be removed during the winter weeks while the unit for the casement window necessitates the removal of a pane of glass for permanent installation.

Through the Wall Air Conditioner. Through the wall models are furnished with a wall sleeve which is permanently installed in an exterior wall. The wall sleeve can be installed flush with the outside wall and covered with a decorative outside grille, **Figure 17.**

Temperature Control by Heating All Rooms

Space heaters must be regarded as supplementary heaters to a central heating system in dwellings located in areas where an efficient heating plant is required when

Fig. 17. Through-the-wall room air conditioner (top) has separate chassis (center) and outside grille (bottom).

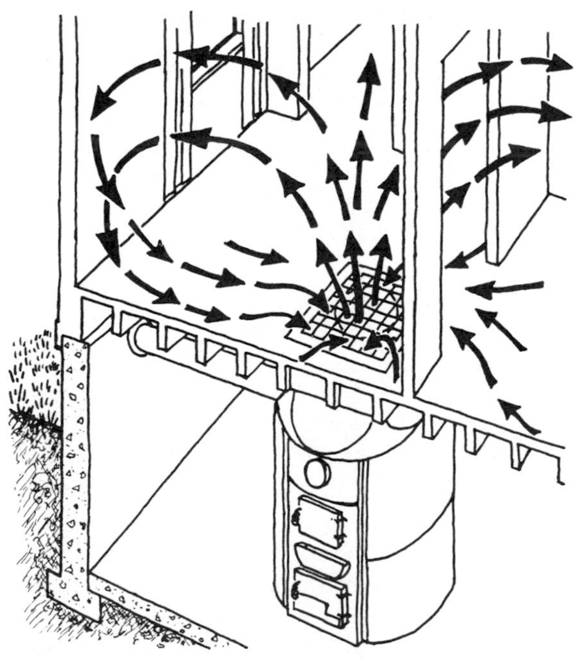

Fig. 18. Large pipeless furnace, located in full basement of house or cottage, provides central heating by circulation.

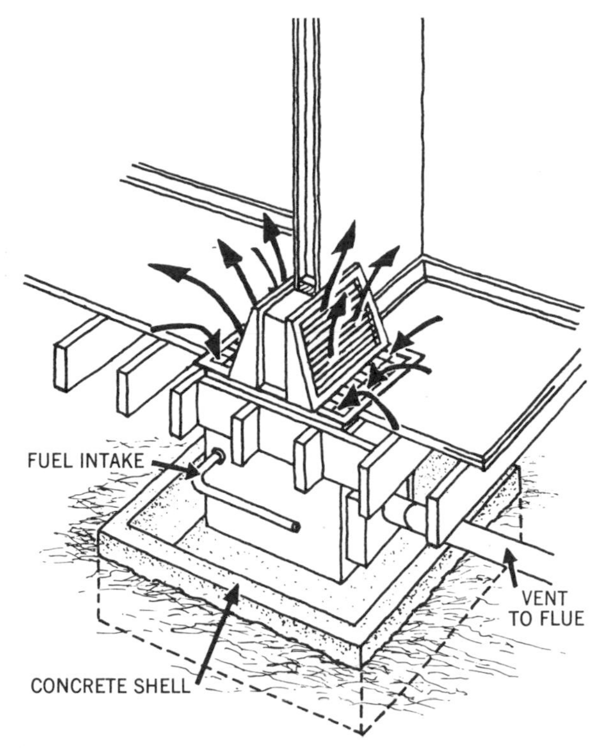

Fig. 19. Small pipeless furnace, located in the crawl space of a cottage or summer home, blows heat in two directions.

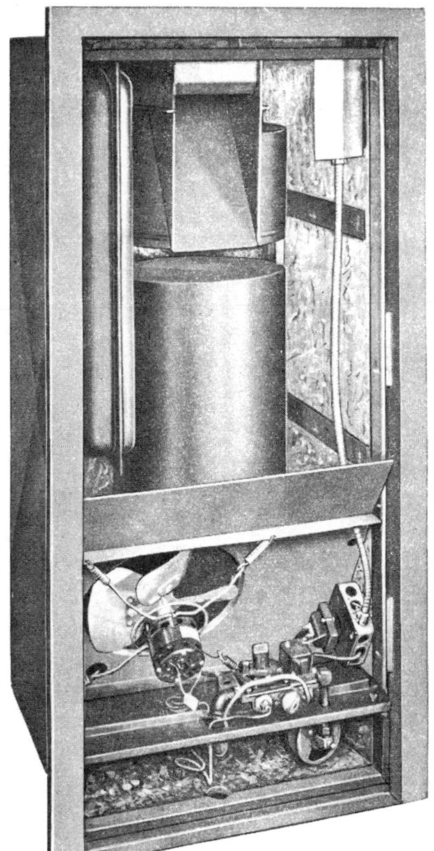

Fig. 20. Compact, vented gas furnace designed for installation in a wall.

outside temperatures become uncomfortably cool or cold.

Space heaters are needed during winter weeks in most any home as a supplement to the central heating system for the bath, the senior citizen, the ill or an infant who might need a particular room temperature 'just a bit warmer' at times.

One other 'space heater' or heating appliance is the electric blanket now in use in many bedrooms. The use of an electric blanket permits lower house temperatures at night.

The Central Heating System. Heating the entire house during the heating season is now accomplished by a central heating system located in convenient places, such as the attic, utility room, the garage, basement or crawl space. 'Central' heating systems are also located in the walls around the baseboards, in ceilings and in concrete floors. This type of heating may be controlled in individual rooms as well as a central control of the entire system.

The Pipeless Furnace. One of the oldest style heating systems, now obsolete, is the gas, oil or coal burning pipeless furnace placed in a full basement, **Figure 18,** or a small pipeless furnace placed in a crawl space under the house, **Figure 19.** The heated air rises at the center of the grille or register circulated through the house and the cooler air returns at the sides of the same register. One disadvantage of this type of furnace is that the room where the register is located is always the warmest and the rooms farthermost from the register are the coldest. Another disadvantage is that doors

Fig. 21 Wall furance shown in Figure 20 with louvered cover in place. Vertical louvers blow warm air to right and left.

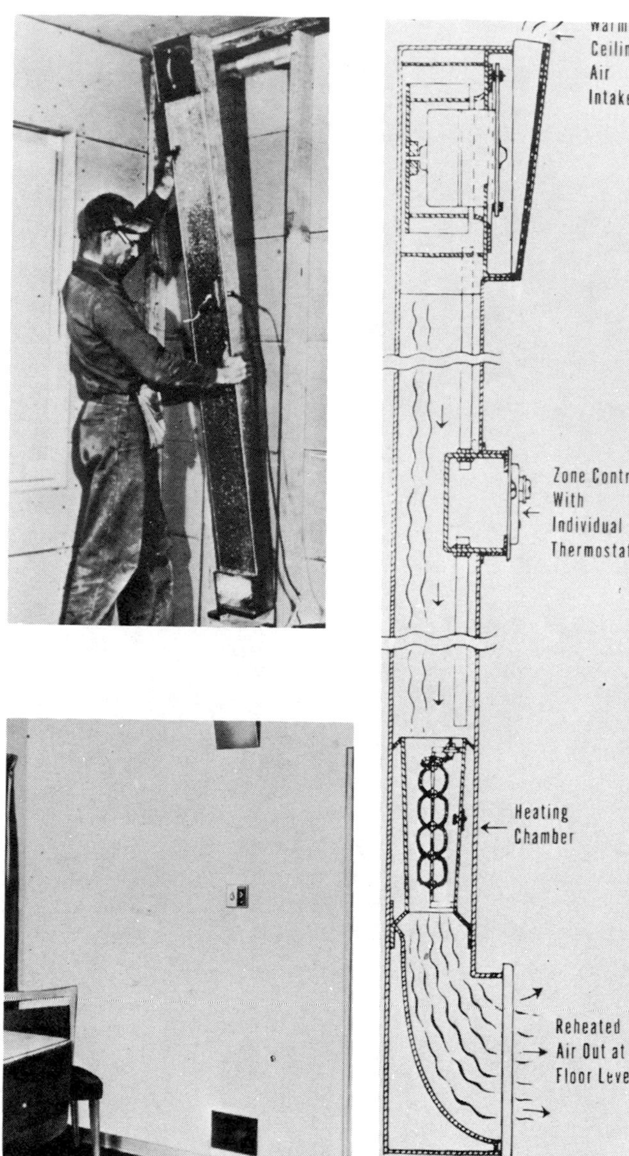

Fig. 22. Electric heater is installed between studs (top). Only intake at ceiling and diffuser at floor are visible; thermostat is at mid-height. Diagram at right illustrates counterflow principle of air distribution, with air pulled in by fan at top, passed through heating chamber to floor level.

must be kept open for complete circulation of air. This system of heating is knows as a warm air-ductless method of temperature control.

The Gas Wall Furnace. A wall furnace is available for conventional or prefabricated houses furnished in a pre-assembled unit. This furnace uses gas as fuel. There are outlets for the heated, forced air, at the front, back or sides of the cabinet, **Figure 20.** Some wall furnaces have sealed combustion chambers which draw air from outside and discharge exhaust gases to the outside.

The furnace is mounted flush with a wall near the center of the house, **Figure 21.** The cabinet is heavily insulated requiring a single vertical flue outlet.

Comfort Control by Electrical Heat

Almost immediate warmth is attained by radiation or electric resistance heating.

Since there is no air movement in most electrically heated rooms, a fan is usually necessary to prevent stratification of smoke or foul air, and there may be need to keep a watch on humidity and to admit dryer outside air in the winter weeks.

Sometimes electric heating units are placed between studding, **Figure 22.** Upper left photograph shows how the unit fits between the studding; lower left shows the finished appearance of the heater; the cut away diagram, right, shows the direction of the air, drawn through the intake at the top of the panel, moving past the heating chamber and out into the room at the floor level.

Temperature Control by Hydronic Heating

Heating with steam or forced hot water has the advantage of freedom from drafts, except those caused by chilled, uninsulated outside walls, which set up a convection current as air, cooled by the wall, drops to the floor and moves into the room, forcing the warmer air there into motion. But since heating is accomplished by radiation and gentle convection, without puffs of warming air, drafts are not found in a well insulated house. Another advantage is a steady heat for a time after the fuel burning has ceased, since radiation units and fluid in them give up their heat slowly.

Hot water distribution is either by gravity or forced circulation. An expansion tank is necessary to make allowance for the different volume of water at different temperatures. In gravity systems, seldom used in modern construction, water circulation is caused by the difference in weight between the lighter heated water in the supply lines and the heavier cooler water in the return lines. Forced circulation systems use a circulator and smaller pipe size, and have more rapid and even heat distribution.

A Modern Electro Hydronic System. This is a combination of an electrically heated boiler and hydronic baseboard panels. In the cut-away views, **Figure 23,** are shown the features of the unit which heats water, generating 81,912 B.T.U. per hour, sufficient to warm eight rooms in a dwelling.

The heated water passes through baseboard units and tubing, **Figure 24.** The floor plan, **Figure 25,** shows a piping layout with arrows indicating the direction of flow of heated water.

New compact, cabinet type electro-hydronic boilers

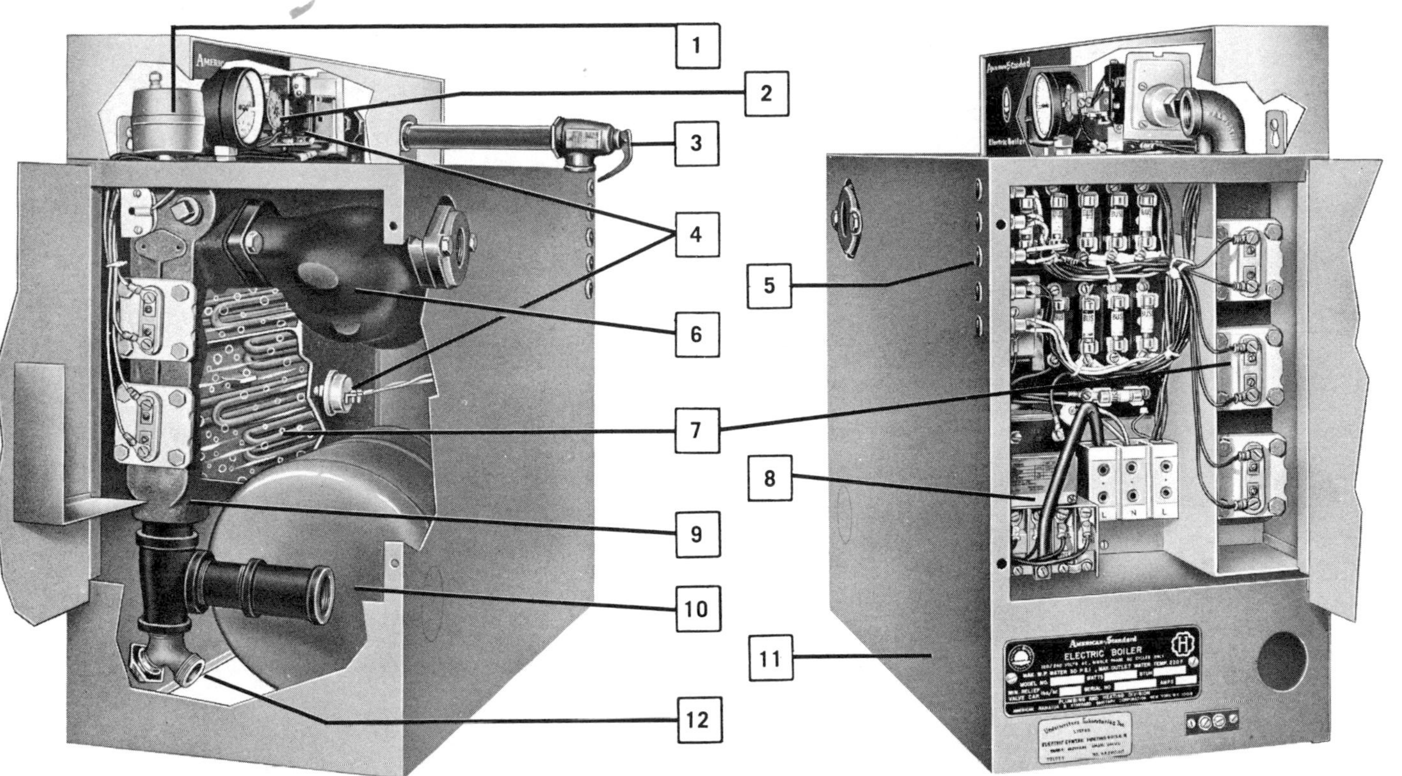

Fig. 23. A modern electro-hydronic cast-iron boiler, showing:
(1) Automatic air vent—bleeds trapped air from water.
(2) Adjustable limit control—sets maximum boiler temperature.
(3) Water pressure relief valve.
(4) Preset pressure controls—guarantee safe operation.
(5) Red indicator lights—show operation of heating elements.
(6) Circulator pump—wired for continuous operation.
(7) Electric heating elements—mounted inside boiler, are low density for quiet operation.
(8) Sequence relay switch—provides incremental loading of electric service line to reduce power surge.
(9) Cast-iron boiler shell—baffled for quiet operation.
(10) Factory-installed expansion tank.
(11) Metal housing—has access doors to controls and elements.
(12) Drain valve—has flat wheel handle and hose fitting.
Boiler of this type generates 81,912 BTU per hour, sufficient to heat an eight room house equipped with hydronic baseboard units or radiant coils in floors or ceilings.

can be mounted on the wall of a workshop or recreation room to heat an entire house, **Figure 26.**

Hydronic Heating by Radiation and Convection. Radiant heat is given off by any warm surface. Convection heat is the result of cool air passing over a warm object and being warmed. Radiant and convection heat are combined in a baseboard type of hydronic heating, **Figure 27.**

The floor plans, **Figure 28,** show the the piping layout for a one pipe system and a series loop one pipe system.

Hydronic heating systems of all types, the one pipe, two pipe and the loop systems are designed for a temperature drop through the system of 20 degrees. This is the temperature difference measured at the point where the water leaves the boiler and the point where the water returns to the boiler.

One-pipe hot water heating systems are arranged so that the heated water flows through each radiator in the circuit successively. Theoretically, then, each successive radiator must have a proportionally larger heating surface to offset the drop in water temperature. Two-pipe hot water heating systems have heated water flowing in a complete circuit to each radiator and back to the boiler through a return pipe line.

In the loop system all of the water in the main passes through each baseboard assembly. Because of this large flow rate, the temperature drop in the radiators in a baseboard loop system is much less than with a one pipe system. Installation is economical, as baseboard sections

Fig. 24. Hot water from hydronic boiler at left is piped through baseboard units such as this one being installed along the wall of a room.

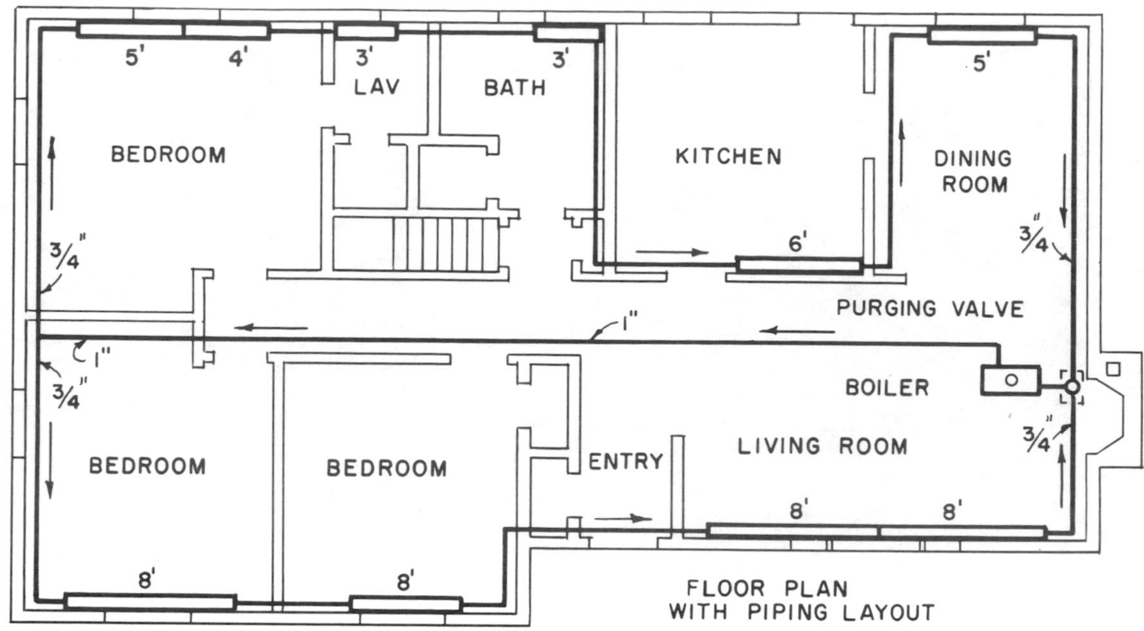

Fig. 25. Layout of hydronic heating system shows boiler, piping, flow of water, and lengths of baseboard for different room sizes.

double as part of the main supply piping.

Figure 29 shows a compact hydronic heater which can be placed in the basement or small utility room. The heater burns fuel oil. The heater may be furnished with an overall deluxe jacket.

A high capacity tankless water heater with double coils is available for installation in the rear section of the boiler to produce an abundant supply of hot faucet water, **Figure 30.**

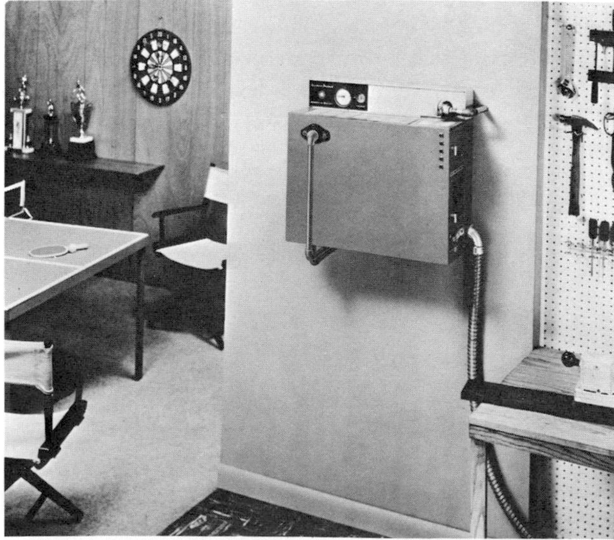
Fig. 26. A small, wall-hung electric boiler has capacity for heating an entire average size home hydronically.

Forced Air Gas Heating Comfort

A wall mounted, multi room gas baseboard heating system is available as shown in **Figure 31.** The insert shows a sectional view of the method used in bringing fresh outside air to support combustion while a vent is used to discharge the fumes to the outside.

A small opening is cut into the outside wall to admit the horizontal direct vent pipe. No chimney or vertical flue is needed.

Baseboard extensions or warm air channels are placed around the room as shown. A four speed blower, housed in the cabinet, diffuses warm air at a selected speed and comfort control. Washable filters, located in the cabinet, screen out dust and dirt from the combustion chamber.

This system is adaptable to any heating situation; one or two extensions (warm air channels) can be used in varying lengths with corner angle accessories permitting peripheral heating of rooms. Open and closed baseboard channel extensions permit full installation flexibility for room heating and for direct warm air to adjoining rooms without heat loss.

This heating system may use gas from the city mains or L-P gas consumed in suburban or rural sections. L-P (liquified petroleum) gas is extracted from so called "wet" natural gas where it occurs as a vapor-like moisture. L-P gas, when used, changes from a liquid to a gas becoming a source of energy for heat and power. L-P gas is available in tanks in small or large containers.

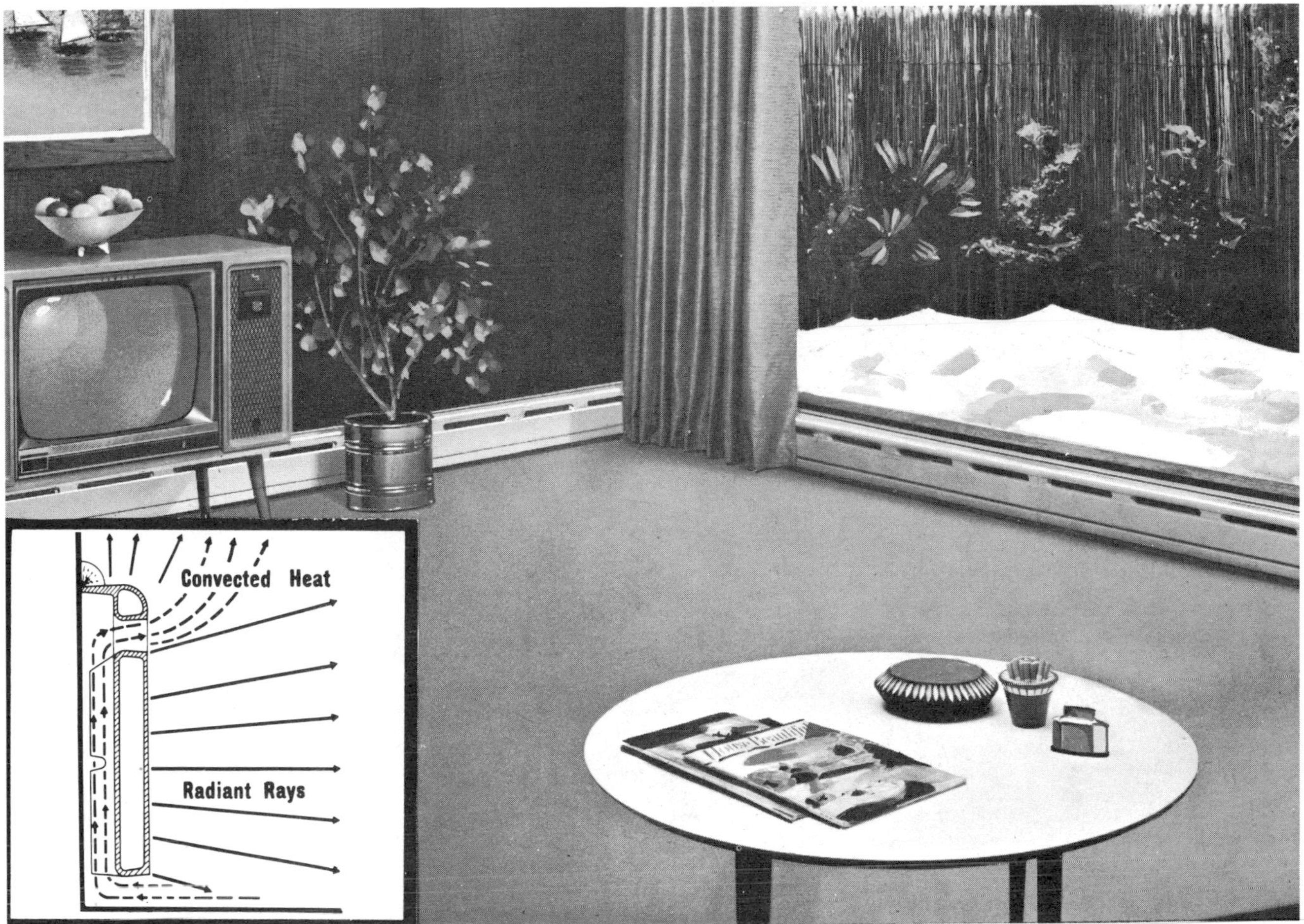

Fig. 27. Radiant-convection baseboard installed along wall and beneath large picture window. Air is heated by convection as it flows under the fixture and emerges through slots at the top (insert dwg.). Radiant heat flows horizontally into room.

The Gas Fired Warm Air Furnace

The gas fired furnace is a unit which can be used in a full basement where overhead space permits plenum and duct work clearance, **Figures 32 and 33.** Ducts beneath the floor and between joists distribute warm air to registers in all the rooms, **Figure 34.**

The gas fired furnace can also be used in a closet or utility room with the duct work overhead, **Figure 35.**

The gas fired furnace can be used in a house without a basement. The duct work is placed in crawl spaces under the floor or imbedded in a slab foundation, **Figure 36.** The furnace is located in a utility room or closet.

The gas fired furnace can also be installed in an attic, ceiling or hallway with adequate headroom or in a crawl space. This installation is ideal for cramped quarters, **Figure 37.**

The gas fired furnace is the basic unit around which all related comfort equipment is installed, such as the air conditioner, electronic air cleaner and the power humidifier.

The gas fired furnace may share space in a full basement with a water heater and incinerator using one common chimney stack, **Figure 38.**

An oil burning furnace is much like a gas fired furnace, except fuel oil is used instead of gas.

Temperature Control by Heating and Cooling

Controlling indoor environment in all seasons is made possible by the use of a combined heating and cooling plant. Illustrated in **Figure 39** is an upflow, gas burning furnace, placed in a full basement, with a heating capacity of 34,000 BTU with a slant type cooling coil of 1-1/2 ton capacity. A compact condensing unit located outside the basement.

The control center is designed to bring both heating (during cool and cold weeks) and cooling (during warm weeks) to all rooms in a house.

The thermostat may be set for a continuous supply of warm or cool dehumidified air at variable blower speeds.

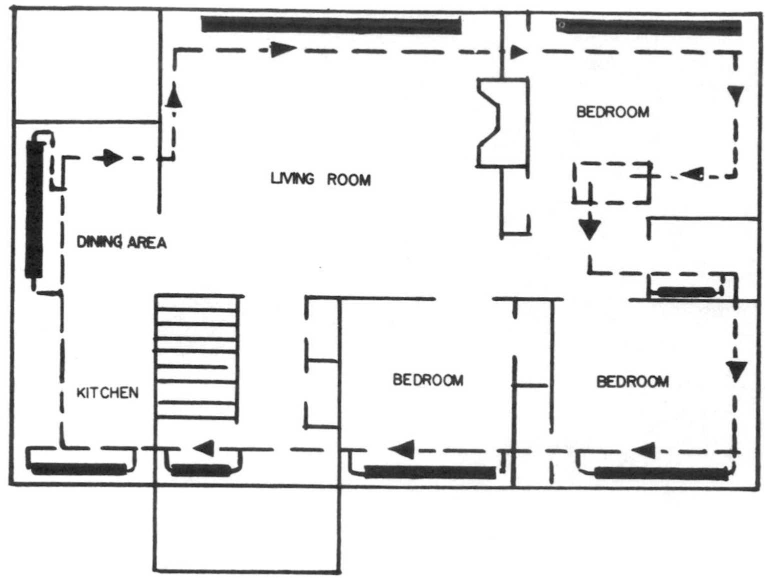

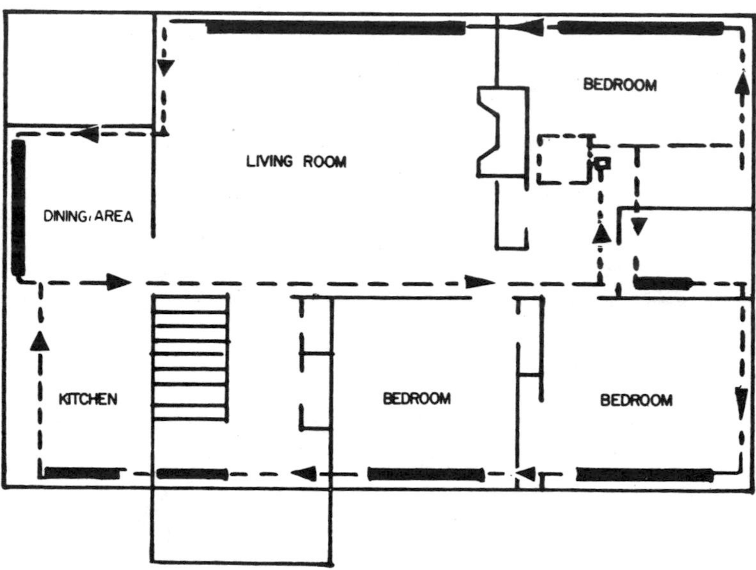

Fig. 28. Floor plan layout of a one-pipe system with stub connectors to baseboard units (above), and a series loop system in which the baseboard also serves as part of the distribution loop for greater economy of installation.

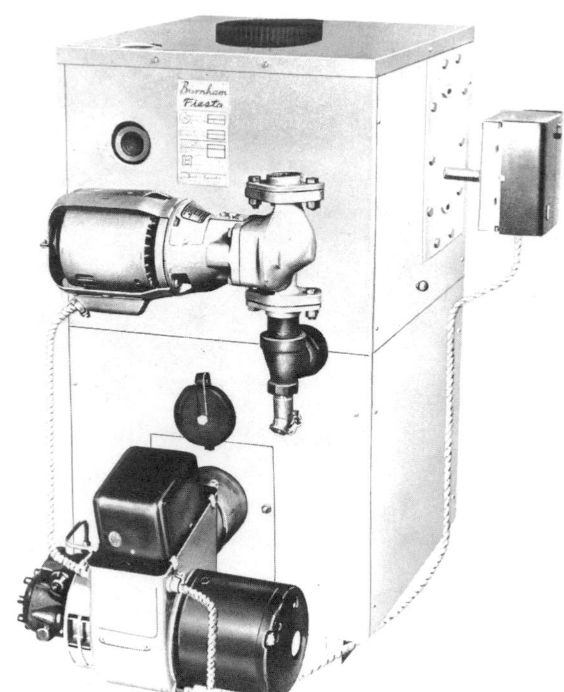

Fig. 29. Oil-burning boiler for a hydronic heating system is packaged at factory, arrives at job ready to connect to system.

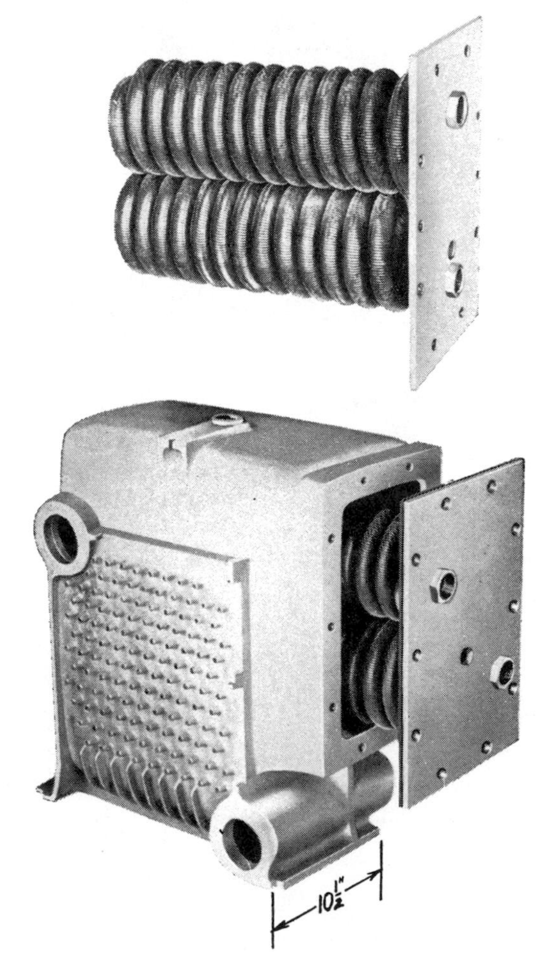

Fig. 30. Heat exchanger coils (above) for domestic hot water supply are installed inside hydronic heating boiler (below).

These plants are also available in the horizontal furnace for burning natural, manufactured, mixed or L-P gases with a cooling fan relay, **Figure 40.** An oil fired furnace is also available in an upright model using a gun type oil burner, **Figure 41.** Electric and gas furnaces are made in upflow, counterflow or horizontal models, equipped with cooling coils for air conditioning, **Figure 42.** When heating and cooling are combined in a single system, ducts must be sized for distribution of cool air. Ducts sized for heating will not provide efficient cooling.

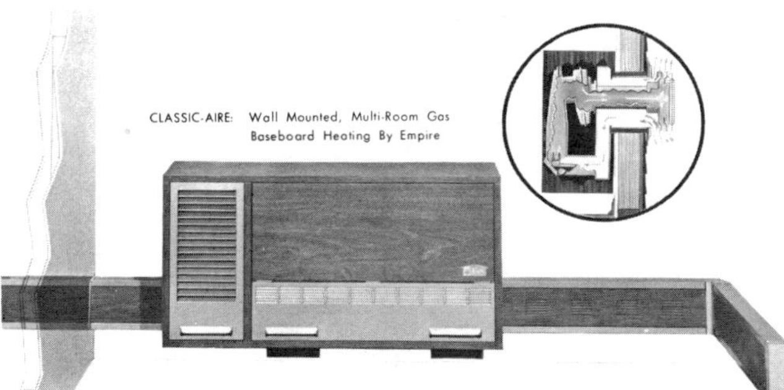

Fig. 31. Wall mounted, multi-room gas baseboard heating system has built-in air inlet and vent through wall to outside.

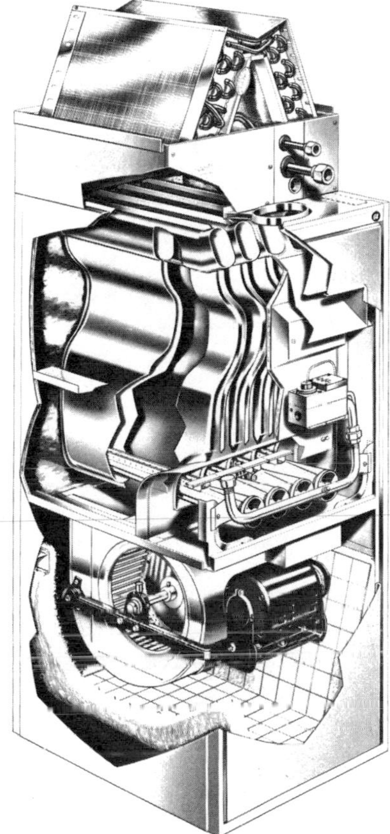

Fig. 32. Gas fired warm air furnace has coils for cooling in plenum chamber at top. A/C compressor is mounted outside house.

Fig. 33. Attractive cabinet occupies little space, need not be hidden if basement is finished. Power humidifier is at top.

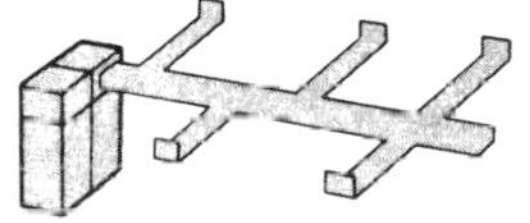

Fig. 34. Ductwork for warm-air furnace located in a full basement. Registers are in floor or walls of main level.

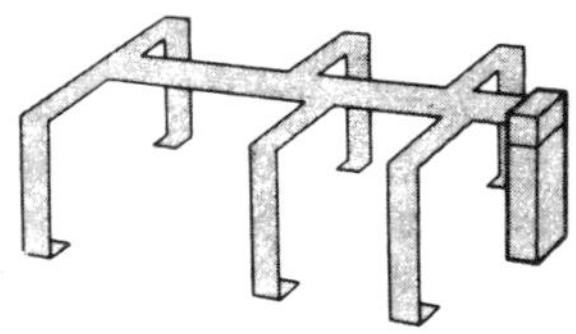

Fig. 35. If furnace is in utility room on main level of house on slab, overhead ducts run down to floor level.

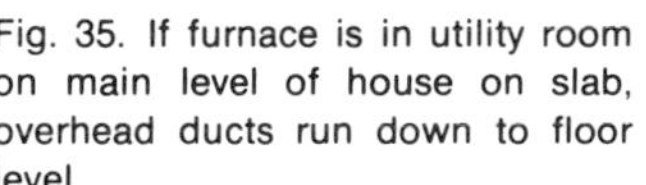

Fig. 36. Downflow furnace in utility room on main level must have ducts embedded in concrete slab floor when poured.

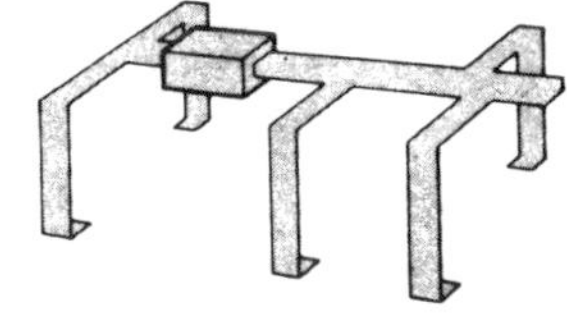

Fig. 37. Furnace can be installed in attic to save space. Ducts run downward through walls to floor level.

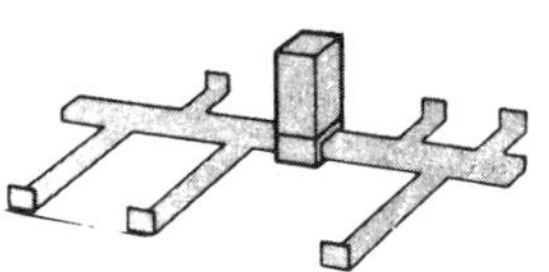

Complete, Constant and Year Round Climate Control in a Home

Modern technology has now made it possible to bring heating, humidification, cooling, dehumidification and electronic air cleaning to a home now completely enclosed in one casing, **Figure 45.**

The several components of this complete climate control comfort center are placed in a full basement. The air condensing unit is located outdoors adjacent to the center.

Fig. 38. Gas-fired furnace, hot water heater and incinerator share common chimney. Return air duct is at rear of furnace.

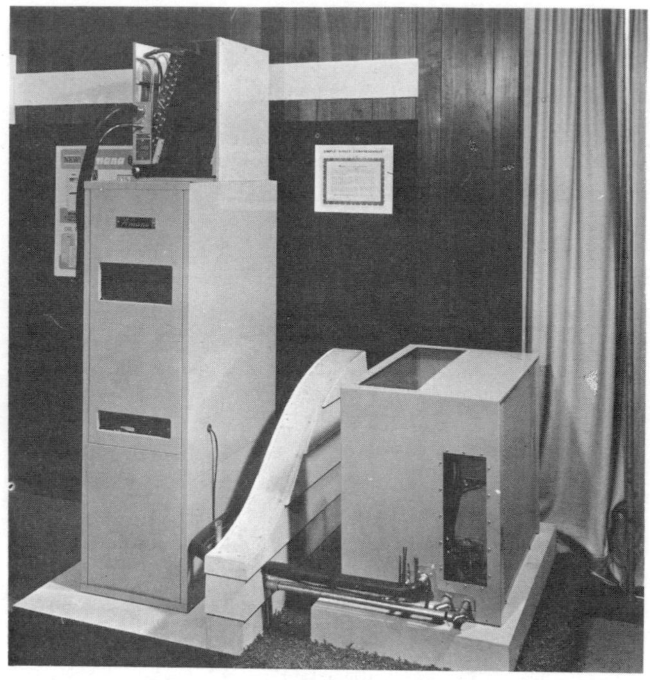

Fig. 39. Cutaway of exterior wall shows how compressor, mounted on pad outside house, connects to A/C coils in furnace plenum.

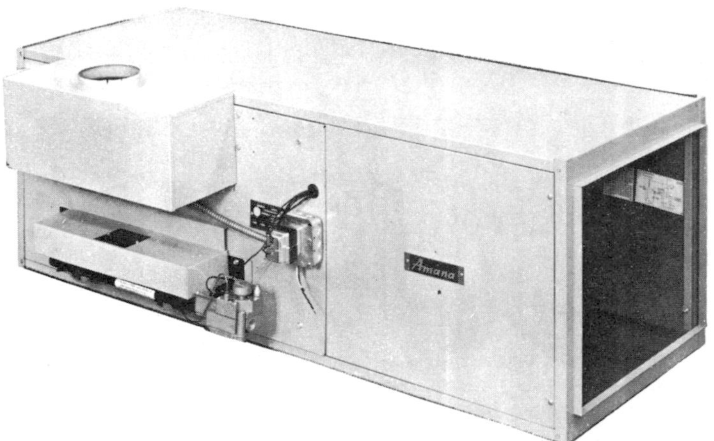

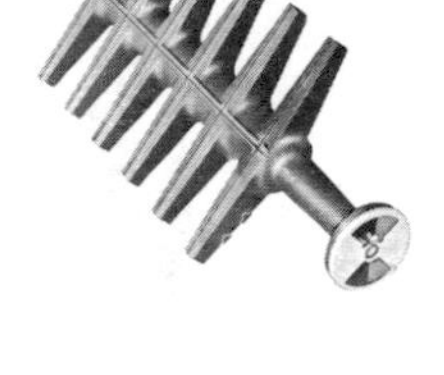

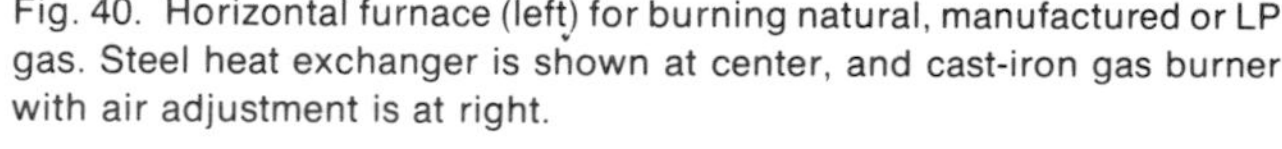

Fig. 40. Horizontal furnace (left) for burning natural, manufactured or LP gas. Steel heat exchanger is shown at center, and cast-iron gas burner with air adjustment is at right.

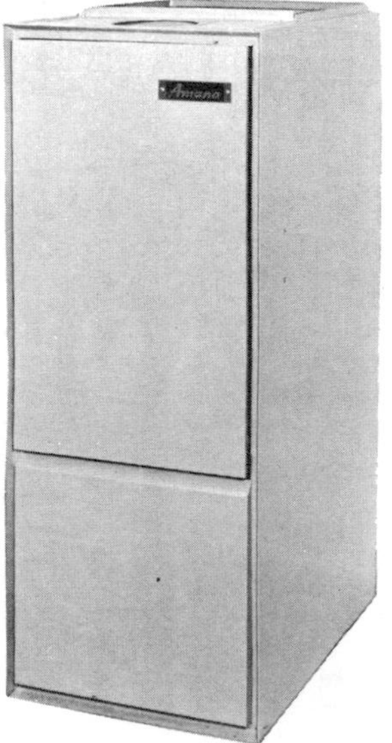

Fig. 41. Warm air upflow furnace shown with cover and with oil burner and blower fan exposed (left). Gun type oil burner is in center. Fan motor is mounted on blower (right).

Fig. 42. Upflow, counterflow and horizontal types of electric furnaces showing placement of blower and motor.

The complete climate control comfort center should be installed only by competent engineers and craftsmen who have the know-how to tailor the equipment to a specific house while it is under construction.

If any additional equipment is to be added to an existing central air heating unit it should be done only after competent engineering advice based on a survey, a cost estimate, the use of adequate equipment and the employment of trained personnel.

Heat Pump is Two-Way Air Conditioner

More costly to the builder, but economical in operation, the heat pump is gradually replacing conventional heating-cooling systems. In winter, the heat pump extracts heat from outside air and transfers it to the house much the same as a refrigerator removes heat from the cooling compartments and dissipates it through the coils at the rear. In summer, the heat pump reverses itself automatically to remove heat from the house, **Figure 44.**

Although its use as a combination heating-cooling system has been somewhat limited by its ability to produce heat, new developments are making the heat pump adequate for northern states. Resistance heating in the main duct can be cut in to increase heat output when outside temperatures drop sharply.

Chilled Water in Hydronic System Cools House

The installation of baseboard type hydronic heating along the ceiling as a valance permits dual use of the heating pipes for cooling by pumping chilled water through them, **Figure 45 and 46.** Water pipes and fins are similar to those used for heating alone, but a trough and drain pipe must be added to carry off condensation which drips from the fins during the cooling cycle.

Solar Heating

As the costs of fossil fuels such as coal, oil, and natural gas continue to rise, and the outlook for the future of these non-renewable fuels is for constantly increasing costs, architects, engineers, and builders have been experimenting with solar heat as a means of lowering the cost to the homeowner of heating his house and obtaining cheaper domestic hot water. Despite the skepticism of many builders, practical systems of home and water heating involving solar heat are being produced and installed in new and existing homes, although not in large quantities.

It is true that the solar heat systems now in use must be supplemented by conventional fuels to some degree, depending on how cold the climate is. Nevertheless, they can effect a considerable saving of fossil fuels. The reason why solar heat reduces the cost of home heating is that its sunlight collectors preheat the water or air in the heat system which therefore require much less conventional fuel to bring them up to the thermostat setting. There are, of course, times when no or almost no conventional fuels are needed to bring the fluids or air in a heating system of the solar type up to a desired thermostat setting. This may happen in southern areas where the sun shines most of the time.

Solar Collectors. The heart of any solar heating system is the collector on the roof. The solar collector in **Figure 47** is designed for heating or pre-heaing fluid temperatures below 140° F. and is made specifically to work with a heat pump for space heating, direct space heating with air distributing sys-

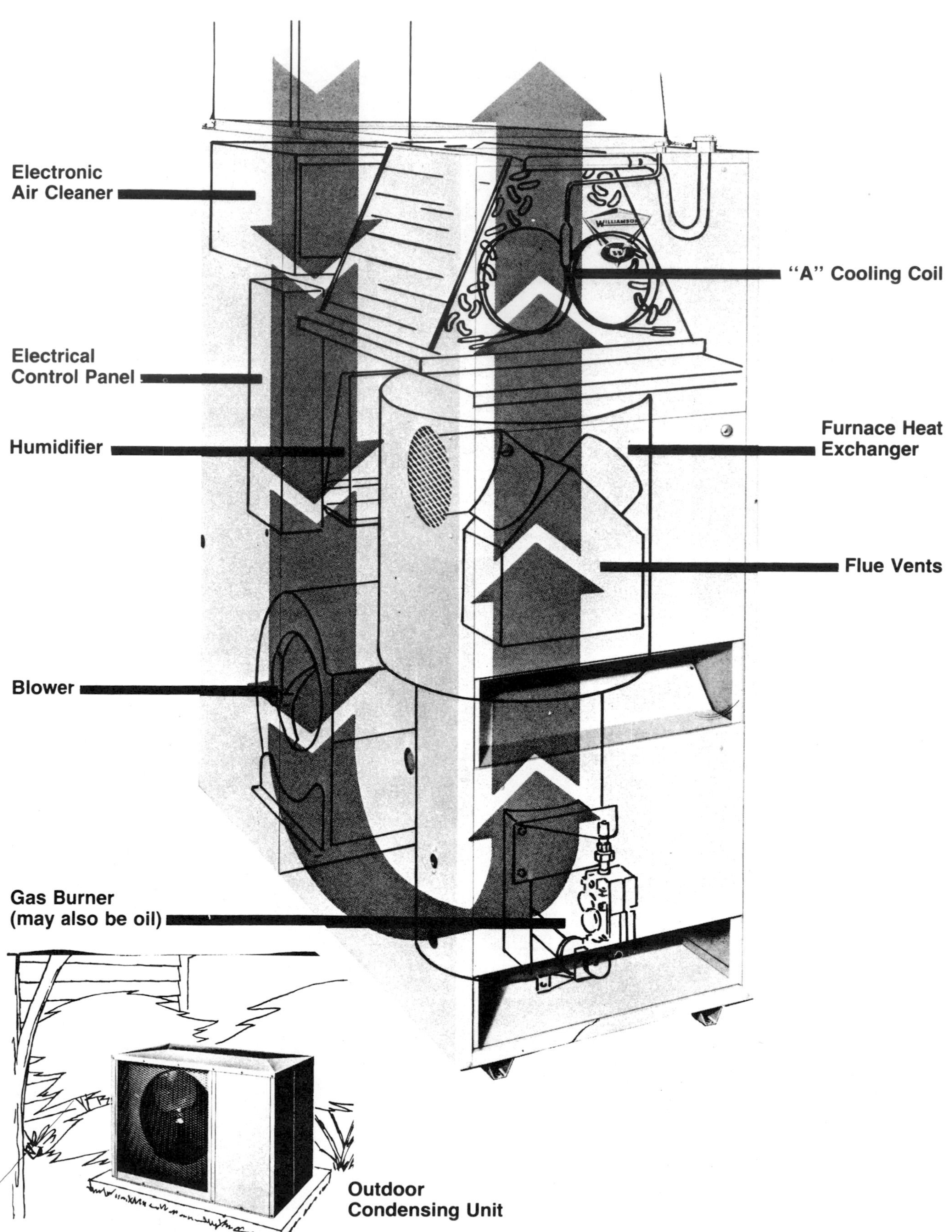

Fig. 43. A complete, year 'round climate control center for heating, cooling, humidification, and air cleaning, packaged in a single, compact housing. Furnace can be fired by gas or oil. Condensing unit for air conditioning is outside house.

tems and heating of domestic hot water and swimming pools.

The collector is a shallow box of white painted aluminum with an arched acrylic plastic cover ⅛″ thick. The sun's rays shining through the arched plastic shield strike a black painted aluminum plate with many copper flow tubes beneath it. The tubes are connected to each other at the top and bottom of the plate so that an electric pump may force a continuous flow of fluid through the collector. The sun's rays heat the air in the enclosed space under the plastic arch and black absorber plate which, in turn, heats the water (or anti-freeze fluid) flowing through its copper tubes. Two inches of fiberglass insulation under the absorber plate minimizes heat loss in the tubes. While some heat from the absorber plate does escape into the atmosphere, most of it is trapped under the acrylic shield in a sort of "greenhouse" effect.

Another type of collector made by the same company, Grumman Energy Systems, Inc., Ronkonkoma, N.Y., **Figure 48,** is made for temperatures below 200° F. and has two flat glass covers each 3/16″ thick directly above the black enameled aluminum absorber plate. The regular size of this collector is 3′ x 9′ but the size can be modified to suit any special needs.

Solar Heating of Domestic Hot Water. One of the most successful aspects of solar heating is domestic hot water. Very simply, a fluid with anti-freeze in it is pumped up to collectors on the roof by an electric pump and then flows down to a heat exchanger in an auxiliary tank which is connected to the existing hot water heater. The fluid in the closed loop never touches the house water but merely passes its heat to the heat exchanger that pre-heats the cold house water which then flows into the regular hot water tank. On sunny days the water in the storage tank may get hot enough so that the existing hot water tank doesn't have to come on at all. At other times the regular hot water heater may only have to go into operation for a short time to bring the pre-heated water up to the thermostat setting, **Figure 49.**

Solar water heaters can be used in any part of the country. In very cold areas the anti-freeze in the closed loop prevents freezing of the fluid that goes to the heat exchanger. Grumman uses a specially formulated anti-freeze in very cold areas and the heated collector fluid goes into a double-walled heat exchanger which surrounds the potable water in the storage tank. In areas of the country where freezing rarely occurs, Grumman uses a system that has no heat exchanger or anti-freeze solution. Instead potable water is circulated directly from the storage tank to the collectors. To prevent freezing, a circulator is activated when the collector temperature reaches 38° F. thereby admitting warm storage water through the collector loop.

A similar double-walled heat exchanger is used in the Lennox Industries solar heat storage tank. Lennox also offers an unusual proportional flow control which senses the difference in temperature between the double-walled storage tank and the fluid in the collectors. When sunlight diminishes, the control slows down the flow transfer fluid from the collectors and finally shuts down the system. An added safety feature in the Lennox system is a tempering valve which ad-

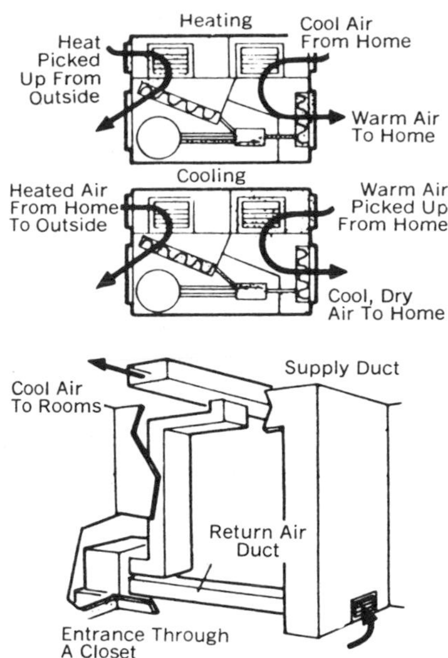

Fig. 44. Heat pump heats and cools from outside. Ducts can be concealed in a closet, run through floors and walls.

Fig. 45. Valance cooler, mounted a few inches below the ceiling, uses chilled water to draw heat from rooms.

Fig. 46. Finned cooling coil enclosed in valance has chilled water circulated through tubes. Room thermostat controls flow of chilled water. Condensation on coils drips into collector trough, is carried off by small tube at bottom of valance.

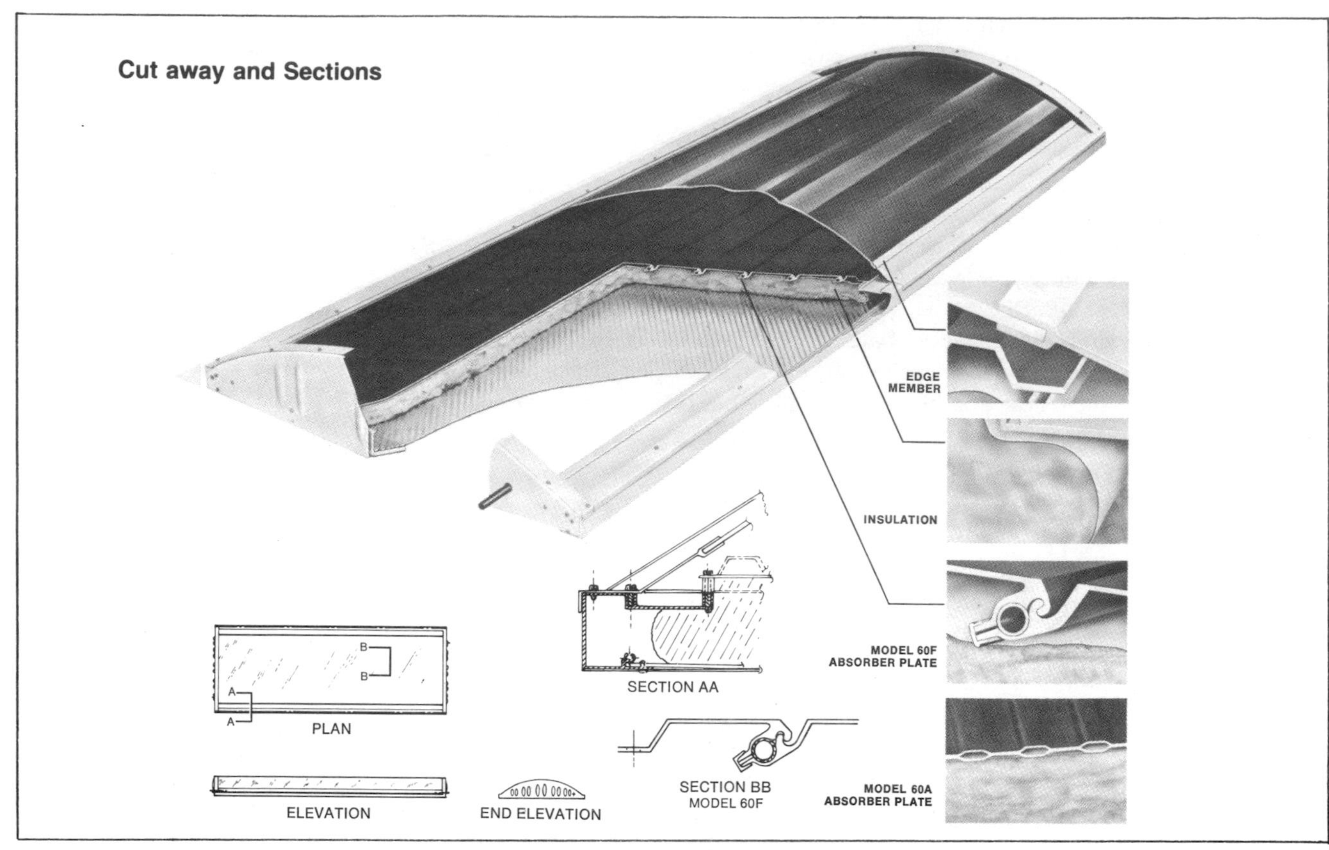

Grumman Energy Systems, Inc.

Fig. 47. Model 60 series Grumman Sunstream solar collectors.

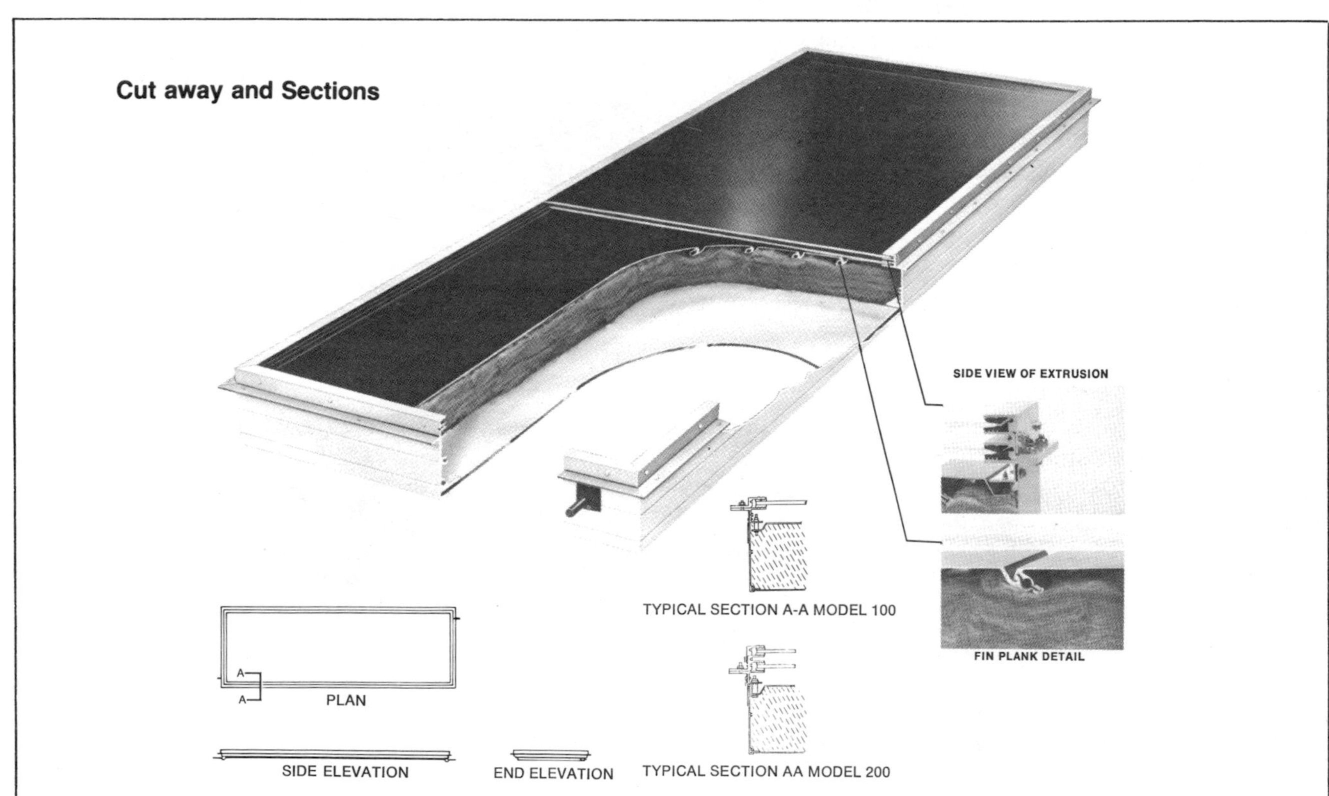

Grumman Energy Systems, Inc.

Fig. 48. Model 100/200 Grumman solar collectors is made for applications requiring temperatures below 200° F.

mits cold water to the solar storage tank when its water gets too hot for the pre-set temperature of the water in the regular hot water tank, **Figure 50.**

Swimming Pools. Compared to other applications, swimming pools require a relatively low operating temperature and make moderate solar collectors particularly attractive. Grumman's 60F Sunstream collector, **Figure 47,** can be combined with an existing pool filtration system to provide a durable installation whose savings should increase as alternative energy sources continue their upward rise. The required number of collectors depends on the size of the pool, the climate, and planned usage. A simplified plan of a typical swimming pool installation is shown in **Figure 51.**

Solar Space Heating. Although there is still much experimentation with solar heat, its practical application for space and hot water heating is typical in the case of four homes in English Tudor style built in Plano, Texas by Design Builders, Inc. The builders used the Solarmate system of Lennox Industries, Inc., Marshalltown, Iowa, to reduce heating costs. Fourteen flat-plate collectors of Lennox's own special design were mounted on the rear roof facing south. Twelve of the collectors are used for space heating and the remaining two are used for domestic hot water. The collectors work together with a 2-speed heat pump, a 500-gallon hot water storage tank, an electric furnace, an electric domestic hot water tank, solar domestic hot water storage tank, and various monitoring controls all linked in an elaborate 3-stage heating system.

In the first stage, on clear sunny days, the panels on the roof heat the circulating anti-freeze solution to some 250° F. and pump it to a heat exchanger in the 500-gallon storage tank where the heat is transferred to the water until needed. The stored heat is transferred again by means of a second heat exchanger in the tank to a sealed hydronic coil in the furnace. A blower in the furnace sends air over the hot coil and the heated air is then distributed through warm air ducts.

Only on fairly cold or cloudy days does the second stage take over. In this stage the temperature of the water in the storage tank may not be warm enough to provide sufficient heat for the warm air system of the house and the anti-freeze fluid goes through a heat pump before it is sent on to the hydronic coil.

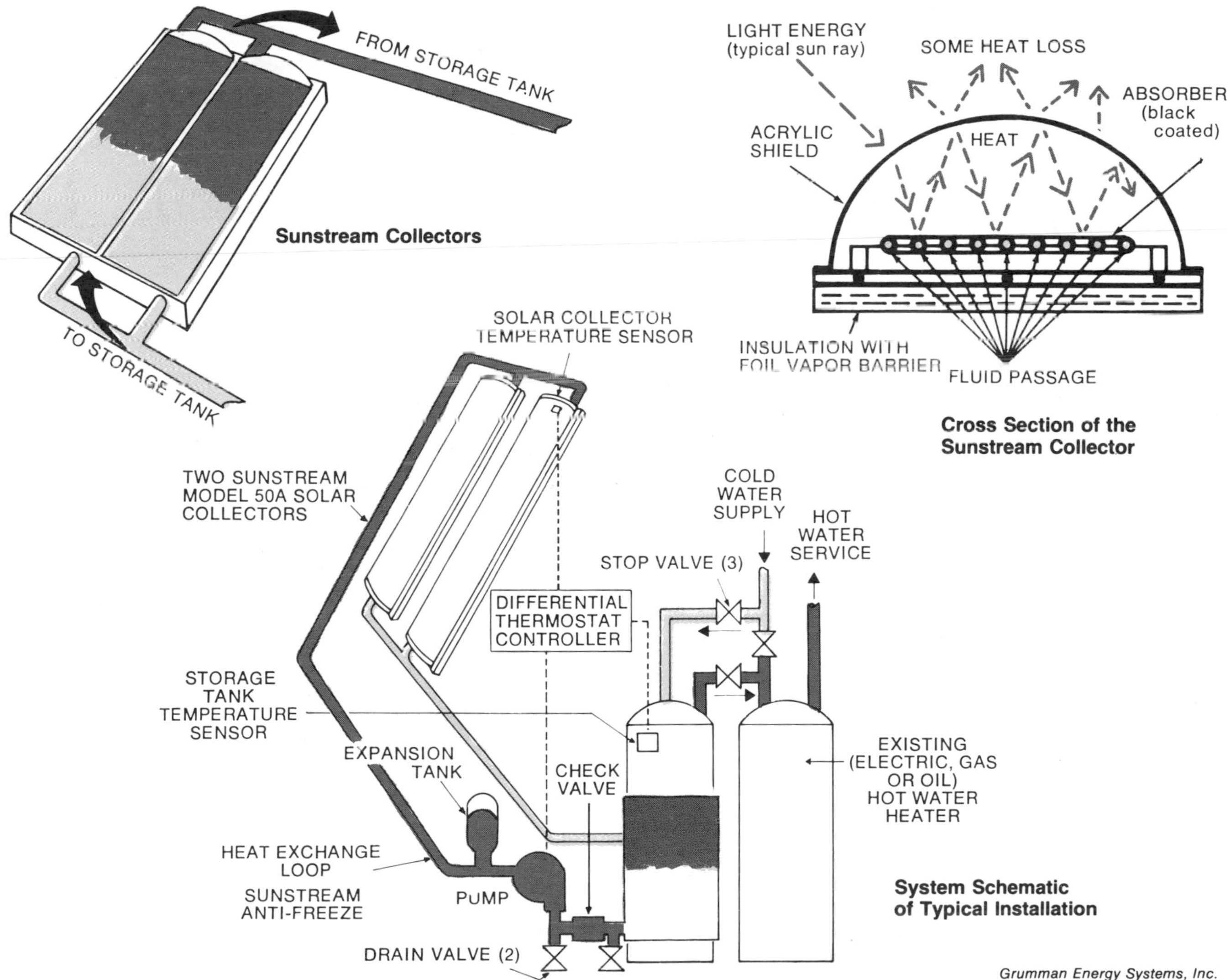

Fig. 49. Solar domestic hot water system requires 2 solar collectors and hot water storage tank with heat exchangers in it in addition to conventional hot water tank.

The energy conservation features of the Vermont home which permitted the greater savings by solar heat as reported by Kenneth Speiser and Edward Diamond in the Journal of the American Society of Heating, Refrigeration, and Air Conditioning Engineers are as follows:

Location of the solar heat collectors on the roof facing south.

Fiberglass insulation—6″ in the walls, 9″ in the roof, 3″ in the outer basement walls.

Triple-glazed windows with very small crack dimensions.

Entrance alcove to prevent escape of warm air during opening of the outside door.

Solarium with insulated shutters to accept solar input during the day and prevent loss at night.

Heatilator type fireplace with glass screen to provide net heat gain to house.

Obviously maximum gains with solar heat requires the fullest

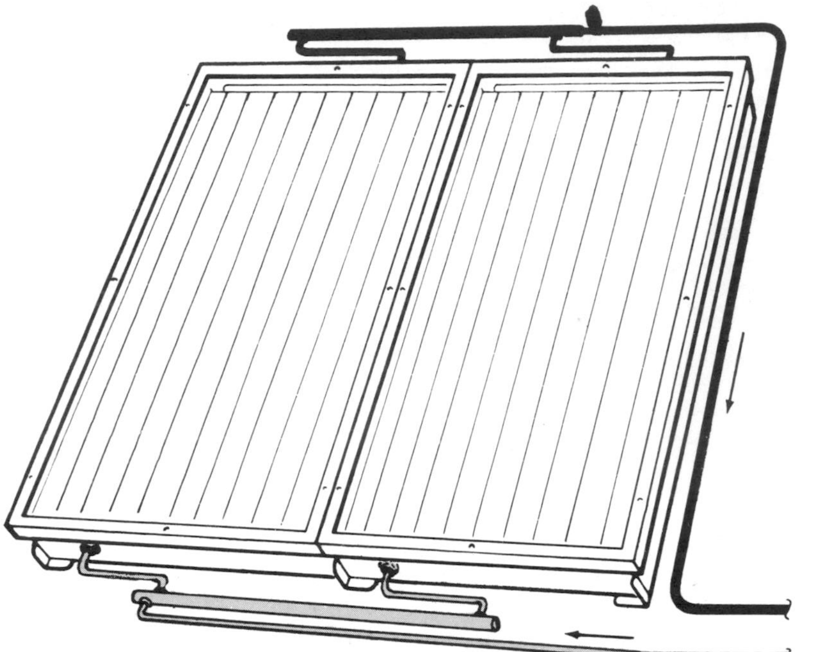

Lennox Industries, Inc.

Fig. 50. Lennox hot water system uses 2 solar collectors, a double-walled heat exchanger storage tank and conventionally heated water tank.

Mounted on a concrete slab outside the house and working like a refrigerator in reverse, the heat pump extracts heat from the outside air and pumps it into the system. Highly efficient, the heat pump does not generate its own heat but simply moves it from outside to inside the house, a procedure made possible because there is always a certain amount of heat even in moderately cold air.

In the third stage during very cold weather, the electric resistance unit of the furnace takes over and provides warm air for space heating.

The domestic hot water system is the same as described in **Figure 50.** A diagram of the Lennox Industries 3-stage system of space heating is shown in **Figure 52.** The 500-gallon storage tank, monitoring devices, and the domestic hot water tanks may be located in the basement, utility room, or, as in the case of the houses in Plano, Tex., in a corner of the garage, **Figure 53.**

While highly efficient, the collector panels are not an attractive element on the roof of the modern home. **Figure 55 and 56,** although the flat type with their black absorber plates close to their glass covers are much less conspicuous, and in this sense, more desirable, **Figures 54 and 57.**

Maximizing Solar Heat Savings. It is now widely recognized that whatever heating system is used in a house, adequate insulation, weatherstripping, and sealing can reduce fuel costs. In an experimental "energy house" built at Quechee Lake, Vermont, it was estimated that in a heating system that included solar collector panels, heat pump, and oil burner, the investment in the solar heat equipment would reduce heating costs by 40% to 50%, whereas in an ordinary house the same solar heat equipment would result in only 20% savings.

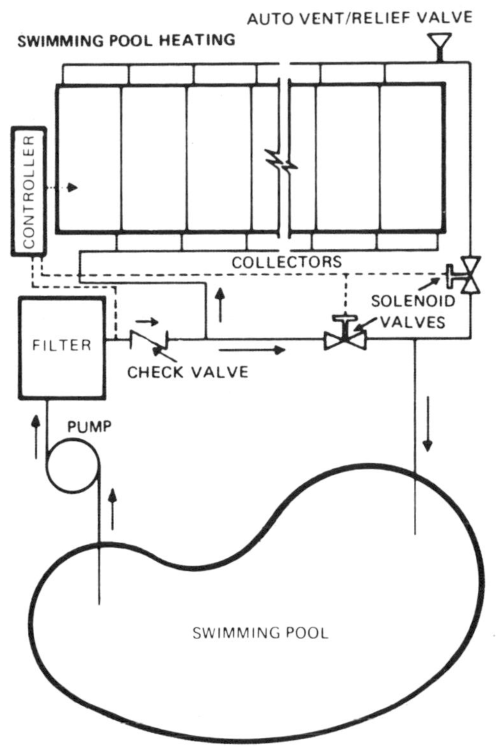

Grumman Energy Systems, Inc.

Fig. 51. Simplified diagram shows how solar collectors can be connected to filtration system of swimming pool to heat water. Number of collectors depends on pool size, climate and frequency of use.

possible energy conservation measures which, of course, is true of any house today even with conventional heating.

As the rough diagram of the Quechee Lake house heating and hot water system indicates, **Figure 58,** a 3-stage system was used including 16 Grumman solar collectors, a heat pump, and an oil burner. The 2400-gallon water storage tank and the oil burner were considered necessary because of the extreme cold of Vermont winters.

Controlling Moisture and Contaminants as well as Temperature

Comfort, as a year round condition in a dwelling, is a more descriptive term than heating and cooling to describe the effect of controlled indoor environment. Total comfort not alone involves the constant circulation of air in heating or cooling but a constant air change in humidification and in air filtering.

Moisture from Within, Moisture from Without. Most people are well aware of the pleasant, exhilarating effect of breathing moist, spring air. Most people are also well aware of the discomforts of a damp, muggy room or the parched, irritated feeling of trying to sleep in an extremely dry room. We have long recognized the important role of humidity (relative humidity is the ratio of water vapor actually present in the air to the greatest

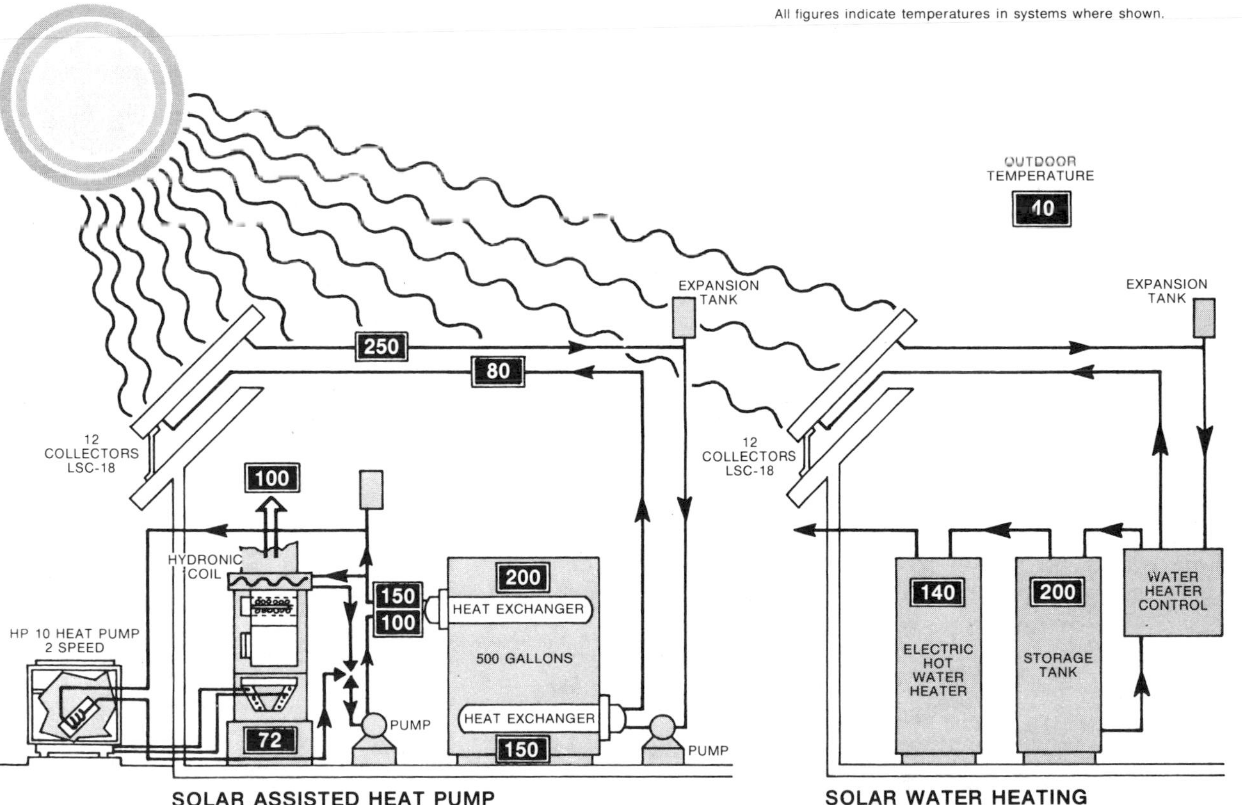

Better Homes and Gardens

Fig. 52. Diagram of the Lennox Industries solar heating system used in Plano, Texas which includes solar collectors, heat pump, and an electric furnace. Figures in boxes indicate temperatures at various points in the system.

Lennox Industries, Inc.

Fig. 53. Lennox solar system in garage of homes in Plano, Texas. At left is large 500-gallon storage tank. Conventional hot water tank is at far right and behind it is solar domestic hot water storage tank. Electric furnace with blower and hydronic coil is to the left of the hot water tank.

Lennox Industries, Inc.

Fig. 54. South roof of home in Plano, Texas showing 12 Lennox flat plate solar collectors.

amount possible at the same temperature) on our comfort, yet until recent years, the true control of humidity in the home has not always received proper attention.

Unbelievable as it may seem, the air in the average home, in winter, is often drier than the air in the middle of a desert. For example, the Sahara Desert has an average humidity of 20%, while many homes in cold weather have humidities of only 10% to 15%. In other homes, the walls and windows may be dripping with moisture at times. Even in homes not plagued by these extremes, every day comfort may be lacking due to too much or too little humidity.

The Effects of Low Humidity. If there is too little moisture in the air of our home, one may suffer from itchy skin or dry, irritated nasal and throat passages. Many medical authorities relate low humidity to increased susceptibility to colds and respiratory infections.

Low humidity can cause woodwork to dry and shrink, furniture joints to loosen and leather goods and furnishings to become brittle. At humidities under 30%, static electricity may cause irritating shocks, and room dust may increase.

When the air is dry, one feels colder because of an excess moisture evaporation from the body. As a result, a higher room temperature is necessary to feel comfortable. Studies show that for a drop of 30% in humidity, the room temperature must be raised five degrees to maintain the same body comfort balance as before. Such an increase in the indoor temperature could result in a 10 to 15% increase in the fuel bill.

Heating costs usually increase 2 to 3% for each degree the room temperature is raised above 70 degrees. In addition, a dry house often shrinks. This tends to open cracks around the doors and windows of the house allowing even more heat loss.

The Effects of High Humidity. If there is too much moisture in the air, as often occurs in summer, it reduces

Grumman Energy Systems, Inc.

Fig. 55. New building development in Princeton, W. Va. showing model 60 Grumman solar collectors on roof. System was used for space heating and domestic hot water along with water-to-air heat pump.

Grumman Energy Systems, Inc.

Fig. 56. Single family residence in Boulder, Colorado which uses 2 Grumman Model 60 solar collectors of hot water heating.

Grumman Energy Systems, Inc.

Fig. 57. One-family residence in Hilton Head, S. Carolina uses Grumman's Model 200 flat plate solar collectors for hot water and space heating along with heat pump.

the evaporation rate of one's body so that one feels hot and sticky, and clothing gets damp and wrinkled. Woodwork in the house swells and wood drawers stick. Mold and mildew may form.

In cold weather, if the indoor humidity is too high, moisture collects on windows and walls. Moisture may drip down and stain and damage sills, draperies or furniture. Sometimes the moisture may form inside the walls and freeze as it collects, and later drip out to blister the outside paint. Such damage can be very costly over the years. There are ways to control moisture and vent it to maintain humidity levels.

The Air We Breathe. Every day we breathe in some 15,000 quarts of air, ten times our daily intake of food and water combined. Man lives, on the earth, at the bottom of a sea of air that is about 100 miles in depth and where gravitational pressure at sea level is 14 pounds per square inch. Air is invisible, odorless and tasteless, a combination of gases. These gases must remain in the proper combination or man and all land animals, and most vegetation, on which they live would wither and die.

The destiny of mankind is bound up with clean air. Life, in all of its form, depends upon the vast ocean of air which surrounds the earth. This transparent ocean of air tempers our climates so that we neither get too cold in the dark of night nor too hot under the heat of the sun. The restless currents and convections of air bring moisture to water the land. But most important in this gaseous envelope surrounding us is its oxygen, necessary to almost all life processes.

For aeons, this atmosphere of ours contained little more than oxygen, nitrogen (about 78%), carbon dioxide, water vapor and traces of other gases. It was relatively pure. Its contaminants were fog, salt spray from the sea, pollen, dust smoke from forest fires and from those fires that warmed mankind, dusts and gases spewed up by volcanoes and little else.

But, in the relatively short period since the start of the industrial revolution, man has displayed an amazing capacity to drastically alter his natural environment. Today his mills and factories, chemical plants, transportation, heating equipment, power generators, and refuse incinerators have added huge amounts of pollutants to the air, some of which are new and strange to the earth's environment. These pollutants are minute particles of dust, smoke or fume, droplets or vapors of liquid chemical compounds or gases, some stable, others rare and transitory.

Why the Air is Polluted. Since 1930, the population of the United States has increased well over 50%, and the standards of living have gone up even faster. This rapid expansion has meant more homes to heat, more automobiles to drive, more factories to operate, more refuse to dispose, and more power to generate. All of these activities of our expanding population will result in more and more air pollution.

Far too many of our towns and cities employ open dumps for the disposal of refuse. The open burning of refuse on domestic, commercial and industrial premises is a common practice that too often cause pollution problems of significant proportions. Waste disposal methods minimizing emissions of air pollutants must be more widely used.

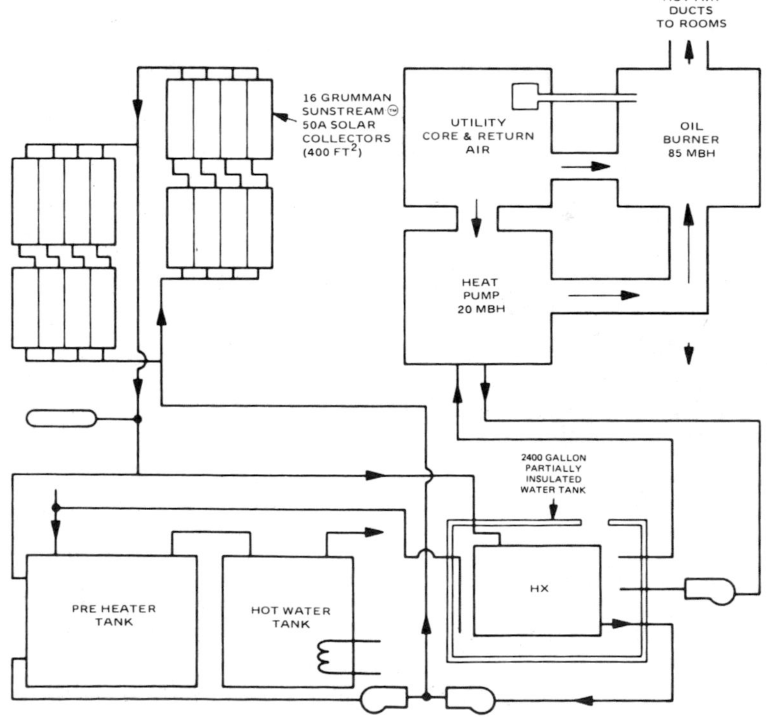

Grumman Energy Systems, Inc.

Fig. 58. Diagram of 3-stage heating system in Quechee Lake, Vermont, of experimental "energy house" involving solar collectors, heat pump, oil burner and 2400-gallon hot water storage tank.

Chapter 21

PAINTING AND FINISHING TODAY'S HOUSE

In recent years there have been remarkable changes in methods and materials used in house construction. Some changes have resulted in economies of labor and material; others have improved comfort conditions; and, still others have reduced maintenance and operational expenses. All have led to the development of better, more durable finishes, which can be applied quickly in the field, **Figure 1.**

The average house has grown tighter thus providing more livable areas in less space. Ceilings are lower. Basements are often omitted, requiring heating plants, laundry equipment, and other utilities to be installed on the first floor level. Architectural designs have also changed. Many houses do not provide the overhang at the eaves and gable ends, which they once did, to protect the walls.

Today's homeowners make more extensive use of water and appliances which discharge moisture into the living areas. Daily activities, such as bathing or cooking- even breathing and perspiring- can add several gallons of water vapor to the air inside a house.

The use of sheet materials for sheathing, plaster base and interior wall finish, has materially reduced air exchange through walls, roofs, and floors, and has helped to seal houses more tightly. Weather stripped doors and windows, and the modern window frame, while increasing the comfort within the house, are also trapping air and moisture in the living areas.

The results of these changes, and many others, make for today's tighter and more compact house which contains higher levels of moisture vapor and less space to hold it.

Moisture is the Biggest Problem

One of the most common and widespread types of damage, for which condensation is often responsible, is in exterior painting. Condensed water, in the form of free water or ice, often collects behind the siding of a building. This excess moisture may absorb extractives from the wood and result in stains as it runs out over the surface of the siding.

In some cases, the siding is thoroughly wetted. If an excessive amount of moisture reaches the painted surface, a loss of paint adhesion and the formation of water filled blisters under the paint film can occur. Rain water, entering the wall at siding laps through capillary action, may also increase moisture in the wood beneath the coating of paint. Adequate protection of exterior walls should be provided against moisture generated within the home and from outside sources. There are a number of effective methods for preventing condensation.

Vapor Barriers in Walls. A properly installed vapor barrier will protect exterior walls and paint from moisture originating within the house. Such barriers should be carefully applied to provide a complete envelope around living areas, thus preventing moisture vapor from entering enclosed wall spaces where condensation may occur. The principle of vapor protection is to make the cold side permeable enough to permit passage of moisture vapor to the outside.

In cold climates, where winter temperatures customarily fall below zero degrees Fahrenheit, walls of all dwellings, which are insulated to produce a U value of less than 0.25, should have a vapor barrier, with a permeability rating no greater than 1 perm, applied to the warm side of the walls. U value is the coefficient of heat transmission and equals the number of BTU's transmitted through a wall per hour per square foot per degree Fahrenheit difference in temperature. Perm is a contraction of the word permeability and is defined as a rating for the amount of water vapor transmitted through a material per square foot per hour per inch of mercury vapor difference. Where a water resistant building paper is used on the cold side, between the sheathing and the siding, or a water resistant material is incorporated into the sheathing, it should be a "breathing" type, having a vapor permeability of at least 5 perms, otherwise moisture trapped inside the walls will condense and saturate the insulation, reducing its effectiveness.

Dull surfaced asphalt or tar saturated sheet products, commonly used as sheathing papers, are water repellent, but not moisture-vapor proof. Such serve adaquately on cold-side, vapor resistant materials. Good warm-side barriers, having a transmission rate of 1 perm or less, are available in many forms. It is now a common practice of insulation manufacturers to provide a vapor barrier on one face of insulation rolls or batts. Many interior wall facing materials, used by builders, are backed with vapor barriers. If not present in one of these forms, however, a separate barrier is necessary.

Sheet metals, metal foils, asphalt laminated papers

Fig. 1. New, more durable finishes are easy to apply. Note dropcloth spread over patio to protect concrete.

and foil laminates are highly vapor resistant and perform well as vapor barriers. Polyethylene films and certain other plastic sheet materials are also satisfactory. To prevent accidental puncturing, the warm side vapor barrier should be installed after heat ducts, plumbing and wiring are in place. Where possible, barrier faced batts and blankets should be compressed and tucked behind conduit, piping and ducts. Where outlet boxes, electrical receptacles and wall projections are encountered, place the insulation behind the obstruction, carefully cutting and fitting the vapor barrier to insure maximum vapor protection.

The barrier is then nailed or stapled over the inside edges of the studs, immediately beneath the interior wall, **Figure 2.** To avoid holes or gaps, apply sufficient fasteners, spaced no more than 6″ apart. The entire stud space, from floor to ceiling, should be covered in this manner. In houses constructed without vapor barriers, good vapor protection can be achieved by applying, to inside walls and ceilings, two coats of a low permeability paint system recommended for this purpose by paint manufacturers. Such measures do not, however, provide complete protection; paints of this type are intended to supplement the action of vapor barriers in new houses.

Vapor Barriers in Ceilings, Crawl Spaces, Basement Floors and Walls. Vapor barriers, in the ceilings of dwellings, are considered necessary only where winter temperatures fall below 20 degrees, Fahrenheit, or where roof slopes are less than 3 inches in 12, or 1/4 pitch. In these instances, the vapor barrier should be installed in the ceilings under attics and flat roofs in the same manner as in walls.

Except in very dry climates, the soils in crawl spaces should be covered with a layer of vapor resistant and durable material, such as asphalt saturated felt roll roofing, weighing at least 55 pounds per 100 square feet, or 4-mil polyethylene film. The ground surface should be leveled and the cover material turned at walls and piers and lapped at least 2″, but need not be sealed. Some codes require 2″ of concrete over crawl space soil.

Where a concrete slab is placed on the ground, a vapor barrier should be installed directly under the slab, to prevent movement of moisture from the damp or wet ground up through the slab and into the dwelling. The barrier must be sufficiently strong to resist puncturing when the concrete is poured, and a type that will not deteriorate with age. Unit masonry basement walls should be thoroughly damp-proofed by applying mortar parging and asphalt coating to the outside surfaces.

Ventilators Protect Against Condensation. Additional protection from interior moisture is provided by crawl space, soffit and roof ventilators which permit moisture laden air to be replaced by dry air from outside. Crawl space vents should provide two square feet of opening for every twenty five feet of exterior wall, and be so placed to allow for the cross currents of air for best ventilation. Adequate roof vents require one square foot of opening for every 300 square feet of attic floor space. Miniature vents can be used for problem areas or as a supplement to larger vents, **Figure 3.** The larger vents for the roof, attic, gable and soffit are also shown, together with brick and cement block vents.

Remarkable Changes in House Paints

The changes in house paints and finishes have been as remarkable as the changes in methods and materials in the construction of a house. Today's tighter and more compact house must receive, both inside and out, better protective and decorative films of paint, stain, enamel or varnish. Through the constant efforts of paint technologists, improvement in house paints is continuing. New products are being developed continuously and existing products are constantly being modified to improve their performance. The ideal house paint is yet to be developed, but today's paints are approaching the ideal.

There is now a competitive battle between traditional paint vehicles (the liquid part of paint) and a host of synthetic materials. Many of these newer materials, the vehicles, have been adapted from other industries. The paint, plastics, rubber and synthetic fiber industries have borrowed ideas and processes from one another. A paint resin may have the same chemical composition as a synthetic textile fiber, but the molecular structure accounts for the physical differences between the two.

Linseed oil used to be the workhorse of the paint industry but today the man with the brush is using alkyd resin paint instead of linseed oil paint. Alkyd resin was originally created as a shellac substitute, but has actually become a linseed oil replacement. The word

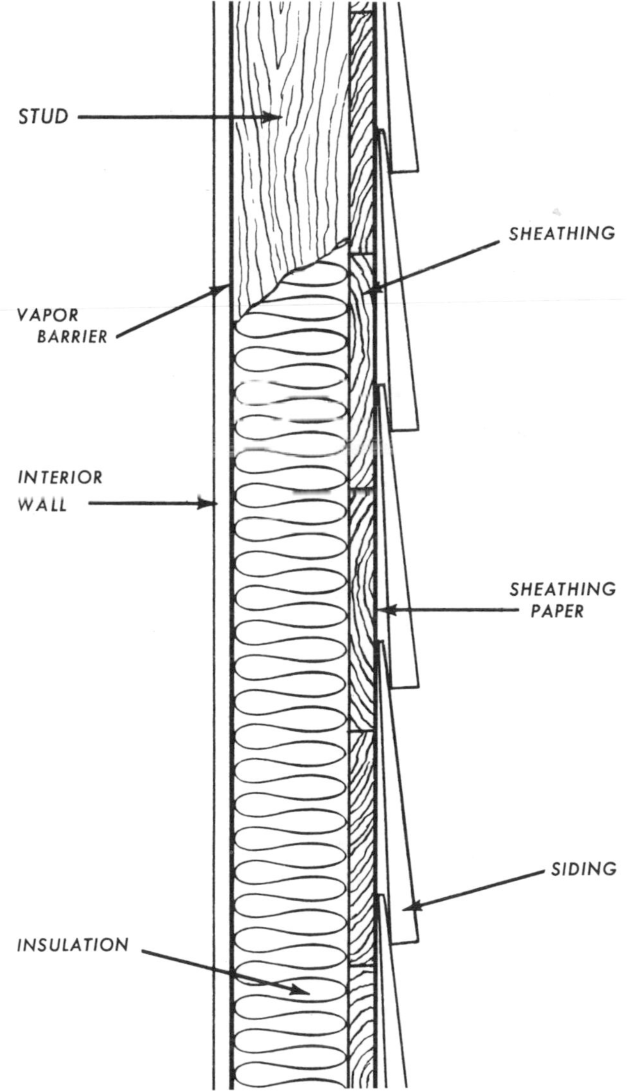

Fig. 2. Typical wood-framed exterior wall, showing vapor barrier applied on warm side of insulation. Building paper on sheathing seals out weather while permitting wall to breathe.

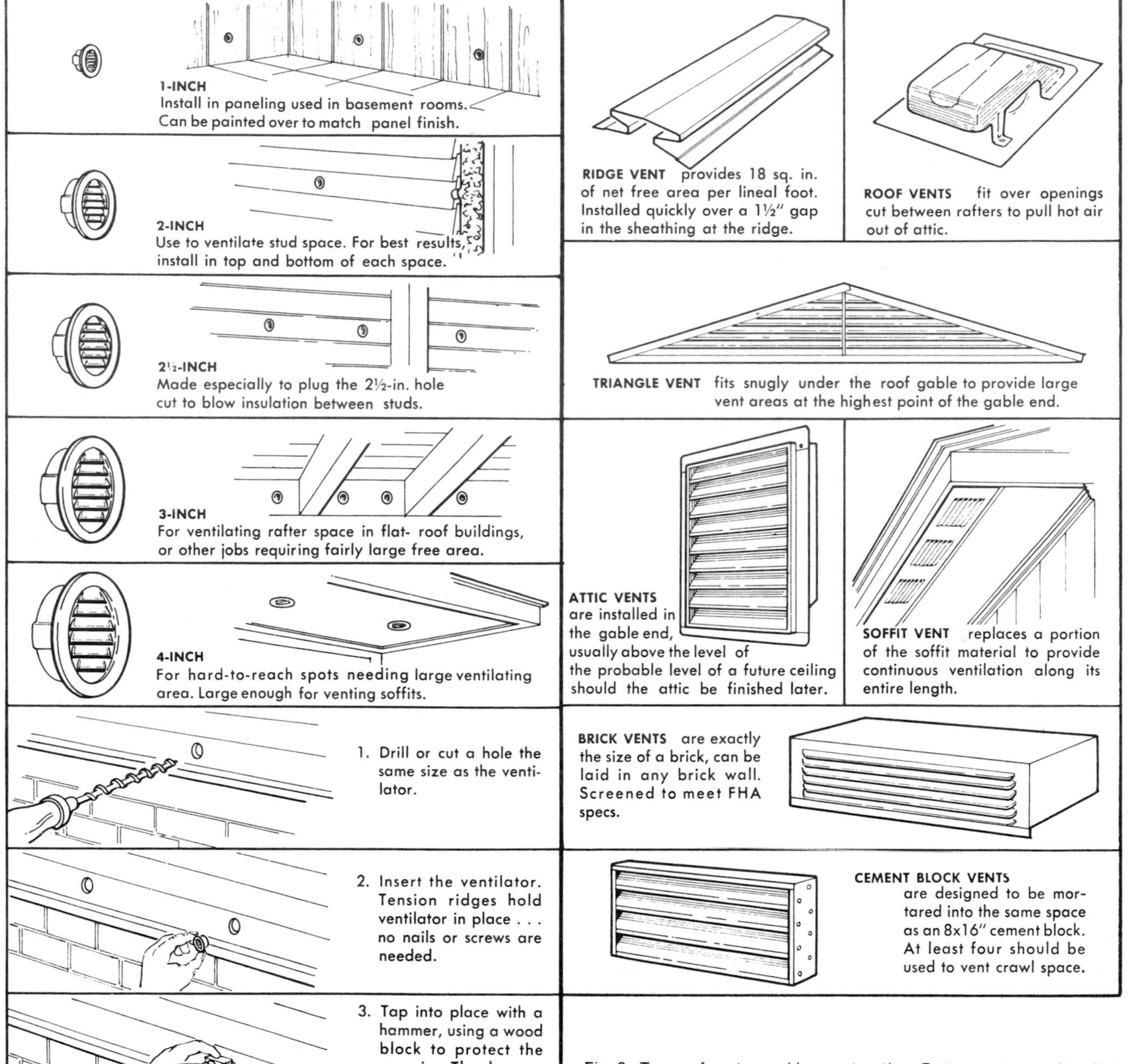

Fig. 3. Types of vents used in construction: Button vents permit walls to release trapped moisture; ridge, roof, triangle, square, and soffit vents rid attic space of heat and moisture; brick and cement block vents fit into masonry coursing of foundation walls.

alkyd was coined from the letters AL in alcohol and CID in acid. Since ALCID is difficult to pronounce, the more euphonious spelling of ALKYD was adopted. The alkyd story dates back to World War I when periodic price squeezes in the Indian shellac market stimulated the General Electric Company to search for a substitute. General Electric used shellac as an adhesive in making insulation. Alkyd resin was developed to serve the same purpose. The first major paint use of alkyd resin was in baked-on automobile enamels. These enamels had to be sprayed and they were hardly adapted for either architectural use or the do-it-yourself market. Air-dry brushing alkyd enamels were not evolved until the late 1930's and became commonplace after World War II.

Alkyd resin has revolutionized the interior white enamel business. In the past, interior enamel, made from linseed oil, always yellowed. Bright sunlight bleaches linseed oil, but the light that filters indoors through window glass is deprived of its bleaching rays and yellowing occurs. Linolenic acid, the largest single constituent in linseed oil, was long recognized as the cause of this yellowing. The problem was how to get rid of the linolenic acid. Soya oil contains very little linolenic acid and its substitution for linseed oil would have been an obvious solution, except that soya oil takes forever to dry. By chemically wedding soya oil to alkyd resin, chemists created a new interior enamel vehicle that is virtually free of linolenic acid and that also dries positively.

About 20 years ago, alkyd resin was first successfully used in flat wall paint. It has now largely displaced linseed oil as a wall paint vehicle, but it in turn faced competition from emulsion vehicles. Wall paint represents about a third of the paint that is sold through trade sales channels.

Until the end of World War II, only two types of wall paint were available. They were oil paint and water thinned paint. Oil paint was used on most architectural jobs because it was serviceable and washable, but it had certain deficiencies. It sometimes reacted with the alkali in fresh plaster and saponified (or turned into soap). It always had to be applied over a special primer, even then darker colors sometimes dried with a smoky or ghosted appearance to the annoyance of everyone concerned. Pre-war water thinned paints, such as calcimine, casein paint and resin emulsion paint, were not noted for quality and shared one big disadvantage. They were water sensitive and would stand little or no washing. This background set the stage for a lot of feverish research. The first new post war (World War II) paint was a water thinned product called latex paint, or rubber latex paint. This nomenclature was unfortunate, for the paints contained neither latex nor natural rubber.

Compositions of Modern Paints. The dictionary defines latex as a milky white liquid found in certain seed plants, notably the rubber plant. The juice in dandelion stems is another latex. Early latex paints were made from an emulsion of synthetic rubber resembling GRS, the material from which tires were produced during World War II. The chemical name for this synthetic rubber is styrene-butadiene, which is derived from gases produced during the cracking of petroleum. Most paint technologists would prefer to identify water thinned paint vehicles as emulsion rather than latices or latexes. Purists like the word "latices"; the paint industry uses the word "latexes."

An emulsion is a mixture of two liquids that do not dissolve in one another. These two liquids do, however, remain in suspension. One of the most familiar emulsions is French dressing used on salads. This emulsion is a mixture of oil and vinegar. In a paint emulsion, the mixture consists of resin globules in water. Styrene-butadiene latex paint proved to be an immediate success with householders and do-it-yourselfers. It worked easily, dried fast and could be washed out of brushes and rollers with soap and water. It had a mild odor. Until the advent of emulsion finishes, paint always had the unpleasant odor that is associated with mineral spirits. Styrene-butadiene paint was moderately washable after it cured for a month. It also had disadvantages. It was rather thin and sloppy, it had only fair hiding qualities and it produced a skimpy appearance. Professional painters were less than enthusiastic about it.

No sooner had styrene-butadiene paint established itself than two new emulsions; PVA and acrylic; became available for paint use. The initials, PVA, stand for polyvinyl acetate. PVA emulsions are also known as vinyl emulsions. PVA is the reaction product of acetylene and acetic acid. Acetylene is used in welding. Acetic acid is the constituent that makes vinegar taste sour. Polyvinyl acetate is not the vinyl compound that is used to make garden hose, plastic shower curtains or phonograph records. Vinyl plastics contain a related material, polyvinyl chloride, not polyvinyl acetate. PVA emulsions were first used as adhesives by the paper industry.

Acrylic resin is closely allied to vinyl resin in chemical structure and it is also derived from acetylene. Acrylic resin, in plastic form, is known as Lucite or Plexiglas and is used to make aircraft windows and automobile tail lights. In filament form, acrylic resin is woven into Dynel or Orlon fabrics. In emulsion form, it is a paint base.

All three emulsion bases; styrene-butadiene, PVA, and acrylic; are milky liquids. They look alike and paints made from them share a number of things in common. All thin with water, work easily, dry fast, are relatively odorless, alkali-resistant, non-flammable, can be cleaned out of brushes and rollers with soapy water, and may freeze.

Alternate freezing and thawing or a protracted freeze will cause an emulsion paint to "break" or to disintegrate. Once an emulsion paint separates, or breaks, it cannot be used or reclaimed. Low temperatures are also bad for application. Most emulsion paints fail to coalesce or fuse into a suitable film if applied at a temperature below 50 degrees F. An emulsion paint produces a different kind of film than a conventional oil or alkyd finish in which the vehicle is dissolved in thinner. After application and drying, the conventional oil or alkyd finishes have a continuous film. In emulsion paint, the resin globules draw together as the water evaporates, and the film that ultimately forms is like a mat of ping pong balls cemented together. The tiny interstices between adjacent resin globules permit the film to breathe. This is probably one reason why emulsion paint works safely over green plaster.

Emulsion paint has the advantage of not penetrating porous substrates. This makes it fine for finishing surfaces like gypsum board, but poor penetration can also result in poor adhesion. If emulsion paint is applied over a chalky or dusty surface, it will not penetrate the chalk or dust and wet the underlying substrate. If paint does not wet a surface, it will not adhere to it.

Surface Must Be Properly Prepared. Adhesion may also be a problem if emulsion paint is applied over old enamel on which there is a residual film of oil or grease. In any house or building area where food is prepared or where oil is burned for heat, grease or oil may deposit on the walls. Water thinned paint does not wet greasy or oily surfaces although it may emulsify these contaminants so they become part of the dried paint film. The solution to adhesion problems is proper surface preparation. In the case of old work, this means washing the surface before it is painted. Proper surface preparation is also important prior to applying other paints, but emulsion paint is more sensitive to surface contaminates than oil and alkyd paints.

In the discussion of emulsion wall paint only low gloss finishes have been considered. No one has yet succeeded in developing a commercially acceptable high gloss emulsion enamel. Many semi-gloss emulsion paints are on the market. It seems likely, however, that good emulsion enamels will ultimately be developed. Probably every paint laboratory is trying to perfect such a product.

Emulsion Versus Alkyd. Which is better, an emulsion wall paint or an alkyd flat? There is no firm answer to this question, for each type fills a somewhat different need. Alkyd resin has stood the test of time. It has been used in paint and enamel for many years. Alkyd flat wall paint was developed primarily for use by the professional painter, and it has qualities that appeal to him. Alkyd resin paint has the consistency that the painter likes. It produces rich, full-bodied, smooth feeling films that attract the decorator and homeowner. It has an advantage in scrubbability and washability over most, if not all, emulsion paints now on the market. It also has an advantage in adhesion and toughness.

Emulsion paint is ideally suited to the do-it-yourself market because it is simple to apply. It brushes easily. Emulsion paint can now be made to hide surface blemishes as well as an alkyd flat. Emulsion paint, however, still freezes, and the water in it may corrode metal and swell wood unless suitable priming coats are used. The primary advantages of emulsion paint for architectural work are its ability to work on new or green plaster, on extremely porous surfaces and to dry fast. It can be used on rush jobs and on work that has to be done under less than ideal conditions.

Fig. 4. Vital areas of house to be inspected carefully before beginning to paint exterior.

The use of alkyd resin in general, and of alkyd flat wall paint in particular, is not shrinking despite the popularity of emulsion paint. Emulsion paint usage, however, is expanding at a rapid rate, whereas alkyd resin demand seems to have reached a plateau. It is commonly assumed that water thinned paint should be relatively inexpensive since water is also inexpensive. It actually costs as much or more to produce a gallon of emulsion paint than it does to produce a gallon of an equivalent solvent thinned oil or alkyd paint. The reason is that water can be incorporated into paint only with the aid of costly wetting agents and emulsifiers. Furthermore, special and generally more expensive pigments have to be used in emulsion paint than in conventional paint. The cost of these additives and special pigments offsets the saving that results from the use of water as a diluent.

Emulsion paints were initially developed for use on interior walls. Later, emulsion type exterior masonry paints were created. Concrete floor versions have also come on the market. It was inevitable that an effort would be made to develop emulsion house paint for use on wood siding. This latter usage has probably caused more controversy than any other paint development in the last decade. Emulsion house paint dates back to 1953 when a leading producer of acrylic emulsion first suggested it as a base for house paint. Two large paint companies began to test market an emulsion house paint shortly after that time. Today it is the most common house paint.

Painting the Exterior With Emulsion Paint. The major appeal of emulsion house paint to the householder is its usual freedom from blistering and flaking. It does produce a breathing type, blister resistant film if it is applied on bare wood over suitable primer. It does not have any better blister resistance than oil base house paint if it is applied over old paint. The harassed householder, with a problem house, may not read the fine print in the promotional literature on emulsion house paint. If he fails to remove the old paint, the performance of an emulsion house paint may be disappointing.

Emulsion house paint, like other emulsion paints, works easily and dries fast, thereby avoiding insect and dust damage to the fresh paint film. The best quality emulsion house paints have excellent color permanence, but colors are available only in a low sheen. Unfortunately, most brands of emulsion house paint must be applied over an oil base primer on new work and must be applied in two coats. Emulsion house paint simply will not survive the weather if it is applied directly over bare wood.

Another problem is performance over heavily chalked old paint. Emulsion paint tends to produce "papery" film over chalk, and scaling may occur. This can be avoided by first applying a coat of oil primer. Bear in mind, however, that oil primer defeats some of the reasons for using an emulsion house paint. Oil primer dries slowly, does not work easily when compared with emulsion paint and does not wash out of brushes with soap and water.

Classification of Paints. A few years ago, most paint fitted into a few neat classifications. Lacquer, for example, dried fast; oil vehicles dried slowly but made good house paint. Most new paint types, however, cannot be pushed into convenient pigeon holes. Alkyd resin is an example. It does not have the fixed composition of a five-grain aspirin tablet which always possesses the same medicinal qualities regardless of the brand name it bears.

Alkyd is a generic name that covers a whole family of resins. Variations are produced by altering the ingredients, changing the cooking cycle or blending with other materials. The name alkyd, or vinyl or acrylic does not signify any specific paint quality or degree of performance. There are good finishes in each category, and there are mediocre finishes in each category.

Good paint is not obtained simply by specifying a single ingredient. Modern paint contains from 10 to 20 different ingredients, each of which contributes to the effectiveness of the paint. The ingredients can be converted into quality paint only with the aid of scientific skill, manufacturing know-how and adequate control techniques. These intangibles may contribute more to paint quality than any single ingredient. There is nothing permanent about any new paint development in the sense that it is the last word. Each, however, represents a significant forward step. New paints always possess some attractive feature or features. They also possess compensating and sometimes unrevealed disadvantages.

Painting Exterior Surfaces. Painting is the act of applying protective and decorative coatings of paint on a surface. It is well to inspect and study the exterior of a house quite thoroughly at least once a year. If one is familiar with properly placing a ladder against a house and climbing up to inspect the roofing of his dwelling, and walking on slanting roof surfaces, a great deal can be learned about the condition of the roofing material, gutters, valleys, chimneys and vents. If one is not familiar with the use of a ladder (see chapter on scaffolding), a roofing contractor should be called to give expert help and advice.

If the chimney is of brick and in need of tuckpointing this work should be done at once. If shingles need replacing or repairing, or if the roofing material is in need of painting it should be done as quickly as possible to avoid more costly repairs at a later time.

The owner of a dwelling can determine the condition of other materials by a careful inspection from the grade level. Are the exterior materials of wood, brick, metal or stucco? Where is tin, iron, copper, steel or aluminum used? In what condition are the painted surfaces? Wood that is not fully protected against the weather can rot

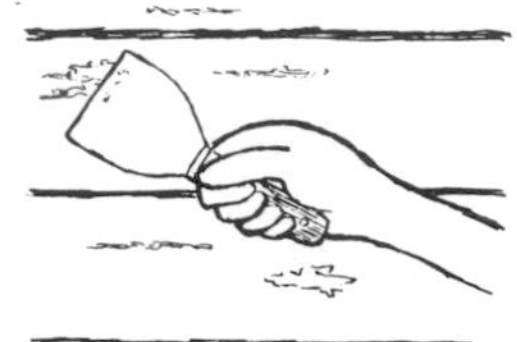

Remove loose paint with a putty knife

Brush dust from surfaces

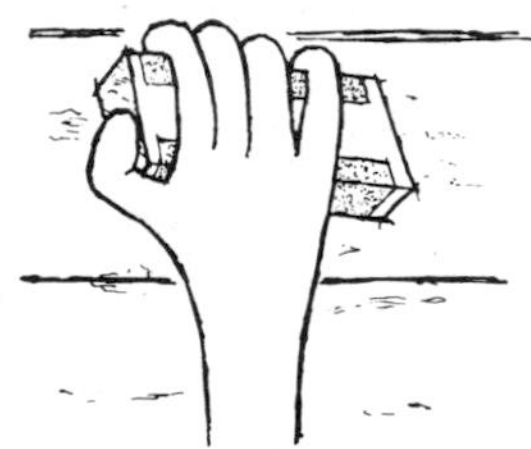

Sand rough spots and edges of siding

Wash dirt from soiled surfaces

Fig. 5. Four methods of preparing old exterior surfaces prior to repainting over old finish.

away, making costly repairs a necessity. Where bare wood is exposed to the weather, ruinous moisture has a chance to cause additional damage beneath the surface. Invading moisture can also force its way to the inner walls of a house and injure side walls and ceilings.

Uncoated brick, stone, stucco and cinder blocks play host to moisture, necessitating more heat from the heating system to overcome clammy dampness which creeps into walls. Metals that are not safe-guarded with paint will rust. Even copper, although it will not rust, develops a corrosive wash when it is uncoated and will stain surrounding surfaces. There are certain places about the exterior of a house where trouble is most likely to start. Examine, with special care, the window frames, siding, the bases of columns on porches or entrance ways, outside steps, downspouts, areas under the eaves, and porch floors and supports, **Figure 4.**

Preparing Exterior Surfaces for Paint. The job of repainting a house does not begin with the application of the paint but with the proper preparation of surfaces. The surfaces on which the new paint is to be applied must be sound, smooth, and if metal, free of rust. To make sure that all surfaces are absolutely clean and smooth, all loose dirt should be completely eliminated with the aid of steel wool and a wire brush. **Figure 5** points out ways to prepare soiled and dirty surfaces for painting. If putty around window panes has cracked or loosened, it should be removed or replaced. If there are cracks or holes around the window or door frames, especially where different materials come together such as in the joints between flashing and brickwork, careful caulking is vital to seal out moisture.

On old areas where old paint has scaled, cracked or blistered, it is often necessary to remove the old coating entirely. Sometimes this can be done with a liquid paint remover or an electric paint remover. It is never wise for an amateur painter to use a blow torch to burn off old paint due to the fire hazards involved.

Where wood replacements have been made or where rust of metal has been removed, it is of course, necessary to apply a priming coat as soon as possible. The primer will block the entrance of moisture, or in the case of metal, prevent further rusting. Zinc chromate, oxide of iron or zincdust-zinc oxide should be used on metal. Outside house primer is used for wood.

Where there are knots or sappy streaks in new wood, a special knot sealer should be brushed on before the priming coat is applied. This will prevent the resin from bleeding through into the final coatings.

Before actual painting is begun with an oil base paint, be sure that the surfaces to be painted are absolutely dry. Be sure, too, to have all tools and materials at hand, **Figure 7.** In addition to the paint, extra cans for mixing the paint and paddles for stirring the paint will be needed. Strainers will be needed for removing lumps which might mar a smooth surface. Rags will be needed for cleaning up spatters, drop cloths to protect porch roofs, floors, steps, shrubbery and plants.

For the one-story house, and many two-story houses, firmly placed ladders and cross planking on brackets will serve for the painter to reach the higher places. On large houses scaffolding will be needed and must be erected by the workmen who know how. In many cities it is now possible to engage skilled painting contractors to paint all of the hard to reach places while the do-it-yourselfer prepares and paints the lower, easy to reach areas.

How to Brush Paint on Exterior Surfaces. For painting relatively large exterior areas a 4-1/2″ or 5″ brush should be used. In painting exterior trim areas a 1-1/2″ to 2″ brush should be used while a sash brush is needed in painting around windows. The top sketch in **Figure 6** shows a beveled brush or sash tool for painting windows and mouldings. The second sketch is a large, 4″ brush, for extensive areas. The round shaped brush is for sash and trim. At the bottom is a smaller, 3″ brush, for smaller areas. Dip the brush in the paint taking care not to overload the brush with an excessive amount of paint. Use a smooth, back and forth stroke that works the paint into the surface.

The addition of an anti-freeze to paint is not recommended. While certain solvent-type paints can be used in freezing weather, it is almost impossible to get a good even film because a paint does not adhere well to a cold surface. In starting to paint a wall, begin in the upper left hand corner of a surface and swing the brush to the right as far as one can comfortably reach. Paint from top to bottom, then begin again at the top. Complete one side wall before starting another. Remember to do the sash, trim and doors first. Then one will not have to rest his ladders against a newly painted wall. The best time to paint a house is during a dry spell when the humidity is low. Painting should be done when the temperature is above 40 degrees F. The chart, **Figure 9,** suggests color combinations for the exterior of a house. Painting should not be started in the morning until the dew has dried from the surfaces. If at all possible follow the sun around the house.

How to Clean a Brush After Painting. If a brush has been used in oil paint it must be cleaned after each day of work or after each job is completed. Pour a small quantity of paint thinner, usually turpentine, in a clean container. Let the brush soak in the thinner for a while, then work the brush against the bottom of the container to release the pigment. To loosen the paint in the center of the brush squeeze the bristles between the thumb and forefinger or press against the brush with a mixing paddle.

Rinse the brush again in the thinner, and if necessary, wash the brush in mild soap suds, and rinse in clear water. Press out the liquids with the fingers or with a mixing paddle. After all of the oil and pigment are out of the brush twirl the handle of the brush between the

Fig. 6. Most common brushes used for house painting: Beveled brush for sash and trim (top); 4″ brush for large areas (second); oval brush for sash and trim (third) 3″ brush for narrow areas.

hands, keeping the bristles below the top of a container to avoid splashing liquid on one's clothing. To store a brush for any length of time it should be wrapped in heavy paper, then laid flat, never standing on the bristles. If an exterior latex paint, instead of an oil base paint is used, the brushes or rollers need but be cleaned in water before the paint hardens.

Painting Porch Floors and Exterior Steps. Finishes for porch floors and steps, constructed of wood or cement, are designed to take the wear and tear of foot travel as well as the weather. For a finish coat, tough, hard wearing, porch and deck paint is usually applied to either wood or cement. A slightly thinned coating of deck paint can be applied to wood, as a primer or undercoat. An alkali-resistant primer is often used on cement as a primer. Muriatic acid is often used to roughen hard and glossy concrete or cement surfaces before a finishing coating of a rubber base or other latex type paint is applied.

Painting Outdoor Furniture, Trellises and Fences. Porch furniture as well as garden furniture should be given protective coatings of paint since both are exposed to the weather. When a colorful finish is desired, a sturdy exterior enamel should be used. House paint is not suitable since it tends to oxidize and the chalking would rub off on one's clothing. An enamel undercoater should be used if the furniture is wood or wicker. If of metal, a special primer for metal should be used. If the finish is relatively good it may be touched up with a coating of spar varnish, which is transparent and weather resisting. The correct material to use for different jobs is listed in the chart, **Figure 10.**

For arbors, trellises and fence posts a wood preservative should be used to retard the rotting of wood that must be placed below grade. Regular house paints can be used on the portions of the arbors, trellises and fences above grade.

Using Latex Paints on Exteriors. Latex paints, de-

Fig. 7. Many helpful accessories which speed and simplify exterior painting while making the job easier.

signed for exterior use, have been used on exterior masonry for about twenty years. They have given excellent results and are among the established coatings for concrete, stucco, brick, cinder block and similar surfaces.

Since the products of different latex paint manufacturers may differ substantially in some of their properties and best conditions for use, it is important that the manufacturer's directions for the use of these products as given on the label be followed implicity. In particular, most manufacturers recommend the use of a specific primer as part of the "system," and the entire system be used to obtain the results desired.

The performance of latex paints, and other paints too, when used for repainting, depends in a large measure upon the condition of the old paint over which it is applied. The performance of these systems, over the wide variety and conditions of old paint which may be encountered in the field, is under intensive study by paint manufacturers but it will be some time before comprehensive information is available on all the conditions which may be encountered.

The products of different manufacturers differ in a number of ways but all latex paints designed for exterior wood may be expected to exhibit the following properties when used as part of the system recommended by the manufacturer:

1. Latex paints are easy to apply and they spread readily. They may be applied by brush, spray or roller, Figure 11. Because latex paints brush so easily, care must be taken to follow the manufacturer's recommendations as to the spreading rate or else too thin a film, with poor hiding power and poor durability will result.

2. Latex paints dry in from one half to one and one half hours, minimizing dirt and bug collection during the drying period. Recoating can be done after a relatively short drying period and often without moving scaffolding.

3. Latex paints often give cleaner and brighter colors than do oil paints. The sheen is low and the uniformity of sheen is generally better than oil paints.

4. Tools such as brushes and rollers used for applying latex paints may be cleaned with water, if this is done before the paint hardens, and spills may be wiped up with a damp cloth.

5. Since most latex paint films are somewhat permeable to moisture, they are less likely to blister because of excess interior moisture than are conventional oil paints.

6. No paint will give as satisfactory a coating when applied at low temperatures as it will when the weather is warmer. While oil paints may be applied with some success in almost any weather the painter can work, the probability of unsatisfactory jobs increases with decreasing temperatures. Latex paints, however, should not be applied at temperatures below 45 degrees F, because satisfactory films are not formed in weather substantially below this temperature.

7. In general, properly formulated emulsion paints will show tint retention superior to properly formulated oil or alkyd paints. While the evidence is not yet complete, indications are that latex paint systems, when used as recommended, can be expected to last at least as long as high-quality oil paints.

8. Since latex paints are thinned with water, it is not necessary that the surface to which they are applied be absolutely dry. This does not apply, of course, to solvent-thinned primers which may be part of the recommended system. These must be applied to dry surfaces.

9. No paint will adhere well to an excessively chalky or dirty surface. Many surfaces to which oil paints adhere readily are not suitable for latex paints. For this reason, most manufacturers recommend priming chalking surfaces with a coating selected for its ability to wet and adhere.

10. Many woods such as red cedar and redwood contain water soluble stains which bleed through latex paints when these are applied directly to the wood. A suitable primer, specifically recommended by the manufacturer for these woods will seal the wood and prevent bleeding.

11. Latex paints perform well over iron which has been adequately primed, but they offer little protection to unprimed iron, and, in many cases, actually stimulate

STEP 1. Load your brush, then apply two or three dabs of paint along the joint of the siding. This helps to distribute the paint quickly and easily.

STEP 2. Next, brush the paint out well, being sure to coat the clapboard under-edge.

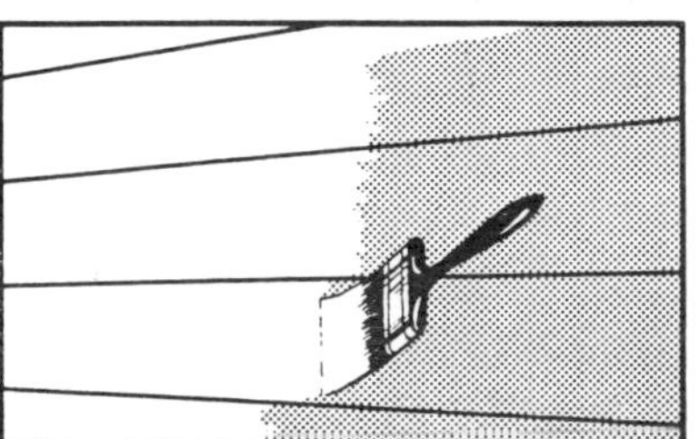

STEP 3. "Feather" the ends of your brush strokes so the coat will be smooth where one section joins another.

National Paint, Varnish and Lacquer Association

Fig. 8. Brushing technique for painting a clapboard wall.

corrosion, which results in unsightly stains on the paint. Aluminum and galvanized steel do not, usually, require a special primer.

12. Specifically recommended latex paint systems can be used on a wide variety of surfaces. In particular, latex paints suitable for use on exterior wood will give satisfactory performance on areas of masonry, asbestos cement shingles, asphalt, and metal which may be adjacent to the wood surface being painted. Some of these surfaces, particularly iron or steel, should be especially primed in advance.

Painting Masonry Surfaces. An important European builder, visiting these shores for the first time, was asked the usual question, "What impresses you most about America?." The gentleman gave the question a few moments thought, then replied, "Colored Concrete." He probably meant painted concrete, which was an unusual sight to him but one that we are apt to overlook or consider commonplace. American builders, landscape architects, farmers and home owners have long since looked to the paint manufacturing industry for the magic that converts dull gray concrete and other masonry into colorful surfaces.

Of course, beautification is just one of the reasons for using paint on concrete, stucco and other masonry surfaces. Paint is also used to change light reflection characteristics, make a masonry surface smoother and easier to clean, or protect the surface from deleterious chemicals and abrasives.

A simple enough matter, the painting of masonry, but as in most simple tasks there are right ways and wrong ways. The purpose of the painting, the condition of the surface to be painted, and the expected exposure all combine to dictate the selection of the correct paint.

If the roof of your house is	You can paint the body	...and the trim or shutters and doors															
		Pink	Bright red	Red-orange	Tile red	Cream	Bright yellow	Light green	Dark green	Gray-green	Blue-green	Light blue	Dark blue	Blue-gray	Violet	Brown	White
GRAY	White	X	X	X	X	X	X	X	X	X	X	X	X	X	X		
	Gray	X	X	X	X		X	X	X	X	X	X	X	X	X		X
	Cream-yellow		X		X		X		X	X							X
	Pale green				X		X		X	X							X
	Dark green	X				X	X	X									X
	Putty			X	X				X	X			X	X		X	
	Dull red	X				X		X						X			X
GREEN	White	X	X	X	X	X	X	X	X	X	X	X	X	X	X	X	
	Gray			X		X	X	X									X
	Cream-yellow		X		X			X	X	X						X	X
	Pale green			X	X		X		X								X
	Dark green	X		X		X	X	X									X
	Beige				X				X	X	X		X	X			
	Brown	X				X	X	X									X
	Dull red					X		X		X							X
RED	White		X		X				X		X			X			
	Light gray		X		X				X								X
	Cream-yellow		X		X						X		X	X			
	Pale green		X		X												X
	Dull red					X		X		X	X						X
BROWN	White			X	X		X	X	X	X	X		X	X	X	X	
	Buff				X				X	X	X					X	
	Pink-beige				X				X	X						X	X
	Cream-yellow				X				X	X	X					X	
	Pale green								X	X						X	
	Brown			X		X	X										X
BLUE	White			X	X		X					X	X				
	Gray			X		X						X	X				X
	Cream-yellow			X	X								X	X			
	Blue			X		X	X					X					X

Fig. 9. Chart suggests color combinations for house exteriors. If house has shutters, paint trim the same color as walls, or white. If there are no shutters, follow chart.

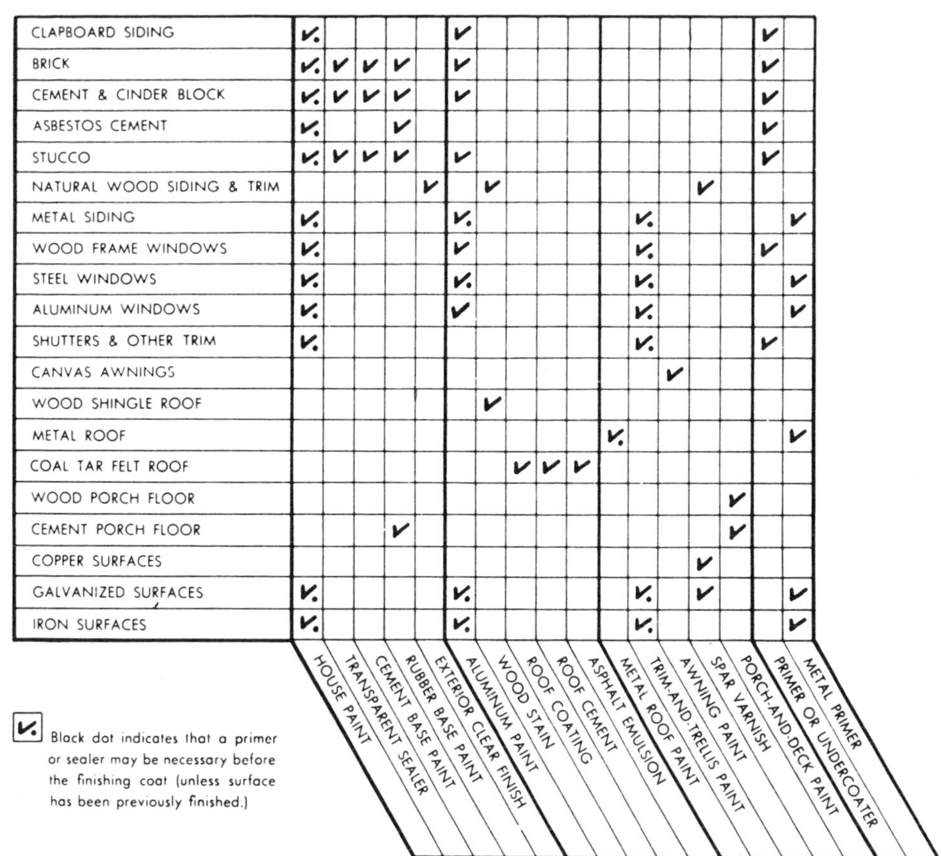

Surface	House Paint	Transparent Sealer	Cement Base Paint	Rubber Base Paint	Exterior Clear Finish	Aluminum Paint	Wood Stain	Roof Coating	Roof Cement	Asphalt Emulsion	Metal Roof Paint	Trim-and-Trellis Paint	Awning Paint	Spar Varnish	Porch-and-Deck Paint	Primer or Undercoater	Metal Primer
CLAPBOARD SIDING	✓•					✓										✓	
BRICK	✓•	✓	✓	✓		✓										✓	
CEMENT & CINDER BLOCK	✓•	✓	✓	✓		✓										✓	
ASBESTOS CEMENT	✓•			✓												✓	
STUCCO	✓•	✓	✓	✓		✓										✓	
NATURAL WOOD SIDING & TRIM					✓		✓							✓			
METAL SIDING	✓•					✓•						✓•					✓
WOOD FRAME WINDOWS	✓•					✓						✓•				✓	
STEEL WINDOWS	✓•					✓•						✓•					✓
ALUMINUM WINDOWS	✓•					✓						✓•					✓
SHUTTERS & OTHER TRIM	✓•											✓•				✓	
CANVAS AWNINGS													✓				
WOOD SHINGLE ROOF							✓										
METAL ROOF											✓•						✓
COAL TAR FELT ROOF								✓	✓	✓							
WOOD PORCH FLOOR															✓		
CEMENT PORCH FLOOR				✓											✓		
COPPER SURFACES														✓			
GALVANIZED SURFACES	✓•					✓•						✓•		✓			✓
IRON SURFACES	✓•					✓•						✓•					✓

Fig. 10. Various surfaces and materials on the exterior of a house which require protective coatings (left side of chart). Checks indicate the proper coating to use as listed along the bottom of the chart. Dot next to check indicates sealer is needed.

Characteristics of Masonry Surfaces. Concrete is quite alkaline, and alkali (ability to neutralize acids) is anything but a friend of many paints. Portland cement is not only in concrete, but in brick and stone mortar, concrete, cinder blocks and stucco. While the alkalinity at the surface decreases with weathering and age, moisture in the concrete may bring more alkali to the surface. Well aged and dry masonry surfaces may be considered comparatively free of this alkali, but one must be sure there is no water present.

In this latter condition, almost any suitable house paint can be employed with good results. But in case of doubt, a paint with built in resistance to alkali should be used, no matter what the age of the concrete.

Alkali Resistant Paints. Of the many alkali resistant paints on the market today, the one selected should be for the expected exposure. For example, if protection is required rather than beauty, select a type of paint designed for the protection needed. Floor paints should be selected for floors and exterior paints for exterior surfaces. In general, a careful study of the instructions on the labels should be a guide. There are six types of alkaline resistant paints available:

1. Latex paints, inherently resistant to alkali, are made from emulsions of resinous materials in water. While other ingredients used in the paint may effect the degree of resistance, any latex paint designed for use on masonry may be used with confidence. Just be sure that the latex paint selected is intended for the surface to be painted, since latex paints come in a wide variety of colors and types for interior or exterior walls, ceilings, floors and other surfaces. Alkyd emulsions (always read the label on the container) are not alkali resistant, so avoid their use on new masonry surfaces. Since latex paints are water thinned, the surface to be painted need not be absolutely dry, but should be dampened before applying the first coat.

2. Portland cement paints contain little organic material and are not subject to attack by alkali. These paints have a long history of success in painting masonry, and are marketed as powders and must be mixed with water before use. Such paints set by the hydration of the cement surface to be covered and must be applied to damp surfaces which must be kept damp until the cement matures, usually a period of two or three days.

Portland cement paints are not suitable for floors and other surfaces where abrasion is anticipated. If other types of paints are applied over a surface painted with portland cement paint the surface may have to be treated or the paint removed completely if peeling has developed.

3. Oil base stucco and masonry paints are similar to conventional house paints in most respects, but are usually reinforced by certain resins to improve their alkali resistance. These paints have the effectiveness and general behavior of conventional house paints and are applied similarly but are least resistant to alkali hence should not be used on fresh cement surfaces. Good results are obtained when concrete and adjacent wooden trim are to be painted the same color, usually a flat or low sheen, which tends to hide surface irregularities.

4. Solvent thinned rubber base paints containing a synthetic rubber type resin and thinned with an organic solvent provide an excellent resistance to alkali and water penetration and resist abrasion. These paints are ideal for basement and porch floors, swimming pools, walks exposed to the weather and are available in a wide range of colors.

5. Alkali resistant paints to which are added catalysts to hasten action can be based on epoxy or urethane resins. They have extraordinary resistance to wear and to chemical attacks and are widely used in chemical plant maintenance. These alkali resistant paints are expensive and difficult to apply and must be applied by professional painters.

6. Due to the low cost and excellent resistance to water penetration, bituminous paints are usually employed for coating the exterior sides of below grade masonry foundations and walls in direct contact with water or wet soils. This covering is not recommended where appearance is a factor and it has a tendency to bleed when coated with solvent thinned paints. The color is most always black.

How to Prepare Masonry Surfaces for Painting. Regardless of the type of paint chosen the surface must be clean. This rule applies whether the surface is new or old. First of all, use a wire brush to remove all dirt, loose particles or other extraneous materials which might interfere with paint adhesion. If there is a white salt-like material adhering to concrete, stucco, or mortar, it is probably efflorescence.

Efflorescence is caused by moisture which dissolves salts in the interior of alkaline materials and carries them to the surface. Efflorescence must be removed before any painting or re-painting is done. Preventing moisture from entering the masonry material will eliminate efflorescence and also prevent damp walls.

An additional step in preparation is to remove all grease or oil by washing with a proprietary cleaner or detergent and water. Poured or precast concrete may have a form release agent on the surface which must be removed by solvent or by several months of weathering. Finally, in preparing masonry surfaces for painting, wash off or hose off surfaces unless efflorescence is present. Allow the surface to dry unless a water thinned paint (latex or portland cement) is to be used. If efflorescence is present, dry-brush it away along with all dirt, dust or other extraneous materials. In some forms of concrete masonry the surfaces are porous, hence a fill coat, wash or grout coat of cement is necessary before painting. Whether or not the fill coat dries with hairline cracks, the finish coat of paint will provide a good seal. To prevent the cracks showing through the finish paint, allow the fill coat to mature before painting.

Before Starting the Painting of Masonry Check These Points. If the concrete, other masonry material or the mortar is well aged (one to two years exposed to the weather except for unusually massive concrete), the chances are that most of the alkali has been washed out or neutralized by the atmosphere. Age increases, however, the likeihood of contamination by dirt, oil or grease, depending on the location. Always be doubly sure that the cleaning job is thorough.

If the surface has been painted, and the paint is in fair condition, use a wire brush to remove loose materials. If the old paint is loose, peeling or heavily chalked, it must be removed, either by sandblasting or other means.

If the old paint is moderately chalked but is otherwise "tight" and non-flaking, and re-painting is contemplated with a water thinned coating (latex or portland cement), a surface conditioner should be used before repainting. A penetrating sealer or masonry conditioner is available from a paint dealer. The paint manufacturer may even recommend the addition of a conditioner to the first coat of latex or portland cement which must be followed

Fig. 11. Spraying paint on exterior of house speeds the job, especially over handsplit shakes, striated shingles or panels, or other heavily textured or very porous materials.

carefully. When using water thinned paint stains may develop when in contact with metals. Thus iron, copper or other metallic objects imbedded in the masonry surface, or adjacent to it, should be primed with a good anti-corrosive primer before painting. Look for nails and other metallic objects, and prime them before applying a water thinned paint.

Painting Interiors. Not all interior walls and ceilings are finished in paneling of wood nor covered with wallpaper or other applied finish. Some owners like a painted wall and ceiling. Properly applied plaster or gypsum board offers an ideal decorative base for the application of paint and an unbroken sweep of surfaces and permits a wide range in the choice of decorative coatings.

Drywall can be painted almost as soon as joints are finished, but up to a generation ago, a painter would wait from six months to two years after a building was plastered before applying paint. He thought rightly, at the time, that plaster had to be cured for this length of time before paint could be successfully applied. Today, a shortened waiting period has been made possible by the use of improved paints and plaster materials. Certain conditions, however, can cause trouble if not recognized, but if proper procedures are followed, a completely satisfactory job will resut.

The first step is to learn the condition of the wall. With this information the painter can determine whether or not it is necessary to apply pretreatment and what types of paint will be necessary. Large volumes of water are necessarily included in wet plaster. Much of this water will not stay in the wall. Some is used in crystallization, but the rest of it will evaporate. The only way it can evaporate is through the surface of the plaster. If paint impedes evaporation damage to the paint may result. It is quite necessary to know just how dry the wall has become. The amount of moisture in a wall can be determined by one of the several moisture meters on the market, and if there is any doubt, a reading should be taken. One way to check the moisture, without the use of a meter, is to loosely cement, over a small area of the wall, a clear plastic film. The presence of moisture will be indicated by condensation on the back of the film after 24 hours. Under normal conditions when the relative humidity is not high, and the temperature is not below 50 degrees F, present day plaster will dry in three to four weeks on a furred wall. A furred wall is one provided with thin wooden strips placed between the plaster and the wall as to form a dead air space. If such a condition is not prevalent a longer time is necessary. A special case is one where plaster is applied to an unfurred wall, preventing evaporation of the water from the back

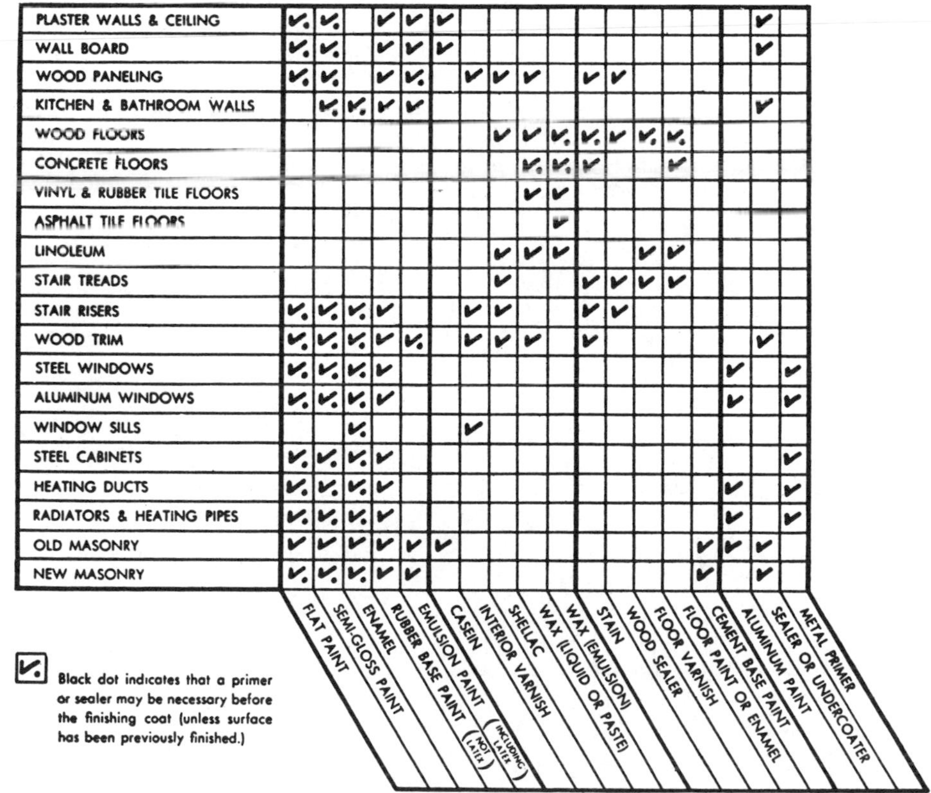

	FLAT PAINT	SEMI-GLOSS PAINT	ENAMEL	RUBBER BASE PAINT (NOT LATEX)	EMULSION PAINT (INCLUDING LATEX)	CASEIN	INTERIOR VARNISH	SHELLAC	WAX (LIQUID OR PASTE)	WAX (EMULSION)	STAIN	WOOD SEALER	FLOOR VARNISH	FLOOR PAINT OR ENAMEL	CEMENT BASE PAINT	ALUMINUM PAINT	SEALER OR UNDERCOATER	METAL PRIMER
PLASTER WALLS & CEILING	✓•	✓•		✓	✓	✓											✓	
WALL BOARD	✓•	✓•		✓	✓	✓											✓	
WOOD PANELING	✓•	✓•		✓	✓•		✓	✓	✓		✓	✓						
KITCHEN & BATHROOM WALLS		✓•	✓•	✓	✓												✓	
WOOD FLOORS								✓	✓	✓•	✓•	✓	✓•	✓•				
CONCRETE FLOORS									✓•	✓•	✓			✓				
VINYL & RUBBER TILE FLOORS									✓	✓								
ASPHALT TILE FLOORS										✓								
LINOLEUM								✓	✓	✓			✓	✓				
STAIR TREADS								✓			✓	✓	✓	✓				
STAIR RISERS	✓•	✓•	✓•	✓			✓	✓			✓	✓						
WOOD TRIM	✓•	✓•	✓•	✓	✓•		✓	✓	✓		✓						✓	
STEEL WINDOWS	✓•	✓•	✓•	✓												✓		✓
ALUMINUM WINDOWS	✓•	✓•	✓•	✓												✓		✓
WINDOW SILLS			✓•				✓											
STEEL CABINETS	✓•	✓•	✓•	✓														✓
HEATING DUCTS	✓•	✓•	✓•	✓												✓		✓
RADIATORS & HEATING PIPES	✓•	✓•	✓•	✓												✓		✓
OLD MASONRY	✓	✓	✓	✓	✓	✓									✓	✓	✓	
NEW MASONRY	✓•	✓•	✓•	✓	✓										✓		✓	

Fig. 12. Various surfaces and materials in the interior of a house which require protective coatings (left side of chart). Checks indicate the proper coating to use a listed along the bottom of the chart. A dot next to a check indicates sealer is needed.

Fig. 13. Many helpful accessories which speed and simplify interior painting while making the job easier.

of the plaster. If but one surface of the plaster is open, the drying time is greatly lengthened. A meter is the safest device to check the progress of drying such walls.

If it is absolutely necessary to paint the plaster while water is still present, the proper selection of paint can minimize the possibility of trouble, but it is best to wait until the drying time is complete. Plaster should be examined for firmness since there may be areas that have become soft from "sweat-outs" and "dry-outs." To correct a "sweat-out" it is necessary to remove the affected area and replaster. A "dry-out" is due to the evaporation of the water before the gypsum crystallizes. To correct a "dry-out" it is necessary to spray the affected area with an alum and water solution. About two pounds of alum per gallon of water. This provides the water necessary for crystalization, and should be done until the affected area has hardened to a condition which will receive paint satisfactorily.

All plastered walls and ceilings to be painted should be examined for uniform density. If the finish trowelling is not uniform over the wall, there may be spots which have non-uniform suction. These spots are usually hard and have a dead appearance. Such spots will cause uneven gloss and color in the paint if this is not prevented. Dead spots can also be caused by the lack of thorough mixing of lime and of gauging plaster. If the plaster is uniformly dull, it should be examined for "chalk." This is a white powder which shows on the hand or a cloth rubbed over the wall.

Sometimes an examination will reveal tiny crystals on the wall. Evaporating water brings these tiny crystals to the surface. These are crystals of soluble salts called efflorescence, and are found infrequently in plastered walls today. If found, these crystals must be brushed off repeatedly until no more are formed.

Just prior to painting, the walls and ceiling must be examined for any grease or oil. If grease or oil is on any portion of the surfaces, such must be removed with a detergent. The plaster must then be allowed sufficient time to dry before paint is applied.

Painting Plastered Surfaces. When it is reasonably certain that plastered walls are dry and if none of the imperfections previously stated are present, any of the standard interior finishes for walls and ceilings, when applied, should give complete satisfaction. Conditions causing difficulties are few and far between,with present day building materials, and by far painting jobs on new plaster do not involve many problems. These paints will last for years. The chart, **Figure 12,** suggests the right paint for various interior jobs.

When an oil or alkyd primer is used, it is wise to allow at least a week for drying before the top coat is applied. Sometimes when the second coat is applied too soon after the primer, peeling develops. One test of a well dried paint job is the lighting of an ordinary kitchen match by striking it in an inconspicuous place. If the match does not light, it indicates a damp surface caused by too much moisture evaporation through the paint. Further drying is required before a second coat of paint. In addition, the pigments in the paint must be insensitive to alkali. Some small quantities of alkaline chemical may remain in the plaster and if this is in solution in water, it will react adversely with certain color pigments.

Latex paints are best for incompletely dried walls. These paints have satisfactory adhesion to damp plaster, and if only one coat is applied the water can continue to evaporate. One coat of latex paint is not usually sufficient to achieve complete coverage, but if one is acceptable and satisfactory to the owner, it will allow evaporation to take place until the plastered surfaces are in condition to receive a permanent decorative finish.

In considering the type of paint to choose the owner should keep in mind the light reflectance of various colors. Does the owner wish to make the most of natural and artificial light within a room or does he desire to soften the skyglare that sometimes enters through large glass areas? Dark colors absorb light, while light colors reflect light. The following list of 14 colors show the percentage of light reflectance when used on interior walls or ceilings; white 80%, ivory (light) 71%, apricot beige 66%, lemon yellow 65%, ivory 59%, light buff 56%, peach 53%, salmon 53%, pale apple green 51%, medium gray 43%, light green 41%, pale blue 41%, deep rose 12% and dark green 9%.

A soft area caused by a sweat-out or a dry-out can sometimes be hardened with a coat of shellac or lacquer if it is not large. If the shellac binds the wall, further finishing can be applied, otherwise the plaster will have to be replaced. If the wall is not uniform due to the presence of uneven suction, it is essential to allow it to dry before painting begins. Then apply an oil type primer-sealer after which any desired paint system can be used.

Fig. 14. Thick texture paints, applied with a roller over drywall construction, cover large areas quickly and resemble stucco. These paints require no sealer and hide imperfections.

Fig. 15. Wallpaper, such as this floral print with a simulated railing, enhances the architecture and character of an interior.

Chalky plaster is quite common and a frequent cause of paint peeling. Chalk must be removed before painting. Sometimes, in a subgrade plaster, there may be loose sand, which must be removed. Vigorous brushing will remove chalk and sand but it is often difficult to determine that all has been removed. Latex paints will not easily wet chalk on a plaster surface and should be avoided. Use instead, an alkyd flat or an oil vehicle primer-sealer, or a proprietary surface hardener prior to any paint system.

At times it may be necessary to paint a wall on which there is efflorescence, before it is certain that the crystals have ceased to form. If the plaster is on a furred wall so that it can dry from the back as well as the front, it is usually possible to stop crystal growth on the inside wall. In this instance the wall is sealed and the crystal formation is transferred to the back of the plaster. The sealing can be accomplished by two coats of a latex paint or any other system of paint that is impervious to water vapor. If the plaster is on a solid wall without furring it must not be sealed before it is dry.

In isolated instances very fine, hairline cracks develop in finish plaster. These are hard to detect until the wall is painted. The paint is mistakenly blamed for this condition. Painted walls should be inspected by viewing from a low angle and the eyes close to the wall to determine if any cracks are present. If hairline cracks are found brushing over them with an emulsion of the paint will frequently obscure them. The secret of a good painting lies in good workmanship when the primer is applied to the wall. The bond between the plaster and the first coat of paint determines the adhesion of future paint coats.

Painting must be done by an experienced and qualified painter. In applying paint to new plaster, it is very important that the painting contractor first consider carefully such variable factors as the condition of the surface to be painted, the type of paint to be used and the existing atmospheric conditions. Suggested equipment for interior painting is shown in **Figure 13.**

Texture Paints. In addition to color, decorative effects can be produced on walls by the use of thick texture paints. Applied with a roller, **Figure 14,** these paints create ripple or spatter finishes which resemble stucco. A textured finish usually requires no sealer or prime coat and is heavy enough to conceal imperfections in the wall surface.

Wallpaper. The renewed popularity of colonial and other types of traditional decor has brought with it an increase in the application of wallpaper. Pre-trimmed, strippable, and pre-pasted materials simplify installation to the point where the homeowner has no difficulty in applying it plumb, without trapped bubbles, and with perfectly matched panels.

Besides carrying out and enhancing the architecture of a house, **Figure 15,** wallpaper also has the advantage of concealing wall defects. Some types of waterproof material, usually applied with special glues instead of paste, provide durable wall surfaces for kitchens and bathrooms.

Chapter 22

MODERN BUILDING MATERIALS

Modern building materials and methods are being created and manufactured in an industry which is turning to new technology for better and faster construction of houses and more comfortable living.

The cost of labor has risen to the point where costs of building can be reduced or held at present levels only through new products, materials and methods, which speed construction and reduce the amount of labor needed.

Buyers' demands and building codes put additional pressure on the manufacturers of building products, and as a result, research labs have sprung up in all parts of the country to bring scientific precision to the development of fire resistant and soundproof materials, durable finishes, maintenance-free materials for interiors and exteriors, and other time and labor saving aids for the builder.

Fig. 1. Built-in gutter of aluminum creates a crisp eave line while eliminating much hard-to-reach trim painting.

Fig. 3. Red cedar shingles are a popular roof covering which requires no finishing. Workman is fitting shingles along valley.

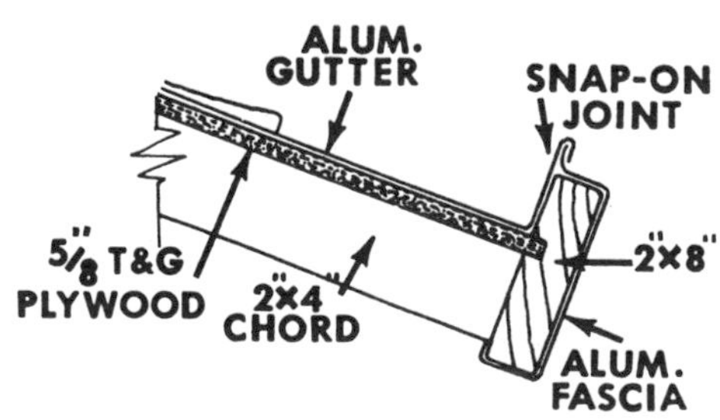

Fig. 2. Gutter extends well up under the wood shingles. Face metal snaps over wood fascia, locks onto gutter section.

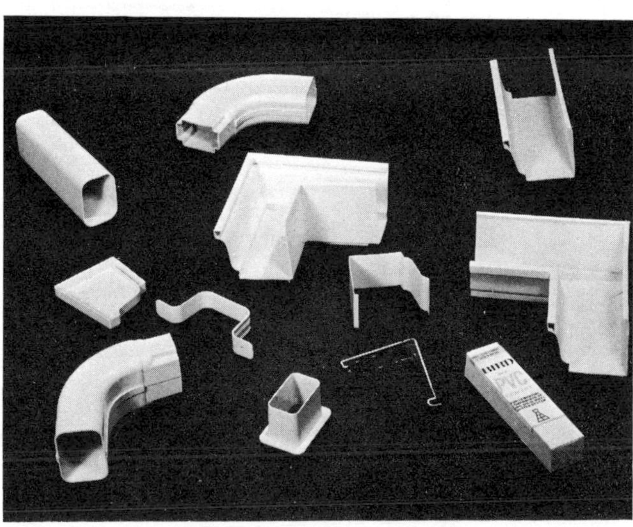

Fig. 4. A complete rain carrying system of solid vinyl includes all parts needed to join, close, hang and drain vinyl gutters.

Throughout the pages of this book, many of these new methods and materials have been used to supplant the methods and materials described in previous editions. This chapter only highlights some of the important developments not mentioned earlier in the various categories of house construction. To endeavor to cover all the work done by manufacturers serving the building industry would require a separate book by itself.

Roofing/Roof Construction

Built-in Gutter. Painting along eaves is eliminated by the prefinished built in gutter shown in **Figure 1.** Note how the aluminum gutter extends well up under the shingles, **Figure 2.** The aluminum is hooked around the 2″ × 8″ wooden fascia strip attached to the rafters and is snapped on to the aluminum fascia strip at the top of the gutter system.

Shingling a Roof. Red cedar shingles, **Figure 3,** are often applied as a finish covering on a roof. Note the aluminum gutter, the aluminum valley nailed in place, the building paper under the shingles and the plywood sheathing laid over trusses spaced 4′ o.c.

Solid Vinyl Gutter System. A complete rain carrying system, of white solid vinyl, consists of gutters, downspouts, inside and outside corners, regular and side elbows, end caps, outlet tubes, joint strips and downspout straps is available, **Figure 4.** The fittings are joined by a cement and a solvent to seal the joints. Specially designed aluminum hangers are used allowing the gutters freedom to float with temperature changes. Vinyl will not blister, dent, warp, corrode, conduct electricity nor support combustion.

Aluminum Shingle Shakes. A roofing of aluminum is made up in deeply embossed, wood textured sheets, 3/4″ × 12″ × 36″ to resemble shakes or shingles of wood, **Figure 5.** The shingle shakes of aluminum are locked on all four sides. Aluminum reflects the heat of the sun and holds in the heat of the house. The textured sheets are available in aluminum, sandalwood, green, white or charcoal colors.

Modern Rafters. The present day rafter is shop built and known as a truss. A truss is made up of several members of wood held rigidly in place by metal or wood gussets. The truss covers twice the length of a pitched rafter, extending from side wall to side wall. A truss supports the roof sheathing and finish roofing, forms the eaves at the exterior walls, and, the horizontal member or lower chord becomes the ceiling joist. This truss, **Figure 6,** is made of 2″ × 4″ stock for a span of 28′ and is placed 4′ on center on the plates of the side wall framing or panels.

The various members are held in place by plywood

Fig. 5. Embossed aluminum shingles, 36″ long, resemble deep textured shingles of wood while reflecting the sun's heat.

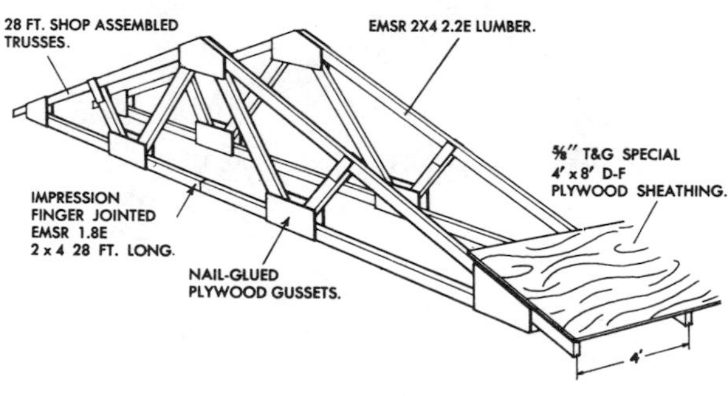

Fig. 6. Workman nails temporary spacer strips (above) to shop assembled truss of 2×4s joined with nail-glued plywood gussets.

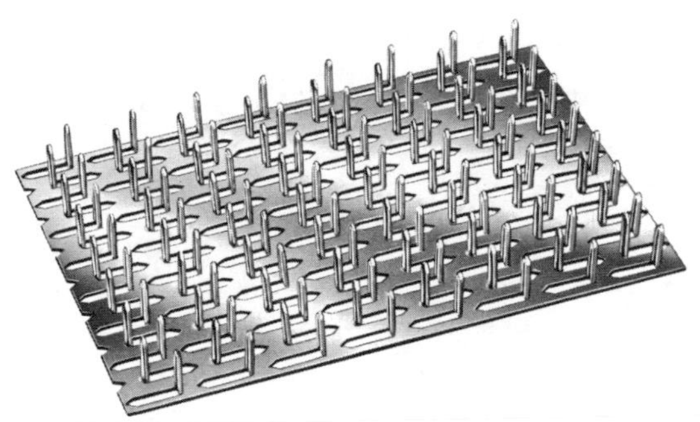

Fig. 7. Metal is partially punched out of truss connector plate and bent upward to form prongs which are pressed into wood.

Fig. 8. Nails are driven through perforations in this plate to join truss members.

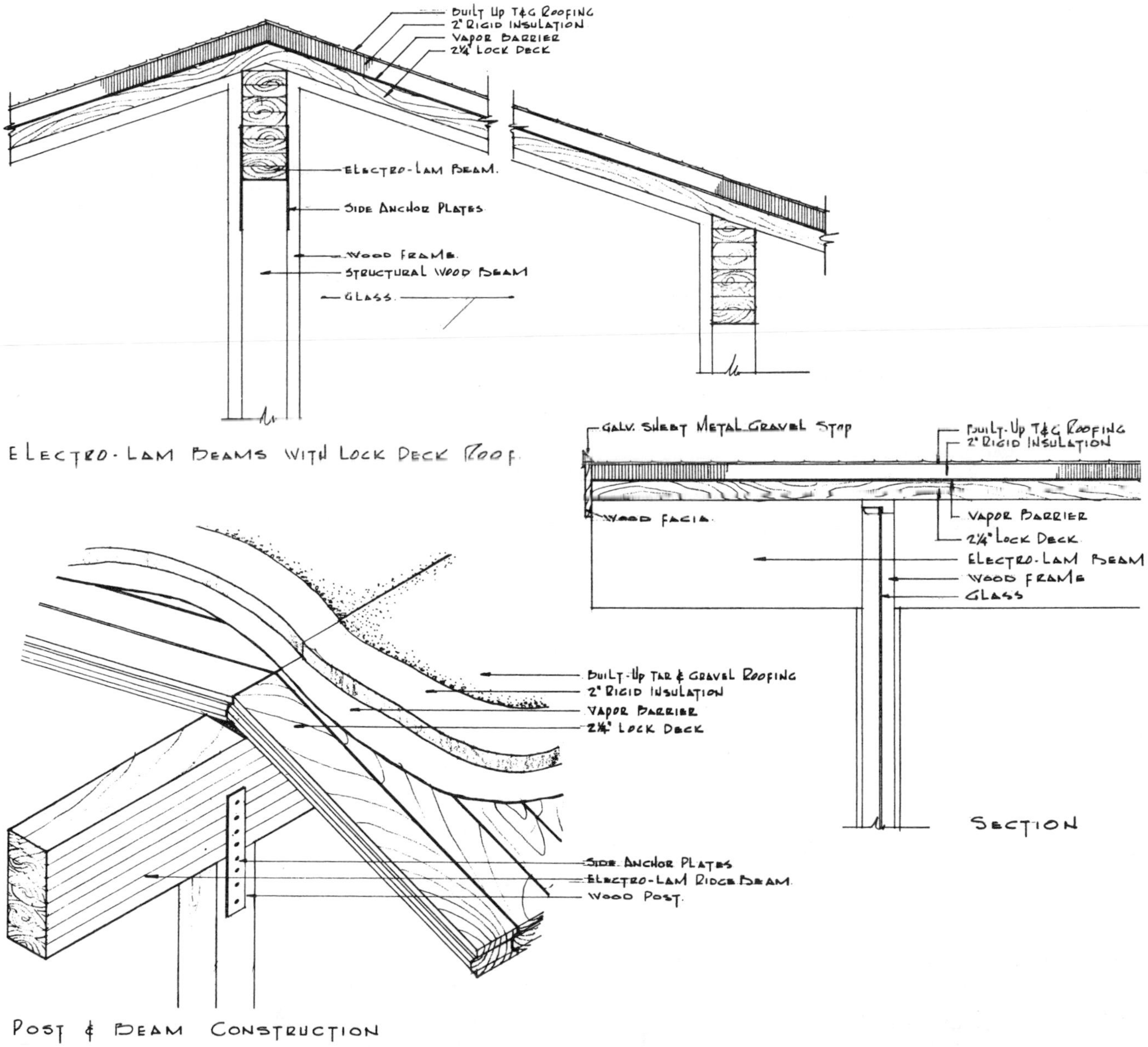

Built Up T&G Roofing
2" Rigid Insulation
Vapor Barrier
2¼" Lock Deck
Electro-Lam Beam.
Side Anchor Plates
Wood Frame.
Structural Wood Beam
Glass.

Electro-Lam Beams With Lock Deck Roof.

Galv. Sheet Metal Gravel Stop
Built-Up T&G Roofing
2" Rigid Insulation
Wood Facia.
Vapor Barrier
2¼" Lock Deck.
Electro-Lam Beam
Wood Frame
Glass

Section

Built-Up Tar & Gravel Roofing
2" Rigid Insulation
Vapor Barrier
2¼" Lock Deck
Side Anchor Plates
Electro-Lam Ridge Beam.
Wood Post.

Post & Beam Construction

Fig. 9. Details of post and laminated beam construction covered with 2-1/4" tongue-and-groove plank decking and built-up roofing.

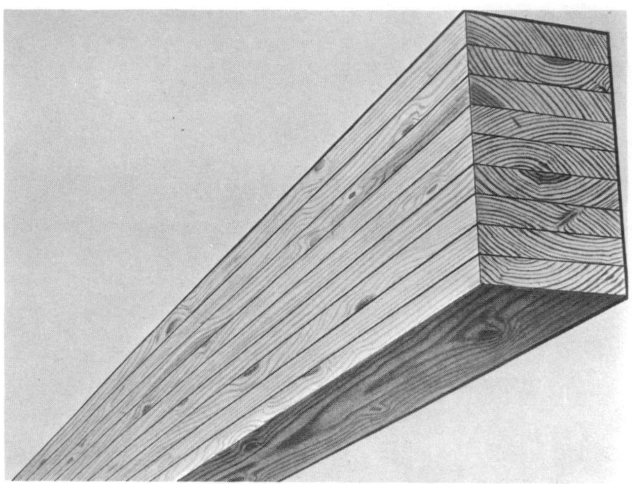

Fig. 10. Electro-laminated beam is built up of select lumber. New beams are built of two or more planks laminated vertically.

Fig. 11. Workman joins end-matched laminated T&G roof decking (above). Staggering laminations (below) creates tongue and groove.

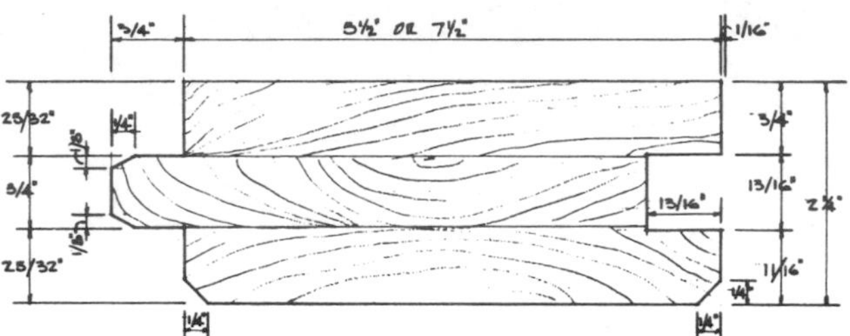

Fig. 12. Trusses are assembled accurately without skill on jig table.

gussets, glued and nailed to the chords.

Metal Truss Plates. Instead of plywood gussets, some trusses are assembled with metal plates, **Figures 7, 8.**

Laminated Beam Construction of Roof. In post and beam construction, the roof deck rests, at the ridge and plate, on built up electro-laminated beams, **Figure 9.**

The beam, **Figure 10,** is built up of select framing stock, glued and pressed together. New laminated beams, joined vertically, resemble solid lumber. Laminated T & G decking, **Figure 11,** completes the roof system.

Jig for Roof Trusses. The wooden table-top jig is completely flexible to accommodate all size trusses. It can be quickly and simply adjusted, **Figure 12,** to form different slopes or spans. It contains an angle iron frame with a channel iron guide for the top cord, and tightening devices, such as screws or cams to hold the truss in alignment while plates are applied to the top surface. The rigid truss is then flipped over onto the second table where the plates are fastened to the opposite face of truss members. Sometimes a roller press is used, **Figure 13,** for the final locking of plates into wood members. The roller press is 12′ in length. Once the end of the truss enters the press, the rollers pull it through without the need for pushing or guiding by workmen. A simple roller conveyor or a mechanical belt is used to convey the trusses to the press.

Fig. 13. Metal connector plates are forced into truss members by feeding truss through massive rollers of this press.

Fig. 14. Large acoustical ceiling tiles and translucent panels drop quickly into suspended grid of aluminum or redwood.

Fig. 15. Wood paneling is available in a wide range of durable factory applied finishes which can be cleaned wth a damp cloth.

Ceiling/Wall Coverings

Suspended Ceiling. An acoustical ceiling system, **Figure 14,** is suspended on aluminum grids resembling redwood to harmonize with the wall paneling. Translucent panels are inserted beneath lighting fixtures for soft, diffused illumination.

Interior Walls. Walls of a dining room, den or foyer are often finished with vertical paneling, **Figure 15,** to harmonize with a suspended ceiling with darker grids. New, factory applied finishes are durable and easy to clean, can be used in kitchens, **Figure 16.**

Interior Wall Covering of Vinyl. The walls of a bedroom, **Figure 17,** may be covered with an interior wall covering of Gypsum panel plasterboard, then finished with a vinyl "wallpaper" with concealed joints created by overlapping flaps of the vinyl surface after filling the joint flush.

Acoustical Tile. A non-combustible, mineral fiber, acoustical tile, **Figure 18,** can be applied with a piggy back stapler to provide sound absorption for rooms. With the piggy back stapler two staples are driven one directly over another without moving the tacker, at midpoints and at corners of the edge flange of the tile joints. The first staple spreads the second staple apart for a more secure grip.

Filigree Panels. Filigree panels (ornamental openwork) are available in hardboard (a pressed wood product) to build folding screens, cabinet grilles, room dividers, suspended ceiling panels, decorative ceilings, etc. The panel material may be cut and worked with ordinary carpenter's tools. The panels should be supported by a frame molding, **Figure 19.**

The panels may be finished with oil or water paint, enamel, stain, lacquer, etc. A suspended ceiling unit is

Fig. 16. New finishes on cabinets and wall paneling in this modern kitchen won't absorb grease, are easy to keep clean.

Fig. 17. Overlapping flaps on edges of vinyl surfaced gypsum board are cemented and trimmed for invisible joint.

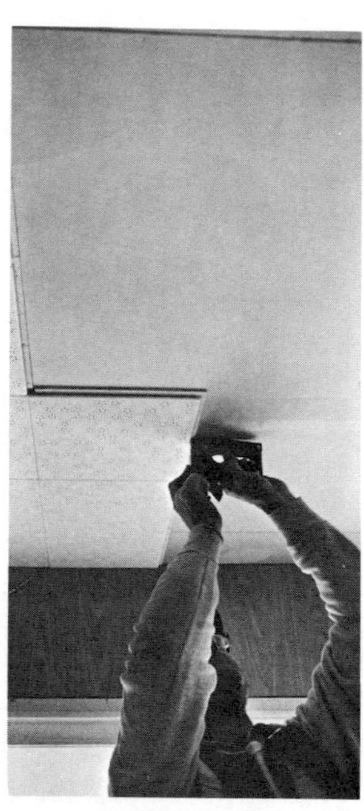

Fig. 18. "Piggy back" staples—one driven over the other-spread to fasten ceiling tile securely to ceiling board.

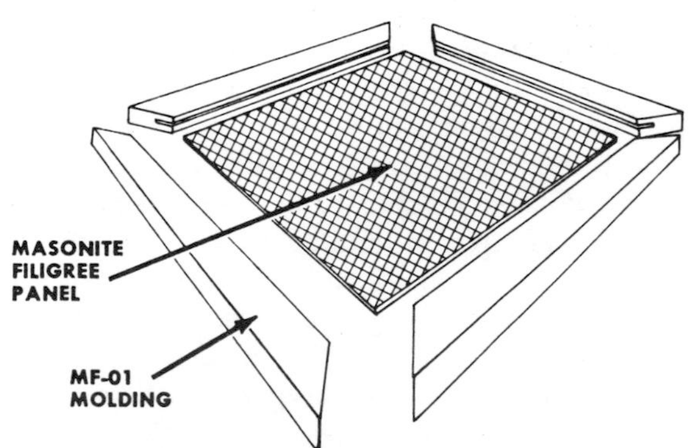

Fig. 19. Filigree panels are stamped from hardboard to create open patterns which can be used for decorative screens, etc.

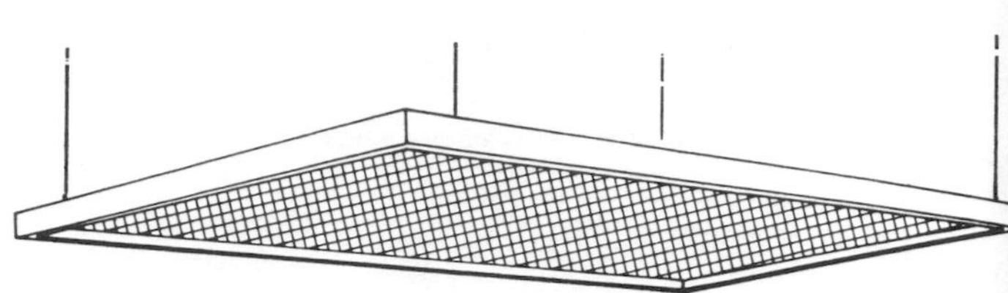

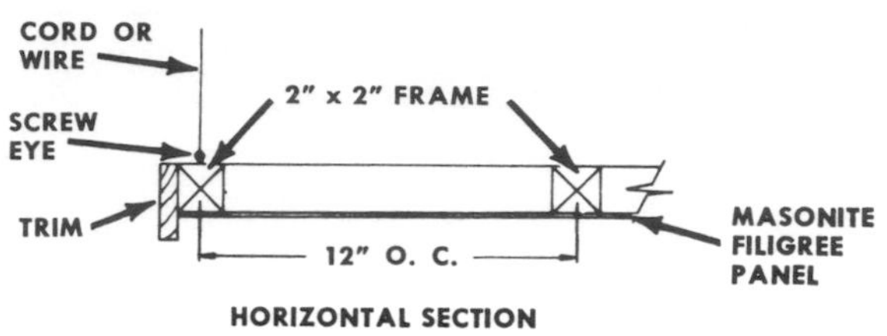

Fig. 20. Filigree panel used as a suspended ceiling.

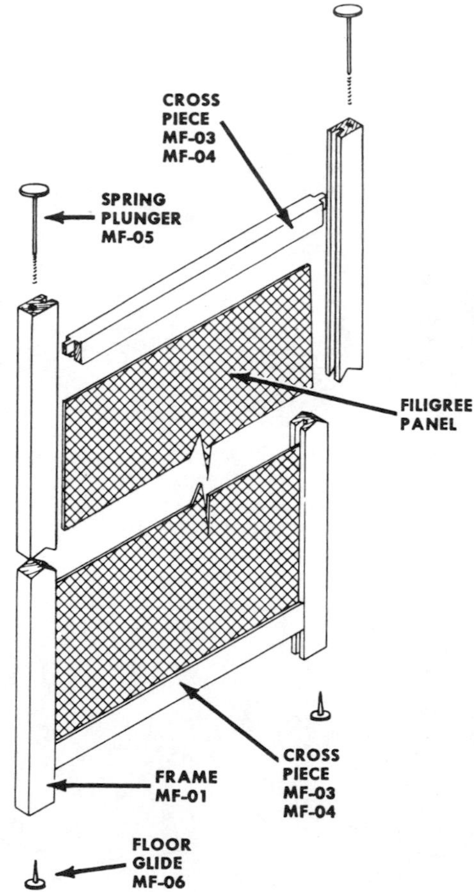

Fig. 21. Framed filigree panel is a room divider.

Fig. 22. Plastic bricks, ¼" thick, are applied to walls with special mastic which simulates real mortar joints.

Fig. 24. Cast-iron drain, waste and vent system joined by stainless steel sleeves clamped over neoprene gaskets.

Fig. 23. Walls and floor of this cheerful kitchen are surfaced with plastic brick that's difficult to tell from real brick.

shown in **Figure 20** and a section of a room divider in **Figure 21.**

Simulated Brick. Plastic bricks 1/4″ thick, in fast color, light weight, and non-porous are available to be pressed on a surface coated with a special mastic which dries to give the appearance of real mortar. Application can be made over painted walls or plywood, **Figure 22,** or on wood flooring, **Figure 23.**

Plumbing

Cast Iron Pipe. Cast iron pipe and joint system, **Figure 24,** for plumbing drain waste and vent piping is designed to permit a 3″ diameter vent stack to fit between 2″ × 4″ wall studding. The overall diameter of the joint is 3-1/2″. The waste pipe diameter is 3-5/8″.

Fig. 25. One-piece shower stall of molded fiberglass is quickly installed and connected to water and drain piping.

All that is needed to install this system is a wrench, screwdriver and a snips. There is no need for a lead pot, lead, oakum, yarning tool, caulking iron or hammer. The cast iron pipe is held in place with Neoprene gaskets bound to the pipe by stainless steel bands as shown.

Shower Stall of Fiber Glass. The shower stall in **Figure 25,** may be nailed directly to studding. The stall is made in modules, with holding clips, making for watertight joints. This stall is made of four matched, fitted modular sections, factory pre-drilled for drain and strainer, mixing valve, shower head, gooseneck and flange. Grouting is not needed.

Plastic Fittings. Fittings for drain and vent pipe, as well as cold water supply pipe, are now available for house plumbing systems. The plastic pipe is cut square with a hand saw in a miter box, **Figure 26,** and any burrs are removed with a pocket knife. The solvent cement is brushed inside the fitting socket and the end portion of the pipe entering the fitting. The pipe is pushed all the way into the socket and rotated, forming a permanent, leak proof joint.

Insulation

Polystyrene Insulation. Rigid foam plastic insulation board is available in a variety of thicknesses, widths and lengths. The polystyrene beads fuse together in the molding process forming millions of small, individual, closed cells that unite in a rigid foam plastic structure, which is waterproof.

These insulation boards are ideal for insulating concrete slabs on grade, requiring the minimum of cutting and fitting and are readily installed in large sections, **Figure 27.** As an insulation for cavity walls this insulating board reduces heat loss and moisture penetration, **Figure 28.**

Woven Pile Weatherstripping. A dense, flexible wood fiber having a backing strip impregnated with a specially formulated polyester resin set in various shaped metal or plastic backing is used for weatherstripping. The pile is set in various backings, such as U-shaped glass channels, flat shape, offset shape, snap on shape or plastic backed

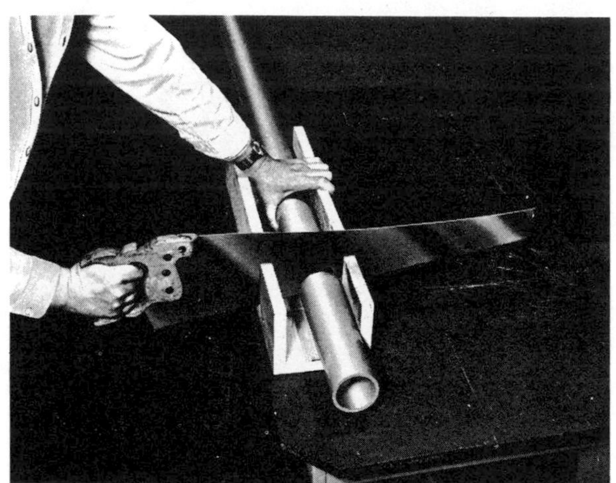

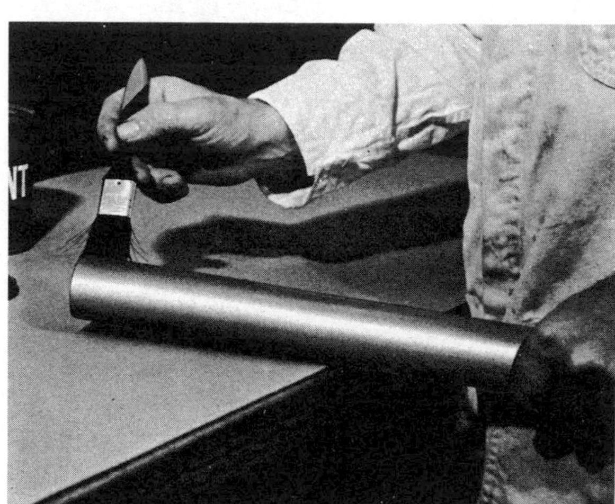

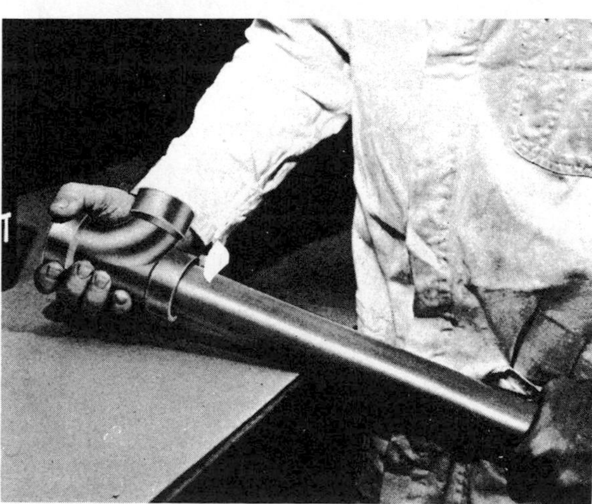

Fig. 26. Four steps in assembling plastic plumbing: Cutting pipe to required length (top left); brushing solvent on end of pipe (top right); brushing solvent inside meeting joint of fitting (bottom left); pressing pipe into fitting (bottom right).

Fig. 27. Rigid foam plastic insulation boards reduce perimeter heat loss in concrete slab floors on grade.

Fig. 28. Installed in 1″ air space between cavity walls, plastic foam insulation reduces heat loss and moisture penetration.

shapes, **Figure 29.** The pile weatherstripping reduces air infiltration, repels water and slides silently and easily.

Electrical Devices

Receptacle Circuits. Convenient receptacle circuits are provided in a surface mounted baseboard system, **Figure 30.** With this system it is possible to install electrical wiring faster and easier than with conventional methods. The surface mounted electrical system makes it easy to add more outlets without breaking into the wall. The need for drilling studs to string conductors in the wall is eliminated, and the number of junction boxes required in a room is reduced from one per wiring device to one or two per room depending on the wiring layout.

Once the vinyl baseway is installed, additional outlets can be added without the necessity of breaking into or through walls. The cover plate can be snapped off, additional outlets installed and the cover plate snapped back on. The fishing of conductors through walls is eliminated. Electrical inspection can take place at any time after installation since the system is readily exposed by removing cover plates. The vinyl baseway is a rigid vinyl that will not support combustion, has a high impact strength, and is resilient enough to permit easy snapon assembly. It cannot be dented and is virtually marproof. It is easily cleaned with soap and water and can be painted.

Intercom System. An intercommunication system which will provide FM or AM music to every room in the house, even to the patio, is shown in **Figure 31.** The intercom can also be used to talk and listen from room to room, and to screen callers at the front door. There is also a telephone jack built into the chassis. The 5″ × 7″ speaker reproduces normal voices clearly, may be installed at a convenient place in the room. It has a rotary volume control, and can be set for privacy with no eavesdropping. The system should be installed as the house is constructed, since wiring runs through walls. The system operates on 33 watts and 120 volts electrical current. **Figure 32** shows the intercom master panel installed in a kitchen.

Fig. 29. Five basic shapes of pile weatherstripping used to reduce air infiltration at doors and windows. Pile is bonded to metal or plastic backing strips.

Door Chimes. Door chimes are available in a variety of styles and colors to match almost any decor. Two notes are usually used for the front door signal and one note for the rear door signal. The chimes operate on a transformer connected to the house current, **Figure 33.**

Fig. 30. Surface mounted electrical system eliminates running wiring through walls. Receptacles are installed in vinyl baseway.

Fig. 31. An intercom system with Am-FM radio provides convenience, entertainment, and protection throughout the house.

Fig. 32. Intercom master panel, usually installed in kitchen, provides instant communication with any room in the house.

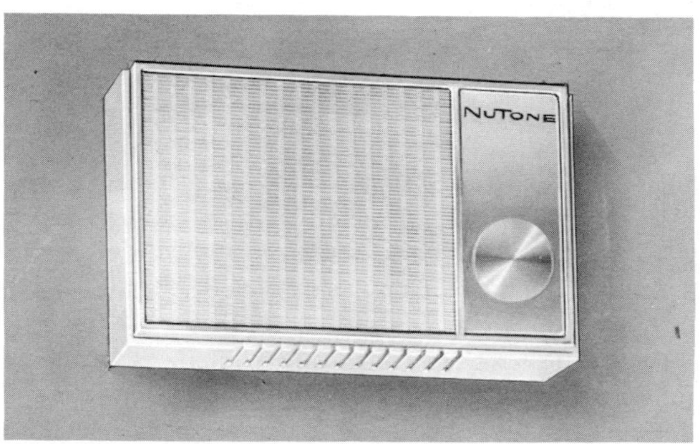

Fig. 33. Modern door chimes are made to sound two notes for the front door, one for the rear door, operate on transformer.

Electric Heating Mats to Melt Snow. Electric heating mats melt snow and ice on sidewalks, steps and driveways to save backbreaking shoveling. The mats are pre-assembled, pre-spaced heater wire anchored in galvanized steel mesh on 1′′ centers. The mats are placed in driveways and finish concretepoured over them.

Heating wires can be quickly and easily installed on black top driveways, too. Spacers for cables are placed on 2″ centers as shown in **Figure 34.** Cables are laced around the spacers, then connected to an electrical outlet as shown in **Figure 35.** Units can be placed in driveways to melt snow or ice in two 18′′ wide tracks. Cables can also be placed in sidewalks, under roofs near the edge of gutters, in gutters and in downspouts. A 120′ length of cable operates on 230 volts at 600 watts.

A Central Vacuum System for Cleaning. A central vacuum system is now available for every room in the house. This system is made up of the power unit and tank, **Figure 36,** the air conveyor system, a 25′ lightweight vinyl hose and the cleaning tools. The vacuum conveying system consists of rigid tubing in 8′ lengths, flexible tubing in 6′ lengths, branch fittings, slip couplings, wall inlet valves, utility inlet valves, and an exhaust vent at the tank. All parts are designed for slip fit, using plastic tape or liquid adhesive.

Low voltage switch leads are built into each wall valve, **Figure 37,** to turn the power unit on when the cleaning hose is inserted. One advantage of this system is that it is not necessary to carry a vacuum cleaner up and down stairs. Even more important, it produces cleaner results. The ordinary vacuum cleaner has a porous bag to hold dirt and to let out sucked in air. The bag holds the larger particles of dust and dirt but the very small ones escape through the fine pores of the bagand are redistributed around the room. In the central system the bag of the power unit is vented to the outside so that the small particles are not retained in the house.

Windows/Doors

Prehung Windows. Workmen are truing up a factory built window frame of wood. **Figure 38,** inserted in the rough opening of an exterior wall. The window is completely factory built with doubled hung glazed sash, saving much time and labor at the building site.

Aluminum Replacement Window. When windows deteriorate or become inoperable, a replacement window designed with an aluminum frame can be fitted easily into the existing opening to cover all of the old exterior window framing, **Figure 39.** The sectional view, **Figure 40,** shows how the replacement window fits into the window opening.

Louvered Screen Shutter Door. A louvered screen

Fig. 34. Before placing final layer of blacktop driveway, Anchor spacers are placed on 2" centers for electric heating wire.

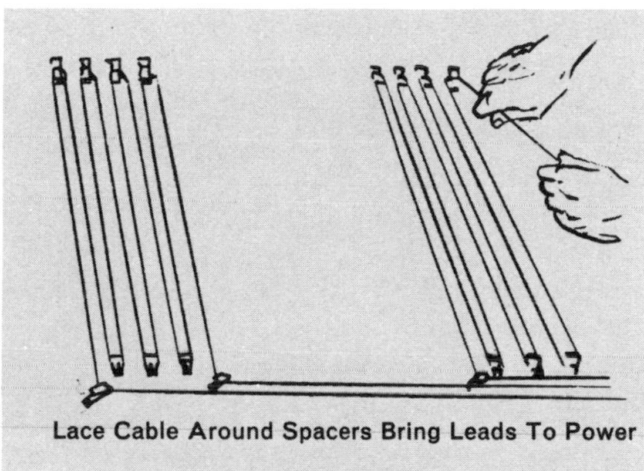

Fig. 35. Lace wire around spacers and bring leads to power source. Test before rolling on finish life of blacktop.

Fig. 36. Central vacuum system deposits dirt in cannister placed in basement or garage, where it can be emptied easily.

Fig. 37. Wall inlets for a central vacuum system are conveniently located throughout the house.

Fig. 38. Workmen plumb factory built window frame in opening.

Fig. 39. Aluminum replacement window can be fitted into the existing opening to cover the old window frame and trim. New sash is aluminum framed.

shutter door made in authentic Colonial style, **Figure 41,** allows full ventilation while keeping out insects, rain or passing glances. Used for the main entry, the 3′-0″ × 6′-8″, doors are made of vertical grain west coast hemlock.

The screens have white enameled frames with a 14 × 18 mesh, gun metal aluminum screen cloth. The screens are recessed into the louvered doors on one side, covering slats, but can be removed for cleaning or painting. When the doors are in an open position, screens are concealed. Louvers face outward when doors are closed so that rain is deflected to the outside.

Pole Type Building

Pole Type Buildings. Pole type construction utilizes pressure treated poles set in the ground as main structural members. This type of construction was used in pioneer days. Because preservation treatments were not available, however, the life of the buildings were comparatively short.

Now, however, with the availability of pressure treatment which resists decay and termite attack, pole type construction over the past decade has been phenomenal. Because of its many inherent advantages, this sturdy low cost construction has rapidly spread from the farm, and today tens of thousands of fine residential, commercial and industrial pole type buildings are built annually.

Poles are sunk from 5′ to 7′ into the ground, in auger bored holes, into which a concrete pad from 6″ to 12″ thick is poured for seating. Depending on the steepness of the terrain, the height of the exposed pole up to the floor joists runs from 4′ to 14′. Some houses are of two-story construction, and in all cases the poles continue straight up supporting the floors as well as the roofs.

Pole type construction is established on sites having steep terrain or over water where conventional construction costs would be very high. **Figure 42** shows a house being constructed on poles. **Figure 43,** shows the completed dwelling.

Exterior Wall Covering

Overlaid Fir Plywood Siding. A medium density, resin-impregnated fiber is permanently bonded to an exterior type of fir plywood. The tough, rugged overlay resists grain raise and checking and retains a smooth, unbroken surface. This siding is available in pre-cut panels or in beveled siding. The beveled siding comes in 12″, 16″ to 24″ widths and lengths 8′ to 9′-4″. The panels are available in 4′ ×8′ size or smaller or larger. Special pre-cut siding widths are available for wide lapped exterior wall covering, **Figure 44.**

Solid Vinyl Siding. Vinyl siding is an extruded, rigid, polyvinyl, chloride compound manufactured for the construction industry. The panel thickness for siding,

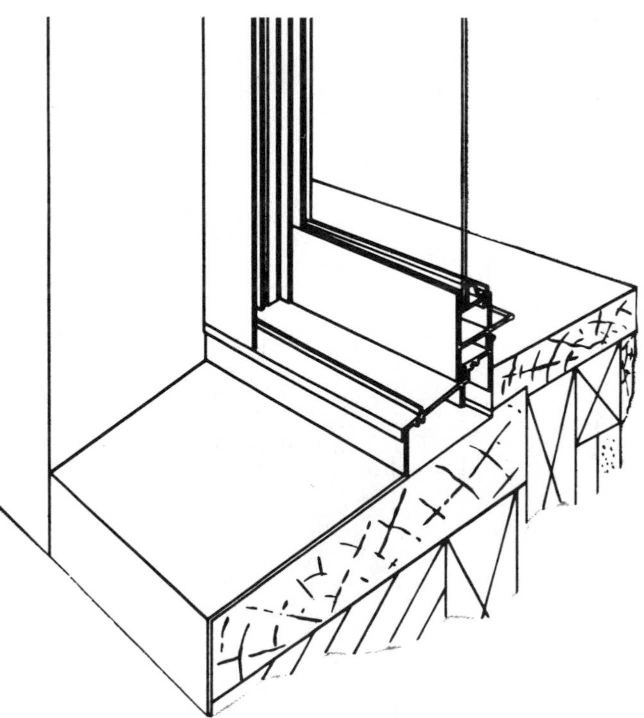

Fig. 40. Sectional view shows how replacement window fits into window opening.

Fig. 41. Louvered shutters have screen installed in side facing wall. When closed across doorway, screen is on outside.

Fig. 42. Floor joists are bolted to pressure treated poles sunk 5′ to 7′ into the ground. Decking is laid over joists.

Fig. 43. Completed dwelling constructed on poles erected on steep sloping site. Tricky foundation problems were eliminated.

Fig. 44. Overlaid plywood siding has tough resin-impregnated surface that resist checking and takes paint well.

Fig. 45. Solid extruded vinyl siding has color molded in, is made in several exposures. Siding never needs painting.

Figure 45, is 0.045″, available in double widths of 4″ and 6′′ and single 8′′ exposures. Each panel is furnished with breather holes to allow the walls to release moisture. Solid vinyl siding accessories are available in the shapes shown in **Figure 46.** Insulation may be placed as shown in the cut-away views, or the siding may be applied direct to sheathing.

Masonry Construction

Masonry. Sandstone for fireplaces, planters, interior walls and retaining walls is sawed with a smooth top and bottom and with broken edges. One face and both edges are straight, permitting the stone to be laid rapidly. A retaining wall and planter, **Figure 47,** is laid dry and a retaining wall for a driveway, **Figure 48,** is made up of 3" sandstone with random lengths under 18".

Masonry Walls, Exterior and Interior. Impressive patterns can be achieved with modular 8′′ × 8′′ × 16′′ concrete block. It is easy to work. The size and shape contributes to handsome structures in every phase of modern architecture. **Figure 49.** The versatility of design makes this concrete masonry unit ideal for exposed walls indoors as well, **Figure 50.**

Pre-Stressed Concrete. Precast T-beams, **Figure 51** strengthened by embedded steel cables under tension, make short work of roofing a masonry structure. The beams are placed edge-to-edge, as shown, anchored and waterproofed with built-up roofing.

Screeding. Concrete or cement may be screeded by a vibrating screed rapidly, by one operator using a power driven machine, **Figure 52.** The vibration is an up-down-forward motion, which brings the fines (thin) and fats (solid) concrete or cement to the surface, making finishing possible sooner.

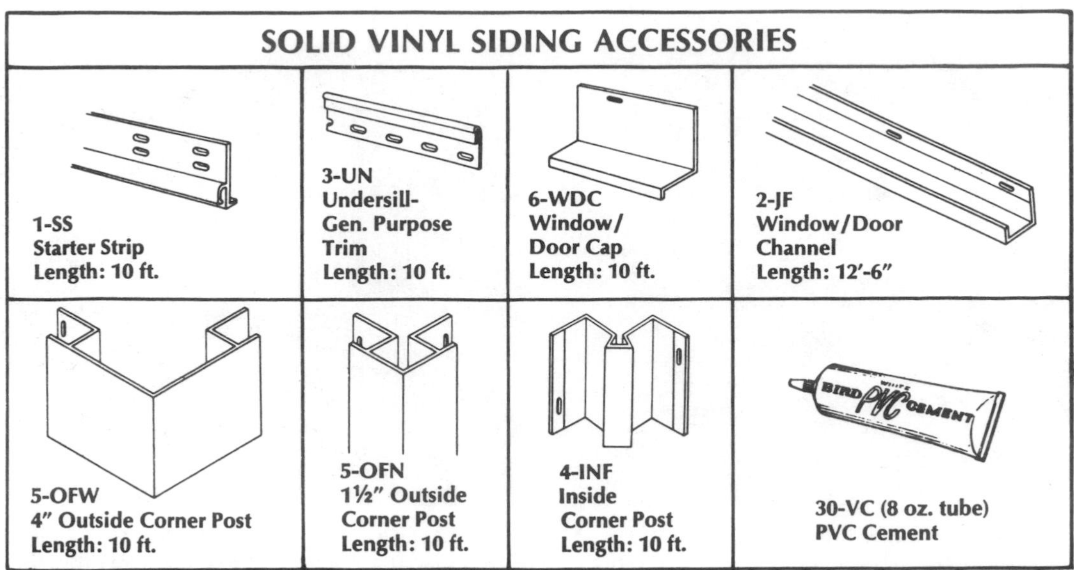

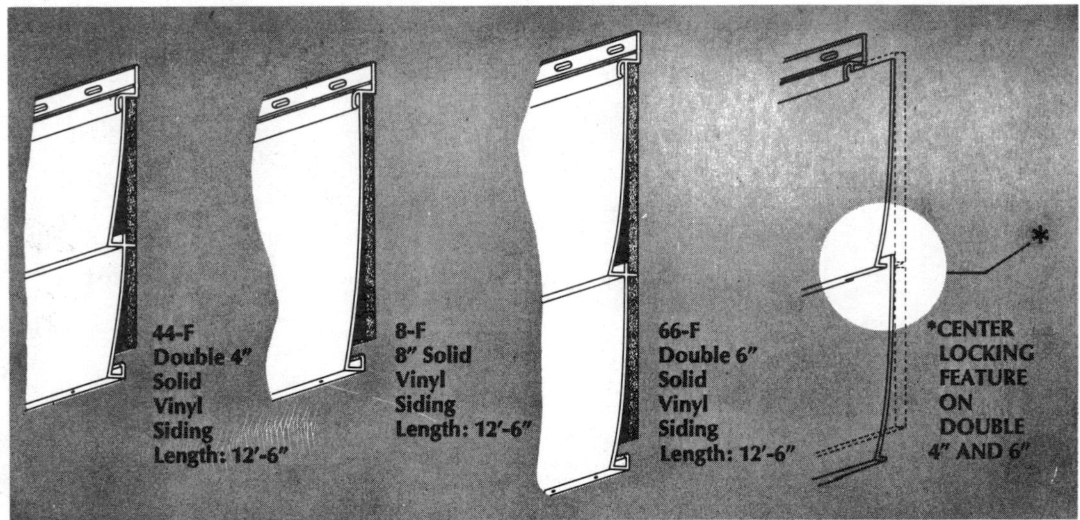

Fig. 46. Extruded vinyl accessories make the siding system complete. Optional backer boards add insulation value.

Fig. 47. Sawed sandstone has smooth top and bottom edges and broken face to add texture as shown in this planter wall.

Fig. 48. Low retaining wall along driveway is built with sawed sandstone blocks laid dry against embankment.

Fig. 50. Interior walls of concrete block are in perfect harmony with wood paneled walls and flush wood doors.

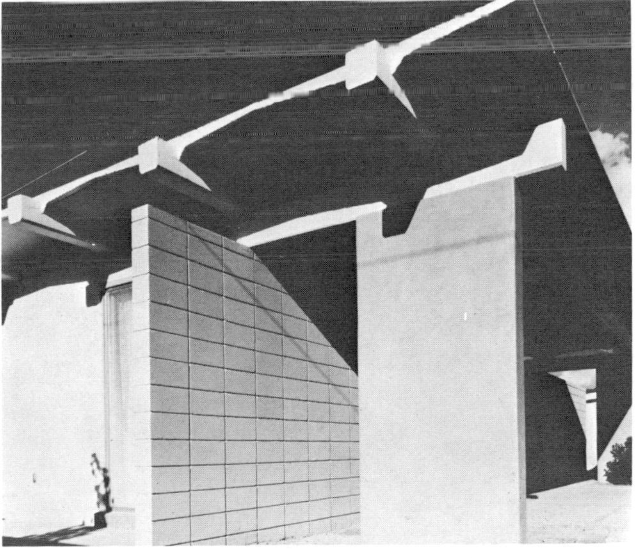

Fig. 51. Precast T-beams are strengthened by embedding steel cables under tension when beams are poured.

Fig. 49. Tooled mortar joints in fireplace built of concrete block create interesting shadow lines under side lighting (above).

The heavily textured wall of concrete block laid at right angles appears to be monolithic because joints are flush.

Fig. 52. Vibrating screed, driven by light gas engine, can be fitted with screeds of different widths to speed finishing.

Fig. 53. Stairguides speed construction of stairs by eliminating notching of stringers, and add strength. Guides are nailed to stringers (bottom left), then stringers are trimmed at top and bottom (top left). Risers and treads are nailed in place (right).

Fig. 54. Open steel staircase was prefabricated for NAHB Research House. Treads were covered with oak parquet squares.

The vibrator compacts the coarse material, dropping the rocks approximately 1/4′′ below the surface. A three horse power motor is used having a recoil starter.

Stairway Construction

Stair Guides. Hot dipped galvanized 22 gauge steel stairguides speed construction and add strength to any stairs, **Figure 53.** They are ideal for basement to first floor stairs or for temporary use during construction. Stairguides are nailed through pre-punched holes to wood stringers. Only one cut at the top of each stringer

Fig. 55. Manufactured slate floor looks like the real thing.

Fig. 56. Finish flooring is laminated to subflooring.

and two at the bottom are needed. Wood treads and risers are cut to required width.

After nailing risers to treads, clips are nailed to under side of treads thus eliminating the problem of spreading stringers.

Stairguides are available with an 8′′ rise and a 9′′ run, plus stair nosing. The guides are for floor heights from an extreme maximum of 8′-4′′ to an extreme minimum of 7′-7″. The top riser remains at 8″ and all treads pitch unnoticeably at either of these extremes. For floor heights from an extreme maximum of 7′-7″ to an extreme minimum of 6′-11″ the top stairguide step unit may be cut off with a tin snips or hacksaw before assembling. Stairguides may be cut off at any point for shorter runs in split level framing, or units for an extra length can be added for longer runs.

Steel Stairway. An open stairway of steel treads, stringers, balusters and hand rail, **Figure 54,** curves slightly in extending from the upper, mid and lower levels of a split house. The steel treads are covered with prefinished oak parquet squares.

Flooring

Slate Flooring. A manufactured slate flooring, **Figure 55,** is often installed as a finish flooring when subjected to constant wear and walking. Modular units are set in a mortar bed and grouted. Vinyl simulated slate in 12″ × 12″ tile is also available in most slate colors.

Plylumber Flooring. A laminated sub-floor-finish floor, **Figure 56,** is factory produced with tongue and groove side and ends. Flooring at left is 7/8″ thick, at right, 1-5/16″ thick. Large panels speed installation, minimize warping.

Chapter 23

PREFABRICATION

Prefabrication is often thought of as a new or revolutionary way of building. Through the centuries it has been man's way of getting the building job done easier and better with the most economical use of his time and materials. Today, prefabrication ranges from assembly of simple components to the manufacture of an entire house.

In its simplest form prefabrication dates back to the time when primitive man cut and trimmed the wood and tanned the skins he needed to put up a crude shelter for his family.

A Short History of Prefabrication

Abodes of the Past. In primeval times shelter was sought in rock caves, arbors of trees or raised on posts to form tents. In caves, huts and tents are found the three types of dwellings forming the bases of subsequent architectural developments. Nature's caves having rough openings of walls and roofs of rock suggested the use of stone by which to build walls to support slabs of rocks for roofs.

Huts with tree trunks for walls and closely laid branches covered with reeds for roofs were suggested by natural arbours. In Jericho, an ancient city in Palestine, huts of two stories, with exterior stairs, were built. One need only to trace the revolutionary changes in building to associate this method of assembling materials with our present day system of wall framing.

Houses of Ancient Times. In building his structures man has experienced one change after another in methods and materials. Many houses of crude masonry have been excavated in Egypt. These houses, built as early as 1400 B.C., were one, two or three stories in height, each story reached by outside steps. The roof was flat or domed and made of puddled clay or sun baked bricks.

Excavations at Pompeii show that Roman houses differed slightly in plan from the Greek dwellings which preceded them. Pompeian houses had plain facades to the street. The frontage on either side of the entrance was let out as shops. The absence of windows facing the street was probably due to the desire for privacy, and glazed windows, even if known, were little used. The rooms were lighted by openings on internal courts, as used in medieval times in England and France.

Prefabricated Houses of the Eighteenth Century. The assembly of fabricated parts for an entire house is not new.

About a century before the Revolutionary War (1775–1783) there occurred another revolutionary change in dwelling construction. A house was manufactured in England, then dismantled, and the parts shipped to the Massachusetts Bay Colony to develop the fishing industry in New England. This is believed to have been the first prefabricated house to have been assembled in America. Today this method of prefabrication of parts for assembly is known as controlled construction.

The "Great House," as it was called, was constructed of precut oak timbers, which, when fitted together, formed the frame of the building. This particular house was dismantled and moved several times to different building sites.

About the year 1725, two houses, "all cut to be erected" were shipped to the West Indies from New Orleans. This is the first recorded instance of a prefabricated house being manufactured in America and shipped out of the country.

While wood up to this time had been the mainstay of the house manufacturing industry, some strides were being made in the use of metals for framing and for exteriors. The firm of Watt and Bolton, of England, in 1801, began to fabricate and erect buildings of cast iron. An all metal, prefabricated house was erected, about 1830, at Tipton Green, Staffordshire, England for the dwelling of a lock keeper on a local canal. The walls of this metal house were of flanged, cast iron panels, placed vertically and bolted together. The interior walls were finished with lath and plaster. When, in 1925, the building was torn down, it was found to be in good condition.

Prefabrication and the California Gold Rush of 1848

The influx of prospectors and adventurers into California during the Gold Rush was so great that existing housing was very soon over-taxed. Lumbermen from the Eastern States, and even from England, Germany, France and Belgium were eager to supply the demand for housing in California. Builders, too, from as far away as New Zealand and China began exporting frame houses made up in panels which could be assembled by semi-skilled workmen. A prefabricated house that sold for $400.00 in New York would bring $5,000.00 in the California region. By 1850 it has been estimated that over 5,000 prefabricated houses had been made and exported from the Eastern States.

The Civil War and the Prefabricated House. The decade of the 1850's witnessed a decline in the activities of the 'Manufactured House' but with the advent of the Civil War in the early 1860's the prefabricated house made a strong comeback. When the Federal troops were mobilized in 1861 training camps were established and the men housed, not alone in tents, but in prefabricated, panelized wooden camp buildings which were easily and quickly built. After the Civil War other revolutionary changes began in the housing industry. In 1882 the Christoph and Unmack Company of Germany began the manufacture of load bearing prefabricated walls.

Manufactured Houses of the Twentieth Century

In the spring of 1918, many concerns began to manufacture houses and building materials in factories. One of the leaders in this field was the world famous mail order firm of Sears, Roebuck & Company. Instead of manufacturing panels and assembling parts of a house this concern instituted a revolutionary change by making pre-cut parts. These pre-cut parts, marketed under the brand name of "Honor-Built Homes" were, in the main, structural members such as sills, girders, shoe plates, studding, plates, floor joists, roof rafters, etc., cut and numbered for assembly by nailing or by the use of holding devices.

In addition to the "pre-cut" parts used in the initial framing, Sears standardized and prefabricated such items as medicine chests, cupboards, china closets, colonnades, ironing boards, breakfast alcoves, door and window frames and doors mortised for lock sets. This method of pre-cutting and standardizing parts materially reduced the time of building a dwelling from almost 600 hours required by conventional methods to about 350 hours in pre-cut assembly. It must be kept in mind that the time factor in erecting a house does not include the time needed to survey the building site,to lay out the location of the basement excavation, to excavate, to erect the basement walls, nor to lay out and place the foundation slab.

Sears manufactured cut and fitted parts for three models: the Homeville which sold for $1,854.00; the Josephine, at $1,405.00; the Springwood, $2,089.00.

In 1934 the accounts of the Modern Homes Department were liquidated.

World War Two and the Prefabricated House

Social changes, economic changes, natural disasters and wars, over the centuries have had a direct influence on the dwellings of man. The greatest war struggle of the present century, World War II, 1939–1946, was the cause of a nation wide demand for temporary housing for defense workers. In order to provide the most rapid development and erection of housing units, building standards were arbitrarily lowered. Good designs were simplified and standardized resulting in many inartistic structures. As a consequence, at the end of the war, when there was a nation wide demand for permanent, private housing, the house manufacturing industries had to overcome the national concept of "prefab" which was based on some very poor construction done during the war years.

Even before the war ended, fabricators were beginning to work on and turn out housing of a more artistic and permanent nature. In the early 1940's the Pease Woodwork Company, an old line building supply firm, entered the housing field. The nation's largest manufacturer of prefabricated houses, the National Homes Corporation, of Lafayette, Indiana, began, before the end of war, to produce a more acceptable house.

Modern Manufacturing Methods Applied in Fabricating Houses. In 1940, a young builder, James R. Price, started making housing history as he supervised the construction of a new dwelling at 812 South Twenty Ninth Street in Lafayette, Indiana. This was a special house, the first National House, the first dwelling produced by a fledgling company, destined to become the leader. This house was built on a building lot costing $250.00 and the purchase price of the house was $3,250.00.

Since the first dwelling was built, National has applied the precision and efficiency of modern manufacturing to the building of houses. One of the latest developments is an amazing nailing machine that can nail into place as many as 22 nails at a single stroke. This rapid nailing is used in fastening sheathing materials to structural framing.

An Electronic Sorting Machine. Another efficient method in the manufacturing plant is the use of an electronic sorting machine. When a builder sends in an order to the plant for a particular house, an electric sorting machine selects, from tickets representing the thousands of panels and parts, those specifically required in the home owner's selections. As soon as these tickets are received in the production department, operations begin in several parts of the plant. In the mill the lumber required for the production of a particular dwelling is cut and shaped to exact specifications. This work is done by finely engineered machines. One of

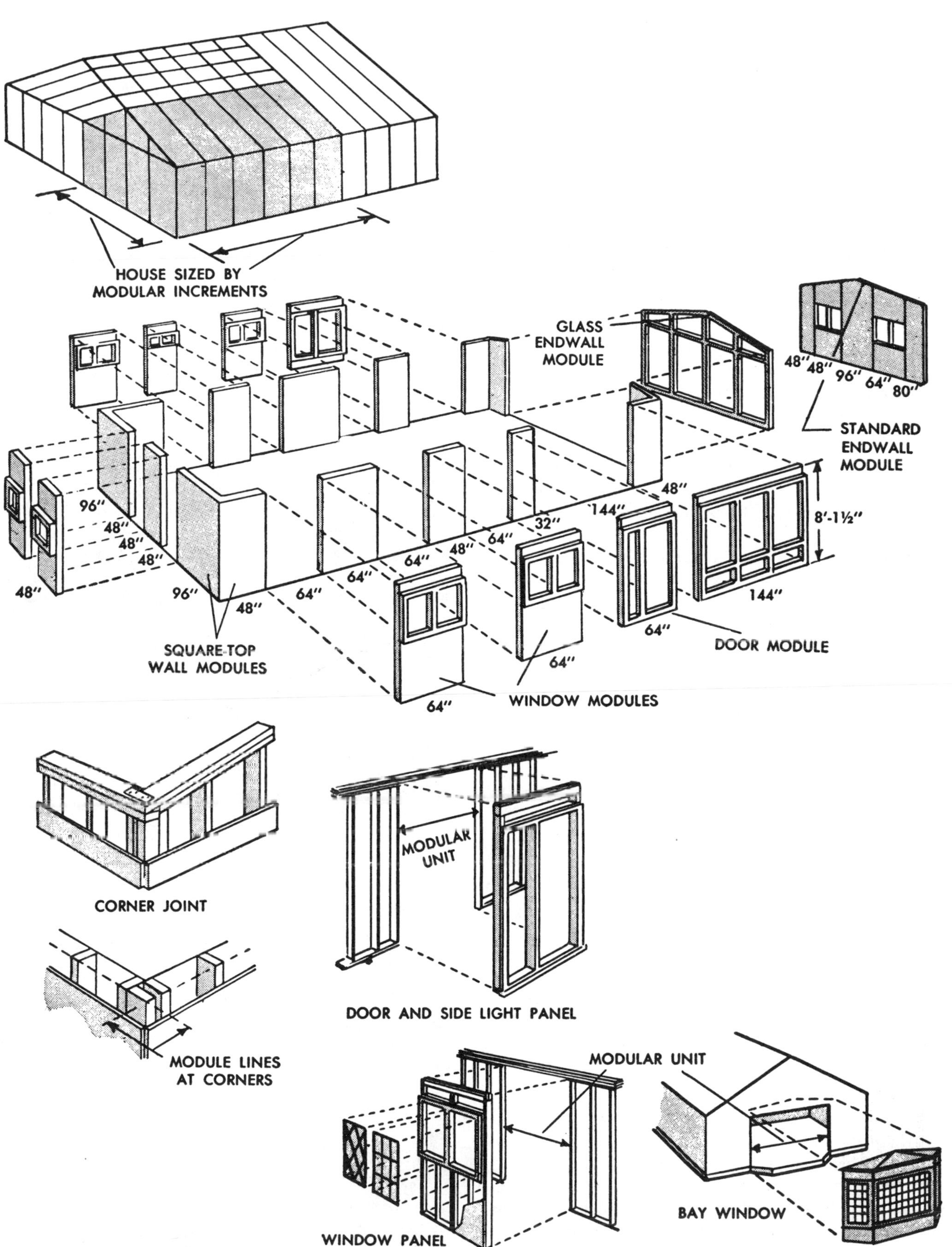

Fig. 1. Details of a modular construction system based on standardized modules for doors, windows, walls and gables.

these machines is a double end tenoner saw adjustable to 23 different kinds of cuts, depending on the job to be done. This tenoner saw can produce the proper cuts on a carload of lumber in an hour.

The Making of Panels. Each panel is made up of precisely cut parts, fitted together tightly in a single, strong section. The interior wall surfacing material is then bonded to the framework with powerful, waterproof adhesives. To reinforce this bond the panels are nailed at the top and bottom. Halfway down the production line the wall panels are turned over mechanically by a giant machine. At this stage a blanket of insulation is applied in the exterior wall panels to insure year-round comfort inside the home. A vapor barrier is applied to the back of the interior wall surfacing material to exclude moistire from the inside.

Placing Windows and Doors. While the panels are still on the assembly line, the preassembled windows are installed, complete with flashing and exterior trim. Likewise the doors and door jambs are prepared and moved to the panels for installation. The exterior sheathing is nailed to the structural framing, enclosing the panels.

Placing Exterior Wall Covering. The exterior wall covering, whether it is wood or masonry, is installed at the building site. If the exterior wall covering is of aluminum, National has developed an exclusive process of applying either shiplap or board and batten with strong, fast drying adhesives. As each section of the particular dwelling is completed it is lifted mechanically to a conveyor system which connects the 22 assembly lines in the plant. There is a final inspection to check on all factory operations. As the panels and parts are moved to the loading dock, the union label is stamped on each one. These panels and parts come off the conveyor system according to a scheduled sequence and bear a number designating each particular part of the buyer's house. The loading of these completed items takes, on the average, one hour and twenty minutes, and the structure is on its way to the building site.

Early House Manufacturers Using Material Other Than Wood

Wood has always been the mainstay of the conventionally built dwelling as well as the factory built house. From the early years of the Twentieth Century persons in the housing field have sought to use materials other than wood. Failure has resulted in most instances. One of the first attempts to use concrete was conceived by a New York architect named Grosvenor Atterbury.

Precast Concrete. The Atterbury system involved the use of precast hollow core panels for floors, walls and roofs. From 1910 to 1918 several hundred houses of this design were erected in a suburb of New York City. Each individual unit was transported from the factory by truck, and put into position on the building site by derricks. While suited for large scale projects it was found that this system of mass manufacture did not easily lend itself to the exclusively individual dwelling, hence it eventually failed.

The Edison System. No less a person than Thomas Alva Edison, native of Ohio, and a professional inventor in the fields of electrical science, in 1908, proposed to pour an entire two or three story house of concrete. Even a poured-in-place concrete bath tub was considered. Cast iron forms, bolted together at the building site, were to be used but proved too costly. Forms of wood were substituted for metal but proved impractical so the scheme was abandoned by Mr. Edison. Others continued

Fig. 2. Casting a pre-stressed concrete grade beam, which will become part of the foundation of a house.

Fig. 3. Grade beam is set in position on the previously leveled piers formed in fibre form tubes. Note vent opening in beam.

with pre-formed concrete "parts" but with no marked success.

Other Materials. During the decade from 1920 to 1930 a large number of experiments and some construction in materials other than wood, were being promoted in Europe. In England, dwellings were constructed of sheet steel, expanded metal sprayed with concrete, rolled steel frames, concrete masonry and precast concrete components. None of these methods proved too successful and were eventually discarded.

In Germany, too, where steel was available in surplus quantities, several methods of manufacturing houses of metal were introduced. The builders in France fabricated houses using steel on both exterior and interior surfaces, then applying stucco on metal lath.

By way of contrast, Sweden, having an abundance of timber, played an important role in the erection of manufactured houses of wood.

In the United States, two concerns utilized materials other than wood during this decade and shortly thereafter. One concern, the Lustron Corporation of Akron, Ohio used porcelain enameled steel as the basic structural unit in manufacturing houses. The product was superior, but this concern did not place sufficient emphasis on marketing and distribution, hence it ceased operation after a few years.

Modular Component Construction

The United States Bureau of Standards defines prefabrication as a house having floors, walls, ceilings, or roof composed of sections or panels, of varying sizes which have been fabricated prior to erection on the building foundation.

Structural Component Construction. In contrast with the time honored piece-by-piece or conventional construction method, there has come into the house building industry the factory fabricated module or part based on dimensional standards to provide a uniform basis of manufacture. This method of manufacture is known as structural component construction. Nearly all factory made components are based on the size of some one part taken as a unit of measure for regulating proportions. Its concept is applicable to the total house, including floors, walls, roofs, partitions and stairways. Since this method is not based on one size of panel, its unlimited flexibility challenges the imagination of the architect, builder and the owner. Modular construction may be applied to a job site assembly of materials or the factory fabrication of components.

Modular Coordination. Exterior walls, doors and windows require major coordination. These three component categories require extremely flexible application in a workable system of construction. Design and structural considerations must be incorporated into each of the components. They must be integrated with all other house segments to provide a completely modular component structure, **Figure 1.**

The 64″ module (multiple of) door and side light panel is integrally designed to become a part of the 16″ module (multiple of) system for wall, door and window components. The pre-assembled component with built-in header fits the 64″ wall opening.

Standard 32″ casement window units are shown mullioned to become a 64″ window unit to fit a 64″ wall opening. The structural jambs of the window panel combine with adjacent blank wall studs to provide double framing at openings. Angular and square bay windows utilize the standard window inventory in all module bay widths.

If a sloping interior ceiling is constructed the front and rear wall plates are given a beveled top with a shim placed between the top and lower plates of the outside partition framing shown in the corner joint detail. Flat top plates are used for flat or horizontal ceilings.

Typical Steps in Component Construction

A building component is an assembly of units of building material fabricated under controlled and efficient shop conditions for erection at the construction site using a minimum of site labor. Savings begin as the component is made. Assembly or manufacture is accomplished under conditions created for convenience and quality control, with power tools and an organized work flow. The designs and working drawings are worked out by engineers, architects and draftsmen.

Modern Building Materials. The reading of a catalog of modern building materials and products show the availability of many components of a dwelling. These components, such as doors, windows, kitchen cabinets, bathroom vanities, china cases, fireplaces, and scores of other manufactured products have become so well constructed and standardized that one need not consider on-site production. It is in the structural phases of building, the phases or operations which were once conventionally erected, that component construction has been developing as rapidly as the factory prefabrication of an entire house.

Constructing Components. The actual construction of a dwelling built of components begins, after the designs and drawings are made, at the building site or building lot. If the dwelling is to be placed over a full basement the complete foundation can be made of concrete block or poured concrete cast in plywood forms. These forms may be considered the first components used. If the dwelling is to be placed on a concrete slab the concrete is poured into forms resting on the ground.

Fig. 4. Concrete piers rising from a sloping site provide a level base for heavy concrete beams hoisted in place by a crane.

Concrete Grade Beams. If the dwelling is to be placed on a prestressed grade beam the beam can be cast or poured under factory conditions, then hauled to the building site. A pre-stressed concrete grade beam, **Figure 2,** is cast in a bed located at or near a local ready mix plant. Two cables under tension run through each beam. The box down the center, and the box in the foreground are dividers between the beams. The form to the right of the first workman creates a vent opening in the beam. After the concrete has set and dried, tension is released on the cables, by being cut at the end of beam, and the beams lifted from the casting bed by loops embedded during the casting process. Beams can be as long as can be handily transported. By using this method, site work can proceed during freezing weather.

Leveling a Concrete Grade Beam. A concrete grade beam is leveled up on a pre-cast footing pier, **Figure 3,** set in holes drilled in the earth. Often these footing piers are made by using fibre tube forms for poured concrete, **Figure 4.**

Plywood Box Beam. Box beams of plywood are made of 2 × 4s with plywood sides, about 10″ to 12″ in width, glued and nailed. This component beam rests on the foundation sill, spaced on 4′ centers.

Another type of support for the floor load is the floor truss made up of top and bottom plates and vertical members of 2 × 4's, strengthened by metal bridging.

The Floor Panel Component. The floor panel is made of seven ply, 1-1/8″ thick, by 4′ × 8′ plywood panels placed over the 48″ floor beam span. Often the panels have tongue and groove edges for firmer support, and to eliminate blocking at joints, **Figure 5.**

A more sturdy component is the 4′ × 8′ panel of 1/2″ plywood, stiffened or braced with 2 × 4 stringers nailed at 16″ spacing across 4′ widths of the plywood. A ledger nailed to the sides of the supporting beams carries much of the load from the 2 × 4 stringers. The plywood panels meet in a butt joint on top of the beams and along stringers.

The Stressed Skin Panel Component. A floor component can be built up of plywood with a finish floor surface of oak or elm. These panel sizes are limited to handling and trucking facilities. A stressed skin flooring component, **Figure 6,** combines the finish floor with the sub floor spanning a 48″ beam.

Exterior Wall, Roof and Partition Components

The exterior wall panel component is somewhat less extensive in horizontal length than the prefabricated full length wall unit. This smaller component panel, as the smaller component floor panel, will lend itself more readily to modular construction. One system of exterior wall components is a factory built, 4′ × 8′ panel, framed with 2 × 4 studding, a top plate and shoe or sole plate, power nailed or stapled.

One side of this component has plywood or insulating board sheathing attached. Often blanket insulation is tacked in between the studding at the factory. Larger units are framed to include the complete window or door set. Sometimes these units, housing garage doors or picture windows are made up of a box beam glued and nailed with plywood and light framing lumber. Walls are tied together with a continuous top plate nailed over a series of small exterior wall components, which have complete window sets framed in at the factory, **Figure 7.**

Roof Components. After exterior wall panels have been erected, the roof structure is put in place. The full potential of the component is a roof structure which spans the building and throws its entire weight on the exterior walls. With the roof structure complete, the house is closed in so that all interior work, including partitioning, can be done without concern for weather conditions.

Manufactured roof trusses were the first true component units introduced in dwelling construction. Today, roof trusses are the most widely adopted and used fabricated component. Due to the well engineered distribution of roof weight, trusses can be fashioned from relatively light and inexpensive dimension stock. Nailed metal plates and combination nailing pressure application systems are the most commonly used methods of joining the framing members after they have been precision aligned in a jig. Nailing plates are pre-punched metal gussets attached with spiral or threaded nails, **Figure 8.** Simple jigs, made of lumber, and space to turn the truss plates to the reverse side, are all that is needed. A roller press, **Figure 9,** is used to embed truss plates, forming a roof truss. Nailed and glued roof truss components, using plywood gussets, are made, but require fabrication under controlled atmospheric conditions most favorable to glue work.

Laminated Beams and Box Beams. The use of roof trusses is not the only method used to span a roof space. Laminated beams and plywood box beams are often used in flat roof construction. Box beams are not unlike the components used in floor construction. Box beams may be of such length as to extend over the outside wall and the exposed portion of the roof beam may be made of fine hardwood. Laminated roof beams present an attractive appearance when exposed.

Stressed Skin Panels. Roof panels, factory made components, **Figure 10,** are used to speed roof construction. These panels are made of large sheets of plywood, glued and tacked over 2 $\times$ 4 inch stock. Insulation batting is inserted between the panels.

Room Partition Components. Partition components, **Figure 11,** need not necessarily be designated as interior load bearing walls dividing the floor area into rooms. Since the roof structure carries the roof load to the exterior walls, the interior components may be storage cabinet room dividers or light weight partitions with ceiling height doors already hung.

Fig. 5. Panels of 1-1/8″ T&G plywood are nailed across wood beams spaced 4′ o.c. T&G edges eliminate blocking across beams.

Fig. 6. Prefabricated stressed-skin flooring panel is toenailed to supporting beam.

Fig. 7. Continuous plate nailed along top edge of wall assembled with component sections ties sections together.

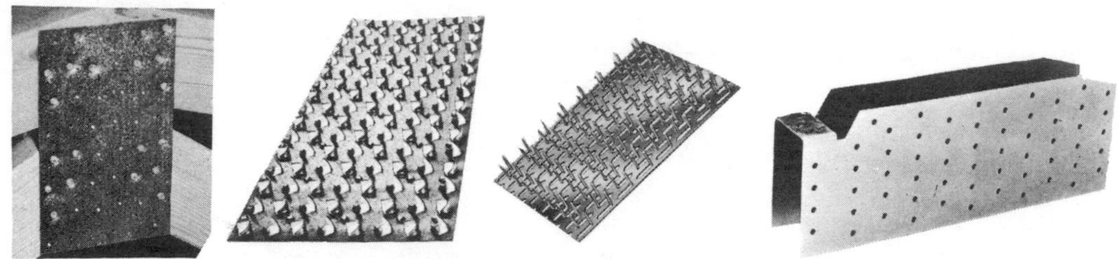

Fig. 8. Truss plates used in prefabrication. Perforated nailing plate (left), connector plates (center), and H-brace plate (right).

Mechanical Unit Components

The modular component constructed house lends itself handily to the installation of the mechanical core, or three dimensional unit, encompassing the most expensive parts of the house.

The making up of any mechanical core involves the work of many skilled tradesmen. Skilled cabinetmakers and carpenters are involved in the factory, in building the various components, heretofore described, and erecting these components at the building site.

Skilled electricians, plumbers, tile setters, and others are involved in the factory making of a mechanical core, which in one unit may include the furnace, water heater and electrical panels. Another core may include the kitchen cabinets, sink, bathroom fixtures and other accessories.

Fig. 9. Workman feeding lightweight truss into rollers to press connector plates securely into meeting truss members.

Fig. 10. Stressed-skin panels with insulation installed between 2″ × 4″ stringers are set in place to close in building quickly.

The core is a pre-planned, finished section of a house that is factory made and then transported to the dwelling and put in place, **Figure 12.** This is a complete kitchen and bathroom core being set in position at the first floor level of a dwelling.

The development of the manufacture of the structural component has now reached an efficient stage in dwelling construction. It is now comparable to the manufacture of the prefabricated house. Mechanical unit component manufacture is being extended into the making of electrical heating panels, factory wired furnace systems, plumbing wall components, etc. In fact, the mechanical parts of a house offer some of the greatest opportunities to benefit from component methods of construction.

Modern Prefabricated House of Wood

Yesterday's trend of mass production appearance of the factory built house of wood has now given way to a dwelling of good design having the latest products and

Fig. 11. Storage walls such as this are often used instead of partitions in prefabricated houses roofed with trusses.

Fig. 12. The ultimate in prefabrication efficiency: A complete kitchen and bathroom core being set in position on the main floor framing. Core includes fixtures, plumbing, wiring, and tile on floors and walls. Final hookup is simple and fast.

Fig. 13. Today's prefabricated house has charm and individuality, and may be modified to suit the preferences of the owner.

top workmanship. Today's prefabricated house is new, has charm and individuality built exclusively for the owner, **Figure 13.**

The Floor Plans. A study of the floor plans, **Figure 14,** will indicate this house to be an ideal home for a family of five or more. The total floor area is 2,334 square feet. At the upper floor level there are three bedrooms, two bathrooms, and a spacious walk-in closet. At the lower level, a large recreation room, a laundry, a small bathroom, and a hobby room which may be used as a fourth bedroom.

At the ground level are located the dining room, a large living room, a walk-through kitchen, a breakfast nook and a low window, a spacious foyer and an inviting fireplace, as well as a two car garage and storage cabinet.

Fabricating Machines. Framing lumber, received from the lumber manufacturer, is fed into a double-end

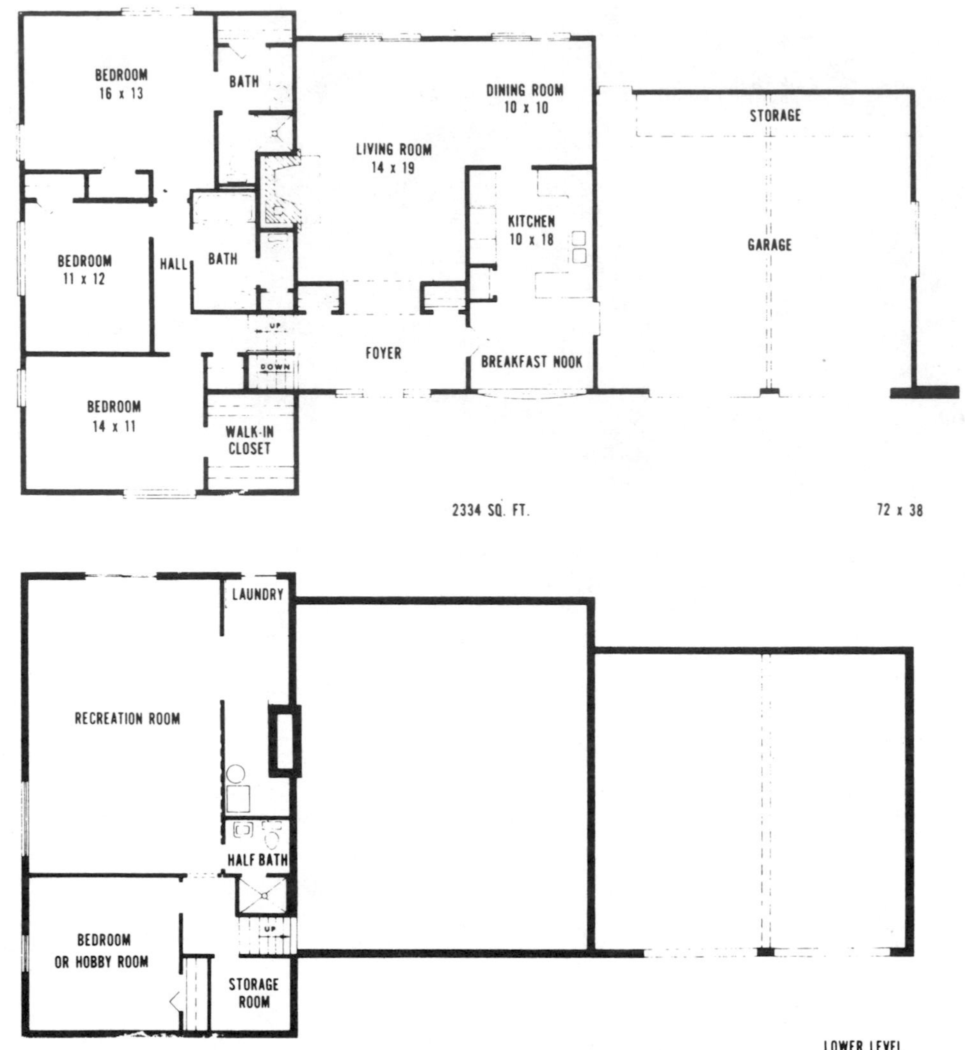

Fig. 14. Floor plan is identical to the plan used in the house illustrated in Figure 13, except that it is flopped left to right.

tenoner saw to produce structural framing members of a panel, **Figure 15.**

Pre-cut structural lumber is received at a framing jig table which has been pre-set for specific sizes of panels for the buyer's order. In **Figure 16,** workmen are framing an outside wall to receive a large picture window.

A framed panel is turned mechanically for work on the opposite side. This panel, **Figure 17,** has been framed on the framing jig and has received the interior wall covering. In this operation the workmen are placing insulation batting between the studding. The large opening in the center is for a window.

An exterior wall panel, having had the interior wall covering applied and the insulating batten inserted, **Figure 18,** is moved up to the jig table where the aluminum exterior wall covering is put in place ready for tacking down.

Installing Glazed Sash. Glazed windows are installed in the window openings, **Figure 19,** as the exterior panel moves along the assembly line.

A Gable Roof Truss Assembly. One of the final structural members to be placed in position in the erection of a prefabricated dwelling is the roof truss. To save time, money, material, and to attain a high degree of accuracy and workmanship, the roof truss was the first structural member to be manufactured in a factory. From the supply of pre-cut parts of a truss, the members, **Figure 20,** are pressed together and held in place by metal gussets.

Prefabricated Parts Loaded for Shipment. The prefabricated parts of a dwelling are loaded on to a truck to be transported to the building site. Note the glazed end wall panel, **Figure 21,** several roof trusses and two gable end roof panels.

The Charm of Furnished Interiors. The charm and individuality of interiors of prefabricated houses are shown in **Figure 22, 23 and 24.** The living room can be readily adapted to modern styles of furniture and dec-

Fig. 15. A double-ended tenoner saw cuts structural framing members for a panel in a home manufacturing plant.

Fig. 16. Precut structural lumber is assembled on a jig table. Jig shown is for a wall component which includes a picture window.

Fig. 17. Overhead crane is used to turn partly completed panel upside down on table so that work can proceed on opposite side.

Fig. 18. Exterior wall section, with interior wall covering and insulation applied, is covered with exterior aluminum siding.

Fig. 19. A completely assembled and glazed window is inserted in opening as exterior wall section moves down assembly line.

Fig. 20. Pre-cut chords and diagonal members are quickly assembled with metal gusset plates to fabricate a truss.

orations. The bedroom is snug, warm and quiet, and the inviting dining room alcove is spacious enough for several large pieces of furniture.

Prefabricated House of Steel

The modern prefabricated house of steel is built with 3-5/8″ galvanized steel framing members. The steel framing **Figure 25** is sheathed with 1/2″ exterior nail base fiberboard attached by steel barbs. Aside from the framing, assembly of the house is similar to the wood-framed prefab.

The inner surface of the outside walls is covered with gypsum board attached to the steel framing panels by a permanent adhesive bond. Insulation is placed between the interior and exterior wall panels.

Slab Foundation or Floor Platform. The foundation slab is poured, or the first floor is framed on a foundation wall, **Figure 26.** If the panels are erected on a foundation slab a sill plate of 2″ × 6″ stock is placed around the perimeter. The sill plate is attached to the foundation slab, a sill plate of 2″ × 6″ stock is placed intervals, **Figure 27.**

The First Steel Framed Panel In Place. The first component of an exterior wall is set in place on the sill plate, **Figure 28.** The interior and exterior gypsum board is already attached. A foreman is standing nearby with construction details in hand.

A Second Panel Forms a Corner. The second component or panel is brought into position, **Figure 29,** to form a corner of the outer walls. Note the workman in the foreground nailing the exterior nail base to the wooden sill member, while a workman at the top of the panels trues the corner.

Nailing Pre-Punched Steel Tabs to Sill Plate. After

Fig. 22. Spacious living room with mural on endwall is more like a custom home than one built on an assembly line.

Fig. 23. Handsome bedroom is snug, warm and quiet—the end result of precision manufacturing methods.

Fig. 21. Prefabricated component sections are loaded on truck for delivery to building site and fast erection.

Fig. 24. A dining alcove, attractively decorated and furnished, is set off by the large picture window in the end wall.

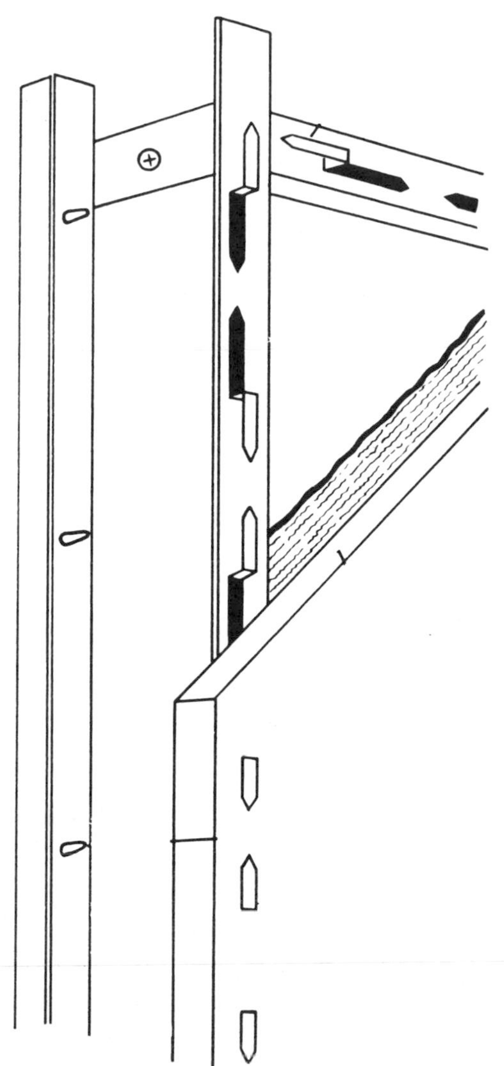

Fig. 25. Steel structural system designed for residential buildings: Exterior wall panel is held in place by steel barbs which pierce material and are bent over or clinched.

the first exterior wall panels have been erected to form the corner of the structure additional panels are erected around the perimeter of the foundation slab. A workman, **Figure 30,** is nailing the base of the panels to the sill plate. Note the small steel tabs which fit down over the plate for nailing.

Erecting Outside Wall Panels. Outside wall panels are erected along the sill plate and secured at the top by a rafter plate member. Note the workman in the foreground moving the panel into position along the sill plate, **Figure 31.** The workman on the stepladder is guiding the top of the panel into position. The third panel from the right has a completely glazed double hung window sash built in between the steel channel rails. Note notches cut into the plate to receive the trusses.

Gable Panels. The gable end panels are set in place after all exterior wall panels or components have been erected. Each gable end is made of two panels, **Figure 32** and framed for a quarter pitch roof. The cripple studs are placed 24″ on center. The first roof truss is being raised in place and fitted into the notches on the continuous plate.

All Roof Trusses Raised and Braced. After the placing of the first gable end panels, the several roof trusses, **Figure 33** are readily set in place on the plates attached atop the side wall panels. The truss used in this structure is the standard Kingpost truss. Shoe strip in the foreground is for the interior wall panels. Plywood sheets form the rough flooring over the concrete slab. The horizontal members of the roof trusses become the joists to carry the ceiling panels.

The Second Gable in Place. The second and final gable is raised in two sections to complete the roof and gable end framing, **Figure 34.** Notched cut in the

Fig. 26. Truck is backed up to first floor platform so that exterior wall panels can be unloaded with minimum handling.

Fig. 27. When house is erected on slab floor, continuous sill plate is fastened to perimeter by shooting drive pins into slab.

Fig. 28. The first steel-framed wall section is set in place on sill plate. Note small nailing tab at bottom of panel.

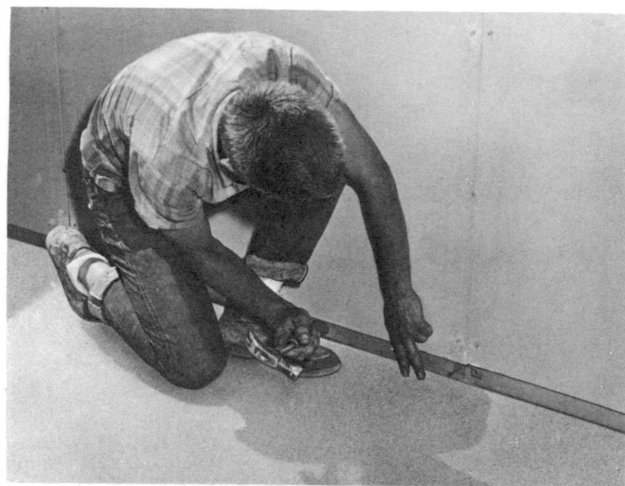

Fig. 30. Pre-punched tabs, installed at factory along inside edge of panels, are nailed to inner edge of sill plate.

Fig. 29. A second steel-framed panel, with exterior sheathing applied, is set in position to form an outside corner.

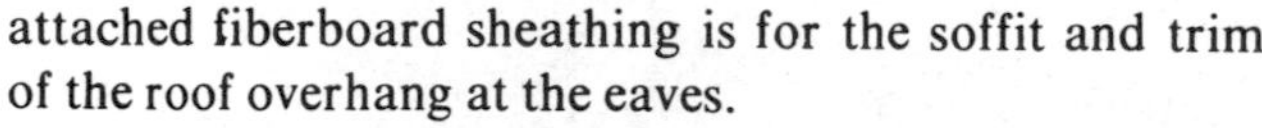

Fig. 31. Additional panels are erected on sill to complete the exterior walls. Overlapping edge of sheathing is nailed to sill.

attached fiberboard sheathing is for the soffit and trim of the roof overhang at the eaves.

Nailing Up the Eave. After the roof trusses are in place and the gable ends of the structure are closed the eave is built up with soffit and fascia trim, **Figure 35.** Note the ventilation slot in the soffit.

Applying Roof Decking and Roofing Felt. The roof decking, **Figure 36,** depends, to some degree, on the kind of roofing to be used. Shingles of wood are often applied to spaced roofing boards, while asphalt shingles are

Fig. 32. Sheathed gable end is prefabricated in two halves, which are joined when halves are positioned on endwall.

applied over solid plywood decking. The decking is being covered with 15-pound roofing felt insulation.

Asphalt Shingle Roofing. Asphalt shingle roofing, due to its resistance to fire, its economy and durability is widely used for pitched roofs. Most asphalt shingles are available in single tabs or in tabs of three. In **Figure 37,** workmen are applying heavy, three tab asphalt shingles. Work is begun at the left as the three men can tack down several rows as they work from the eave to the ridge. The tabs must be nailed securely and staggered to avoid water and wind damage.

Applying the Ceiling. Installing the ceiling and the interior partitions follow the completion of the work on the roof and the eaves. Large sheets of 4′ × 12′ gypsum board are tacked to the lower members of the roof trusses, **Figure 38.** These lower or horizontal members of the roof trusses become the joists to which the ceiling and upper rail of the room partitions are attached. Insulation battens are often placed on top of the ceiling, between the joists, to provide a completely insulated house and assure lower heating and cooling costs.

Sealing the Joints in Wall and Ceiling. In order to prepare the interior walls and the ceiling for plastering or painting, all joints must be taped and spackled with joint cement to provide a smooth surface for plastering or painting, **Figure 39.**

Placing Interior Walls Into Position. The final step in placing factory built panels in the fabricated steel house is the setting of interior partitions according to specifications provided with each dwelling, Figure 40. These panels have a steel framework covered with 1/2″ gypsum board applied to both sides with a strong adhesive. This adhesive eliminates nails, gives a smoother wall and provides a much stronger bond. The partitions are nailed to the floor and the ceiling.

Fig. 33. Roof trusses are set in place in notches precut in wall sections. Temporary braces hold trusses erect.

Fig. 35. Fascia is nailed to truss ends after roof decking and soffit panels are installed. Note screened vents in soffit.

Fig. 34. Two halves of second gable end are erected and joined after all interior trusses have been placed and secured.

Fig. 36. Overlapping strips of 15 pound roofing felt are unrolled and nailed to plywood roof decking to seal out weather.

Fig. 37. Three-tab asphalt shingles are laid over roofing felt. Once edge is started, three men can lay staggered rows in sequence.

Fig. 39. All joints in wall and ceiling panels are taped, cemented and sanded to provide a smooth surface for painting.

Fig. 38. Large 4′ × 12′ panels of gypsum board are nailed to lower truss chords to finish ceiling before erecting partitions.

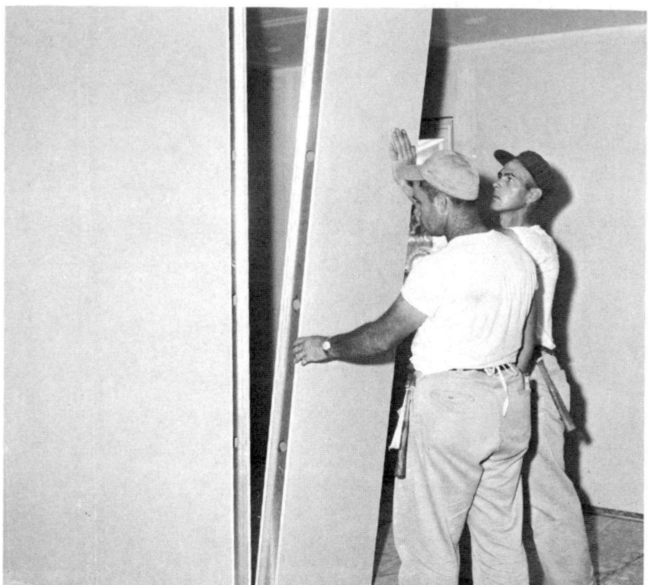

Fig. 40. Interior partitions of gypsum board adhered to steel studs are erected according to the floor plan.

Fig. 41. Steel-framed split-level house has exterior finish of stone veneer and vertical wood siding with batten joints.

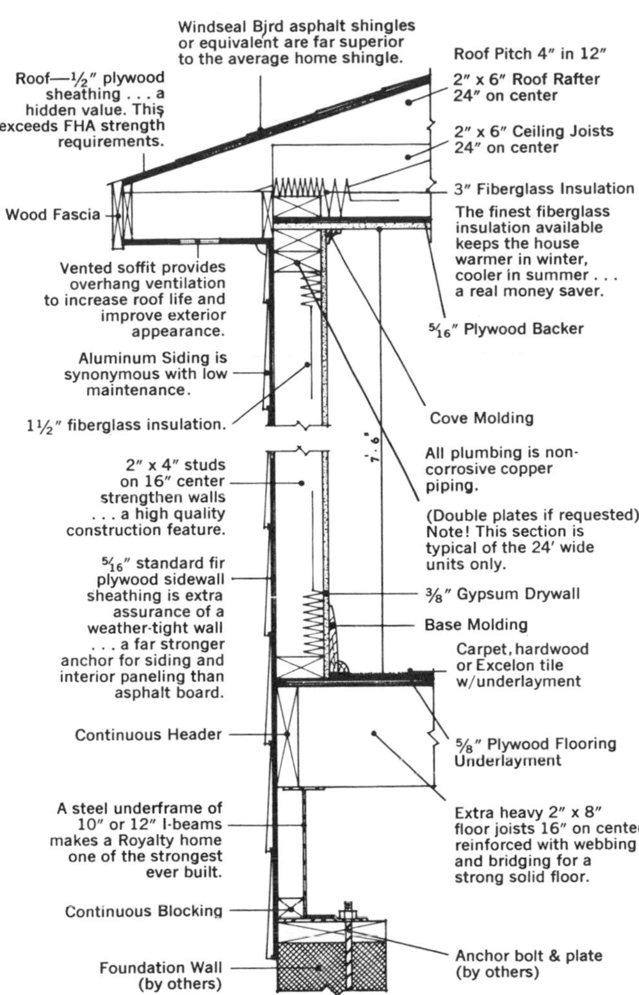

Fig. 42. Exterior wall section of a sectional house built in two half sections and shipped to the site for final joining.

Fig. 43. This shop-welded steel frame provides a rigid foundation which makes it possible to ship half sections without damage.

Fig. 44. Exterior walls are framed and insulated as complete units, then moved to another area for final assembly.

A Bi-Level House. The house illustrated in **Figure 41,** is a steel-framed split level with 1,746 square feet of floor area. In addition to the living room and dining room on the upper level there is a kitchen, three bedrooms and two baths.

The lower level consists of a large bedroom, a spare bedroom or den and a bath with shower. A two car garage is at the ground level. The exterior walls are finished with a combination stone veneer and vertical wood siding with batten joints. The garage door is of aluminum clapboard style opening overhead. The walk from the front entrance is made of flagstone.

The Sectional House

A shop fabricated wood frame house consisting of two complete half sections is designated as a sectional house. The fabricated units are shipped completely finished. A typical exterior wall section is shown in **Figure 42.** The units are from 10′ to 12′ wide in lengths up to 57′, are joined together at the site.

Plumbing, heating and electrical work may be shop installed in accordance with minimum property standards. Foundations are conventional and field installed.

The Floors of a Sectional House. A shop welded steel frame is provided for support of the wood floor system. Framing consists of 10″ junior I-beams at the sides of each unit and 10″ deep, 11 gage channels at the ends of the unit. One X-shaped 2″ channel brace is installed 4′ o.c. between the beams. All joints are welded, and 2″ × 1/8″ clip angles are welded to the top of the frame at 4′ intervals for attaching the wood framing, **Figure 43.**

Insulation board and blanket type insulating material may be shop installed over the cross members of the steel frame.

Floor framing consists of 2″ × 6″ joists, spaced 16″ o.c. with clear spans up to 9′ - 4-1/2″ or 2″ × 8″ joists spaced 16″ o.c. with clear spans up to 11′ - 4-1/2″. A header member corresponding to the joist size is nailed on the ends of the joists with three 16d nails evenly spaced 12″ o.c. in a staggered pattern.

Solid bridging of the same size as the joist is installed between the joists at the midpoint of the joist span. Two 1″ × 2″ continuous rails are dadoed into the joists at the halfway point between the bridging and the outside headers.

The floor framing is covered with 5/8″ plugged and touch-sanded underlayment grade fir plywood with the plywood sheet running parallel to the joists. The plywood is glued to each joist, the rails and the bridging. The plywood is also secured with 2″, #10 sheet metal screws spaced 12″ o.c. at the edge of each sheet and 30″ at the intermediate supports.

Exterior Walls. Wall framing, **Figure 44,** consists of 2″ × 4″ studs spaced 16″ o.c. with a single 2″ × 4″ bottom plate and a single 2″ × 4″ top plate. Each joint is secured with two 16d nails. Joints in the top and bottom plates are fastened together with a 4″ × 6″ × 1/4″ fir plywood splice which is glued and stapled to the plate using five staples on each side. Splices are located near the center of the plywood panels covering the wall framing.

The bottom plate is secured to the floor framing with 16d nails spaced 12″ o.c. Door openings are double framed with double 2″ × 4″ headers set on edge and blocking above. The headers are end nailed through the full length jamb stud with four 16d nails at each end. The jamb studs are nailed together with 16d nails spaced 24″ o.c.

Window openings are single framed with a single 2″ × 4″ header and a single 2″ × 4″ subsill. The top header is dadoed into the jamb stud 1/2″. The header, subsill and jamb stud are securely glued and stapled to the inside wood paneling and the exterior plywood sheathing.

The maximum span used on a single framed window is 36″. Openings in excess of 36″ are double framed with a 2″ × 4″ subsill and 2″ × 4″ jack studs below, spaced not over 16″ o.c. Window headers consist of double 2″ × 4″ members on edge for spaces up to 4′ and double 2″ × 6″ members for spans up to 6′. Headers are end nailed through the full length jamb stud with four 16d nails at each end. Jamb studs are nailed together with 16d nails spaced 24″ o.c.

The outside face of the wall framing, **Figure 44,** left, is covered with 5/16″ fir plywood glued around all door and window openings and along the top plate and at the floor line where the plywood laps. The exterior sheathing is further secured with staples. As an alternate, 1/2″ dense type fiberboard sheathing may be specified, in which case window openings will conform to the construction outlined for openings over 36″.

Fig. 45. Roof over each half section of house consists of preassembled rafter units bearing on wall plate and ridge beam.

Fig. 46. In-place scaffolding at factory saves hours of labor when finishing the half sections and installing equipment.

The interior finish consists of 1/4″ plywood glued and stapled to the studs, sill and plate.

Various types of siding materials may be shop installed in conventional manner, or the siding may be omitted when brick veneer is specified.

Three-inch fiberglass or rock wool insulation, with the vapor barrier on the warm side, is shop installed between the studs.

Gable Walls. Framing for the gable wall consists of a standard half truss. The bottom cord is secured to the top wall plate with 10d nails spaced 12″ o.c. or, if brick veneer is used, to 1″ × 4″ filler blocks attached to the wall framing.

Partitions. A bearing partition is installed beneath

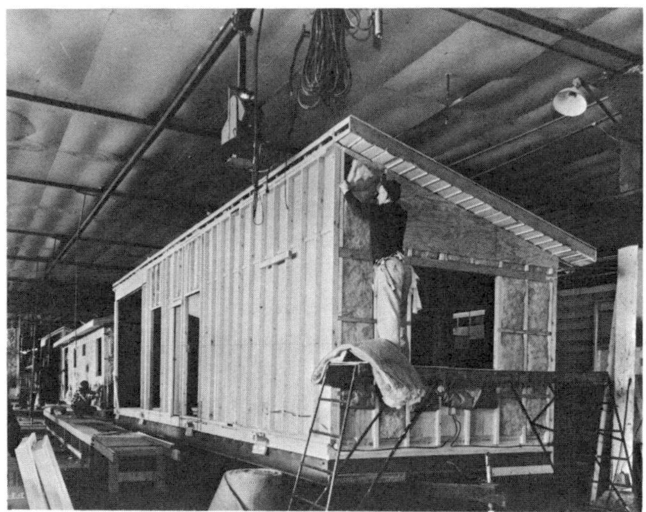

Fig. 47. Insulation is installed in end wall of sectional house. Note rigid steel frame beneath wood framing.

Fig. 50. One half of the house is already in place on full basement foundation while second half is moved into position.

Fig. 48. A completed half of a sectional house is wrapped in a protective covering of polyethylene film for shipment.

Fig. 51. Attractive interior of a sectional house.

Fig. 49. Two sections of a sectional house loaded on trailer for transporting to the building site.

Fig. 52. Completed sectional house has brick veneer laid up at site. Endwall siding is applied after joining sections.

the ridge of each half house section. Framing consists of 2″ × 2″ studs spaced 16″ o.c. with a single 2″ × 2″ top and bottom plate. Each joint in the framing is secured with two 12d nails and all plates are spliced with 2″ × 6-1/4 × 2 plywood glued and stapled as outlined under the section relative to Exterior Walls.

Door openings are double framed with a 2″ × 4″ header resting on the inner jamb stud. The header is secured by end nailing through the full length jamb stud with two 16d nails.

All other non-bearing partitions are framed, as described heretofore, except that both sides of the wall are stress skinned by gluing and stapling 1/4″ plywood to the studs.

Roof Framing. Roof framing consists of preassembled rafter units, **Figure 45,** bearing on the ridge partitions and outer wall plates. Ceiling joists run from the ridge partitions to outer wall plates to tie the rafters.

Roof Sheathing. Plywood roof sheathing, 3/8", is secured to the roof framing with staples. Asphalt shingle roofing is installed in accordance with applicable minimum property standards.

Ceiling Finish. The ceiling consists of 5/16" plywood applied to joists, over which acoustical tile is installed in conventional manner. Insulation, with integral vapor barrier on the warm side, is shop installed between joists.

Equipment. All component parts of the dwelling unit including plumbing, heating and electrical work and both interior and exterior finish material are factory installed to form the completely prefinished half sections of a dwelling unit, **Figures 46 and 47.** Sections are wrapped in polyethylene film for shipment, **Figure 48.**

Field Erection. During transportation, adequate support is provided to avoid deflection in the unit, **Figure 49.** At the site, the units are placed on conventional foundations, **Figure 50.** Two sections are bolted together through the 10″ Junior I-beam with 1/2″ × 6-1/2″ bolts spaced 8′ o.c. Metal strap ties at 32″ o.c. are placed across the ridge and secured to each roof framing member with three 8d nails. The ridge cap is installed in a conventional manner. The end wall sections are attached together by the use of metal ties at 24″ o.c. secured to each section with two 8d nails. The two bearing partitions running the length of the units at approximate center are attached at all openings in the same manner as the end wall sections. The bearing walls are further attached together by 3″ #8 screws, which are spaced vertically on stud location at 24″ o.c. and longitudinally every 6′. Openings between units are covered with appropriate door stops or arch trim. Exterior joints are covered with siding or brick veneer.

The Sectional House in Place. A typical interior is shown in **Figure 51.** Note the wall finish and the beam ceiling of this attractive kitchen-dining alcove. A completed sectional house, with an exterior of brick veneer and aluminum horizontal siding is shown in **Figure 52.**

Chapter 24

TIME AND MONEY SAVING DETAILS

No matter how comprehensive a book on the subject of house construction may appear to be as each phase of the art is covered in chapter after chapter, there is usually a wealth of information left over after fitting and editing the material. To discard such material would be wasteful, so the authors have assembled the following sixty pages of details which should be useful to all builders in their day-to-day activities. A single idea that might be gleaned from these pages would more than pay the cost of this book!

Reference is made in the index to each topic covered, but the list of details by category at right may be helpful in referring to them.

DETAIL SHEET .... PAGE

Interior Moldings 390
Exterior Moldings 391
Construction Joints 392
Framing Hardware 393
Masonry Details 394
Brickwork at Entry 395
SCR Brick Panel 396
Decorative Concrete Block 397
Texture with 4 Basic Blocks 398
How To Build Dry Basements 399
Rigid Frames for Wide Spans 400
Roof Construction Details 401
Wetern Cornice Details 402
Stressed Skin Panels 403
Dormers 404
Eaves, Gravel Stops, Gutters 405
Plastic Flashing 406
Porch Details 407
Carport Details 408
Bay Windows 409
Corner Windows 410
Louver Vents 411
Glazing with Insulating Glass 412
Window & Door Details 413
Locksets 414
Unusual Entrance and Bay 415
Entrance and Entrance Hall 416
Entrance and Window Wall 417
Colonial Stairway 418
Stair Stringers 419
Roughing In Stairways 420
Attic Stair Construction 421
Attic Stair Winders 422
Attic Stairs, Finish Carpentry 423
Attic Alteration, Boy's Room 424
Good Kitchen Planning 425
Built-Ins In Kitchens 426
Kitchen Cabinetwork 427
How To Build A Corner Sink 428
Ventilation for Kitchens 429
Family Bathrooms 430
Master Dressing Room and Bath 431
Planning Plumbing 432
Septic Tanks and Absorption Fields 433
Laundry Room as Part of House 434
Linen and Towel Cabinets 435
Hanging Closets 436
Free Standing Closet 437
Distinctive Closet Details 438
Multiple Use Sliding Table 439
Modern Fireplace 440
Inside-Outside Fireplace 441
Open Hearth Fireplace 442
Circular Fireplace 443
Small Commercial Building Details 444
Office Building Entrance 445
Commercial Building Stairway 446
Garden Boxes 447
Garden Fences 448

·Nº D-25 · INTERIOR · MOULDINGS · AND · APPLICATION ·

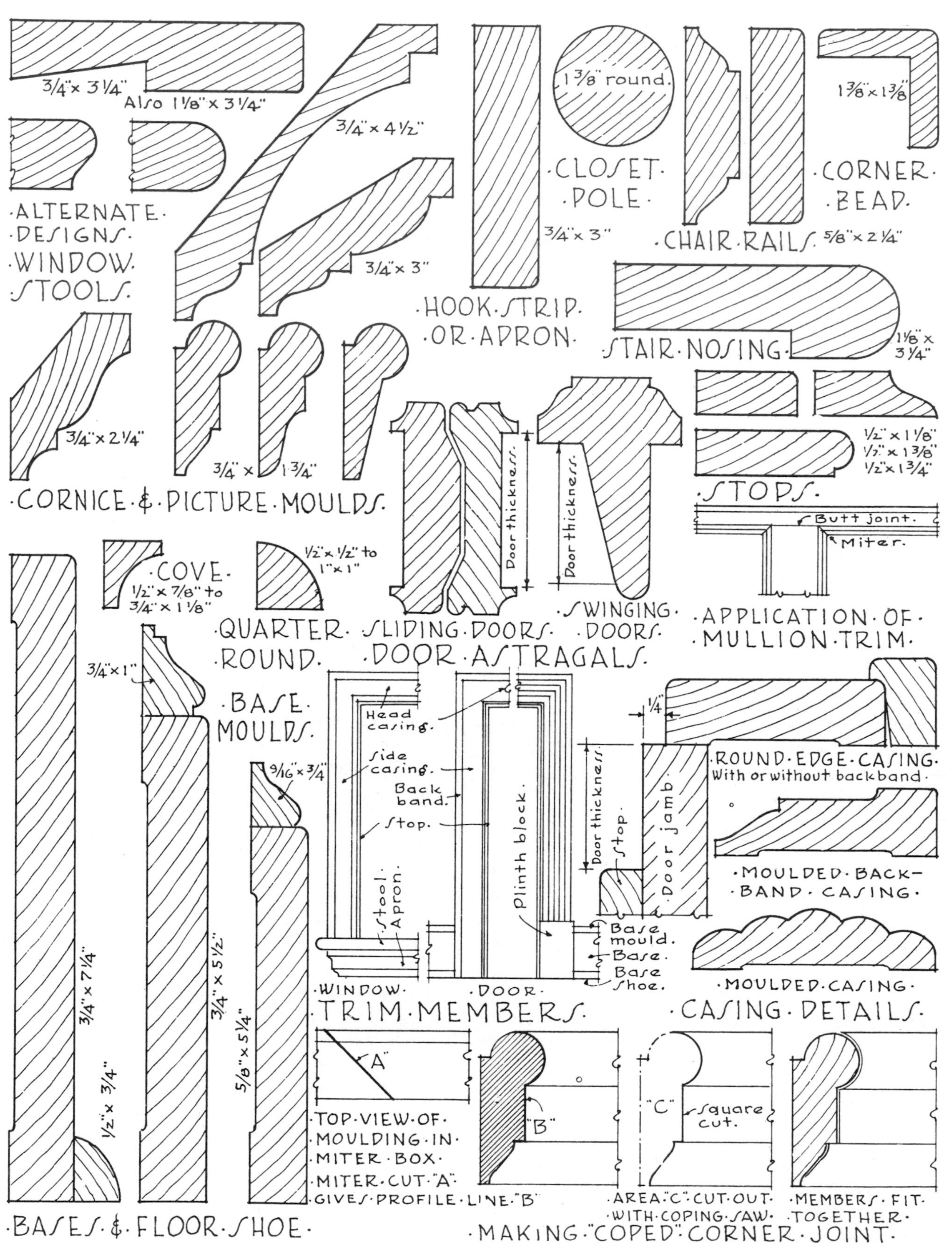

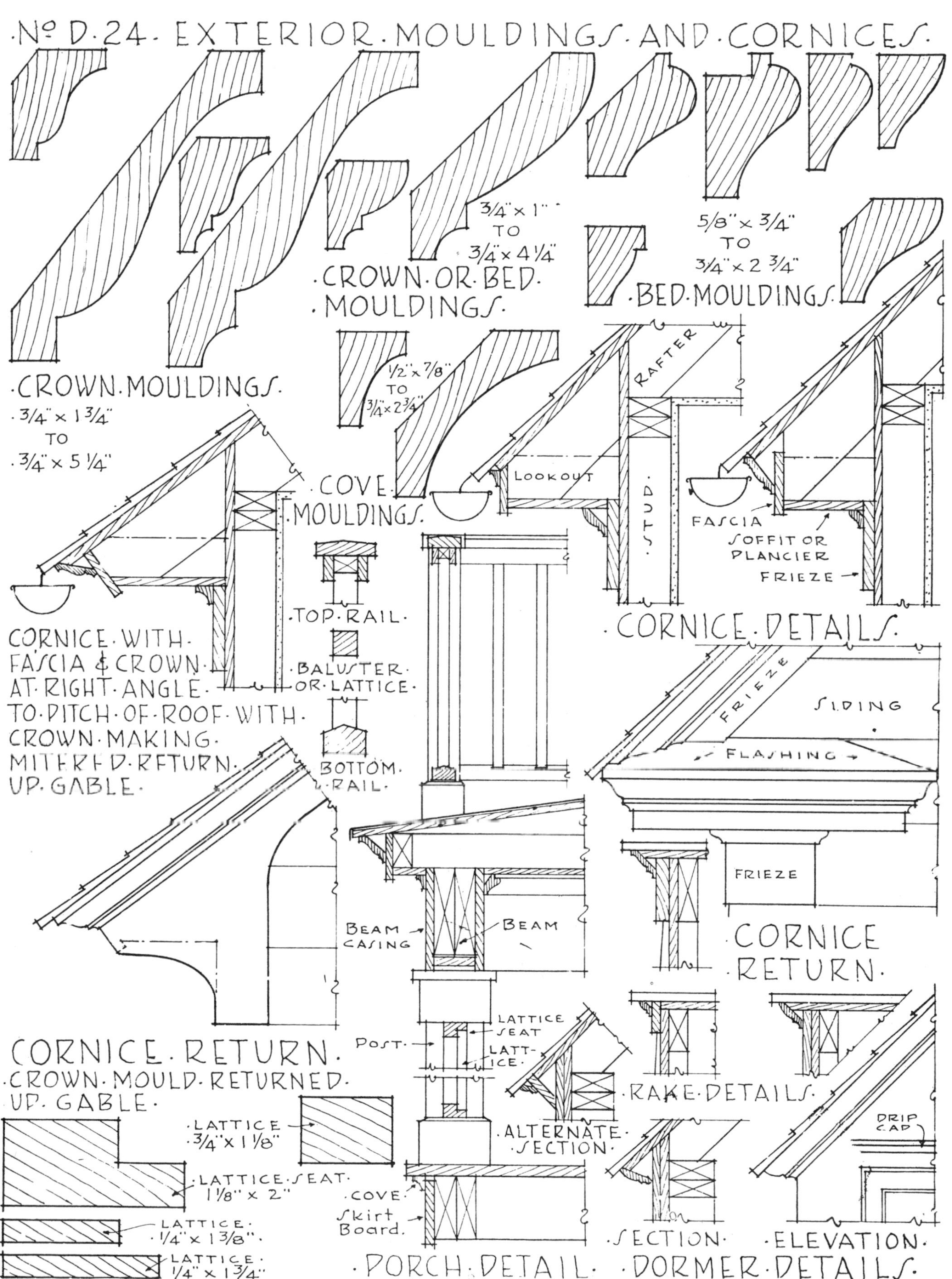

TIME AND MONEY SAVING DETAILS

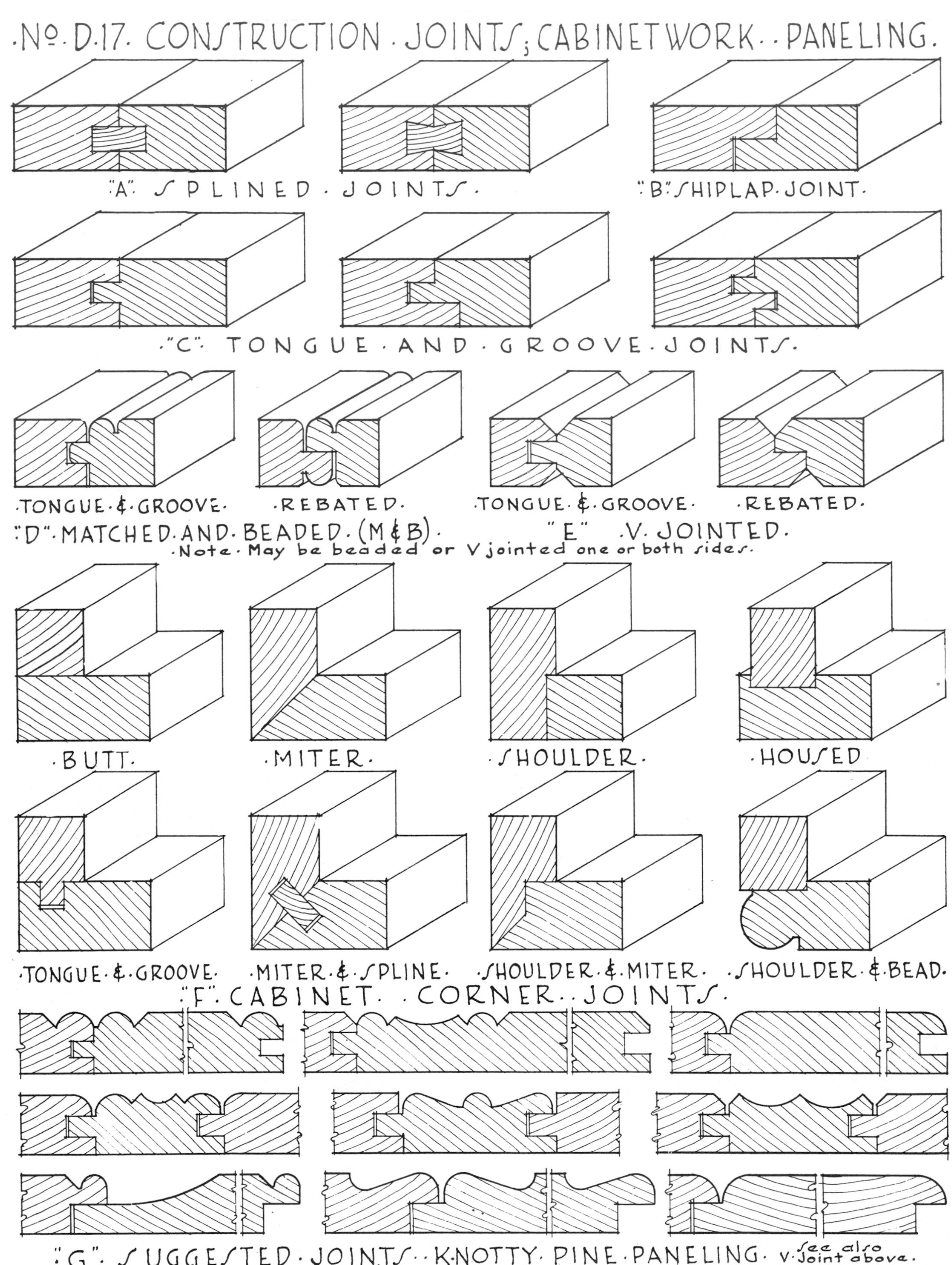
·Nº·D.17· CONSTRUCTION · JOINTS; CABINETWORK · · PANELING.
"A" SPLINED · JOINTS ·
"B" SHIPLAP · JOINT ·
·"C"· TONGUE · AND · GROOVE · JOINTS ·
·TONGUE · & · GROOVE ·
·REBATED ·
·TONGUE · & · GROOVE ·
·REBATED ·
"D" MATCHED · AND · BEADED · (M & B) ·
"E" ·V· JOINTED ·
·Note · May be beaded or V jointed one or both sides ·
·BUTT ·
·MITER ·
·SHOULDER ·
·HOUSED
·TONGUE · & · GROOVE ·
·MITER · & · SPLINE ·
·SHOULDER · & · MITER ·
·SHOULDER · & · BEAD ·
"F" · CABINET · · CORNER · · JOINTS ·
"G" · SUGGESTED · JOINTS · · KNOTTY · PINE · PANELING · V · See also Joint above ·

Framing Hardware Speeds House Construction, Saves Labor

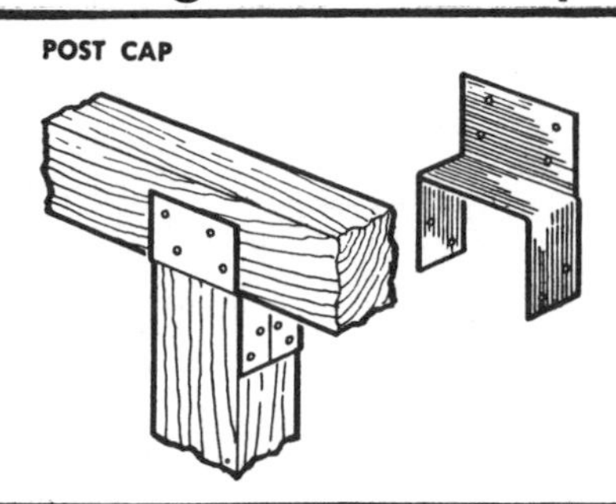

Post caps connect beams to posts efficiently and economically, assuring perfect alignment.

The old and not so dependable toenail is being replaced by a wide variety of efficient hardware designed specifically for fastening wood structural members together. Some, like the post cap and plywood clip shown at left and right, align mating parts. Others, like the anchors, tie-downs, truss plates, and joist connectors are installed to fasten structural members after they are brought together in correct alignment. Joist and beam hangers serve as supports to transfer heavy loads from one member to another. Split ring truss connectors prevent movement in shear by supplementing the bolts used to fasten chords together. The slight extra cost of this hardware is more than offset by the savings in time and labor it offers.

Plywood clips align and stiffen joints between panels, are available to fit ⅜", ½", ⅝" and ¾" plywood.

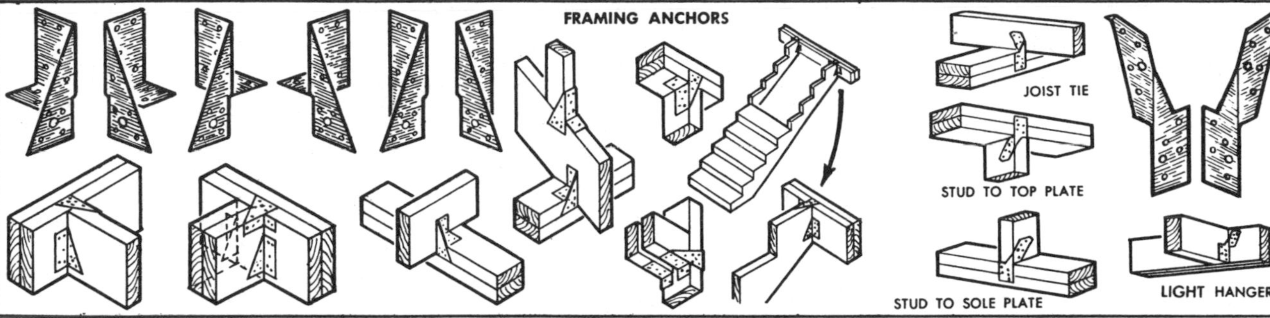

Framing anchors are made in a wide variety of sizes and shapes to fasten structural members securely together. Nails used are equal in thickness to 8d common, but are about half as long. The anchors develop the full strength of the nails in shear. FHA has approved the use of all hardware shown on this page when used as recommended.

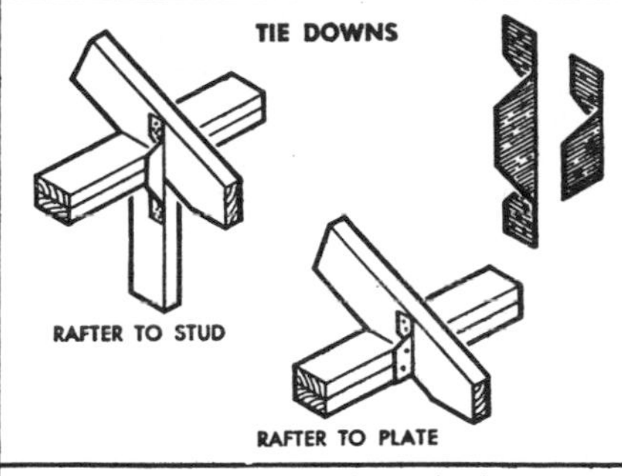

Tie-downs secure rafters to wall plates and studs to resist uplift resulting from high winds.

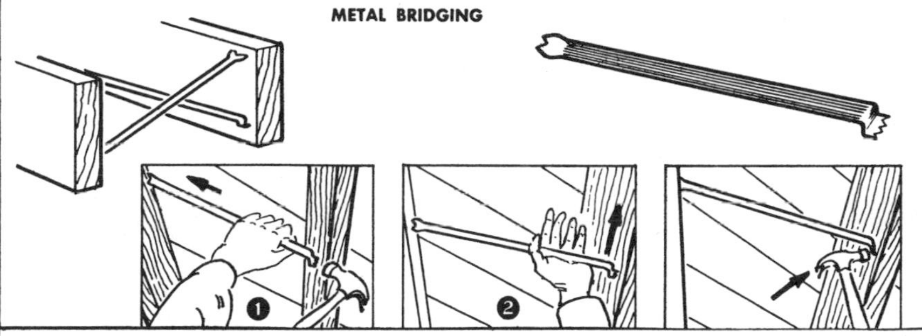

Metal bridging can be installed in seconds, saves cutting and beveling hundreds of short pieces of wood. To install, hammer one end into joist (1), push lower end into position (2), and drive end into joist (3).

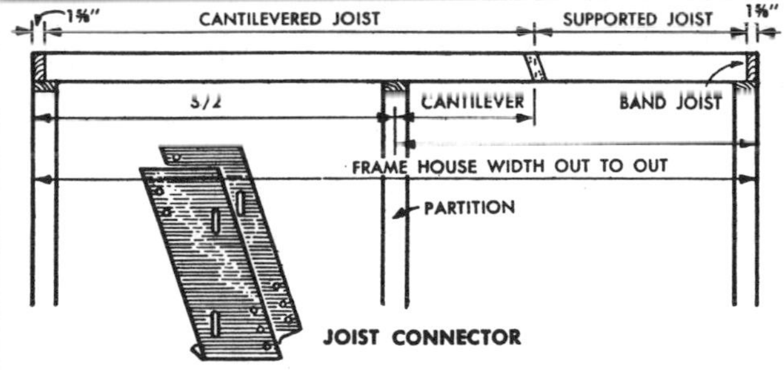

Joist connectors reduce subflooring labor by assembling joists in line instead of lapped or staggered. Their use can save $30 in materials per house by eliminating overlapping.

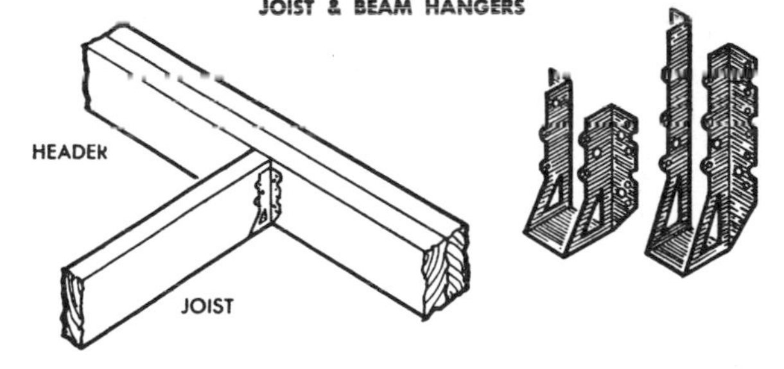

Joist and beam hangers are formed of heavy gauge zinc coated steel to support wood joists and beams from 2x4 to 4x14, including double 2x6s and 2x14s.

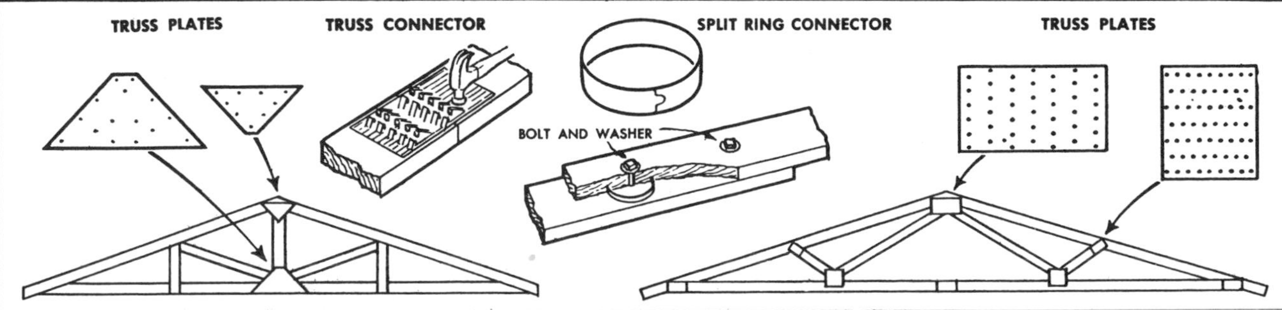

Truss plates are made in many sizes, shapes and types; some simply drilled for nailing, others punctured so that the protruding metal bites into the wood when installed. Some even have nails formed in the plate. Ring-type truss connectors fit into grooves bored in the wood to supplement the shear strength of the bolt and distribute stress.

MASONRY · CONSTRUCTION · DETAILS

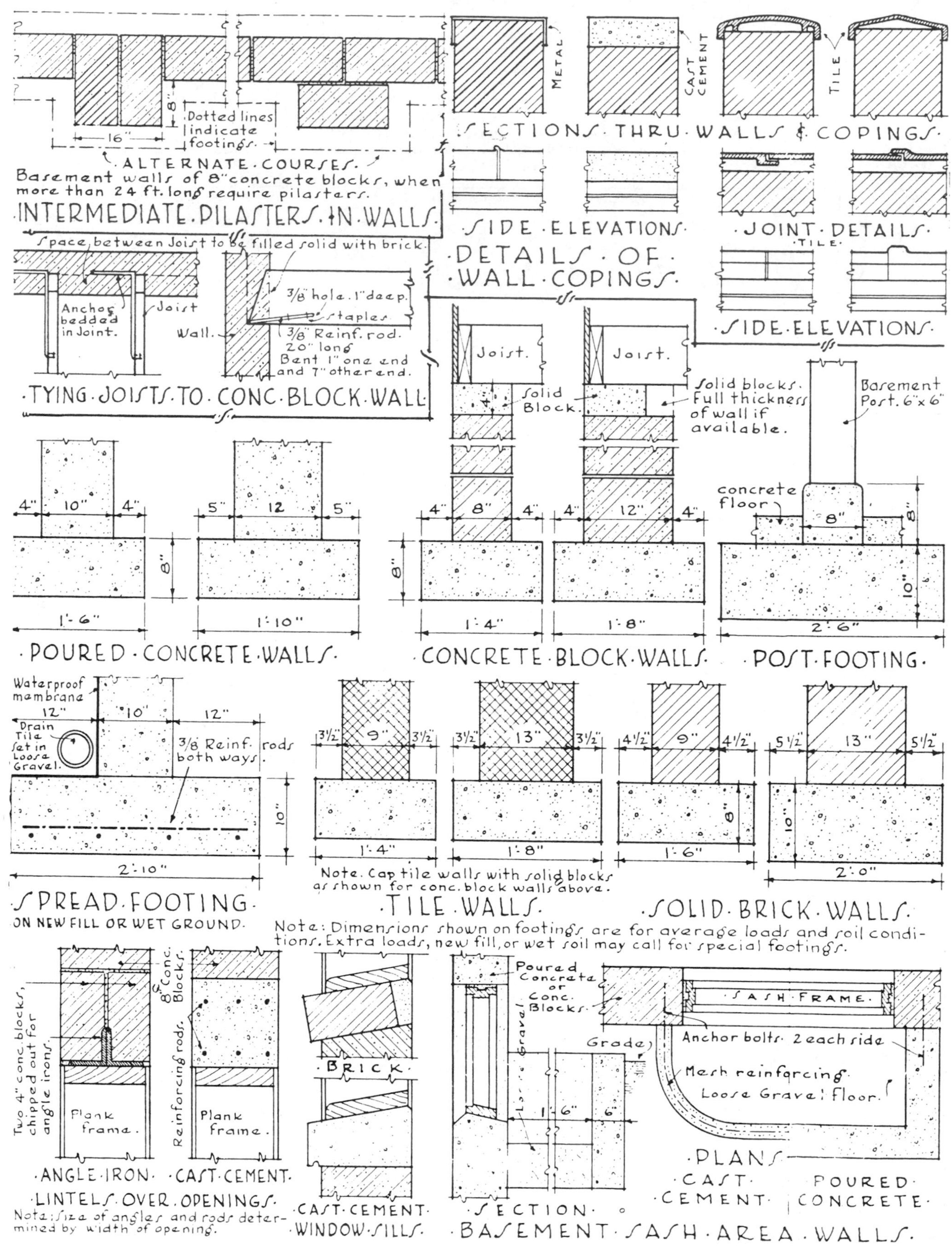

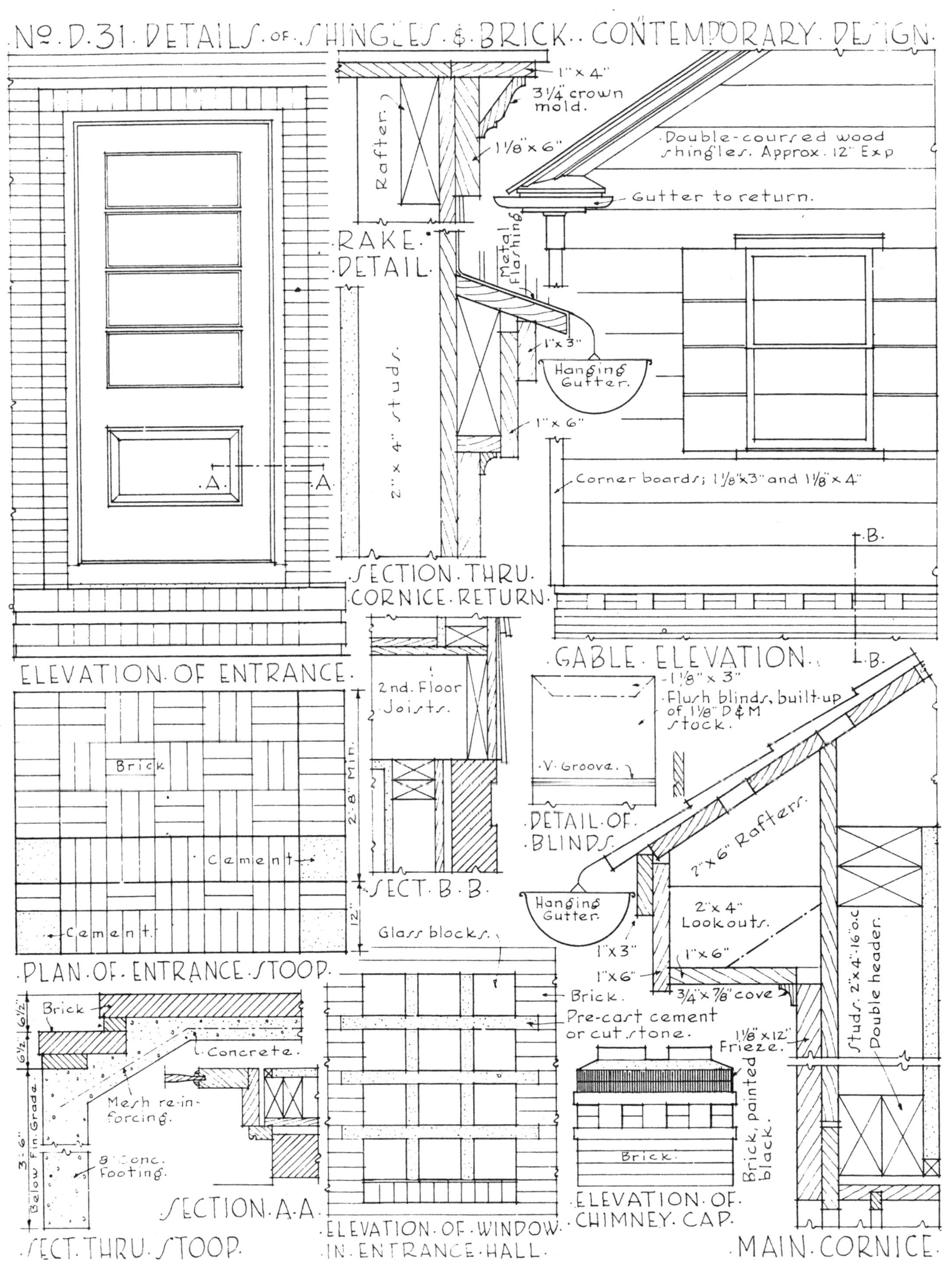

No. D.31. DETAILS of SHINGLES & BRICK. CONTEMPORARY DESIGN.
1" x 4"
3 1/4" crown mold.
Rafter.
1 1/8" x 6"
Double-coursed wood shingles. Approx. 12" Exp
Gutter to return.
RAKE. DETAIL.
Metal Flashing
1" x 3"
Hanging Gutter.
2" x 4" studs.
1" x 6"
A.
A
Corner boards; 1 1/8" x 3" and 1 1/8" x 4"
B.
SECTION. THRU. CORNICE. RETURN.
ELEVATION. OF ENTRANCE.
GABLE. ELEVATION.
B.
1 1/8" x 3"
Flush blinds, built-up of 1 1/8" D & M stock.
2nd. Floor Joists.
Brick
2' 8" Min.
V. Groove.
DETAIL. OF. BLINDS.
2" x 6" Rafters.
Cement
SECT. B. B.
Hanging Gutter
2" x 4" Lookouts.
12"
Cement.
Glass blocks.
1" x 3"
1" x 6"
1" x 6"
PLAN. OF. ENTRANCE. STOOP.
Brick.
3/4" x 7/8" cove
Brick
6 1/2"
Pre-cast cement or cut stone.
6 1/2"
Concrete.
1 1/8" x 12" Frieze.
Studs. 2" x 4". 16" o.c.
Double header.
Mesh re-in-forcing.
3' 6" Below Fin. Grade
8" Conc. Footing.
Brick, painted black.
Brick.
SECTION. A.A.
ELEVATION. OF. CHIMNEY. CAP.
SECT. THRU. STOOP.
ELEVATION. OF. WINDOW IN. ENTRANCE. HALL.
MAIN. CORNICE.

New SCR Brick Building Panel: How It Works

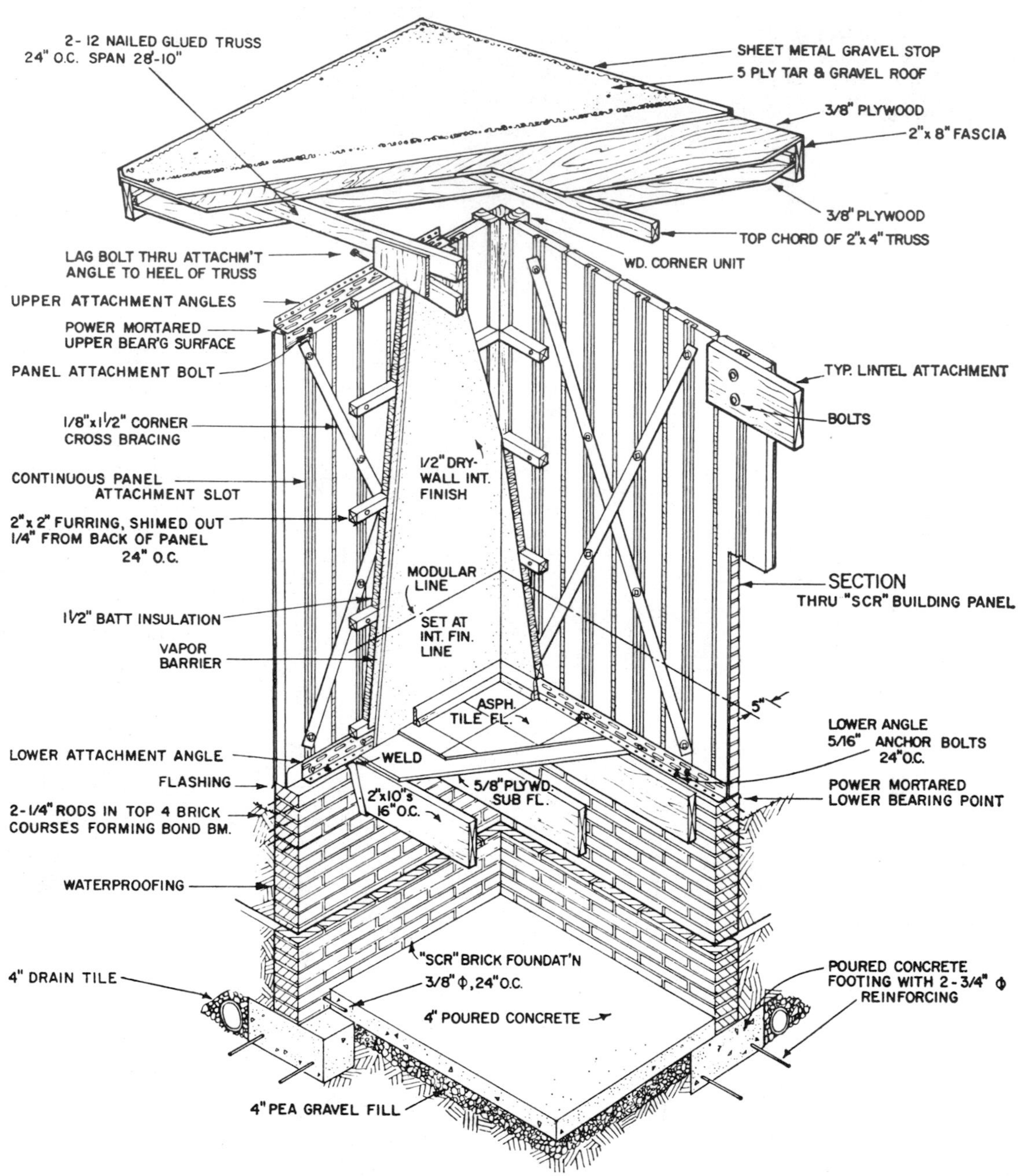

Direct from the Structural Clay Products research laboratory is this new load-bearing brick panel. The perspective view illustrates the construction of the first successful brick-panel house. The house was built for \$12 per sq. ft. in Geneva, Ill., utilizing the 1x8′ by 2½″ norman-faced panel, but it may be a while before it's available commercially.

Decorative Concrete Block Adds Pleasing Texture to Exposed Walls

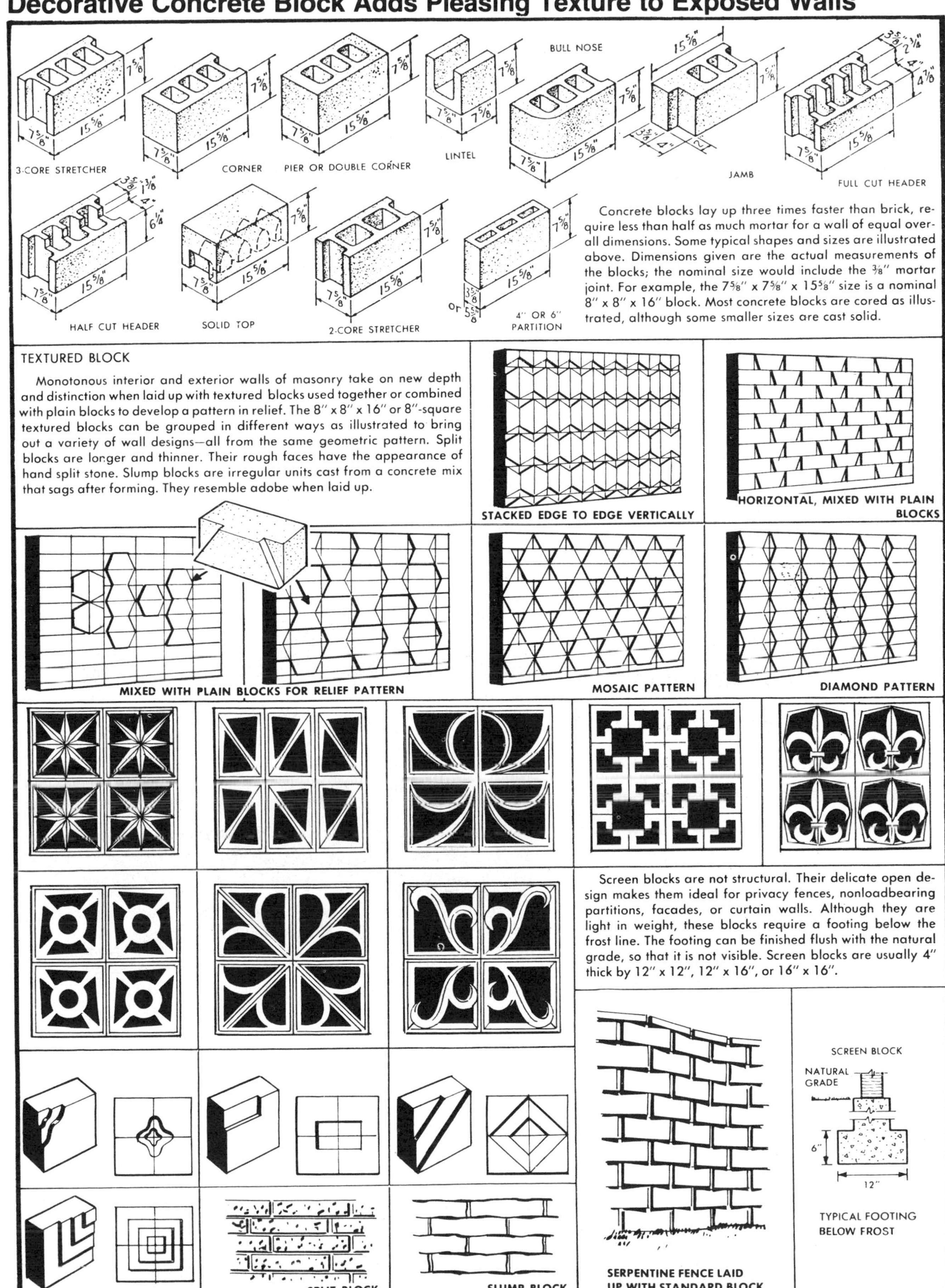

Concrete blocks lay up three times faster than brick, require less than half as much mortar for a wall of equal overall dimensions. Some typical shapes and sizes are illustrated above. Dimensions given are the actual measurements of the blocks; the nominal size would include the 3/8" mortar joint. For example, the 7 5/8" x 7 5/8" x 15 5/8" size is a nominal 8" x 8" x 16" block. Most concrete blocks are cored as illustrated, although some smaller sizes are cast solid.

TEXTURED BLOCK

Monotonous interior and exterior walls of masonry take on new depth and distinction when laid up with textured blocks used together or combined with plain blocks to develop a pattern in relief. The 8" x 8" x 16" or 8"-square textured blocks can be grouped in different ways as illustrated to bring out a variety of wall designs—all from the same geometric pattern. Split blocks are longer and thinner. Their rough faces have the appearance of hand split stone. Slump blocks are irregular units cast from a concrete mix that sags after forming. They resemble adobe when laid up.

Screen blocks are not structural. Their delicate open design makes them ideal for privacy fences, nonloadbearing partitions, facades, or curtain walls. Although they are light in weight, these blocks require a footing below the frost line. The footing can be finished flush with the natural grade, so that it is not visible. Screen blocks are usually 4" thick by 12" x 12", 12" x 16", or 16" x 16".

Unusual design and texture is achieved in commercial...

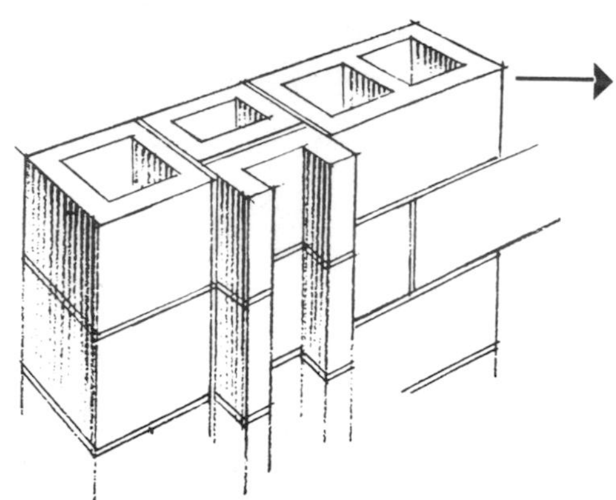

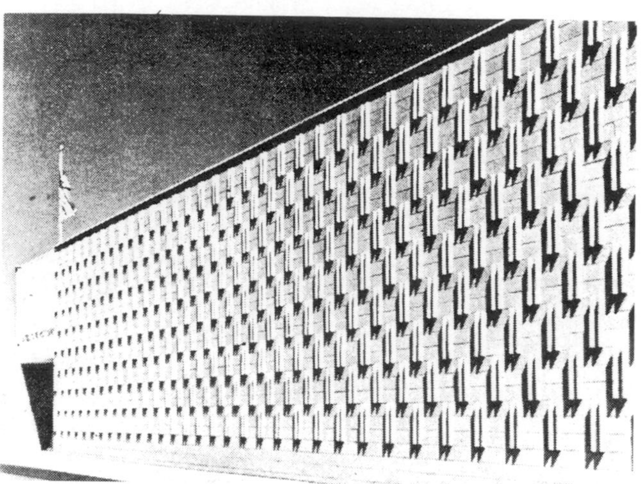

STANDARD LINTEL block, three high, adds beauty and drama to post office wall in La Mirada, Cal. Builder: Devon Constr. Co.; Architect: Victor Gruen & Assoc.

...and residential construction with 4 basic concrete blocks

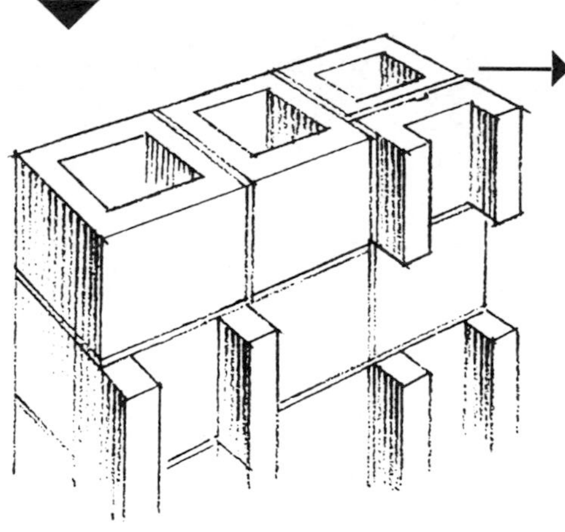

IN LIVING AREA of contemporary home, block wall is used effectively to contrast with the plain fireplace wall. Builder: Sanford Adler; Architect: Palmer & Krisel.

ONLY four standard concrete-block units were used to add originality and freshness of design to the post office wall and the contemporary house, shown above. These units are: 8″x8″x8″ half; 8″x4″x8″ "U" lintel; 6″x8″x8″ half; 8″x8″x16″ standard. In the post office wall, three Rocklite lintel units are used, one on top of the other in staggered rows. In the house, same staggered row pattern is used, but with a single lintel unit in the rows. The 6″x8″x8″ backs up lintel units.

Dry Basements: How to Build Them to Satisfy Today's Buyer

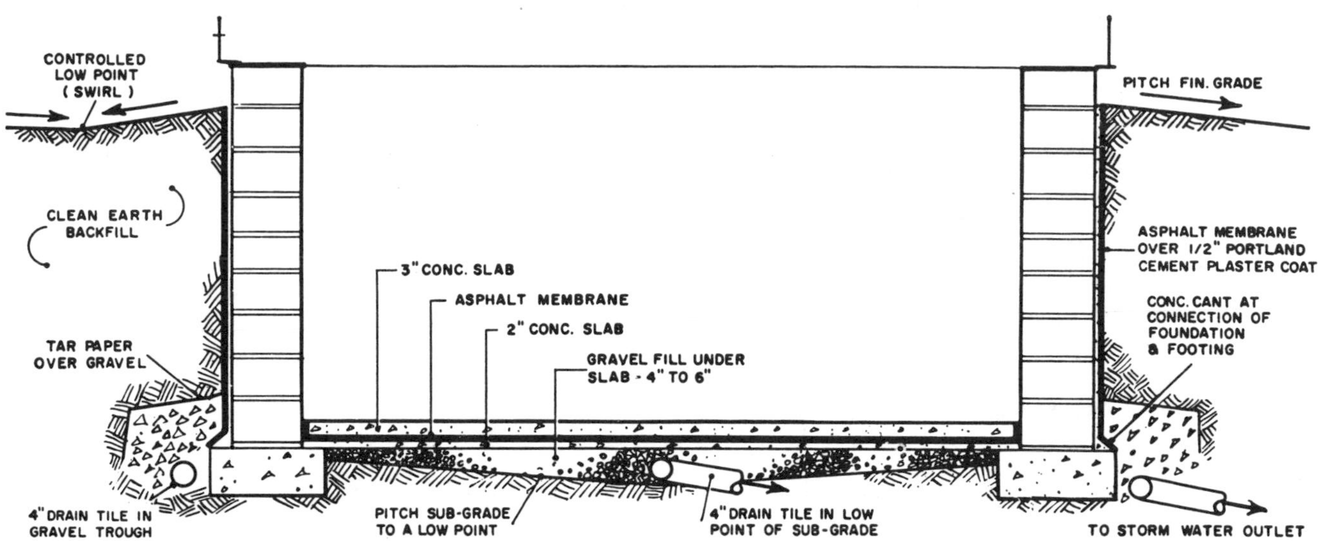

Every buyer expects a dry basement in his home, and rightfully—he's paid for it. And when a puddle of water appears on the floor, the buyer goes after the builder.

To avoid both the aggravation and the expense of making good on a faulty job, the smart builder tackles the problem at the start when preparing the site, and follows through with all necessary precautions until the foundation is completed.

The drawing at the top of the page illustrates all structural, water-proofing and drainage elements that should be considered. The slight additional cost of combining all of them in one foundation will insure a dry basement under the most extreme water conditions. But there is more to waterproofing than simply installing drain tiles. A builder should study each site to determine the nature of sub-surface water and take whatever steps may be necessary to control it.

Below are six common causes of seepage, both natural and man made, along with the most effective steps to prevent it.

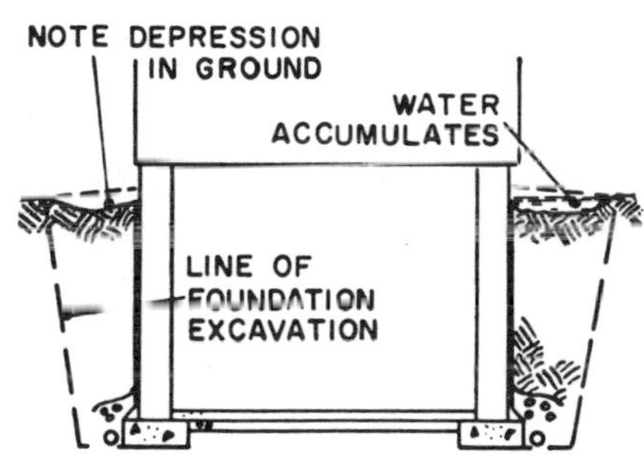

BACKFILL THAT HAS SETTLED will form a continuous depression that will allow water to accumulate around the house. *Backfill above grade and tamp.*

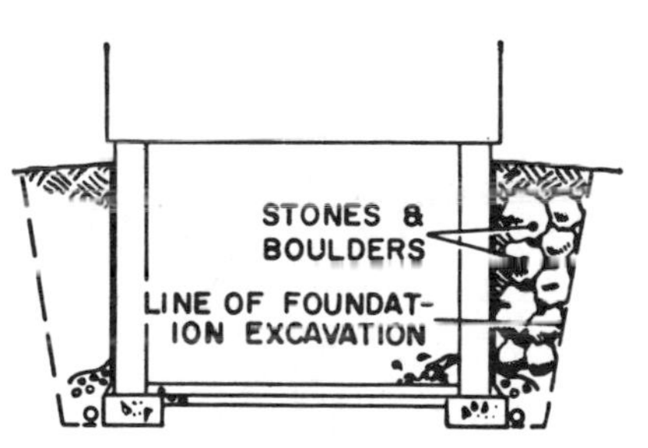

DUMPING LARGE ROCKS in with the backfill causes large quantities of water to be drawn into the voids between them, creating great pressure. *Use clean fill.*

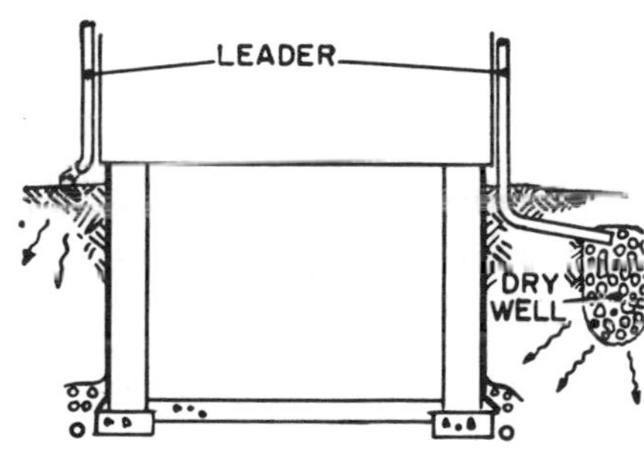

DOWNSPOUTS that drain on the ground or into drywells too close to the house can cause excessive saturation of the earth. *Dig drywells in ground beyond backfill.*

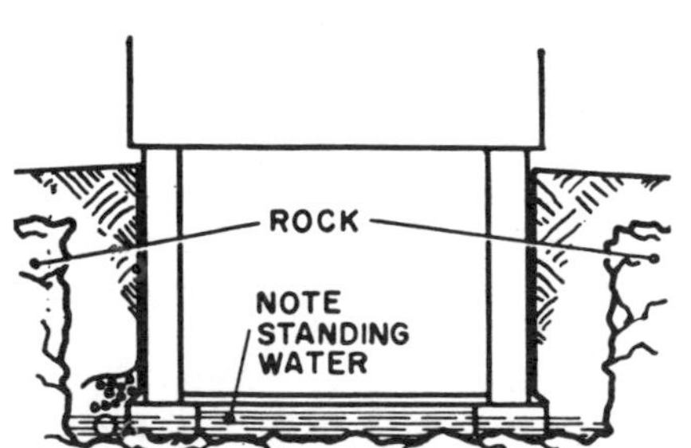

STANDING WATER trapped by rock around the foundation can exert upward pressure against the slab. *Drain water off to a low point and build a two-layer slab.*

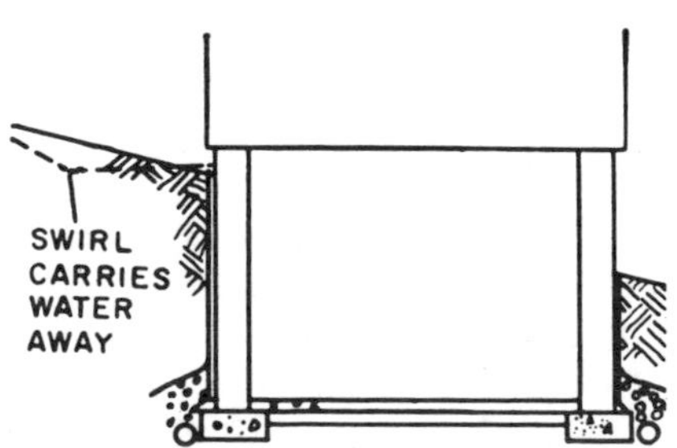

HILLSIDE LOTS that slope toward house will flood basement with surface water running down from higher elevations. *Create a swirl to divert this water.*

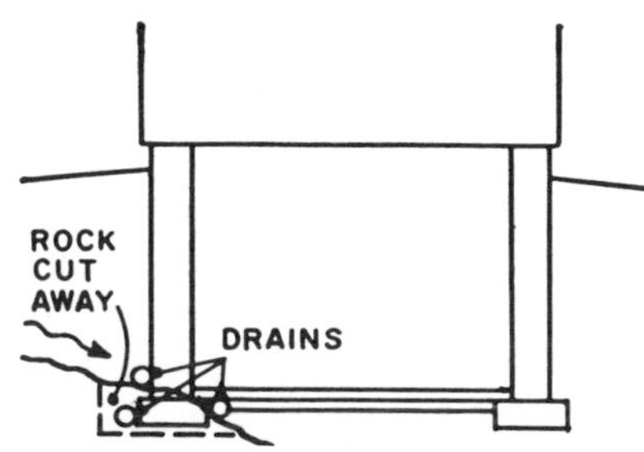

UNDERGROUND LEDGE ROCK can carry unseen water to foundation that's built on it. *Rock should be cut away and drains installed on both sides of footing.*

Rigid Frames: One Way to Frame Wide Clear Spans

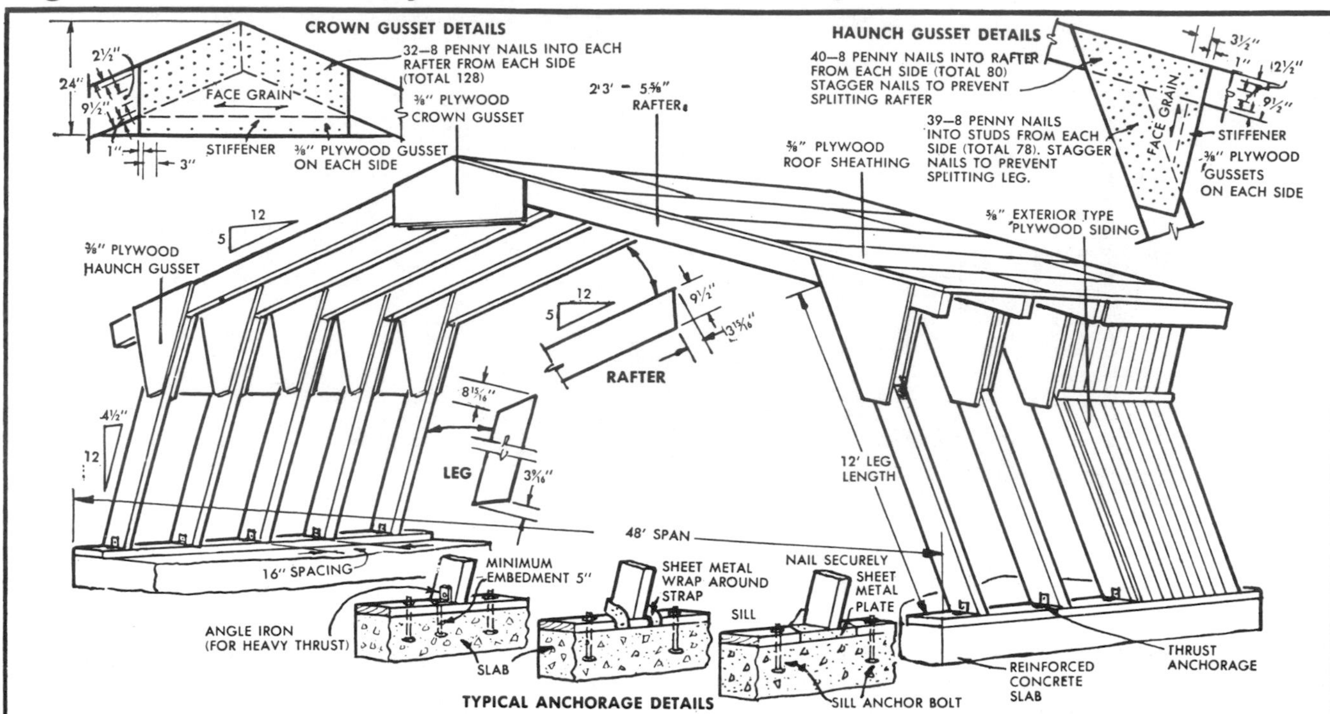

Plywood rigid frames are structural arches formed by joining four straight lengths of lumber with nailed plywood gussets. Once assembled, they become rigid load-carrying units suitable for farm and industrial buildings, churches, airplane hangars and recreational buildings.

Rigid frames are fabricated with slanted legs as illustrated, or with vertical legs, similar in appearance to conventional wall framing. The slant-leg type is more popular because it makes the most efficient and economical use of materials—a vertical-leg design built for the same loads as the slant-leg version would require heavier framing to meet the load specifications.

A 48′ span slant-leg structure designed by the Douglas Fir Plywood Assn. is detailed above. It requires 3 x 10 legs and rafters spaced 16″ o.c., 3/8″ interior grade plywood haunches and crown gussets, 5/8″ exterior grade siding and roof sheathing, and 3/8″ interior grade inside finish. The 3 x 10s for the rafters must be ordered in 24′ lengths, as splicing these members is not considered practical. DFPA cautions that the design is based on the use of DFPA plywood, because the sheathing and roofing provide diaphragm strength and eliminate all bracing other than the plywood gussets.

A reinforced slab floor makes the most effective foundation. It ties the walls securely together. Where no slab is laid, buttresses must be formed to transfer the thrust into the earth. This, combined with secure leg anchorage, results in a sturdy structure.

Steps in rigid frame erection are illustrated below.

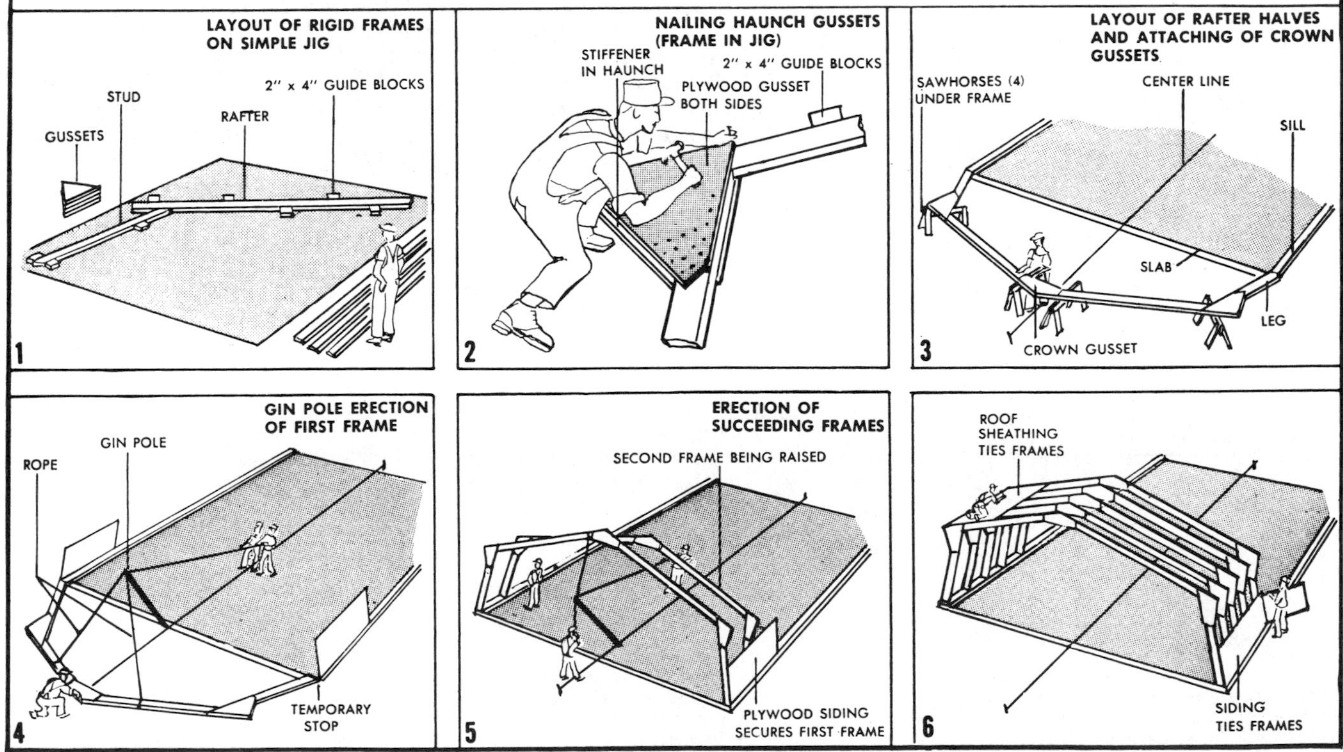

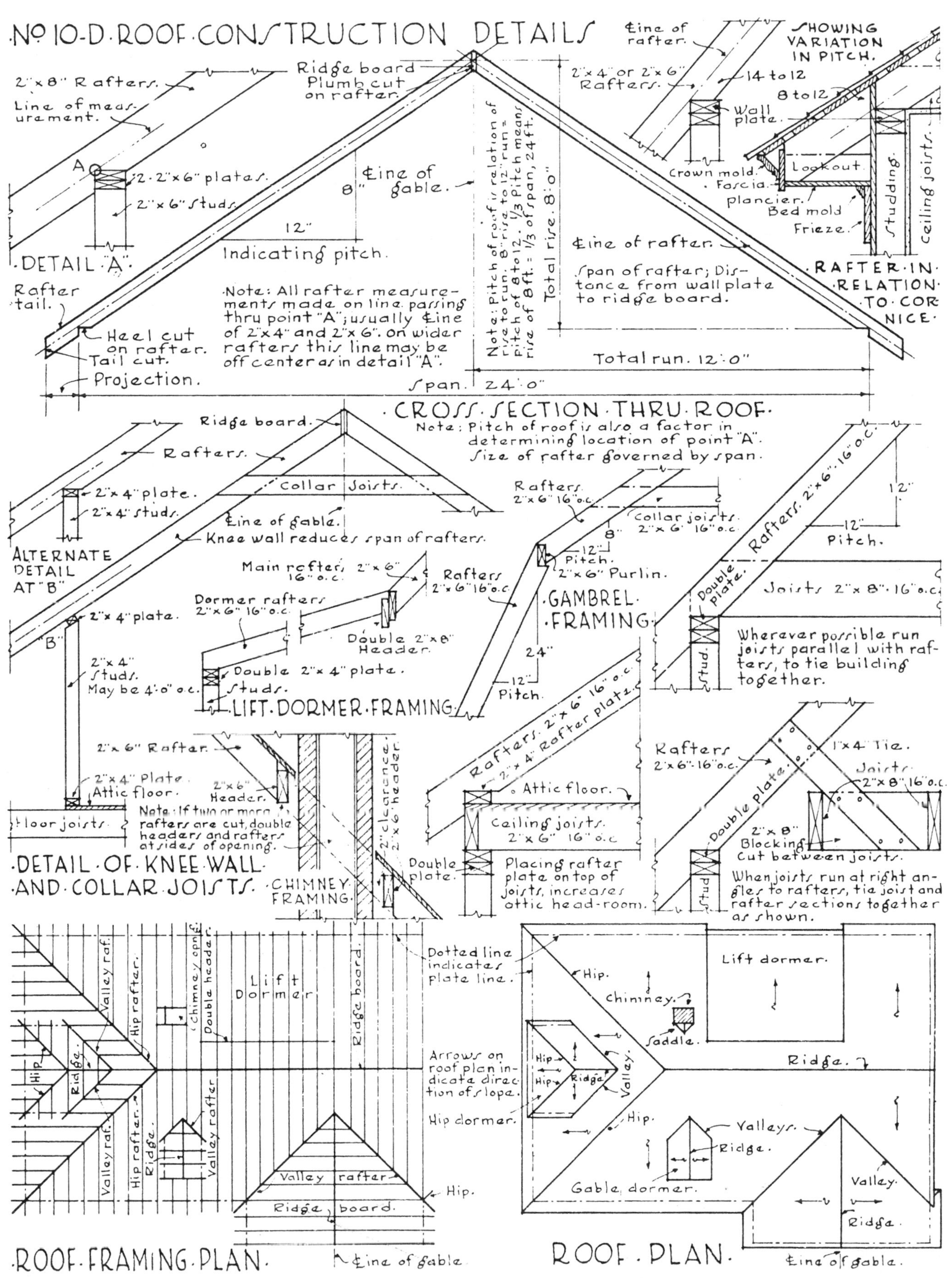

TIME AND MONEY SAVING DETAILS

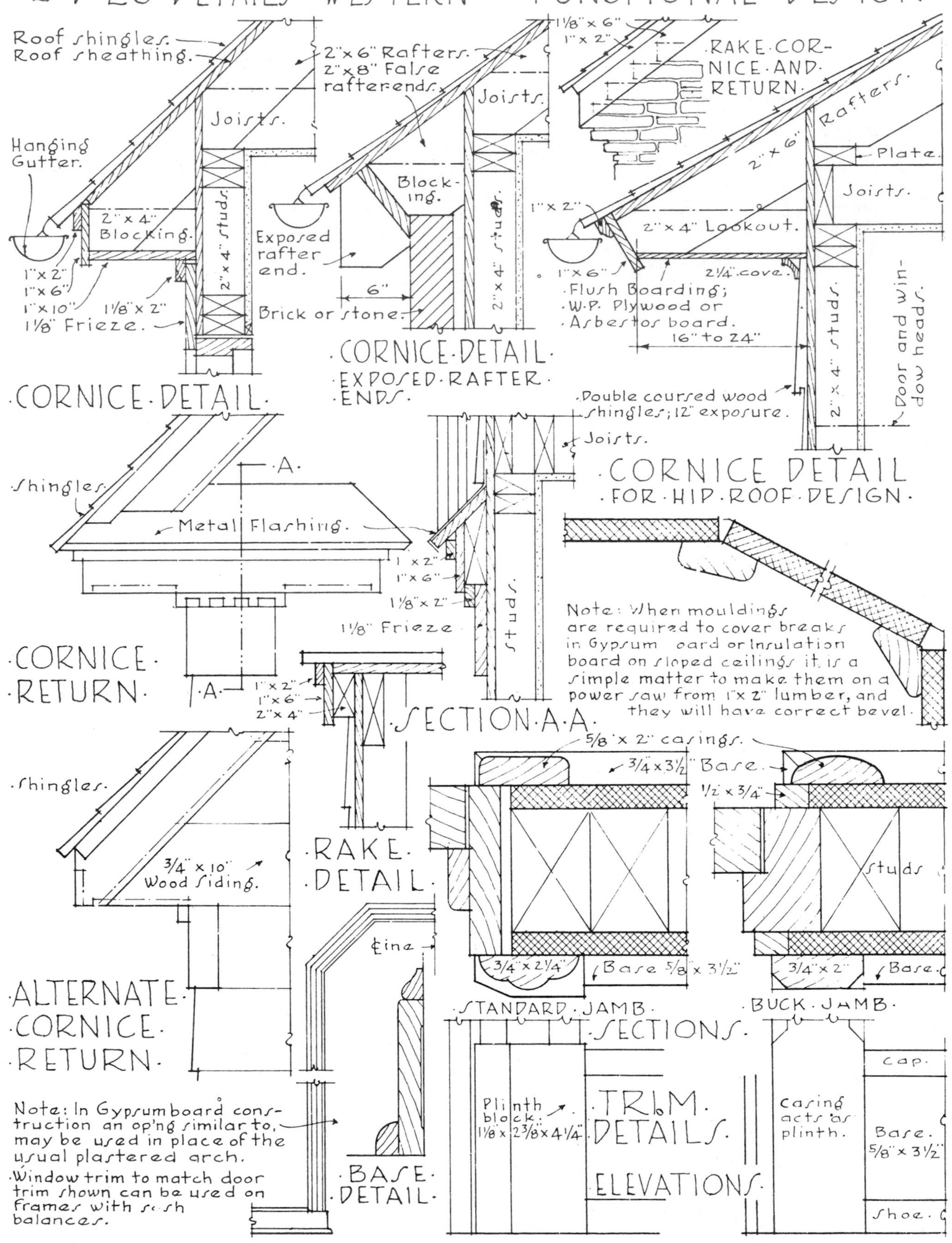
NO. D. 26. DETAILS "WESTERN" OR "FUNCTIONAL" DESIGN.
Roof shingles.
Roof sheathing.
2"x6" Rafters.
2"x8" False rafter-ends.
Joists.
Hanging Gutter.
2"x4" Blocking.
2"x4" studs.
1"x2"
1"x6"
1"x10"
1 1/8"x2"
1 1/8" Frieze.
CORNICE DETAIL.
Blocking.
Exposed rafter end.
6"
Brick or stone.
CORNICE DETAIL. EXPOSED RAFTER ENDS.
1 1/8"x6"
1"x2"
RAKE CORNICE AND RETURN.
2"x6" Rafters.
Plate.
Joists.
2"x4" Lookout.
1"x6"
2 1/4" cove.
Flush Boarding; W.P. Plywood or Asbestos board.
16" to 24"
2"x4" studs.
Door and window heads.
Double coursed wood shingles; 12" exposure.
Joists.
CORNICE DETAIL FOR HIP ROOF DESIGN.
Shingles.
A
Metal Flashing.
1"x2"
1"x6"
1 1/8"x2"
1 1/8" Frieze
studs
CORNICE RETURN.
A
1"x2"
1"x6"
2"x4"
SECTION A.A.
Note: When mouldings are required to cover breaks in Gypsum board or Insulation board on sloped ceilings it is a simple matter to make them on a power saw from 1"x2" lumber, and they will have correct bevel.
Shingles.
3/4"x10" Wood siding.
RAKE DETAIL.
5/8"x2" casings.
3/4x3 1/2" Base.
1/2"x3/4"
studs
3/4"x2 1/4"
Base 5/8x3 1/2"
3/4"x2"
Base.
ALTERNATE CORNICE RETURN.
Line
STANDARD JAMB SECTIONS.
BUCK JAMB.
Cap.
Note: In Gypsumboard construction an op'ng similar to, may be used in place of the usual plastered arch.
Window trim to match door trim shown can be used on frames with sash balances.
Plinth block. 1 1/8x2 3/8x4 1/4"
TRIM DETAILS.
ELEVATIONS.
Casing acts as plinth.
Base. 5/8"x3 1/2"
BASE DETAIL.
Shoe.

Stressed Skin Panels—Stronger but Lighter

Plywood stressed skin panels consist of plywood sheets glued to the top, and in most cases to the bottom edges of longitudinal framing members, so that the assembly will resist bending stresses. Structurally, the plywood acts as the flanges of an I-beam, the framing members as webs. A typical stressed skin panel is detailed at left.

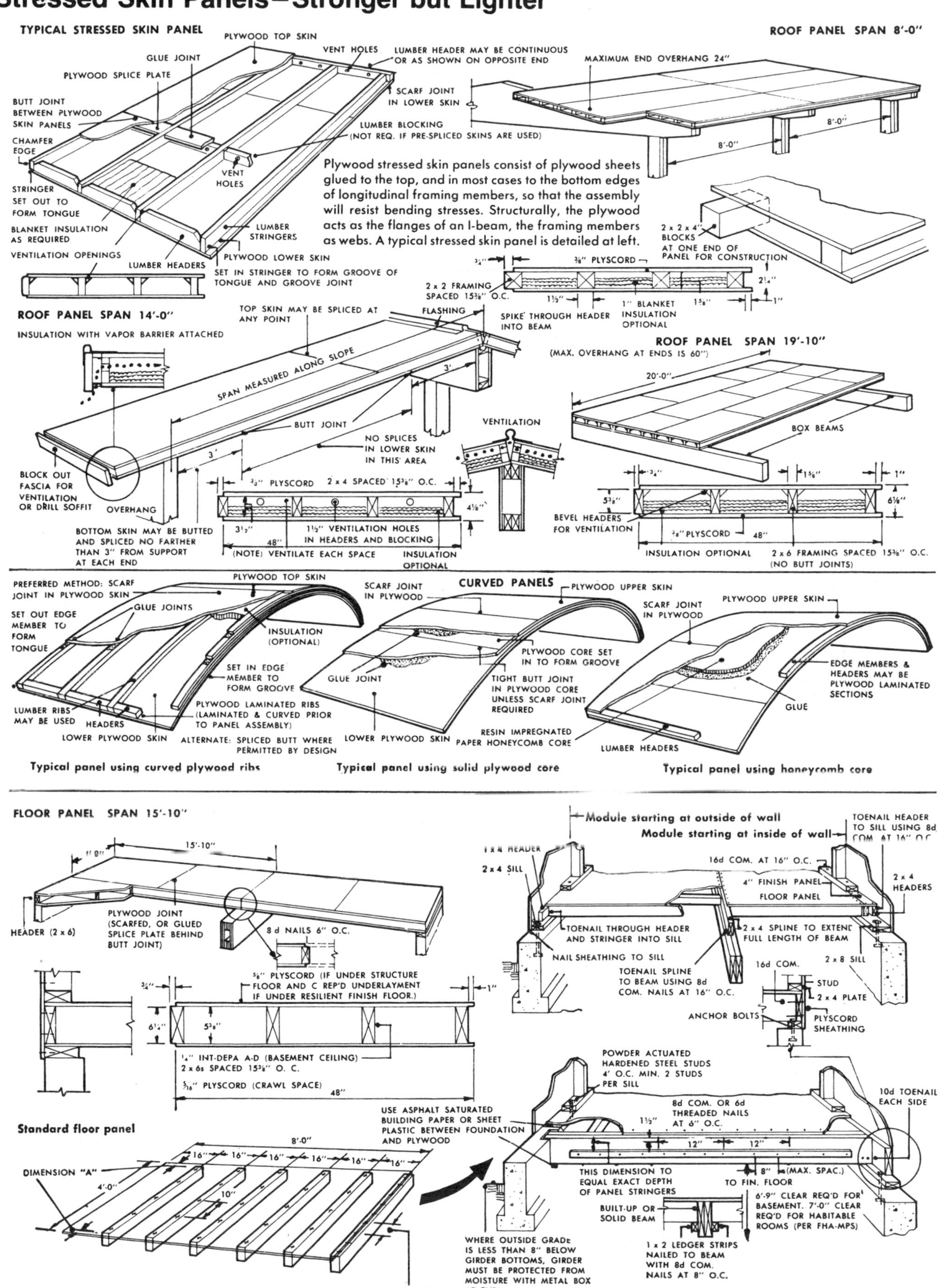

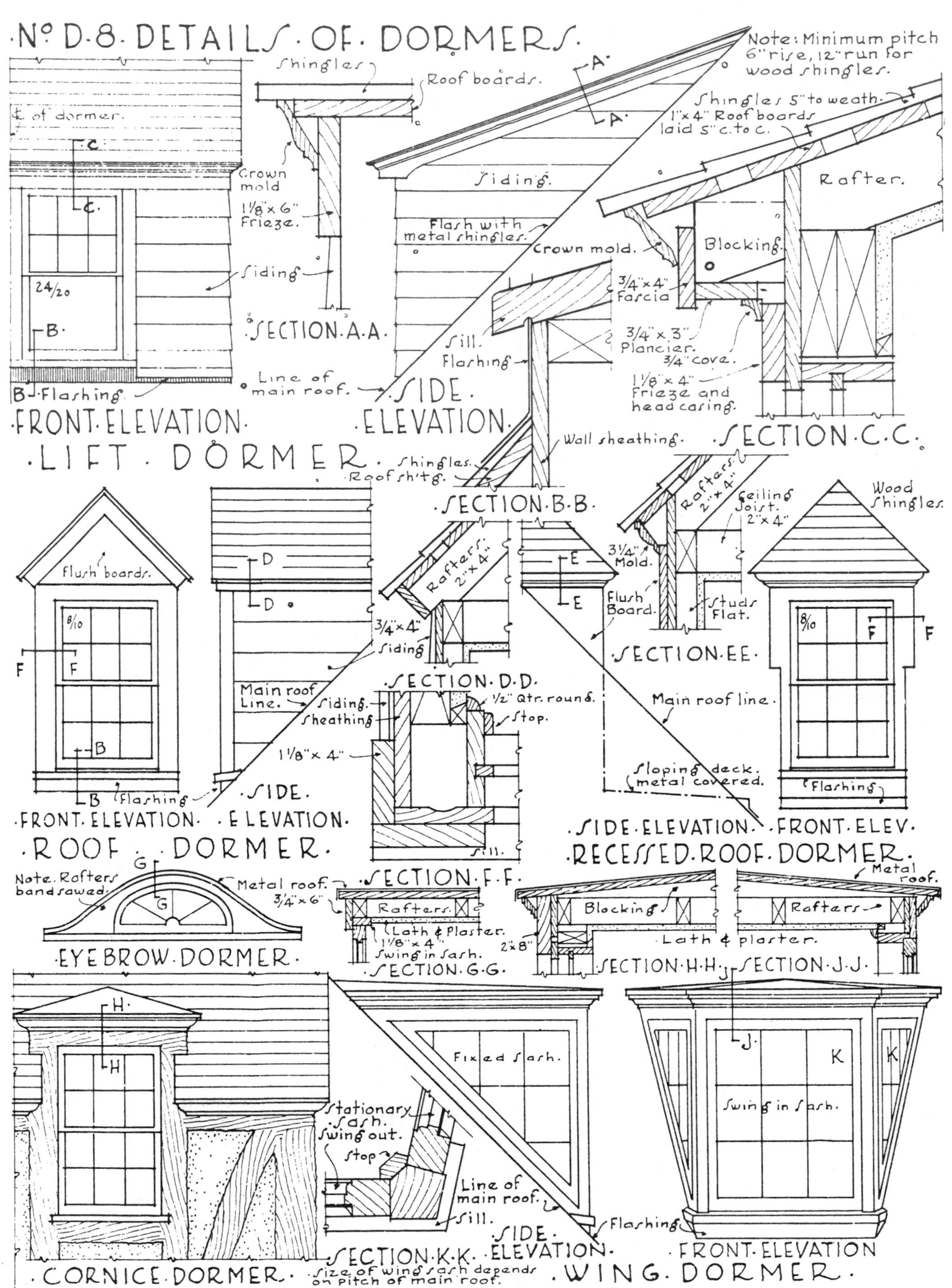
No D-8. DETAILS OF DORMERS.
Note: Minimum pitch 6" rise, 12" run for wood shingles.
LIFT DORMER
FRONT ELEVATION
SIDE ELEVATION
SECTION A-A
SECTION B-B
SECTION C-C
ROOF DORMER
FRONT ELEVATION
SIDE ELEVATION
SECTION D-D
SECTION E-E
SECTION F-F
RECESSED ROOF DORMER
SIDE ELEVATION
FRONT ELEV.
EYEBROW DORMER
Note. Rafters bandsawed.
SECTION G-G
SECTION H-H
SECTION J-J
CORNICE DORMER
SECTION K-K
Size of wing sash depends on pitch of main roof.
WING DORMER
SIDE ELEVATION
FRONT ELEVATION

NO. D-75 EAVES . . GRAVEL STOP . . GUTTER . . .

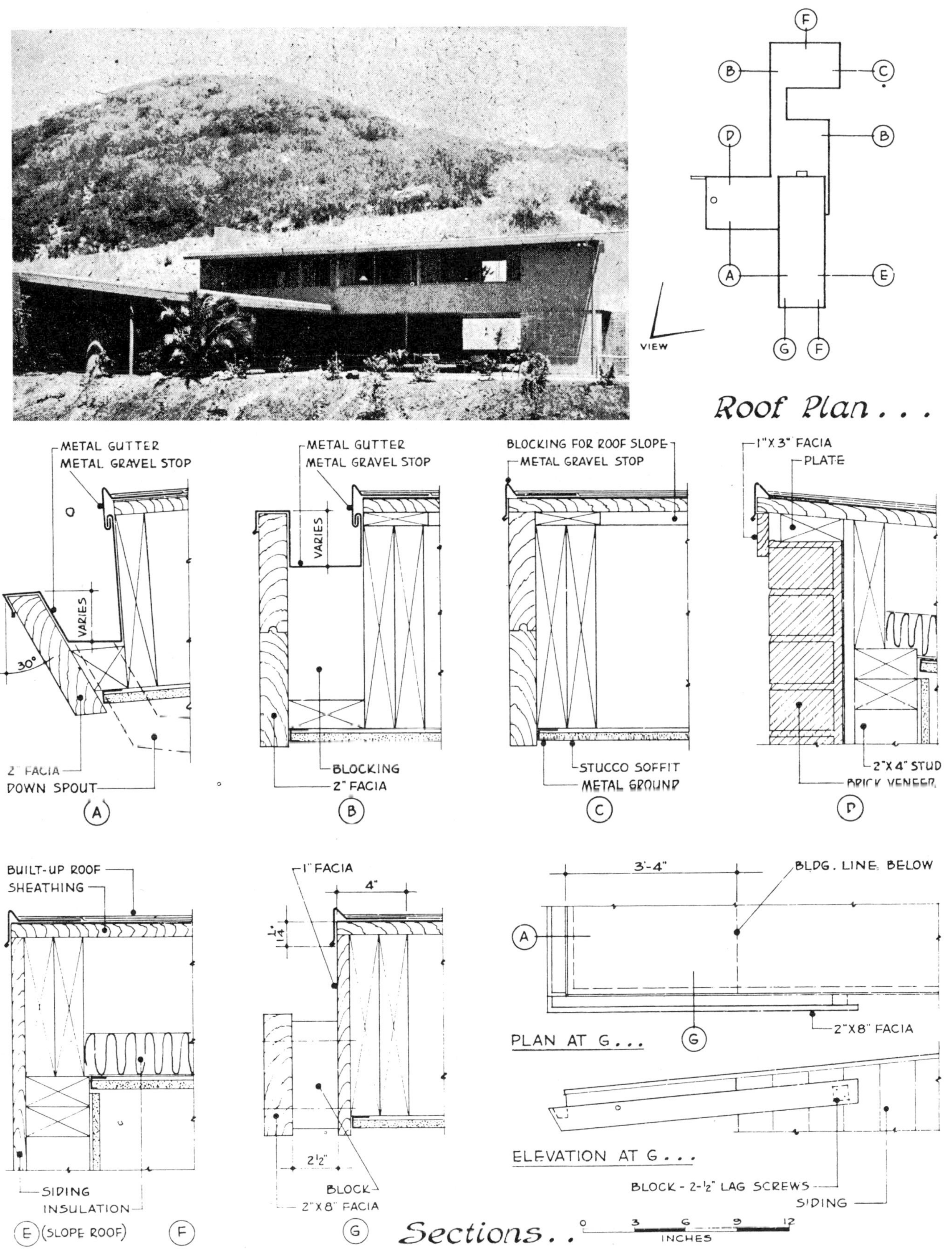

TIME AND MONEY SAVING DETAILS

Plastic Flashing Solves Difficult Sealing Jobs

CHIMNEYS

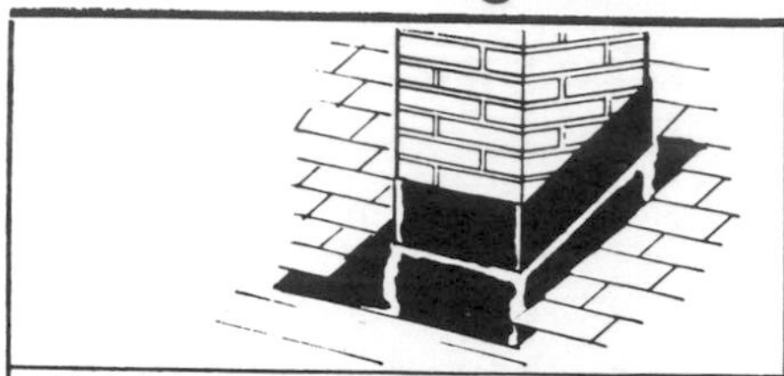

Plastic flashing is especially effective around a chimney. It can be applied after the chimney is built and need not be embedded in the mortar.

After cutting the flashing to size and bonding it to the roof and masonry with flashing cement, heat it gently with a torch to soften it.

Press the plastic firmly into the mortar joints, reheating it if necessary to keep it soft. Wear cotton gloves to rub the plastic smooth.

SHINGLE ROOFS

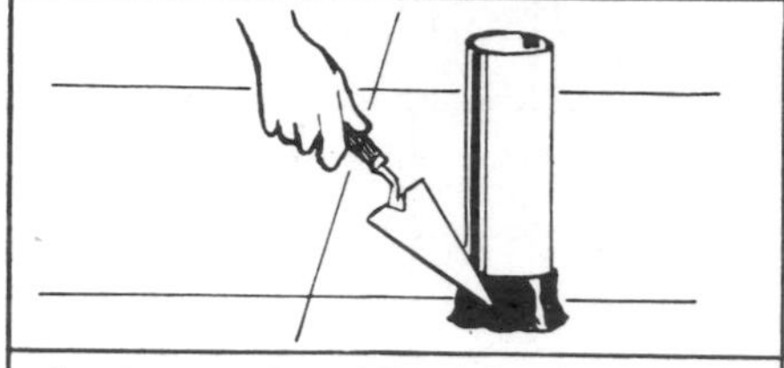

Trowel a generous layer of fibrated plastic flashing cement around base of vent where the flashing collar will seat.

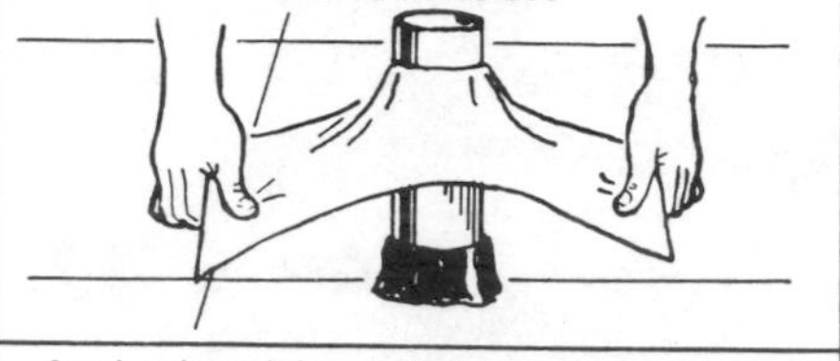

Stretch undersize hole cut in flashing over vent and pull the collar down so underside of sheet lies flat against the roof.

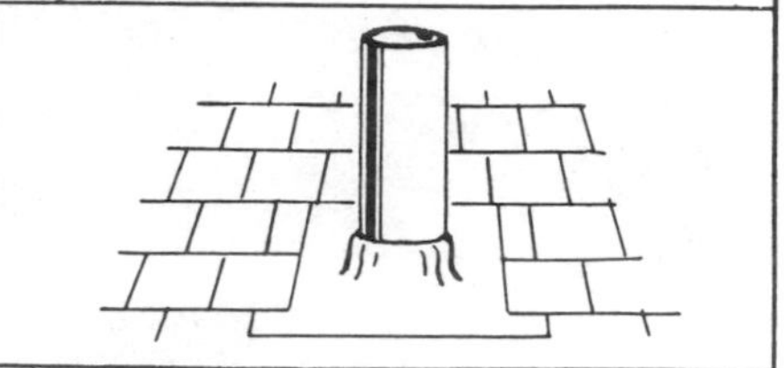

Nail shingle courses so that they overlap flashing at top and sides. Drive nails through plastic, it will seal itself against them.

CORRUGATED DECKS

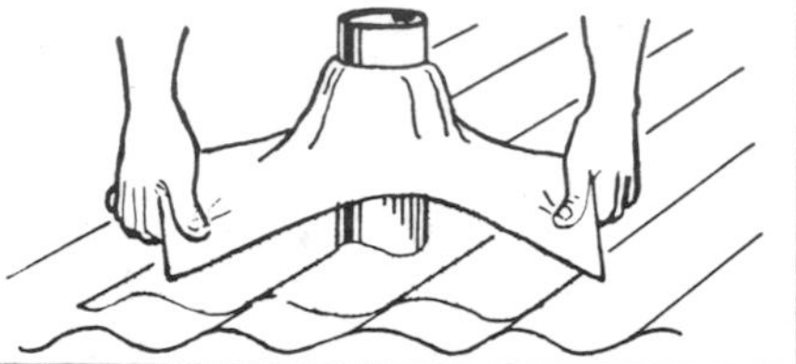

Apply adhesive to roof around base of vent. Stretch flashing over vent and slide it down until it lies flat on top of corrugations.

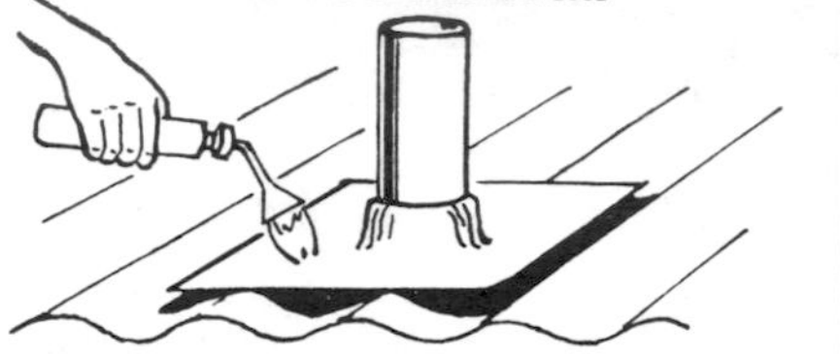

Heat flashing gently with a torch, hot air gun or heat lamp until it begins to drape itself into valleys of the corrugations.

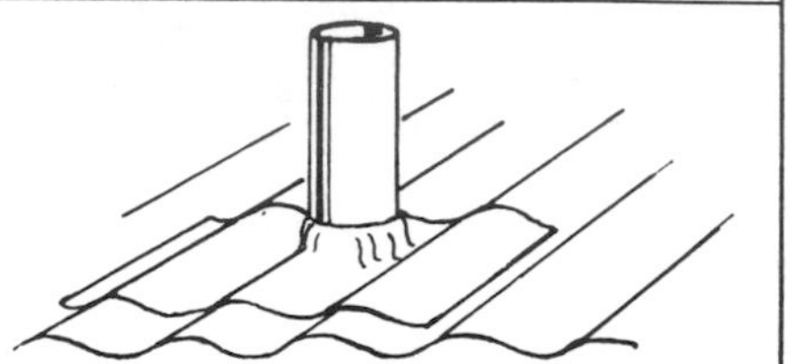

Press the flashing firmly into the valleys, reheating it, if necessary, to keep it at the 180-200 degree conforming temperature.

BUILT-UP ROOFS

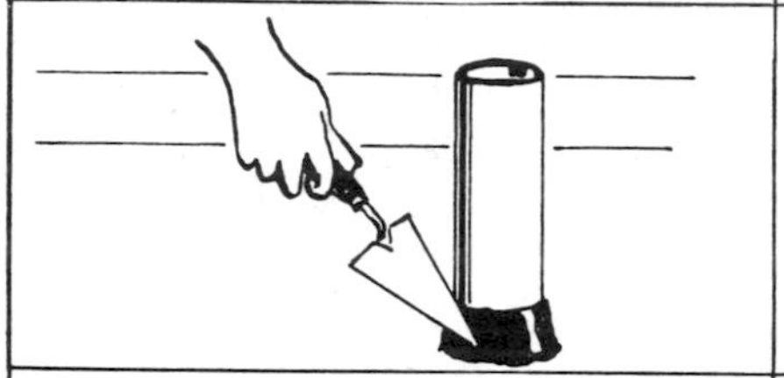

Trowel a generous layer of fibrated plastic flashing cement around base of vent, filling the gap between the vent and roof.

Stretch flashing over vent and slide it down until it lies flat on roofing felt. It can be cemented in place, or you can . . .

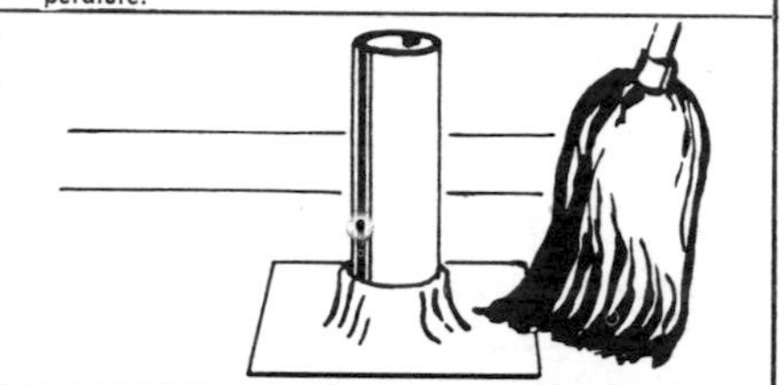

. . . mop the flashing in with hot asphalt or pitch. If gravel is applied to roofing, spread it over flashing, too.

The use of plastic flashing can solve many sealing problems and speed ordinary flashing jobs. The 1/32″ to 1/16″ thick material can be shaped to fit most contours with minimum effort, and it can be softened to fit irregular surfaces or flow around sharp corners by warming it to 180-200 degrees with a torch or heat lamp.

The plastic is applied to standard construction materials with the same flashing cements or bituminous coatings used for roofing. It also bonds readily and permanently to itself when solvent-softened wth methyl ethyl ketone. Once placed, the elastic material will flex with the expansion and contraction of building sections caused by temperature changes or settling. The plastic also acts as a gasket, sealing itself around nails driven through it.

Although it costs more to install than many materials commonly used for flashing, the plastic has excellent weathering qualities and resists chemicals and airborne fumes, resulting in a longer useful life and a low cost per year of service.

The installation techniques illustrated on this page have been developed for residential and light construction by technicians at Dow Chemical Co.

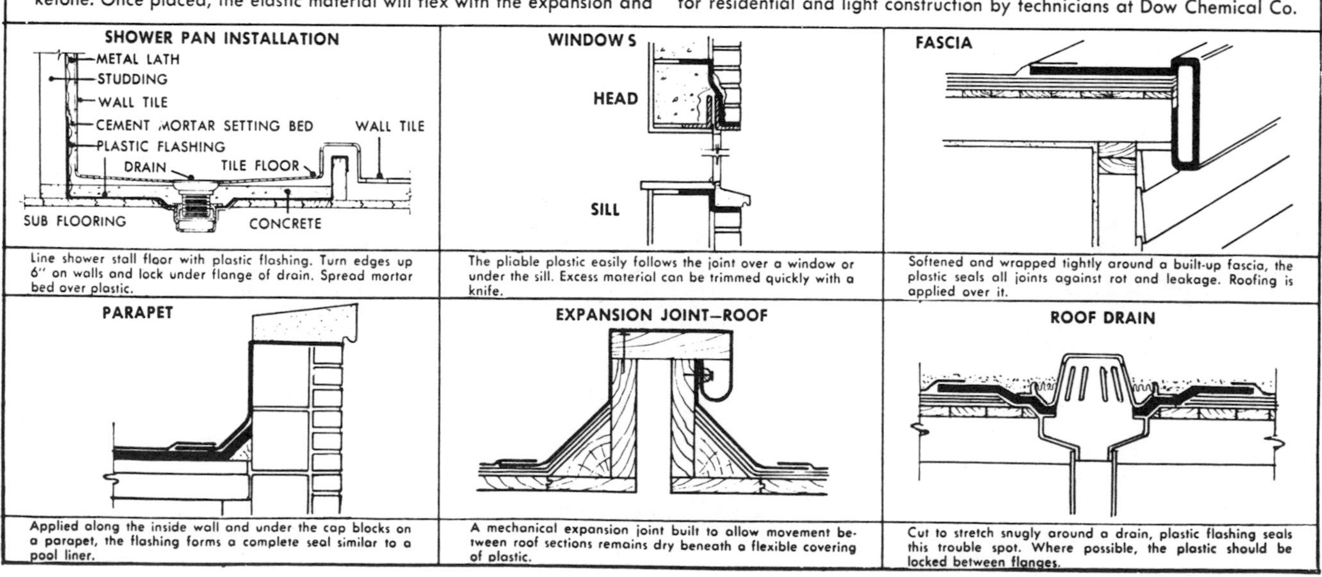

Line shower stall floor with plastic flashing. Turn edges up 6″ on walls and lock under flange of drain. Spread mortar bed over plastic.

The pliable plastic easily follows the joint over a window or under the sill. Excess material can be trimmed quickly with a knife.

Softened and wrapped tightly around a built-up fascia, the plastic seals all joints against rot and leakage. Roofing is applied over it.

Applied along the inside wall and under the cap blocks on a parapet, the flashing forms a complete seal similar to a pool liner.

A mechanical expansion joint built to allow movement between roof sections remains dry beneath a flexible covering of plastic.

Cut to stretch snugly around a drain, plastic flashing seals this trouble spot. Where possible, the plastic should be locked between flanges.

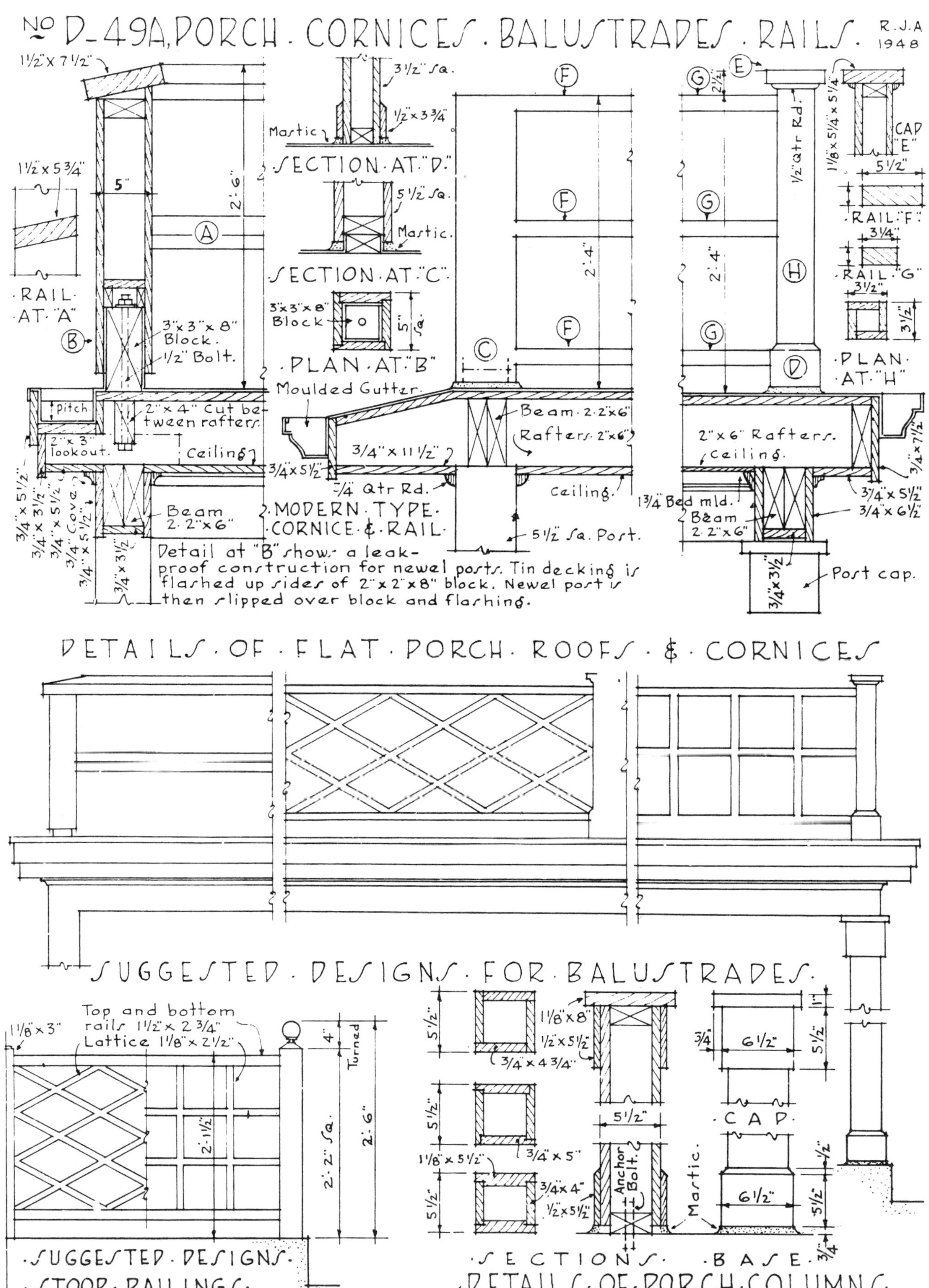
No D-49A, PORCH · CORNICES · BALUSTRADES · RAILS ·
R.J.A 1948
1½" x 7½"
1½" x 5¾"
5"
2'-6"
A
· RAIL · AT "A"
B
3" x 3" x 8" Block.
½" Bolt.
3½" Sq.
Mastic
½" x 3¾"
· SECTION · AT · "D."
5½" Sq.
Mastic.
· SECTION · AT · "C".
3" x 3" x 8" Block
5" Sq.
· PLAN · AT · "B"
pitch
2" x 3" lookout.
2" x 4" Cut between rafters.
Ceiling
Beam 2. 2" x 6"
¾" x 5½"
¾" x 3½"
¾" x 5½"
¾" Cove
¾" x 5½"
¾" x 3½"
Detail at "B" shows a leakproof construction for newel posts. Tin decking is flashed up sides of 2" x 2" x 8" block. Newel post is then slipped over block and flashing.
F
2'-4"
C
Moulded Gutter.
Beam · 2. 2" x 6"
Rafters 2" x 6"
¾" x 11½"
¾ x 5½"
¼ Qtr Rd.
Ceiling.
· MODERN · TYPE · CORNICE · & · RAIL ·
5½ Sq. Post.
G
E
2¼"
½" Qtr Rd.
H
D
1⅛ x 5¼ x 5¼
CAP "E"
5½"
· RAIL · "F"
3¼"
· RAIL · "G"
3½"
· PLAN · AT · "H"
2" x 6" Rafters.
Ceiling.
1¾ Bed mld.
Beam 2. 2" x 6"
¾ x 7½"
¾" x 5½"
¾ x 6½"
¾" x 3½"
Post cap.
DETAILS · OF · FLAT · PORCH · ROOFS · & · CORNICES
SUGGESTED · DESIGNS · FOR · BALUSTRADES.
1⅛" x 3"
Top and bottom rails 1½" x 2¾"
Lattice 1⅛" x 2½"
4"
Turned
2'-1½"
2'-2" Sq.
2'-6"
· SUGGESTED · DESIGNS · STOOP · RAILINGS ·
5½"
¾" x 4¾"
¾" x 5"
1⅛" x 5½"
¾" x 4"
½" x 5½"
1⅛" x 8"
½" x 5½"
Anchor Bolt.
Mastic.
1"
¾"
6½"
CAP
½"
¾
· SECTIONS · · BASE ·
· DETAILS · OF · PORCH · COLUMNS ·

NO. D-66. DETAILS OF OPEN CARPORT.

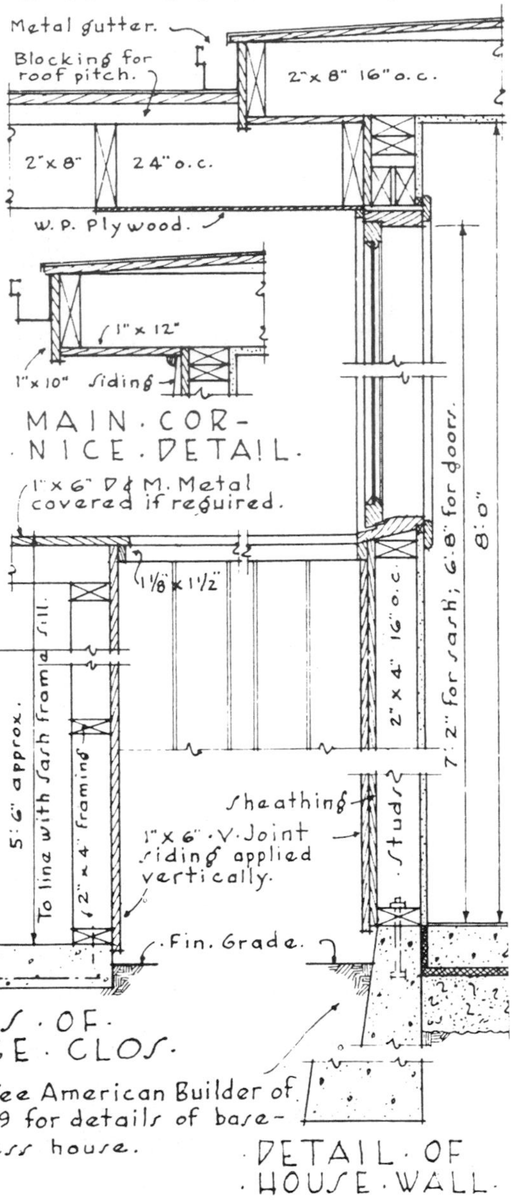

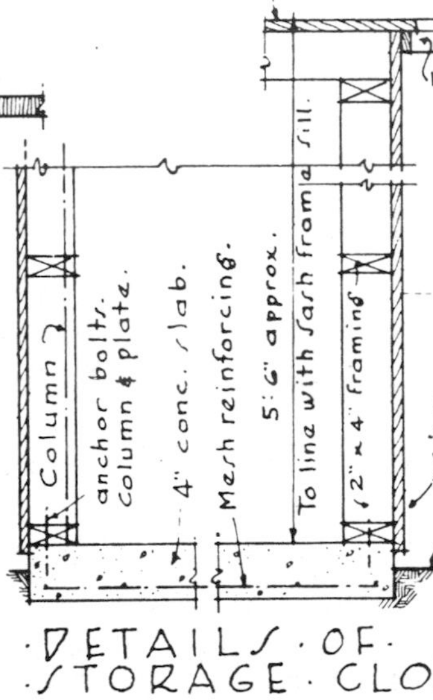

Metal gutter.
Blocking for roof pitch.
2"x 8" 16" o.c.
2"x 8" 24" o.c.
W. P. Plywood.
1"x 12"
1"x 10" siding

MAIN CORNICE DETAIL.

1"x 6" D & M. Metal covered if required.
1 1/8"x 1 1/2"
7' 2" for sash; 6' 8" for doors.
8' 0"
2"x 4" 16" o.c.
Sheathing
Studs
1"x 6" V-Joint siding applied vertically.
Fin. Grade.

Metal or wood sash.
Entrance.
Line of cornice overhang.
columns.
STORAGE
Doors.
Doors.
Joist above exposed.
CARPORT
crushed rock floor.
10' 6"
2' 6"
5' 6"
columns.
Conc. or flagstone walk.
2' 0" | 3' 0" | 9' 6" | 2' 0" | 4' 0"

PLAN OF CARPORT.

Column
anchor bolts. column & plate.
4" conc. slab.
Mesh reinforcing.
5' 6" approx.
To line with sash frame sill.
2"x 4" framing

DETAILS OF STORAGE CLOS.

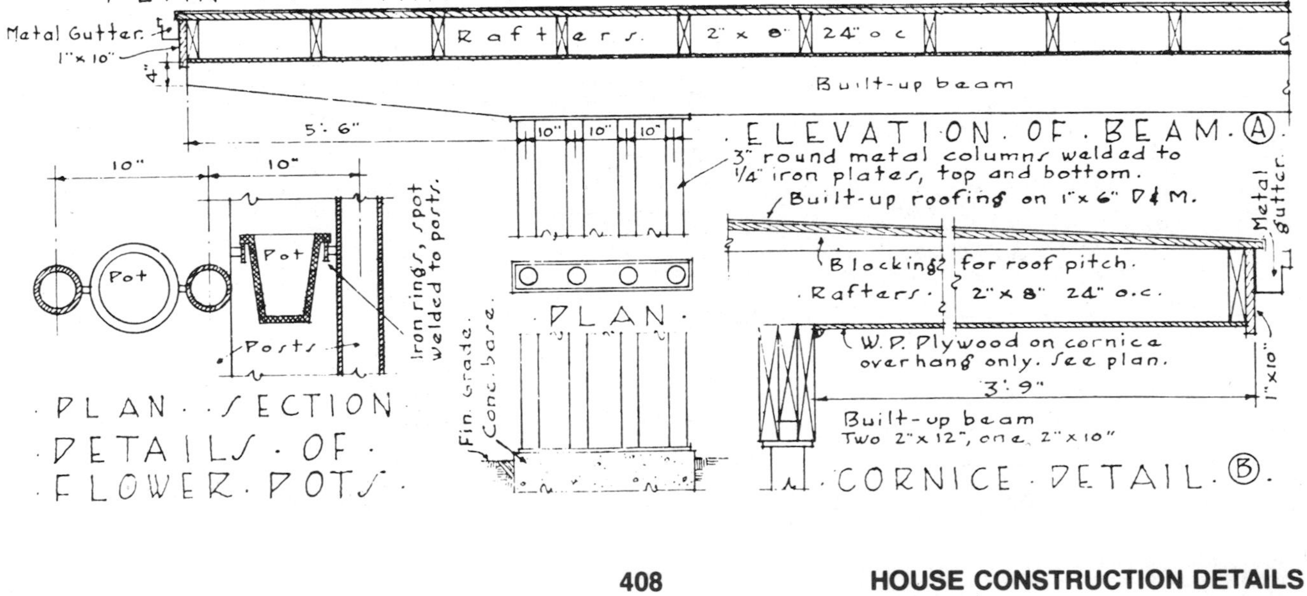

Note: See American Builder of Oct. 1949 for details of basementless house.

DETAIL OF HOUSE WALL.

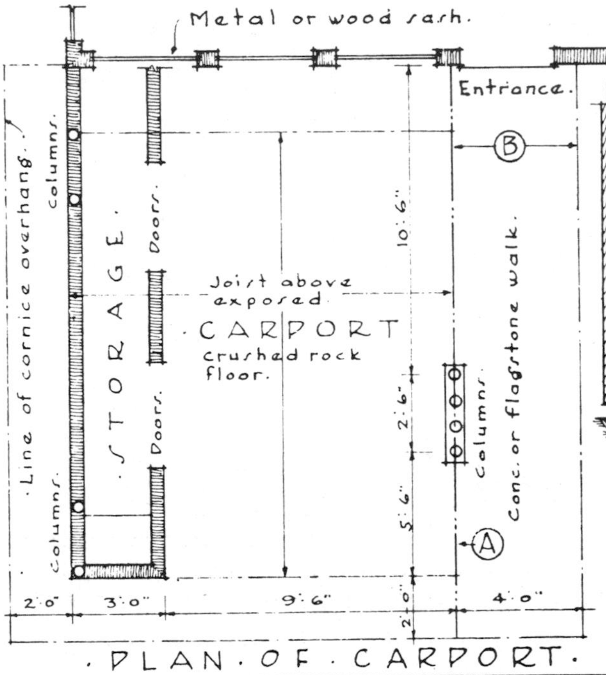

Metal Gutter.
1"x 10"
4"
Rafters. 2" x 8" 24" o.c.
Built-up beam
5' 6"
10" 10" 10"

ELEVATION OF BEAM (A).

3" round metal columns welded to 1/4" iron plates, top and bottom.
Built-up roofing on 1"x 6" D & M.

PLAN.

Fin. Grade.
Conc. base.

10" 10"
Pot
Pot
Posts
Iron rings, spot welded to posts.

PLAN. SECTION DETAILS OF FLOWER POTS.

Blocking for roof pitch.
Rafters. 2"x 8" 24" o.c.
Metal gutter.
W. P. Plywood on cornice overhang only. See plan.
3' 9"
1"x 10"
Built-up beam Two 2"x 12", one 2"x 10"

CORNICE DETAIL (B).

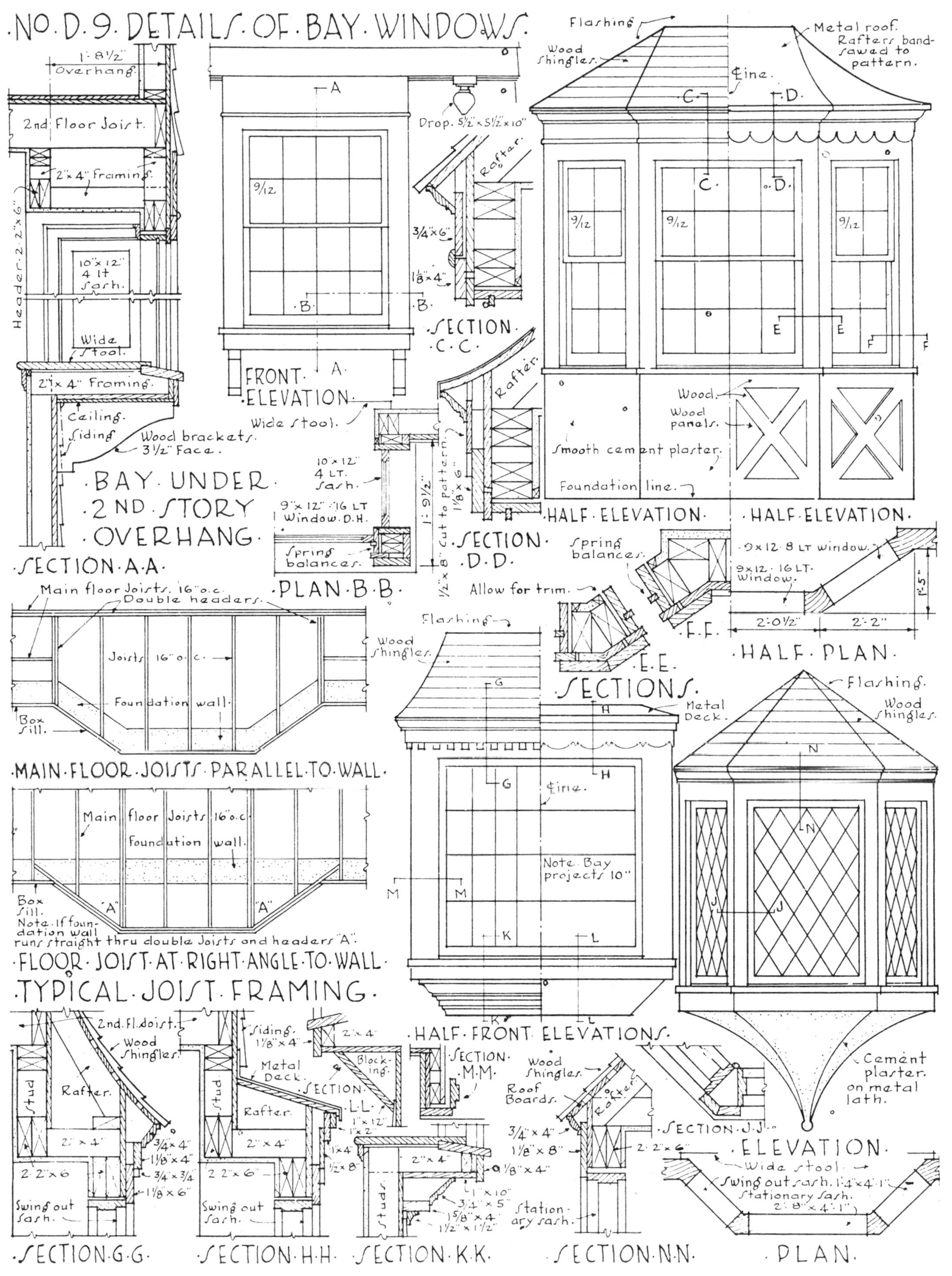

TIME AND MONEY SAVING DETAILS

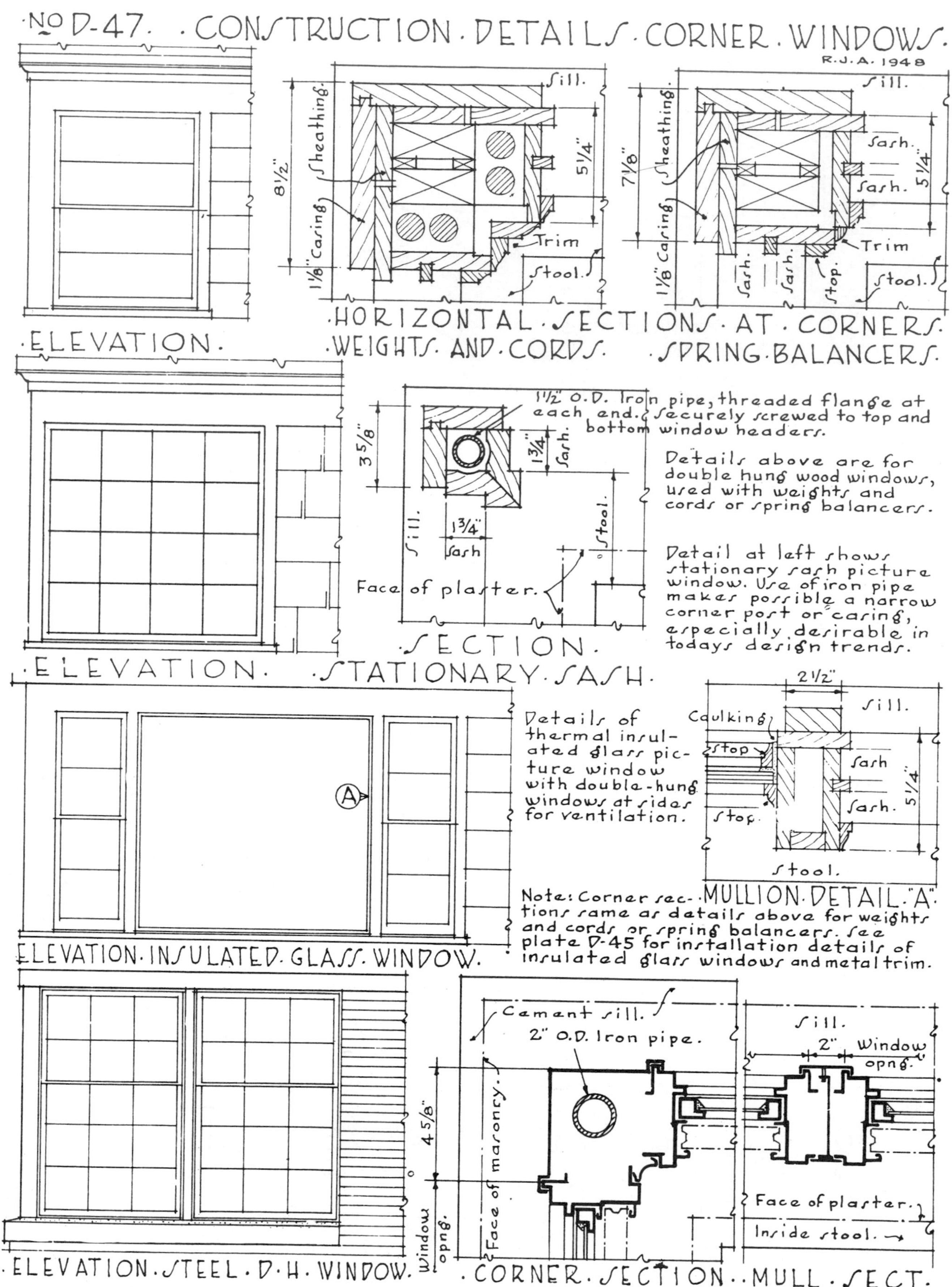

No D-47. CONSTRUCTION · DETAILS · CORNER · WINDOWS ·
R.J.A. 1948
ELEVATION ·
HORIZONTAL · SECTIONS · AT · CORNERS ·
WEIGHTS · AND · CORDS ·
SPRING · BALANCERS ·
Sill.
Sheathing.
8½"
1⅛" Casing
5¼
Trim
Stool.
7⅛"
Sash.
Stop.
1½" O.D. Iron pipe, threaded flange at each end. Securely screwed to top and bottom window headers.
Details above are for double hung wood windows, used with weights and cords or spring balancers.
Detail at left shows stationary sash picture window. Use of iron pipe makes possible a narrow corner post or casing, especially desirable in todays design trends.
3⅝
1¾
1¾"
Face of plaster.
ELEVATION · STATIONARY · SASH ·
SECTION ·
Details of thermal insulated glass picture window with double-hung windows at sides for ventilation.
2½"
Caulking
Stop
MULLION · DETAIL · "A" ·
Note: Corner sections same as details above for weights and cords or spring balancers. See plate D-45 for installation details of insulated glass windows and metal trim.
ELEVATION · INSULATED · GLASS · WINDOW.
Cement sill.
2" O.D. Iron pipe.
Window opng.
2"
4⅝"
Face of masonry.
Face of plaster.
Inside stool.
ELEVATION · STEEL · D · H · WINDOW.
CORNER · SECTION ·
MULL · SECT ·

NO. D-57 - DOUBLE · GLAZING · WITH · LOUVER · VENTS ·

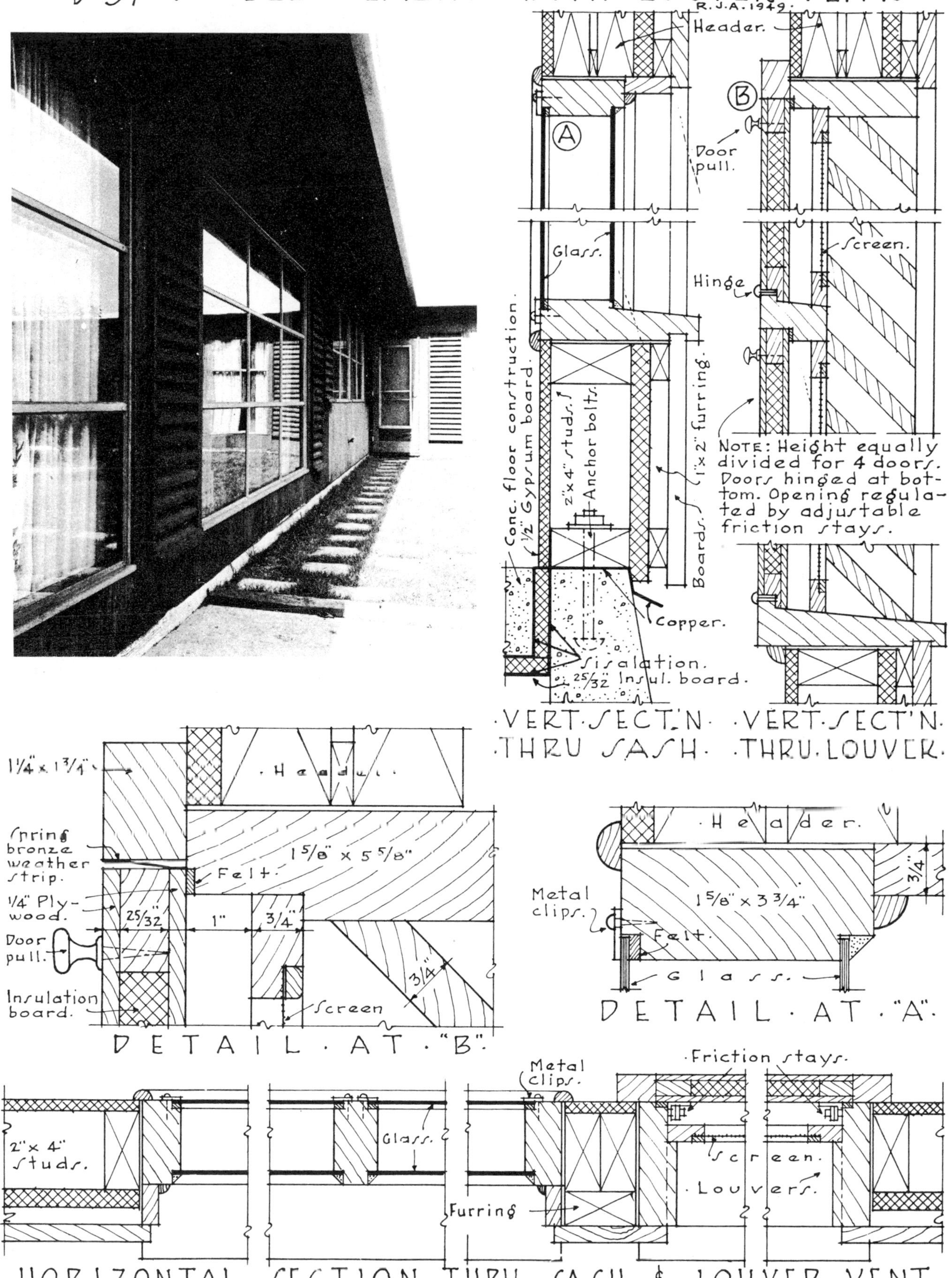

Glazing With Insulating Glass Is Easy, but Must Be Done Right

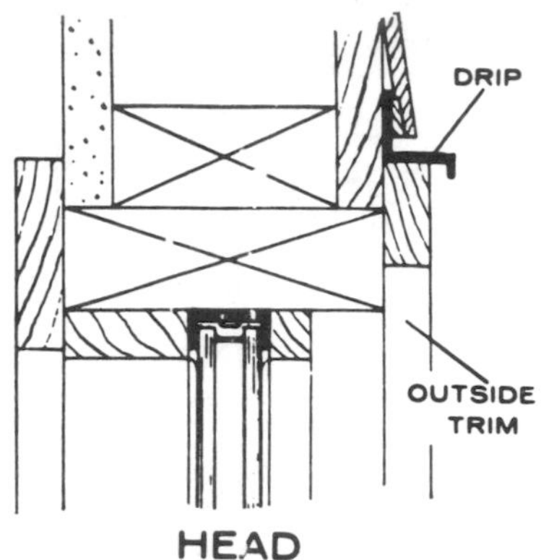
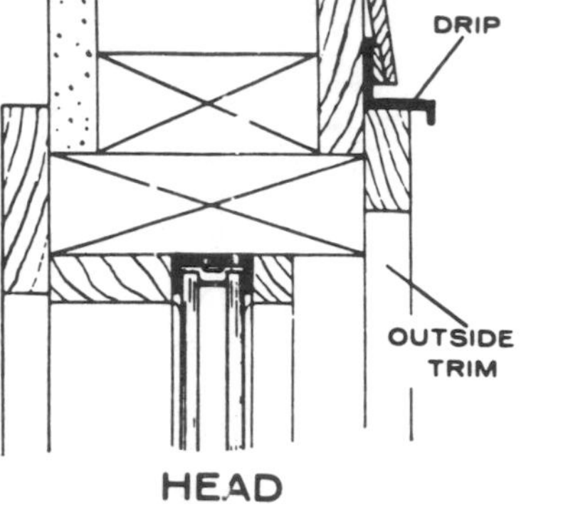

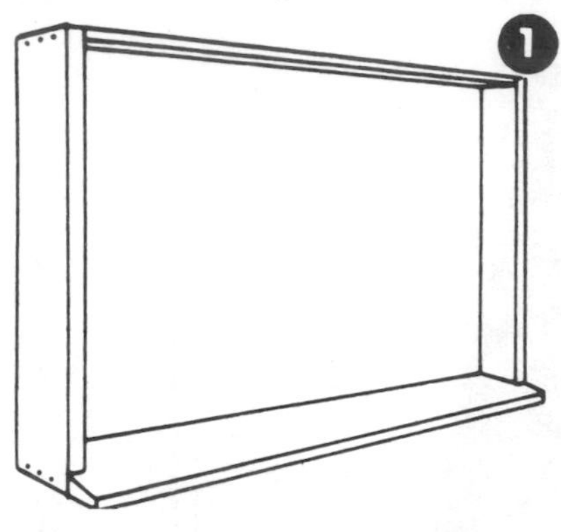

FRAME OPENING with a 2 x 8 sill and 2 x 6 sides and top, making the opening ½" wider than the window.

CHECK SILL to be sure it's flat and level, then square the rest of the frame with the sill. Nail securely.

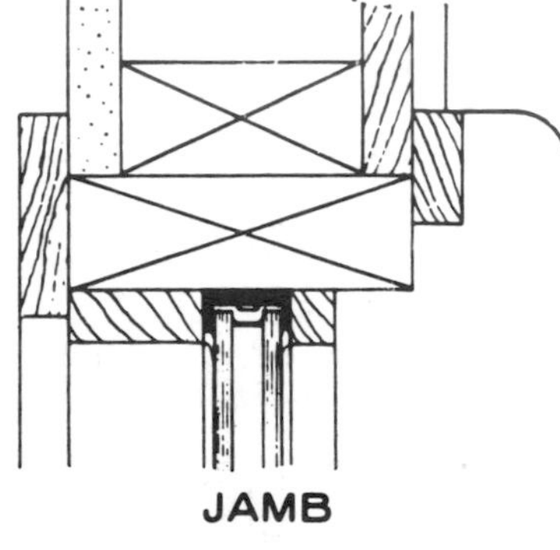

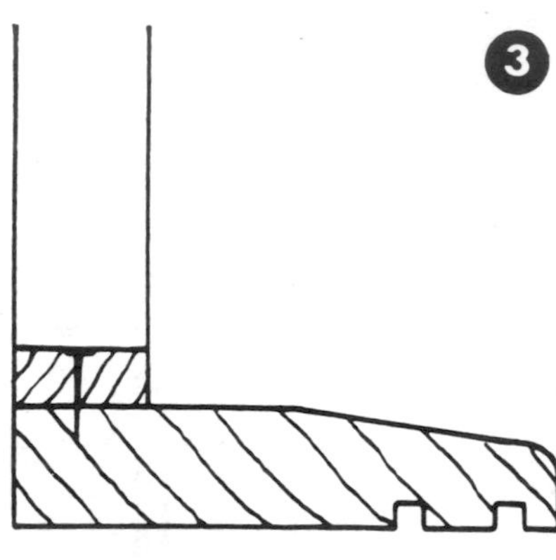

NAIL BACK STOP in place all around frame. If stool is used, it becomes the lower stop (see sill details at left).

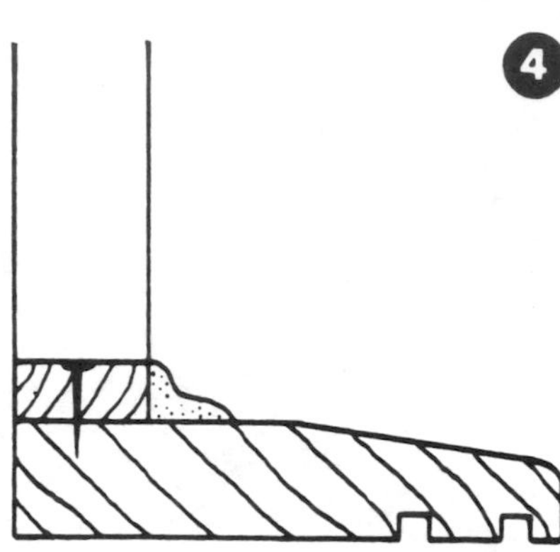

APPLY knife-grade glazing compound (not putty) along the edge of the stop. This will seal and cushion the unit.

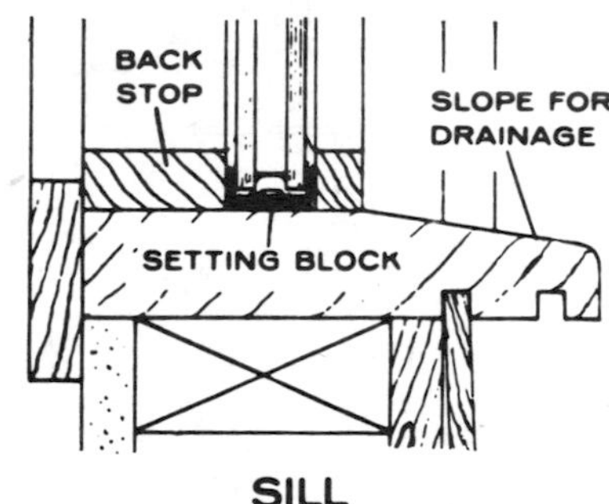

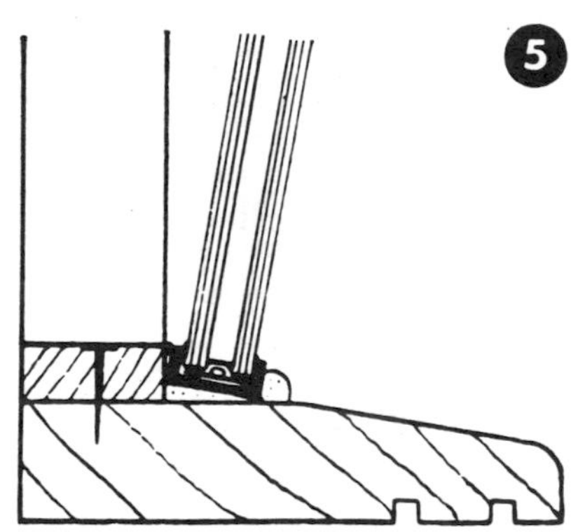

PLACE BOTTOM edge of window on sill and press into glazing compound until neoprene setting blocks touch stops.

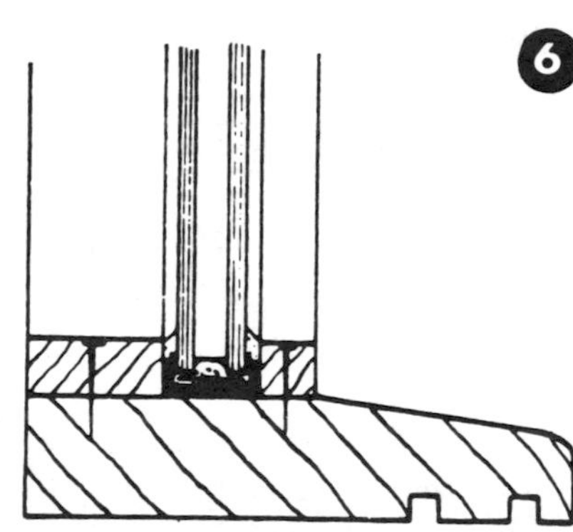

USE FULL BED of glazing compound at top, bottom and sides to make weather-tight seal, then nail on face stops.

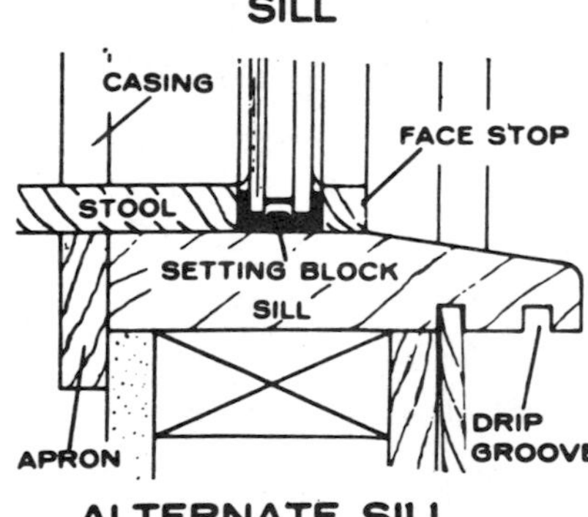

Costly callbacks for leaks, drafts, or replacement of cracked glass can be eliminated completely by simply knowing the right way to install double-glazed windows . . . and doing the job the right way takes no longer than the wrong way.

The most important single consideration in installing insulating glass is to allow sufficient edge clearance for expansion and contraction. Forcing a unit into a tight-fitting opening is almost certain to cause the glass to crack when it expands or if the header sags.

Total or partial covering of the glass with paint, decals or signs should be avoided. Glass will trap the sun's heat, causing failure due to an unnatural and concentrated increase in temperature and resulting expansion around areas that are covered in any way.

NO. D-99 WINDOW & DOOR DETAILS . . .

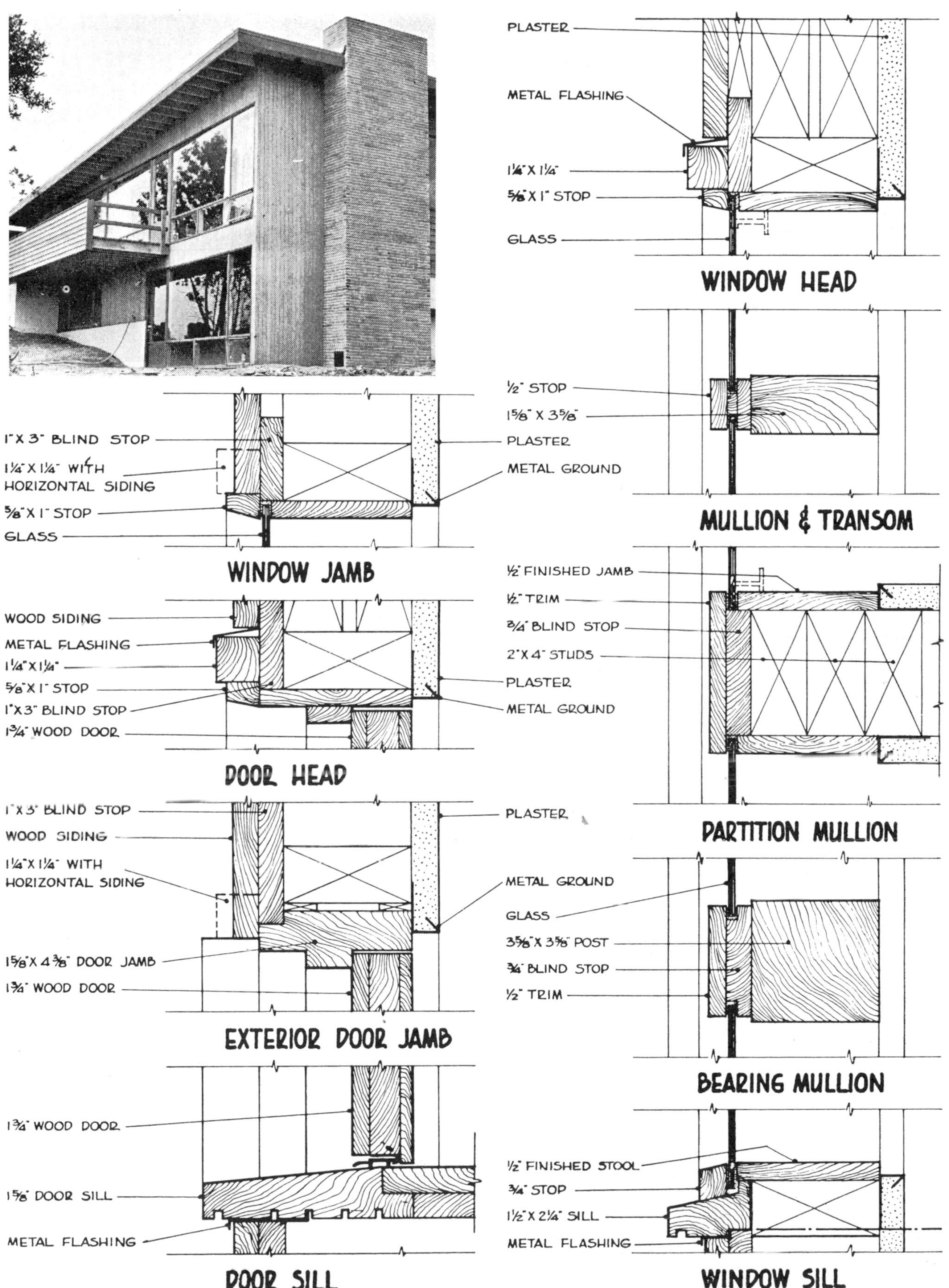

Smart Selection of Locksets Can Save You Money

Tubular type locks and latches are suitable for the lowest cost housing, yet offer the latest styling and security at competitive prices. In addition, there are low-cost cylindrical type locks and latches to match the tubulars for installations requiring cylindricals on the exterior doors, tubulars on the interior. This permits a great variety of decorative escutcheon styling on exterior doors, while retaining the economy of tubular locks inside the house.

TUBULAR LOCKSET

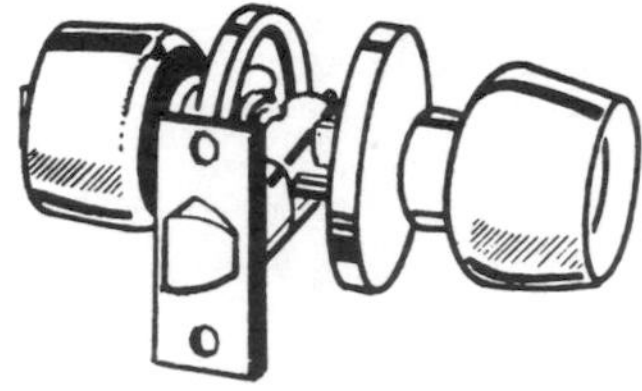

CYLINDRICAL LOCKSET

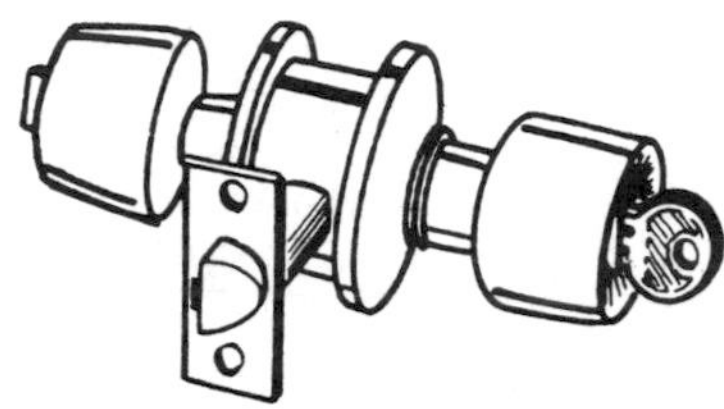

OUTSIDE LOCK
Pintumbler lock for maximum security, turnbutton in inside knob locks outside knob. Latch automatic deadlock.

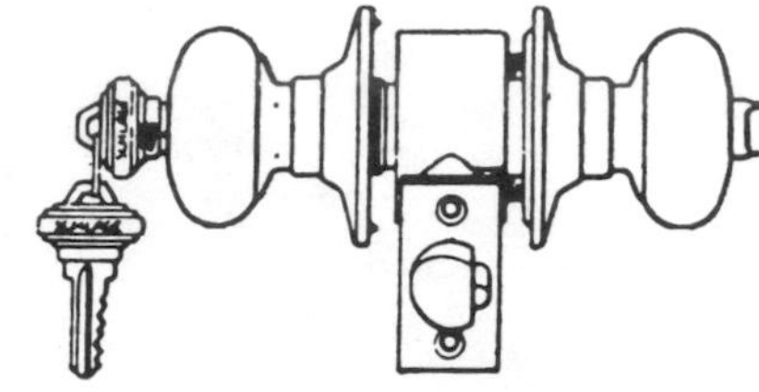

EXIT LOCK
For patio, porch and basement doors. Push-button locking on inside, cannot be opened from outside.

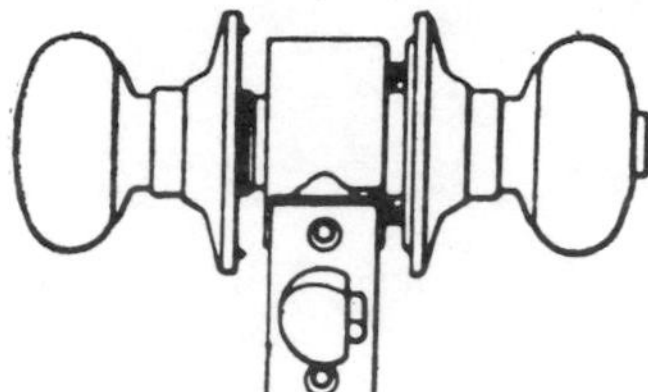

PRIVACY LOCK
For bath and bedroom doors. Push button in inside knob with emergency key for outside knob. Unlocks when inside knob is turned.

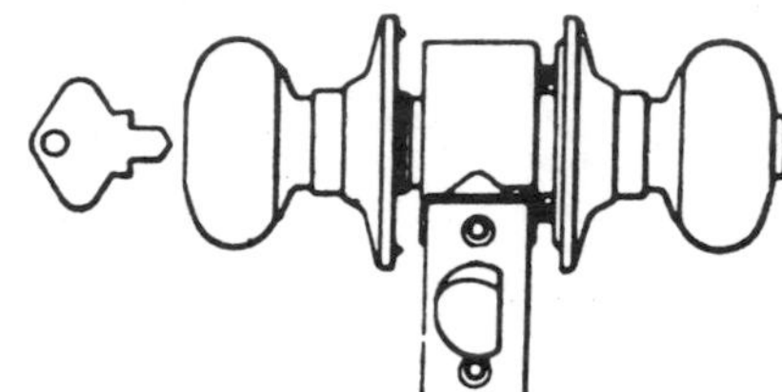

THREE METHODS OF KEYING TO MEET ALL NEEDS.

All alike for convenience. Using this system, two or more locks can be operated by the same key.

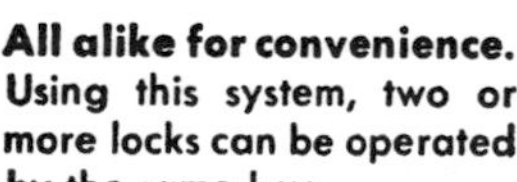

All different for security. Each lock is set to a different key change and no other key will operate it.

Master-keyed for control. Two or more locks operated by a different key known as a change key, and all operated by a master key.

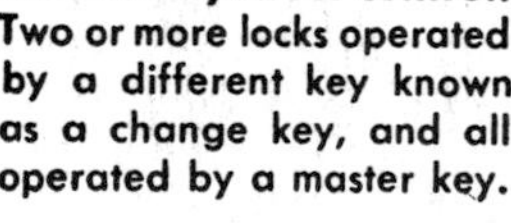

Pin tumbler action assures maximum security and smooth operation with unlimited key changes described above.

KNOB LATCH
For passage and closet doors. Turning knob on either side will retract latch bolt.

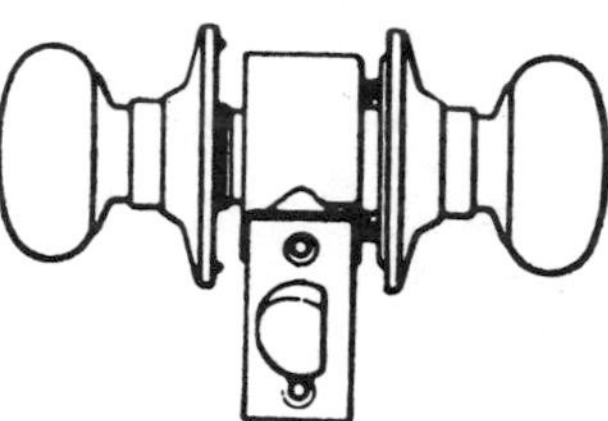

DUMMY KNOB
For doors where only a pull is required.

EXTENSION LINKS
For longer backset from edge of door, or for popular installation of lock in center of door.

JIG SPEEDS INSTALLATION
Combination jig is available for 2⅜" and 5" backset installations. Assures accurate right-angle holes.

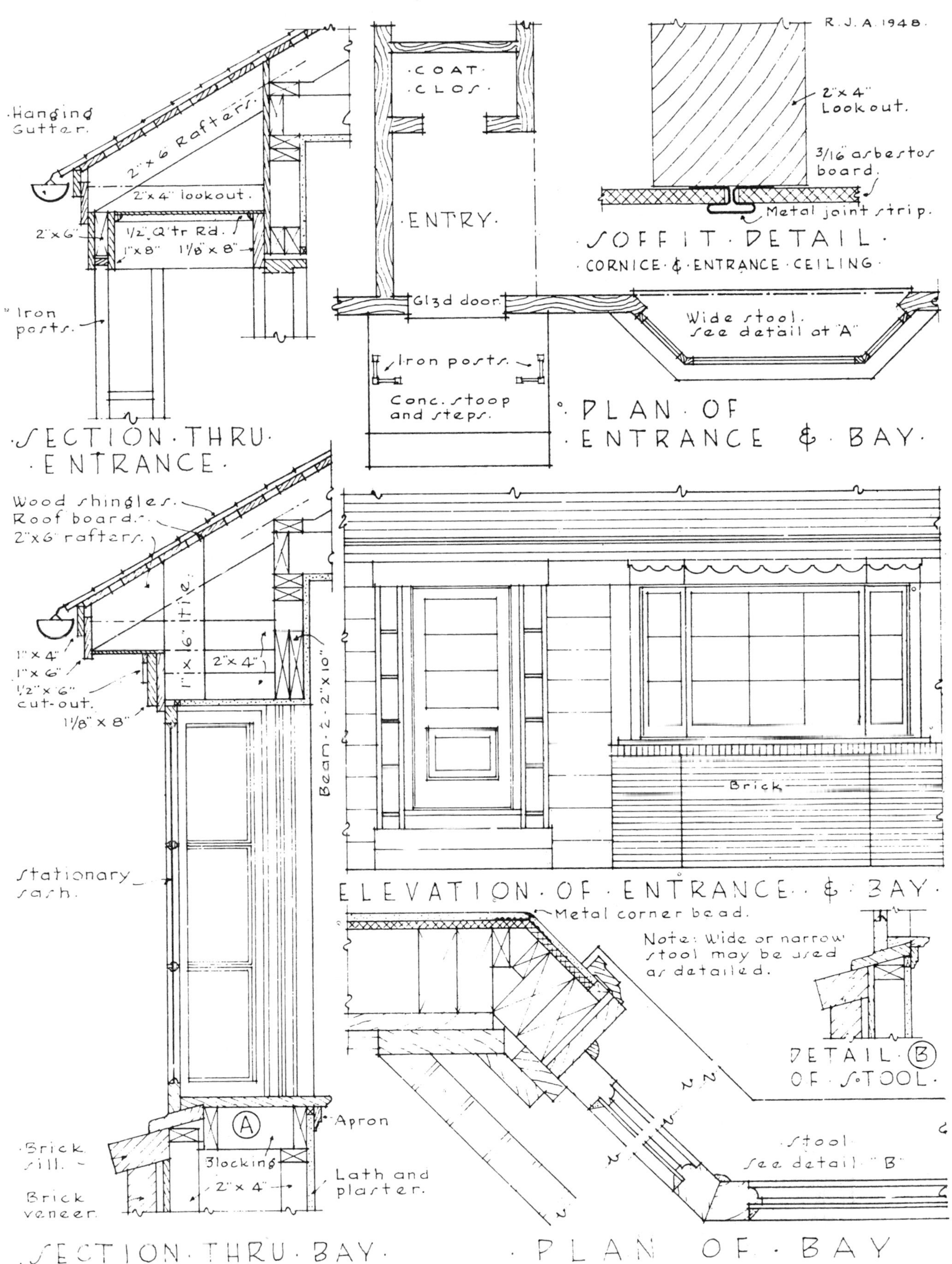
No. D.52. ENTRANCE & BAY OF UNUSUAL DESIGN.
R.J.A. 1948.
Hanging Gutter.
2"x6" Rafters.
2"x4" lookout.
2"x6"
1/2" Q'tr Rd.
1"x8"
1 1/8" x 8"
Iron posts.
SECTION THRU ENTRANCE.
COAT CLOS.
ENTRY.
Gl3d door.
Iron posts.
Conc. stoop and steps.
Wide stool. see detail at "A"
PLAN OF ENTRANCE & BAY.
2"x4" Lookout.
3/16" asbestos board.
Metal joint strip.
SOFFIT DETAIL.
CORNICE & ENTRANCE CEILING.
Wood shingles.
Roof boards.
2"x6" rafters.
1"x6" tie.
2"x4"
Beam 2-2"x10"
1"x4"
1"x6"
1/2"x6" cut-out.
1 1/8" x 8"
Stationary sash.
A
Apron
Brick sill.
Blocking
2"x4"
Lath and plaster.
Brick veneer.
SECTION THRU BAY.
Brick
ELEVATION OF ENTRANCE & BAY.
Metal corner bead.
Note: Wide or narrow stool may be used as detailed.
DETAIL B OF STOOL.
Stool. See detail "B"
PLAN OF BAY

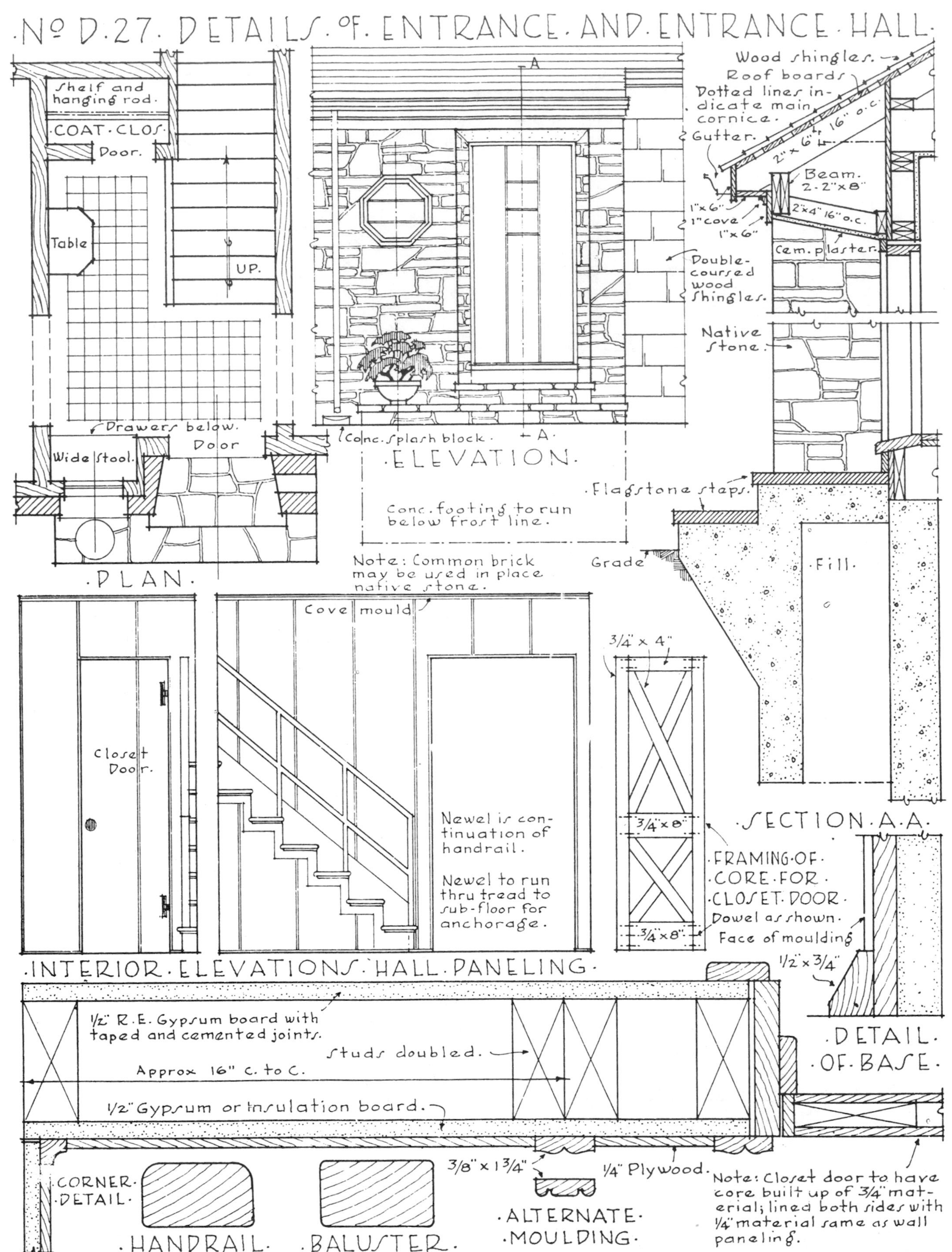

Nº D.27. DETAILS OF ENTRANCE AND ENTRANCE HALL.
Shelf and hanging rod.
COAT CLOS.
Door.
Table
UP.
Drawers below.
Door
Wide Stool.
PLAN.
Conc. splash block.
A
ELEVATION.
Conc. footing to run below frost line.
Note: Common brick may be used in place native stone.
Wood shingles.
Roof boards
Dotted lines indicate main cornice.
Gutter.
2"x6" 16" o.c.
Beam. 2-2"x8"
1"x6"
1" cove
1"x6"
2"x4" 16" o.c.
Cem. plaster.
Double-coursed wood shingles.
Native stone.
Flagstone steps.
Grade
Fill.
SECTION A.A.
Cove mould
Closet Door.
Newel is continuation of handrail.
Newel to run thru tread to sub-floor for anchorage.
3/4" x 4"
3/4"x8"
3/4"x8"
FRAMING OF CORE FOR CLOSET DOOR.
Dowel as shown.
Face of moulding
1/2"x3/4"
DETAIL OF BASE.
INTERIOR ELEVATIONS HALL PANELING.
1/2" R.E. Gypsum board with taped and cemented joints.
Studs doubled.
Approx 16" c. to c.
1/2" Gypsum or insulation board.
3/8" x 1 3/4"
1/4" Plywood.
CORNER DETAIL.
HANDRAIL.
BALUSTER.
ALTERNATE MOULDING.
Note: Closet door to have core built up of 3/4" material; lined both sides with 1/4" material same as wall paneling.

No. D-67 Details of Entrance and Window Wall

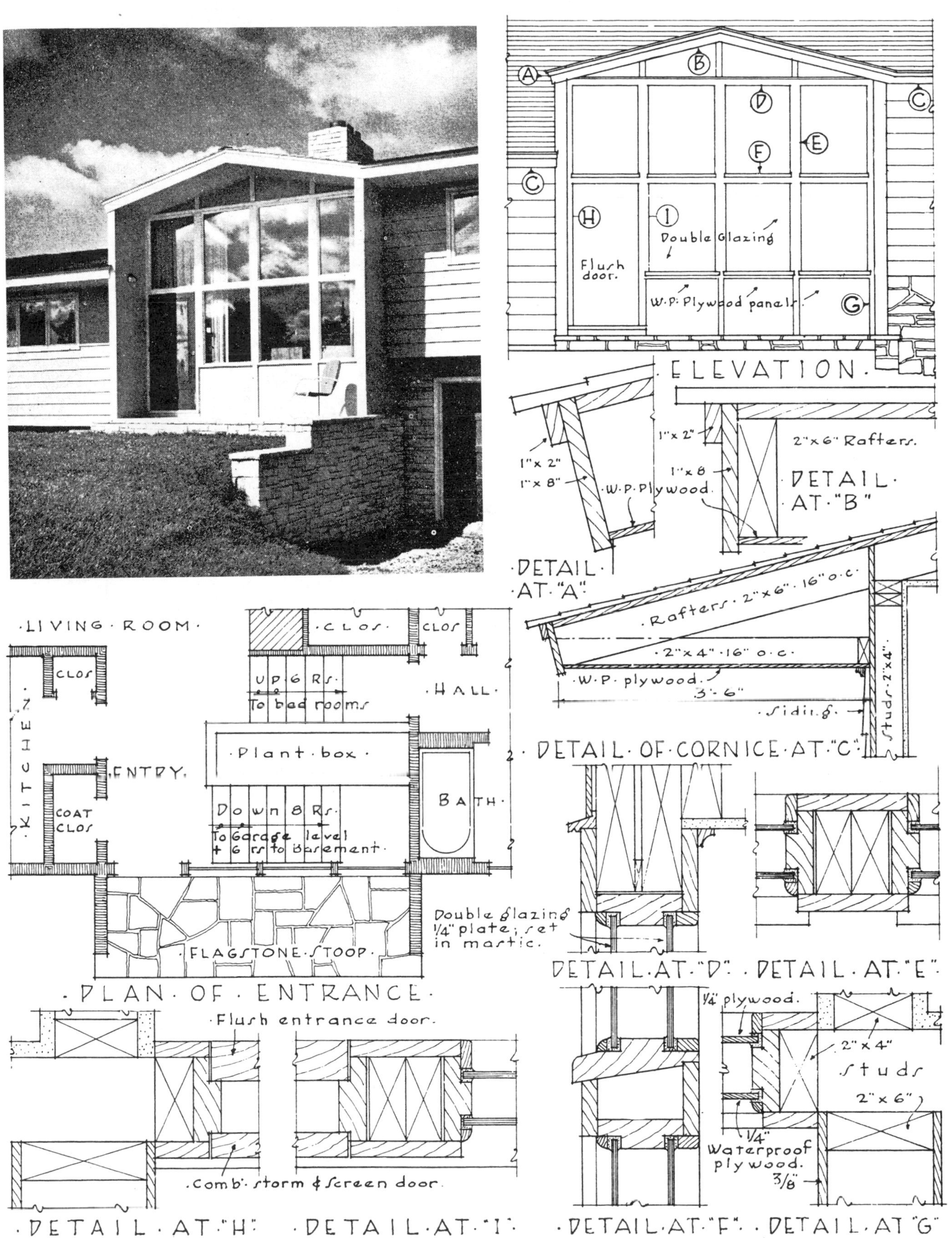

TIME AND MONEY SAVING DETAILS

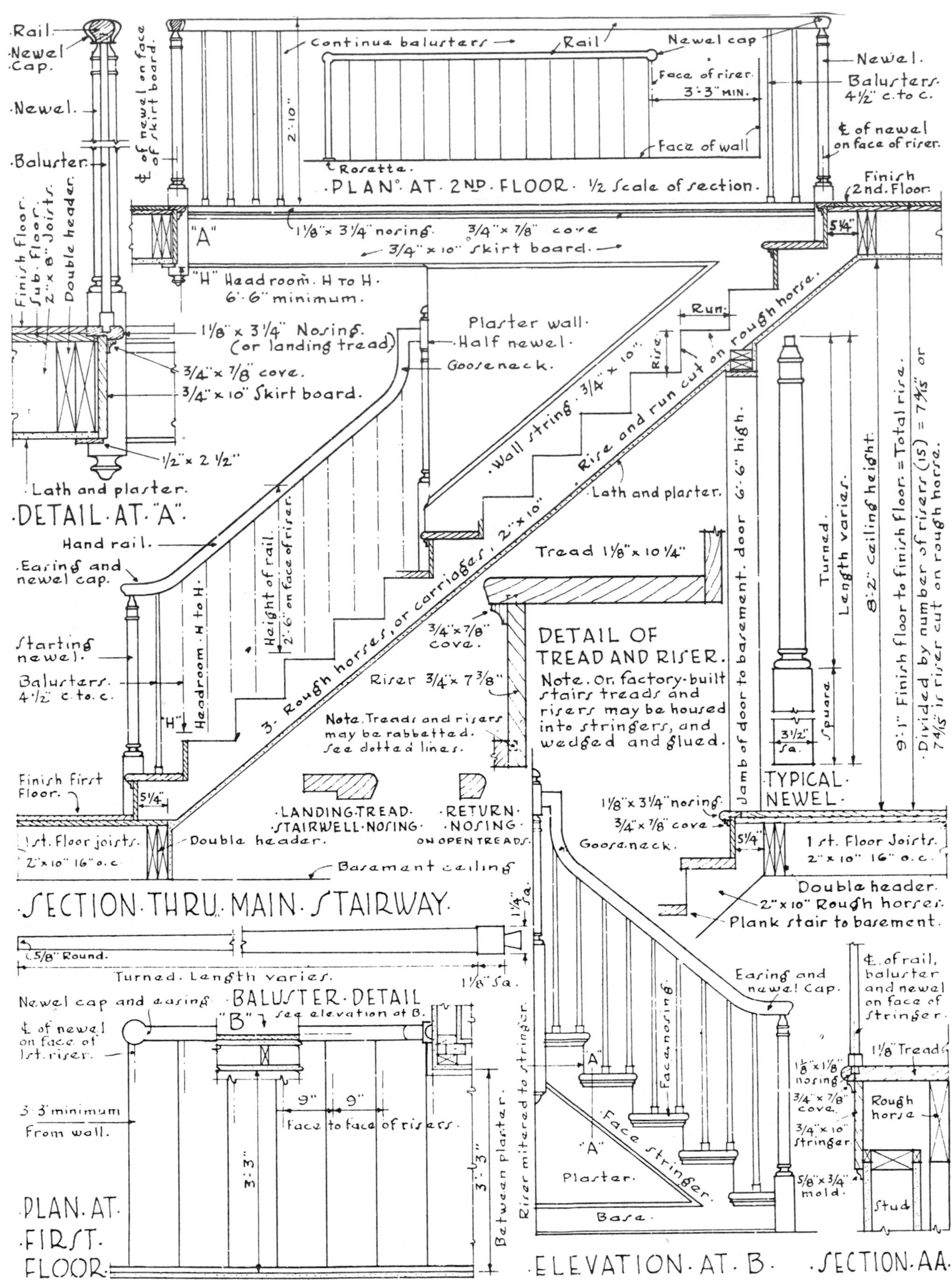
NO. D.5.. DETAILS. OF. MAIN. STAIRWAY. . TWO. STORY. COLONIAL. HOUSE
PLAN AT 2ND FLOOR. 1/2 Scale of section.
DETAIL. AT. "A".
SECTION. THRU. MAIN. STAIRWAY.
DETAIL OF TREAD AND RISER.
Note. On factory-built stairs treads and risers may be housed into stringers, and wedged and glued.
TYPICAL. NEWEL.
LANDING TREAD. STAIRWELL NOSING.
RETURN. NOSING. ON OPEN TREADS.
BALUSTER. DETAIL
PLAN. AT. FIRST. FLOOR.
ELEVATION. AT. B.
SECTION. AA.

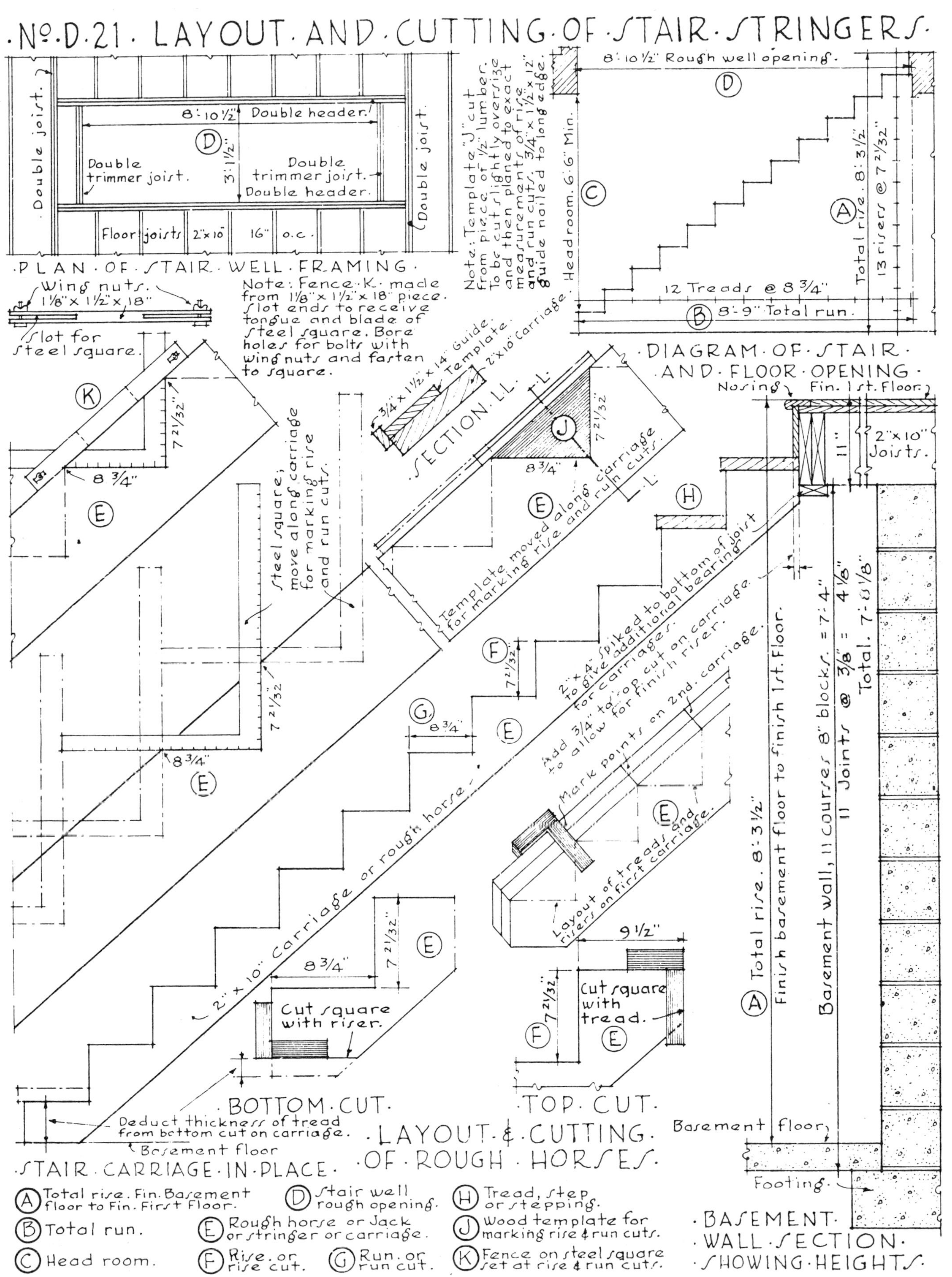

·No.D.21· LAYOUT·AND·CUTTING·OF·STAIR·STRINGERS·
8'-10½" Double header.
D
Double trimmer joist.
3'-1½"
Double trimmer joist.
Double header.
Double joist.
Double joist.
Floor joists 2"x10" 16" o.c.
·PLAN·OF·STAIR·WELL·FRAMING·
Wing nuts.
1⅛" x 1½" x 18"
Slot for steel square.
Note: Fence K made from 1⅛" x 1½" x 18" piece. Slot ends to receive tongue and blade of steel square. Bore holes for bolts with wing nuts and fasten to square.
Note: Template "J" cut from piece of ½" lumber. To be cut slightly oversize and then planed to exact measurements of rise and run cuts. ¾" x 1½" x 12" guide nailed to long edge.
Headroom. 6'-6" Min.
8'-10½" Rough well opening.
12 Treads @ 8¾"
8'-9" Total run.
Total rise. 8'-3½"
13 risers @ 7 21/32"
·DIAGRAM·OF·STAIR·AND·FLOOR·OPENING·
K
7 21/32"
8¾"
E
Steel square, move along carriage for marking rise and run cuts.
¾" x 1½" x 14" Guide.
Template.
2"x10" Carriage.
SECTION LL.
L
J
7 21/32"
8¾"
Template moved along carriage for marking rise and run cuts.
7 21/32"
8¾"
H
F
G
2" x 4" spiked to bottom of joist to give additional bearing for carriages.
Add ¾" to top cut on carriage to allow for finish riser.
Mark points on 2nd. carriage.
Layout of treads and risers on first carriage.
2" x 10" carriage or rough horse.
8¾"
7 21/32"
Cut square with riser.
9½"
7 21/32"
Cut square with tread.
Deduct thickness of tread from bottom cut on carriage.
Basement floor
·STAIR·CARRIAGE·IN·PLACE·
·BOTTOM·CUT·
·TOP·CUT·
·LAYOUT·&·CUTTING·OF·ROUGH·HORSES·
Nosing
Fin. 1st. Floor.
11"
2"x10" Joists.
Total rise. 8'-3½"
Finish basement floor to finish 1st. Floor.
Basement wall, 11 courses 8" blocks = 7'-4"
11 Joints @ ⅜" = 4⅛"
Total. 7'-8⅛"
Basement floor
Footing.
·BASEMENT·WALL·SECTION·SHOWING·HEIGHTS·
A Total rise. Fin. Basement floor to Fin. First Floor.
B Total run.
C Head room.
D Stair well rough opening.
E Rough horse or Jack or stringer or carriage.
F Rise or rise cut.
G Run or run cut.
H Tread, step or stepping.
J Wood template for marking rise & run cuts.
K Fence on steel square set at rise & run cuts.

No. D-51 · ROUGHING-IN · DETAILS · STAIRWAYS

·R·J·A·1948·

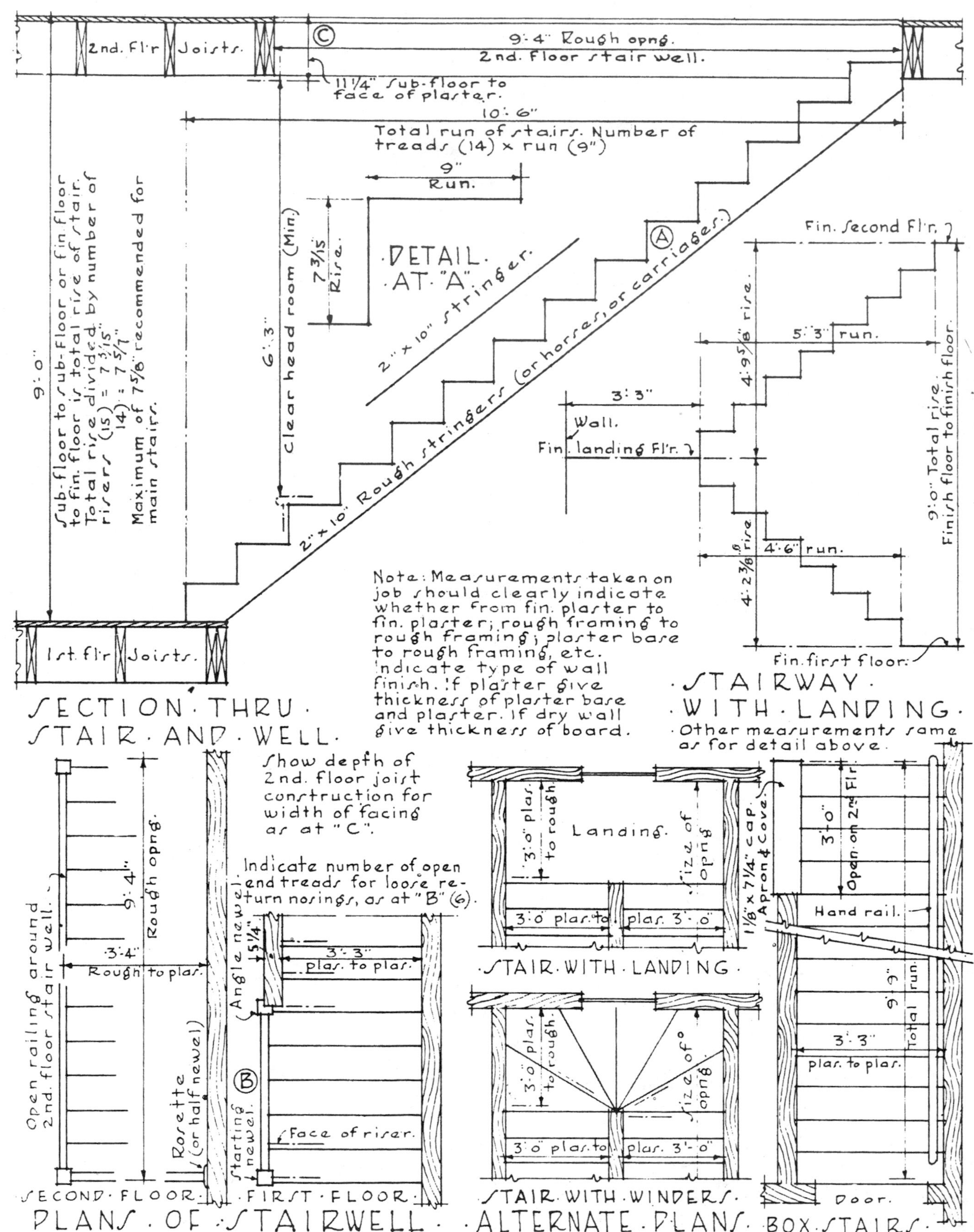

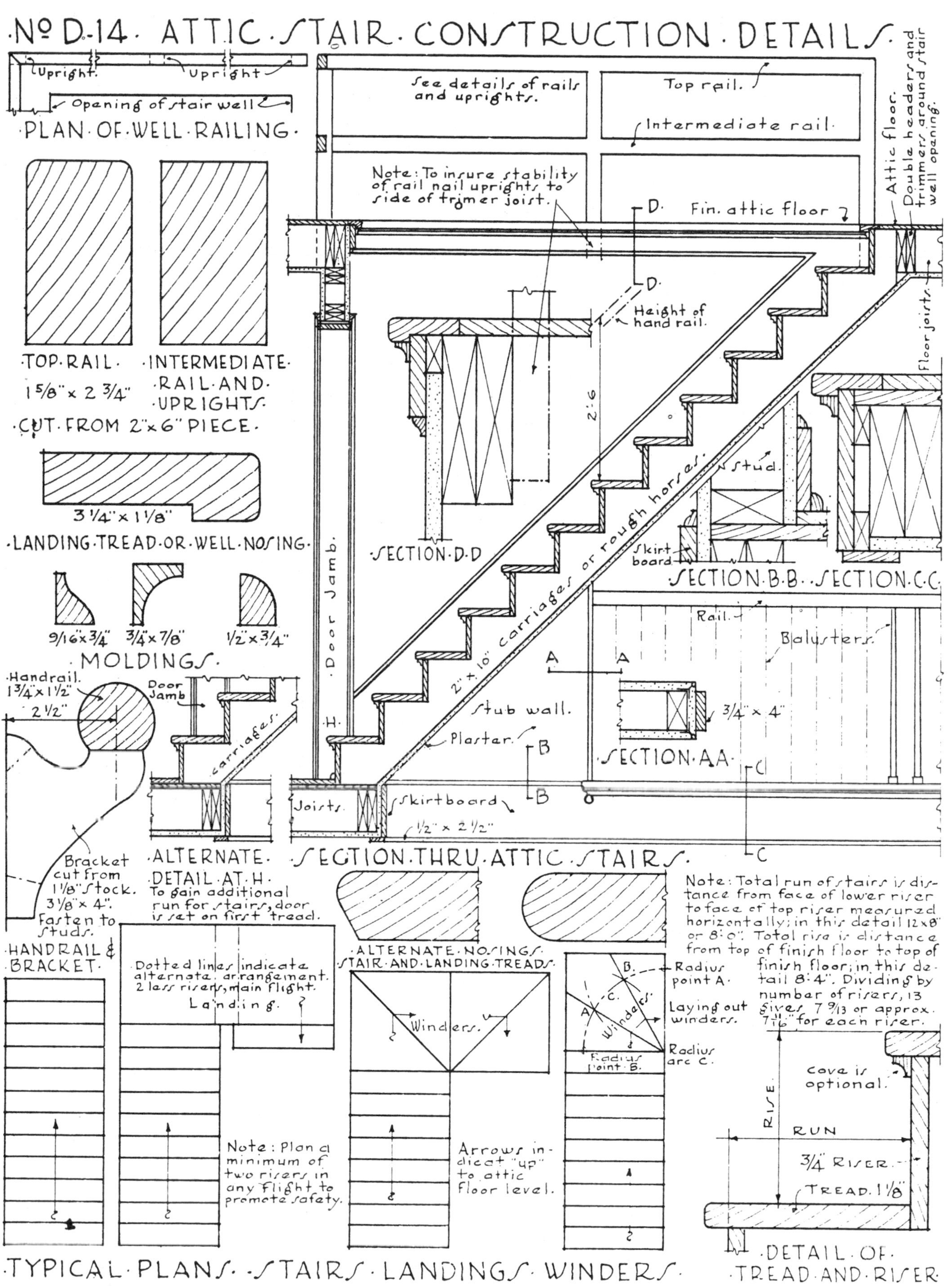

TIME AND MONEY SAVING DETAILS

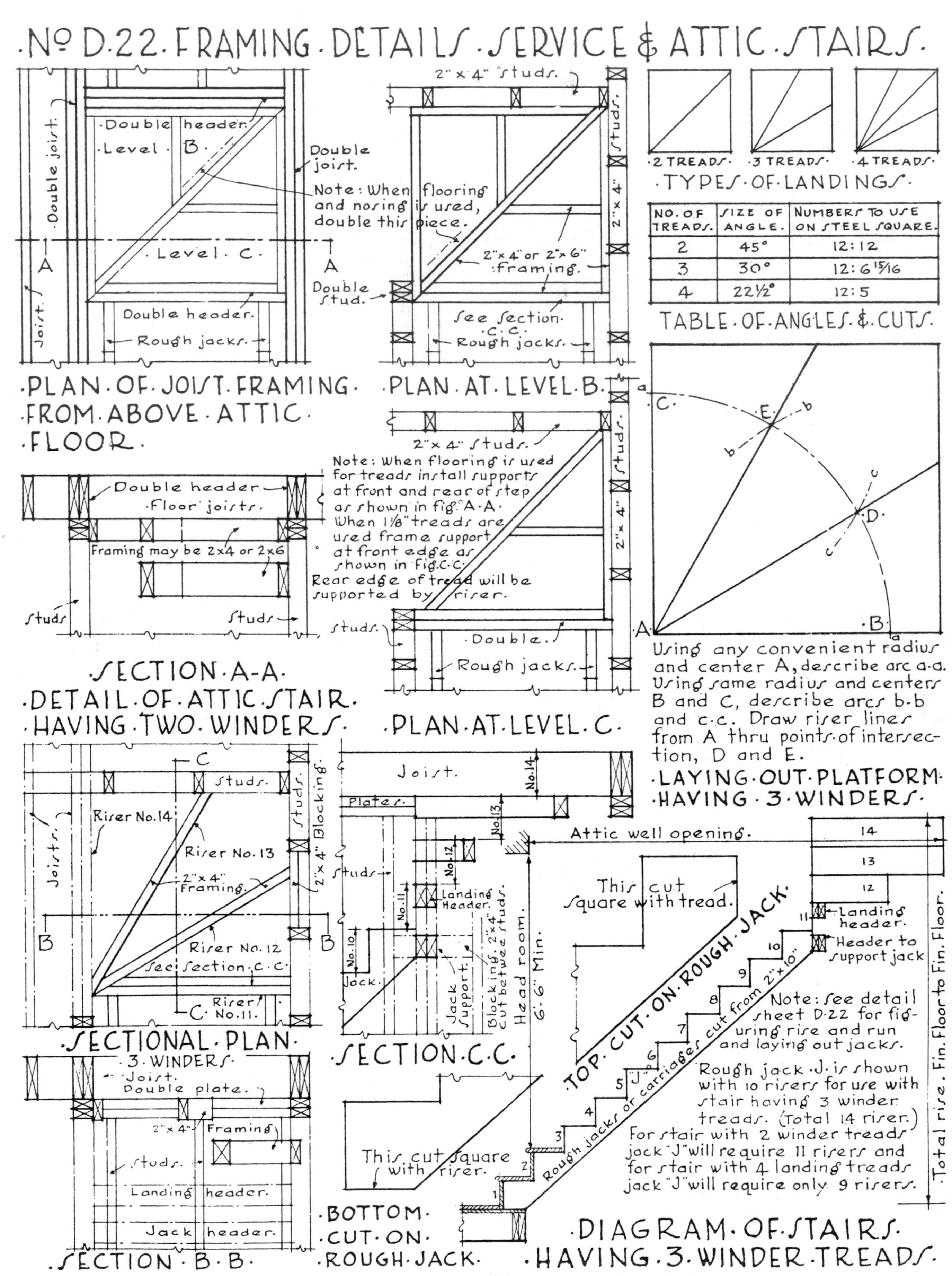

NO. OF TREADS.	SIZE OF ANGLE.	NUMBERS TO USE ON STEEL SQUARE.
2	45°	12:12
3	30°	12:6 15/16
4	22½°	12:5

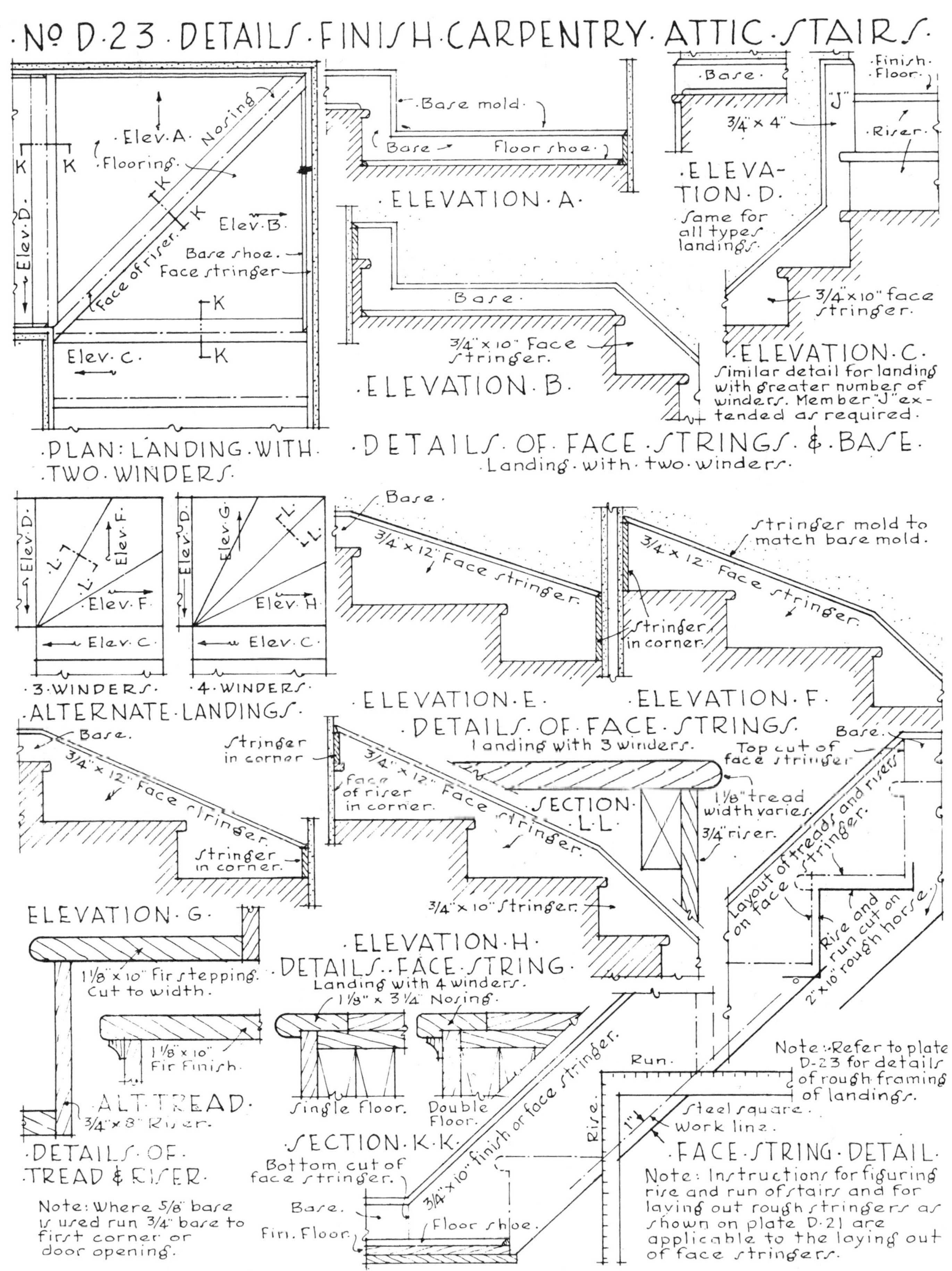
·No D·23·DETAILS·FINISH·CARPENTRY·ATTIC·STAIRS·
Elev·A·
Nosing
Flooring·
Elev·D·
Face of riser
Elev·B·
Base shoe·
Face stringer
Elev·C·
·PLAN: LANDING·WITH·TWO·WINDERS·
·Base mold·
Base
Floor shoe·
·ELEVATION·A·
Base·
3/4"x10" Face stringer.
·ELEVATION·B·
·Base·
·Finish·Floor·
"J"
3/4"x4"
·Riser·
·ELEVATION·D·
Same for all types landings·
3/4"x10" face stringer·
·ELEVATION·C·
Similar detail for landing with greater number of winders. Member "J" extended as required·
·DETAILS·OF·FACE·STRINGS·&·BASE·
·Landing·with·two·winders·
Elev·F·
Elev·G·
Elev·H·
·3·WINDERS·
·4·WINDERS·
·ALTERNATE·LANDINGS·
Base·
3/4"x12" Face stringer.
stringer mold to match base mold.
Stringer in corner.
·ELEVATION·E·
·ELEVATION·F·
·DETAILS·OF·FACE·STRINGS·
landing with 3 winders.
3/4"x12" face stringer.
Stringer in corner
face of riser in corner.
SECTION·L·L·
Top cut of face stringer
1 1/8" tread width varies.
3/4" riser.
3/4"x10" Stringer.
ELEVATION·G·
·ELEVATION·H·
DETAILS··FACE·STRING·
Landing with 4 winders.
1 1/8"x3 1/4" Nosing·
Single Floor.
Double Floor.
·SECTION·K·K·
Layout of treads and risers on face stringer.
Rise and run cut on 2"x10" rough horse.
1 1/8"x10" Fir stepping. Cut to width.
1 1/8"x10" Fir Finish.
·ALT·TREAD·
3/4"x8" Riser.
·DETAILS·OF·TREAD & RISER·
Note: Where 5/8" base is used run 3/4" base to first corner or door opening.
3/4"x10" finish or face stringer.
Bottom cut of face stringer.
Base.
Fin. Floor.
Floor shoe.
Run.
Rise.
1"
Steel square·
Work line.
Note:·Refer to plate D-23 for details of rough framing of landings.
·FACE·STRING·DETAIL·
Note: Instructions for figuring rise and run of stairs and for laying out rough stringers as shown on plate D·21 are applicable to the laying out of face stringers.

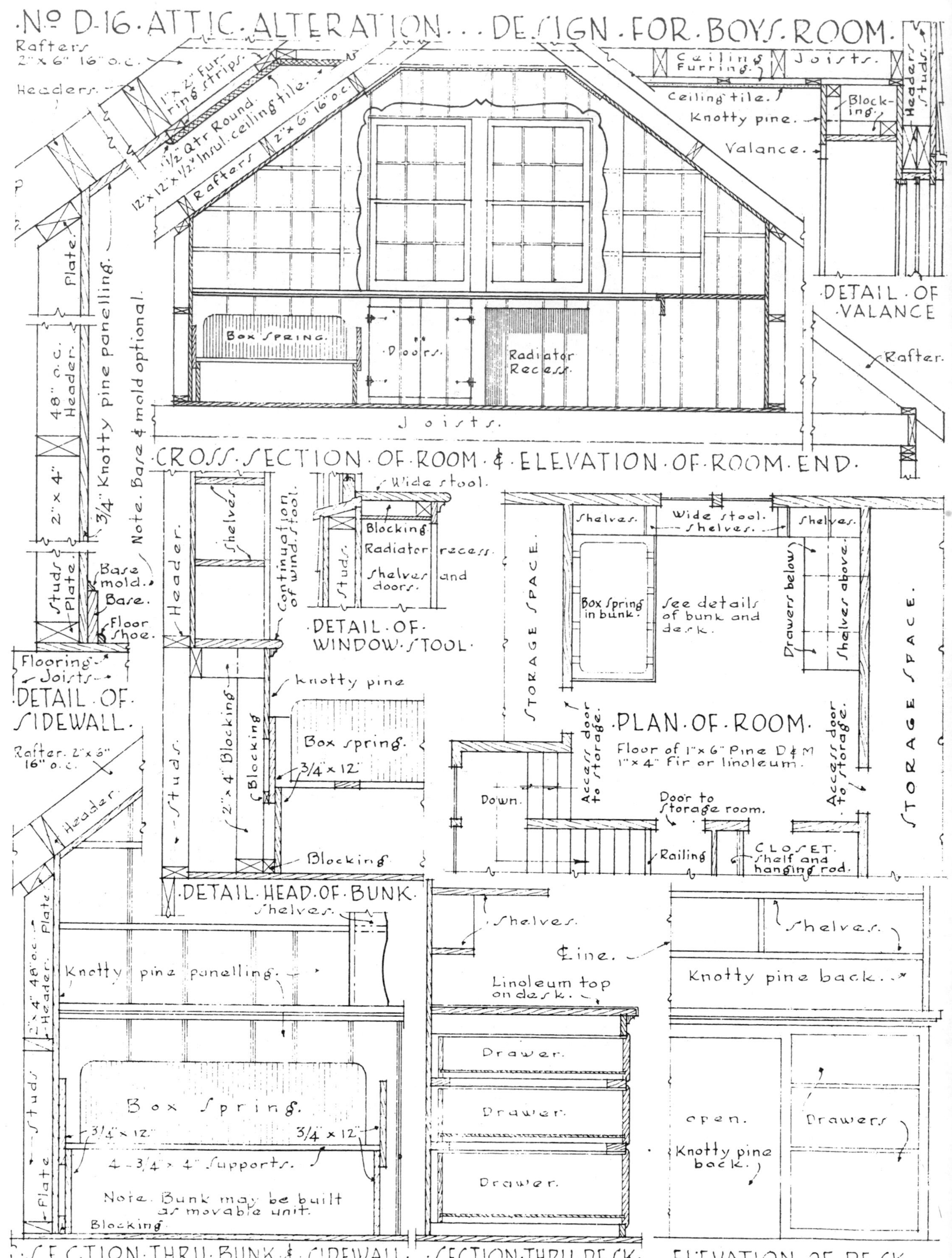
No D-16 · ATTIC · ALTERATION . . . DESIGN · FOR · BOYS · ROOM ·
Rafters 2" x 6" 16" o.c.
Headers.
1" x 2" Furring strips.
1/2" Qtr Round.
12" x 12" x 1/2" Insul. ceiling tile.
2" x 6" 16" o.c.
Rafters.
Plate.
3/4" Knotty pine panelling.
Note. Base & mold optional.
48" o.c. Header.
2" x 4"
Studs Plate.
Base mold.
Base.
Floor shoe.
Flooring.
Joists.
DETAIL · OF · SIDEWALL.
Ceiling Furring.
Joists.
Ceiling tile.
Knotty pine.
Valance.
Blocking.
Headers Studs.
DETAIL · OF · VALANCE
Rafter.
Box Spring.
Doors.
Radiator Recess.
Joists.
CROSS · SECTION · OF · ROOM · & · ELEVATION · OF · ROOM · END ·
Header.
Shelves.
Continuation of wind. stool.
Studs.
2" x 4" Blocking.
Blocking.
Knotty pine
Box spring.
3/4 x 12"
Blocking.
DETAIL · HEAD · OF · BUNK.
Wide stool.
Blocking.
Radiator recess.
Shelves and doors.
Studs.
DETAIL · OF · WINDOW · STOOL ·
Shelves.
Wide stool.
Shelves.
Shelves.
STORAGE SPACE
Box Spring in bunk.
See details of bunk and desk.
Drawers below.
Shelves above.
STORAGE SPACE
PLAN · OF · ROOM ·
Floor of 1" x 6" Pine D&M 1" x 4" fir or linoleum.
Access door to storage.
Access door to storage.
Down.
Door to Storage room.
Railing.
CLOSET. Shelf and hanging rod.
Rafter. 2" x 6" 16" o.c.
Header.
Plate.
2" x 4" 48" o.c. Header.
Studs.
Plate.
Shelves.
Knotty pine panelling.
Box Spring.
3/4" x 12"
3/4" x 12"
4-3/4" x 4" Supports.
Note. Bunk may be built as movable unit.
Blocking.
SECTION · THRU · BUNK · & · SIDEWALL ·
Shelves.
Line.
Linoleum top on desk.
Drawer.
Drawer.
Drawer.
SECTION · THRU · DESK ·
Shelves.
Knotty pine back.
open.
Knotty pine back.
Drawers.
ELEVATION · OF · DESK ·

Good Kitchens Are Planned Before They're Built

FIVE BASIC KITCHEN PLANS

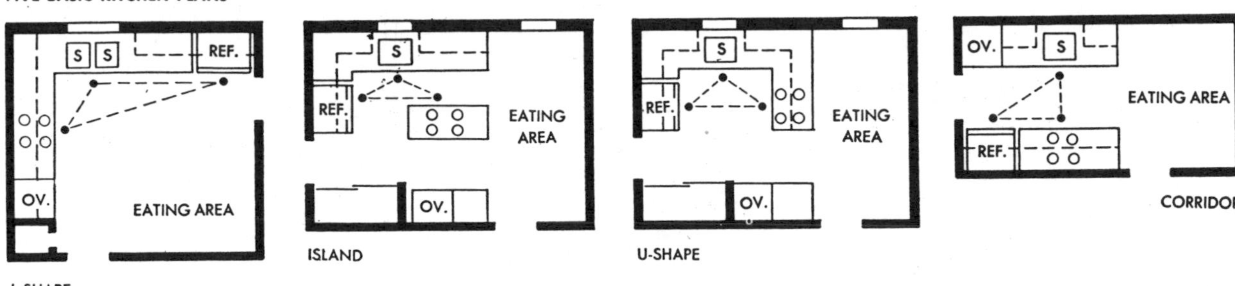

The overall floor plan affects the shape, location and arrangement of the kitchen, so careful thought must go into its layout to insure sufficient wall space for the installation of cabinets and all necessary appliances. Five basic plans for kitchens of different shapes are shown above and at right.

Ideally, the kitchen should provide a continuous line of appliances linked by cabinets and counter tops. The U-shape does this most effectively. An L-shaped kitchen usually occupies two walls of the room and results in a triangular work area. If this triangle becomes too large, consider adding an island to bring the activity centers closer together. Single wall and corridor kitchens work well in long, narrow spaces, are best suited to small homes or apartments.

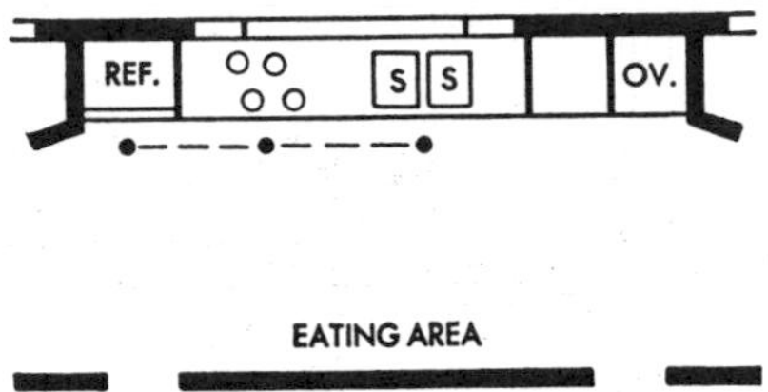

SINGLE WALL

LOCATION OF DOORS AND WINDOWS AFFECTS PLANNING

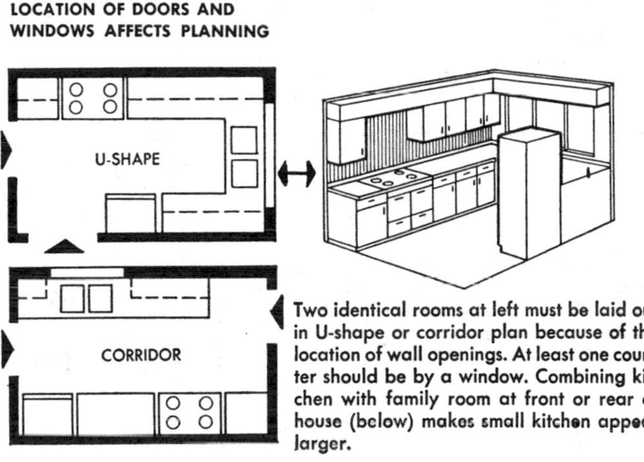

Two identical rooms at left must be laid out in U-shape or corridor plan because of the location of wall openings. At least one counter should be by a window. Combining kitchen with family room at front or rear of house (below) makes small kitchen appear larger.

KITCHEN FOR A SPLIT LEVEL

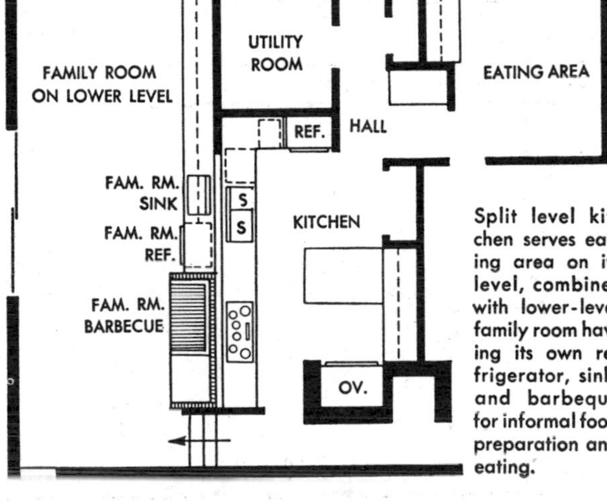

Split level kitchen serves eating area on its level, combines with lower-level family room having its own refrigerator, sink, and barbeque for informal food preparation and eating.

KITCHEN-FAMILY ROOM COMBINATIONS

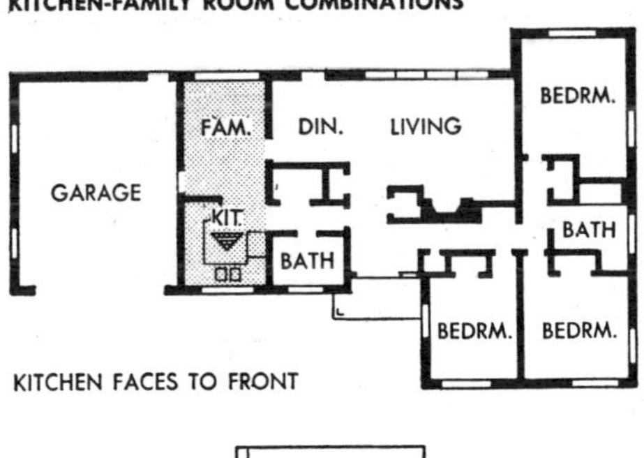

KITCHEN FACES TO FRONT

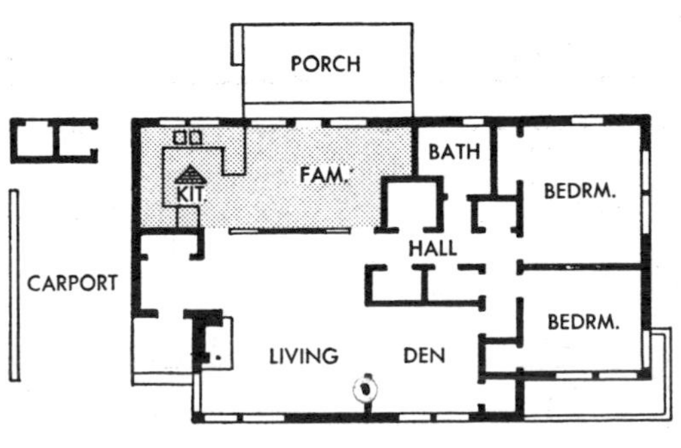

KITCHEN FACES TO REAR

PLAN ON DROP-INS FOR A BUILT IN LOOK

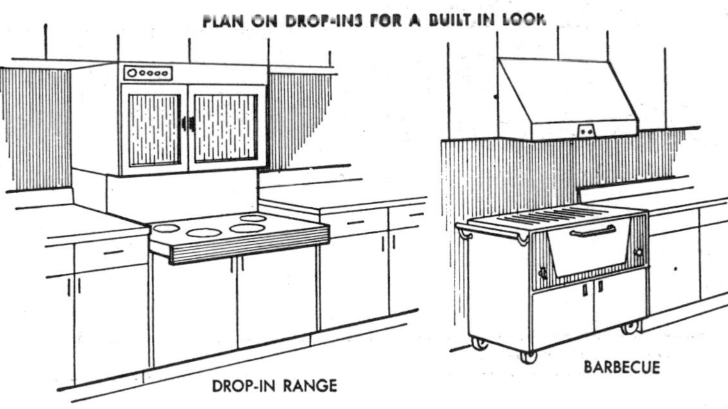

DROP-IN RANGE

BARBECUE

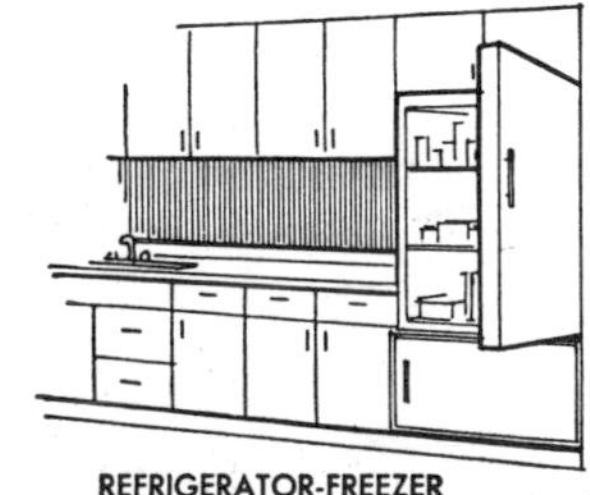

REFRIGERATOR-FREEZER

Latest trend: free-standing appliances that provide appearance of built-ins at lower installation cost.

Electric range has eye-level oven. Unit sets on a cabinet, requires no special installation. Refrigerator-freezer can be installed flush against wall. Doors swing within the dimensions of the unit, eliminating need for space between cabinet on hinged side. Hood makes portable barbeque look built in.

Cabinets and Built-ins Combined for Efficient Kitchens

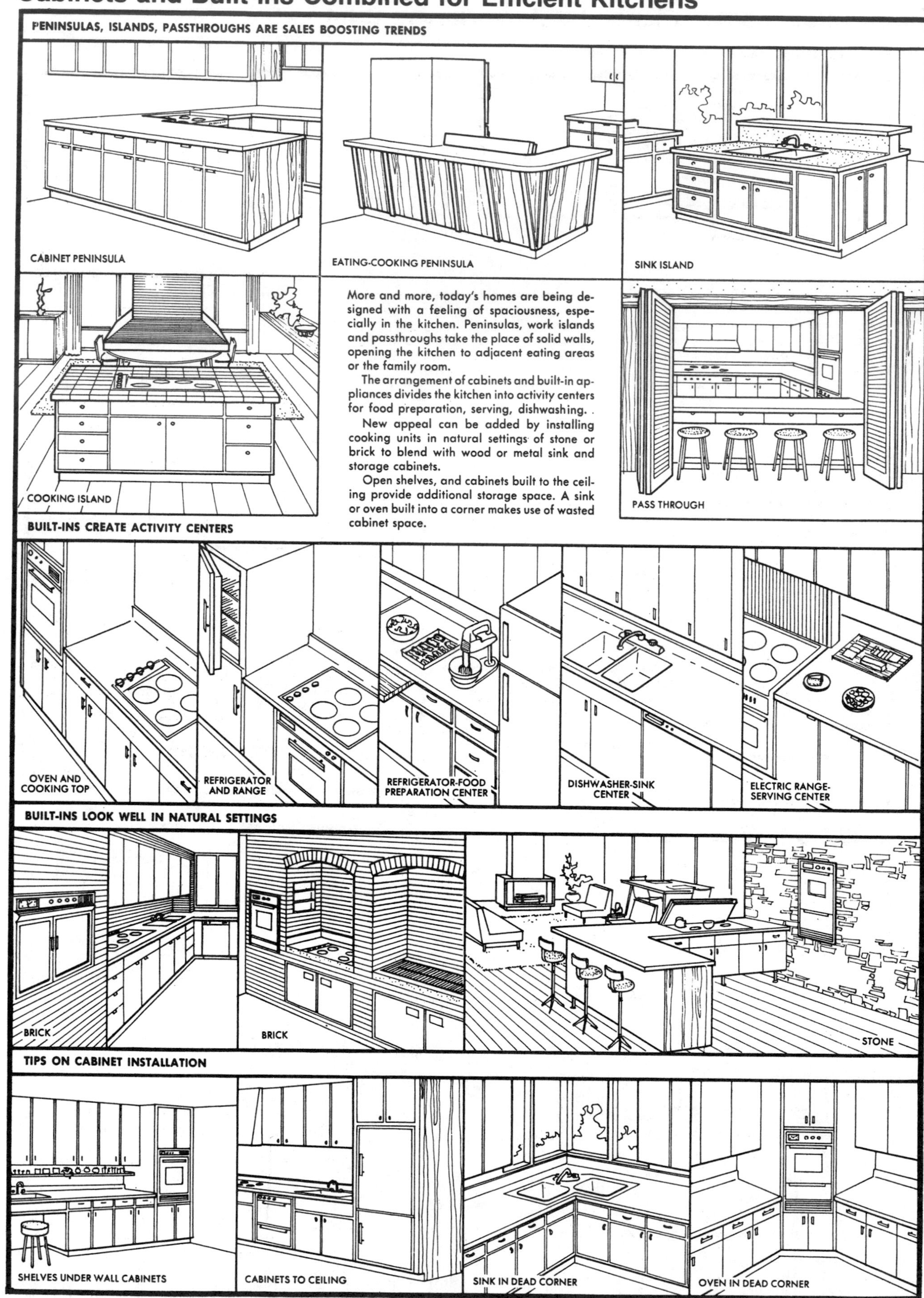

More and more, today's homes are being designed with a feeling of spaciousness, especially in the kitchen. Peninsulas, work islands and passthroughs take the place of solid walls, opening the kitchen to adjacent eating areas or the family room.

The arrangement of cabinets and built-in appliances divides the kitchen into activity centers for food preparation, serving, dishwashing. .

New appeal can be added by installing cooking units in natural settings of stone or brick to blend with wood or metal sink and storage cabinets.

Open shelves, and cabinets built to the ceiling provide additional storage space. A sink or oven built into a corner makes use of wasted cabinet space.

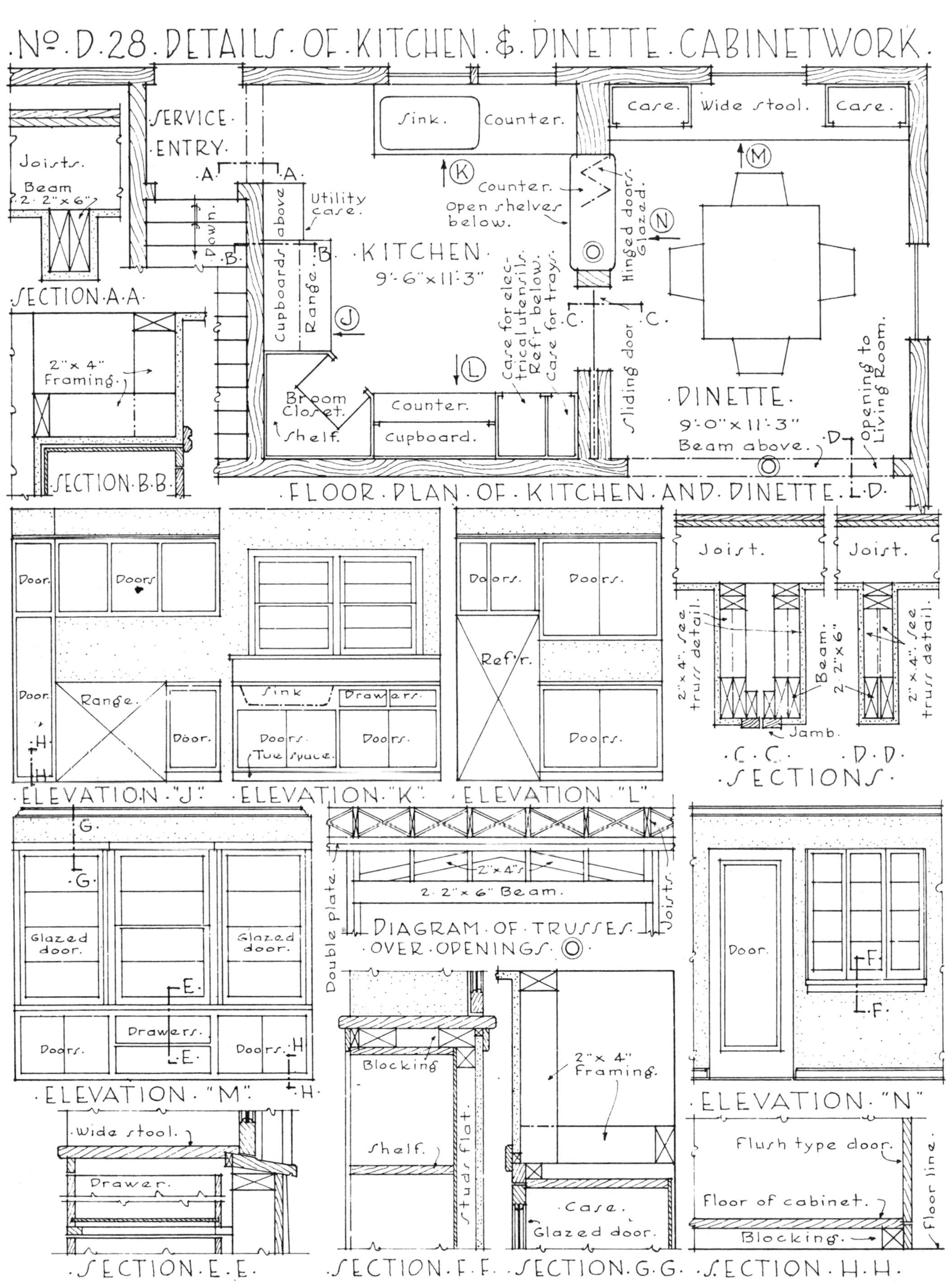
No. D.28. DETAILS OF KITCHEN & DINETTE CABINETWORK.
SERVICE ENTRY.
Joists.
Beam 2. 2"x6"
SECTION A.A.
2"x 4" Framing.
SECTION B.B.
Sink.
Counter.
Case.
Wide stool.
Case.
Utility case.
Cupboards above
Range
Down
Counter.
Open shelves below.
Hinged doors. Glazed.
KITCHEN.
9'-6"x11'-3"
Case for electrical utensils. Ref'r below. Case for trays.
Broom Closet.
Shelf.
Counter.
Cupboard.
Sliding door
DINETTE.
9'-0"x11'-3"
Beam above.
Opening to Living Room.
FLOOR PLAN OF KITCHEN AND DINETTE.
Door.
Doors
Door.
Range.
Door.
ELEVATION "J".
Sink
Drawers.
Doors.
Tub space.
Doors.
ELEVATION "K".
Doors.
Doors.
Ref'r.
Doors.
ELEVATION "L".
Joist.
Joist.
2"x4". See truss detail.
Beam. 2. 2"x6"
Jamb.
2"x4". See truss detail.
C.C.
D.D.
SECTIONS.
Glazed door.
Glazed door.
Drawers.
Doors.
Doors.
ELEVATION "M".
2"x4"
2. 2"x6" Beam.
Double plate.
Joists.
DIAGRAM OF TRUSSES OVER OPENINGS.
Door.
ELEVATION "N".
Wide stool.
Drawer.
SECTION E.E.
Blocking
Shelf.
Studs flat.
SECTION F.F.
2"x 4" Framing.
Case.
Glazed door.
SECTION G.G.
Flush type door.
Floor of cabinet.
Blocking.
Floor line.
SECTION H.H.

How to Build in a Corner Sink and Make Most of the Space

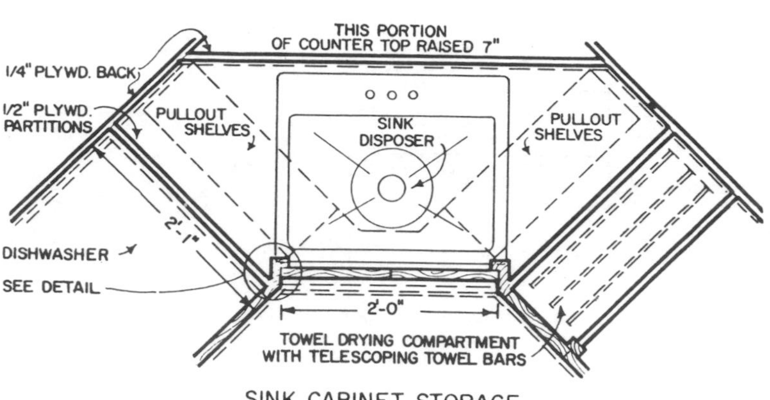

NEATNESS is no struggle with the specially designed shelves nestled around the food disposer for soaps and cleaners.

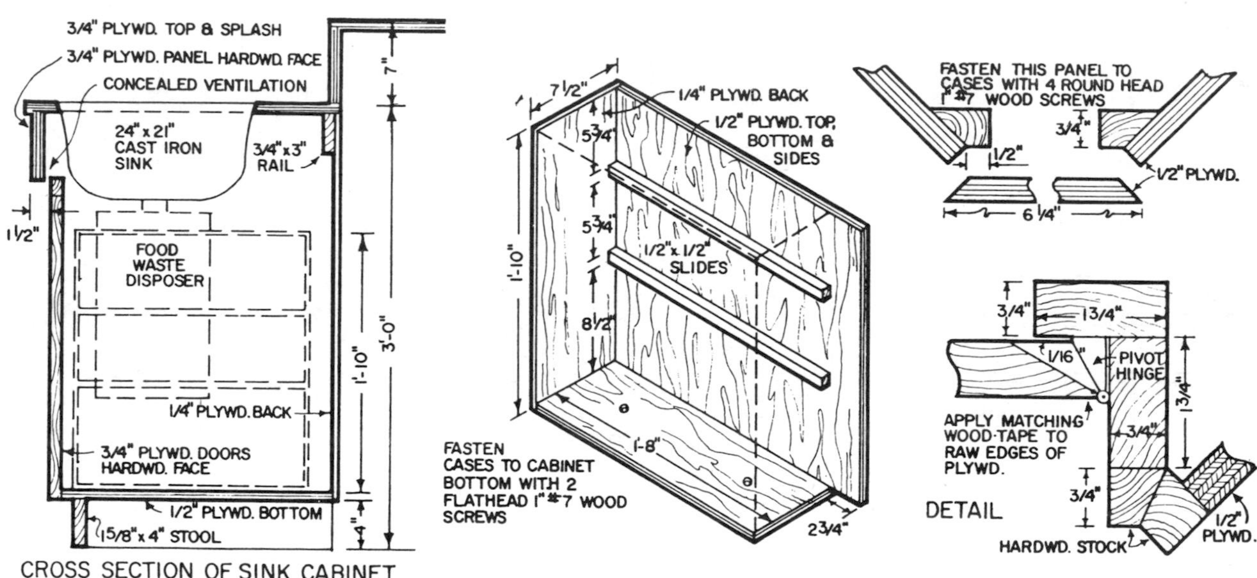

DETAILS show how to construct the corner sink cabinet assembly. Refer to page 208 for details dealing with the shelves.

A corner-sink arranged in the manner of Seattle Builder Elmer V. Moss (as shown in the above photo and plan view) would make a highly efficient addition to any kitchen plan. Aside from the pull-out feature for soaps and cleaners, a dishwasher is just a step to the left. To the right is a towel-drying compartment with its telescoping bars for dish towels and cloths. The broken lines in the plan and sectional drawings above refer to the position of cases for the sliding shelves. Should the plumbing or food disposer require servicing, the compact assembly can easily be dismantled. This can be accomplished by first removing the shelves and then unscrewing the two 1" flat-head No. 7 wood screws that fasten the sliding-shelf cases to the floor of the plywood cabinet. One right and one left case are required.

Ventilation Heads List of Kitchen Equipment

To rid the kitchen of the by-products of cooking—grease, smoke, heat, moisture and odors—an efficient ventilating fan should be installed as near to the range or cooking unit as possible. A wall or ceiling fan is the most economical way to vent the room, but these are not necessarily the most effective. A broad hood above the cooking unit will trap air and prevent its escape from a wall fan, but a hood equipped with a blower and ducting to outside will remove cooking fumes most effectively. The latest development in kitchen air purification is a filtering hood. Grease-laden air moves through aluminum mesh, fiberglass and a layer of activated charcoal, then returns to the room grease and odor free.

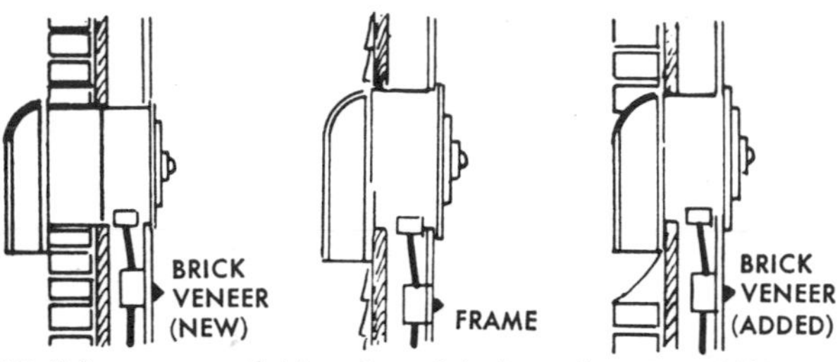

Wall fans are available to fit wall thickness from 5″ to 10⅛″. A pull chain opens a shutter on the outside and turns on the fan. Can be had for wall-switch control.

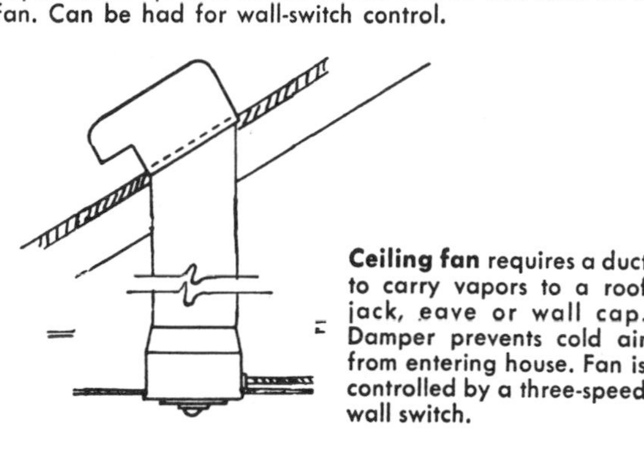

Ceiling fan requires a duct to carry vapors to a roof jack, eave or wall cap. Damper prevents cold air from entering house. Fan is controlled by a three-speed wall switch.

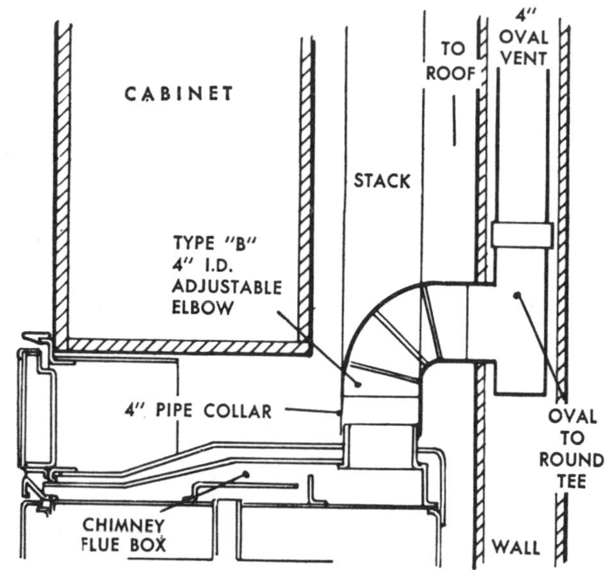

Oven can be fitted with a hood or vented direct to outside through ducts running behind cabinet or in the wall to the roof or eave. No blower is needed.

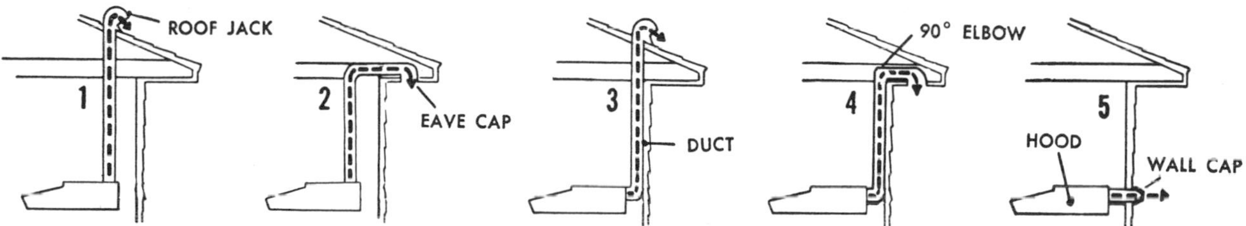

Power range hoods have openings at top and rear, can be fitted with ducts to exhaust air in a variety of ways: (1) through the cabinet to roof, (2) through the cabinet to eave, (3) through the wall to roof, (4) through the wall to eave, (5) directly through wall to outside. Cover plate is left over unused opening in hood.

ROOF JACK

WALL CAP

FLUSH EAVE CAP

WALL CAP

FLUSH WALL CAP

A variety of vent fittings provide flexibility in assembling ducts. Roof jacks, wall and eave caps can be had with built-in dampers to keep out cold air. Adapters (not shown) permit round ducts to be connected to flat in-the-wall ducts.

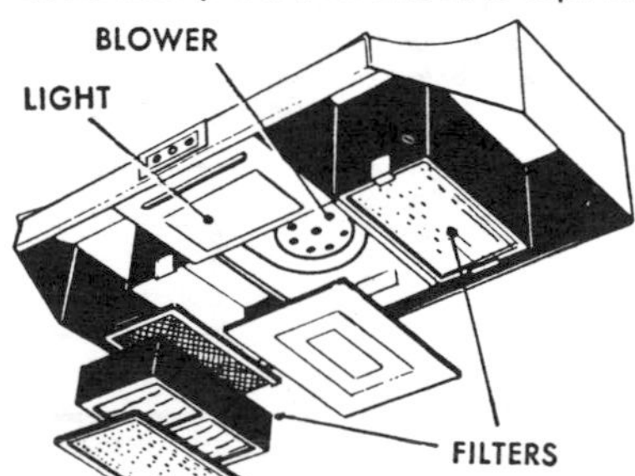

Non-ducted filtering hood is self contained, requires no exhaust passage to outside. Filters clean air.

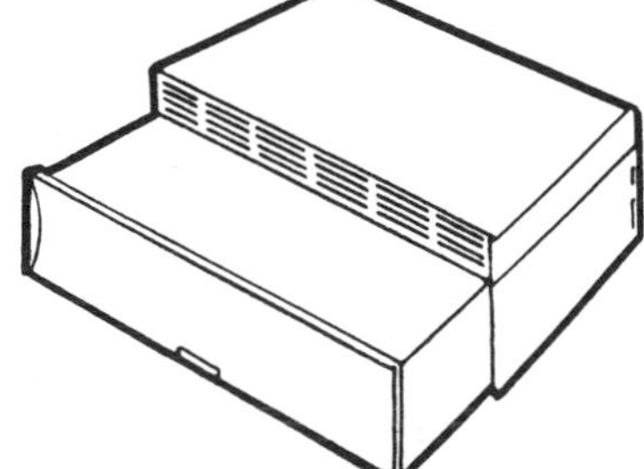

Roll-out hood retracts into main housing when not in use. It can be had with filters as shown or for hookup to duct.

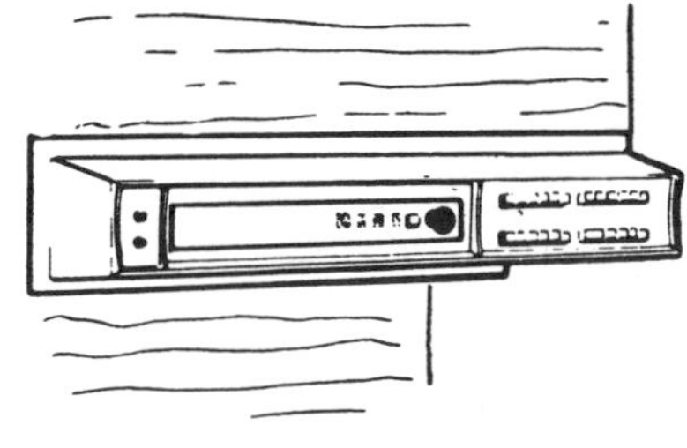

Combination hood puts cooking top controls at eye level where they're easy to see and reach.

Bathrooms Planned for the Family, Not the Individual

TWO STORY

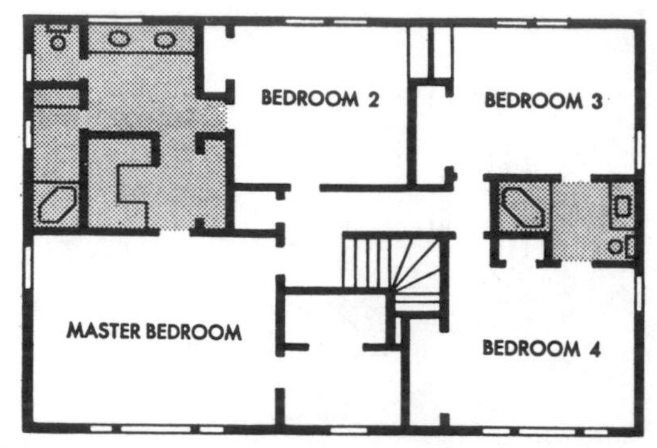

Economy in construction often conflicts with good bathroom planning; placing two baths back to back saves on plumbing costs, but seldom provides maximum convenience and privacy. Three good arrangements are illustrated. In the two-story plan above, the bathroom adjacent to the master bedroom also serves bedroom 2, while occupants of bedrooms 3 and 4 share a smaller bathroom without going out into the hall. The idea is carried out in the one-story ranch and split level with equal effectiveness.

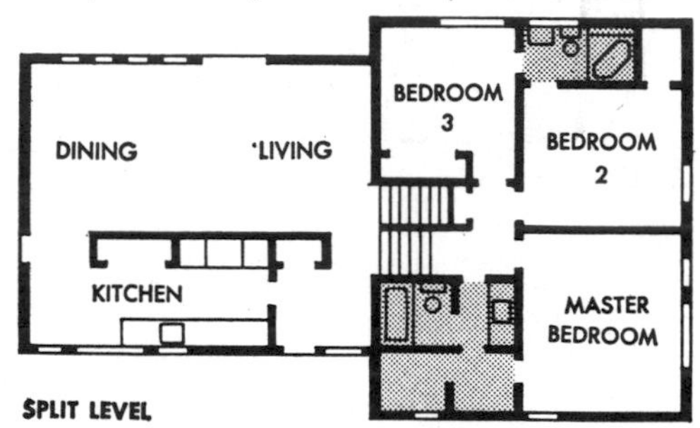

ONE STORY RANCH

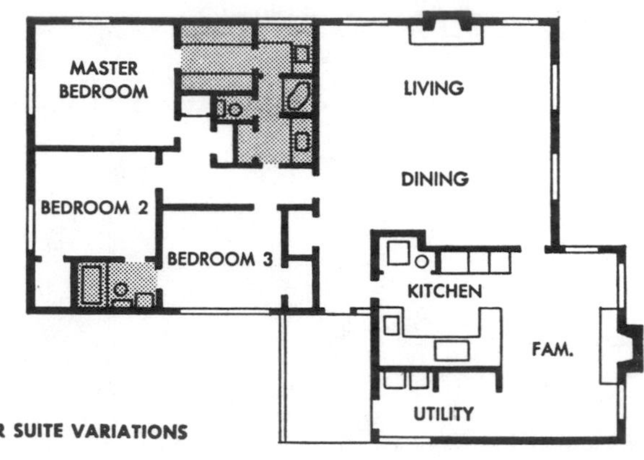

MASTER SUITE VARIATIONS

Master suites are gaining in popularity because they provide the utmost in privacy and convenience. The long plan below combines four rooms in one suite having three entrances. The separate shower is available to all without interrupting activities in the dressing room and adjacent baths. The smaller plans illustrate variations of the suite idea.

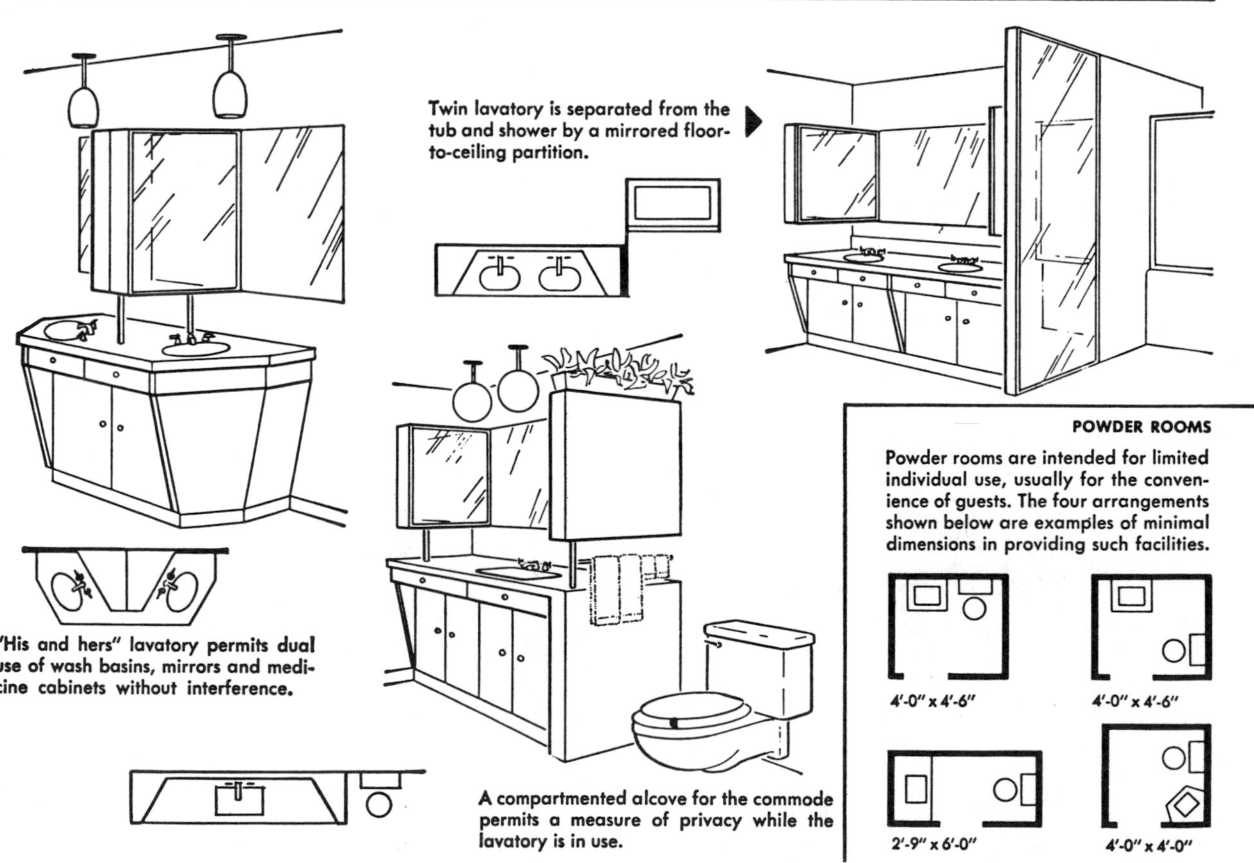

Twin lavatory is separated from the tub and shower by a mirrored floor-to-ceiling partition.

"His and hers" lavatory permits dual use of wash basins, mirrors and medicine cabinets without interference.

A compartmented alcove for the commode permits a measure of privacy while the lavatory is in use.

POWDER ROOMS

Powder rooms are intended for limited individual use, usually for the convenience of guests. The four arrangements shown below are examples of minimal dimensions in providing such facilities.

No D 41 · DRESSING · ROOM · & · BATH · OFF · MASTER · BED · ROOM.

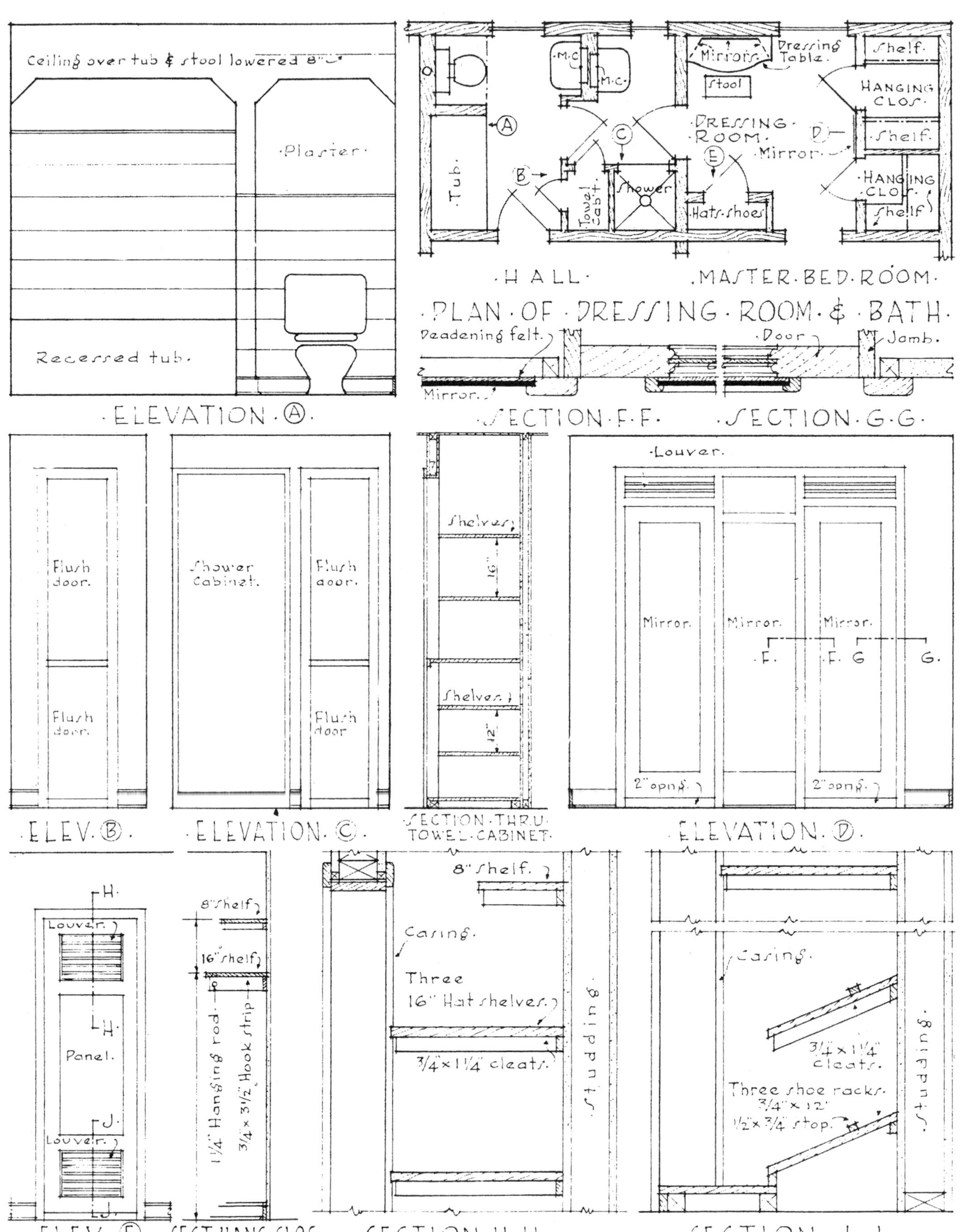

TIME AND MONEY SAVING DETAILS

Good Planning Cuts Plumbing Costs

The drainage system combines the use of TRAPS, VENTS, WASTE LINES, and SOIL STACKS

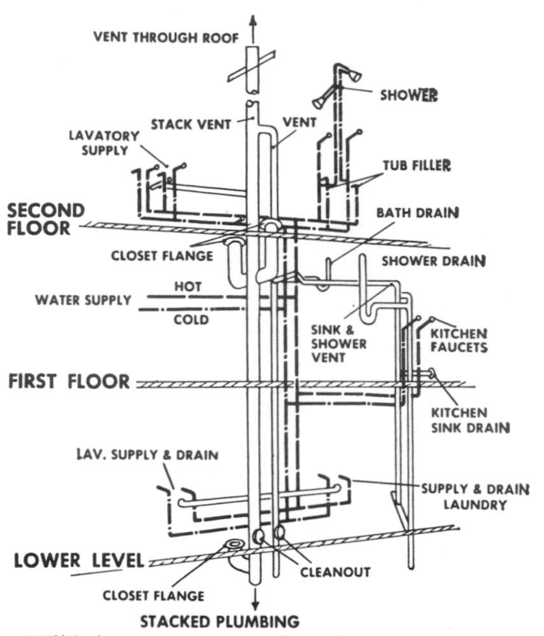

STACKED PLUMBING

A 2½ bath economy arrangement for a split or two-story house is comprised of a "stack on" type supply and drainage assembly which accommodates two baths and kitchen on the upper levels and a laundry with half bath on the lower level.

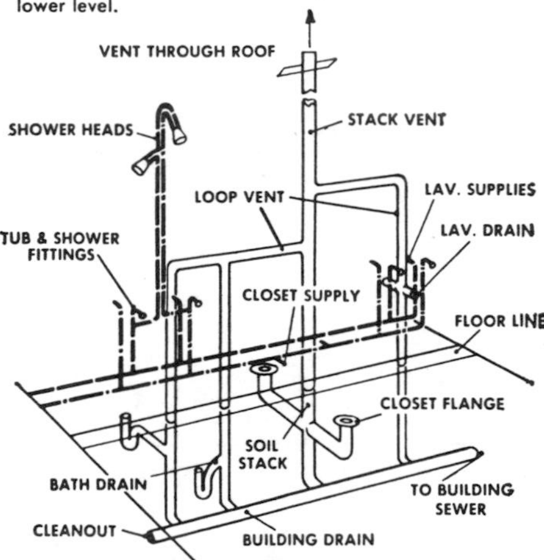

BACK TO BACK PLUMBING

A back to back 2 bath arrangement on one story would be comprised of a supply and drainage assembly as shown at right.

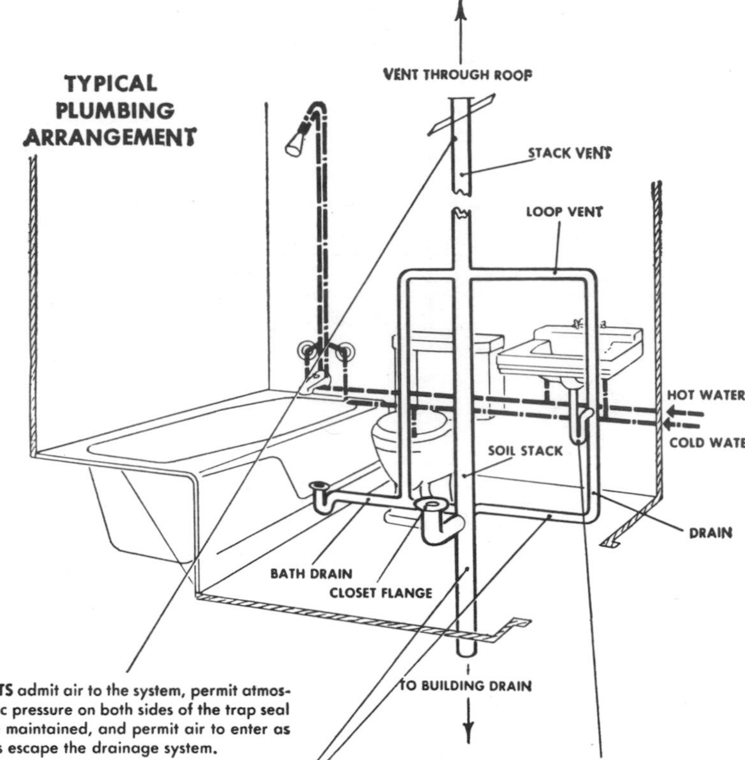

VENTS admit air to the system, permit atmospheric pressure on both sides of the trap seal to be maintained, and permit air to enter as gases escape the drainage system.

TRAPS permit waste and waste water to enter the drainage system and prevent any sewer gases from entering the house. The water seal utilizes a portion of the waste water to act as the barrier.

WASTE LINES AND SOIL STACKS connect the plumbing fixtures to the traps and vents and eventually to the main disposal system. Waste piping is smaller in diameter than the main soil pipe. Cleanouts should be located so that the entire system can be opened up if necessary.

- Wherever possible, it is best to locate rooms containing fixtures near the point of entrance of the water and sewer services. This will cut the length of service lines.

- Many plans today include compartmented baths; toilet and sink separate from tub and shower, or toilet separate from sink and tub and shower. Piping should be planned so that fixtures drain and are vented with a single stack.

- The suggestions on this page are not meant to be absolute. Local conditions and codes sometimes prevent installations as shown here. However, when they can be used, savings mount — especially in multi-home projects.

WATER-HEATER AND CHIMNEY LOCATED FOR ECONOMICAL HOT-WATER PIPING

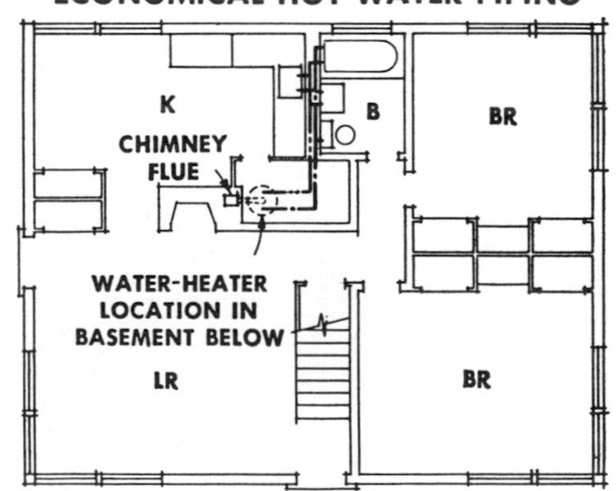

EQUIPMENT LOCATION: When water-heater is located too far from fixtures using hot-water, there is extra cost for material and labor to install the piping, and heat loss through the extra run of pipe makes the heater less efficient. Note in the plan above that fireplace was located so that water-heater could be properly vented and still be close to those fixtures needing hot-water.

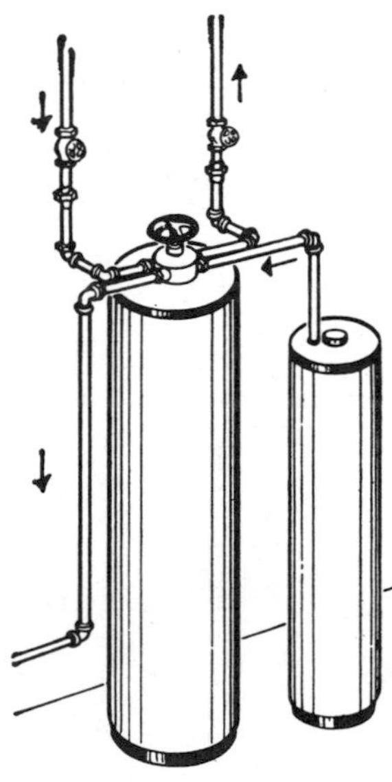

Soft water is often supplied to the water heater only, all cold-water lines being left with hard water. When both cold and hot water are softened, three pipes (the third for cold hard water) are needed for each sink and lavatory as many people object to drinking the soft cold water.

Septic Tanks and Absorption Fields Solve Sewage Problems

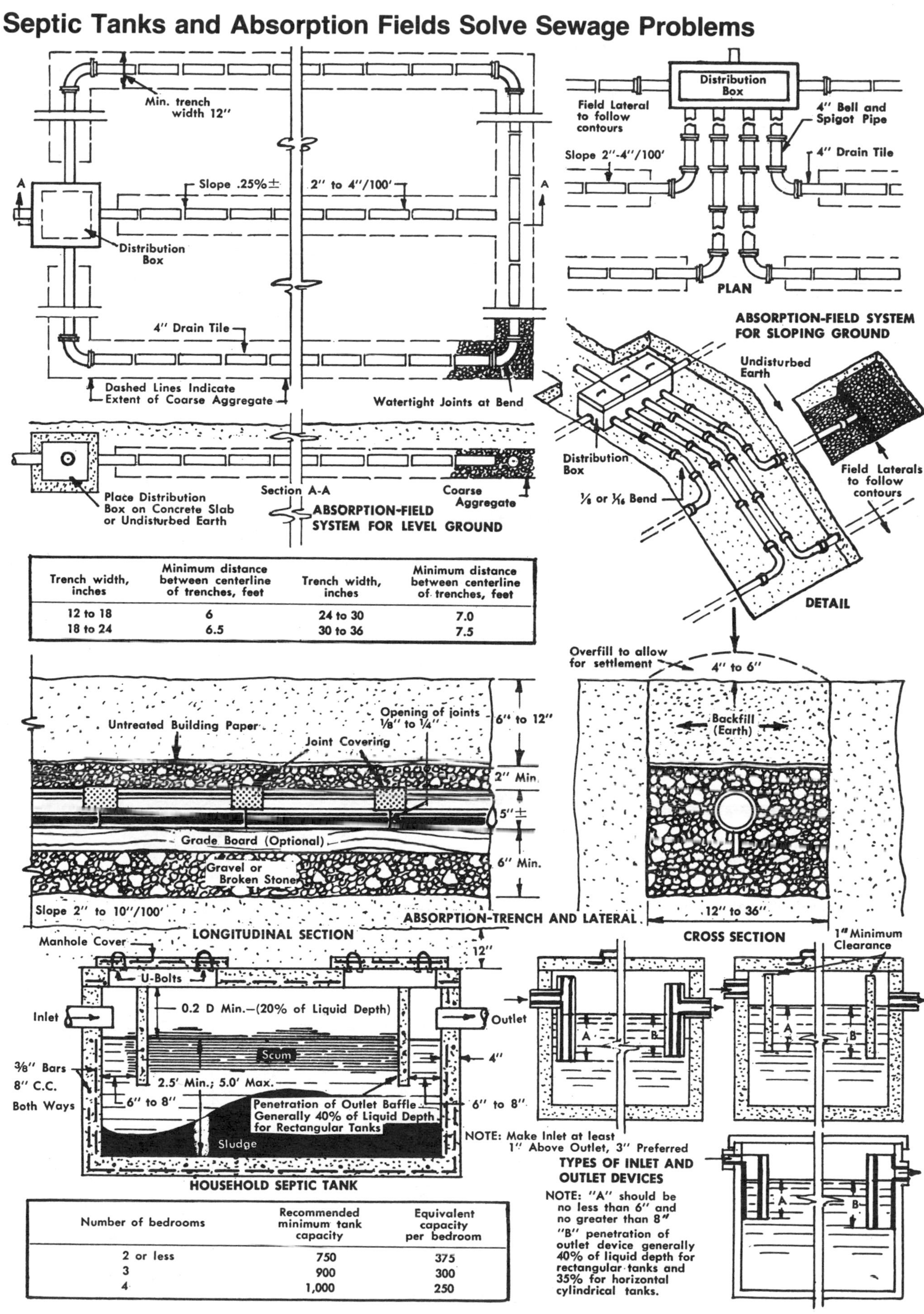

Trench width, inches	Minimum distance between centerline of trenches, feet	Trench width, inches	Minimum distance between centerline of trenches, feet
12 to 18	6	24 to 30	7.0
18 to 24	6.5	30 to 36	7.5

Number of bedrooms	Recommended minimum tank capacity	Equivalent capacity per bedroom
2 or less	750	375
3	900	300
4	1,000	250

Laundry Room Should Be Part of the House

Today's laundry appliances have turned the ugly duckling laundry room into a smart work center that need not be hidden in the basement or in a closed-off utility room. Convenience now is the first thought in planning the arrangement and location of the laundry room. The latest idea is to put the laundry room in a rear hall near the bath and bedrooms, where most soiled clothes originate. Eight convenient locations are illustrated. Note that a screen or folding louvered doors can be used to conceal a laundry center installed in the kitchen or family room.

Provision must be made to vent a dryer, otherwise moisture removed from the wash will create a humidity problem inside the house. The most common method is to run a duct through the floor and out the crawl space or basement wall as shown at the bottom of the page. The duct should be no longer than 30′, with 4′ subtracted for each elbow.

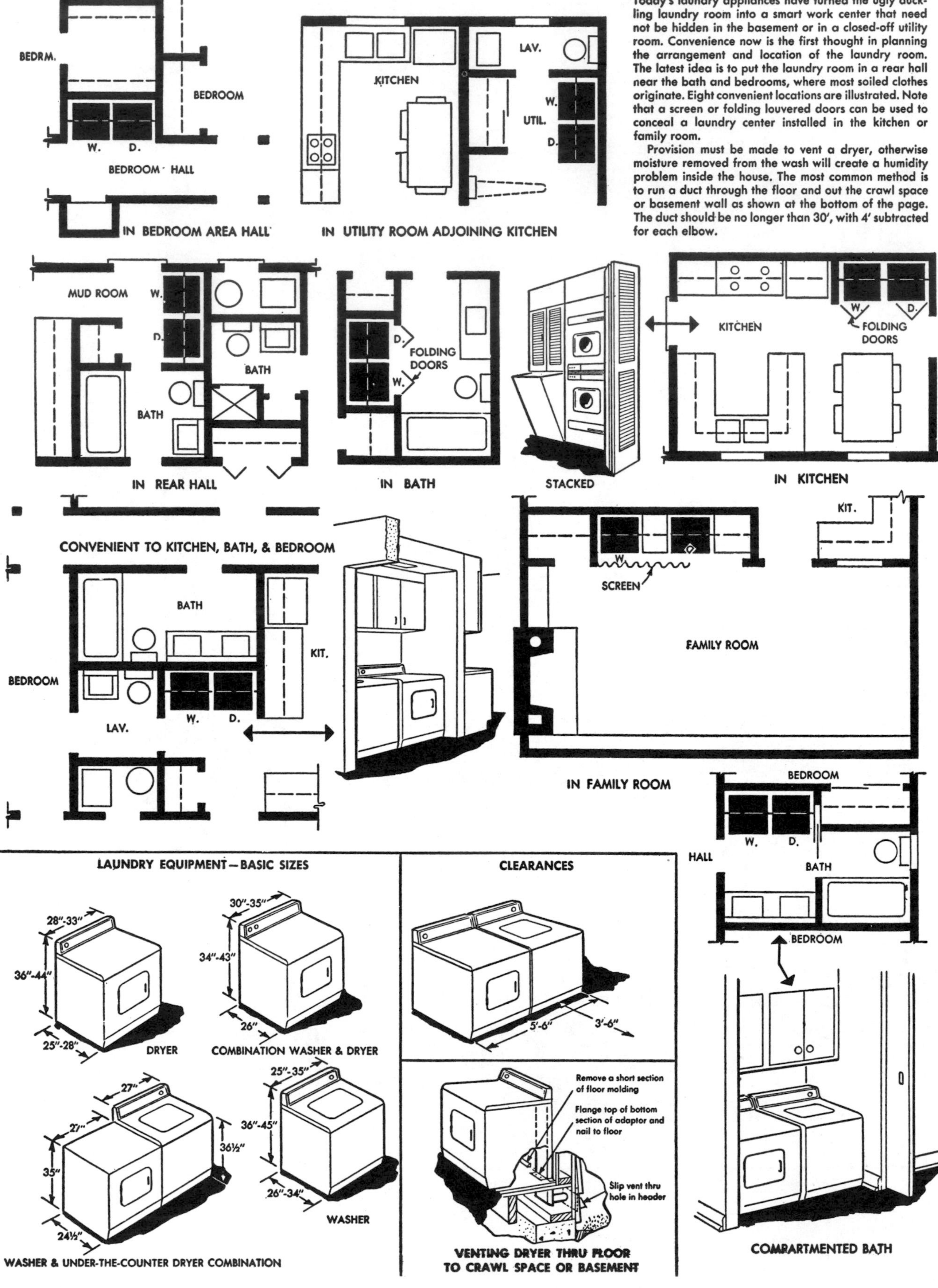

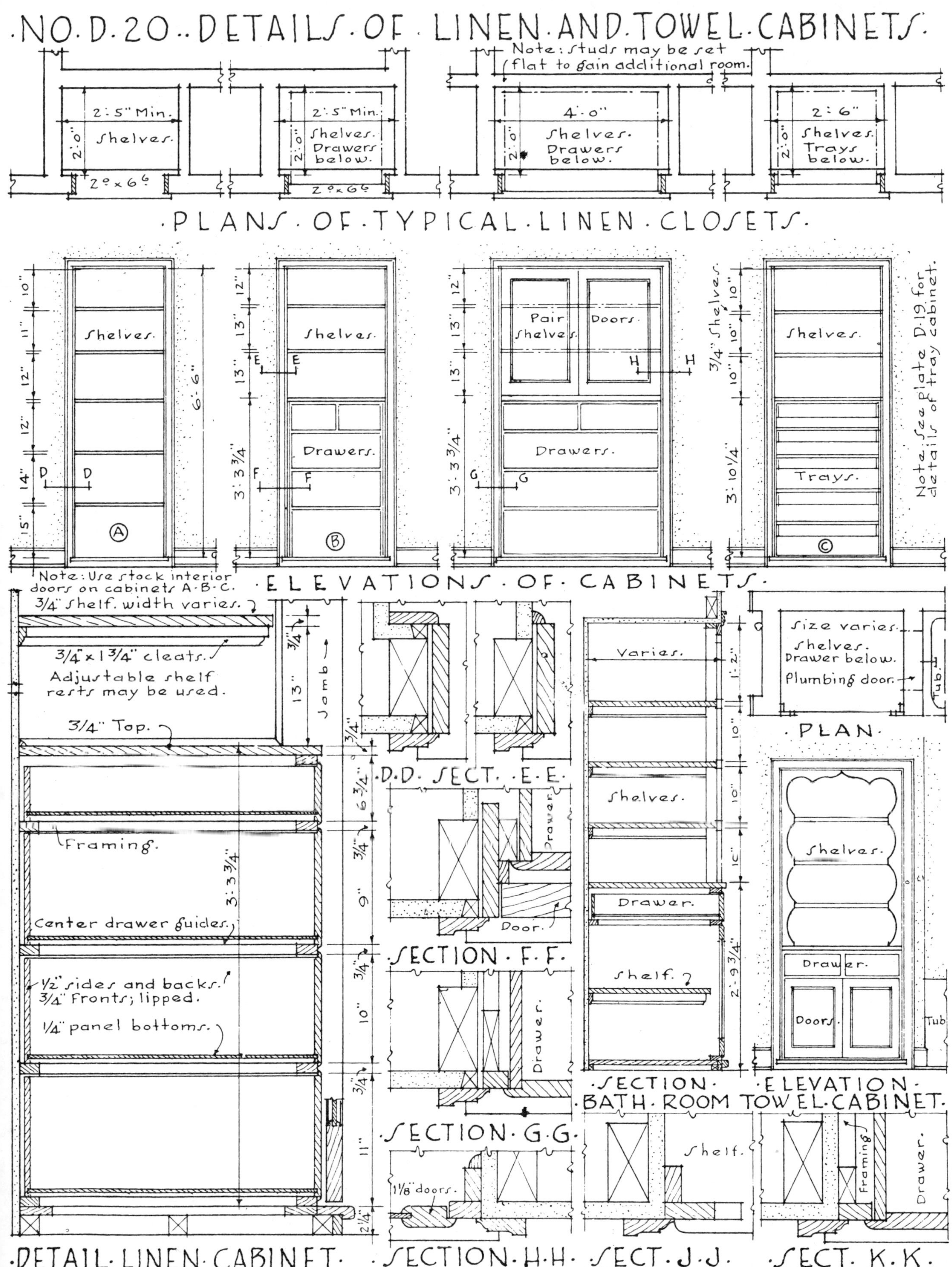

TIME AND MONEY SAVING DETAILS

No D.19 . DETAILS OF HANGING CLOSETS; WARDROBE TRAYS.

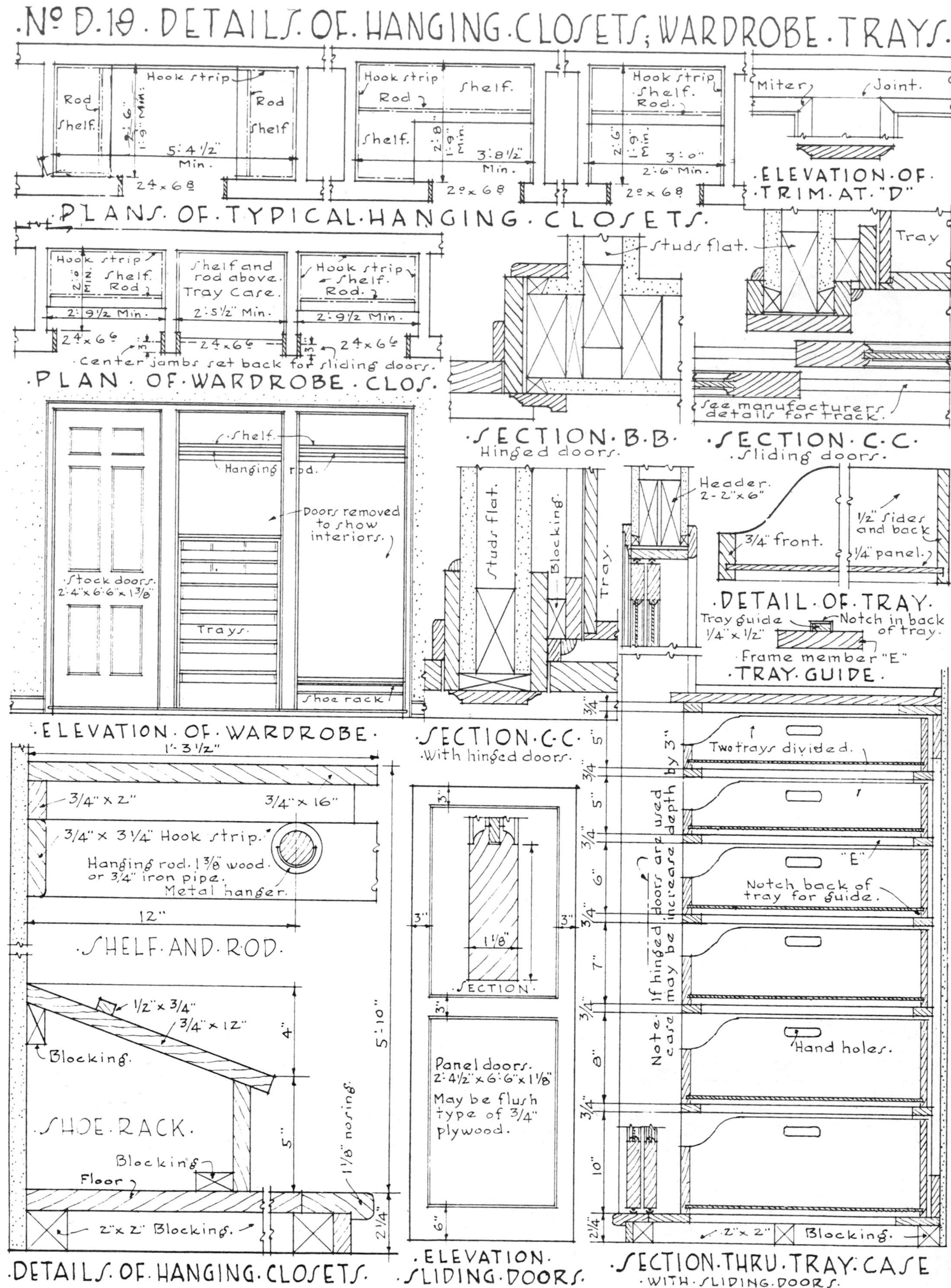

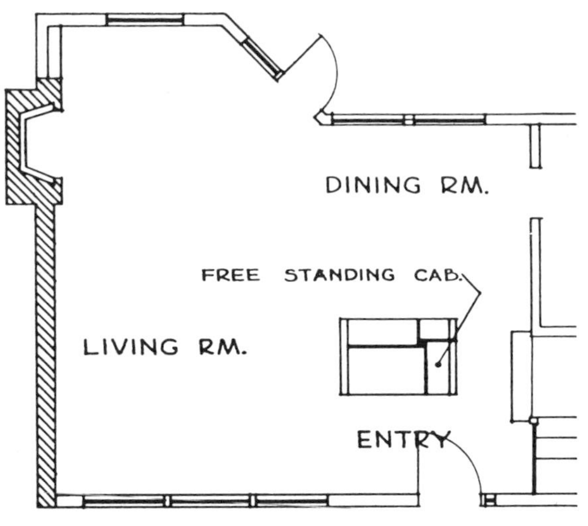

Plan . . .

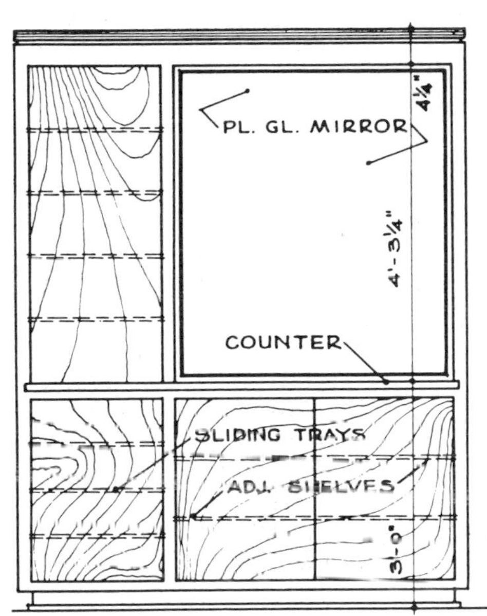

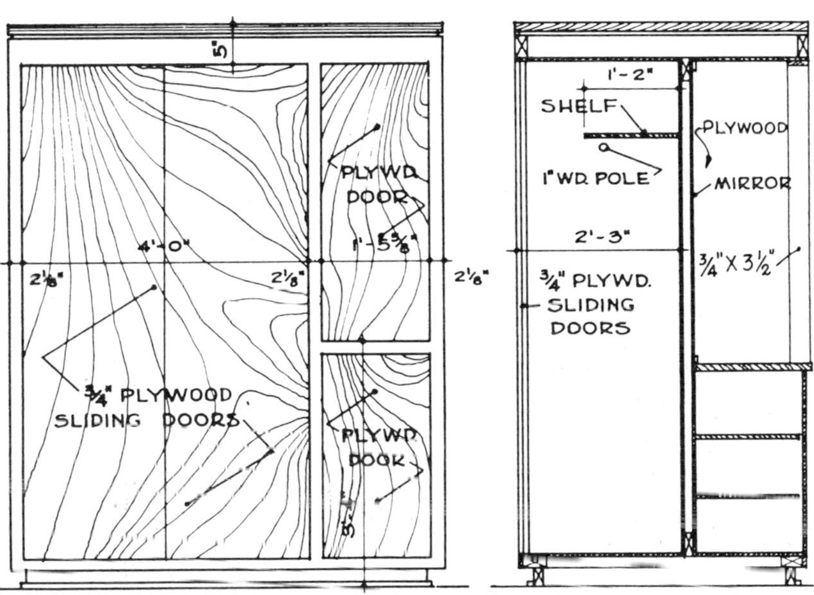

Elevations . . .

Section . . .

1'-9⅞" 4'-2⅛"

CLOSET

15" HANGING ROD SHELF

3'-11"

MIRROR

BUFFET

6'-0"

Plan . . .

NO. D 104

FREE STANDING CABINET

No. D.63. . DISTINCTIVE. DETAILS: CLOSETS. &. DRAWERS.

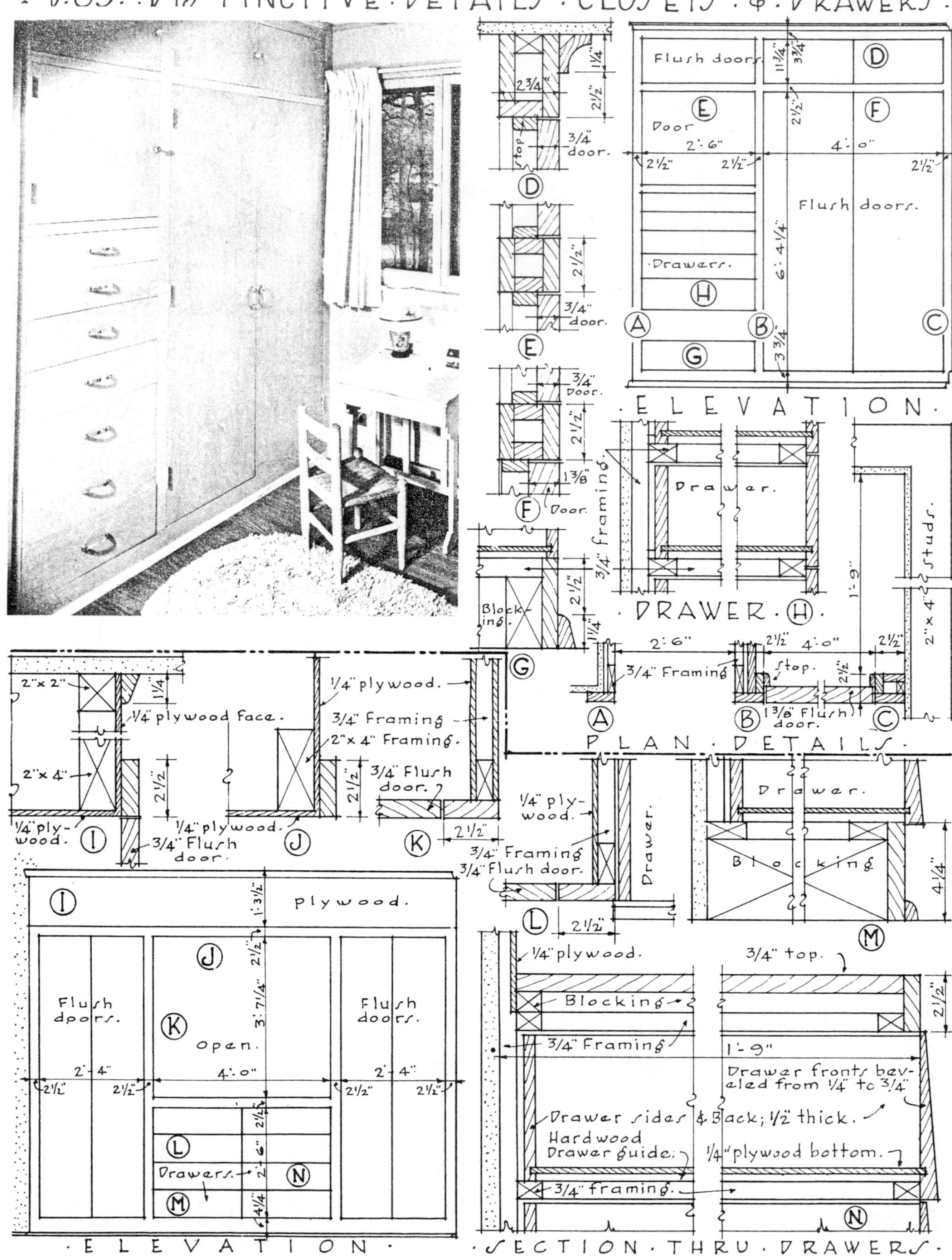

NO. D-64. MULTIPLE USE SLIDING TABLE.

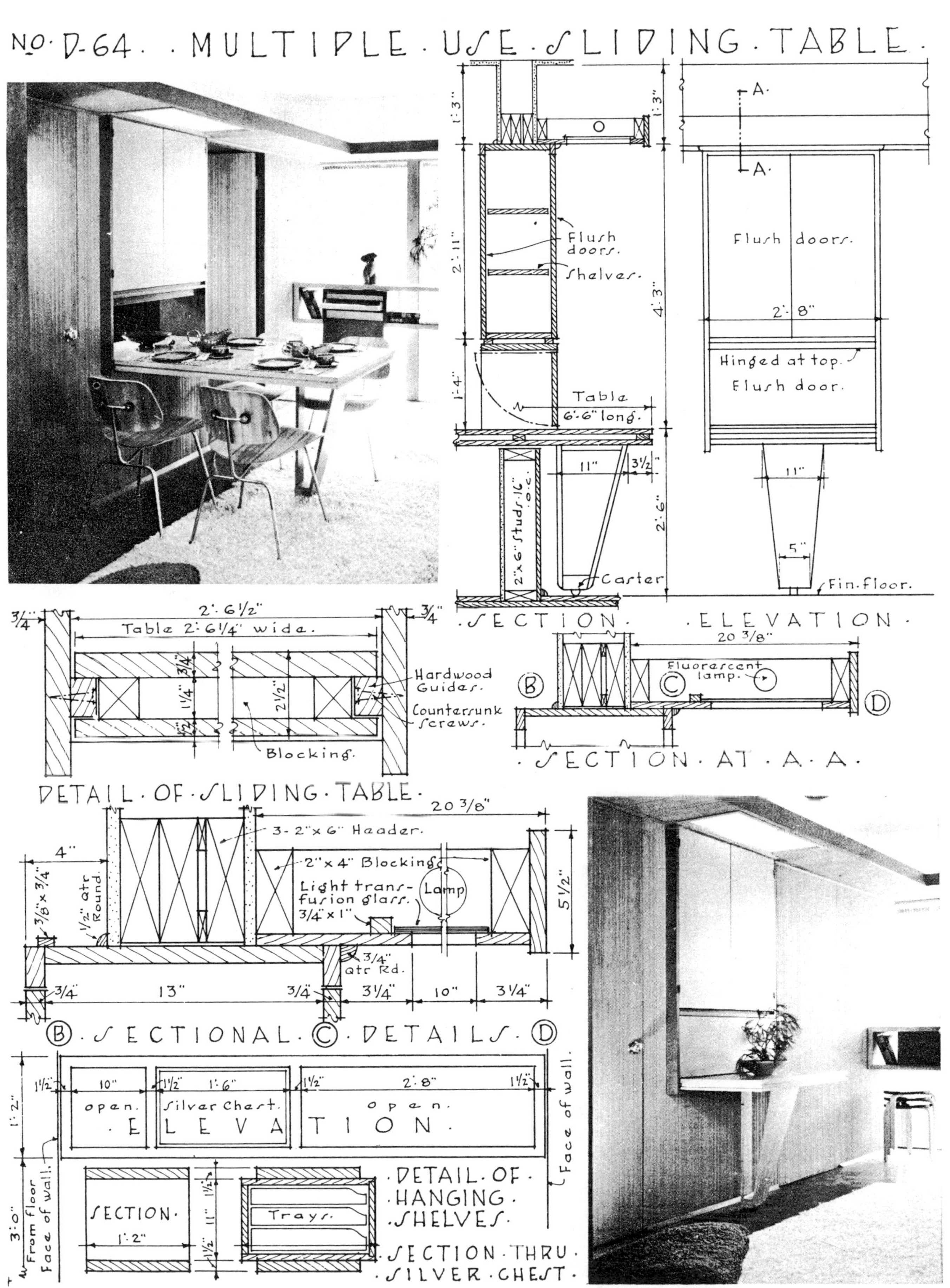

NO. D-105 FIREPLACE . . .

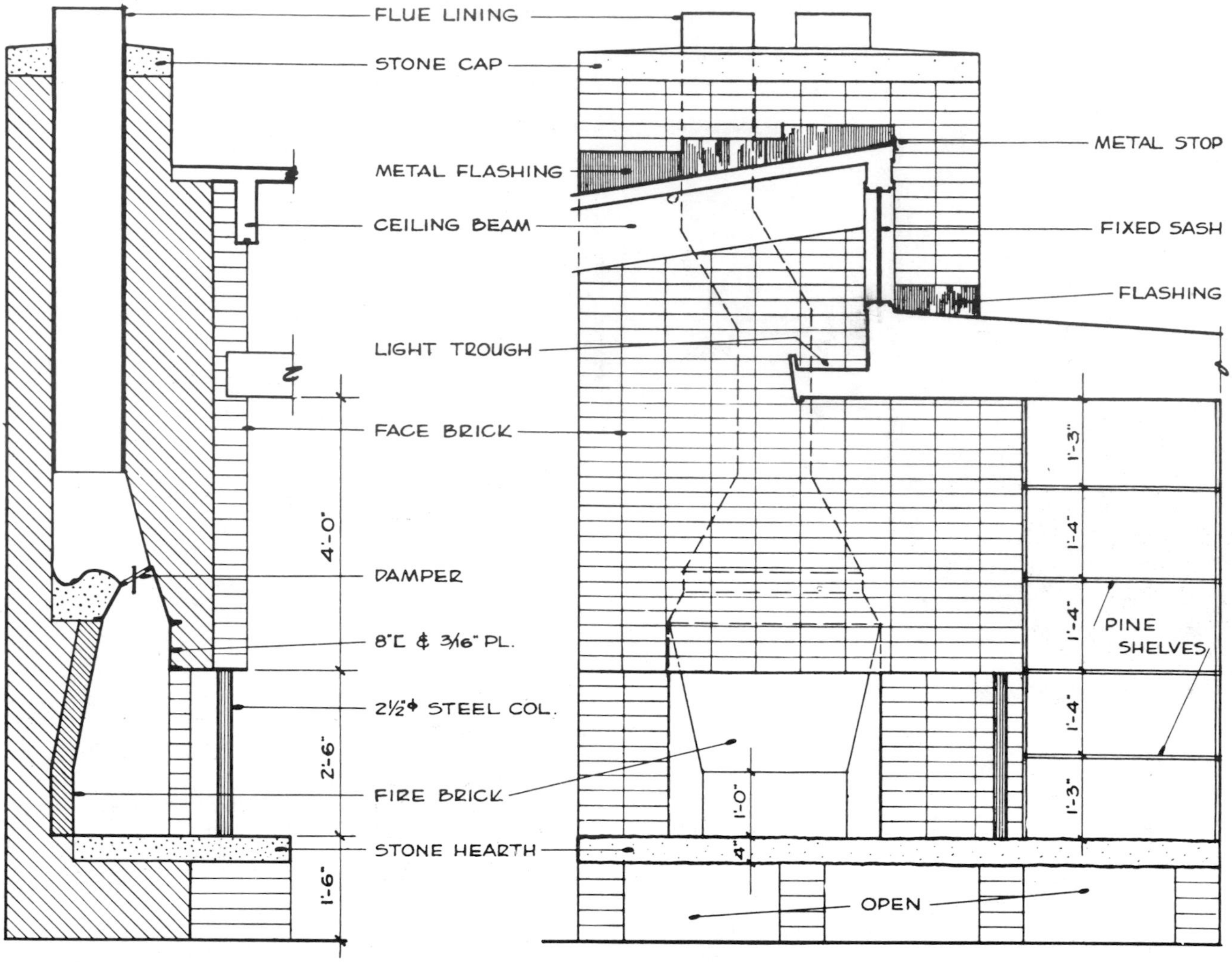

FLUE LINING
STONE CAP
METAL FLASHING
CEILING BEAM
LIGHT TROUGH
FACE BRICK
DAMPER
8"⊏ & 3/16" PL.
2½"φ STEEL COL.
FIRE BRICK
STONE HEARTH
4'-0"
2'-6"
1'-6"
METAL STOP
FIXED SASH
FLASHING
1'-3"
1'-4"
1'-4"
1'-4"
1'-3"
PINE SHELVES
1'-0"
4"
OPEN

Section . . .

Elevation . . .

12"
1'-9"
SHELVES
BR. LINE ABOVE
COL.
6'-0"
1'-4"
3'-2"
2'-0"
6'-6"
2'-11"

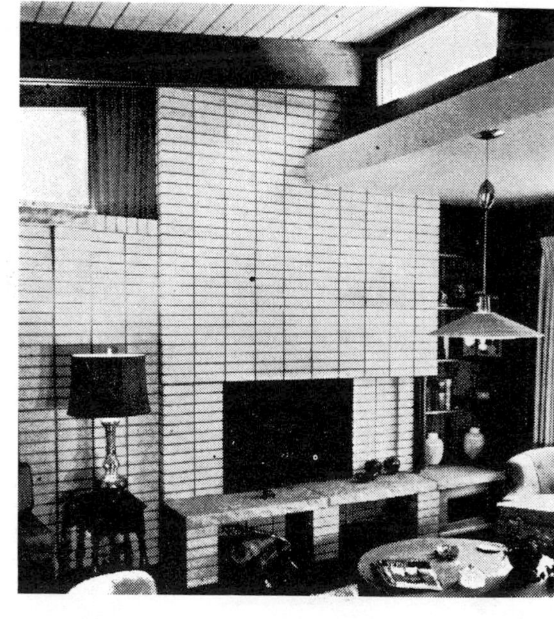

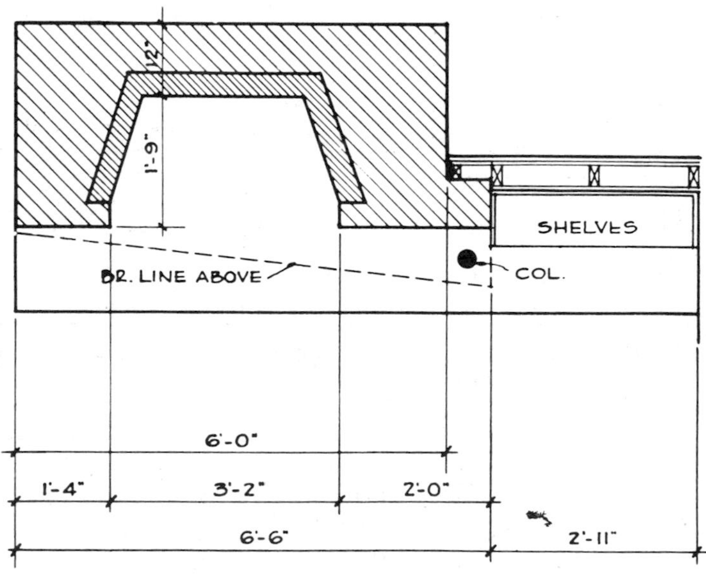

Plan . . .

NO. D-113 INSIDE-OUTSIDE FIREPLACE

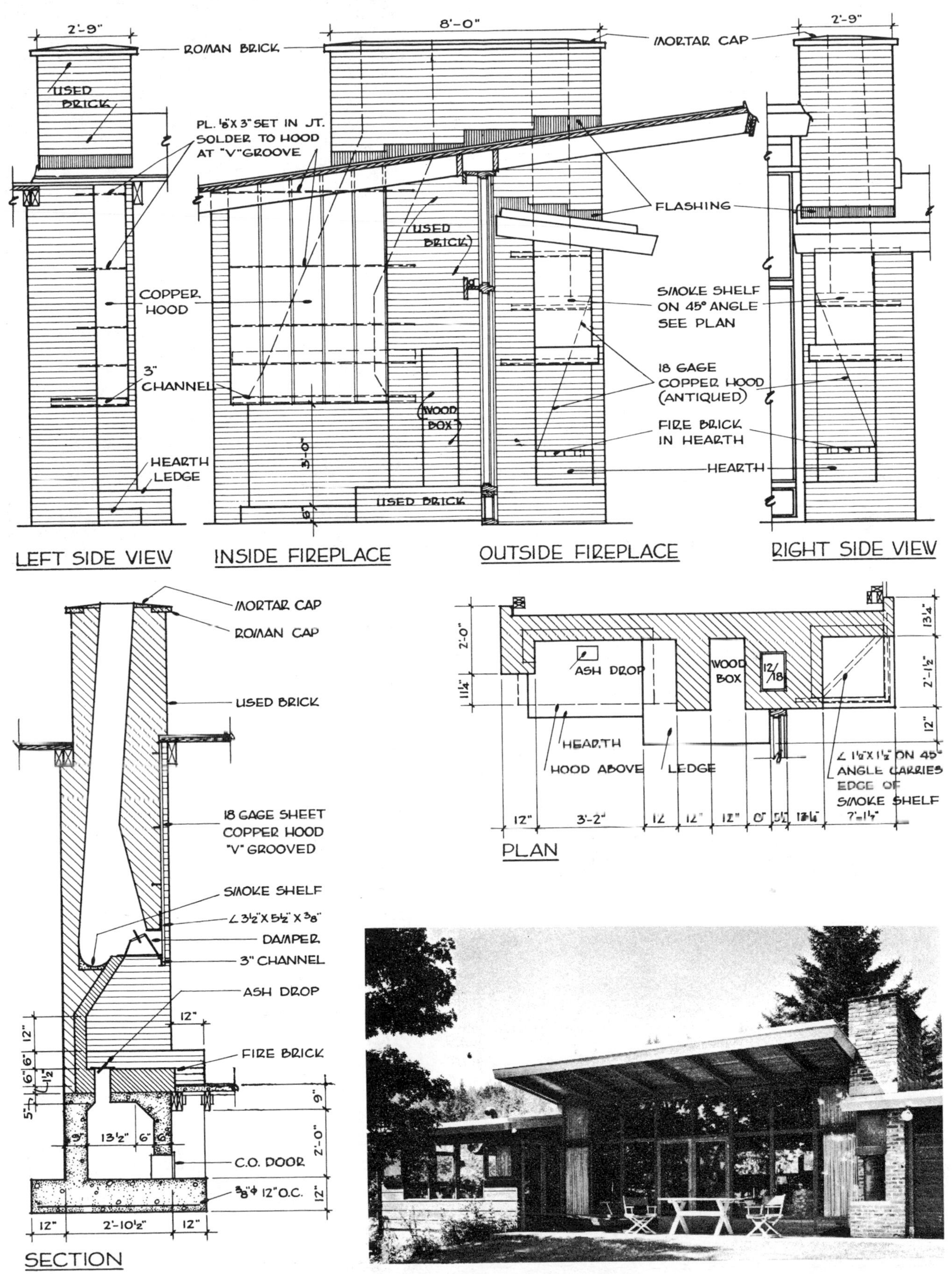

NO. D-107 FIREPLACE

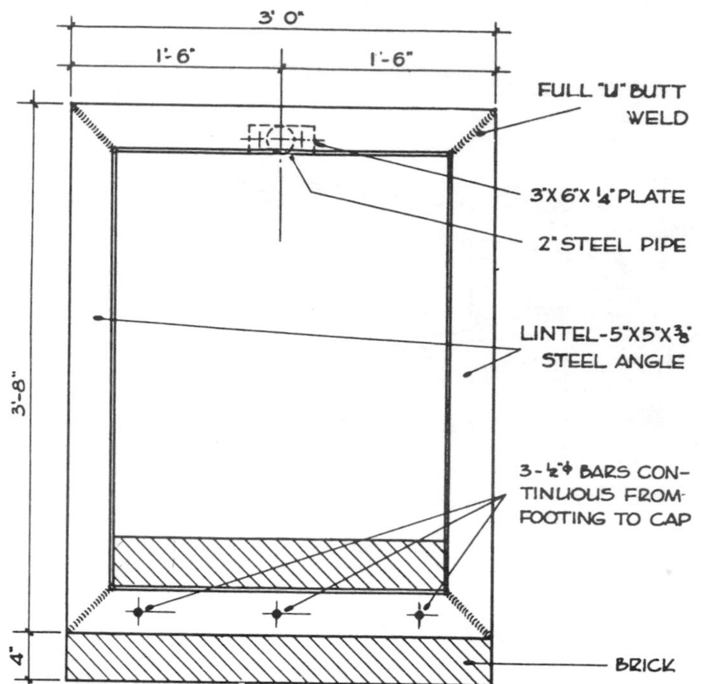

Plan (AT LINTEL) . . .

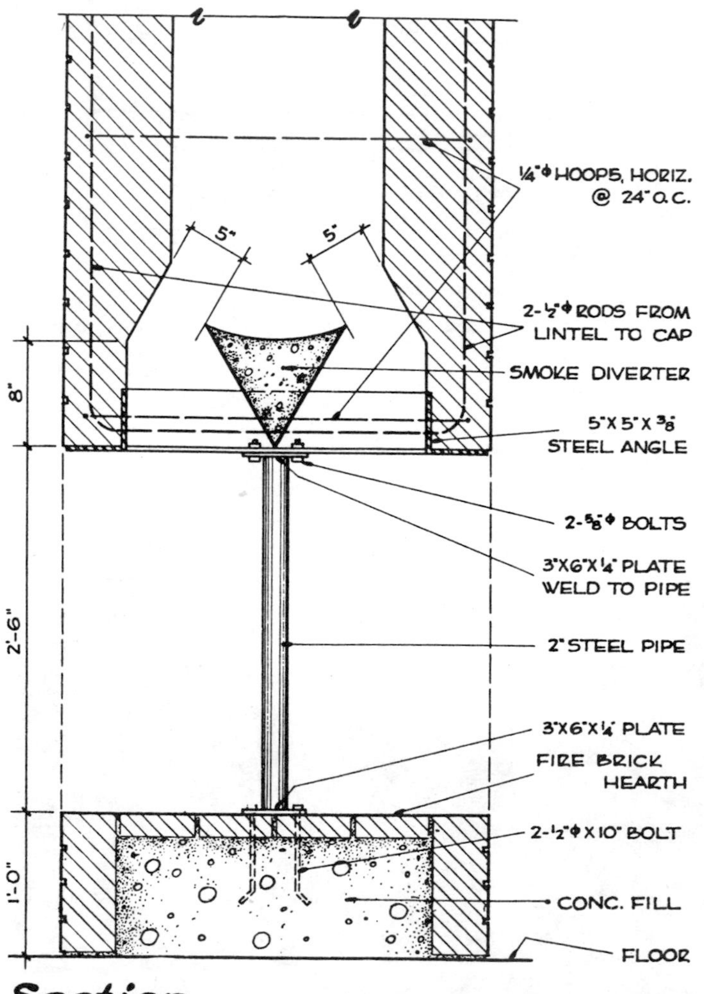

Section . . .

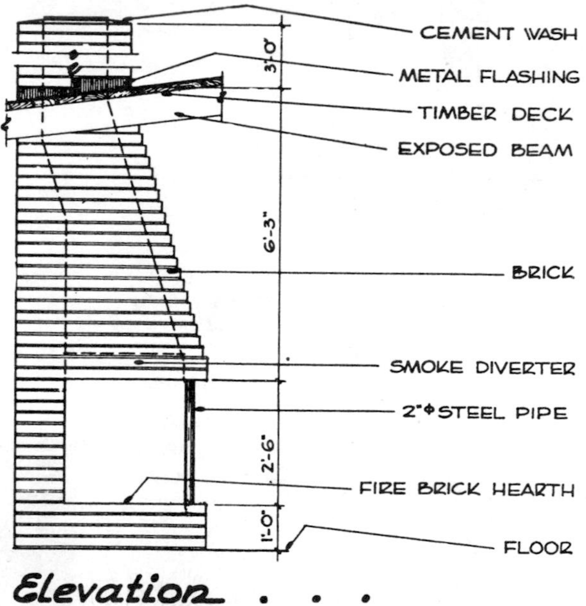

Elevation . . .

Plan (AT HEARTH) . . .

A Cast Cylindrical Foundation Supports an Unusual Fireplace

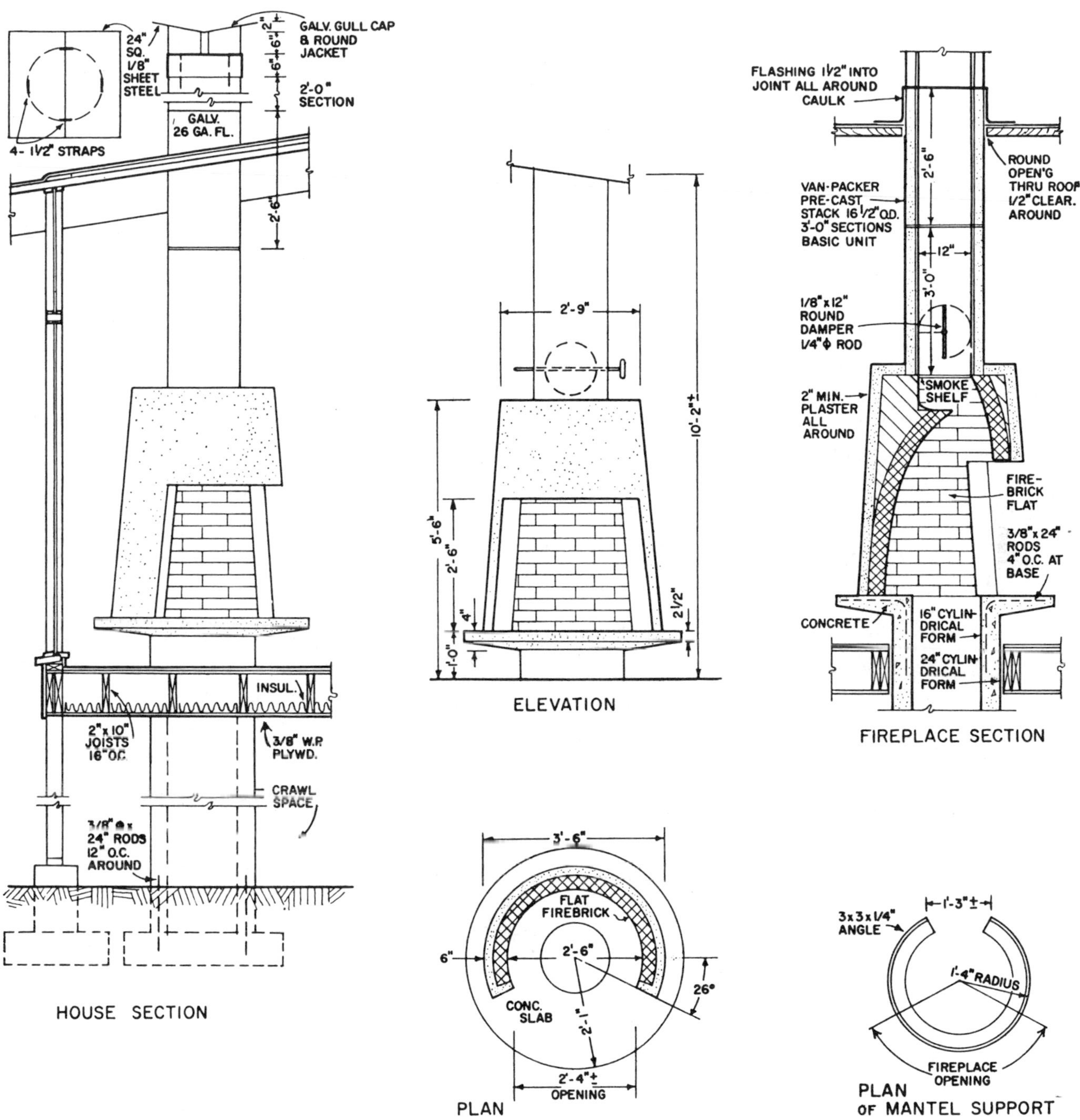

With the availability of packaged chimneys and fireplaces, the design, construction and placement of today's fireplace are limited only to the imagination of the designer. Earlier this year in the Better Detail of the Month we showed how a Chicago architect, Marvin Fitch, customized the outward appearance of a pre-fab fireplace. This month we show a very unusual fireplace, in both appearance and construction, which utilizes a pre-cast stack. Designed by Seattle Architect Zema Bumgardner, the fireplace features a circular reinforced concrete foundation. Two concentric Sonotube forms were used in construction. Support for the fireplace extends up through the insulated crawl space.

NO. 13. C. CONSTRUCTION DETAILS. A. SMALL. COMMERCIAL BUILDING

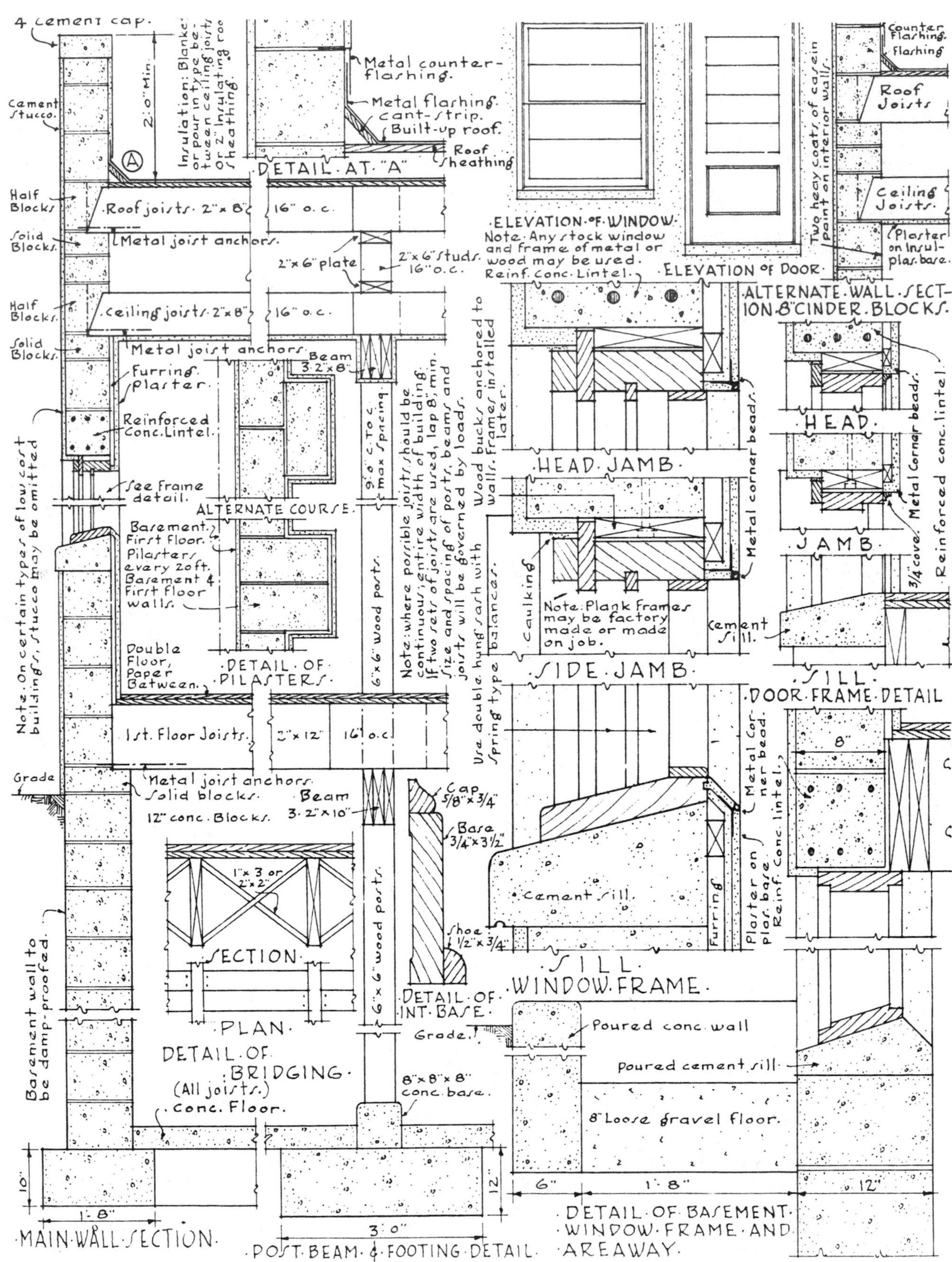

NO. G-47 OFFICE ENTRANCE DETAILS . . .

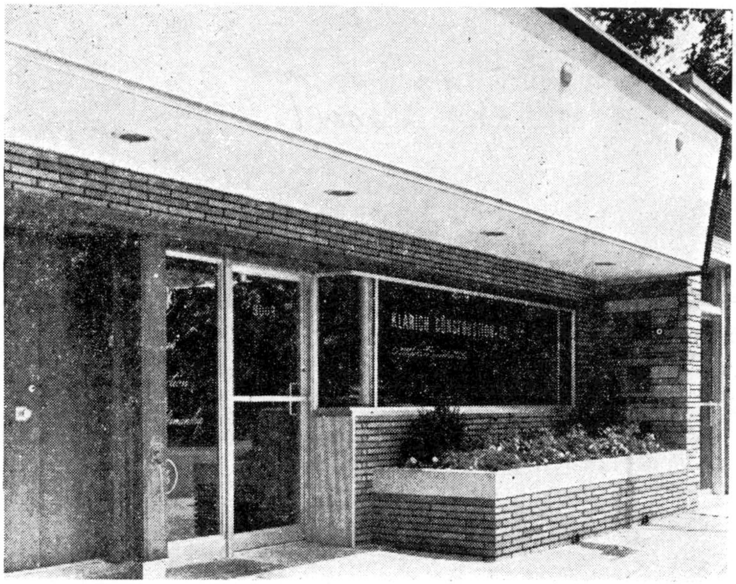

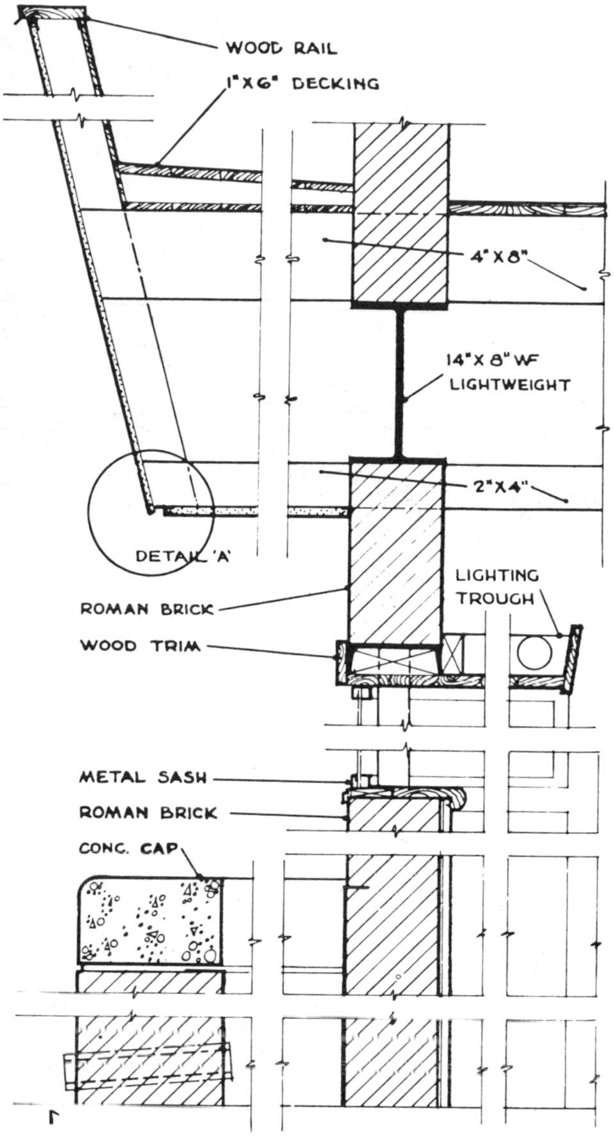

WOOD RAIL
1"X6" DECKING
4"X8"
14"X8" WF LIGHTWEIGHT
2"X4"
DETAIL 'A'
LIGHTING TROUGH
ROMAN BRICK
WOOD TRIM
METAL SASH
ROMAN BRICK
CONC. CAP

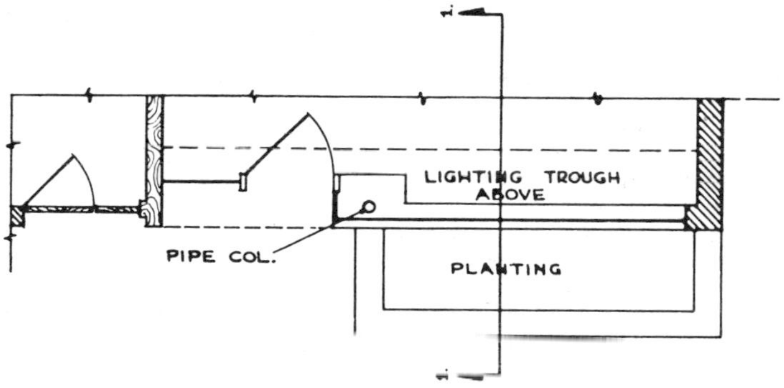

Section 1-1 . . .

LIGHTING TROUGH ABOVE
PIPE COL.
PLANTING

Plan . . .

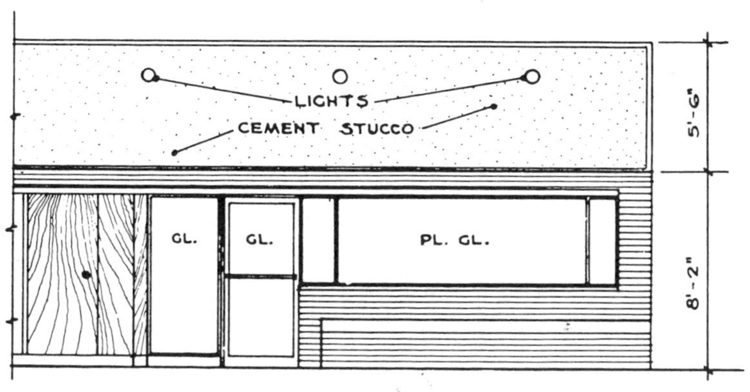

Elevation . . .

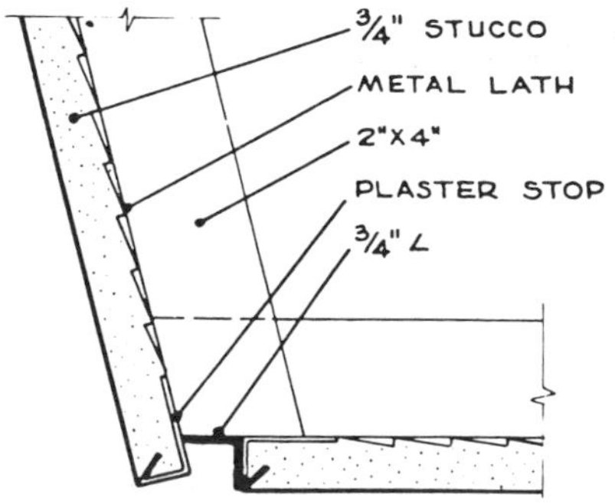

3/4" STUCCO
METAL LATH
2"X4"
PLASTER STOP
3/4" ∠

Detail A . . .

NO. 118 COMMERCIAL STAIR....

DOOR
1" STD. PIPE RAIL
SEE STAIR DETAIL
VAULT DOOR

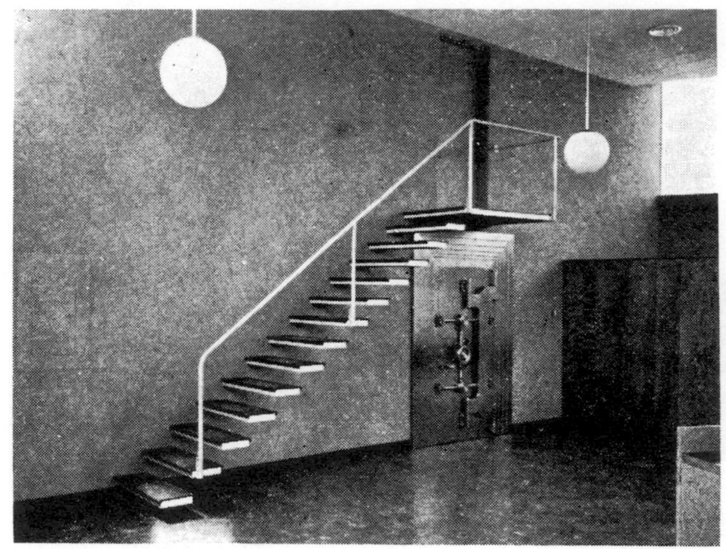

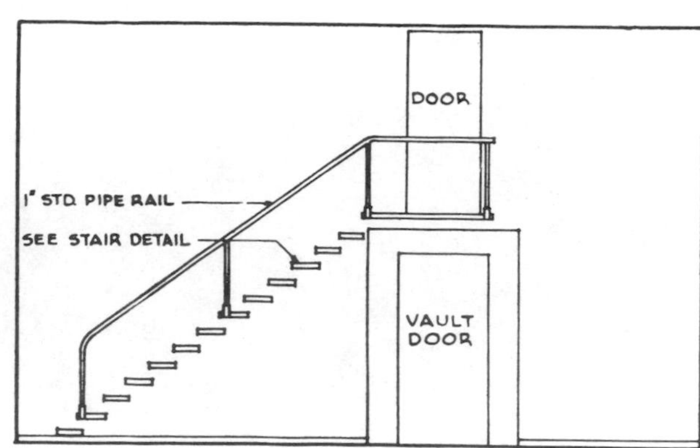

Elevation....

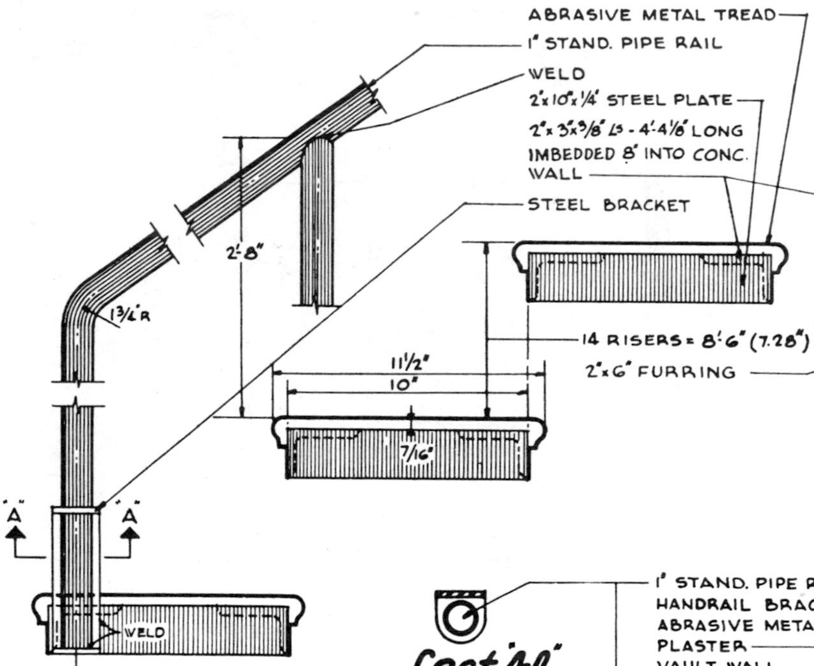

ABRASIVE METAL TREAD
1" STAND. PIPE RAIL
WELD
2"x10"x1/4" STEEL PLATE
2"x 3"x3/8" Ls - 4'-4 1/8" LONG
IMBEDDED 8" INTO CONC.
WALL
STEEL BRACKET
2'-8"
1 3/4" R
11 1/2"
10"
7/16"
14 RISERS = 8'-6" (7.28")
2"x6" FURRING
"A"
"A"
WELD

End Elevation....

Sect."A-A"....

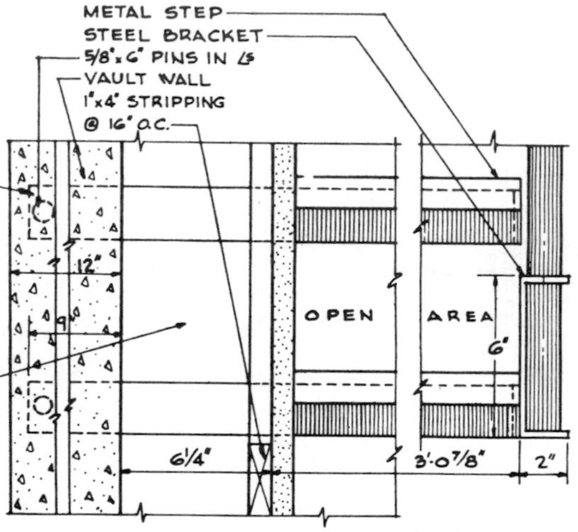

METAL STEP
STEEL BRACKET
5/8"x 6" PINS IN Ls
VAULT WALL
1"x4" STRIPPING
@ 16" O.C.
12"
9"
OPEN
AREA
6"
6 1/4"
3'-0 7/8"
2"

Elevation & Section....

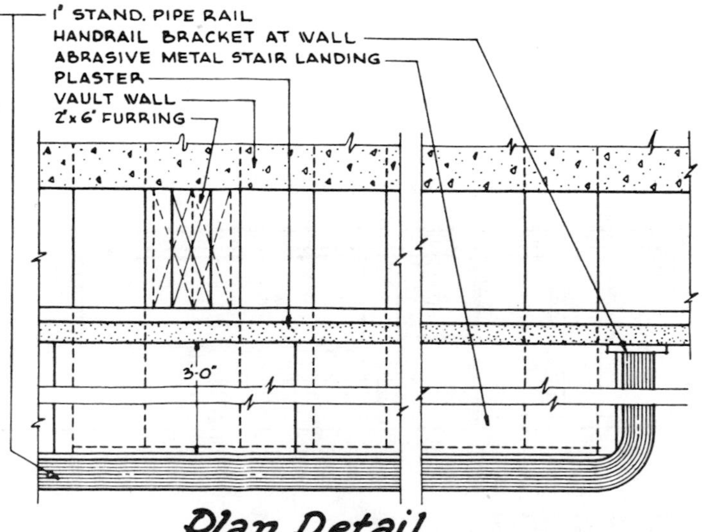

1" STAND. PIPE RAIL
HANDRAIL BRACKET AT WALL
ABRASIVE METAL STAIR LANDING
PLASTER
VAULT WALL
2"x 6" FURRING
3'-0"

Plan Detail....

NO. D-77 GARDEN BOXES . . .

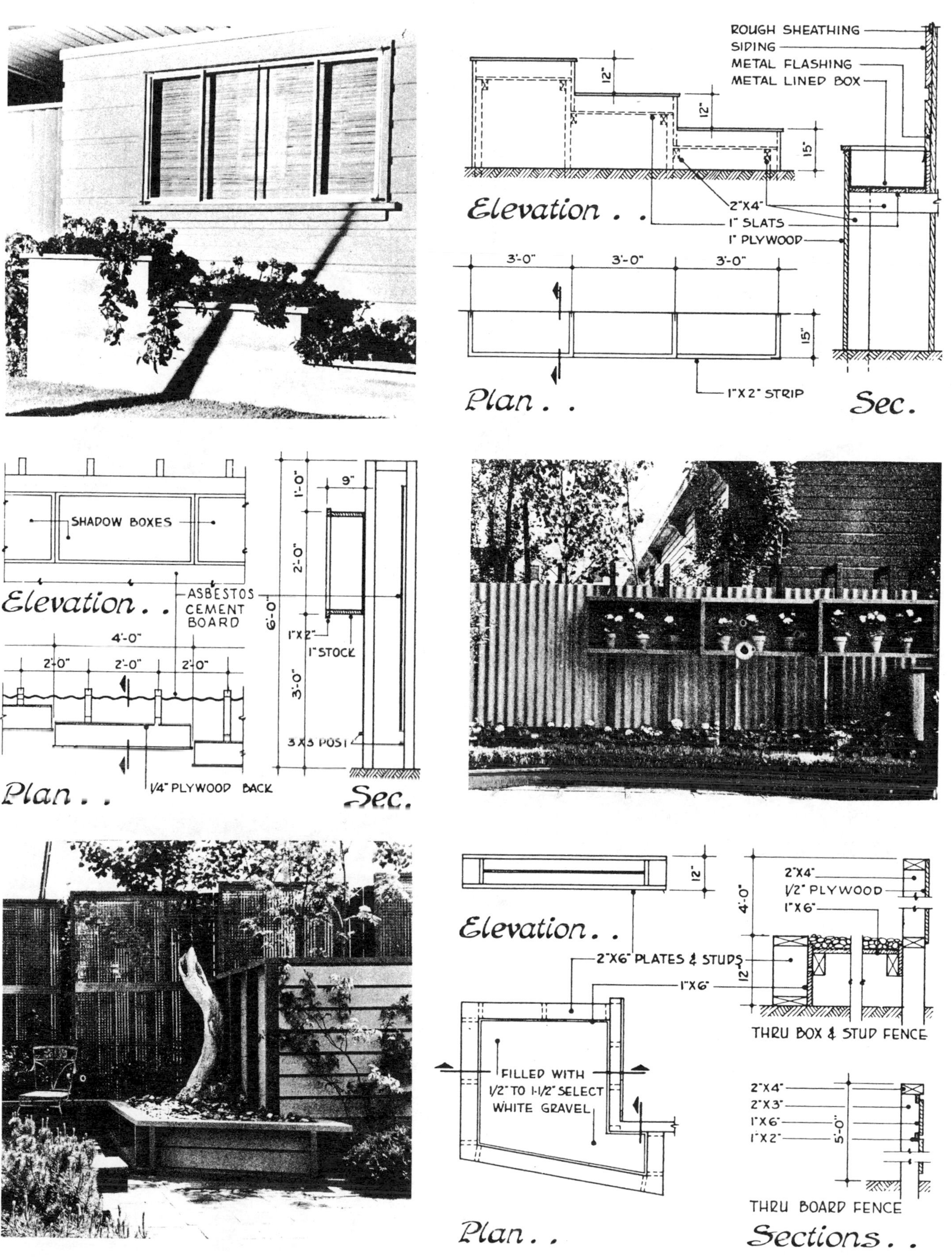

TIME AND MONEY SAVING DETAILS

NO. D-76 GARDEN FENCES . . .

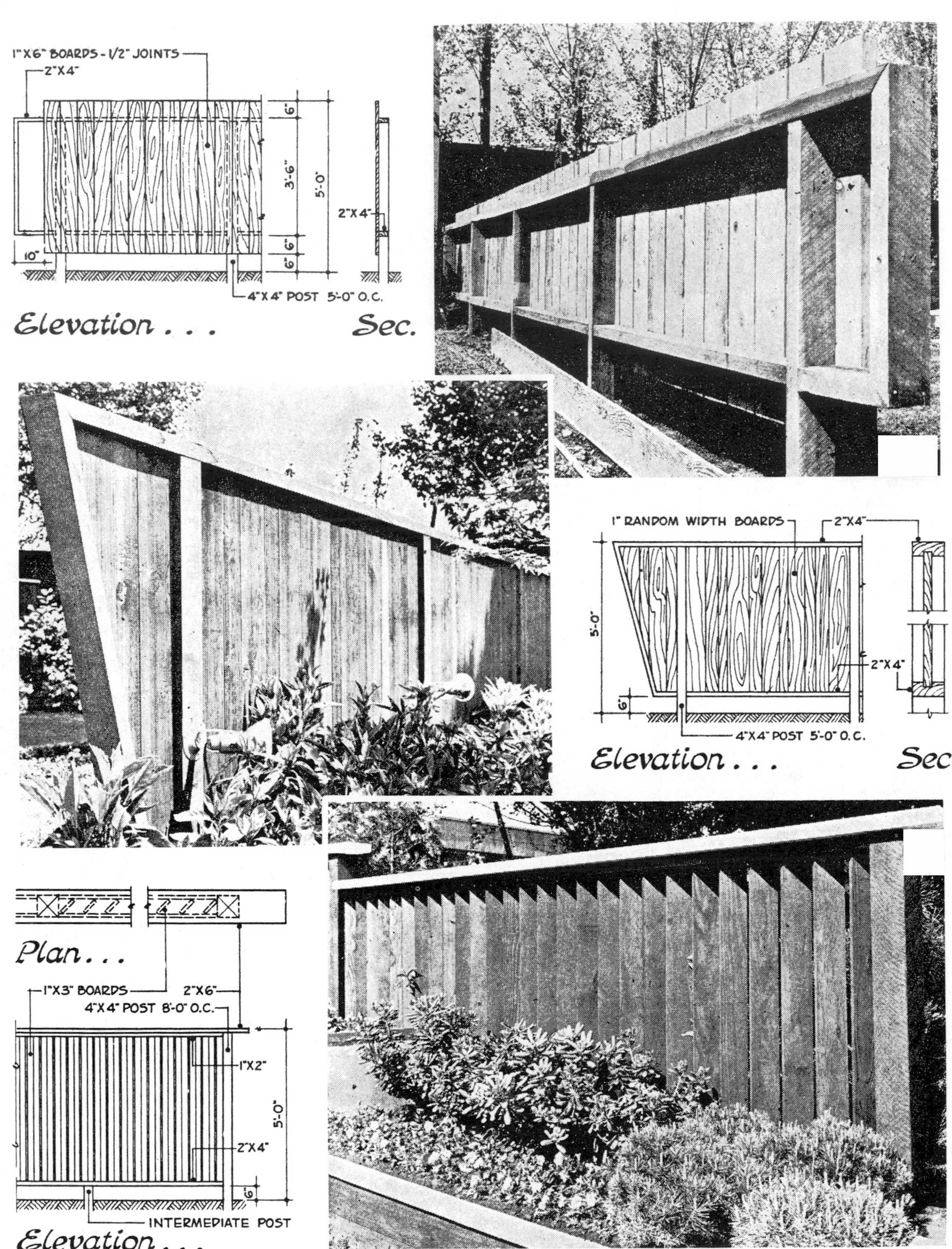

GLOSSARY

Aggregate – Crushed stone added to sand and cement of certain types of concrete.

Anchor Nails – Nails set at right angles to perforated squares of metal which can be glued to concrete with mastic or other special glue.

Anodized – A treatment of aluminum surfaces which makes it weatherproof and comes in a variety of colors including gold and silver.

Apron – Molding under interior window sill.

Ashlar Stone – Precut rectangular stone laid with all joints offset.

Auger, Earth – A power driven auger about 6" in diameter used to bore holes in earth for fence posts.

Awning Window – See Window, Awning.

Back Fill – Earth and other material previously removed to permit construction of a foundation which is then returned and pressed against the exterior wall of the foundation.

Balloon Framing – Type of framing in which studs in outer walls run continuously from foundation sill to roof.

Baluster – Vertical support on which a stair railing rests.

Batten – Narrow strip of wood used to cover a joint in vertical siding.

Batter Boards – Sticks and cords used as guides by masons in making footings and foundations.

Bay Window – A framed structure housing window, ceiling, and floor in a semi-octagonal plan.

Beveled Siding – See Siding.

Bifold Doors – Doors with 2 folding panels hinged to each other and to one jamb of the doorway. The door may have an upper track or an upper and lower track.

Bi-passing Doors – Sliding doors that hang from an upper track on which they roll back and forth. There are 2 tracks at the top and on the floor. Doors overlap by about 1" and can pass each other.

Birds Mouths and Tails – Shape of cuts at the ends of rafters under shed roofs.

Blast Furnace Slag – A waste product in the process of making steel which is crushed and used for fill or aggregate.

Blind Stop – Outside wood strip that holds in the outer sash of double hung windows.

Board and Batten – Vertical siding in the form of boards joined side by side with vertical battens covering the joints.

Bow Window – Same as a bay window but on a semicircular plan. (See Bay Window)

Bridging – The X-shaped double pieces of wood or steel which brace joists to prevent lateral movement.

Brown Plaster Coat – Coarse brown plaster that is applied over a rough scratch coat before the last coat of fine white plaster is applied.

Buck – The welded steel sides and top that frames the opening for a steel or wooden door.

Building Paper – Black, permeable paper which is water-repellent but not waterproof. Used between finish flooring and subflooring and between exterior siding and sheathing.

Casings – The moldings used around the top and sides of windows and doors on the inside and exterior of the house.

Cavity Wall – A double wall with a hollow space between outer wall sections. Mostly these are brick walls.

Check Rails – Also called meeting rails. The bottom rail of the upper sash of a double hung window and the top rail of the lower sash. They are usually notched or weather-stripped so that they can interlock and keep out drafts.

Chipboard – A building material made of wood chips and plastic available in 4′ × 8′ sheets and in thicknesses from 3/8" to 3/4" or more.

Clapboards – Boards used as siding. Applied horizontally, they have a 1" overlap and are 1/2" thick and 12" wide.

Clerestory – Narrow, horizontal windows connected end to end in ribbon form. Often used in bedrooms to assure privacy yet provide ventilation.

Collar Beams – Light horizontal beams joining rafters on opposite sides of a roof to stiffen rafter and roof structure.

Compacter – A tool driven by gasoline, compressed air, or electricity used to compact earth or aggregates.

Composition Board – Hardboard sheet material.

Corbel – Projection from the face of a wall that supports a bay, bow, or oriel window.

Cornice – Open or closed roof overhang.

Cripple Studs – See Jack Studs.

Cross Banding – The veneer layer immediately below and at right angles to the outermost layer of veneer of plywood.

Curtain Walls – The metal or masonry walls that form the exterior of a steel structure.

Cut Nails – Flat, wedge-shaped nails with square tips and slight heads that are stamped out of a sheet of hardened steel. Used in strip hardwood flooring and hard enough to penetrate concrete.

Dado – Groove cut across a board to receive the end of a board or shelf.

Dentil – A type of molding.

Door, Dutch – Exterior door divided horizontally along the middle so that either half can be moved independently of the other. Usually a back door.

Dormer – Outer section of attic room that projects through the slant of the roof.

Double Acting Hinges – Hinges that permit cafe doors to turn in both directions.

Double Glazing – Windows with double panes of glass with sealed edges so no air can penetrate between windows.

Double Hung Windows – Type of window with 2 sashes which slide vertically and are counter balanced by springs.

Dowels – Round pegs in diameters from 1/8" to 3/4" to join pieces of wood with matching holes.

Drip Cap – Narrow, projecting horizontal member usually above a window or door to prevent dripping from rain.

Drop Siding – Decorative wood siding in the form of milled boards which may have T&G edges or be rabbeted at the top or bottom.

Drywall – Also called Sheetrock, gypsum board, and wallboard. It is made in the form of 4′ × 8′ sheets of gypsum plaster with a special paper on both sides and is nailed to studs to form wall surfaces.

D-W-V – Term used in plumbing to identify drain, waste, and vent pipes.

Eaves – Rafter end edges.

Efflorescence – Natural salts which show up on brickwork exposed to moisture in the form of hard, white crystals.

Escutcheon – A metal plate of brass or aluminum, often decorative, around the knob of a door.

Espagnolette – Bolts with hooks at the ends that run the full vertical length of a door and engage strikes at the top and bottom.

Extrusion, Aluminum – Lengths of aluminum, often with an intricate cross section, made by softening a round billet of this metal and forcing it through a die by a giant hydraulic press.

Fan Lights – Wedge-shaped window panes arranged in a semicircle, usually above a door.

Fascia Board – A board nailed to the ends of rafters at the lower edge of the roof.

Felt, Asphalt – A heavy, asphalt impregnated paper which is water repellent.

Fenestration – A general term for windows.

Fire stop – Wood block between studs to prevent up draft and minimize progress of fire.

Flashing – Copper, zinc, aluminum, or plastic sheet used at exterior joints to keep out moisture.

Flitch Plate – Steel plate bolted between beams for added strength and stiffness in forming a joist.

Float – Flat tool with a wood surface used for rough slip-proof finish of wet concrete.

Floating Jamb – Spring loaded metal side of a window frame which can be pushed back to permit sash to be withdrawn.

Flush Door – Door which is smooth and continuously flat on both sides.

Fluting – Shallow, round bottomed, decorative vertical grooves. Fluted pillars are common on Greek temples and classical architecture.

Footings – Concrete shapes or walls that support foundation walls.

Frieze – Highest board directly above siding and immediately below roof deck.

Gable – Angled wall section at the end of a peaked roof.

Gable Roof – Pitched roof with equal angles from ridge board.

Galvanized – Coated with zinc.

Gambrel Roof – A roof with 2 planes on each side of the ridge board. Sometimes called a Dutch roof.

Glass, Insulated – Double sheets of glass with sealed edges and evacuated air space between them.

Grade – The degree of earth surface slope around a house.

Grout – A thin cement that fills in the spaces between tiles and other masonry parts set close together.

Gusset – Flat piece of metal or plywood nailed over a joint to reinforce it. Commonly used to strengthen joints of roof trusses.

Gypsum – A white mineral which forms the main ingredient of plaster and gypsum board.

Hardboard – Exploded wood fibers formed into brown sheets under great heat and pressure and held together by the natural adhesive, lignin, which is retained in the wood. The product is grainless, long lasting and widely used in building for paneling.

Header – Any beam or structural member at the top of an opening above windows and doors or other openings.

HIP Roof – Four-slant roof.

Hopper Window – Sash hinged at bottom that slants inward and directs air toward ceiling. Usually placed under large, fixed windows. Also used as basement window.

Hydrated Lime – Also called slaked lime. A dry white powder treated with water to form calcium hydroxide used in making plaster.

Hydronic – Heating or cooling system that employs hot or cold fluids moving through closed circuit pipe systems. Hot water heating and air conditioning are examples.

Insert – The 2 movable glass panes and the half screen of an aluminum combination storm sash are called inserts.

Jack Studs – Short 2×4 studs above and below a window opening in wall frame.

Jalousie – Aluminum-framed window with glass louvers which can be turned up or down by geared operating mechanism.

Jamb – Sides of a window or door opening.

Joists – Beams that support floors and ceilings.

Joist Hangers – Metal hangers which support the ends of joists and keep them flush with walls to which they are attached.

Keyway – Rectangular slot or notch in a footing into which another wall section can fit.

Kiln Drying – Drying of wood in an oven-like apparatus to reduce the amount of water in it to a desirable level. A freshly cut stud at the mill is 2″ × 4″ but is reduced to 1½″ × 3½″ after kiln drying.

Knee Wall – A short wall or partition often no more than 4 feet high.

Kraft Paper – A tough brown paper often impregnated with asphalt which makes it black and water repellent.

Lally Column or Post – A steel pipe with a flat square section at each end which is used to support a steel I-beam or joist in a cellar. Some have a threaded section at the top which permits them to be used to gradually raise sagging joists.

Laminated Beams – Large beams made up of many layers of boards formed under heat and pressure into a single unit.

Lath – Wood or gypsum strips designed to be nailed to a wall to hold plaster. Gypsum lath has tiny holes to increase holding power of the lath. Metal lath is expanded metal, a mesh made by stretching slotted sheet metal.

Ledger – A strip or section of wood fastened to a vertical surface on which the ends of joists, rafters, or beams may rest.

Let-In – Wood structural members or braces that rest in notches so that they are flush with the studs they cross diagonally. Let-in braces are commonly used to strengthen the corners of stud walls.

Light – Individual window pane.

Lintel – Wood or metal member that bridges the space above a window or door opening to provide support for the wall structure above such open space.

Load Bearing Partition – An interior wall that is essential to the support of the structure above it.

Lock Rail – Center door rail that meets door stile on which lock is installed.

Lookouts – Short projecting beams that overhang a wall.

Medium Density Overlaid Plywood – Plywood overlaid with medium density phenolic-impregnated brown kraft paper to assure a continuously smooth surface for painting.

Meeting Rails – See Check Rails.

Mortise and Tenon – A joint in carpentry in which a rectangular tongue (tenon) fits into a similar shaped hole or mortise.

Mullions – Vertical wood or metal members that stand between and join a row of windows. Sometimes the sides of the mullions form the jambs of the windows.

Muntin – Wood or metal sash members that divide sash into separate glass panes.

Newel Post – A post at the start of a stairway.

Nosing – Rounded front edge of a stair tread.

Operator – Geared mechanism that opens and closes jalousies and casement windows.

Oriel Window – A 3-sided window projecting from the face of an upper wall supported by a bracket or corbel.

Panel Door – A door with raised panels.

Parapet – Low wall section that rises above level of flat roof.

Parging – Thin coat of mortar spread on a rough stone or brick wall to smooth or waterproof it.

Parting Bead – Wood strip between inner and outer sashes of double-hung window.

Pediment – Arched, triangular, or curved decorative shape above a door common in Colonial and Georgian architecture.

Perm – A rating for the amount of water vapor transmitted through a material per sq. ft., per hour, per inch of mercury vapor.

Pier – Wall section between windows or doors. Massive vertical support.

Pilaster – A square column built as an integral part of a wall foundation to support the ends of beams or joists.

Pitch – Rate of rise or slope of a roof. A rise of 6″ per foot is ¼ pitch.

Plate – Horizontal single or double beams on which vertical members rest. (See Sole and Sill plate).

Platform Framing – A type of construction in which subfloors extend to outside edges of building and provide platform upon which exterior walls and inside partitions can be erected.

Points – Triangular zinc pieces that can be forced into the sash of a wood window to hold the glass pane in place. Points are concealed by putty or glazing compound.

Polyvinyl Fluoride – An exceedingly tough, durable film which is cemented to the more expensive grades of aluminum siding.

Purlin – Horizontal roof member to which upper and lower rafters of gambrel roof are attached.

Rabbet – Step or L-shaped recess in the edge of a frame member to receive another member or panel edge.

Radiant Heating – A form of heating in which electric heating cables or copper pipes bearing hot water are buried in a concrete floor.

Rafters – Slanting or horizontal beams that support the roof deck of a house.

Rafter Plate – Horizontal member at top of wall on which notched rafters rest.

Rail – Horizontal members of a door or sash frame.

Raised Panel – A panel beveled on all four sides. Usually seen in doors.

Rake – Outermost rafters of a gable roof.

Ridge Beam – Central beam at the top of a pitched roof to which all rafters are fastened.

Riser – The vertical pieces that connect the treads of steps.

Roll Roofing – Asphalt impregnated felt supplied in rolls that weigh 55 pounds when spread over 100 square feet. Commonly used on flat roofs.

Roof Truss – Triangular wooden shape with internal bracing which takes the place of rafters and ceiling joists and permits non-load bearing partitions.

Run (of stairs) – The horizontal distance between the bottom riser and a perpendicular drop from the top riser.

Sash – Movable part of a window that contains the glass panes.

Scratch Coat – Preliminary brown plaster coat that forms base for all other plastering. It is usually criss-crossed with scratches so that following plaster coats will hold better.

Screws, Sheet Metal – A type of screw with sharp threads that go all the way up to the head and are hard enough to cut their own threads in preliminary holes in sheet metal, aluminum, etc.

Shake Shingles – Hand split red cedar shingles which have a rough surface and are split with the grain.

Sheathing – The wooden boards, plywood, or insulating sheets nailed to the exterior of the stud walls before the siding is applied.

Shed Roof – Flat, single roof with very limited slant.

Sheet Rock – The copyrighted named by the U.S. Gypsum Co. for its product otherwise known as plasterboard, wallboard and gypsum board. It is a sheet of plaster in 4′ × 8′ size with specially treated paper on both sides.

Shiplapped – A term often applied to overlapping wood shingles in which a rabbeted cut at the bottom of the shingle fits into a similar cut at the top of the shingle below it.

Siding – The exterior cover applied over sheathing. It may be overlapping wooden boards, wood or asbestos shingles, aluminum or vinyl siding in the shape of boards.

Sill – Bottom part of a window or doorway opening.

Sill Plate – Horizontal 2 × 4 framing member that rests on floor or foundation top to which vertical studs are nailed 16″ o.c.

Slaked Lime – See Hydrated lime.

Soffit – Hollow space above kitchen wall cabinets that do not reach to ceiling. Also hollow space above roof overhang.

Sole Plate – Single or double horizontal 2 × 4's nailed to the subflooring and to which, in turn, the studs are nailed. The horizontal 2 × 4 at the top of the studs is called a top or ceiling plate.

Spiral Spring Balance – A spiral spring which counterbalances the sashes in a double hung window.

Stile – Vertical parts of a door's frame.

Stool – Bottom part of a window frame that connects to interior window sill and slopes downward and outward.

Storm Window – An added exterior window which provides a dead air space between it and the regular house window. The dead air space serves as insulation.

Striated – Having the appearance of stripes.

Strike – Hardware on door jamb that receives the latch or bolt of the lock.

Stringers – Heavy 2″ × 10″ or 2″ × 12″ boards with right angle cutouts to receive risers and treads of stairs.

Stud – Vertical 2 × 4 timber used 16″ o.c. in framing of walls.

T & G – Boards or plywood sheets that have tongues and grooves in their edges to provide a close fit.

Toggle Bolt – Threaded bolt with hinged wings at the end that open out into a hollow space behind the wall.

Treads – The horizontal part or top of each step in a stairway.

Trimmer – A supporting stud. Where studs are doubled around a window or door opening in house framing, the inner stud is called a "trimmer."

Tuck Pointing – Repair of mortar lines around brick work by removing and replacing part of the mortar.

Volute – Curved stair rail end that rests on the newel post.

Wallboard – Gypsum Board – See Sheet Rock.

Weep Holes – Openings in masonry walls or at the bottom of combination storm windows to permit rain water to drain away.

Winders – Triangular treads used where a stairway makes a right angle turn.

Windows, Awning – Windows hinged at the top which swing outward and may or may not be operated by a rotary screw mechanism.

Windows, Casement – Windows hinged on a vertical axis that swing outward.

Windows, Clerestory – Fixed, narrow panes usually high up on a wall.

Windows, Double Hung – A sliding window in which two sashes slide up and down vertically in separate tracks.

Windows, Hopper – A sash hinged at the bottom which projects outward and has metal arms at the sides to limit its outward movement.

Wythes – Sections of a cavity wall.

Yoke – Head jamb of a window.

INDEX

Acoustical ceiling tile, 355
Acrylic paint, 337, 339
Air conditioning:
  central A/C in heating system 321
  chilled water system, 325, 323
  orientation of house for, 309
  room air conditioners, 314
  temperature control, 309, 311, 312, 314
Air pollution, 331, 332
Alkyd resin paint, 335, 336, 337, 339
Anchor bolts, 36, 225
Asbestos-cement shingles, 101
Asbestos-cement siding, 160, 161
Asphalt shingles, 95, 96
Atrium, 110
Attic fans, 311
Attic remodeling, 424
Attic ventilation, 105
Automatic garage door, 129, 130, 131, 132
Awning windows, 115

Backhoe, 19
Balconies, 106, 107
Balloon framing, 46, 48
Balustrade, 205, 208
Basements, 18, 399
Bathroom:
  cabinets, 250, 251
  dressing room, 432
  family layouts, 431
Bathtub installation, 59, 60
Batter boards, 20
Bay window, 86, 88, 122, 409
Bi-level house, 3
Bolster, 38
Box sill, 37
Breezeways, 307
Brick flooring, 261, 265
Brick and tile walls, 165
Brick veneer, 59, 162, 163, 164
Bridging, 40
Built-ins, 232
Bulldozer, 19

Cabinets:
  bathroom, 250, 251
  hardware, 216, 217
  kitchen, 232, 233, 234, 235, 238, 239, 240, 241
Cape Cod house, 2
Carpeting, 266, 267
Carporch, 297
Carports, 308, 408
Casement:
  doors, 139
  windows, 117
Cast iron pipe, 357
Cavity walls, 59, 165
Ceilings:
  stapled, 175, 176, 179
  suspended, 175
Ceramic tile, 176
Chimneys:
  construction, 268, 270, 271, 273
  flashing and capping, 99, 100, 272, 275
  flues, 274, 275
  height, 271, 272
  prefabricated, 276, 277, 278, 279
  two-flue, 276, 277
  waterproofing top, 275
Clerestory window, 111, 122, 124
Climate control, 321, 323
Closets, 232, 233, 234, 235, 238, 243, 244, 246, 247, 250, 437, 438, 439
Collar beams, 74, 75
Columns, concrete, 21
Combination doors, 141, 142
Commercial buildings:
  details, 445
  entrance, 446
  stairway, 447
Common rafters, 75
Composition board walls, 172
Concrete:
  air entrained, 24
  coloring, 25
  columns, 21
  construction, 24
  curing, 25
  estimating quantity, 25
  mixing, 24
  placing in cold weather, 25
  placing in hot weather, 25
  slab, 29
  troweling, 32
  wall forms, 28
Concrete block:
  decorative, 397, 398
  foundation, 27
  walls, 363
Control joints, 32
Conveyor belt, 96, 295, 296
Cornices, 101, 102, 103, 104, 402

Corrugated fasteners, 225, 229
Corrugated aluminum roofing, 98
Crawl space foundation, 18
Cricket (see chimneys)
Cripples (see studs)
Cripple jacks (see rafters)

Deadlock, 219, 220
Dining room, 242
Disappearing stairs, 211, 212
Doors:
- basement access, 146, 147
- bi-fold, 143
- by-passing, 144, 145
- cafe, 143
- casement doors, 139
- checks, 216
- chimes, 359
- combination, 141, 142
- details, 413
- Dutch, 142
- entry, 132, 133, 134
- flush, 136, 138, 139, 140
- folding 146, 147
- framing, 140, 141
- garage, 129, 130, 131, 132, 301, 303, 304, 305
- garage door operators, 131
- in house design, 142
- jalousie, 146
- jambs, 133
- louvered, 143
- louvered screen door, 360, 362
- panel, 136, 137, 138
- shower and tub, 145
- sliding glass, 142, 143
- steel, insulated, 135

Dormers, 78, 79, 404
Double-glazed sash, 127
Double-hung windows, 112, 113, 114
Double studding, 66, 68
Drip cap, 155
Drywall:
- Board-See Gypsum, Plasterboard
- trestle, 290

Ducts, insulated, 196

Earth moving equipment, 19, 20
Electric baseboard raceway, 359
Electronic air cleaner, 321
Elevator for stairway, 212, 213
Emulsion paints, 337, 338, 339
Entrances, 132, 133, 134, 415, 416, 417
Escutcheons, 221, 223
Excavation, laying out, 19, 20
Exterior wall covering, 148

Fascia, 6, 94, 102, 104
Fences, 449
Fink truss, 88, 89
Fireplaces:
- as room dividers, 268
- barbeques, 268, 282, 283
- chimneys, 268, 270, 271, 272, 273
- circular, 444
- damper, 273
- flues, 274
- framing for, 268, 273
- inside-outside, 442
- mantle, 270
- modern, 441
- open hearth, 443
- prefabricated, 276, 278, 279, 280

Firestopping, 47, 48, 50, 52, 58
Flashing:
- chimney, 99, 100
- plastic, 406
- roofs, 98, 99, 100
- valley, 99

Flat plane truss, 91
Flat roof, 73
Flitch plate, 70
Floor plans:
- one-story house, 4, 10
- prefabricated split level, 378
- split level, 14, 15

Flooring:
- brick, 261, 265
- carpeting 266, 267
- ceramic tile, 264
- component panels, 43
- construction, 252, 253, 254
- finishes, 259, 261
- hardwood, 253, 254, 255
- laminated hardwood, 383
- linoleum, 262
- nailing, 255, 256
- nails, 257, 258
- parquet block, 258, 259
- plank, 256, 258
- resilient, 262
- shrinkage, 256
- subflooring, 252, 253, 254, 256

Footing, 23
Foundations:
- concrete block, 27
- concrete grade beams, 372, 374
- concrete piers, 374
- drains, 26, 27
- forms, 23, 28, 29
- plan, 11
- types, 17
- walls, 27

Framing:

balloon, 46, 48
bay window, 86
brackets and joist hangers, 244, 411
collar beams, 74, 75
component wall panels, 50
firestopping, 47, 48, 50, 52, 58
ladder overhang at gable, 75, 76
overhang, 50, 52
partitions, 61
plank and beam, 63, 64, 65, 68
platform, 46, 47
plumbing wall, 63
prefabrication, 69, 369
split-level, 49
stud spacing, 46
Frieze board, 13, 101, 102, 103
Front-to-back split level, 3
Furring 59, 60, 170, 171

Gable end truss, 90
Gable roof, 74, 75
Gambrel roof, 78, 81
Garages:
attached, 298, 299, 300, 301
breezeways, 307
construction, 304, 305, 306, 307
doors, 129, 130, 131, 132, 301, 303, 304, 305
prefabricated, 302
storage areas, 308
stressed-skin panel, 304
Girders:
wood, 38, 39, 47
steel, 37, 38
Glass block, 111, 122, 176, 177
Gooseneck, 207
Gravel stop, 405
Gussets, 352
Gutters:
aluminum, built-in, 351
extruded vinyl, 351
standard, 405
Gypsum board:
drywall wallboard, 169
pre-decorated, 170
vinyl-surfaced, 356
walls, 169, 170

Hand of doors, 214
Hand-split shakes, 158, 159
Hardboard:
filigree panels, 356
paneling, 172, 173, 174
perforated, 174, 175, 176
Hardware:
anchor bolts, 36, 228
bifold door, 225
corrugated fasteners, 225, 228
door checks, 219
door locks, 219, 220, 221
door viewer, 223
framing hardware, 231, 393
garage door, 129, 130
hinges, 215, 216, 217, 218
nails and screws, 227, 228
sliding door, 224, 225, 226
truss connectors, 229, 230, 231
Heating:
central heaters, 315
electric furnaces, 316, 323
heat pump, 323
hydronic, 316, 317, 318
orientation of house, 309
portable heaters, 312
solar heat, 323, 325, 326, 328, 329, 330, 331, 332, 337
space heaters, 312
temperature control, 309, 310, 311
warm air, 319
Hinges, 215, 216, 217, 218
Hip jacks, 80
Hip roof, 80
Hoists, 295, 296
Hot water heaters, 319, 320, 322
Hubless cast-iron DWV pipe, 357

Incinerator, 282
Insulation:
batt, 191, 192, 193, 194
blanket, 190, 191, 192, 198, 203
comfort, 194
controlling heat loss, 187, 188
loose fill, 190, 191, 200, 201, 202
plastic foam, 358, 359
rigid foam, 30, 190
sound control with 197, 198
stapling to studs, 191, 192
vapor barrier, 195, 197
Intercom, 360
Interior trim, 181, 182, 184, 185, 186, 390, 391
Interior walls, 169
Interior wall treatments, 7, 8
Intersections in gable roofs, 78, 80

Jalousie windows, 118, 120, 121, 122, 125
Jambs, 132, 133
Joints, brick, 57
Joints used in construction, 392
Joists:
lookout, 52
steel, 37, 38
wood, 39, 40
hangers, 40, 231, 393

Keyway in footing, 22
King post truss, 89

Kitchen:
appliances, 235, 238, 240, 241
built-ins, 427
cabinets, 234, 235, 236, 237, 238, 239, 240, 241, 428
corner sink, 429
planning, 239, 240, 241, 426
ventilation, 430
Knee wall, 82

Ladders, 287, 288, 289
Ladder framing at gable, 75, 76
Lally columns, 8, 37
Laminated beams, 354
Laminated roof decking, 353, 354
Latches, 216, 219
Latex paint, 337
Lath, metal, 72
Laundry room, 251, 435
Lean-to rack for sheathing, 94, 291
Ledger strip, 39, 47, 48
Locksets, 232, 233, 432
Lookout joists, 52
Lookout rafters, 73
Lookouts for soffit, 76
Luminous ceiling, 355

Magnetic catch, 220, 222
Mansard roof, 78
Masonry:
details, 394, 395, 396
reinforcing, 55, 56
tooling joints, 55, 57
Medicine cabinet, 251
Metal tiles, 177
Modular wall system, 52, 53
Moisture and condensation, 333, 334, 335
Moldings, 181, 184, 185, 186, 390, 391
Mullion, 113
Muntins, 126, 127

Nailing:
flooring, 270
framing, 49
machines, 93, 95
Nails used in house construction, 223, 227, 228
Nail-on brick, 165, 166
Newel, 206, 207
Night latch 221, 222
Nosing, 204, 206

Octagon window, 123
One-story house, 2
One-and-one-half-story house, 2
One-and-one-half-story truss, 90
Open riser stairs, 201
Oriel window, 86
Orientation of house on site, 6, 9, 309

Painting:
brush technique, 341, 343
changes in paints, 335, 336, 337, 338, 339, 340.
color chart, 344
equipment, 357, 365
masonry surfaces, 345, 346
moisture problems, 333
preparing surfaces, 337, 340, 346
rollers, 349, 350
texture, 350
types of paint to use, 345, 347
Paneling:
hardboard, 172, 173, 174, 175, 176
perforated hardboard, 177
plywood, 172
wood, 170, 171
Parapet, 108
Partitions:
framing, 61
split-level, 69, 70
steel stud, 72
storage walls, 232, 233
Patios, 12, 107, 109
Perimeter insulation, 30
Permeability 333
Placing house on site, 6, 9, 309
Plancier (see soffit)
Plank and beam roof, 84
Planters, 448
Plasterboard, 169, 170
Plastic pipe, 358
Platform framing, 46, 47
Platform trestle, 291
Plumbing:
core, 63, 64
planning, 433
septic tanks, 434
walls, 63, 64
Pocket door, 225
Pole-type construction, 362, 363
Porches, 106, 107, 109, 407
Prefabricated houses:
history of, 369, 370, 372, 373, 384, 385, 386
split-level, 66
sectional house, 385, 386, 388
steel-framed house, 380, 381, 382, 383, 384
Prefabrication:
concrete grade beams, 372, 387
construction details, 376, 377, 378, 380
mechanical cores, 376, 377
modular components, 373
wall and roof components, 374, 375
Purlin, 81
Rafters, 74, 75, 76, 77, 78, 79, 80, 81, 82, 83, 84, 86, 87
Rake molding, 102
Ranch house, 8

Remodeling for outdoor living, 110
Re-roofing over old shingles, 97, 98
Ribbon, 43
Ridge board, 74, 76, 78, 79
Rigid frames, 400
Riser, 200, 201, 202, 204, 206
Roof saddle, 290
Roofing:
  aluminum roll roofing, 97, 98
  asphalt shingles, 95, 96
  embossed aluminum shakes, 352
  roll roofing, 97
  terne, 99, 100
  wood shingles, 94, 95
Roofs:
  anchoring against wind, 80, 83
  combination roof-ceiling decking, 94
  construction details, 73-84, 401
  flashing, 98, 99, 100
  flat roof, 73
  gable roof, 75, 76
  gambrel roof, 81
  gravel stops, 405
  gutters, 351, 405
  hip roof framing, 80
  mansard, 78
  plank and beam, 84, 85
  rafters at right angle to joists, 77
  sheathing, 93, 94
  shed roof, 73
  stressed-skin panels, 95
  trusses, 87, 88, 89, 90, 91
  valley framing, 78
Room dividers, 65

Sash frame, 116
Sawhorses, 284, 285, 286
Scaffolds, 285, 286, 287, 292, 293, 294, 295
Scarifier, 19
Screen walls, 108, 109
Second floor framing, 61
Sheathing:
  board, 54
  insulating, 55
  plywood, 54
Shed roof, 73, 75
Shingles:
  asphalt, 95, 96, 97
  wood, 95, 98
Side-to-side split-level, 14, 15
Siding:
  aluminum and steel, 166, 176
  asbestos-cement, 160, 161
  beveled, 148, 149
  board and batten, 155, 156, 157
  brick veneer, 164, 166, 167
  clapboard, 150, 151
  corner trim, 151
  drop, 150, 151
  horizontal application, 151, 154, 155
  nailing, 151, 152
  plywood, 153, 156, 157, 158, 159, 362
  simulated stone and brick, 160, 161, 162, 165
  vertical 152, 153, 154, 155
  vinyl, 166, 168, 362, 364
  wood shakes and shingles, 159
Sills, 36, 37
Simulated stone and brick, 161, 162, 357
Skylight, 122, 125
Slab:
  compacting sub-base, 30
  construction on grade, 29
  foundation, 22, 27, 28, 30, 31, 32
  reinforcing, 31
Sliding table, 440
Snow melting cable, 361
Soffits, 101, 102, 103, 104
Soil:
  analysis, 17
  compaction, 30
Solar Heating, 323, 325, 326, 328, 329, 330, 331, 332, 337
Sole plate, 41
Sound control, 197, 198
Split-level house, 3
Split-ring timber connectors, 87, 88
Stairs:
  attic, 421, 422, 423
  balustrades, 205, 206, 207
  Colonial, 418
  concrete, 210, 211
  disappearing, 211, 212
  elevator for, 212, 213
  gooseneck, 207
  rise and run, 203
  roughing in, 420
  safety factors, 201, 202
  shop-fabricated, 202
  spiral, 208
  stairwell construction, 202, 203, 204
  standards, 200, 201, 202
  steel, 212, 213, 268
  steel stair guides, 361
  stringers, 202, 203, 204, 419
  suspended, 208
  volute, 207
  winders, 209
Stilts, 310
Stone, simulated, 161
Storage walls:
  as room dividers, 65, 232, 233
  in bedrooms, 245, 246, 247, 248, 249
  in living and family rooms, 243, 244
Storm windows, 118, 121

Stoops, 210, 211
Stressed-skin panels for floors and roofs, 322, 421
Stucco, 159, 160
Studs:
  cripple, 46
  double, 68
  gable, 51
  size and spacing, 46
Subflooring, 41

Termite shields, 41
Terne roofing, 99, 100
Ties in masonry, 55, 56, 59
Tooling masonry, 55, 57
Trowel, power, 32
Troweling, 32
Thread, 204, 205, 206
Trimmers, 79, 277
Trusses:
  connectors, 87, 89, 229, 230, 231, 353
  construction, 87
  jig table for, 354
  plywood gussets, 352
  steel, 89
T-sill, 37
Two-story house, 2

Vacuum cleaner system, 361
Valley:
  flashing, 99
  framing, 78, 79, 80
  jacks, 80
Vapor barrier, 30, 153, 333, 335
Ventilation, 105, 108, 310, 336
Volute, 207

Walers, 23
Wall construction:
  brick veneer, 60
  cavity, 59
  masonry, 55, 72
  wood, 53
Wallboard, 169
Wallpaper, 350
Weatherstripping, 359
Winders, 209
Windows:
  awning, 115, 119
  bay, 120, 409
  basement, 118, 120
  casement, 115, 117
  clerestory, 111, 122, 124
  construction details, 112, 113,
  double-hung, 112, 113, 114
  glass block, 122, 124
  hopper, 118, 120
  installation, 125
  insulating glass, 412
  jalousie, 118, 120, 121, 122, 125
  picture, 120, 122
  planning principles, 127
  pre-hung, 360
  removable muntins, 126, 127
  replacement, 360
  skylight, 122, 125
  sliding, 115, 116, 123, 125, 127
  special purpose, 122, 123
  spring balances, 112, 113
  storm sash, 118
  types of, 111, 112
  u-shaped channels, 118
Wythe, 59, 60, 165, 168